红树林研究论文集

(2005—2009)

主编／林光辉 林 鹏

林光辉教授（1962—），福建省莆田市人，1983年7月毕业于厦门大学生物学系，1986年6月获得厦门大学生物学系硕士学位，1992年12月获得美国迈阿密大学博士学位，1992—1994年在犹他大学从事博士后研究。先后担任过美国生物圈研究所研究员、哥伦比亚大学地球研究院副研究员、中国科学院植物研究所首席研究员（入选“百人计划”）和中国科学院研究生院教授、博士生导师。现为厦门大学“闽江学者”特聘教授、博士生导师和生命科学学院副院长，兼任滨海湿地生态系统教育部重点实验室主任、Journal of Plant Ecology-UK 共同主编、中国生态学会红树林学组执委会主席、湿地科学专业委员会副主任委员、海洋生态学专业委员会常务委员和中国自然资源学会湿地资源保护专业委员会副主任委员。

林光辉教授长期从事红树林生态学、稳定同位素生态学、全球变化生态学和环境科学的研究及教学工作，先后主持过美国自然科学基金会、环保署和中国科学院、科技部、农业部、自然科学基金会以及深港科技创新圈资助的课题多项；已发表90多篇学术论文，其中SCI论文50多篇；参加编写的专著4部。

林鹏教授（1931—2007），福建省龙岩市人，1955年毕业于厦门大学生物系植物学专业，先后任助教、讲师、副教授、教授、博士生导师，湿地与生态研究中心主任、中国工程院院士。兼任中国生态学会顾问、福建省生态学会名誉理事长、福建省科学技术协会第二、三、四届委员和国家教委首届环境科学教学指导委员会委员、第二届副主任，国际红树林生态系统学会（ISME）理事。

林鹏教授主要从事植物生态学和群落学，环境生物学和红树林生态学研究，曾在国内外发表论文400多篇，专著有《植物群落学》（1986）、《福建植被》（1990）、《武夷山研究—森林生态系统（Ⅰ）》（1998）、《南药栽培》（1980）、《红树林》（1984，中文版；1988，英文版）、《红树林研究论文集（第一集）（1980—1989）》（1990）、《红树林研究论文集（第二集）（1990—1992）》（1993）、《红树林研究论文集（第三集）（1993—1996）》（1999）、《红树林研究论文集（第四集）（1997—1999）》（2000）、《红树林研究论文集（第五集）（2000—2001）》（2002）、《红树林研究论文集（第六集）（2002—2004）》（2005）、《中国红树林环境生态和经济利用》（1995，中文版；2000，英文版）和《中国红树林生态系》（1997，中文版；1999，英文版）。

2002 年 12 月 18 日林鹏院士生日和部分学生的合影

2006 年 4 月至九龙江口红树林调研

2006 年 5 月闽港两地就漳江口红树林保护与利用开展系列研讨

2006 年 6 月至翔安山亭红树林人工林调研

漳江口红树林生态系统定位站通量塔

湛江红树林生态系统定位研究站

NR01 辐射仪

三维超声风速仪和红外气体分析仪

湛江红树林生态系统定位研究站

2009年4月2日科技部办公厅一行至漳江口红树林生态定位站调研

2008年9月27日美国托莱多大学陈吉泉教授至漳江口红树林定位站考察

2008年12月21日谢联辉院士、唐崇惕院士漳江口红树林定位站考察

2010年1月华东师范大学陆健健教授至漳江口红树林定位站考察

2010年1月台湾“中央研究院”周昌弘院士至漳江口红树林定位站考察

2010 年 1 月 17 日海洋公益性项目“新兴经济区滨海湿地生态系统修复技术研究与工程示范”项目启动会

2010 年 1 月国家自然科学基金重点项目采样

2010 年 7 月美国俄勒冈州立大学研究生 Julie Doumbia 和中国地质大学李海龙教授至漳江口红树林定位站参加 973 前期专项研究

林光辉教授指导学生野外工作

2008 年 11 月香港城市大学谭凤仪教授至湛江高桥红树林定位站考察

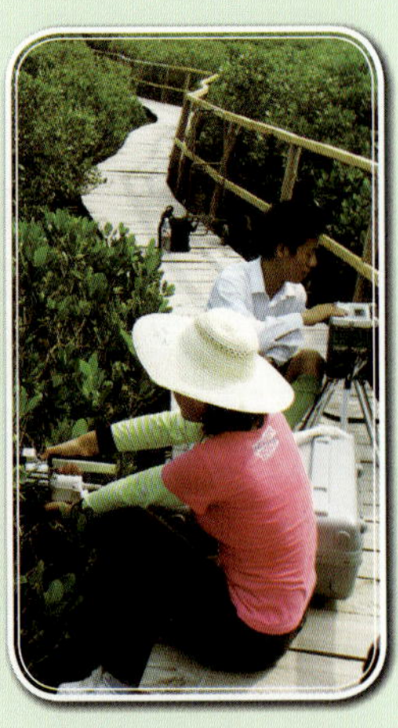

红树植物光合测定

序 言

红树林是热带海岸潮间带的木本植物群落。由于温暖洋流的影响，有的可以分布到亚热带；由于受潮汐影响，有的也可分布于河口海岸和水陆交叠的地方。因此，红树植物包括了具有每日浸润的潮间带生长的真红树和只有高洪潮方可浸润的高潮带以上的两栖性的半红树植物。红树林具有以下 6 个方面生态系统服务功能：(1)促进土壤的形成、抵抗潮汐和洪水的冲击；(2)过滤陆地径流和内陆排放的有机物质和污染物；(3)为许多海洋动物（包括渔业、水产生物）提供栖息和觅食的理想生境；(4)为滨海湿地和近海海洋生态系统次级生产力提供有机碎屑；(5)提供木材、薪炭、食物、药材和其他化工原料等产品；(6)为环境科普教育和生态旅游提供自然和人文景观。开展红树林研究不仅具有重要的理论意义，还有现实的经济意义。

本书是继林鹏院士的《红树林研究论文集》(1980－1989)和第二集(1990－1992)、第三集(1993－1996)、第四集(1997－1999)、第五集(2000－2001)、第六集(2002－2004)之后的续集。林鹏院士因车祸于 2007 年 5 月逝世，为了纪念他对红树林科研事业所做出的毕生贡献，我和其他同仁认真整理了 2005 年以来所有林鹏院士署名的红树林相关论文以及我们课题组 2005－2009 年发表的其他相关论文，编辑《红树林研究论文集》(第七辑，2005－2009)。全书共分七个部分：前言部分收录纪念林鹏院士的论文 1 篇：《情系红树、献身中国生态事业——记“中国红树林之父”林鹏院士》，概括了林鹏院士研究红树林、献身中国生态事业的一生；第一部分收录论文 22 篇，论述红树植物生态生理学特性，如抗盐特性、淹水胁迫响应等；第二部分收录论文 18 篇，论述了红树林中的鸟类、底栖动物、微生物和藻类等类群的生物多样性；第三部分收录论文 8 篇，论述了红树林植物的活性物质的特性、丹宁代谢以及植物激素作用等；第四部分收录论文 4 篇，论述红树植物形态学特性；第五部分收录论文 4 篇，论述了红树植物种群遗传变异和分化；第六部分收录论文 6 篇，论述了红树林的保护和生态恢复。

全书收录了 62 篇(2005－2009)正式发表的论文，是林鹏院士及其领导下的厦门大学红树林科研组的集体成果之一。本书可供国内外有关学者参阅，并作为进一步开展此项研究的参考资料。本书汇编过程中得到王新丽、杨志伟等的帮助，谨此感谢。本论文集出版以及涉及的部分研究得到国家自然科学基金重点项目(30930017)、国家自然科学基金面上项目(40776046、30700092、30600077、30972334)、科技部 973 预演研究专项(2009CB426306)、国家海洋局海洋公益性科研专项(200905009)科技支撑计划项目(2009BADB2B0605)以及近海海洋与环境国家重点实验室自主课题经费等资助。

由于编写匆促，书中错漏或不足之处，敬请同行专家和读者批评指正。

林光辉

2010 年 2 月 23 日于厦门

序

Preface

Mangroves are woody plants in the intertidal zones of tropical coasts. They can occur in some subtropical areas because of ocean warm currents, and others may grow into the inland areas of river estuaries or land-sea ecotones due to tidal effects. Mangroves are divided into true mangroves that subject to daily tidal inundations and semi-mangroves that occur in high tide zones and are flooded occasionally only by high tides. Mangroves in general can offer the following six categories of ecosystem services: (1) facilitating sedimentation and reducing erosions of tidal currents and floods; (2) filtering organic matter and pollutants derived from surface runoffs and river discharges; (3) providing ideal feeding habitats for many marine animals including some important fishery species; (4) supplying organic detritus for secondary producers in coastal wetlands and near-shore oceans; (5) producing woods, charcoals, foods, medicines and other chemicals; and (6) providing natural and cultural landscapes for scientific outreach and eco-tourism. Thus, mangroves are very important study materials not only in theoretical research but also in application research.

This book is a continuation of Mangrove Research Papers Series edited by Academician Peng Lin, which included "Mangrove Research Papers (1980—1989)", "Mangrove Research Papers II (1990—1992)", "Mangrove Research Papers III (1993—1996)", "Mangrove Research Papers IV (1997—1999)", "Mangrove Research Papers V (2000—2001)" and "Mangrove Research Papers VI (2002—2004)". Unfortunately, Prof. Peng Lin passed away in May 2007 due to a car accident. In recognizing his whole-life contributions to mangrove research, my colleagues and I decided to continue publishing Mangrove Research Papers series. The resulted "Mangrove Research Papers VII" included all mangrove papers co-authored by Prof. Peng Lin since 2005. Also included in this book are other mangrove papers authored by me and my colleagues in Xiamen University between 2005 and 2009.

The book has seven sections in total. Introduction includes a eulogy article titled "Fond with mangroves and devotion to China ecological research: in memory of 'Father of China Mangroves' Prof. Peng Lin", which praised his life-long contributions to mangrove and ecological research in China. Section I consists of 22 papers on mangrove ecophysiological characteristics such as salt tolerance, waterlogging resistance, etc. In Section II, total of 17 papers described the biodiversity of birds, benthos, microorganisms and algae in mangrove wetlands. Section III is composed of 8 papers on bio-active compounds, tannin metabolisms and plant hormone functions. The 4 papers in Section IV focused on morphological characteristics of selected mangrove species. Section V included 4 papers on polygenetic variations and differentiation of mangroves. Section VI had 6 papers on mangrove conservation and ecological restoration.

In total, the book contains 62 mangrove papers published previously in scientific journals between 2005 and 2009, which resulted from the research efforts of Prof. Peng Lin and his co-workers at Xiamen

University. It can serve as a reference for professionals who are in the field of mangrove and other coastal wetland research. We acknowledged help from Ms. Xinli Wang and Mr. Zhiwei Yang in editing this book. The financial supports for publishing this book and for part of research reported here came from Chinese National Natural Science Foundation (30930017, 40776046, 30700092, 30600077, 30972334), National Basic Research Program (2009CB426306), State Administration of Oceanography (200905009), State Administration of Forestry (2009BADB2B0605) and State Key Laboratory of Marine Environment.

Guanghui Lin
February 23, 2010

前言

情系红树　献身中国生态事业*

——记“中国红树林之父”林鹏院士

撰文・供图/林光辉　陈鹭真

在中国甚至国外许多国家，只要一提起红树林研究，人们就会自然想起著名的红树林专家林鹏教授。林鹏教授出生于福建省晋江市，原籍福建省龙岩市。1955 年毕业于厦门大学。林鹏长期从事河口海岸红树林和陆地植被生态学研究，率先对中国六省区（包括台湾）红树林进行了系统调查和研究，是中国红树林生物量、生产力、物流能流等生态系统研究的开拓者。专著《中国红树林生态系统》填补了中国红树林生态系统学科的空白，为中国红树林的研究和生态恢复工程起到奠基作用。2001 年，他因在这些方面的突出贡献当选为中国工程院院士。

弃学贩盐

鲜为人知的是，这位因红树林生态学研究享誉海内外的工程院院士早年有段当学徒、贩盐谋生的经历。

由于家庭困苦，林鹏 17 岁就辍学当学徒，半年后在亲友支持下才得以回校继续念书。然而报考大学时，又因家里出不起赴厦门大学的路费，他只得放弃。自此开始挑担赶集的生活，每天天没亮就和弟弟挑着百多斤盐担子，到 40 多公里外的镇上贩卖。1951 年，挑着盐担子的林鹏在龙岩街头布告栏得知厦门大学因金门炮击将迁到龙岩，并补招 30 名学生的消息。他不肯放弃这宝贵的机会，在获得家人同意后参加了招生考试，并以第三名的优异成绩被厦门大学生物系录取。

艰苦的生活不但磨砺了他坚韧的性格，也使他明白在通往成功的路上，有两个代价总是形影相随：牺牲和困苦。这种坚韧的性格伴随了林鹏教授一生，也体现在他对红树林的研究过程中。

---

* 原载于生命世界，2008，8：98－101

结缘红树林

其实,林鹏院士最早并不是专门研究红树林的。他在科研道路上最终与红树林结缘还有一段故事。

早在1953年,当时还是厦门大学生物系学生的林鹏在导师何景教授带领下第一次在野外见到了红树林。他曾经回忆道,当时觉得真有意思,竟然有树长在海里。

可是到了1977年,林鹏教授在翻阅《湿地海岸生态系统》时,发现这本书把中国的红树林列为空白。1980年,有美国学者说"中国红树林已经消失"。林鹏教授当然知道外国人的观点并不能反映中国红树林资源的真实情况,但要打破这种观点则需要确实的证据。于是,林鹏教授决心对中国红树林进行系统研究。

林鹏院士指导厦门市翔安区的红树林生态恢复工程。

他带领自己的科研组深入闽、浙、两广等我国沿海各省份,开展红树林实地调查,足迹遍布几乎所有的红树林区。随后,林鹏教授又带领他的科研组在中国的三大红树林基地——海南东寨港、广西英罗湾和福建九龙江口,分别进行了为期6年、5年和11年的定点生态系统研究。

红树林给人们的印象通常是美丽的,漫步于浪漫的红树林中更使人向往,但对红树林的野外考察就是另外一回事了。长期的野外工作异常艰苦,寒来暑往、风餐露宿不说,有时候为了观察海滩上一棵树木的情况,不得不在齐膝深的海泥里泡上三四天,走出来的时候,腿都泡得肿胀了。更多的时候,为了一个群落的红树林得花上七八年乃至十几年的时间。后来,林鹏教授笑言:当时就只想着要为国争光,为民造福。

所有的辛苦终于没有白费。1980年,林鹏教授赴美参加第三届红树林学术会议。在会议上,林鹏教授以确凿的事实、精确的研究、完善的数据向世人展示了中国丰富的红树林资源,改变了国际学术届对中国大陆红树林及其研究的错误认识。随后,林鹏教授在1985年澳大利亚海洋研究所举行的一个国际红树林会议上,以报告的形式详细地介绍了中国的红树林生态系统,赢得了与会各国专家的认可。报告结束后,会议执行主席、著名的红树林专家费德尔教授拉住他的手连声赞叹。这个报告的研究成果打破了"中国除台湾外没有红树林"的偏见。

从这以后,中国的红树林研究迅速发展,很快进入国际先进行列。林鹏教授也成为了中国当代红树林湿地研究权威和学术带头人,甚至被一些人誉为"中国红树林之父"。

硕果累累

林鹏院士带领他的团队先后在我国沿海省份建立了3个红树林基地,进行了红树林生态系统长期定位研究，揭示了海莲林、红海榄林和秋茄林的现存生物量、生产力、凋落物量和落叶半分解期。在此基础上提出了红树林的“三高”特性（高生产率、高归还率、高分解率），为发展海岸河口湿地水产渔业，选择鱼虾亲本苗饵料基地奠定了理论基础，为红树林资源的保护和可持续利用提供了科学依据。

林鹏院士视察红树林新造林的情况。

红树林在沿海防风护堤、保护人民生命财产方面起着重要作用。20世纪80年代以来，林鹏院士开始着手建立红树林恢复生态工程体系。在海南、广西、福建各保护区应用生态恢复原理，提出生态适应条件、林地选择和育种等栽培技术规范。在厦门海沧台商投资区滨海大道外侧滩涂红树林营造工程中，建立潮滩、潮流、盐度等可行性技术指标以及抗盐、抗潮的方案。这些技术在应用上取得重大效益。

结合实践研究的同时，在林鹏院士的指导下，厦门大学的红树林研究也在向纵深发展。首先，应用分子生态学技术揭示了中国红树植物的种群遗传中心。通过对盐胁迫和红树植物器官衰老机制的研究，揭示了抗盐胁迫对抗衰老物质形成的正效应，并进一步向筛选抗盐基因的方向迈进。通过木材结构的研究发现了海生红树植物远比陆生红树植物原始，纠正了国外学者提出是陆生红树被挤入海滩的进化途径假说。

由于红树植物生长于海岸潮间带，在种类的界定上容易带来诸多不便，为此，林鹏院士在曼谷参与讨论联合国教科文组织和国际红树林生态系统学会的“红树林宪章”的基础上，结合20多次参加国际会议和长期的工作积累和论证经验，严格界定了红树种类标准，确定了“真红树”和“半红树”种类的科学界限，成为全国广大红树林研究、保护和管理人员引用的依据。他的这些研究成果，打开了中国红树林研究走向世界的通道，填补了国内在这个方面的空白，使中国红树林生态系统和保护工作进入国际先进行列并占有一席之地。

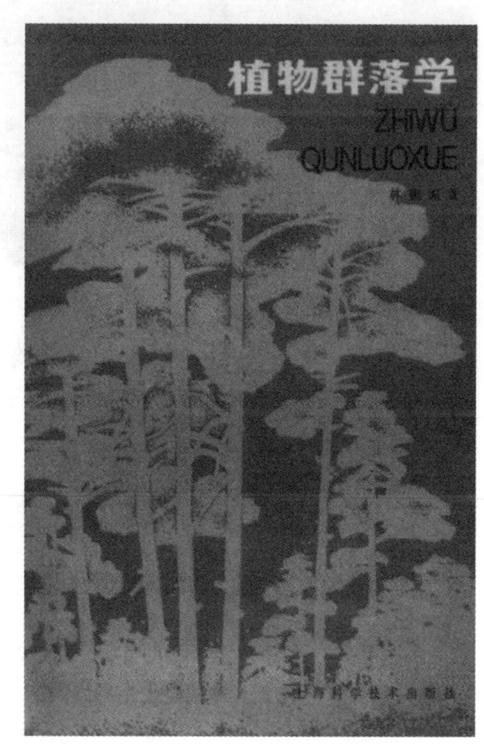

林鹏院士所著的44万字的《植物群落学》，被许多专家称之为中国第一部植物群落学著作。

执着追求

为了红树林和我国的生态事业，他曾经5次在野外工作时发生意外，全身上下遍布伤疤，大腿里至逝世时还镶着根钢条。他也曾在外出研究途中遭遇车祸昏死两个小时后转醒。1987年的一次野外考察，让林鹏教授与死神擦身而过。考察工作接近尾声时，他在赶回学校给研究生上课的路上不幸发生了车祸，身负重伤，四肢仅剩左手完好，当时就被交警宣布死亡。两个小时后他竟醒了过来,之后做了5次手术、全身开了7个切口,住院长达一年半。在住院的400多天里,林鹏教授仍坚持撰写科研论文,抓紧点滴时间修改《海洋高等植物学》这部书,并做些翻译和校对工作。病情减轻时,他坐在病床上,为参加博士生入学考试的考生进行复试。出院第二天,他撑着双拐,在客厅里为研究生主持论文答辩。答辩会上林鹏教授特别兴奋,甚至忘记自己是个刚出院的病人,完全沉浸在浓郁的学术氛围中。

作为一个科研工作者,林鹏教授取得的成果之多令人称奇!多年来他独立撰写和参加编写的书有14部。除此还在国内外发表论文300多篇,获国家自然科学二等奖一项,国家科技进步奖三等奖一项,部(省)级二等奖四项,省部级三等奖八项。林鹏教授在红树林生态学领域取得了系统的创新性成果,使他成为国内外同仁公认的中国红树林生态系统研究的权威和学术带头人,被台湾学者推崇为中国红树林生态系统"研究大师"。

注重培养"造血型"人才

林鹏教授不仅在科研上取得丰硕成果,在教学岗位上也培养了众多优秀学生。他主张培养自立自强、具创新意识的开拓型人才。他对研究生的培养,不是采取"输血型"教育,而是注重培养他们成为"造血型"人才。一有机会,他就让研究生参加学术会议和国内外专家学术讲座,增长学生独立工作的才干。他常对学生讲"一本书主义"和"尽信书不如无书",就是说既要尊重老师认真读书,又要不怕权威敢提新观点。他说:"我真诚地希望,你们都能胜过我,长江后浪推前浪,这样科学事业才能不断发展。"数十年来,他先后指导培养了博士后研究人员8名,博士研究生28名,硕士研究生25名,桃李满天下。学生遍布国内外,成为生态学领域的高级人才。

林鹏院士与学生在一起。

林鹏教授生前曾有三句座右铭:"方向明、干劲大、及时总结"、"顺境时更加谦虚、逆境时更加自强"、"有志者事竟成、有恒者业必兴"。他说过:"过去我是这样的,以后也将如此。我要工作,直到最后一息"。他的一生,就是这些座右铭的最好写照!

作者简介

林光辉,教授,厦门大学生命科学学院副院长,主要从事红树林生态系统生态学、全球变化生态学和稳定同位素生态学等方面的研究。

目 录

序言

前言

二、红树林生物多样性

三、红树林植物化学

四、红树林植物形态学

五、红树林分子生态学

六、红树林保护与生态恢复

Mangrove Research Paper (Ⅷ) (2005—2009)

Preface

Introduction

Part Ⅰ. Mangrove Eco-physiology

Part Ⅱ. Mangrove biodiversity

Part Ⅲ. Mangrove Phytochemistry

Part Ⅳ. Mangrove plant morphology

Part Ⅴ. Mangrove molecular ecology

Part Ⅵ. Mangrove conservation and ecological rehabilitation

一 红树林生理生态

PART I MANGROVE ECO-PHYSIOLOGY

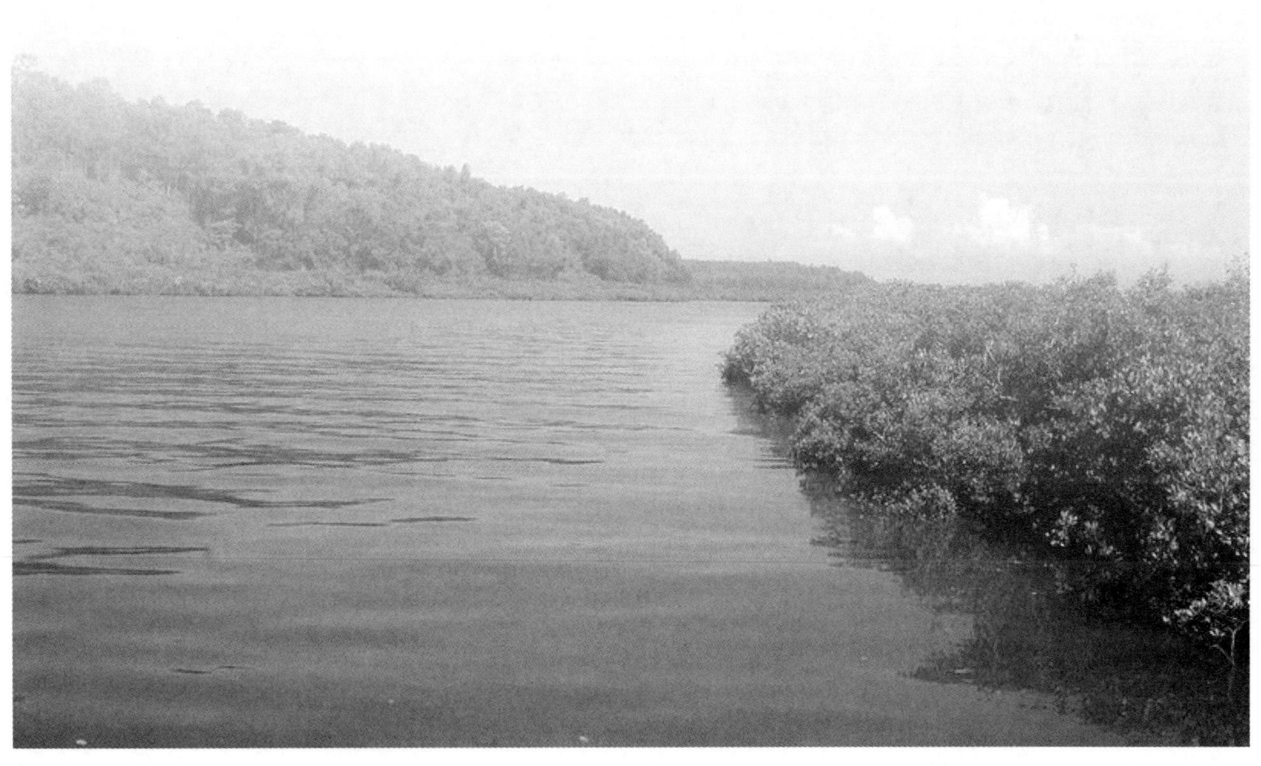

红树植物淹水胁迫响应研究进展*

陈鹭真[1,2] 林 鹏[1] 王文卿[1]
（1. 厦门大学生命科学学院，厦门 361005；
2. 中国科学院植物研究所植被数量生态学重点实验室，北京 100093）

摘要:潮汐淹水是红树植物面临的主要环境胁迫之一,也是导致目前红树林造林成活率低的一个关键因子。由于长期适应于水淹生境,红树植物发育出一套适应于潮间带生长的抗淹水机制。综述了与红树植物相关的抗淹水胁迫响应机制,包括了形态结构、生长、水分和光合作用、膜脂过氧化系统和根系脱氢酶系统、内源激素和胁迫多肽等5个方面。提出应用人工潮汐系统研究红树植物的淹水抗性机理是确定不同种类红树植物的耐淹水能力的有效手段。并指出生长的研究是淹水胁迫响应研究的基础,而与分子手段相结合的激素水平的研究将在红树植物抗性胁迫研究中得到重视。

关键词:红树林;潮汐;淹水胁迫;抗性机制
文章编号:1000-0933(2006)02-0586-08 **中图分类号**:Q143,Q178.1,Q948.8 **文献标识码**:A

Mechanisms of mangroves waterlogging resistance

CHEN Lu-Zhen[1,2], LIN Peng[1], WANG Wen-Qing[1] (1. *School of Life Sciences, Xiamen University, Xiamen* 361005, *China*; 2. *Laboratory of Quantitative Vegetation Ecology, Institute of Botany, Chinese Academy of Sciences, Beijing* 100093, *China*). Acta Ecologica Sinica, **2006, 26(2): 586～593.**

Abstract: Tidal waterlogging is one of the most important stresses to mangroves. The duration of waterlogging is a limiting factor for the survival of mangrove seedlings. On adapting to waterlogging in the intertidal zones, mangroves have developed a set of mechanisms of waterlogging resistance, such as the growth of aerial roots and air space in the roots, special photosynthesis rates and nutrient circle. Scientists have shown great interesting in mangroves waterlogging resistance. The present paper reviews five aspects of the mechanisms of mangroves in responses to waterlogging, which are morphological and anatomical factors, growth, water use efficiency and photosynthesis, activities of alcohol dehydrogenase and enzymes processing reactive oxygen species, and plant growth regulators. Aerial roots and aerenchyma in cortex are important for mangroves' resistance to waterlogging, by allowing oxygen to be replenished. Under tidal flooding, oxygen transported in the aerenchyma can maintain the oxygen demand of the roots. Like other stress, under waterlogging conditions the growth of mangroves falls, the photosynthetic rate declines, and leaves pigment contents change. In addition, the activities of enzymes processing reactive oxygen species and of dehydrogenase change. These physiological and photosynthetic responses of mangroves facilitate their tolerance to waterlogging. The plant growth regulator abscisic acid is also greatly induced by waterlogging. However, no reports about the relationship between content of ethylene or polyamine and waterlogging in mangroves have been reported.

To investigate waterlogging resistance mechanisms of mangroves, artificial tidal equipment is useful. It simplifies the conditions, and simulates different kinds of tidal cycles, including different durations of waterlogging. We find it necessary to study further the growth of mangroves to waterlogging, because it is the base for plant resistance research. Similarly, studies on plant hormone combined with molecular techniques will be of great interest in research on mangroves waterlogging. Further studies

* 国家基金资助项目(30200031);教育部博士点基金资助项目(20030384007);厦门市科技基金资助项目(350272021046)
原载于生态学报,2006,26(2):586－593

should be focused on the molecular mechanisms of hormone changes in mangroves during waterlogging. Studies on the mechanisms of mangroves waterlogging resistance will provide theoretical guidance for mangroves rehabilitation.

Key words:mangroves; tide; waterlogging stress; resistance mechanisms

自然状态下,红树植物及其幼苗的生长受到各种不利因素的干扰,除了人为干扰外,海水盐度、滩面高程、淹水时间,以及藤壶附着和啮齿类动物的啃食等都对其生长具有显著的影响。自然状态下,不同种类的红树植物根据其耐淹水能力不同,分布在不同高程的滩涂上[1]。在大部分的潮汐周期中,红树植物幼苗经常被潮水淹没,处于没顶淹水的状态[2]。由于生境的恶化,幼苗没顶淹水的时间延长,淹水深度加深[3~7],将成为导致红树林造林成活率过低的一个关键因素[2]。

国内外学者对红树植物的淹水耐性进行了大量的研究,其研究表明:红树植物由于长期适应于水淹生境,发育出一套适应于潮间带生长的抗淹水机制[8~13],如特异的形态特征、生理响应以及营养循环。在不同高程的滩涂上,这些生理上和结构上的适应特性具有显著的梯度变化[14~16]。本文将从形态结构、生长、水分和光合作用、膜脂过氧化系统和根系脱氢酶系统、内源激素和胁迫多胺等 5 个方面对红树植物的淹水抗性研究进行综述。

1 红树植物与淹水相关的形态结构

1.1 根系的形态学和通气组织的解剖学特征

和许多湿地植物一样,红树植物由于长期适应于水淹生境,发育出了发达的不定根,植株内存在的发达的气体交换系统,能将氧气源源不断的运送到根系,躲避根系厌氧[17]。

红树林林下的土壤是由细质颗粒的沉积物组成的富含有机质的无结构土壤,高水分、高盐度、并缺乏氧气[1],仅在表层沉积物中的一个薄层有氧气。假根或次生根的数量增多是红树植物缓解根系缺氧的一种机制[18]。在白骨壤属(*Avicennia*)和海桑属(*Sonneratia*)的种类中,背地性生长气生假根能有效地在大气和红树植物根系内部进行气体交换[19];木榄属(*Bruguiera*)和榄李属(*Lumnitzera*)中的膝状呼吸根和表面根起到气体交换的作用[19];红树属(*Rhizophora*)的植物中的支柱根能进行气体交换[20]。

淹水改变了根系的解剖学构造。根系表面的皮孔是气体交换的门户,例如,萌芽白骨壤(*Avicennia germinans*)和大红树(*Rhizophora mangle*)通过皮孔可以从地上向地下部分运送空气[8]。在许多红树植物的地上部分的根系皮孔周围,可见剥落的组织,它们构成了内部组织与外间环境进行气体交换的门户[1]。白骨壤(*A. marina*)具有皮孔的假根在淹水后 3min 后可以完全恢复氧气的运送,而没有皮孔的假根至少需要 10min 才能恢复根内氧气的运送[21]。皮孔不仅有利于氧气的进入,也有利于挥发性物质如乙醇和乙烯等的排出[22]。同时,皮孔在气生根上的分布也很有规律,白骨壤(*A. marina*)的每个气生根上大约有 25 个皮孔,而在气生根与地下水平根系的交界处,皮孔多于 25 个[21]。

一般而言,植物缺氧刺激乙烯合成,进而刺激纤维素酶的活性增强,细胞壁中果胶质分离,导致细胞分离,逐步形成由小到大的细胞间隙,在皮层中形成发达的通气组织,这是皮层气室形成的裂生方式[23];若由于皮层中某些细胞的整个细胞崩溃而形成的通气组织,为溶生性通气组织[23]。研究表明:红树植物根系中的几乎所有的气室都是通过裂生方式产生的,而只有小根中的部分气室是溶生方式产生的[29]。在通气组织形成过程中,乙烯作为一种厌氧信号,激活 Ca^{2+} 和磷酸肌醇(phosphoinositides)的信号传导途径,最后导致细胞程序性死亡[25]。根系皮层中产生的连续气室构成了气体通路,氧气可以在这个通道中运送[26~28]。厌氧条件下,红树植物根系发达的通气组织能够把氧气从植株的地上部分运送到根部,保障缺氧条件下正常的生理活动[1]。同时,红树植物茎的皮层中也有发达的气道起着输送氧气的作用[1]。

大多数红树植物在生长 1a 内的幼苗期,植株未长出气生根,例如,白骨壤(*A. marina*)幼苗通常在 1a 后长出气生根[30]。对于幼小的植株而言,当地上部分如茎、叶和胚轴暴露于空气中时,气体通过气孔进入叶片的海绵组织,经皮孔进入茎和胚轴,运送到根系,供给地下的根系[31]。在 10 个月大小的白骨壤(*A. marina*)植株中有连续的气室;叶柄、根、茎节间和胚轴这些较长的组织中,皮层的气室形成彼此连接的宽而长的通道,而茎节处则有大量的海绵组织连接,从而构成一个气体运送通道[31];当植株长出呼吸根后,高潮时被海水覆盖,而退潮时呼吸根露出水面,进行气体交换,此时,气体通过呼吸根进入根系是红树植物的主要呼吸方式。Curran等认为在正常的低潮时通过植株呼吸根进入根系的氧气就足以提供根内的气体需求[32]。

通气组织的发达程度与耐淹水的能力呈正相关。通常根系皮层中的气室数量最多,气室体积最大,占根系体积40%~50%[33];而从根系切片上看,在白骨壤(*A. marina*)气生根和粗根(Cable roots)的通气组织分别占根系横截面积的69%~80%和81%~85%[19]。因此,研究根系皮层及其气道面积占根系横截面积的比例、或通气组织在根系所占的比例是衡量不同红树植物不同根系贮气能力的有效手段[19,29,31]。

1.2 红树植物的氧气运送

测定根系中氧气含量的变化是确定根系内贮气能力的另一有效手段。Scholander等证明了涨潮时根系中的氧气含量下降,而退潮时氧气的含量恢复正常;在整个淹水的过程中,气生根内部的气压呈梯度下降,当根系露出水面后气压迅速恢复,而淹水后的低压能促使大量的空气在假根恢复呼吸时流入假根[8]。Allway等测定出潮水淹没白骨壤(*A. marina*)的气生根后,气生根中的气压下降到1.7kPa,其中氧气的含量也下降,约为3mol·m^{-3};但潮水退去,气生根干燥时,气压迅速回升到与大气相当,根系氧气含量也缓慢升高至与大气中的含量相当[29]。

氧气是通过气体扩散进入气生根的皮孔[32,34],因为在退潮时根系氧气浓度的变化与通过扩散作用进入气生根的氧气基本一致[34]。Andersen和Kristensen的研究表明气生根中的氧气含量占空气饱和度的63%~88%时,地下根系中的氧气含量只占62%~73%,从地上部分到地下部分氧气含量呈下降趋势[19]。Hovenden等通过计算证明:当持续淹水超过3.5h以上,白骨壤(*A. marina*)根系中的氧气就会下降至0[35]。通过氧气的储存和交换能力的计算发现:具有3~9个气生根的较大的白骨壤(*A. marina*)植株根系中贮存的氧气至少可以在淹水条件下保持6h的有氧呼吸[32]。

2 红树植物的生长分析

淹水能减缓植物的生长,促进根系分蘖、不定根增生、根系气腔形成[36~38]。但红树植物是一类生长在海岸潮间带的植被类型,生长过程需要一定的潮汐作用[17],适当的淹水对其生长有利[2]。但其植株淹水过深、淹水时间过长也会导致生长发育减慢、叶面积减少、生存率下降[2,5,35,39]。

淹水处理下的植株生物量及其分配和相对生长率(*RGR*)可以反映淹水胁迫响应的能力[5,40,41]。土壤表面淹水使白骨壤(*A. marina*)的总生物量和根系生物量显著下降[34];而萌芽白骨壤(*A. germinans*),拉关木(*Laguncularia racemasa*)和大红树(*R. mangle*)等3种红树植物的根系和叶片的生长都受到显著抑制,总叶面积减少,碳同化率下降[39]。

以大红树(*R. mangle*)为例,其幼苗在高潮带或低潮带的生长速率减缓,均低于在中潮带上的幼苗生长速率[42]。在水深较深的滩涂上种植的大红树(*R. mangle*)幼苗,胚轴萌发初期生长最快,当植株成苗(sapling)后其生长下降[5],同时地上部分的比例(root/shoot)升高,生长率下降[43];反之,水深较浅处的大红树(*R. mangle*)的植株矮小,分枝和叶片少,叶片C∶N比下降;而在正常潮位生长的植株,其生长指标均比水深较深和较浅的两个处理的植株高10%~20%,相对生长率(*RGR*)比其它两个处理高3%~23%[5]。种于低潮位的红海榄(*R. stylosa*)幼苗成活率低,生长不良[44]。淹水时间延长导致秋茄(*Kandelia candel*)植株趋向于减少根系的生物量的分配而增加幼枝的生物量的分配,秋茄根系将通过提高每单位根系生物量的营养吸收来维持较高的相对生长率(*RGR*)[45]。这与耐淹水植物中存在一种机制相一致,即当根系中生物量的分配降低,植株可以提高每单位根系的营养吸收来维持较高的相对生长率(*RGR*)[46]。

3 红树植物水分关系与光合呼吸作用

淹水导致红树植物叶片气孔关闭,RuBP羧化酶活性受到抑制,光合速率(P_n)下降[5,41,47]。叶片色素含量发生变化[45,47]。蒸腾速率(T_r)和气孔导度(g_s)也随淹水而发生变化,但对其水分利用率(*WUE*)没有显著影响[47,48]。Ellison 和 Farnsworth 对大红树(*R. mangle*)幼苗的研究结果表明:淹水较深的处理植株的光合速率(*Pn*)最低,而生长在中潮带的幼苗,气孔导度下降,光合速率极大增加,净光合速率比淹水较深的处理高 30 %[5]。对萌芽白骨壤(*A. germinans*),拉关木(*L. racemosa*)和大红树(*R. mangle*)[39],秋茄(*K. candel*)[47],白骨壤(*A. marina*)[48]和木榄(*Bruguiera gymnorriza*)[49]的研究表明:淹水使叶片的水势和气孔导度减少,光合作用受到极显著抑制。

与其他环境胁迫相似,淹水胁迫使红树植物植株的光合特性降低,CO_2 同化速率减缓,最终减缓植株生长[14,50,51]。大红树(*R. mangle*)受到高于正常潮位 16cm 的水深浸淹时,苗木的最大光合同化率下降,生长减缓[5]。红海榄(*R. stylosa*)在胚轴淹水的生境里,叶片光合同化速率减慢,幼苗生长缓慢[44]。

4 红树植物的膜脂过氧化系统和根系脱氢酶系统

4.1 红树植物的膜脂过氧化系统

当植物处于逆境时,活性氧等自由基能引起膜脂过氧化作用,从而破坏质膜,影响细胞的正常生理过程。但植物体内的抗氧化酶保护系统,能够防止自由基破坏的膜保护系统[52]。抗淹水耐性较强的植物,过氧化酶系统活性提高,保护植物免受氧化损伤[52,53]。

和湿地植物一样,红树植物的活性氧系统酶类有较高的活性,超氧化物歧化酶(SOD)和过氧化物酶(POD)的活性随淹水的增加而提高[17,36]。Takemura 等报道了木榄(*B. gymnorriza*)在环境胁迫下 SOD 酶活性升高[54];Ye 等发现随土壤淹水时间延长,秋茄(*K. candel*)叶片中的 POD 和 SOD 活性显著增高,而木榄(*B. gymnorriza*)叶片中仅 POD 活性有显著变化,并认为秋茄(*K. candel*)的抗淹水能力高于木榄(*B. gymnorriza*)[45]。Chen 等发现当周期性没顶淹水时间超过 8h,秋茄(*K. candel*)幼苗叶片的 POD 和 SOD 活性显著提高,幼苗抗性提高[47]。Youssef 和 Saenger 研究表明抗淹水能力强的红树幼苗气生根的抗氧化能力强[16]。与其他环境胁迫相似[36],膜脂过氧化性可以作为红树植物淹水胁迫的指示剂。

丙二醛(MDA)是膜脂过氧化的重要产物,会引起叶片的损伤[17,36]。淹水胁迫导致 MDA 含量升高,质膜受到损伤,膜透性升高,特别是根系细胞质膜透性增高,细胞内含物外渗[36],这一胁迫特性在秋茄(*K. candel*)幼苗叶片中也同样存在[47]。

4.2 红树植物的根系脱氢酶系统

对于旱生植物而言,缺氧迅速触发糖酵解过程,乳酸脱氢酶(LDH)活性迅速升高,乳酸发酵;当细胞质中的 pH 值达到 6.8 时,乙醇脱氢酶(ADH)与丙酮酸脱羧酶(PDC)被激活,进入乙醇发酵途径[17,23]。植物在缺氧时,将启动脱氢酶系统降解植物体内的毒性物质,如乳酸、乙醇以及苹果酸等毒害物质[10]。其中,ADH 是根系在厌氧条件下产生的主要酶,ADH 活性的增加可以将毒害物质乙醇转化成乙醛,帮助植物躲避缺氧根系的主要毒害物质——乙醇的损伤[10,41,55]。抗淹水植物的淹水耐性与 ADH 活性的变化是成正比的[10]。湿生植物,特别是红树植物的无氧呼吸功能强,酶活性高,在淹水条件下,根系迅速厌氧呼吸,ADH 活性迅速升高[17]。Pezeshki 等研究表明萌芽白骨壤(*A. germinans*)和大红树(*R. mangle*)的幼苗根系的 ADH 在淹水后有较大地提高[41];Chen 等研究表明秋茄(*K. candel*)幼苗在每个潮水周期没顶淹水 4～6h,根系的 ADH 活性较其他淹水处理高[47];但这也说明当幼苗期气生根还未形成时,幼苗根系中的充氧作用系统还不足以应付淹水时土壤条件的极度变化。另外,ADH 能通过提高巴斯德效应以维持较高的能荷,有利于保护膜结构功能的完整性,延长根系寿命[36]。

同时,缺氧引起根系呼吸速率和根系活力变化显著降低[36,37]。红树植物淹水胁迫的研究结果也证明了根系缺氧时,根系活力显著降低[45,47,56]。

5 红树植物的内源激素和胁迫多胺

淹水导致植物的一些生长发育特性改变,诸如气孔关闭、叶柄偏上性生长、叶片衰老、胚轴膨胀、通气组织形成、生长减缓,而这些变化主要受到内源激素的影响[10]。淹水胁迫诱导的几种激素中,乙烯是研究得比较透彻的激素之一,也是植物对淹水胁迫反应最敏感的激素之一[38]。但在红树植物淹水胁迫的研究中,胁迫乙烯的含量变化还未见报导。

厌氧条件下,根系中柱中的Met(甲硫氨酸)转化为SAM(S-腺苷蛋氨酸),在ACC(1-氨基环丙烷-1-羧酸)合成酶作用下生成ACC,扩散到低氧的根系皮层组织中,在需氧的ACC氧化酶作用下,ACC被氧化成乙烯。乙烯再通过木质部运送到地上部分而行使功效[10,38]。淹水胁迫导致植株体内乙烯含量增加,主要有3个来源途径:(1)淹水使乙烯合成能力加强;(2)淹水使体内乙烯扩散减少;(3)淹水增加土壤微生物产生外源乙烯[36]。乙烯含量增加使植株出现不定根增生、气腔形成、生长减慢、器官脱落以及衰老加速等症状[57],特别是能激活纤维素酶活性,溶解细胞壁,破坏细胞结构,导致细胞死亡,促进根系气腔形成[23],加速茎的韧皮部和木质部分化[58]。

淹水胁迫诱导的另一类重要激素就是脱落酸(ABA)。作为植物的生长调节剂,在正常植株中,ABA基本上是由地上部分合成的,特别是在老叶中合成,并向根系运送[17]。植株淹水后,地上部分ABA合成加速,并减少了向根系运输的数量[38]。ABA经过韧皮部运送到茎中,在茎中从韧皮部运送到木质部,再运送到根系[36],但淹水胁迫导致ABA合成增加的原因尚不清楚[10,38]。植株淹水后,ABA首先在成熟叶和老叶中合成,再转运到幼叶中[10],叶片中ABA含量发生变化,导致气孔关闭,蒸腾作用和呼吸速率均下降[18,23,36,59]。植物体内的ABA含量变化可以反映植物的抗涝性[59]。但与红树植物淹水胁迫相关的ABA的研究鲜见报道[47]。Chen等研究表明:淹水时间的延长显著促进了秋茄(*K. candel*)幼苗叶片中的ABA含量提高[47]。

多胺是一种低分子量含氮碱,一般认为它可作为广义上的植物激素,调节植物的生长发育,有助于提高植物的抗逆性[60]。植物中常见的多胺有腐胺(Put)、尸胺(Cad)、亚精胺(Spd)、精胺(Spm)以及其他胺类,Put是多胺生物合成途径的中心产物[61]。植物体内的多胺常以游离态、结合态和束缚态形式存在。结合态的多胺能形成分子屏障,抵御外界不良因素的侵染;束缚态多胺则通过大分子的交联稳定细胞内成分。由于多胺带多个正电荷,易与多价阴离子的核酸和质膜的磷脂相结合,改变膜透性。多胺还能转化成生物碱达到解毒作用[62]。

多胺代谢与胁迫乙烯的产生有一定的相关性,由于具有抑制衰老作用的亚精胺和精胺与促进衰老的乙烯具有相同的生物合成前体,即SAM(S-腺苷蛋氨酸),因此有研究认为乙烯和多胺在胁迫条件下竞争同一底物[63],是具有相反作用的调节物质[64]。多胺可能通过降低ACC合成酶的合成和清除自由基而抑制乙烯产生,达到抑制衰老的作用[65]。

在分生组织和生长细胞中,多胺的含量及多胺合成酶的活性最高,而在衰老组织中则最低;多胺的氨基越多,延缓衰老的活性越高,一般表现为Spm > Spd > Put[66]。同时在稳定膜结构、抑制核酸酶和蛋白酶活性的升高、抑制叶绿体的降解等方面,Spd和Spm的功能远大于Put[67]。多胺不仅可以通过抑制内源乙烯的合成而延缓衰老,还能抑制外源乙烯所诱导的衰老[68]。

在酸胁迫、渗透胁迫、盐胁迫和水分胁迫下,对植物内源多胺的含量变化及植物抗逆性有较多研究[70],还未见淹水胁迫下多胺的变化研究;特别是与红树植物淹水胁迫相关的多胺研究也未见报道。

6 总结与展望

由于红树植物的自然生境受到自然气候条件和人为干扰的影响,红树植物的淹水抗性的研究是关系到红树林造林成活率低的重要问题,越来越多地受到国内外学者的关注,因此,此项研究也将继续得到深远发展。通过本文的综述,对红树植物的淹水胁迫响应提出以下展望:

(1) 植物生长的研究 植物的生长研究是研究的基础,能直观的反映出红树植物在淹水胁迫下的生长状况。在红树植物的抗淹水研究中,生长的研究是不容忽视的。

(2) 激素水平的研究 基于对各种激素了解的不断加深,在红树植物胁迫研究的激素水平响应也应得到重视。特别是与淹水胁迫相关的内源激素调节和作为广义激素的内源多胺的研究,也将逐步得到重视。还可以将其与分子和蛋白质技术相结合,从激素水平进一步揭示红树植物的淹水胁迫响应的分子机理。

(3) 红树植物长期适应于水淹生境,不同红树种类在潮间带的分布受周期性淹水频率的影响,当淹水频率超过正常水平时,红树林将退化甚至死亡[76]。因此,红树植物特别是其幼苗在潮间带的浸淹频率将引起国内外学者的关注。Waston、de Haan 和 Chapman 都从淹水等级的角度研究了不同红树植物种类的淹水耐性,对红树植物耐淹水能力和在潮间带的自然分布做了较好的总结[77]。应用潮汐模拟装置,有效模拟自然潮汐作用下的红树植物的生长特征[2, 47],是研究红树植物的淹水胁迫响应的一种新的思路。

因此,在探讨红树植物幼苗的抗淹水机制,确定红树植物在潮间带的浸淹频率和临界高程,划定不同红树植物种类的宜林临界线具有一定的指导意义,将对红树林造林具有重要的实践意义。

References:

[1] Lin P. Mangrove ecosystem in China. Beijing: Science Press, 1999.

[2] Chen L Z, Wang W Q, Lin P. Influence of waterlogging time on the growth of *Kandelia candel* seedlings. Acta Oceano. Sinica, 2004, 23(1): 149～158.

[3] Snedaker S C, Meeder J F, Ross M S, *et al*. Mangrove ecosystem collapse during predicted sea-level rise — Holocene analogues and implications — discussion. J Coast Res., 1994, 10: 497～498.

[4] Field C D. Impact of expected climate change on mangroves. Hydrobiologia, 1995, 295: 75～82.

[5] Ellison A M, Farnsworth E J. Simulated sea level change alters anatomy, physiology, growth, and reproduction of red mangrove (*Rhizophora mangle* L.). Oecologia., 1997, 112: 435～446.

[6] Chen X Y, Lin P. Responses and roles of mangroves in China to global climate changes. Transactions of Oceanology and Limnology, 1999, 2: 11～16.

[7] Fan H Q, Li G Z. Effect of sea dike on the quantity, community characteristic and restoration of mangroves forest along Guangxi coast. Chinese J. of Applied Ecol, 1997, 8: 240～244.

[8] Scholander P F, Van Dam L, Scholander S I. Gas exchange in the roots of mangroves. Amer. J. Bot., 1995, 42: 92～98.

[9] Scholander P F, Hammerl H T, Hemmingen E, *et al*. Salt balance in mangroves. Plant Physiol, 1962, 37: 722～729.

[10] Kozlowski T T. Flooding and plant growth. London: Academic Press, INC, 1984.

[11] Tomlinson P B. The botany of mangroves. Cambridge: Cambridge University Press, 1986.

[12] Armstrong W, Brändle R, Jackson M B. Mechanisms of flood tolerance in plants. Acta. Bot. Neerl., 1994, 43: 307～358.

[13] Ball M C. Ecophysiology of mangroves. Trees, 1988a, 2: 129～142.

[14] Ball M C. Salinity tolerance in the mangroves *Aegiceras corniculatum* and *Avicennia marina* I. Water use in relation to growth, carbon partitioning and salt balance. Aust. J. Plant Physiol, 1988, 15: 447～464

[15] Pezeshki S R, DeLaune R D, Patrick W H. Differential response of selected mangroves to soil flooding and salinity: Gas exchange and biomass partitioning. Can. J. For. Res., 1990, 20: 869～874.

[16] Youssef T, Saenger P. Anatomical adaptive strategies to flooding and rhizophere oxidation in mangrove seedlings. Aust. J. Bot., 1996, 44, 297～313.

[17] Wang W Q, Zhang F S. The physiological and molecular mechanism of adaptation to anaerobiosis in higher plants. Plant Physiology Communications, 2001, 37, 63～70.

[18] Lin P. Mangrove vegetation. Beijing: Ocean Press, 1988.

[19] Andersen F Ø, Kristensen E. Oxygen microgradients in the rhizosphere of the mangrove *Avicennia marina*. Mar. Ecol. Prog. Ser., 1988, 44: 201～204.

[20] Chapman V J. Mangrove vegetation. Cramer. Vaduz, 1976.

[21] Hovenden M J, Allaway W G. Horizontal structures on pneumatophores of *Avicennia marina* (Forsk.) Vierh — a new site of oxygen conductance. Ann. Bot., 1994, 73: 377～383.

[22] Yu S W, Tang Z C. Plant physiology and molecular biology. Beijing: Science Press, 1998. 739～751.

[23] Pan R Z. Plant physiology. Beijing: Chinese High Education Press, 2001.

[24] Zhao K F, Wang S T. Crops resistance physiology. Beijing: Agriculture Press, 1900. 226～248.

[25] Drew M C, He C J, Morgan P W. Programmed cell death and aerenchyma formation in roots. Trens in Plant Sci., 2000, 5(3): 123～127.

[26] Armstrong W. Rhizosphere oxidation in rice and other species: a mathematical model based on the oxygen flux component. Physiol. Plant., 1970, 23: 623～630.

[27] Smirnoff N, Crawford R M M. Variation in the structure and response to flooding of root aerenchyma in some wetland plants. Ann. Bot., 1983, 51: 237～249.

[28] Armstrong W. Justin S H F W, Beckett P M, *et al*. Root adaptation to soil waterlogging. Aquatic Botany, 1991, 39: 57～73.

[29] Allaway W G, Curran M, Hollington L M, Ricketts M C, Skelton N J. Gas space and oxygen exchange in roots of *Avicennia marina* (Forssk.) Vierh. var. australasica (Walp.) Moldenke ex N. C. Duke, the Grey mangrove. Welands Eco. & Management, 2001, 9: 211～218.

[30] Curran M. Gas movements in the roots of *Avicennia marina* (Forsk.) Vierh. Aust. J. Plant Physiol, 1985, 9: 519～528.

[31] Ashord A E, Allaway W G. There is a continuum of gas space in young plants of *Avicennia marina*. Mar. Ecol. Progr. Ser., 1995, 295, 5～11.

[32] Curran M, Cole M, Allaway W G. Root aeration and respiration in young mangrove plants (*Avicennia marina* (Forsk.) Vierh.. J. Exp. Bot., 1986, 37: 1225～1233.

[33] Curran M, James P, Allaway W G. The measurement of gas spaces in the roots of aquatic plants ——Archimedes revistited. Aqua. Bot., 1996, 54: 255～261.

[34] Skelton N J, Allaway W G. Thermo-osmotic gas supply not detected in *Avicennia marina* seedlings. Hydrobiologia, 1995, 295: 1～4.

[35] Hovenden M J, Curran M, Cole M A, Coulter P F E, *et al*. Ventilation and respiration in roots of one-year-old seedlings of grey mangrove *Avicennia marina* (Forsk.) Vierh. Hydrobiologia, 1995, 295, 23～29.

[36] Guan Z H. Introduction of plant iatrology. Beijing: China Agriculture University Press, 1996.

[37] Liu YL. Physiology of plant water stress. Beijing: Agriculture Press, 1992.

[38] Li R Q, Wang J P. Plant stress physiology. Wuhan: Wuhan University Press, 2002.

[39] Pezeshki S R, Delaune R D, Patrick W H Jr. Differential response of selected mangroves to soil flooding and salinity: gas exchanges and biomass partitioning. Can. J. For. Res., 1989, 20, 869～874.

[40] Naidoo G. Effects of waterlogging and salinity on plant-water relations and on the accumulation of solutes in three mangrove species. Aquat. Bot., 1985, 22: 133～143.

[41] Pezeshki S R, Delaune R D, Meeder J F. Carbon assimilation and biomass partitioning in *Avicennia germinans* and *Rhizophora mangle* seedlings in response to soil redox conditions. Environ. Exp. Bot., 1997, 37: 161～171.

[42] Ellison A M, Farnsworth E J. Seedling survivorship, growth, and response to disturbance in *Belizean mangal*. Am. J. Bot., 1993, 80: 1137～1145.

[43] Ellison A M, Farnsworth E J. Spatial and temporal variability in growth of *Rhizophora mangle* saplings on coral cays: links with variation in insolation, herbivory, and local sedimentation rate. J. Ecol., 1996, 84: 717～731.

[44] Kitaya Y, Jintana V, Piriyayotha S, *et al*. Early growth of seven mangrove species planted at different elevations in a Thai estuary. Trees, 2002, 16, 150～154.

[45] Ye Y, Tam N F Y, Wong Y S, *et al*. Growth and physiological responses of two mangrove species (*Bruguiera gymnorrhiza* and *Kandelia candel*) to waterlogging. Environ. Exp. Bot., 2003, 49: 209～221.

[46] Rubio G, Oesterheld M, Alvarez C R, Lavado R S. Mechanisms for the increase in phosphorus uptake of waterlogged plants: Soil phosphorus availability, root morphology and uptake kinetics. Oecologia, 1997, 112: 150～155.

[47] Chen L Z, Wang W Q, Lin P. Photosynthetic and physiological responses of *Kandelia candel* (L.) Druce seedlings to duration of tidal immersion in artificial seawater. Environ. Exp. Bot., 2005, 54: 256～266.

[48] Naidoo G, Rogalla H, Von-Willert D J. Gas exchange responses of a mangrove species, *Avicennia marina*, to waterlogged and drained conditions. Hydrobiologia, 1997, 352: 39～47.

[49] Naidoo G. Effects of flooding on leaf water potential and stomatal resistance in *Bruguiera gymnorrhiza* (L.) Lam. New Phytol., 1983, 93: 369～376.

[50] Ball M C, Farquhar G D. Photosynthetic and stomatal responses of two mangrove species, *Aegiceras corniculatum* and *Avicennia marina*, to long term salinity and humidity conditions. Plant Physiol, 1984, 74: 1～6.

[51] Lin G, Sternberg L da S L. Effects of salinity fluctuation on photosynthetic gas exchange and plant growth of the red mangrove (*Rhizophora mangle* L.). J. exp. Bot., 1993, 90: 9～16.

[52] Bowler C, Montagu M V, Inze D. Superoxide dismutase and stress tolerance. Annu. Rev. Plant Physiol. Plant Mol. Biol., 1992, 43: 83～116.

[53] Monk L S, Fagerstedt K V, Crawford R M M. Superoxide dismutase as anaerobic polypeptide. A key factor in recovery from oxygen deprivation in Iris pseudacorus. Plant Physiol., 1987, 85: 1016～1020.

[54] Takemura T, Hanagata N, Sugihara K, *et al*. Physiological and biochemical responses to salt stress in the mangrove, *Bruguiera gymnorrhiza*. Aquat. Bot., 2000, 68:15～28.

[55] Akhtar J, Gorham J, Qureshi R H, *et al*. Does tolerance of wheat to salinity and hypoxia correlate with root dehydrogenase activities of aerenchyma formation. Plant and Soil, 1998, 201: 275～284.

[56] McKee KL. Growth and physiological respones of neotropical mangroves seedlings to root zone hypoxia. Tree Physiol, 1996, 16: 883～889.

[57] Dong J G, Yu Z W, Yu S W. Effect of increased ethylene production during different periods on the resistance of wheat plants to waterlogging. Acta Phytophysiologia Sinica, 1983, 9(4): 383～389.

[58] Shen H J. Plant hormone and wood formation. Sci. Silvae Sinicae, 1996, 32(2): 165～170.

[59] Li Z T, Zhou X. Plant hormone and immunoassay. Nanjing: Jiangsu Science and Technology Press, 1996.

[60] Galston A W. Polyamines as modulaters of plant development. Bioscience, 1983, 33: 382～388.

[61] Shen H J, Xie Y F. Polyamine and plant stress. J. Nanjing Forestry University, 1997, 21(4): 26~30.

[62] Zhao F G, Liu YL. Metabolism and regulation of uncommon polyamines in high plants. Plant Physiology Communications, 2000, 36(1): 1~6.

[63] Wang S Y, Steffens GL. Effect of paclobutrazol on water stress —induced ethylene biosynthesis and polyamine accumulation in apple seedling leaves. Phytochemistry, 1985, 24: 2185~2190.

[64] Kaur-Sawhney R, Shin L M, Flores H E. Relation of polyamine synthesis and titer to aging and senescence in oat leaves. Plant Physiol, 1982, 69: 405~415.

[65] Drolet G, Dumbroff E B, Legg R, *et al*. Radical scavenging properties of polyamines. Phytochem, 1986, 25: 367~371.

[66] Wang X Y, Zou Q. Advances in studies on relation ship between polyamines and plant senescence. Chinese Bulletin Botany, 2002, 19(1): 11~20.

[67] Ting D T. Stress ethylene production —A measure of plant response to stress. Hortscience, 1980, 15(5): 16~19.

[68] Fuhrer J, Kaur-Sawhney R, Shih L M, *et al*. Inhibition of ethylene biosynthesis by aminoethoxyvinylglycine and by polyamines shunts lavel from Cl_4-methionine into spermidine in aged orange peel discs. Plant Physiol., 1982, 70: 1597~1600.

[69] Zhao F G, He L F, Luo Q Y. Plant stress physiological ecology. Beijing: Chemical Industry Press and Environmental Science and Ecology Press, 2004. 193~216.

[70] Wasinger V C, Cordwell S J, Cerpa-Poljak A, *et al*. Progress with gene-product mapping the mollicutes: mycoplasma genitalium. Electrophoresis, 1995, 16: 1090~1094.

[71] Sachs M M, Freeling M, Okimoto R. The anaerobic proteins of maize. Cell, 1980, 20:761~767.

[72] Andrews D L, Cobb B G, Johnson J R. Hypoxicandanoxic induction of alcohol dehydrogenase in root sand Zea may. Plant Physiol., 1993, 101: 407~414.

[73] Zeng Y, Wu Y, Avigne W T. Differential regulation of sugar sensitive sucrose synthase by hypoxia and anoxia indicate complementary transcriptional and post translational response. Plant Physiol., 1998, 116: 1573~1583.

[74] Mujer C V, Rumpho M E, Lin J J. Constitiutive and inducible aerobic and anaerobic stress proteins in the Echinochloa complex and rice. Plant Physiol., 1993, 101:217~226.

[75] Chang W W, Huang L, Shen M. Patterns of synthesis and tolerance of anoxia in root tips of maize seedlings acclimated to alow oxygen environment, and identification of proteins by mass spectrometry. Plant Physiology, 2000, 122: 3517~3526.

[76] Zhang Q M, Yu H B, Chen X S, *et al*. The relationship between mangrove zone on tidal flats and tidal levels. Acta Ecologica Sinica, 1997, 17: 258~265.

[77] Snedaker S C, Snedaker J G. The mangrove ecosystem: research methods. UK: Unesco Press, 1984.

参考文献:

[6] 陈小勇,林鹏.我国红树林对全球气候变化的响应及其作用.海洋湖沼通报,1999,2:11~16.

[7] 范航清,黎广钊.海堤对广西沿海红树林的数量、群落特征和恢复的影响.应用生态学报,1997,8(3):240~244.

[17] 王文泉,张福锁.高等植物厌氧适应的生理及分子机制.植物生理学通讯,2001,37(1):63~70.

[22] 余叔文,汤章城主编.植物生理与分子生物学.北京:科学出版社,1998.739~751.

[23] 潘瑞炽主编.植物生理学.北京:高等教育出版社,2001.

[24] 赵可夫.王韶唐主编.作物抗性生理.北京:农业出版社,1990.226~248.

[36] 管致和编.植物医学导论.北京:中国农业大学出版社,1996.100~104.

[37] 刘友良.植物水分逆境生理.北京:农业出版社,1992.144~187.

[38] 利容千,王建波主编.植物逆境细胞及生理学.武汉:武汉大学出版社,2002.

[57] 董建国,俞子文,余叔文.在渍水前后的不同时期增加体内乙烯产生对小麦抗渍性的影响.植物生理学报,1983,9(4):383~389.

[58] 沈惠娟.植物激素与木材形成.林业科学,1996,32(2):165~170.

[59] 李宗庭,周燮著.植物激素及其免疫检测技术.南京:江苏科学技术出版社,1996.114~203.

[61] 沈惠娟,谢寅峰.多胺(Pas)与植物的几种胁迫反应.南京林业大学学报,1997,21(4):26~30.

[62] 赵福庚,刘友良.高等植物体内特殊形态多胺的代谢及调节.植物生理学通讯,2000,36(1):1~6.

[66] 王晓云,邹琦.多胺与植物衰老关系研究进展.植物学通报,2002,19(1):11~20.

[69] 赵福庚,何龙飞,罗庆云.植物逆境生理生态学.北京:化学工业出版社环境科学与工程出版中心,2004.193~216.

[76] 张乔民,于红兵,陈欣树,郑德璋.红树林生长带与潮汐水位关系的研究.生态学报,1997,17(3):258~265.

Leaf anatomical responses to periodical waterlogging in simulated semidiurnal tides in mangrove *Bruguiera gymnorrhiza* seedlings*

Wenqing Wang[a,b] Yan Xiao[a] Luzhen Chen[a] Peng Lin[a]

(a. *School of Life Sciences, Xiamen University, Xiamen* 361005, *PR China*;

b. *State Key Laboratory of Marine Environmental Science (Xiamen University), Xiamen* 361005, *PR China*)

Abstract

Leaf anatomical changes of *Bruguiera gymnorrhiza* (L.) Lamk seedlings grown in experimental equipment that simulated semidiurnal tides with salinities of 15‰ under greenhouse conditions were studied. Compared with the 0 h treatments, leaf thickness, palisade parenchyma thickness, spongy parenchyma thickness, palisade–spongy thickness ratio, xylem length of the vascular system and number of vessels and vessel lines under the 12 h treatments declined 31.9%, 59.1%, 21.7%, 47.1%, 48.9%, 67.1% and 51.6%, respectively. However, the upper and lower epidermis to leaf thickness ratio, upper and lower hypodermis to leaf thickness ratio and stomatal density of 12 h treatments showed increases of 47.9%, 50.9%, 14.3%, 21.4% and 104.3% over those of 0 h treatments, respectively. The cuticle to leaf thickness ratio (inundated for 0–6 h) decreased significantly with waterlogging duration at first and then increased. Moreover, the percentage of intercellular spaces in spongy tissue decreased from 4 to 10 h treatment and then tended to increase by nearly 20% in the 12 h treatment. Tannin cells that were distributed in the vascular tissue, crystalliferous cells and phloem fibers were more abundant in the short-duration waterlogging treatments than in the long-duration waterlogging treatment. It was concluded that significant changes in the leaf anatomical features as a result of periods of immersion would have come at the cost of reduction of photosynthesis and water transport when waterlogging duration was longer than 2 h. These anatomical characteristics further proved that *B. gymnorrhiza* had a relatively low tolerance to waterlogging at the seedling stage.

Keywords: *Bruguiera gymnorrhiza*; Inundation; Mangrove; Leaf; Anatomy; Waterlogging

1. Introduction

Mangroves are likely to be one of the first ecosystems to be affected by global changes because of their location at the interface between land and sea. As sea level rises, accompanying the rapid climate changes (Field, 1995; Ellison and Farnsworth, 1997), mangroves tend to retreat landwards. On the other hand, the prolonged waterlogging duration caused by land enclosure and jetty construction leads to a low survival rate of mangrove seedlings in China (Fan and Li, 1997). Waterlogging is one of the most serious problems that confront mangroves in their habitats. Consequently, the growth, physiology, morphology, anatomy and reproduction of mangrove under waterlogged conditions are widely observed (Naidoo et al., 1997; Ellison and Farnsworth, 1997; Kitaya et al., 2002; Ye et al., 2003; Chen et al., 2004, 2005). Generally, mangroves are highly adaptable to aquatic environments in terms of their anatomy and morphology; for example, they have specialized roots for gas exchange when living in an anaerobic environment, and the wood anatomical features associated with mangrove forests in a coastal environment have been studied extensively (Tomlinson, 1986; Hovenden and Allaway, 1994; Youssef and Saenger, 1996; Allaway et al., 2001; Yáñez-Espinosa et al., 2001; Angeles et al., 2002; Deng et al., 2004). Morphologically and anatomically, the leaf is the most variable plant organ (Fahn, 1983); leaves of most mangroves exhibit a range of xeromorphic features that have probably developed in response to the physiological drought conditions of the mangrove environment, including thick cuticle; wax coatings; hairs or epidermal outgrowths; sunken stomata; the distribution of cutinized and sclerenchymatous cells throughout the leaf; well-developed, large celled, water-storing hypodermis; strongly developed palisade mesophyll; and, generally, small intercellular volumes (Rao and Tan, 1984; Saenger, 2002). Unfortunately, very little is known about the anatomical

* From Aquatic Botany, 2007, 86: 223—228

responses of mangrove seedling leaves under waterlogging conditions. *Bruguiera gymnorrhiza*, of the family Rhizophoraceae, is distributed naturally in high tidal zones, and is one of mangrove types of the later succession stage of mangroves (Lin, 1999). It is also one of the main species used for mangrove silviculture in China. The experimental equipment that simulates semidiurnal tide is applied to obtain the mature leaf anatomy features of *B. gymnorrhiza* in order to study the tendencies and mechanisms of tolerance to waterlogging.

2. Materials and methods

2.1. Experimental design

The experimental design is referred to by Chen et al. (2004). Seven plastic tanks acting as artificial-tidal tanks simulating the semidiurnal tides were arranged, and inundation periods in the these tanks were 12, 10, 8, 6, 4, 2 and 0 h, respectively. There were two tide cycles each day. There were four pots in each tank, each being 25 cm tall and 25 cm in diameter, with a small hole at the bottom to allow rapid drainage while water in the tanks was drained away. Each pot was filled with washed river sand (diameter = 1 mm). Three sets of equipment acted as three replicates.

2.2. Plant materials and culture conditions

Mature hypocotyls of *B. gymnorrhiza* were collected from the Jiulong River Estuary in Fugong Town, Longhai County, Fujian Province of China (24°29′N, 117°55′E). The average seawater salinity there was 17‰ (Lin, 1999). Five hypocotyls were planted in each pot and the seedlings were periodically submerged under diluted natural seawater with a salinity of 15‰ (seawater from the west coast of Xiamen of 22–28‰ in salinity was diluted by tap water). Tap water was added daily to compensate for evaporation losses and the seawater was renewed weekly. All seedlings were grown in a greenhouse with an air temperature of 27–32 °C, where irradiance level was approximate 1000–1200 μmol m^{-2} s^{-1} at full sun. Seedlings were flooded at 'high tide' to a maximum depth of 60 cm above the bottom of the tank, and at 'low tide' the water was slightly below the sand level.

2.3. Measurement of anatomical features of leaves

After 100 days in culture, mature leaves (first pair from the shoot top) were fixed in formalin–alcohol–glacial acetic acid. Samples were dehydrated in an alcohol series, cleared in xylene and embedded in paraffin. Transverse sections of 10 μm thickness were obtained using a rotary microtome, and stained with safranin-fast green. Sections of leaf samples were photographed under light microscope (OLYMPUS BX41) and digitized (OLYSIA BioReport software) to determine the thickness of the total leaf, cuticle, epidermis, hypodermis, palisade parenchyma, spongy parenchyma, vessel wall, and the tangential and radial diameters of vessels (maximum in each line), xylem length of main vein (maximum in each field), palisade cell length and width and stomatal density. The number of crystals, vessels and vessel lines distributed in the main vein were counted under a light microscope. The number of crystals was calculated in the field of 456 μm × 341 μm. The percentage of intercellular spaces in spongy tissue was quantified by using the photographs of each treatment. Ten measurements per sample were carried out in 10 different fields, respectively. All the measurements were made on sections that were taken from the main vein of the lamina midway between the leaf tip and base.

2.4. Statistical analysis

Mean and standard deviation (S.D.) values of three replicates were calculated. The least significant differences at $P \leq 0.05$, 0.01 and 0.001 were determined by one-way ANOVA. The Tukey multiple comparison method was used to analyze the difference among the seven treatments and Pearson correlation analysis was applied to evaluate association between anatomical features and waterlogging duration. All statistical analysis was performed with SPSS 11.0 software.

3. Results

3.1. Epidermal tissue system

There was a positive correlation between stomatal density and duration of waterlogging ($P = 0.000$, $R = 0.737$). A

Table 1
Leaf anatomical characteristics of epidermal tissue system of *Bruguiera gymnorrhiza* seedlings under periodical waterlogging (mean ± S.D.)

Waterlogging time (h)	Upper epidermis thickness (μm)	Lower epidermis thickness (μm)	Upper hypodermis thickness (μm)	Lower hypodermis thickness (μm)	Upper cuticle thickness (μm)	Lower cuticle thickness (μm)	Stomatal density (mm^{-2})
0	15.2 ± 2.2	15.7 ± 2.7	17.6 ± 2.7	16.3 ± 2.5	5.1 ± 1.0	3.9 ± 0.6	184.8 ± 52.3
2	17.4 ± 2.6	16.2 ± 1.5	21.5 ± 3.3	18.0 ± 2.7	3.4 ± 0.5	2.2 ± 0.4	234.6 ± 41.7
4	17.3 ± 1.6	14.6 ± 1.8	19.1 ± 2.3	15.3 ± 2.5	3.2 ± 0.8	2.1 ± 0.5	278.3 ± 42.2
6	18.2 ± 2.6	15.7 ± 1.5	21.8 ± 3.6	17.2 ± 1.9	2.4 ± 0.3	1.4 ± 0.4	367.9 ± 52.6
8	17.1 ± 2.3	15.6 ± 2.1	26.8 ± 5.9	17.5 ± 3.0	3.5 ± 0.5	2.0 ± 0.3	397.7 ± 61.2
10	15.10 ± 2.0	15.1 ± 2.0	22.7 ± 2.9	18.1 ± 2.9	3.6 ± 1.0	2.5 ± 0.4	340.4 ± 48.7
12	15.1 ± 2.1	15.3 ± 1.7	22.5 ± 7.5	19.7 ± 2.8	3.2 ± 0.7	3.0 ± 0.6	377.5 ± 30.4
F-value	10.355***	2.206 (ns)	13.394***	8.285***	39.524***	87.587***	84.682***

ns, not significant.
*** $P \leq 0.001$.

stomatal density of 377.5 ± 30.4 mm^{-2} occurred in the plants that had been treated for 12 h and was considerably lower in the plants that had been treated for 0 h (Table 1). The upper epidermis thickness varied over the range 0–12 h immersion without significant regularity, but the lower epidermis thickness showed no significant differences among the treatments (Table 1). However, the upper and lower epidermis to leaf thickness ratio increased significantly with prolonged waterlogging duration ($P = 0.000$ and 0.000) (Fig. 1A). The hypodermis thickness changed without significant regularity too (Table 1), but the upper and lower hypodermis to leaf thickness ratio increased significantly with prolonged waterlogging duration ($P = 0.000$ and 0.000) (Fig. 1B). The upper and lower cuticle thickness of the 0–6 h waterlogging treatments declined significantly with increasing duration of waterlogging ($P = 0.000$ and 0.000) at first and then increased significantly from the 6 h to the 12 h treatments ($P = 0.000$ and 0.000) (Table 1). The upper and lower cuticle thicknesses of plants under the 6 h treatments were 52.8% and 65.1% lower than those of the 0 h treatments, and were 24.2% and 54.9% lower than those of the 12 h treatments, respectively. Moreover, cuticle to leaf thickness ratio exhibited a similar trend to cuticle thickness (Fig. 1C).

3.2. Mesophyll

Negative correlations were found between leaf thickness and palisade–spongy ratio and waterlogging duration ($P = 0.000$, $R = -0.828$; $P = 0.000$, $R = -0.771$) (Table 2). Palisade and spongy parenchyma thickness decreased significantly with prolonged waterlogging duration (Table 2); leaf thickness, palisade parenchyma thickness, spongy parenchyma thickness and palisade–spongy ratio of plants under the 0 h treatment were 20.1%, 47.2%, 10.6% and 32.8% thicker than those of the 2 h treatments, respectively. Palisade cell width remained fairly constant in all treatments ($P = 0.175$), but palisade cell length declined significantly in response to inundation (Table 2), and that of plants under the 2 h treatment was 43.2% lower than that of those under the 0 h treatments. In the 12 h treatment, the palisade cells were elliptically shaped and arranged in 2–3 layers, whereas those of the 0–6 h treatments were rod-shaped and appeared to be arranged in 3–4 layers relatively tightly. The percentage of intercellular spaces in the spongy tissues of the 0–4 h treatments showed no significant difference ($P = 0.338$). In the 4–10 h treatment, intercellular spaces declined with prolonged waterlogging duration at first but then tended to increase by nearly 20% in the 12 h treatment (Table 2).

3.3. Vascular system

Significant negative linear correlations were found between the xylem length of the vascular system, the number of vessels and vessel lines, and waterlogging duration ($P = 0.000$, $R = -0.863$; $P = 0.000$, $R = -0.953$; $P = 0.000$, $R = -0.872$) (Table 3). All of the minimum values occurred in the 12 h treatments, and increased by 95.7%, 203.5% and 106.7% in the

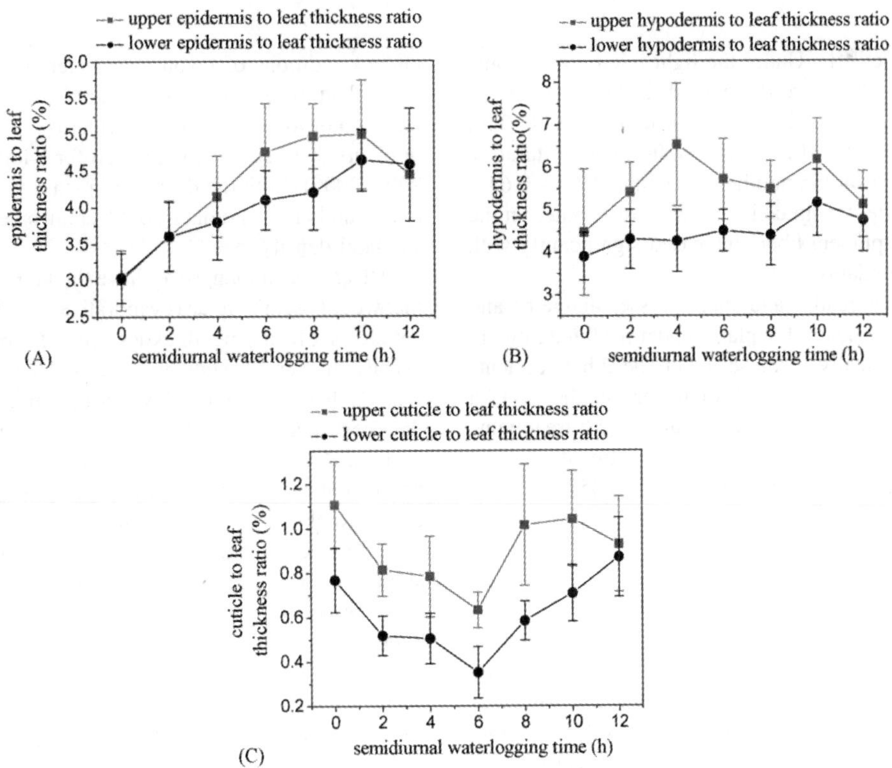

Fig. 1. Anatomical features of *Bruguiera gymnorrhiza* seedlings under periodical waterlogging grown for 100 days. (A) Epidermis to leaf thickness ratio; (B) hypodermis to leaf thickness ratio; (C) cuticle to leaf thickness ratio.

Table 2
Leaf anatomical characteristics of mesophyll tissue system of *Bruguiera gymnorrhiza* seedlings under periodical waterlogging (mean ± S.D.)

Waterlogging time (h)	Leaf thickness (μm)	Palisade parenchyma thickness (μm)	Spongy parenchyma thickness (μm)	Palisade–spongy ratio (%)	Palisade cell length (μm)	Palisade cell width (μm)	Intercellular spaces in spongy tissue (%)
0	502.8 ± 56.0	150.9 ± 20.5	280.3 ± 33.8	54.0 ± 5.2	55.8 ± 10.8	19.7 ± 3.5	29.4 ± 2.4
2	418.6 ± 15.5	102.5 ± 7.0	253.4 ± 17.6	40.6 ± 4.2	31.7 ± 9.1	20.1 ± 3.8	30.3 ± 2.2
4	410.1 ± 10.9	103.4 ± 7.6	240.7 ± 12.5	43.1 ± 3.9	33.7 ± 5.9	20.3 ± 2.8	29.4 ± 3.2
6	381.9 ± 21.3	87.8 ± 15.2	230.4 ± 13.7	38.2 ± 7.0	30.5 ± 10.8	18.6 ± 3.3	25.4 ± 3.8
8	347.8 ± 21.1	78.7 ± 5.0	252.3 ± 37.0	32.1 ± 6.5	27.8 ± 9.0	18.5 ± 3.2	21.0 ± 2.3
10	348.3 ± 11.4	76.5 ± 7.0	228.4 ± 13.4	33.7 ± 4.2	28.8 ± 6.9	19.2 ± 3.7	20.9 ± 2.8
12	342.8 ± 15.7	61.7 ± 6.0	219.5 ± 23.3	28.6 ± 4.9	22.3 ± 5.6	18.5 ± 3.1	25.0 ± 3.5
F-value	144.485***	204.651***	22.621***	77.583***	46.872***	1.515 (ns)	53.614***

ns, not significant.
*** $P \leq 0.001$.

Table 3
Leaf anatomical characteristics of vascular system of *Bruguiera gymnorrhiza* seedlings under periodical waterlogging (mean ± S.D.)

Waterlogging time (h)	Tangential vessel diameter (μm)	Radial vessel diameter (μm)	Vessel wall thickness (μm)	Xylem length (μm)	Number of vessels	Number of vessel lines	Number of crystals
0	16.5 ± 2.6	17.7 ± 1.8	2.6 ± 0.3	117.8 ± 6.2	142.8 ± 7.6	20.7 ± 1.0	7.9 ± 4.2
2	12.9 ± 1.8	13.6 ± 1.9	1.9 ± 0.2	89.7 ± 7.8	132.9 ± 13.1	21.3 ± 1.3	8.2 ± 2.6
4	12.8 ± 1.9	13.4 ± 2.4	1.8 ± 0.3	81.5 ± 1.03	109.3 ± 11.9	17.7 ± 1.3	4.9 ± 3.1
6	11.4 ± 1.3	12.4 ± 1.8	1.8 ± 0.3	76.5 ± 8.3	102.3 ± 9.4	16.3 ± 2.5	3.8 ± 2.6
8	10.8 ± 1.9	11.7 ± 1.2	1.7 ± 0.3	66.4 ± 7.2	76.7 ± 10.0	15.3 ± 2.4	3.6 ± 3.6
10	12.0 ± 2.0	12.3 ± 1.6	1.8 ± 0.3	64.4 ± 5.8	66.7 ± 8.7	12.7 ± 2.1	2.3 ± 2.8
12	11.2 ± 2.1	10.7 ± 2.2	1.7 ± 0.3	60.2 ± 5.3	47.0 ± 6.0	10.0 ± 0.8	0.6 ± 1.0
F-value	28.193***	42.908***	38.027***	212.065***	384.3975***	163.866***	26.587***

*** $P \leq 0.001$.

0 h treatment, respectively. Under the light microscope, some crystalliferous cells that were distributed in the midrib were observed, especially in the short-duration treatments (0–2 h). Correlation analysis showed prolonged waterlogging duration had a negative effect on the number of crystals ($R = -0.641$) (Table 3). Meanwhile, tannin cells that were distributed in the vascular tissue and phloem fibers increased significantly with the reduction of inundation.

The maximum tangential and radial vessel diameter and vessel wall thickness occurred in plants under the 0 h treatment and declined dramatically in those under the 2 h treatments (Table 3). The tangential vessel diameter of the 6–12 h treatments, the radial vessel diameter of the 8–12 h treatments and the vessel wall thickness of the 4–12 h treatments showed no significant differences, respectively ($P > 0.05$, 0.05 and 0.05).

4. Discussion

In the present research, there was a positive correlation between stomatal density and waterlogging duration (Table 1). Similar to previous findings, in *B. gymnorrhiza*, stomatal densities significantly increased with a water level rise in fine soils (Ye et al., 2004). Plants growing in permanently damp environments have more, but smaller, stomata than those growing in dry habitats (Stephen et al., 1983). The increases in the number of stomata allow a higher rate of gas exchange under conditions of favourable water supply (Fahn, 1983). A parallel study found that stomatal conductance and the mature leaf area of *Kandelia candel* (L.) Druce decreased significantly in plants that had been inundated for a long time (Chen et al., 2004, 2005). Both the decrease in leaf area and the increase in total number of stomata could contribute to the increase in stomatal density.

All species of mangroves have thick, waxy, lamellar cuticles that would seem to retard evaporative loss (Saenger, 2002). The cuticle is of varying thickness in different plants, and it is usually thicker in plants growing in dry habitats. As shown in Fig. 1C, between 0 and 6 h waterlogging duration, both cuticle thickness and cuticle to leaf thickness ratio declined with prolonged inundation, which might result from the relatively dry habitat due to reduction of waterlogging duration. However, cuticle thickness and cuticle thickness ratio increased from 6 to 12 h treatments. According to Fahn (1983), the structure and amount of surfaces wax has an extremely important effect on the degree to which a surface can be wetted. It was hypothesized that the extremely long period of inundation stimulated the thicker cuticle and epicuticular waxes, which decreased the influx of water into submerged leaves; moreover, the plants might improve their capacity to withstand mechanical impact during inundation.

A well-developed hypodermis is a conspicuous xeromorphic feature of many mangrove leaves and various functions have been ascribed to it, including water storage, salt accumulation or osmoregulation, mesophyll protection via light back-scattering and/or heat dissipation, nutrient conservation, and preventing fungal infestations via tannin content (Saenger, 2002). Compared with other mangrove species, the hypodermis is less prominent in *Bruguiera* (Rao and Tan, 1984). Based on the increase of epidermis and hypodermis to leaf thickness ratio in the present study (Fig. 1A and B), it was predicted that these changes might be responsible to the 'physiological drought' that accompanies the prolonged immersion duration.

Leaf thickness, palisade parenchyma thickness, spongy parenchyma thickness, palisade cell length and palisade–spongy ratio decreased significantly with prolonged inundation. All of the maximum values occurred in plants under the 0 h treatments and these values were inhibited significantly when waterlogging duration was longer than 2 h (Table 2). The specialization of the palisade tissue that results in more efficient photosynthesis is brought about not only by the increased number of chloroplasts in the cells but also by the dimensions of its free surface area. Another important factor that increases photosynthesis is the presence of a well-developed system of intercellular spaces in the mesophyll, which facilitates rapid gas exchange (Fahn, 1983). In this experiment, the percentage of intercellular spaces decreased significantly from 4 to 10 h treatment (Table 2). Changes in leaf anatomy probably affect the conductance of CO_2 diffusion (Longstreth and Nobel, 1979; Evans et al., 1994). The high ratio of mesophyll area to leaf area was suggested limit to CO_2 diffusion and photosynthesis in bean and cotton, and a low tissue density was associated with high mesophyll conductance in peach leaves (Longstreth and Nobel, 1979; Syvertsen et al., 1995). In *Bruguiera parviflora* (Roxb.) W. et A., salt-induced decrease of mesophyll thickness might have contributed to the reduction in the mesophyll and in photosynthesis (Parida et al., 2004). Salt and waterlogging stress apparently reduced photosynthesis by similar mechanisms. A parallel study showed that the photosynthetic rate declined progressively with increasing duration of waterlogging in *K. candel* (Chen et al., 2005). These results support the idea that a direct relationship exists between leaf porosity and mesophyll conductance (Loreto et al., 1992). Therefore, we can conclude that the leaf anatomy of these plants that have been waterlogged for a short period enable the utilization of light, depending on not only a higher palisade–spongy ratio but also larger intercellular volumes. Compared with the 10 h treatment, the percentage of intercellular spaces in plants under the 12 h treatment appeared to increase dramatically (Table 2), which might reduce the mechanical pressure in aquatic environments when seedlings were inundated completely in artificial seawater for a long time. However, the detailed mechanisms that are involved still need further study.

The idioblast cells, including crystalliferous cells and tannin cells distributed in vascular tissue of plants that had been waterlogged for a short time (0–2 h), were found to be more abundant in the 0–2 h treatment than in other treatments (Table 3). Many plants deposit excess inorganic materials, consisting mostly of calcium salts and silicon dioxide, in their cells and these first occur as crystals (Fahn, 1983), but the ecological significance of this process in this experiment was not clear.

Xylem length, number of vessels and vessel lines was found to be correlated with waterlogging duration (Table 3). The number of phloem fibers decreased with prolonged inundation, and was hardly observed in the 12 h treatment. Consequently, flooding had a significant effect on the leaf vascular system. Tangential and radial vessel diameter and vessel wall thickness declined markedly in plants under the 2 h treatments compared with the 0 h treatment, and showed a slight difference in plants that had been waterlogged for a long time (Table 3). It was concluded that the leaf vascular system of *B. gymnorrhiza* was sensitive to waterlogging duration, especially from 0 to 2 h immersion. These results were in accordance with the fact that mangroves growing in the low tidal zone have narrower vessels and higher vessel densities than those growing in the high tidal zone (Deng et al., 2004). Generally, mangrove woods have narrow and densely distributed vessels that help in overcoming the high osmotic potential of seawater and transpiration caused by high temperatures (Tomlinson, 1986). Wide vessels have more efficiency of water conduction and less safety, but narrow vessels have more safety and less efficiency of water conduction (Zimmermann, 1983). Short and narrow vessel members with thick walls would increase the strength of a vessel and make it resistant to high negative pressure in xeric habitats (Carlquist, 1975). However, we observed that the number of vessels per unit area and vessel wall thickness declined significantly with prolonged inundation in *B. gymnorrhiza*. It was predicted that the development of the leaf vascular system was influenced by prolonged inundation, which was associated with inhibition of the growth of mangrove seedlings in a parallel experiment (Chen et al., 2004). In addition, previous authors have focused on the secondary xylem in their studies (Deng et al., 2004), but not on the leaf vascular system, which might be responsible for those differences.

These anatomical results agree with previous studies about the growth and physiological responses of *B. gymnorrhiza* to waterlogging (Ye et al., 2003), and match the geographical distribution of *B. gymnorrhiza* in natural mangrove swamps in China. We can conclude that *B. gymnorrhiza* has a relatively low tolerance to waterlogging at the seedling stage. Compared with other mangrove species, *B. gymnorrhiza* may be more easily threatened by prolonged immersion duration and more difficult to be replanted in middle and low tide zones. Nevertheless, to prove the tolerance of mangroves anatomically, further systematical research on stems, hypocotyls and roots not only in the laboratory, but also in the field are warranted.

Acknowledgements

The project was supported by Natural Science Foundation of Fujian Province (No. 2006J0146), National Natural Science Fund of China under contract No. 30200031 Program for New Century Excellent Talents in Xiamen University.

References

Allaway, W.G., Curran, M., Hollington, L.M., Ricketts, M.C., Skelton, N.J., 2001. Gas space and oxygen exchange in roots of *Avicennia marina*

(Forssk.) Vierh. var. *australasica* (Walp.) Moldenke ex N. C. Duke, the Grey Mangrove. Wetlands Ecol. Manage. 9, 211–218.

Angeles, G., López-Portillo, J., Ortega-Escalona, F., 2002. Functional anatomy of the secondary xylem of roots of the mangrove *Laguncularia racemosa* (L.) Gaenrtn. (Combretaceae). Trees 16, 338–345.

Carlquist, S., 1975. Ecological Strategies of Xylem Evolution. University of California Press, Berkeley.

Chen, L.Z., Wang, W.Q., Lin, P., 2004. Influence of waterlogging time on the growth of *Kandelia candel* seedlings. Acta Oceanol. Sin. 23, 149–158.

Chen, L.Z., Wang, W.Q., Lin, P., 2005. Photosynthetic and physiological responses of *Kandelia candel* L. Druce seedlings to duration of tidal immersion in artificial seawater. Environ. Exp. Bot. 54, 256–266.

Deng, C.Y., Lin, P., Guo, S.Z., 2004. Wood structures of some Sonneratia species and their adaptation to intertidal habitats. Acta Phytoecol. Sin. 28, 392–399 (in Chinese).

Ellison, A.M., Farnsworth, E.J., 1997. Simulated sea level change alters anatomy, physiology, growth, and reproduction of red mangrove (*Rhizophora mangle* L.). Oecologia 112, 435–446.

Evans, J.R., von Caemmerer, S., Setchell, B.A., Hudson, G.S., 1994. The relationship between CO_2 transfer conductance and leaf anatomy in transgenic tobacco with a reduced content of Rubisco. Aust. J. Plant Physiol. 21, 475–495.

Fahn, A., 1983. Plant Anatomy. Pergamon Press, New York.

Fan, H.Q., Li, G.Z., 1997. Effect of sea dike on the quantity, community characteristic and restoration of mangroves forest along Guangxi coast. Chin. J. Appl. Ecol. 8, 240–244 (in Chinese).

Field, C.D., 1995. Impact of expected climate change on mangroves. Hydrobiologia 295, 75–81.

Hovenden, M.J., Allaway, W.G., 1994. Horizontal structures on Pneumatophores of *Avicennia marina* (Forsk.) Vierh.—a new site of oxygen conductance. Ann. Bot. 73, 377–383.

Kitaya, Y., Jintana, V., Piriyayotha, S., Jaijing, D., Yabuki, K., Izutani, S., Nishimiya, A., Iwasaki, M., 2002. Early growth of seven mangrove species planted at different elevations in a Thai estuary. Trees 16, 150–154.

Lin, P., 1999. Mangrove Ecosystem in China. Science Press, Beijing.

Longstreth, D.J., Nobel, P.S., 1979. Salinity effects on leaf anatomy. Plant Physiol. 63, 700–703.

Loreto, F., Harley, P.C., Di Marco, G., Sharkey, T.D., 1992. Estimation of mesophyll conductance to CO_2 flux by three different methods. Plant Physiol. 98, 1437–1443.

Naidoo, G., Rogalla, H., Von Willert, D.J., 1997. Gas exchange responses of a mangrove speices, *Avicennia marina*, to waterlogged and drained conditions. Hydrobiologia 352, 39–47.

Parida, A.K., Das, A.B., Mittra, B., 2004. Effects of salt on growth, ion accumulation, photosynthesis and leaf anatomy of mangrove, *Bruguiera parviflora*. Trees 18, 167–174.

Rao, A.N., Tan, H., 1984. Leaf structure and its ecological significance in certain mangrove plants. In: Soepadmo, E., Rao, A.N., Mclntosh, D.J. (Eds.), Proc. As. Symp. Mangr. Environ.—Res. Manage. UNESCO, pp. 183–194.

Saenger, P., 2002. Mangrove Ecology, Silviculture and Conservation. Kluwer Academic Publishers, The Netherlands.

Stephen, M.E., Donkin, M.E., Stevens, R.A., 1983. Stomata. Edward Arnold, London.

Syvertsen, J.P., Lloyd, J., McConchie, C., Kriedemann, P.E., Farquhar, G.D., 1995. On the relationship between leaf anatomy and CO_2 diffusion through the mesophyll of hypostomatous leaves. Plant Cell Environ. 18, 149–157.

Tomlinson, P.B., 1986. The Botany of Mangrove. Cambridge University Press, New York.

Yáñez-Espinosa, L., Terrazas, T., López-Mata, L., 2001. Effects of flooding on wood and bark anatomy of four species in a mangrove forest community. Trees 15, 91–97.

Ye, Y., Tam, N.F.Y., Wong, Y.S., Lu, C.Y., 2003. Growth and physiological responses of two mangrove species (*Bruguiera gymnorrhiza* and *Kandelia candel*) to waterlogging. Environ. Exp. Bot. 49, 209–221.

Ye, Y., Tam, N.F.Y., Wong, Y.S., Lu, C.Y., 2004. Does sea level rise influence propagule establishment, early growth and physiology of *Kandelia candel* and *Bruguiera gymnorrhiza*? J. Exp. Mar. Biol. Ecol. 306, 197–215.

Youssef, T., Saenger, P., 1996. Anatomical adaptive strategies to flooding and rhizophere oxidation in mangrove seedlings. Aust. J. Bot. 44, 297–313.

Zimmermann, M.H., 1983. Xylem Structure and Ascent of Sap. Springer-Verlag, Berlin.

淹水胁迫对秋茄(*Kandelia candel*)幼苗叶片 C、N 及单宁含量的影响*

——一个关于碳素-营养平衡假说的实验

何 缘 张宜辉 于俊义 黄冠闽 林 鹏

(厦门大学生命科学学院,福建 厦门 361005)

摘要:通过在福建厦门同安湾潮间带滩涂种植红树植物秋茄,测定不同滩面高程 (黄海高程 1. 6,1. 0,0. 4 m)下 1年生秋茄幼苗叶片单宁、C、N、叶绿素含量及幼苗的生长指标,研究了淹水胁迫对秋茄幼苗次生代谢物质单宁的影响及作用机理。结果表明,随着滩面高程的降低,淹水胁迫增强,秋茄幼苗生物量、叶片 C/N 及单宁含量显著降低,在滩面高程 0. 4 m处,与 1. 0 m和 1. 6 m相比,幼苗生物量分别降低了 18. 2%和 47. 0%,叶片 C/N比值分别降低了 17. 5%和 20. 0%,相应地,叶片单宁含量也分别降低了 44. 6%和 70. 5%。秋茄幼苗叶片单宁含量与叶片 C/N比值呈显著正相关 (R =0. 8425),表明秋茄幼苗叶片中 C、N含量及单宁含量对淹水胁迫的响应符合碳素/营养平衡假说。

关键词:秋茄;淹水;单宁;C/N比值

文章编号:1000-0933(2008)10-4725-07 **中图分类号**:Q142,Q178,Q948 **文献标识码**:A

Effect of waterlogging on the contents of C, N and tannins of *Kandelia candel* seedlings: a test of the carbon-nutrient balance hypothesis

HE Yuan, ZHANG Yi-Hui , YU Jun-Yi, HUANG Guan-Min, LIN Peng

Acta Ecologica Sinica, **2008**, **28**(**10**): **4725**～**4731**.

Abstract: This study aimed to determine the effect of waterlogging on the contents of C、N、tannins and Chlorophyll in the leaves of *Kandelia candel* (L.) seedlings, which were planted at various topographic sites in an intertidal zone of Tongan Bay, Xiamen, China. The experimental plots were on a slop and showed a maximal elevation difference of 1. 6 m. Three experimental plots were set up in bare areas of intertidal zones. The mean elevations of the plots were 0. 4 m, 1. 0 m and 1. 6 m respectively. The intertidal elevation affects tidal inundation and the inundation period. At 1. 6 m above the zero tidal level of Huang Ocean, *Kandelia candel* seedlings showed best growth and the higest tannin concentrations and C/N ratios. Tannin concentrations and C/N ratios tended to decline with decreasing elevation. The results showed that changes in the C、N and tannin concentrations in response to waterlogging. This indicates that longer waterlogging time resulted in lower C/N ratios. The responses of tannins to the altered C/N ratios were that lower C/N ratios resulted in lower tannin concentrations of leaves, decreased by 44. 6% and 70. 5%, respectively, the biomass of *Kandelia candel* seedlings decreased by 18. 2% and 47. 0%. In fact, there was a significant positive correlation between tannin concentrations and C/

* 国家自然科学基金资助项目(30600077)

原载于生态学报,2008,28(10):4725－4731

N ratio ($R = 0.8425$). The response of C、N and tannins in leaves to waterlogging was consistent with carbon-nutrient balance hypothesis

Key Words: *Kandelia candel*(L.) Druce; waterlogging; tannin; carbon/nitrogen ratio

红树林是热带、亚热带海岸潮间带的木本植物群落,对维持生态平衡和保护环境起着重要的作用。秋茄(*Kandelia candel*)是我国红树植物群落中最为常见的红树植物之一[1]。在大部分的潮汐周期中,秋茄幼苗经常被水淹没,处于没顶淹水的状态[2,3]。随着滩面高程的降低,生长在中低潮带的红树植物幼苗没顶淹水时间延长,淹水深度加深,导致红树植物幼苗成活率降低[4~6]。

在次生代谢物质随环境条件变化的生理机制方面,曾被提出了一些有意义的假说,其中碳素/营养平衡(CNB)假说成功地预测了许多有关营养、光照对植物次生代谢物质的影响[7,8],但是许多研究也得出了一些相反的结论,对CNB假说的可适用性及适用范围一直存在争议[9,10],这可能是因为不同植物之间确实存在着这种差异,另一方面,植物体内次生代谢物质的合成,不只与体内的C、H、O、N等元素的比例有关,它还受到其它因素比如酶活力的影响[11]。本研究对种植在不同滩面高程的红树植物秋茄进行一年的野外跟踪调查,测定其幼苗叶片中C、N含量、单宁含量和叶绿素含量及幼苗生长指标,进一步分析了植物体内C、N含量与单宁积累的相关关系,旨在探讨植物次生代谢物质在水淹胁迫下变化的内在生理机制,研究结果可为红树林的生态恢复和造林工作提供科学依据。

1 研究方法

1.1 样地设置

样地位于厦门东海域同安湾同安西溪河口(24°38′N, 118°11′E)。厦门属南亚热带季风气候区,年均气温20.7℃,厦门港属正规半日潮,海水盐度高、潮差大,多年平均高潮位5.49 m、平均低潮位1.50 m,平均潮差3.98 m,最大潮差6.92 m,最小潮差0.99 m。样地滩面高程最低处为0.09 m,最高处为1.60 m(以黄海平均海平面为潮位基准面)。秋茄幼苗种植于2005年4月,为胚轴插植,株行距30 cm ×30 cm,样地靠堤岸一侧呈直线,长300 m,靠海一侧呈弧形,最宽处为160 m,样地种植面积6.7 ×10⁴m²。

在样地布设3条高程水准测量横断面,沿堤岸垂直方向拉一条样线,从岸边人工堤岸开始到离岸160 m外的红树林带前缘,沿样线每隔5 m为一个测点。退潮时,携带标志杆和已装好水的透明塑料软管,根据连通器原理进行测量,得到离岸0~160 m之间样地滩涂的相对垂直高程(图1)。其中,离岸160 m处红树林生长带前缘高程为0.09 m(以黄海平均海平面为潮位基准面),每个样带设有5 m ×5 m的3个重复的平行样方。3个样带在滩涂上的分布位置如图1所示,样地高程、土壤pH、盐度及基质情况如表1所示。

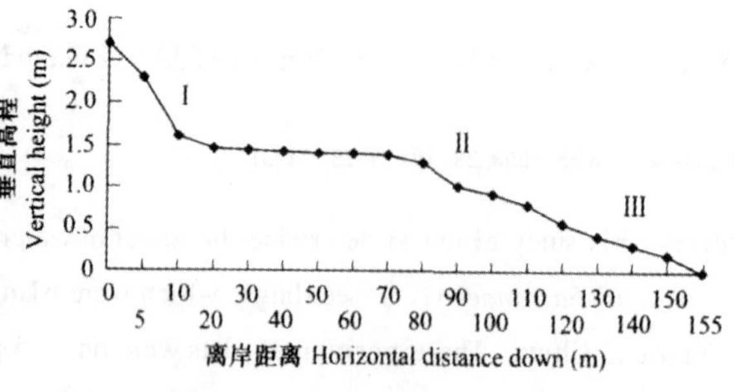

图1 同安湾样地滩面高程剖面特征

Fig 1 Beach profile of the sampling sites in the intertidal zone of Tongan Bay

滩面高程 Tidal level: I 1.6m, II 1.0m, III 0.4m

同时,每次采样过程中,采集各样方内0~20 cm的表层土壤带回实验室测定。土壤样品测定pH值(酸度计法)、盐度(电导法)、有机质含量(重铬酸钾法)以及土壤质地(我国土壤质地分类法)等指标,土壤背景值见表1。

1.2 样品采集和测定

2006年6月在Ⅰ、Ⅱ和Ⅲ样地分别选取成熟叶和幼叶各6对,幼叶为主枝顶端完全展开的第1对叶;成熟叶为主枝顶端完全展开的第3对叶。叶片用蒸馏水冲洗阴干,单宁含量当天测量,剩余叶片105 ℃杀青10

min, 80 ℃烘干至恒重,研磨过筛,保存备用。并在每个样方内随机选取 2株幼苗带回实验室测量生长指标,测定的生长指标包括生物量、基径、株高及全株叶片面积。

表 1 厦门同安湾各秋茄样方的土壤背景值

Table 1 Soil characters in different sample plots in the intertidal zones of Tongan Bay of Xiamen

土壤背景值 Soil characters	样带 Sample plots		
	Ⅰ	Ⅱ	Ⅲ
样地离岸距离 Horizontal distance down shore(m)	20	100	150
滩面高程 (黄海高程) Tidal level, upper the zero tidal level of Huang	1.6	1.0	0.4
每个潮水周期平均浸淹时间 Waterlogging time per-tide-cycle (h)	4	5	5.9
土壤 pH值 Soil pH	7.0a ±0.02	6.7a ±0.1	6.9a ±0.1
土壤盐度 Soil salinity(‰)	20.0b ±2.9	17.6a ±0.6	21.2b ±0.3
土壤有机质 Soil organic matter(%)	4.5a ±1.8	3.4a ±1.9	3.9a ±0.4
土壤质地 Soil texture	粘壤土 Clay loam	粘壤土 Clay loam	粘壤土 Clay loam

*同一行中标有不同字母的数值表示存在显著差异, $P < 0.05$ Means followed by same letters on the same column indicate significant differences at $P < 0.05$

1.3 秋茄幼苗叶片 C、N含量及成熟叶片叶绿素含量的测定

幼苗叶片 C、N含量用 CN元素分析仪 (Elementar Vario EL III)测定。成熟叶片的叶绿素含量采用丙酮提取比色法,测定 645 nm和 663 nm的光吸收值,并计算 Chl a和 Chl b含量。

1.4 秋茄幼苗叶片单宁含量的测定

称取样品 0.1 g,加入 70%丙酮提取液研磨,浸提 3次,每次用 5 ml提取液提取 30 min,离心 (5000 r/min)后收集上清液并定容至 25 ml。可溶性缩合单宁 (ECT)、残渣中结合态缩合单宁 (BCT)含量测定用正丁醇-盐酸法[12]。总缩合单宁为可溶性缩合单宁和结合态缩合单宁的总和。

1.5 数据处理和分析

试验原始数据的处理采用 Excel软件完成,并采用 SPSS11.5软件进行方差分析 (ANOVA)完成。

2 结果与分析

2.1 水淹胁迫下秋茄幼苗生长的变化

随着滩面高程的降低,水淹胁迫程度增强,植株总生物量呈现下降的趋势 (表 2)。滩面高程 0.4 m处,总生物量最小,为 (30.5 ±1.6) g DW,与高程 1.0 m和 1.6 m相比,分别降低了 18.2% ($P < 0.01$)和 47.0% ($P < 0.01$)。随着水淹胁迫程度的增强,茎的伸长生长加剧,茎的径向生长减缓。与高程 1.0 m和 1.6 m处相比,高程 0.4 m处的植株茎高表现为先下降后升高,植株基径分别减小了 20.0%和 42.9%。从茎生物量的比重情况来看,随着淹水胁迫程度的增强,茎生物量的比重表现为先下降后升高,高程 0.4 m处为 29.2%, 1.0 m处为 27.6%, 1.6 m处 30.5%。

表 2 淹水对秋茄幼苗茎和生物量的影响

Table 2 Effect of waterlogging on stems and biomass of *Kandelia candel* seedlings

参数 Parameter	滩面高程 Tidal level(m)		
	0.4	1.0	1.6
茎高 Lengths of stems(cm)	43.3a ±4.9	39.3b ±4.3	48.5a ±4.0
基径 Diameters of stems(cm)	0.8b ±0.1	1.0b ±0.2	1.4a ±0.2
茎干生物量 Biomass of stems(gDW)	8.9b ±0.7	10.3b ±1.3	17.6a ±1.1
总生物量 Biomass of seedlings(gDW)	30.5c ±1.6	37.3b ±4.2	57.6a ±2.8
全株叶面积 Leaf area of a seedling(cm^2)	101.7c ±39.9	406.1b ±104.9	867.2a ±78.2

*同一行中标有不同字母的数值表示存在显著差异, $P < 0.05$ Means followed by same letters on the same column indicate significant differences at $P < 0.05$

2.2 水淹胁迫下秋茄叶片叶绿素的变化

由图2可以看出,随着淹水时间的增长,叶片的叶绿素 a(Chl a)含量显著降低,滩面高程 0.4 m处,叶片叶绿素 a(Chl a)含量显著低于滩面高程 1.0 m和 1.6 m的叶绿素 a含量 ($P<0.05$)。叶绿素 b(Chl b)的含量有所降低,但其变化不显著 ($P>0.05$)。全株叶面积的变化情况见表 1,随着滩面高程的降低,叶片面积显著减小 ($P<0.01$),高程 1.6 m处,总叶面积为 (867.2 ±78.2) cm^2, 1.0m处,总叶面积为 (406.1 ± 104.9) cm^2, 0.4 m处,总叶面积为 (101.7 ±39.9) cm^2。

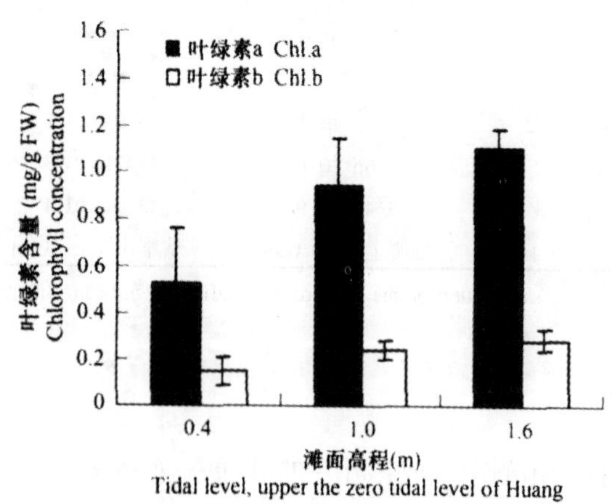

图2 淹水对秋茄叶片叶绿素含量的影响

Fig. 2 Effects of waterlogging on contents of Chl in *Kandelia candel* leaves

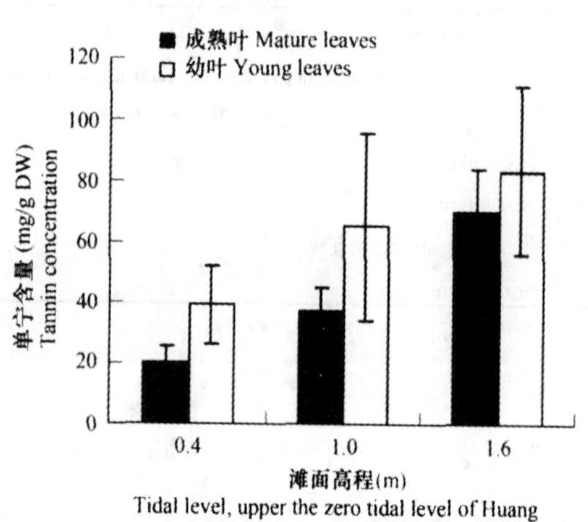

图3 淹水对秋茄叶片单宁含量的影响

Fig. 3 Effects of waterlogging on contents of tannin in *Kandelia candel* leaves

2.3 水淹胁迫下秋茄叶片单宁的变化

实验结果 (图 3)表明:高程 1.6 m处幼苗成熟叶片单宁含量最高,为 (69.5 ±14.1) mg/g DW,随着淹水时间的增长,胁迫增强,叶片单宁含量降低,高程 0.4 m处成熟叶片单宁含量降至最低,为 (20.5 ±5.0) mg/g DW,与高程 1.0 m和 1.6 m处相比,单宁含量分别降低了 44.6% ($P=0.042$)和 70.5% ($P<0.01$)。幼叶单宁含量与成熟叶单宁含量的变化趋势一致,但幼叶单宁含量显著高于成熟叶单宁含量 ($P<0.01$),其变化范围为 39.1～83.2 mg/g DW。

2.4 水淹胁迫下秋茄叶片 C、N含量的变化

淹水对秋茄幼苗成熟叶片 C和 N的影响见表 3。随着淹水时间的增长,叶片 C含量降低,而 N含量有所增高, C/N比值降低,与高程 1.0 m和 1.6 m处相比, 0.4 m处幼苗叶片 C含量分别降低了 15.2% ($P<0.01$)和 17.1% ($P<0.01$), C/N比值分别降低了 17.5% ($P<0.01$)和 20.0% ($P<0.01$), N含量变化差异不显著 ($P>0.05$)。

表 3 淹水对秋茄幼苗叶片 C和 N含量的影响

Table 3 Effect of waterlogging on concentrations of N and C in *Kandelia candel* leaves

滩面相对高程 Tidal level(m)	有机 C (g/kg) Organic C	全 N (g/kg) Total N	C/N比值 C/N ratios
0.4	324.6b ±16.2	24.8a ±2.2	13.2b ±0.6
1.0	382.9a ±10.8	24.0a ±1.4	16.0a ±0.9
1.6	394.6a ±1.2	23.9a ±0.3	16.5a ±0.2

*同一列中标有不同字母的数值表示存在显著差异, $P<0.05$

Means followed by same letters on the same column indicate significant differences at $P<0.05$

3 讨论

3.1 水淹胁迫下对秋茄幼苗生长的影响

以往对植物的淹水胁迫的研究表明,淹水对植物的生长有显著的影响,白骨壤 (*Avicennia marina*)在淹水胁迫下,总生物量显著下降[13],秋茄在长时间淹水条件下幼苗的生长受到显著抑制[14]。

研究样地的不同滩面高程是经过人工填土而成,土壤质地、pH值及有机质含量等不存在显著差异 (表 1)。随着滩面高程的降低,淹水时间逐渐延长,胁迫增大,秋茄幼苗生物量显著下降,幼苗基径有下降的趋势,而幼苗茎高则呈现为先下降后升高的趋势。从茎的生物量、茎高、基径及茎生物量占总生物量的比例可以

看出,在生长过程中,由于受到淹水胁迫,茎的伸长生长加剧,特别是在淹水时间最长的样 Ⅲ,植株通过加大茎生物量的比重、加快茎的伸长生长、减缓茎的径向生长,最大限度的使茎伸出水面。研究认为,淹水水稻将增加内源激素的分泌,进而加快茎、叶的向上生长[15],秋茄幼苗也存在类似的现象,即在淹水时间较长的环境条件下,茎的生物量升高,将茎叶伸出水面,可能是缓解根系缺氧的一种机制[14],而茎生物量的升高可能是以牺牲植物体内的次生代谢物质含量为代价[16,17]。

3.2 水淹胁迫下对秋茄叶片 C、N含量的影响

植物体内的次生代谢物质的产生与体内的 C、N元素的比例有关[18],营养、光照强度、CO_2浓度等因子都会使植物体内的 C/N比值发生变化[19~21]。Lawler的研究表明,高光照强度和高浓度的 CO_2使植物叶片的 C/N比值增高[19]。同样,Cronin证实,低光照强度和高营养分别使植物叶片的 C/N比值降低了 37%和 31%[21]。这可能是由于淹水使气孔关闭,CO_2扩散的气孔阻力增加,随着淹水时间的延长,羧化酶活性逐渐降低,叶绿素含量下降,绿叶面积减少,淹水不仅降低光合速率,光合产物的运输也有所下降[22]。

本实验结果表明,随着淹水时间的延长,植物叶片的叶绿素含量降低、叶片面积减少,植物叶片 C含量也逐渐降低,N含量则出现上升的趋势,与高程 1.0 m和 1.6 m处相比,0.4 m处秋茄幼苗叶片 C/N比值分别降低了 17.5%和 20.0%,差异极显著 ($P<0.01$)。表明不同的淹水时间对秋茄幼苗叶片的 C/N比值产生了显著影响。长时间淹水条件下,由于光照强度的减弱、光照时间的降低及幼苗叶片叶绿素含量的降低,同化作用受到抑制,以 C为基础的能源物质积累减少,从而导致植物体内 C/N比值降低。有研究认为,随着植物体内 C含量的降低,N含量则会上升,植物体内的 C含量的增大会稀释 N的含量[20],而长时间淹水,会减弱植物有效的光合作用和干物质的的合成[14],降低植物体内的 C含量[23]。而在水淹胁迫下叶片的 N含量没有明显改变,推测其原因,随着淹水时间的延长,植物体内 C含量降低,N含量上升,另一方面,水淹胁迫与其它逆境胁迫不同,长时间水淹胁迫明显抑制植物根系对氮素的吸收,加快脱氮和淋洗过程,降低有机氮的矿化速率,从而降低植株地上部的 N含量[22]。

3.3 水淹胁迫下对秋茄叶片单宁含量的影响

单宁作为植物次生代谢物质,主要存在于植物的叶、根、皮、果及花中[24]。植物体内单宁累积水平受生物和非生物等因子影响,如营养、光照、温度、CO_2 浓度、水分、紫外线辐照[25~27]。目前,在次生代谢物质随环境条件变化的生理机制方面,提出了一些有意义的假说[11]。其中碳素/营养平衡 (Carbon-Nutrient Balance Hypothesis, CNB)假说认为,植物体内以 C为基础的次生代谢物质 (如植物多酚等以 C、H、O为主要结构的化合物)与植物体内的 C/N (碳素/营养)比呈正相关,而以 N为基础的次生代谢物质 (如生物碱等含 N化合物)与植物体内 C/N比呈负相关。

本研究表明,高程 1.6 m处秋茄幼苗叶片单宁含量最高,为 (69.5 ±14.1) mg/g DW,随着淹水时间的延长,叶片单宁含量迅速减少,0.4 m处秋茄幼苗叶片单宁含量降低了 70.5%,与 C/N比值的变化趋势相同。秋茄幼苗叶片单宁含量与叶片 C/N比值呈正相关,且相关性显著 ($R=0.8425$),与 CNB假说一致。在本实验中,样 Ⅲ的幼苗通过加大茎生物量的比重、加快茎的伸长生长、减缓茎的径向生长,最大限度的使茎伸出水面,而次生代谢物质单宁含量却显著降低,CNB假说认为,植物光合作用降低,植物体内 C/N比值降低,导致酚类等不含 N次生代谢物质含量降低,且次生代谢物质的产生是不会以减少植物生长为代价的,在淹水胁迫程度足以影响到植物的生命时,植物则可能将更多的能量分配到生长上,进而导致以 C为基础的次生代谢物质单宁含量降低。另外,幼叶和成熟叶比较结果表明,幼叶单宁含量明显高于成熟叶,从植物的角度来看,这种对于次生物质防御功能的分配是合理的,只将最多的防御物质分配到昆虫最可能为害的部位[28]。

4 结论

环境胁迫下,植物必然产生一系列的生理和化学响应,环境、生理和次生代谢物质三者之间必然存在联系。

(1)水淹胁迫下秋茄幼苗叶片中的 N含量没有明显改变,而叶片中的单宁和 C含量均受到水淹的影响,

随着淹水胁迫的增强,光合能力下降,同化作用受到抑制,从而引致缺乏合成单宁的原料和能量的来源,最终引起单宁含量显著降低。

(2) CNB假说成功的解释了秋茄幼苗叶片中的单宁、C、N含量及幼苗生长对淹水胁迫的响应及三者之间的关系:秋茄幼苗叶片中的 C、N含量及单宁含量对淹水胁迫的响应符合 CNB假说。本实验中,只测定分析了一种次生代谢物质单宁,对 CNB假说的适用范围有待进一步研究和验证。

References:

[1] Lin P. Mangrove Research Papers (Ⅲ). Xiamen: Xiamen University Press, 1990. 30 - 40.

[2] Chen L Z, Wang W Q, Lin P. Influence of waterlogging time on the growth of Kandelia candel seedlings. Acta Oceanologica Sinica, 2004, 23 (1): 149 - 158.

[3] Chen L Z, Lin P, Wang W Q. Mechanisms of mangroves waterlogging resistance. Acta Ecologica Sinica, 2006, 26 (2): 586 - 593.

[4] Lin P. A review on the mangrove research in China. Journal of Xiamen University (Natural Science), 2001, 40 (2): 592 - 603

[5] Zhang Q M, Sui S Z, Zhang Y C, *et al*. Marine environmental indexes related to mangrove growth. Acta Ecologica Sinica, 2001, 21 (9): 1427 - 1437.

[6] Kitava Y, Jintana V, Piriyayotha S, *et al*. Early growth of seven mangrove species planted at different elevations in a Thai estuary. Trees, 2002, 16: 150 - 154.

[7] Gershenzon J. Changes in the levels of plant secondary metabolites under water and nutrient stress. Recent Adv. Phytochem. 1984, 18: 273 - 320.

[8] Herms D, Mattson W J. The dilemma of plants: To grow or to defend. Q. Rec Biol, 1992, 67: 283 - 335.

[9] Lerdau M. Benefits of the Carbon-Nutrient Balance Hypothesis. Opinion, 2002, 98 (3): 534 - 536.

[10] Jason G, Hamilton, Arthur R, *et al*. The carbon-nutrient balance hypothesis: its rise and fall. Ecology Letters, 2001, 4: 86 - 95.

[11] Kong C H, Xu T, Hu F, Huang S S. Allelopathy under environmental stress and its induced mechanism. Acta Ecologica Sinica, 2000, 20 (5): 849 - 854.

[12] Porter L J, Hrstich L N, Chan B G. The conversion of procyanidins and prodelphinidins to cyanidin and delphinidin. Phytochemistry, 1986, 25 (1): 223 - 230.

[13] Hovenden M J, Curran M, Col E M A, *et al*. Ventilation and respiration in roots of one-year-old seedlings of grey mangrove Avicennia marina (Forsk) Vierh. Hydrobiologia, 1995, 295: 23 - 29.

[14] Chen L Z, Wang W Q, Lin P. Influence of waterlogging time on the growth of Kandelia candel seedlings. Acta Oceanologica Sinica, 2005, 27 (2): 141 - 147.

[15] Guan Z H, Phytomedicine: An introductory treatise. Beijing: China Agricultural University Press, 1996. 100 - 104.

[16] Bazzaz F A, Nona R C, Coley P D, *et al*. Allocating resources to reproduction and defense. Bioscience, 1987, 37: 58 - 67.

[17] Chapin F S Ⅲ, Arnold J B, Christopher B F, *et al*. Plant response to multiple environmental factors. Bioscience, 1987, 37: 49 - 77

[18] Bryant J P, Chapin III F S, Klein T H. Carbon/nutrient balance of boreal plants in relation to vertebrate herbivory. Oikos, 1983, 40: 357 - 368.

[19] Lawler L R, Foley W J, Woodrow I E, *et al*. The effects of elevated CO2 atmospheres on the nutritional quality of Eucalyptus foliage and its interaction with soil nutrient and light availability. Oecologia, 1997, 109: 59 - 68.

[20] Cronin G, Lodge D M. Effects of light and nutrient availability on the growth, allocation, carbon/nitrogen balance, phenolic chemistry, and resistance to herbivory of two freshwater macrophytes. Oecologia, 2003, 137: 32 - 41.

[21] Booker F L, Maier C A. Atmospheric carbon dioxide, irrigation, and fertilization effects on phenolic and nitrogen concentrations in loblolly pine (*Pinus taeda*) needles. Tree Physiology, 2001, 21: 609 - 616.

[22] Liu Y L. Physiology of plant water stress. Beijing: Agriculture Press, 1992. 163 - 171.

[23] Xiao Q, Zheng H L, Ye W J, *et al*. Effects of waterlogging on growth and physiology of Spartina alterniflora. Chinese Journal of Ecology, 2005, 24 (9): 1025 - 1028.

[24] Kraus T E C, Yu Z, Preston C, Dahlgren R A and Zasoski R J. Linking chemical reactivity and protein precipitation to structural characteristics of

foliar tannins J. Chem. Ecol, 2003, 29: 703 - 730.

[25] Kraus T E C, Zasoski R J, Dahlgren R A. Fertility and pH effects on polyphenol and condensed tannin concemtrations in foliage and roots Plant and Soil, 2004, 262: 95 - 109.

[26] Alonso-Amelot M E, Oliveros A, Calcagno-Pisarelli M P. Phenolics and condensed tannins in relation to altitude in neotropical *Pteridium* spp. A field study in the Venezuelan Andes Biochemical Systematics and Ecology, 2004, 32: 969 - 981.

[27] Hyvarinen M, Walter B, Koopmann R. Impact of fertilisation on phenol content and growth rate of Cladina stellaris: a test of the carbon-nutrient balance hypothesis Oecologia, 2003, 134: 176 - 181.

[28] Yan F M. Chemical Ecology. Beijing: Science Press, 2003. 50 - 59.

参考文献:

[1] 林鹏. 红树林研究论文集(1980～1989年). 厦门:厦门大学出版社,1990. 30～40.

[4] 陈鹭真,林鹏,王文卿. 红树植物淹水胁迫响应研究进展. 生态学报,2006,26(2):586～593.

[4] 林鹏. 中国红树林研究进展. 厦门大学学报(自然科学版),2001,40(2):592～603.

[5] 张乔民,隋淑珍,张叶春,于红兵,孙宗勋,温孝胜. 红树林宜林海洋环境指标研究. 生态学报,2001,21(9):1427～1437.

[11] 孔垂华,徐涛,胡飞,黄寿山. 环境胁迫下植物的化感作用及其诱导机制. 生态学报,2000,20(5):849～854.

[14] 陈鹭真,王文卿,林鹏. 潮汐淹水时间对秋茄幼苗生长的影响. 海洋学报,2005,27(2):141～147.

[15] 管致和. 植物医学导论. 北京:中国农业大学出版社,1996. 100～104.

[22] 刘友良. 植物水分逆境生理. 北京:农业出版社,1992. 163～171.

[23] 肖强,郑海雷,叶文景,陈瑶,朱珠. 水淹对互花米草生长及生理的影响. 生态学杂志,2005,24(9):1025～1028.

[28] 阎凤鸣. 化学生态学. 北京:科学出版社,2003. 50～59.

Photosynthetic and physiological responses of *Kandelia candel* L. Druce seedlings to duration of tidal immersion in artificial seawater*

Chen Luzhen[a,b] Wang Wenqing[a,b] Lin Peng[a,b]

(a. *School of Life Sciences, Xiamen University, Xiamen*, 361005 *Fujian, PRChina*;

b. *Research Centre for Wetlands and Ecological Engineering, Xiamen University, Xiamen*, 361005 *Fujian, PR China*)

Abstract

The present study demonstrates the influence of the duration of periodical waterlogging with artificial seawater on the photosynthetic and physiological responses of *Kandelia candel* L. Druce seedlings, the pre-dominant species of subtropical mangroves in China. Artificial tidal fluctuations applied here closely mimicked the twice daily tidal inundation which mangroves experience in the field. All the seedlings were immersed in artificial seawater during 70-day cultivation. Similar trends with increasing duration of immersion occurred in photosynthetic rate, transpiration rate, stomatal conductance and intercellular CO_2 concentration, where significant decreases occurred only in long time treatments of 10 or 12 h. Water used efficiency and chlorophyll contents showed lower in medium periods and higher in long periods of immersion. This indicates that the increase in pigment contents of leaves was ineffective in promoting P_n under long time immersion. Light saturation points under short time waterlogging (0–4 h) occurred at light intensities of 800–1000 μmol/m^2/s, and at around 400 μmol/m^2/s in long time treatments (8–12 h). Long periods of tidal immersion therefore significantly inhibited photosynthesis of mature leaves. Alcohol dehydrogenase and oxidase activity in roots both increased under longer immersion periods, suggesting that roots are sensitive to anaerobiosis under long term waterlogging. The activities of peroxidase and superoxide dismutase in mature leaves increased in 8 h and 10 h treatments, respectively. The content of malondialdehyde in mature leaves increased under long time treatments. Abscisic acid accumulation in mature leaves also had a sharp increase from 8 h to 12 h inundation. Even though the anti-oxidative enzymes were induced by waterlogging, this was not sufficient to protect the seedlings from senescence. The results suggested that *K. candel* seedlings completely tolerated tidal immersion by seawater up to about 8 h in each cycle, which matches the natural distribution of *K. candel* in inter-tidal zones of China.

Keywords: Mangroves; Semidiurnal tide; Artificial tide; Photosynthesis; Abscisic acid (ABA)

1. Introduction

Mangroves are inter-tidal forests of tropical and subtropical regions. They are adapted to periodical

* From Environmental and Experimental Botany, 2005, 54: 256—266

waterlogging. Due to enclosure for aquaculture and jetty construction, some/many mid-tidal zones of mangroves have been damaged. Mangroves face the challenge of subsiding plantable tidal flats in China (Fan and Li, 1997; Mo and Fan, 2001). Furthermore, the rise of sea level accompanying the rapid changes of climate (Field, 1995; Ellison and Farnsworth, 1997) also prolongs the period of waterlogging of mangroves at high tide. Consequently, the growth of natural mangroves and the survival rate of mangrove plantings have been greatly influenced, and this is of great concern to scientists. Under waterlogged conditions, root respiration rate and oxidase activity declined (McKee, 1996; Ye et al., 2003). Contents of pigments in leaves changed (Ye et al., 2003); stomatal closure, water uptake reduced, transpiration rate (T_r) and photosynthetic rate (P_n) declined (Naidoo, 1984; Ellison and Farnsworth, 1997; Pezeshki et al., 1997). Though the transpiration rate (T_r) and stomatal conductance (g_s) were decreased, the water used efficiency (WUE) appeared lower and less variable under waterlogging (Naidoo et al., 1997). Activities of enzymes processing reactive oxygen species, such as superoxide dismutase (SOD) and peroxidase (POD) activities, increases with prolonged waterlogging (Ye et al., 2003). The plant growth regulator, abscisic acid (ABA) was greatly induced by waterlogging (Kozlowski, 1984; Guan, 1996; Li and Zhou, 1996), and inactivated the PSII complex (Ahmed et al., 2002). No reports about the relation between ABA and waterlogging in mangroves have been published. These physiological and photosynthetic responses of mangrove seedlings would indicate that they tolerate waterlogging.

Kandelia candel (L.) Druce, Rhizophoraceae, is a major mangrove species of the eastern group (Lin, 1999), occurring only along the coastlines of Asia and East–Pacific Archipelagos (Lin, 1999). *K. candel* can be found in all mangrove associations in China, and it is the major mangrove species for forestation on the southeast coast of China (Lin, 1988). It occurs naturally in mid-to-high tidal zones (Lin, 1999). Under field conditions, one-year-old *K. candel* seedlings are usually fully covered in seawater at high tide, and exposed to air at low tide. The duration of seedling fully covered depends on their location in the tidal zone. The duration of periodical waterlogging is a limiting factor for the survival of seedlings (Liu, 1995; Komiyama et al., 1996; Chen et al., 2001; Kitaya et al., 2002). There is an urgent need to ascertain the suitable for this species (Fan and Li, 1997; Chen et al., 2001).

In the current study, we focussed on the physiological responses and resistance of *K. candel* seedlings to periodical waterlogging. We hypothesized that *K. candel* seedlings are tolerant of waterlogging because the distribution of *K. candel* seedlings in the inter-tidal zones was consistent with this. A former study has found out that the maximum period of waterlogging that allows growth of *K. candel* seedlings was 8 h (Chen et al., 2004). We also aimed to seek out the correlation between physiological responses and the growth of *K. candel* seedlings under periodical waterlogging. To achieve this, we used equipment, which simulated the tidal waterlogging at different levels on the tidal flats. In the field work of Watson, the Western Malayan mangrove forest was divided into five gradients of tidal inundation classes, and the frequency of flooding per month was the basis of his inundation classes (Watson, 1928). In our simulation experiment, we simplified the conditions, and averaged the inundation time for a semidiurnal tidal cycle into seven classes, according to the length of waterlogging during a tidal cycle at different locations in the inter-tidal zones. Twelve hours treatment meant that seedlings were waterlogged all the time in a tidal cycle (12 h), which denoted the waterlogged time at the lowest tidal level, while 0 h treatment equalled the highest tidal level where seedlings were not inundated. All plants were inundated twice daily except treatment *G* (0 h). In this paper, we present data showing changes in rates of leaf gas exchange, enzyme activities and hormone levels accompanying the different periodical waterlogging times after 70 days culture.

2. Materials and methods

2.1. Experimental design

A set of artificial-tidal tanks simulating the semidiurnal tide were constructed from seven plastic tanks (65 cm × 50 cm × 50 cm) arranged as shown in Fig. 1. It took 2 h to fill a tank with artificial seawater through pipes. After Tank A was full, the water flowed over into Tank B, and so on. After Tank F was full, all water in Tanks A, B, C, D, E and F was unloaded by timer-controlled valves at the bottom of each tank. It took 5 min to let the water drain out of the tank. Therefore,

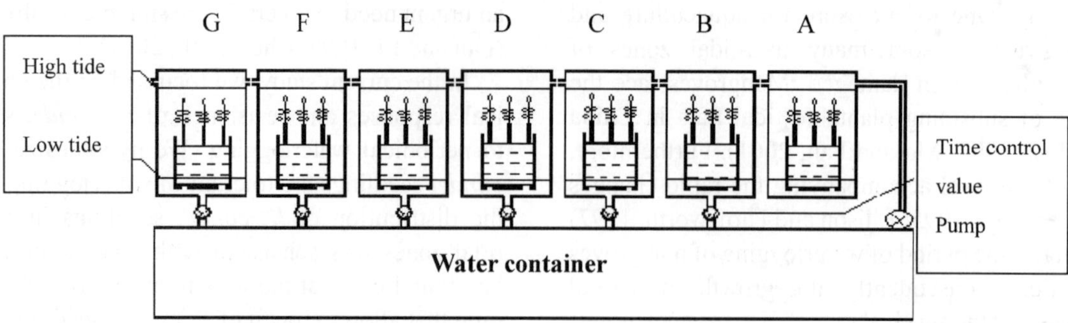

Fig. 1. Arrangement of artificial tidal tanks. It takes 2 h to fill each tank with artificial seawater through the pipe. After Tank A was full it overflowed to fill Tank B, and so on. When Tank F was full, all water in Tanks A, B, C, D, E and F was unloaded by timer-controlled valves at the bottom of each tank. Therefore, the inundated periods of tanks from A to G were 12, 10, 8, 6, 4, 2 and 0 h per-tidal cycle, respectively. There were 2 tide cycles every day.

Tank A was full of water for 12 h per tidal cycle, and the inundation periods in the other tanks were 10, 8, 6, 4, 2 and 0 h, respectively. There were two tide cycles every day. There were four pots in each tank, each 25 cm tall and 25 cm in diameter, with a small hole at the bottom to allow rapid drainage while water in the tanks was drained away. Each pot was filled with washed river sand (diameter = 1 mm). Three sets of equipment acted as three replicates. Tanks were arranged by randomized block design in the same greenhouse and the arrangement of tanks was changed randomly every week.

2.2. Materials and growing conditions

Healthy and mature hypocotyls of *K. candel* were collected from Jiulong River Estuary in Fugong Town, Longhai County, Fujian Province of China (24°29′N, 117°55′E). The average seawater salinity there was 17‰ (Lin, 1988, 1999). Five hypocotyls were planted in each pot at random. The seedlings were periodically submerged under artificial seawater, as in Fig. 1, with a salinity of 15‰ (seawater from the west coast of Xiamen of 22–28‰ in salinity was diluted by tap water). Tap water was added daily to compensate for evaporation losses and the seawater was renewed weekly. All seedlings were grown in a greenhouse with air temperature of 27–32 °C, and under natural sunlight for 70 days (28 April, 2003–3 July, 2003). Seedlings were flooded at 'high tide' to a maximum depth of 60 cm above the bottom of the tanks, and at 'low tide' the water was very slightly below soil level as showed in Fig. 1. After 70 days culture, each seedling had two to three pairs of leaves and the second leaf blades had matured. All the analyses were performed on the second leaf blades from the base.

2.3. Gas exchange measurements and contents of pigments in mature leaves

Photosynthesis and transpiration of mature leaves were measured using a portable photosynthesis system (Model CIRAS-1, UK). Before the measurement, water was drained from all the tanks. Three to five mature leaves for each replicate were chosen for the measurement of photosynthetic rate (P_n), transpiration rate (T_r), stomatal conductance (g_s) and intercellular CO_2 concentration (C_i). Mean values of P_n, T_r, g_s and C_i for each replicate were used for statistics. Gas exchange measurements were made on entire and fully expanded leaves and were carried out between 10 a.m. and 12 p.m. in natural sunlight of 800–1000 μmol/m^2/s photosynthetically active radiation and in the temperature range 29–32 °C. Water used efficiency (WUE) was calculated from the ratio P_n/T_r. Light-response curves were obtained using the same portable photosynthesis system equipped with artificial light. The measurements were carried out at 10 levels of illumination intensity, 1400, 1200, 1000, 800, 600, 400, 200, 100, 50 and 0 μmol photons/m^2/s, in the sequence from the highest to the lowest values.

For the measurement of pigment contents, about 0.1 g of mature leaves were cut and immersed in a mixture of ethanol: acetone: H_2O = 4.5:4.5:1 for 3 days, protected from light by the method of Zhang (1990). The absorbance at 645 nm and 663 nm was measured.

2.4. *Physiological analysis*

2.4.1. *Alcohol dehydrogenase (ADH) activity of roots*

The methods used for ADH extraction and activity determination were those of Liu et al. (1991), and Kato-Noguchi and Saito (2000), with some modifications. Fresh roots were homogenized in four volumes of ice-cold solution containing 100 mM Tris–HCl (pH 8.0), 2 M KCl, 20 mM EDTA, and 1.0% (m/v) polyvinyl pyrrolidone (PVP) to neutralize the effect of phenol from mangrove tissues. The homogenate was centrifuged at 12,000 × *g* and 4 °C for 20 min and the supernatant was used for ADH assay. Activity of ADH was measured by monitoring ethanol-dependent NAD^+ reduction at 340 nm in a spectrophotometer. The reaction mixture with a final volume of 3.0 ml, contained 2.7 ml of 50 mol/l Gly-NaOH buffer (pH 9.0), 0.1 ml NAD^+ solution (2 mg NAD^+), 0.1 ml enzyme extract and 0.1 ml 97% ethanol. The absorbance at 340 nm was measured at 1 min intervals for 3 min. An increase of 0.01 absorbance units per minute equalled one unit of ADH activity. The activity of ADH was expressed as units per mg protein. Protein was determined using Coomassie Brilliant Blue G-250 (Bradford, 1976), using bovine serum albumin (BSA) as a standard.

2.4.2. *Root oxidase activity*

Root oxidase activity was determined according to Zhang (1990). One to two grams of fresh fine roots were washed with distilled water and blotted dry. They were then immersed in 50 ml of solution (1:1 mixture of 0.1 mM phosphate buffer at pH 7.0 and 40 ppm α-naphthylamine) and placed on a shaker table at 25 °C for 5 h. Before and after shaking, 2 ml of solution was taken out and added to 10 ml of distilled water plus 1 ml 1% sulfanilic acid (w/v in 30% acetic acid) and 1 ml 100 ppm sodium nitrite. The mixture was kept at room temperature for 5 min, diluted by adding 6 ml of distilled water, and then the absorbance at 510 nm was measured. During the incubation period, the content of α-naphthylamine decreases, due to the root oxidase activity. Root oxidase activity was expressed as μg α-naphthylamine/h/mg FW.

2.4.3. *Lipid peroxidation of leaf membranes*

The methods of extraction and activity determination were those of Liu and Zhang (1994) and Li (2000) with a little modification. 0.5 g of mature leaves were collected and ground in a mortar with 5 ml of ice-cold 62.5 mM phosphate buffer (pH 7.8) with 1.0% (m/v) polyvinyl pyrrolidone (PVP). The homogenates were centrifuged at 4 °C and 12,000 × *g* for 20 min. The supernatant was stored at 4 °C and then used for assays of enzymatic activities and protein content.

Superoxide dismutase (SOD) activity was measured as the amount of inhibition of photo-reduction of nitroblue tetrazolium (NBT). The reaction mixture, with a final volume of 3.0 ml, contained 1.5 ml 62.5 mM phosphate buffer, 0.3 ml 20 μM riboflavin, 0.3 ml 130 mM methionine, 0.3 ml 100 μM Na_2EDTA, 0.3 ml 750 μM NBT, 0.05 ml enzyme extract and 0.25 ml deionized water. Phosphate buffer was substituted for enzyme to determine the maximum photo-reduction of NBT. The reaction was carried out in 75 μmol/m^2/s illumination for 20 min. One unit of SOD activity was calculated as that inhibiting the maximum photo-reduction of NBT by 50% at 560 nm of optical density, *k* measured with a spectrophotometer.

For the measurement of peroxidase (POD) activity, 2.9 ml of 100 mM phosphate buffer + 20 mM guaiacol, pH 7.0, was mixed with 0.1 ml of enzyme extract and allowed to stand at room temperature for 3 min. Twenty microliters of 2% hydrogen peroxide was added to activate the reaction. The absorbance at 470 nm was measured at 1 min intervals for 5 min, and an increase of 0.01 absorbance units per minute was equated to one unit of peroxidase activity.

The activities of SOD and POD was expressed as unit per mg protein. Protein was determined according to Bradford (1976) by using bovine serum albumin (BSA) as a standard.

Malondialdehyde (MDA) of mature leaves was determined according to Li (2000). Two microliters enzyme extracts were incubated with 10% (w/v) trichloro acetic acid (TCA) plus 0.6% thibarbituric acid solution in boiling water for 10 min. Then the mixture was centrifuged at room temperature and 10,000 × *g* for 20 min. Absorbances in 450, 532 and 600 nm were recorded. The MDA content was calculated as described by Li (2000), as follows:

$$C\ (\mu M) = 6.45\ (A_{532} - A_{600}) - 0.56A_{450}$$

2.5. *Endogenous ABA content of mature leaves*

The endogenous ABA in mature leaves was determined using the ELISA method according to Wu et al. (1988).

3. Data analysis

Mean and standard deviation (S.D.) values of three replicates were calculated. Data for leaf gas exchange; physiological responses and ABA concentration were analyzed by one-way ANOVA using *F*- and *t*-tests. The Student–Newman–Keuls univariate comparison method was used to analyze the difference among the light-response curves of seven treatments. The illumination effects on each treatment were analyzed by one-way ANOVA using a *t*-test.

3.1. *Gas exchange measurement and pigment contents*

With increasing duration of waterlogging, P_n of mature leaves declined progressively (Fig. 2, Table 1). Maximum P_n occurred in the 0 h and 2 h treatments, 23.89 ± 1.13 and 23.78 ± 1.44 μmol/m²/s, respectively. Transpiration rate (T_r) of 0–8 h treatments were not significantly different ($P > 0.05$) (Fig. 2). Waterlogging periods of up to 8 h did not affect T_r, but 10 h and 12 h immersions reduced T_r by 20.5% and 47.0%, respectively. Stomatal conductance (g_s) followed the same trend as P_n (Fig. 2). But g_s under the 0 h treatment was lower than that of the 2 h treatment ($^{***}P < 0.001$). Maximum g_s occurred with 2 h immersion, and was 162.4% higher than that of the minimum of 12 h treatment ($^{**}P = 0.003$). Intercellular CO_2 concentration (C_i) and water use efficiency (WUE) both changed significant over the range 0–12 h immersion (Fig. 2 and Table 1).

The photo-response curves of mature leaves were expressed by P_n in different illumination ranges from 0–1400 μmol/m²/s. The trend of photo-response curves in each treatment was significantly different (Table 2. and Fig. 3). Photosynthetic rate (P_n) under short time treatments was much higher, and the light saturation point occurred at 800–1000 μmol/m²/s of illumination. Otherwise, in 8–12 h treatments, P_n was low and increased smoothly when illumination exceeded 400 μmol/m²/s. At low light levels (0–200 μmol/m²/s), P_n was considerably higher in all treatments. The maximum P_n under 200 μmol/m²/s illumination occurred in the 4 h treatment, at 8.75 ± 0.92 μmol/m²/s (Figs. 4 and 5).

Significant changes in photosynthetic pigments of Chl.a, Chl.b and total chlorophyll (Chl.) were approximately uniform and all followed a U-shape (Fig. 6 and Table 1). The amount of Chl.a, Chl.b and Chl. was

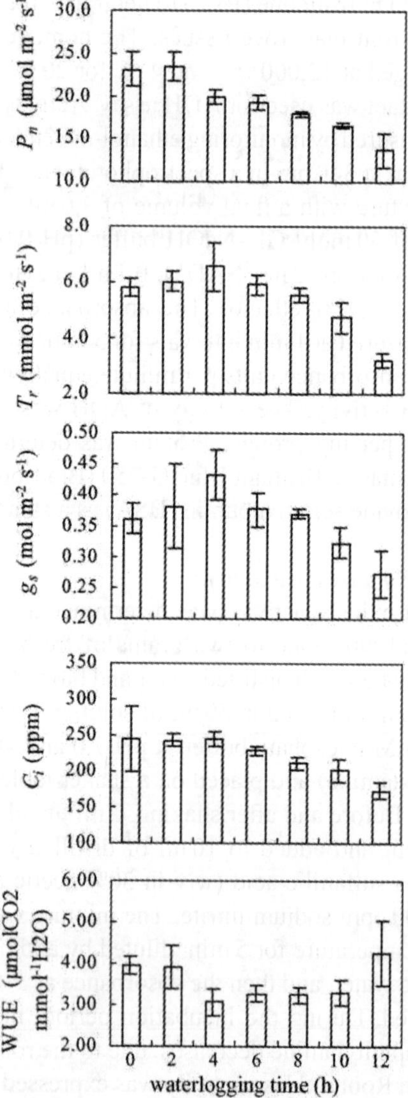

Fig. 2. P_n, T_r, g_s, C_i, and WUE of mature leaves of *Kandelia candel* seedlings under periodical waterlogging. One-way ANOVA showed significant differences between effects of treatments on P_n ($^{***}P < 0.001$), T_r ($^{***}P < 0.001$), g_s ($^{**}P = 0.005$), C_i ($^{**}P = 0.003$) and WUE ($^{*}P = 0.027$).

Table 1
Effect of immersion in seawater on photosynthetic and physiological activities in *Kandelia candel* seedlings

Parameter	Immersion time (h)							One-way ANVOA
	0	2	4	6	8	10	12	
P_n (μmol/m²/s)	23.1 ± 2.0	23.5 ± 1.8	19.9 ± 0.8	19.3 ± 0.8	17.9 ± 0.4	16.6 ± 0.3	13.7 ± 2.0	20.2***
T_r (mmol/m²/s)	5.8 ± 0.32	6.0 ± 0.4	6.6 ± 0.9	5.9 ± 0.3	5.5 ± 0.3	4.8 ± 0.6	3.2 ± 0.3	16.8***
g_s (mol/m²/s)	0.36 ± 0.03	0.38 ± 0.07	0.43 ± 0.04	0.38 ± 0.03	0.37 ± 0.01	0.32 ± 0.02	0.27 ± 0.04	5.32**
C_i (ppm)	246 ± 45	244 ± 9	245 ± 10	229 ± 7	212 ± 9	202 ± 17	174 ± 12	5.70**
WUE (μmol/mmol)	3.96 ± 0.21	3.93 ± 0.52	3.07 ± 0.31	3.27 ± 0.21	3.24 ± 0.21	3.30 ± 0.29	4.28 ± 0.79	3.524*
ADH (U/mg protein)	3.55 ± 0.69	6.78 ± 1.06	8.75 ± 1.29	6.15 ± 1.46	7.53 ± 0.26	6.63 ± 0.44	4.51 ± 0.36	7.49**
Root oxidase-activity (μg/h/g FW)	13.4 ± 2.1	24.0 ± 1.8	27.3 ± 4.3	21.3 ± 3.1	20.5 ± 6.1	17.0 ± 1.8	15.6 ± 3.3	5.25**
POD (U/mg protein)	901 ± 50	605 ± 46	587 ± 67	587 ± 103	894 ± 70	920 ± 105	1159 ± 315	7.98**
SOD (U/mg protein)	71.4 ± 6.4	67.9 ± 10.1	62.5 ± 3.5	73.7 ± 7.1	72.6 ± 9.4	102.8 ± 11.0	140.0 ± 25.2	13.7***
MDA (μmol/g FW)	3.65 ± 0.10	2.13 ± 0.29	2.77 ± 0.22	2.61 ± 0.18	3.18 ± 0.31	3.36 ± 0.18	3.54 ± 0.53	7.21**
Chl.a (mg/g FW)	1.39 ± 0.11	1.24 ± 0.21	0.59 ± 0.13	0.62 ± 0.14	0.85 ± 0.12	0.76 ± 0.05	1.30 ± 0.142	15.7***
Chl.b (mg/g FW)	0.436 ± 0.033	0.283 ± 0.071	0.197 ± 0.049	0.199 ± 0.044	0.272 ± 0.041	0.262 ± 0.014	0.417 ± 0.048	12.5***
Chl. (mg/g FW)	1.84 ± 0.15	1.65 ± 0.29	0.77 ± 0.18	0.83 ± 0.19	1.13 ± 0.16	1.03 ± 0.07	1.36 ± 0.35	10.3**
ABA (fmol/g FW)	851 ± 68	753 ± 49	726 ± 17	717 ± 15	854 ± 113	1233 ± 102	1669 ± 77	48.1***

Values are means of three replicates ±S.D. *F*-values are given and significance is shown as: *0.05, **0.01 and ***0.001.

Table 2
Results of ANOVA (*F*-values) for the illumination effects on P_n of *Kandelia candel* seedlings in different waterlogging times

Parameter	Sources of variation for univariate comparison method			One-way ANOVA: illumination effects on seven treatments						
	Illumination (*I*)	Treatment (*T*)	$I \times T$	0 h	2 h	4 h	6 h	8 h	10 h	12 h
P_n (μmol/m²/s)	745***	88.8***	7.83***	603***	1318***	496***	525***	795***	162***	82.2***

F-values are given and significance is shown as: *0.05, **0.01 and ***0.001.

lower in 4 h and 6 h treatments and higher in 0 h and 12 h treatments.

3.2. Physiological responses

Activities of all the enzymes exhibited significant changes with the prolonging of periodical waterlogging time (Figs. 4 and 5 and Table 1). The maximum ADH activity occurred in the 4 h treatment, 87.45 ± 12.89 U/mg protein. ADH activities under 0 h and 2 h treatments were less and were 59.5% and 22.5% lower than that of the maximum (4 h), respectively (**P=0.001; P=0.067). When the periodical waterlogging time was longer than 4 h, ADH activity decreased up to 12 h treatment showing a reduction by 94.0% of the 4 h treatment (**P=0.002).

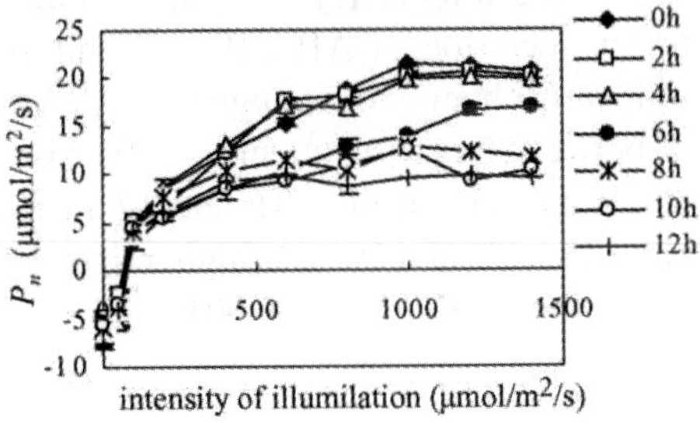

Fig. 3. The photo-response of mature leaves of *Kandelia candel* seedlings to periodical waterlogging. Analysis by Student–Newman–Keuls univariate comparison method showed no significant difference among treatments (***P < 0.001).

Root oxidase activity increased at first and then fell significantly when the periodical waterlogging time was longer than 4 h (Fig. 4 and Table 1). The maximum root oxidase activity was observed in the 4 h treatment at 27.27 ± 4.34 μg/h/g FW, approximately twice the lowest value in the 0 h and 12 h treatments (**P=0.001; **P=0.002). The decrease of root oxidase activity from 4 h treatment to 8 h treatment was not significant (P=0.056; P=0.053).

Changes in SOD activity of mature leaves is shown in Fig. 5 and Table 1. There were insignificant changes

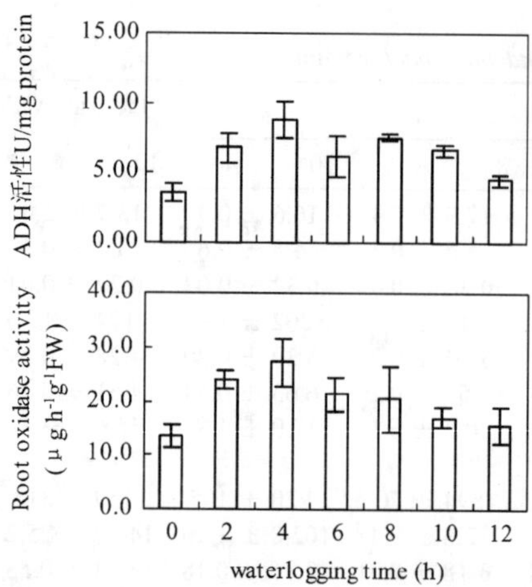

Fig. 4. Activities of ADH and oxidase in roots of *Kandelia candel* seedlings under periodical waterlogging. One-way ANOVA test showed significant differences between effects of treatments on ADH ($^{**}P=0.009$) and root oxidase activity ($^{**}P=0.009$).

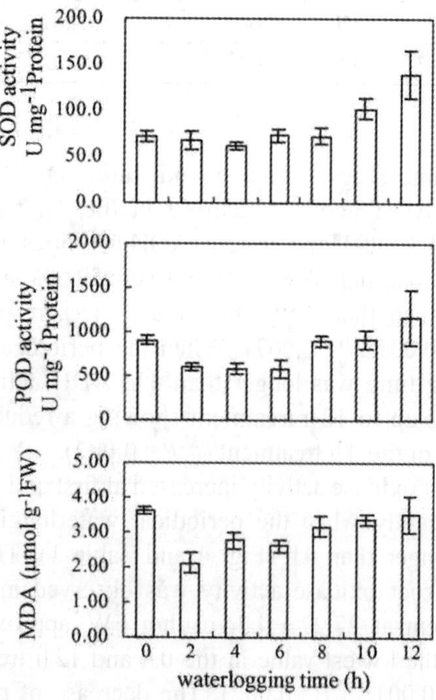

Fig. 5. Activities of SOD and POD and content of MDA in mature leaves of *Kandelia candel* seedlings under periodical waterlogging. One-way ANOVA test showed significant difference between effects of treatments on SOD ($^{***}P<0.001$), POD ($^{**}P=0.002$) and MDA ($^{**}P=0.003$).

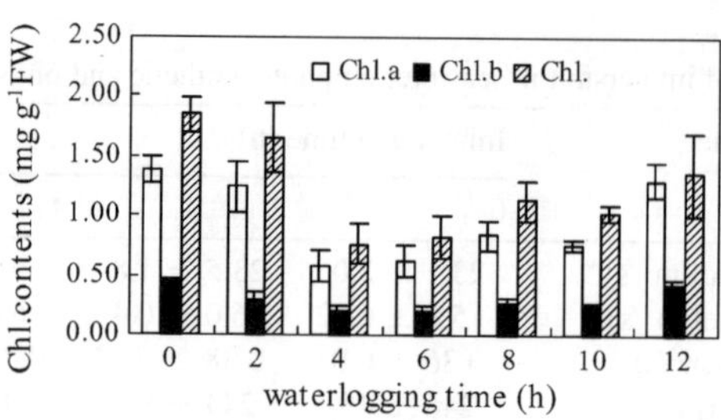

Fig. 6. Contents of Chl.a, Chl.b, and total Chl in mature leaves of *Kandelia candel* seedlings under periodical waterlogging. One-way ANOVA test showed significant difference between effects of treatments on Chl.a ($^{***}P<0.001$), Chl.b ($^{***}P<0.001$) and Chl. ($^{**}P=0.002$).

among the treatments from 0–8 h ($P>0.05$). SOD activity was remarkably activated when periodical waterlogging time was prolonged to 10 h and 12 h, being 164.4% and 223.9% of the 8 h treatment, respectively ($^{*}P=0.013$; $^{*}P=0.04$).

Peroxidase (POD) activity of mature leaves followed the same trend as SOD (Fig. 5 and Table 1). Insignificant differences were found between the 2–6 h treatments ($P>0.05$) and between the 8–12 h treatments ($P>0.05$). POD activity at 8 h was 152.1% of that of the 6 h treatment ($^{*}P=0.019$).

The content of MDA also changed similarly to that of POD and SOD (Fig. 5 and Table 1). A sharp fall occurred in the 2 h treatment ($^{**}P=0.001$), but the content rose significantly when periodical waterlogging time was longer than 2 h ($^{**}P=0.003$).

3.3. Content of ABA in leaves

The content of endogenous ABA in mature leaves was highest in the 12 h treatment followed by 10 h, 8 h and 6 h, showing a tendency for longer immersion to promote the secretion of ABA (Fig. 7 and Table 1). No significantly changes of endogenous ABA contents occurred between 0–8 h treatments ($P>0.05$), but there was a sharp increase in the 10 h and 12 h treatments, which were 144.3% and 195.3% higher than that of 8 h treatment, respectively ($^{***}P=0.001$; $^{***}P<0.001$).

4. Discussion

Mangroves are well adapted to waterlogging either in anatomy or in physiology. Parallel study has proved

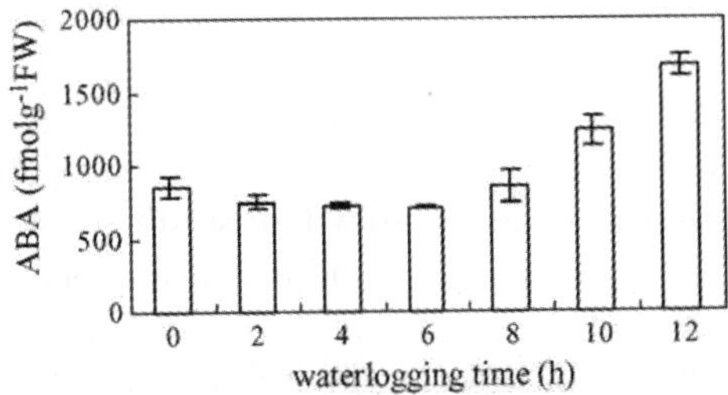

Fig. 7. ABA accumulation in mature leaves of *Kandelia candel* seedlings under periodical waterlogging. One-way ANOVA test showed significant difference between effects of treatments on ABA ($^{***}P < 0.001$).

that *K. candel* seedlings grown in different waterlogging duration when fully covered by artificial seawater and growth was reduced with long immersion. The gas exchange characteristics and physiological responses in current results have proved further that *K. candel* seedlings also have strong tolerance to long time waterlogging.

4.1. Photo-response to waterlogging

Due to stomatal closure and ribulose bisphosphate carboxylase/oxygenase (RUBISCO) activity being inhibited under waterlogging, photosynthetic rate (P_n) declined in mangroves (Ellison and Farnsworth, 1997; Pezeshki et al., 1997). Though the transpiration rate (T_r) and stomatal conductance (g_s) were also reduced, the water used efficiency (WUE) appeared lower and less variable under waterlogging (Naidoo et al., 1997). The chlorophyll contents also changed (Ellison and Farnsworth, 1997; Ye et al., 2003). In the present study, similarly, P_n of mature leaves was maximum under short time inundation, and declined progressively with the prolonging of periodical waterlogging. Transpiration rate (T_r) did not significantly differ between 0 h and 8 h treatments, which lead to a reduction of WUE (P_n/T_r). However, T_r declined sharply when inundation periods were longer than 8 h, which resulted in a rise of WUE under 10–12 h inundations. Changes of g_s and C_i indicate that waterlogging inhibited P_n and T_r by closing the stomata and reducing the intercellular CO_2 concentration. On the other hand, the contents of photosynthetic pigments of Chl.a, Chl.b and total amount of chlorophyll (Chl.), which changed uniformly, all showed a U-shaped response. We concluded that the increase of pigment contents of *K. candel* seedlings under long inundations was ineffective in promoting P_n.

The photo-response curves also showed the inhibition of P_n in long time treatments, indicating that the carbon assimilation rates were reduced. Data here were consistent with the effects of inundation on growth in a parallel study (Chen et al., 2004) where the biomass and area of mature leaves were reduced under long time treatments. Similar conclusions about carbon assimilation in mature leaves have also been reached for the other mangroves species *Avicennia germinans*, *Laguncularia racemosa* and *Rhizophora mangle* (Pezeshki et al., 1989), *Avicennia marina* (Naidoo et al., 1997) and *R. mangle* (Ellison and Farnsworth, 1997) under prolonged waterlogging.

4.2. Physiological response to waterlogging

It has been reported that young *A. marina* seedlings without pneumatophores had a continuum of gas-spaces throughout leaves, stem, hypocotyls and roots (Ashord and Allaways, 1995). For a one-year-old *A. marina* seedling, oxygen in the root system might be reduced to zero, if flooding lasted more than 3.5 h (Hovenden et al., 1995). As indicated by alcohol dehydrogenase (ADH), the anaerobic respiration increased in *A. germinans* seedlings when subjected to hypoxia for 96 h (McKee and Mendelssohn, 1987). During waterlogging, roots may eventually use up all their internally stored air and then begin to respire anaerobically, ethanol accumulates and alcohol dehydrogenase (ADH) activity increases in other hydrophytes (Pazeshki et al., 1993; Akhtar et al., 1998). ADH helped plants avoid the damage caused by ethanol. In this study ADH activity was the highest in the 4 h treatment, and was less in long time treatments. Roots in long time inundation were damaged, as shown by the decrease of the root's oxidase activity (Fig. 4). This suggested ADH could not detoxify that ethanol accumulated in roots of *K. candel* seedlings during long time inundations effectively. This result agrees with previous studies showing that hypoxia leads to a significant decline in respiration rate and oxidase activity of roots (McKee, 1996; Ye et al., 2003). Changes in ADH and root oxidase activities reported here suggest that long time waterlogging significantly inhibited the normal metabolism in roots of *K. candel* seedlings. We conclude that toxicity caused by anaerobiosis was the major inhibiter to roots under waterlogging (Liu, 1992; Guan, 1996).

Lipid peroxidation is an indicator of oxidative damage under environmental stress, (Guan, 1996). Under prolonged waterlogging, superoxide dismutase (SOD) and peroxidase (POD) activities increased in *K. candel*, as have previously been reported for other hygrophytes (Guan, 1996; Ye et al., 2003). Malondialdehyde (MDA) is a major product of lipid peroxidation, and its increase reflects damage to leaves under waterlogging (Guan, 1996). The observation of significant increases in SOD and POD, and the contents of MDA in mature leaves indicated that *K. candel* seedlings have resistance to oxidation damage due to waterlogging, which agrees with the results of Ye et al. (2003).

Abscisic acid (ABA) is one of the most important hormones and indicators of stress caused by flooding (Kozlowski, 1984; Guan, 1996). It is produced in mature or aged organism and transported to the younger. The accumulation of ABA in mature leaves of *K. candel* seedlings was higher under inundations longer than 8 h, which suggests that ABA was induced by longer waterlogging. ABA stimulates stomatal closure, which leads to the decreased in g_s, P_n and T_r of mature leaves (Guan, 1996; Li and Zhou, 1996). The observed increase in ABA reported here is probably the cause of the photosynthetic response we reported. ABA is also known to induce the growth of aerial roots (Guan, 1996; Li and Zhou, 1996), which was also observed in the parallel study (Chen et al., 2004). Senescence is caused by waterlogging (Kozlowski, 1984; Guan, 1996). It therefore seems that even the induction of anti-oxidative enzymes under long time treatments is not sufficient to protect the seedlings from senescence.

4.3. Waterlogging tolerance of K. candel *seedlings*

Previous studies have established that *K. candel* seedlings have a strong resistance to tidal inundation by seawater (Ye et al., 2003; Chen et al., 2004). Data here nevertheless showed that immersion of more than 8 h lead to inhibition of photosynthesis to senescence as reflected by hormone production in leaves and to ADH activity declining in roots. These results matched the immersion times that *K. candel* seedlings tolerate under their natural distribution in mid-to-high tidal zones and some in mid-to-low tidal zone as a pioneering species in the field (Lin, 1988, 1999).

The increase of tidal immersion time caused by the subsidence of plantable tidal flats influences survival rates in mangroves (re)-forestation (Liu, 1995; Komiyama, 1996; Chen et al., 2001; Kitaya et al., 2002) by inhibiting the growth of seedlings (Ye et al., 2003; Chen et al., 2004), and by causing other physiological changes, even resulting in stunting or death of mangroves (Zhang et al., 1997). A parallel study Chen et al. (2004) showed that *K. candel* seedlings had strong tolerance of periodical waterlogging and that 8 h was the maximum, beyond which growth was reduced. Combined with the photosynthetic and physiological responses, this supports the conclusion that regular immersion in seawater for more than 8 h significantly reduces carbon assimilation and speeds up senescence of seedling leaves. Despite the detrimental effects of long term waterlogging, the induction of anti-oxidation enzymes indicates that *K. candel* seedlings have some resistance to waterlogging.

The survival rate of *K. candel* seedlings during afforestation is reduced by zero waterlogging and long time waterlogging (Chen et al., 2004). Action should be taken to raise the tidal flats and shorten waterlogging time in mangrove restoration. Our results provide theoretical guidance for tidal level selection in mangrove rehabilitation.

5. Conclusion

Mangroves are threatened by the subsidence of plantable tidal flats, by rising sea level and the consequent increase of duration of waterlogging by high tides. In order to undertake the most effective mangrove rehabilitation there is an urgent need to determine the appropriate tidal levels for mangrove seedlings. The results of this simulation experiment clearly show the influenced of periodical waterlogging time on the photosynthetic and physiological responses on *K. candel* seedlings, and we concluded that seedlings are threatened by long periods of immersion in seawater. Inundation by sea water for more than 8 h in a tidal cycle leads to increased ABA accumulation and decreased P_n, and induction of anti-oxidative enzymes in leaves, and reduced ADH and oxidase activity in roots. *K. candel* seedlings were tolerant of tidal immersion in sea water for up to 8 h, which matches its natural distribution in the inter-tidal zones of China.

Acknowledgements

The project was supported by National Natural Science Fund of China (NNSFC) under contract No. 30200031, Doctorate Program Foundation of the Education Ministry of China under contact No. 20030384007, and Item of Science and Technology in Xiamen, China under contact No. 3502Z20021046. We greatly appreciate Prof. W.J. Cram for his comments and advices on phraseology and expression on the manuscript. We also thank Dr. Chen C.P., Dr. Zhang Y.H., and Wang L., Jie Z.L., Chi M.J., for their assistance in seedlings cultivation and measurement. We are grateful to two anonymous reviewers for their comments on this work.

References

Ahmed, S., Nawata, E., Hosokawa, M., Domae, Y., Sakuratani, T., 2002. Alterations in photosynthesis and some antioxidant enzymatic activities of mungbean subjected to waterlogging. Plant Sci. 163, 117–123.

Akhtar, J., Gorham, J., Qureshi, R.H., Aslam, M., 1998. Does tolerance of wheat to salinity and hypoxia correlate with root dehydrogenase activities of aerenchyma formation? Plant Soil 201, 275–284.

Ashord, A.E., Allaways, W.G., 1995. There is a continuum of gas space in young plants of *Avicennia marina*. Mar. Ecol. Progr. Ser. 295, 5–11.

Bradford, M.M., 1976. A rapid and sensitive method for the quantitation of microgram quantities of protein using the principle of protein–dye binding. Ann. Biochem. 72, 248–254.

Chen, L.Z., Wang, W.Q., Lin, P., 2004. Influence of waterlogging time on the growth of *Kandelia candel* seedlings. Acta. Oceano. Sin. 23 (1), 149–158.

Chen, Y.J., Chen, W.P., Zheng, S.F., Zheng, D.Z., Liao, B.W., Song, X.Y., 2001. Researches on the mangrove plantation in Panyu, Guangdong. Ecol. Sci. 20, 25–31 (in Chinese).

Ellison, A.M., Farnsworth, E.J., 1997. Simulated sea level change alters anatomy, physiology, growth, and reproduction of red mangrove (*Rhizophora mangle* L.). Oecologia 112, 435–446.

Fan, H.Q., Li, G.Z., 1997. Effect of sea dike on the quantity, community characteristic and restoration of mangroves forest along Guangxi coast. Chin. J. Appl. Ecol. 8, 240–244 (in Chinese).

Field, C.D., 1995. Impact of expected climate change on mangroves. Hydrobiologia 295, 75–81.

Guan, Z.H., 1996. Introduction of Plant Iatrology. China Agriculture University Press, Beijing (in Chinese).

Hovenden, M.J., Curran, M., Cole, M.A., Goulter, P.F.E., Skelton, N.J., Allaway, W.G., 1995. Ventilation and respiration in roots of one-year-old seedlings of grey mangrove *Avicennia marina* (Forsk.) Vierh. Hydrobiologia 295, 23–29.

Kato-Noguchi, H., Saito, H., 2000. Induction of alcohol dehydrogenase in lettuce seedlings by flooding stress. Biologia. Plant. 43, 217–220.

Kitaya, Y., Jintana, V., Piriyayotha, S., Jaijing, D., Yabuki, K., Izutani, S., Nishimiya, A., Iwasaki, M., 2002. Early growth of seven mangrove species planted at different elevations in a Thai estuary. Trees 16, 150–154.

Komiyama, A., Santiean, T., Higo, M., Patanaponpaiboon, P., Kongsangchai, J., Ogino, K., 1996. Microtopography, soil hardness and survival of mangrove (*Thizophora apiculata* BL.) seedlings planted in an abandoned tin-mining area. For. Ecol. Manage. 81, 243–248.

Kozlowski, T.T., 1984. Flooding and Plant Growth. Academic Press. INC, London.

Li, H.S., 2000. Principles and Techniques of Plant Physiological Biochemical Experiment. Higher Education Press, Beijing (in Chinese).

Li, Z.T., Zhou, X., 1996. Plant Hormone and Immunoassay. Jiangsu Science and Technology Press, Nanjing (in Chinese).

Lin, P., 1988. Mangrove Vegetation. Ocean Press, Beijing.

Lin, P., 1999. Mangrove Ecosystem in China. Science Press, Beijing.

Liu, X.Z., Wang, Z.L., Gao, Y.Z., 1991. Relationships between alcohol dehydrogenase activity and flooding tolerance in corn roots under waterlogging stress. Jiangsu J. Agri. Sci. 7, 1–7 (in Chinese).

Liu, Y.L., 1992. Physiology of Plant Water Stress. Agriculture Press, Beijing (in Chinese).

Liu, Z.P., 1995. A study on the methods and technique of mangroves ecological afforestation. Ecol. Sci. 2, 100–104 (in Chinese).

Liu, Z.Q., Zhang, S.C., 1994. Plant Stress-Resistant Physiology. China Agriculture Press, Beijing (in Chinese).

McKee, K.L., Mendelssohn, 1987. Root, metabolism in the black mangrove (*Avicennia germinans* L): responses to hupoxia. Environ. Exp. Bot. 27, 147–158.

McKee, K.L., 1996. Growth and physiological responses of neotropical mangrove seedlings to root zone hypoxia. Tree Physiol. 16, 883–889.

Mo, Z.C., Fan, H.Q., 2001. Comparison of mangrove forestation methods. Guangxi Forestry Sci. 30, 73–75, 81. (in Chinese).

Naidoo, G., 1984. Effects of flooding on leaf water potential and stomatal resistance in *Bruguiera gymnorrhiza* L. Lam. New Phytol. 93, 369–376.

Naidoo, G., Rogalla, H., Von-Willert, D.J., 1997. Gas exchange responses of a mangrove species, *Avicennia marina*, to waterlogged and drained conditions. Hydrobiologia 352, 39–47.

Pazeshki, S.R., Pardue, J.H., DeLaune, R.D., 1993. The influence of oxygen deficiency and redox potential on alcohol dehydrogenase activity, root porosity, ethylene production and photosynthesis in *Spartina patens*. Environ. Exp. Bot. 33, 565–573.

Pezeshki, S.R., Delaune, R.D., Meeder, J.F., 1997. Carbon assimilation and biomass partitioning in *Avicennia germinans* and *Rhizophora mangle* seedlings in response to soil redox conditions. Environ. Exp. Bot. 37, 161–171.

Pezeshki, S.R., Delaune, R.D., Patrick Jr., W.H., 1989. Differential response of selected mangroves to soil flooding and salinity: gas exchanges and biomass partitioning. Can. J. For. Res. 20, 869–874.

Watson, J.G., 1928. Mangrove Forests of the Malay Peninsula. Malayan For. Rec. 6, 275.

Wu, S.R., Chen, W.F., Zhou, X., 1988. Enzyme linked immunosorbent assay for endogenous plant hormones. Plant Physiol. Commun. 5, 53–57 (in Chinese).

Ye, Y., Tam, N.F.Y., Wong, Y.S., Lu, C.Y., 2003. Growth and physiological responses of two mangrove species (*Bruguiera gymnorrhiza* and *Kandelia candel*) to waterlogging. Environ. Exp. Bot. 49, 209–221.

Zhang, Q.M., Yu, H.B., Chen, X.S., Zheng, D.Z., 1997. The relationship between mangrove zone on tidal flats and tidal levels. Acta. Ecol. Sin. 17, 258–265 (in Chinese).

Zhang, Z.L., 1990. Guides to Plant Physiological Experiments. Higher Education Press, Beijing.

潮汐淹水时间对秋茄幼苗生长的影响*

陈鹭真[1,2] 王文卿[1,2] 林 鹏[1,2]
(1. 厦门大学生命科学学院,福建 厦门 361005;
2. 厦门大学湿地与生态工程研究中心,福建 厦门 361005)

摘要: 模拟正规半日潮淹水的条件下,不同淹水时间对秋茄[*Kandelia candel*(L.)Druce]幼苗生长的影响.应用沙和土两种培养基质,制备盐度为15的人工海水栽培幼苗培养时间为70 d.不淹水(0 h)和长时间淹水(>8 h)对秋茄幼苗的生长有一定的抑制作用,而短时间淹水(淹水2～4 h)对幼苗生长有所促进.较长时间的淹水可促进贮气根的数量增加和长度的增长.在当前培养条件下,秋茄幼苗生长的最佳淹水时间是每个潮水循环淹水2～4 h.2 h处理的沙培秋茄幼苗总生物量和成熟叶面积最大,胚轴失重最小,土培幼苗也有相应的变化趋势.秋茄幼苗生长的临界淹水时间是每个潮水循环淹水8 h.秋茄幼苗有较强的抗淹水能力,适当的淹水对其生长有利.

关键词: 秋茄;生长;最佳淹水时间;临界淹水时间
中图分类号: Q945.14 **文献标识码**: A **文章编号**: 0253-4193(2005)02-0141-07

1 引言

被誉为"海上森林"的红树林是亚热带和热带海岸潮间带的一种特殊植物群落,在防风防浪、保护海堤、保护生物多样性等方面具有重要作用.20世纪60年代以来,随着围海造田、围塘养殖和城市建设的开展,红树林这种特殊的森林在全球面临濒危.目前我国红树林总面积仅有2.2万hm^2[1].从红树林资源日趋减少和大量红树林宜林滩涂亟待绿化的现状出发以及沿海防护林体系建设的需要,开展大规模的红树林造林已是迫在眉睫的任务[2].2002年初,国家林业总局在深圳专门召开会议,计划在2010年前在全国营造6万hm^2红树林.上世纪90年代至今,我国红树林造林取得了很大的成绩,但是国内红树林造林成活率低的问题依然没有根本改变.据我们对福建沿海各县市的调查统计,红树林造林成活率不超过50%,许多地方甚至不到20%,全部覆灭的情况也时有发生[3].红树林造林成活率低在国内其他省区也非常普遍[4].由于围海造田使原先适合红树林生长的中高潮带滩涂损失殆尽,目前的红树林造林多在中低潮带滩涂,而且是逆境条件下的特种造林[5].大量的造林实践表明,潮汐浸淹时间是影响红树林造林的重要的限制因子[5~7],红树林宜林临界线的确定是红树林造林成败的关键[4,7,8].

国外研究者已对淹水胁迫下红树植物的生长[9~13]、叶片营养积累[14]、根系发育[14]、根系形态[12,15~18]、气生根的氧浓度[17]、植物水分关系[10]、离子富集[10]和根系气体运送机制[11,19~21]等方面进行了大量研究.上述研究的淹水胁迫方式是土壤表面渍水,此时整个植株基本暴露于空气中.野外的潮汐涨落过程中,由于幼苗植株矮小,因此地上部分暴露于空气中而土壤表面渍水的时间很短,而土表曝露于空气中和整株幼苗被海水浸淹这两个过程占用了绝大部分时间.幼苗受海水的浸淹时间则与植株在潮间带的位置有关.上述研究在阐明红树植物耐

* 国家自然科学基金资助课题(30200031);厦门市重点科技资助项目(3502Z20021046)
原载于海洋学报,2005,27(2):141－147

淹水(周期性潮汐浸淹)机理方面已取得了一系列成果,但没有给出红树幼苗生长的最佳淹水时间1) 和临界淹水时间2) . 秋茄是我国境内天然分布最广且纬度最高的红树植物,最北可以分布到福建福鼎,也是浙江省北移引种惟一成功的种类[22] ,同时秋茄也是我国东南沿海主要的红树林造林树种[23] . 本文采用模拟潮汐的人工控制装置,系统研究了完全淹水胁迫对红树植物秋茄幼苗生长的影响,以确定秋茄的最佳和临界淹水时间,为秋茄引种栽培的宜林地选择、提高秋茄的造林成活率提供科学依据.

2 材料和方法

2.1 实验装置的设计

温室环境下,设计一组模拟潮汐的装置[24] . 由A ,B ,C ,D ,E ,F ,G 七个桶组成,每个桶注满水的时间为2 h ,培养液从水缸中首先泵入 A 桶,注满 A 桶,耗时2 h ,再注入 B 桶,依此类推,当水注满 F 桶时,在时间控制装置控制下,各桶开始同时放水. 因此,桶 A 的淹水时间是12 h ,桶 G淹水时间为 0. 这样由 A 到 G 在淹水时间上构成一个 12 h ,10 h ,8 h ,6 h ,4 h ,2 h ,0 h 的梯度. 每天进行两个完整的循环,模拟了正规半日潮情况下的潮汐作用.

2.2 样品采集、栽培和参数测定

2002 年 4 月从九龙江口的龙海市浮宫镇草埔头村((24°29′N ,117°55′E) 的红树林引种园内,采回无病虫害、发育良好、成熟度接近的秋茄胚轴,选择长度、重量相近[平均鲜重为(12.97 ±3.00) g]的胚轴用于实验. 在 A ,B ,C ,D ,E ,F ,G 七个大桶(桶面积为240 000 mm^2 ,桶高600 mm) ,每桶各放入 6 个小桶(口径为200 mm ,高200 mm ,桶底穿孔,当大桶放水时,小桶中的水可迅速排干) . 培养基质为沙和土. 3 个小桶中装沙(沙取自建筑工地的河沙,粒径约为1 mm ,经自来水洗净) ,3 个小桶中装土(土取自采集胚轴的秋茄林下,和匀) . 将胚轴随机植于小桶中,每小桶种植胚轴 5 棵. 在温室中培养(自然光,温度为 27～32 ℃) . 培养液为海水(取自厦门西海域,经高压沙滤净化,盐度为 22～28) ,用自来水调配至盐度为 15 (接近九龙江口的龙海市浮宫秋茄林的海水平均盐度 17) [22,23] ,每周更换一次,每天补充适量自来水以保持盐度稳定. 人工潮汐的高潮水位是从大桶桶底向上60 cm (桶高) ,低潮水位线略低于培养基质表面. 培养时间为70 d (2002 年 4 月 28 日至 7 月 4 日) . 培养过程中定期记录生长状况.

培养后把植株的根、茎、叶、胚轴分开. 实验过程中,我们发现一类直径一般大于 5 mm ,疏松而表面光滑的根,而现有文献并未报道这类根系[22] ,我们暂且称之贮气根. 将秋茄幼苗的根系分成 3 部分:大根(1 mm < D < 5 mm) 、细根(D < 1 mm) 和贮气根. 测量大根数目、长度,贮气根数目、长度,叶片鲜重、成熟叶片数和叶面积. 在105 ℃杀青后,在60 ℃烘干至恒重. 测定各组分的干重. 叶面积的测定采用剪纸衡重法. 叶片肉质化程度根据多汁度(S) 的计算方法[24] ,为单位面积叶片的饱和水分含量(mg/cm^2) .

多汁度等于饱和水分含量(g) 除以表面积(dm^2) . 所有数据用 SPSS 软件进行 I-way ANOVA 分析.

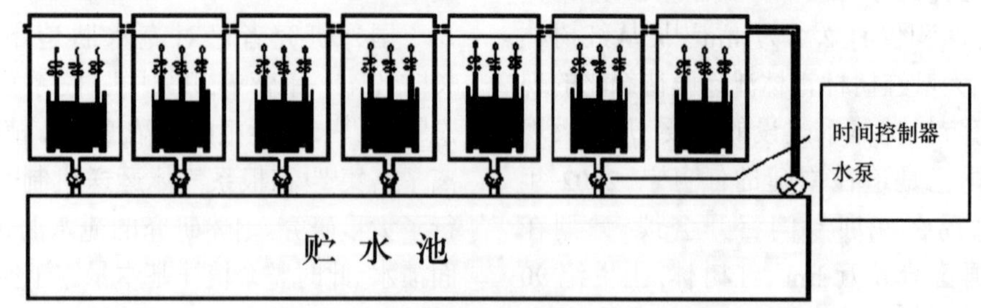

图 1 模拟滩涂淹水实验装置示意图

3 结果

3.1 淹水时间对根系生长的影响

淹水超过2 h ,沙培幼苗大根和细根生物量随淹水时间延长而呈递减趋势(见图 2 ,3) ;淹水2 h ,大根和细根生物量均为最高,比最低的处理(12 h) 分别高 154.2 %和 209.9 %. 土培的大根和细根生物量均在淹水 4 h达到最大,分别比淹水 12 h的处理高

1) 最佳淹水时间指潮汐的最佳淹水周期.

2) 临界淹水时间指对幼苗生长显著抑制的潮汐浸淹的临界值.

200.8 %和 273.8 %. 大根长度在2 h最大，比最低的(12 h)处理高 125.6 %. 不淹水(0 h)和长时间淹水(>8 h)条件下，根系明显生长不良，大根数目少、长度短(表 1). 在淹水 2～12 h，淹水时间显著抑制了大根生物量、大根长度和细根生物量的增加($P=0.000$，$P=0.021$，$P=0.000$，沙培；$P=0.000$，$P=0.000$，$P=0.000$，土培；$df=6$).

不同处理在贮气根长度和数量上有不同的响应. 不淹水和淹水2 h，沙培和土培植株均不出现贮气根. 淹水12 h，沙培处理仅出现 1 条贮气根，长度为9.3 cm，而土培处理不出现贮气根；淹水6 h贮气根数量最多，而土培和沙培贮气根长度的最大值分别出现在 6 和4 h. 总体而言，淹水时间为 6，8，10 h，贮气根数量较多，长度较长(表 1). 适当的淹水对贮气根数量和长度伸长都有一定的促进.

表1　不同淹水时间下每苗秋茄大根和贮气根长度和数量

淹水时间/h	大根				贮气根			
	数量/条		长度/cm		总数量/条		长度/cm	
	沙培	土培	沙培	土培	沙培	土培	沙培	土培
0	22.5 ±0.9	21.3 ±0.4	8.7 ±1.8	10.1 ±1.3	0	0	—	—
2	22.0 ±2.1	19.5 ±3.7	10.8 ±1.91	4.2 ±2.0	0	0	—	—
4	24.8 ±1.7	25.0 ±0.9	10.6 ±1.3	12.2 ±1.2	1	4	0.60 ±0.0	4.33 ±3.72
6	26.3 ±3.1	22.2 ±1.0	10.2 ±1.6	10.7 ±1.7	7	8	6.17 ±2.90	2.90 ±2.69
8	23.1 ±3.0	24.2 ±0.9	10.1 ±0.5	10.4 ±1.7	5	6	2.90 ±2.32	3.32 ±3.21
10	23.4 ±1.6	21.3 ±1.8	9.1 ±0.8	9.3 ±1.4	6	7	1.95 ±1.17	3.54 ±2.87
12	19.9 ±2.1	20.6 ±3.9	8.6 ±0.9	8.1 ±1.6	1	0	9.30 ±0.0	—

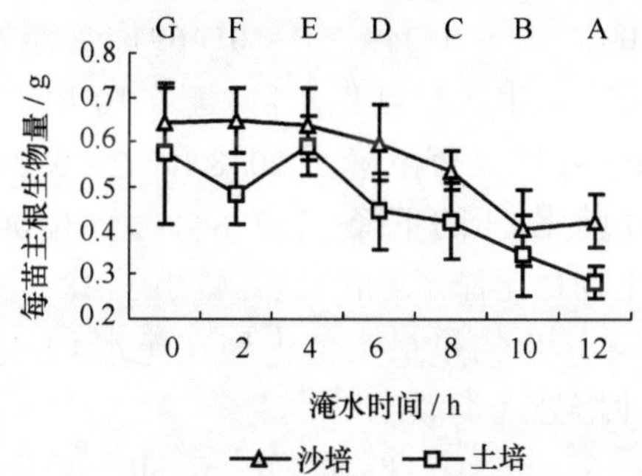

图 2　淹水时间对秋茄幼苗主根生物量的影响

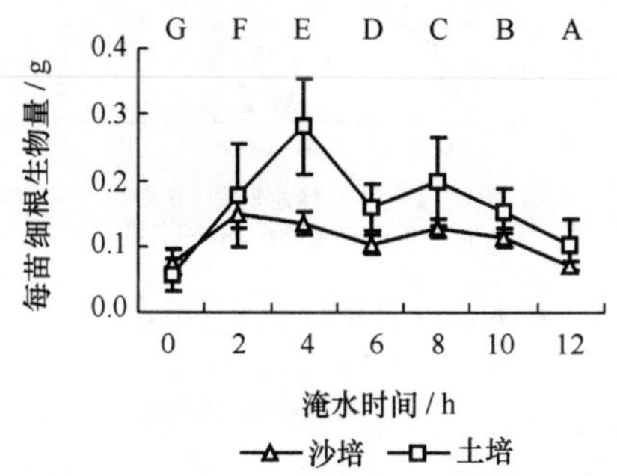

图 3　淹水时间对秋茄幼苗细根生物量的影响

3.2　淹水时间对茎生长的影响

沙培茎生物量经8 h的处理最大，10 h次之，然后依次是 6，4 和2 h. 土培茎生物量具有相似的趋势. 由此可见，一定的淹水时间对茎的生长有促进作用($P=0.000$，$df=6$). 沙培处理不淹水和淹水12 h的生境下，茎生物量很少，生长不良，分别比最大茎生物量低 72.1 %和 64.8 %(图 4). 土培不淹水和淹水12 h的处理，分别比最大茎生物量低 77.7 %和 45.5 %.

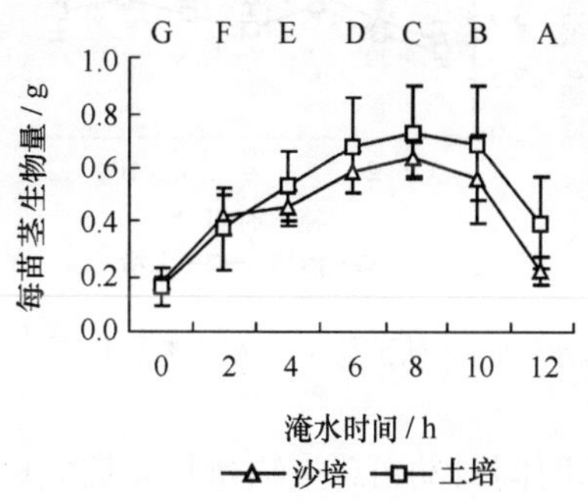

图 4　淹水时间对秋茄幼苗茎生物量的影响

3.3　淹水时间对叶片生长的影响

秋茄幼苗成熟叶的生物量、叶面积在沙培处理下淹水2 h达到最大，分别比淹水时间最长的12 h处理高 595.5 %和 435.3 %，而土培条件下淹水4 h达到最大. 淹水 2～12 h，沙培成熟叶生物量和叶面积与淹水时间成线性负相关关系($P = 0.000$, $df = 6$),土培处理淹水 4～12 h具有相同的趋势($P = 0.000$, $df = 6$)(图 5,6).叶片肉质化程度随淹水时间的延长而极显著增大($P = 0.005$,沙培; $P = 0.031$,土培; $df = 6$)(图 7).淹水12 h,成熟叶肉质化程度均急剧减小.

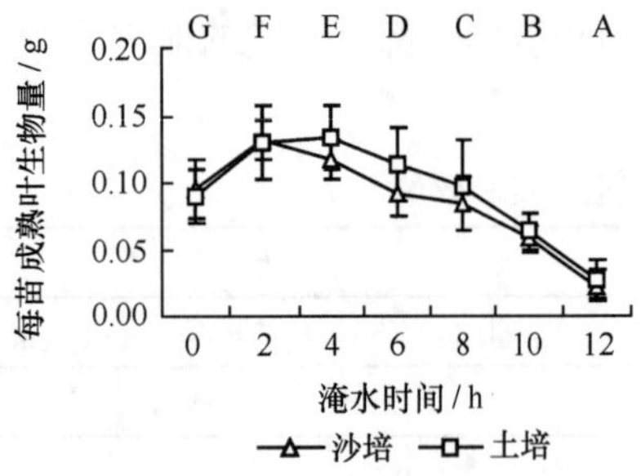

图 5 淹水时间对秋茄成熟叶生物量的影响

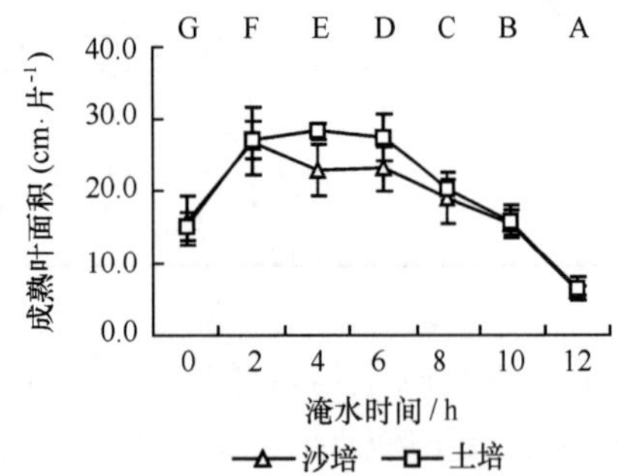

图 6 淹水时间对秋茄幼苗成熟叶面积的影响

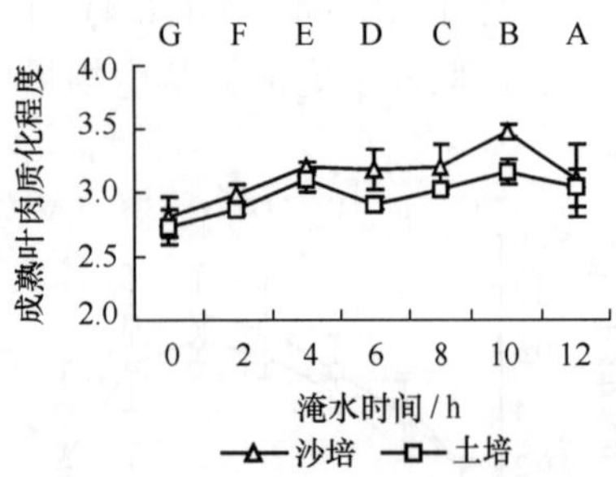

图 7 淹水时间对秋茄幼苗成熟叶肉质化程度的影响

3.4 淹水时间对幼苗总生物量和胚轴失重的影响

秋茄是显胎生红树植物,种子在果实脱离母体前就萌发,形成胚轴.在幼苗早期的生长过程中,胚轴为之提供了大量的营养,因此常表现为负增长.经过 10 周的淹水培养,胚轴重量表现为负增长,比培养前的5.60 g干重明显下降.沙培处理淹水2 h,胚轴失重最小.淹水6和8 h,胚轴重量损失最大,是2 h的 2 倍.当淹水超过8 h,胚轴失重随淹水时间的延长而显著减小($P = 0.005$, $P = 0.001$)(表 2).土培处理在淹水时间较短时,失重最小,而淹水时间 4～8 h时失重最大,差异不显著($P = 0.640$, $P = 0.695$).在淹水时间 2～8 h,发育过程中的胚轴失重有所增大.

表 2 淹水时间对秋茄幼苗胚轴失重的影响

淹水时间/h	胚轴失重/g	
	沙培	土培
0	1.241 ±0.274	1.245 ±0.165
2	1.109 ±0.315	1.863 ±0.075
4	1.292 ±0.279	2.290 ±0.033
6	2.219 ±0.028	2.348 ±0.172
8	2.218 ±0.112	2.300 ±0.257
10	1.696 ±0.238	2.182 ±0.055
12	1.539 ±0.126	1.565 ±0.143

经过 10 周的培养,植株的总生物量变化如图 8 所示.淹水2 h,根、茎、叶生物量总和最大,比总生物量最小的处理(12 h)高 232.2 %.淹水 2～8 h,沙培总生物量受淹水时间的延长而减少,但变化不显著($P = 0.037$, $P = 0.249$, $P = 0.469$).当淹水超过8 h,幼苗的总生物量陡然下降.淹水10 h的总生物量比8 h的处理显著下降了 20.8 %($P = 0.000$),而12 h的处理比8 h的下降了 52.6 %($P = 0.000$).不淹水的生物量比淹水2 h的显著减小了 38.5 %($P = 0.000$).土培生物量最大值出现在淹水4 h,其变化趋势与沙培处理类似.

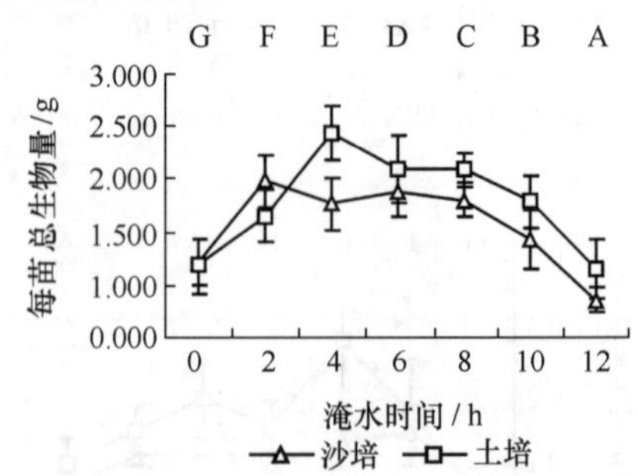

图 8 淹水时间对秋茄幼苗总生物量的影响

4 讨论

4.1 淹水时间对秋茄幼苗生长的影响

以往对其他植物的淹水胁迫的研究表明,淹水对于植物的生长会产生显著影响,特别是促进根系分蘖、不定根增生、根系气腔形成和减缓植株生长[25,26]. 白骨壤(*Avicennia marina*)在土壤表面淹水条件下,总生物量和根系生物量显著下降[12],萌芽白骨壤(*Avicennia germinans*)、拉关木(*Laguncularia racemasa*)和大红树(*Rhizophora mangle*)等3种红树植物在种植过程中表土渍水导致根系和叶片的生长都受到显著抑制[11]. 本研究发现延长淹水时间对秋茄幼苗的生长存在显著影响. 不淹水和长时间淹水(>8 h)均抑制了幼苗的生长. 沙培处理,除单位面积叶片肉质化程度、茎生物量、大根长度外,其他指标都在淹水2 h达到最大值;土培处理中,最大值均出现在淹水4 h的处理. 淹水时间延长,大根和细根的生物量显著减少(见图2,3),特别是与有效吸收营养物质有关的细根的生物量很少. 成熟叶生物量和叶面积也随淹水时间延长而减小,进而减弱了有效的光合作用和干物质的合成.

研究认为,淹水水稻将增加内源激素的分泌,进而加快茎、叶的向上生长[25]. 秋茄幼苗在完全淹水条件下也存在类似的现象,即在淹水时间较长(6~10 h)的处理中,茎的生物量升高. 这可能缓解根系缺氧的另一机制,即通过伸出水面的茎叶,将氧气输送到根系. 成熟叶肉质化程度是通过单位叶面积的饱和水分含量计算得到的. 植物淹水时,叶片通过关闭气孔、减少失水来保持稳定的膨压[26]. 本研究发现,适当的淹水有利于提高成熟叶片的肉质化程度.

4.2 淹水对秋茄贮气根生长的影响

淹水条件下,植物存在增加呼吸根的数量以缓解根系缺氧的机制[25,26],在其他红树植物中,呼吸根也与缓和根系缺氧有关[22]. 我们发现在幼苗胚轴略高于整个根系的着生位置上,有一类具疏松的海绵状结构、表面光滑无侧根、直径一般大于5 mm的根系. 经初步切片观察发现其皮层在横截面上占据较大的面积,且多孔隙,而在幼苗的生长过程中,这类根系并不露出土表. 我们认为这是一类与呼吸相关的根,可能与贮存氧气、在潮汐缺氧环境下提供植株所需的氧气有关. 现有文献也未报道这类根系[22,23],贮气根的具体结构特性正在进一步研究. 本实验中,淹水2~12 h,沙培和土培处理的贮气根的数量和长度均表现为先增后减,这说明在秋茄幼苗能最大限度维持正常生长的条件下,贮气根数量的增多可以缓和根系缺氧的逆境胁迫.

4.3 影响秋茄幼苗生长的最佳周期性淹水时间和临界淹水时间的探讨

由于红树林是一类生长在海岸潮间带的植被类型,其生长过程需要一定的潮汐作用[22]. 张乔民等[1,27]实地调查结果也表明,红树林主要分布在平均海平面稍上与回归潮平均高潮位(相当于正规半日潮型的大潮平均高潮位)之间的潮滩面. Ellison等[13]在人工半日潮条件下栽培的红树幼苗发现中潮带最有利于红树植物的生长. Komiyama等[28]报道土壤高程过低严重影响红树幼苗的成活率. 在自然状态下,秋茄多分布在中高潮带[23],每个潮水周期平均淹水时间大概为2~4 h. 从本实验结果看,当幼苗淹水2 h,沙培处理除茎生物量较少外,大根、细根和成熟叶以及总生物量均最大,同时大根数量最多、成熟叶面积最大,能进行有效的养分吸收和营养物质积累,而且2 h的处理,生长过程中没有出现明显的胁迫特征. 土培处理,除茎生物量较少外,其他指标的最大值基本出现在淹水4 h,其变化趋势与沙培处理基本相同. 在10周的培养过程中,幼苗胚轴均处于负增长状态,沙培2 h的处理胚轴失重最少;土培处理,在不淹水情况下胚轴失重最少. 可见,淹水2~4 h幼苗的生长正常,能进行一定的光合作用和营养吸收,以补充发育过程中胚轴的能量损耗. 当淹水超过8 h,与秋茄幼苗生长相关的各项指标,如大根生物量、长度,成熟叶生物量和叶面积,茎生物量,总生物量等均显著减小(见图2~8),秋茄幼苗生长受到显著抑制,植株的光合同化和营养积累下降. 根据沙培和土培研究结果,我们推测在模拟正规半日潮汐培养条件下,海水盐度为15时,秋茄幼苗完全淹水的最佳淹水时间是每个潮水周期浸淹2~4 h,而淹水临界时间为每个潮水周期淹水8 h左右. 当淹水时间超过这个临界值,秋茄幼苗生长受到显著抑制. 由于本实验的处理是幼苗完全淹水,而以往的研究[1,13,28]主要针对土壤表面渍水,因此相对于其他研究,本实验的滩面高程升高了,淹水时间就相对减少了.

由于红树林在潮滩的带状分布格局地区差异大,而且受到光、温度、水、气和生物因子等的影响. 实际造林中,即使确定了适合秋茄幼苗生长的最佳淹水时间和临界淹水时间,还必须综合以上因子,才能确定造林的宜林地. 与红树林野外造林的相关实验正在进一步进行中.适合秋茄幼苗的最佳淹水时间和临界淹水时间的确定,在红树林造林的宜林地选择和滩涂改造上具有一定的指导作用,对于有效提高造林成活率、推广福建沿海地区红树林造林具有实践意义.

参考文献:

[1] 张乔民,隋淑珍,张叶春,等.红树林宜林海洋环境指标研究[J]. 生态学报,2001,21(9):1 427—1 437.

[2] 莫竹承,范航清.红树林造林方法的比较[J].广西林业科学,2001,30(2):73—75,81.

[3] 王文卿,赵萌莉,邓传远,等.福建沿岸地区红树林的种类与分布[J].台湾海峡,2000,19:534—540.

[4] 廖宝文,郑德璋,郑松发,等.我国华南沿海红树林造林现状及其展望[J].防护林科技,1996,4(29):30—34.

[5] 刘治平.深圳福田红树林生态造林方法技术研究[J].生态科学,1995,2:100—104.

[6] KITAYA Y, JINTANA V, PIRIYAYOTHA S. Early growth of seven mangrove species planted at different elevations in a Thai estuary [J]. Trees, 2002, 16:150—154.

[7] 陈玉军,陈文沛,郑松发,等.广东番禺红树林造林研究[J].生态科学,2001,20(1,2):25—31.

[8] 范航清,黎广钊.海堤对广西沿海红树林的数量、群落特征和恢复的影响[J].应用生态学报,1997,8(3):240—244.

[9] CLARKE L D, HANNON N J. The mangrove and marsh communities of the Sydney district: Ⅲ. Plant growth in relation to salinity and waterlogging[J]. J Ecol, 1970, 58: 351—369.

[10] NAIDOO G. Effects of waterlogging and salinity on plant-water relations and on the accumulation of solutes in three mangrove species [J]. Aquat Bot, 1985, 22: 133—143.

[11] PEZESHKI S R, DELAUNE R D, PATRICK W H Jr. Differential response of selected mangroves to soil flooding and salinity: gas exchanges and biomass partitioning[J]. Can J For Res, 1989, 20: 869—874.

[12] HOVENDEN M J, CURRAN M, COLE M A, et al. Ventilation and respiration in roots of one-year-old seedlings of grey mangrove *Avicennia marina* (Forsk.) Vierh[J]. Hydrobiologia, 1995, 295: 23—29.

[13] ELLISON A M, FARNSWORTH E J. Simulated sea level change alters anatomy, physiology, growth, and reproduction of red mangrove (*Rhizophora mangle* L.)[J]. Oceanographic Literature Review, 1998, 45(6):1 003—1 004.

[14] MISRA S, CHOUDHURY A, GHOSH A. The role of hydrophobic substances in leaves in adaptation of plants to periodic submersion by tidal water in a mangrove ecosystem[J]. J Ecol, 1984, 72(2): 621—625.

[15] ASHORD A E, ALLAWAYS W G. There is a continuum of gas space in young plants of *Avicennia marina*[J]. Mar Ecol Progr Ser, 1995, 295:5—11.

[16] YOUSSEF T, SAENGER P. Anatomical adaptive strategies to flooding and rhizophere oxidation in mangrove seedlings[J]. Aust J Bot, 1996, 44: 297—313.

[17] SKELTON N J, ALLAWAY W G. Oxygen and pressure changes measured in situ during flooding in roots of the grey mangrove *Avicennia marina* (Forssk.) Vierh[J]. Aquat Bot, 1996, 54(2,3): 165—175.

[18] KOCH M S, SNEDAKER S C. Factors influencing *Rhizophora mangle* L. seedling development in everglades carbonate soils[J]. Aquat Bot, 1997, 59(1,2): 87—98.

[19] PEZESHKI S R, DELAUNE R D, MEEDER J F. Carbon assimilation and biomass partitioning in *Avicennia germinans* and *Rhizophora mangle* seedlings in response to soil redox conditions[J]. Environ Exp Bot, 1997, 37(2-3): 161—171.

[20] CHIU C Y, CHOU C H. Oxidation in the Rhizosphere of mangrove *Kandelia candel* seedling, soil sci[J]. Plant Nutr, 1993, 39(4): 725—731.

[21] NAIDOO G, ROGALLA H, VON-WILLERT D J. Gas exchange responses of a mangrove species, *Avicennia marina*, to waterlogged and drained conditions[J]. Hydrobiologia, 1997, 352(0): 39—47.

[22] 林 鹏.中国红树林生态系[M].北京:科学出版社,1997. 85—91.

[23] 林 鹏.红树林[M].北京:海洋出版社,1984.4—16.

[24] 王文卿,林 鹏.不同盐胁迫时间下秋茄幼苗叶片膜脂过氧化作用的研究[J].海洋学报.2000,22(3):49—54.

[25] 管致和.植物医学导论[M].北京:中国农业大学出版社,1996.100—104.

[26] 刘友良.植物水分逆境生理[M].北京:农业出版社,1992.144—187.

[27] 张乔民,于红兵,陈欣树,等.红树林生长带与潮汐水位关系的研究[J]. 生态学报,1997,17(3):258—265.

[28] KOMIYAMA A, SANTIEAN T, HIGO M, et al. Microtopography, soil hardness and survival of mangrove (*Thizophora apiculata* BL.) seedlings planted in an abandoned tin-mining area[J]. For Ecol Mange, 1996, 81:243—248.

Influence of waterlogging time on the growth of *Kandelia candel* seedlings

CHEN Lu-zhen[1,2], WANG Wen-qing[1,2], LIN Peng[1,2]

(1. *School of Life Science, Xiamen University, Xiamen* 361005, *China;* 2. *Research Center for Wetlands and Ecological Engineering, Xiamen University,* 361005, *China*)

Abstract: Influence of waterlogging time on the growth of *Kandelia candel* (L.) Druce seedlings grown for 70 d in artificial-tidal tanks simulated semidiurnal tide under greenhouse was studied. Sand and soil acted as the substrate and artificial seawater with the salinity of 15 were used in cultivation. Shorter waterlogging time (inundated for about 2～4 h) promoted the growth of *K. candel* seedlings, while longer time (inundated more than 8 h) or no waterlogging (0 h) inhibited their growth. The number and length of aerating roots increased with the increase of waterlogging time. In current condition, the optimal waterlogging time for the growth of *K. candel* seedlings was about 2～4 h in every tide cycle. Comparing with other treatments, seedlings in 2 h sanded treatments obtained the highest biomass, made the lowest mass loss of hypocotyl and broadened the photosynthetic area by increasing area per leaf after 70d cultivation. And the soil treatments had the similar tendency. However, waterlogging for 8 h in every tide cycle is critical for normal development of seedlings. *K. candel* seedlins were highly tolerant to waterlogging and a proper waterlogging was beneficial to the growth of *K. candel* seedlings.

Key words: *K. candel*; growth; optimal waterlogging time; critical waterlogging time

福建漳江口红树林区秋茄幼苗生长动态*

张宜辉[1] 王文卿[1] 吴秋城[2] 方柏州[2] 林 鹏[1]
(1. 厦门大学生命科学学院,厦门大学湿地与生态工程研究中心,厦门 361005;
2. 福建漳江口红树林国家级自然保护区管理局,福建 云霄 363300)

摘要:通过福建漳江口红树林自然保护区内8个样方24个小样方人工种植600个秋茄胚轴,在3a时间内对秋茄胚轴建立、幼苗生长以及环境因子进行定期观测。研究结果表明:林缘空地的秋茄生长状况良好,白骨壤林内最不利于秋茄幼苗的生长。潮位、盐度、底质土壤理化因子不是造成该样地各样方间秋茄幼苗生长差异的主要原因。动物取食、光照状况以及种间竞争是限制秋茄生长的主要环境因子。秋茄胚轴在长根前易于随潮水漂走,底质土壤中白骨壤致密的根系抑制了秋茄胚轴的定植,导致白骨壤林内秋茄幼苗漂走的数量最多。昆虫和螃蟹等动物的取食是导致林内已经固着生长的秋茄幼苗大量死亡的最主要原因,而林外被取食的幼苗个体极少。此后秋茄幼苗能否继续成长,主要取决于幼苗所接受到的光照条件。3a后,在荫蔽的树冠下,秋茄幼苗无法存活;而在林外,秋茄幼苗已经长成幼树。在林外滩涂上迅速生长的互花米草,也将影响秋茄幼苗的更新和生长。

关键词:红树林;秋茄;生长;福建(漳江口)
文章编号:1000-0933(2006)06-1648-09 **中图分类号**:Q948.1 **文献标识码**:A

The growth of *Kandelia candel* seedlings in mangrove habitats of the Zhangjiang Estuary in Fujian Province, China

ZHANG Yi-Hui[1], WANG Wen-Qing[1], WU Qiu-Cheng[2], FANG Bai-Zhou[2], LIN Peng[1] (1. *School of Life Sciences, Xiamen University, Research Center for Wetlands and Ecological Engineering, Xiamen* 361005, *China*; 2. *Administrative Bureau of Zhangjiang Estuary Mangrove National Natural Reserve, Fujian, Yunxiao* 363300, *China*). Acta Ecologica Sinica, **2006, 26(6): 1648～1656.**

Abstract: To evaluate the establishment and early growth of the mangrove species *Kandelia candel* in the intertidal zone, and to develop a better understanding of biotic and abiotic factors influencing the regeneration of its seedlings, we conducted a field experiment in Zhangjiang Estuary in Fujian. Different positions along the intertidal gradient were selected from 20 m to 120 m horizontal distance down the shore, including eight sampling sites in the mangrove areas. Equal numbers (75) numbers of mature propagules of *K. candel* were planted in each sampling site. The fates of propagules and growth of seedlings were monitored for 3 years.

The rates of rooting of *K. candel* propagule varied spatially. The lowest rates occurred in sites with an *Avicennia marina*-dominated overstory (69.7%). The rates were higher in sampling sites with a *K. candel*-dominated overstory (90.0%), at the fringe of the mangrove forest (89.3%), and on the bare tidal flat outside the mangrove forest (82.7%). After 1 year, the survival rates of seedlings planted under *A. marina* forest, *K. candel* forest, at the fringe of the mangrove forest, and the bare tidal flat were 13.7%, 54.7%, 76.0%, and 34.7%, respectively. Among the surviving *K. candel* seedlings, those at the fringe of the mangrove forest and on the bare tidal flat had greater height, stem diameter, leaf number, leaf area, and biomass

* 国家自然科学基金资助项目(30200031);福建省自然科学基金资助项目(B0410001)
原载于生态学报,2006,26(6):1648－1656

than did those under the *A. marina* and *K. candel* forests. In general, our experiment demonstrated that establishment and growth of *K. candel* seedlings occurred successfully at the fringe of the mangrove forest but did worst under the *A. marina* forest.

We analyzed the factors which translate *K. candel* seedlings' performance into significant differences in terms of establishment and early growth among sites. The performance of *K. candel* seedlings was not correlated with sediment texture, pH, salinity, organic matter, total N, and total P among the sites. However, interspecies competition, propagule predation by insects and crabs, and the incident light had significant effects on seedling survival and growth.

At the early growth stage, the probability of establishment of *K. candel* propagules planted in the intertidal sediments was influenced by predators and tidal disturbance. Under the *A. marina* forest, the compact root system of *A. marina* prevents *K. candel* propagules from rooting so that the propagules tend to be carried away by tidal currents. Insects and herbivorous crabs can play a considerable role in the predation of mangrove propagules and possibly are a threat to the regeneration of mangroves. We found that rates of insect and crab predation were higher in the intertidal location under intact canopies than at the fringe of the mangrove forest and on the bare tidal flat.

Longterm survival of seedlings and their development into saplings depend on light availability. Analyses showed that correlation between growth parameters of one year old *K. candel* seedlings and light intensity was significant. Shade reduced seedling growth in the field. Only those seedlings at the fringe of the mangrove forest and bare tidal flat established successfully and grew to maturity. These sites afford better growth conditions than the surrounding understory and, as importantly, provide a refuge from predation by insects and crabs.

Our results also indicate that the rapid growth of *Spartina alterniflora* reduces the regeneration of *K. candel*. As a competitive plant to *K. candel* in the mid intertidal zone, *S. alterniflora* may be having a large impact on the mangrove composition of our study forests. It is necessary to search for ways to protect this reserve area of mangrove wetland.

Key words: mangroves; *Kandelia candel*; growth; Fujian (Zhangjiang Estuary)

红树林是热带亚热带海岸潮间带具有重要生态防护功能的植被类型,由于受到海水周期性浸淹,红树植物特化出胎生繁殖的方式来适应潮滩生境[1]。因此,胎生苗(胚轴)→幼苗→幼树是红树林种群更新和发展的重要阶段[2]。对红树植物成熟繁殖体从母树上掉落后幼苗生长阶段的研究,有助于了解成年红树植物种群生长分布的特征[3]。

潮间带红树植物幼苗的生长受到多种因素(生物、非生物因素)的影响。已有许多研究从光照状况[4~7],底质土壤颗粒大小和理化性质[4,6,8~11],潮汐[8,12],盐度[4,7,13],动物取食[14~17],繁殖体大小[17~19],繁殖体传播方式[3,20,21]以及种间竞争[8,22]等因子对红树植物胚轴或种子的发芽和生长的影响进行分析和探讨。

本研究根据福建漳江口红树林自然保护区滩涂和红树林植被的分布状况,在选定的样地中,从林内到林外光滩插植秋茄(*Kandelia candel*)胎生苗(胚轴),定期调查各样方内秋茄幼苗的存活和生长情况,同时测定不同样方生境的土壤、光照等生态因子。通过上述测定工作,力图从光照、底质土壤理化性质、潮汐、种子或幼苗被捕食以及种间竞争等方面分析影响秋茄幼苗存活和生长的主要因素,从而为当地红树林的恢复和红树林造林工作提供科学依据。

1 样地概况与研究方法

1.1 样地概况

样地位于福建云霄县漳江口国家级红树林自然保护区(23°55′N,117°26′E)。属亚热带海洋性季风气候,气候温暖湿润,光、热、水资源丰富。根据 1960～1999 年的气候资料统计,年均气温 21.2 ℃,最高月均温 28.9 ℃(8 月份),最低月均温 13.5 ℃(1 月份);年平均降雨量为 1714.5 mm,降雨量主要集中在 4～9 月份。保护区近岸表层海水温度随季节变化较大,2 月份水温较低,8 月份水温较高,变化范围在 14.9～25.6 ℃;受降雨、江河径流和潮汐的影响,海水盐度在 12～26 之间变化;该海域潮汐属不正规半日潮,最大潮差 4.67 m,最小潮差 0.43 m,平均潮差 2.32 m,最高潮位 2.80 m,最低潮位 - 2.00 m,平均海平面 0.46 m(以黄海平均海平面为潮位高程基准面),平均涨潮历时 397 min,平均落潮历时 315 min[23]。

通过对整个保护区的踏勘,所选定的样地位于云霄县东厦镇竹塔村附近,红树林带宽 105 m,林下滩涂较为平整。并且该样地离码头以及潮沟较远,受人为干扰破坏较少。从堤岸开始,垂直离岸 0～20 m 之间为白骨壤林,处于一个凹地中,地势较低;20～40 m 处为秋茄纯林,地势最高;40～100 m 之间为白骨壤林,105～120 m之间为互花米草盐沼,在白骨壤林缘外侧和互花米草盐沼之间 5～10 m 宽的交界处,有秋茄和桐花树混生,树高在 1.0～2.0 m 之间,林下较多裸露的空地。离岸 120 m 外的滩涂为泥蚶、缢蛏养殖区。在样地布设 3 条高程水准测量横断面,测量距离从岸边人工堤岸开始到离岸 120 m 处的红树林带前缘,平均 10 m 为一个测点。退潮时,携带标志杆和已装好水的透明塑料软管,根据连通器原理进行测量,得到离岸 0～120 m 之间样地滩涂的相对垂直高程(图 1)。其中,离岸 120 m 处红树林生长带前缘高程为 1.03 m(以黄海平均海平面为潮位基准面),在平均海平面稍上,每天 2 次的涨潮均能淹及样地土壤。

根据《中国植被》[24]的划分方法,该样地内有红树林和滨海盐沼两个植被类型。其中红树林植被类型中有秋茄林(Form. *Kandelia candel*)和白骨壤林(Form. *Avicennia marina*)两个群系,滨海盐沼为天然互花米草盐沼群系(Form. *Spartina alterniflora*)。

(1)秋茄林　外貌整齐,青绿色或深绿色,结构简单,郁闭度在 80 %左右。纯林,偶有桐花树(*Aegiceras corniculatum*)混生,树高 3～6 m,平均高 4.5 m,平均胸径 5.7 cm,密度为 30 株/100 m^2。

(2)白骨壤林　外貌整齐,灰绿色,结构简单,郁闭度在 70 %左右。纯林,偶有秋茄、桐花树混生,呈丛生状萌生林,高度 1.5～3.0 m,平均高 2.4 m,平均基径 14.0 cm,密度为 15 丛/100 m^2,地面有从表土伸出的指状呼吸根,(472 ±63)条/m^2。

(3)互花米草盐沼　植被繁茂,外貌整齐,青绿色,结构简单,郁闭度在 80 %～90 %。以互花米草(*Spartina alterniflora*)为单优势种,高 1.5～2.0 m,周边偶有桐花树和秋茄的幼苗生长。

1.2 研究方法

1.2.1 秋茄胚轴人工插植及其幼苗生长状况调查

根据生产实践经验,采集胚轴最好在胚轴脱落初、中期进行,此时采摘的成熟胚轴粗壮,插至海滩后容易生根固定,不易被浪潮漂走[25]。漳江口秋茄胚轴采收时间宜安排在 3 月上旬至 5 月上旬。在 2002 年 4 月 11 日,采集该保护区内的秋茄成熟胚轴,挑选发育良好,成熟度接近且重量、长度大小相近的个体,于 4 月 12 日栽培。栽培前秋茄单个胚轴平均鲜重(14.35 ±2.27) g,长度(22.24 ±1.66) cm。

根据样地滩涂剖面以及植被分布状况,采取如下种植方案:离岸 0～20 m 之间,地势较低且人为干扰相对较大,离岸 120 m 外的滩涂为泥蚶、缢蛏养殖区,受人为影响非常大,因此仅选择离岸 20～120 m 之间的滩涂进行栽培实验。沿堤岸垂直方向拉一条样线,

表 1　样方具体位置和所处林带

Table 1　Horizontal distance down shore and vegetations of sampling sites

样方序号 Sampling site number	离岸距离 (m) Horizontal distance down shore (m)	所处林带 Vegetation zonation
1	25	秋茄林 *Kandelia candel* forest
2	40	秋茄林和白骨壤林交界处,秋茄林下 boundary between *K. candel* and *Avicennia marina* forest, under *K. candel* forest
3	55	白骨壤林 *A. marina* forest
4	70	白骨壤林 *A. marina* forest
5	85	白骨壤林 *A. marina* forest
6	95	白骨壤林(林缘内侧 5 m) *A. marina* forest, 5 m inside the forest fringe
7	105	林缘外侧 Forest fringe
8	120	互花米草盐沼外侧滩涂 Bare tidal flat outside the Spartina alterniflora

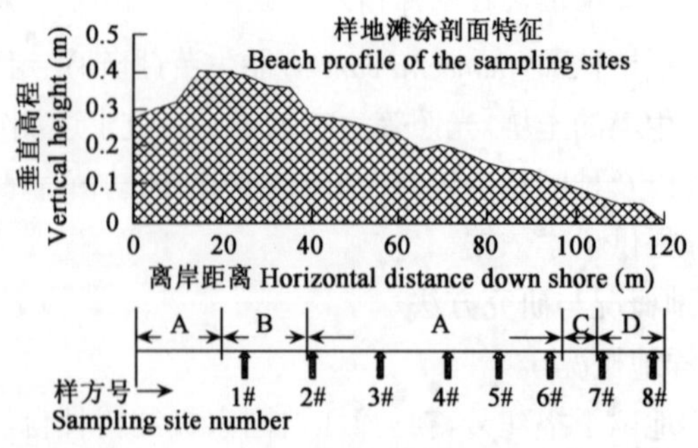

图 1　样地内不同水平离岸距离的滩涂剖面、林带分布及样方位置

Fig. 1　Beach profile, vegetation zonation and sampling sites location horizontal distance down shore at Zhuta of Yunxuao, Fujian

A:白骨壤林 *Avicennia marina* forest;B:秋茄林 *Kandelia candel* forest;C:林缘 Fringe;D:互花米草 *Spartina alterniflora*

从秋茄林内开始(离岸 25 m 处) ,沿样线每隔 10～15 m 为 1 个栽培样方,每个样方中选定 3 个 1 ×1 m^2 小样方(重复) ,小样方之间相隔 5 m,离岸距离相同。每个小样方插植 25 个秋茄成熟胚轴,株距 0.2 m ×0.2 m。总计 8 个样方,24 个小样方,插植 600 个胚轴。各样方具体位置以及所处林带见图 1 和表 1。

从 2002 年 4 月到 2003 年 4 月,定期(2 月/次)调查各样方内固着、漂走、外来这 3 种不同类型秋茄幼苗的数量,其中固着生长包括定居成活和死亡两类,死亡的幼苗进一步分为损伤性和非损伤性两种情况。同时观测定居成活幼苗的生长状况(3 株/小样方),记录幼苗的主茎节数、分枝数、叶片数,量取茎高(胚轴顶端到顶芽之间的长度,不包括胚轴)和基径(第 1 节中部)。

2003 年 4 月,随机挖取 1 年龄的秋茄幼苗(2～3 株/小样方),记录各株幼苗的主茎节数、分枝数、叶片数,剪纸衡重法测定叶面积。量取茎高、基径。烘干法(105 ℃,24 h)测定幼苗各部分的生物量。并在 2004 年和 2005 年跟踪调查各样方中幼苗的生长状况。

1.2.2 林下各样方生态因子的测定 光照强度采用上海市嘉定学联仪表厂生产的数字式照度计,每 2 个月测定 1 次各个小样方的光照强度,测定高度为离地面 0.5～0.8 m。以互花米草盐沼外侧光滩测定的光照强度为 100 %相对光照强度,林内各样方的相对光照强度按以下公式计算:

样方相对光照强度 = 林内光照强度/林外滩涂光照强度 ×100 %

从 2002 年 4 月到 2003 年 4 月,每 3 个月采集各个小样方的表层土壤(0～20 cm),带回室内自然风干、研细、拣去根系,过 35 号筛,贮存备用。土壤质地,有机质,N、P 含量,pH 值等理化性质的测定参照《土壤农业化学分析方法》[26]。

1.3 数据分析

采用 SPSS 软件对各样方间的生态因子、幼苗生长参数进行方差分析,并对生态因子、秋茄幼苗更新数量和幼苗生长参数之间进行相关分析。

2 结果

2.1 样方生态因子

各样方的生态因子见表 2。

秋茄林的郁闭度高于白骨壤林,因此秋茄林下(样方 1)的相对光照强度最小,仅为 7.15 %;而从秋茄林和白骨壤林交界处到白骨壤林林缘(样方 2～6)的相对光照强度在 14.15 %～16.01 %之间,较为一致。林缘外侧空地的光照程度受树冠影响小,其相对光照强度为 83.77 %。

各样方底质土壤均为粉粘土;土壤有机质含量和全氮量均以林外(样方 7、8)相对较低,全磷含量差别不大;土壤盐度在 13.86～21.79 之间,以秋茄林下最高;土壤 pH 值在 6.51～6.60 之间,较为一致。

表 2 各样方的生态因子

Table 2 Ecological factors in sampling sites

样方序号 Sampling site number	相对光照强度(%) Relative light intensity	土壤主要理化性质 Edaphic physical and chemical characters					
		土壤质地 Soil texture	有机质(%) Organic matter	全氮量(%) Total N	全磷量(%) Total P	盐度(%) Salinity	pH
1	7.15a ±2.21	粉粘土 Silty clay	3.34b ±0.33	0.33b ±0.06	0.04a ±0.01	21.79b ±1.35	6.55a ±0.08
2	14.15a ±2.80	粉粘土 Silty clay	3.80c ±0.25	0.33b ±0.08	0.04a ±0.01	19.40b ±2.94	6.60a ±0.09
3	16.01a ±3.54	粉粘土 Silty clay	3.94c ±0.14	0.34b ±0.03	0.03a ±0.01	18.62 ab ±0.33	6.53a ±0.14
4	14.51a ±3.98	粉粘土 Silty clay	3.61c ±0.18	0.32b ±0.08	0.03a ±0.01	16.68 ab ±2.25	6.54a ±0.04
5	14.25a ±3.30	粉粘土 Silty clay	3.41b ±0.35	0.32b ±0.06	0.03a ±0.01	15.97 ab ±1.45	6.55a ±0.08
6	15.68a ±4.11	粉粘土 Silty clay	3.16 ab ±0.33	0.33b ±0.08	0.03a ±0.01	16.59 ab ±1.96	6.51a ±0.07
7	83.77b ±11.27	粉粘土 Silty clay	2.74a ±0.16	0.27 ab ±0.05	0.03a ±0.01	13.86a ±2.26	6.57a ±0.04
8	100.00c ±0.00	粉粘土 Silty clay	2.45a ±0.07	0.20a ±0.02	0.03a ±0.01	15.12a ±1.04	6.60a ±0.05

同一列数据的不同字母表示多重检验结果差异显著,$p < 0.05$ Mean values in the same column having the different letters are significantly different at $p < 0.05$ level

此外,从样方 1 到样方 8,滩涂垂直高程依次下降,高程差为 0.39 m(图 1)。经实测,在样方 1～8 之间,平均涨潮历时 41 min,平均落潮历时 32 min。

2.2 各样方中插植秋茄幼苗的命运

表 3 为栽培 1a 过程中,每 2 个月观测,总计 6 次调查数据的统计结果。

表3 栽培1年内各样方中不同类型秋茄幼苗的数量统计

Table 3 Statistic of different types of *Kandelia candel* seedlings in sampling sites in 1 year. The values were the means of 3 replicates

项目 Item	样方序号 Sampling site number	幼苗类型 Seedling's type						
		固着* Settled	漂走* Carried away by tidal current	定居成活 Survival	死亡 Death	非损伤性死亡 Undamaged -death	损伤性死亡 Damaged -death	外来* Newly recruited naturally
个数 Number	1	22.0 ±1.7	3.0 ±1.7	12.3 ±0.6	9.7 ±1.2	6.0 ±1.0	3.7 ±0.6	5.3 ±0.6
	2	23.0 ±0.0	2.0 ±0.0	15.0 ±2.6	8.0 ±2.6	3.3 ±1.2	4.7 ±1.5	6.7 ±1.5
	3	20.0 ±1.7	5.0 ±1.7	3.0 ±0.0	17.0 ±1.7	2.7 ±0.6	14.3 ±1.5	0.7 ±0.6
	4	16.7 ±0.6	8.3 ±0.6	3.3 ±0.6	13.3 ±0.6	1.0 ±1.0	12.3 ±1.2	0.3 ±0.6
	5	16.3 ±2.9	8.7 ±2.9	2.0 ±2.0	14.3 ±1.5	2.3 ±0.6	12.0 ±2..0	0.0 ±0.0
	6	16.7 ±2.5	8.3 ±2.5	5.3 ±0.6	11.3 ±3.1	1.3 ±0.6	10.0 ±3.6	0.0 ±0.0
	7	22.3 ±0.6	2.7 ±0.6	19.0 ±1.7	3.3 ±1.5	1.3 ±1.5	2.0 ±0.0	3.3 ±1.2
	8	18.7 ±4.2	6.3 ±4.2	6.7 ±11.5	12.0 ±11.1	12.0 ±11.1	0.0 ±0.0	2.3 ±1.2
占栽培总体的百分比(%) Percentage of the total seedings planted in each sampling site	1	88.0 ±6.9	12.0 ±6.9	49.3 ±2.3	38.7 ±4.6	24.0 ±4.0	14.7 ±2.3	
	2	92.0 ±0.0	8.0 ±0.0	60.0 ±10.6	32.0 ±10.6	13.3 ±4.6	18.7 ±6.1	
	3	80.0 ±6.9	20.0 ±6.9	12.0 ±0.0	68.0 ±6.9	10.7 ±2.3	57.3 ±6.1	
	4	66.7 ±2.3	33.3 ±2.3	13.3 ±2.3	53.3 ±2.3	4.0 ±4.0	49.3 ±4.6	
	5	65.3 ±11.5	34.7 ±11.5	8.0 ±8.0	57.3 ±6.1	9.3 ±2.3	48.0 ±8.0	
	6	66.7 ±10.1	33.3 ±10.1	21.3 ±2.3	45.3 ±12.2	5.3 ±2.3	40.0 ±14.4	
	7	89.3 ±2.3	10.7 ±2.3	76.0 ±6.9	13.3 ±6.1	5.3 ±6.1	8.0 ±0.0	
	8	82.7 ±4.6	17.3 ±4.6	34.7 ±41.1	48.0 ±44.5	48.0 ±44.5	0.0 ±0.0	

* 2002 年 6 月测定数据 Data were obtained in June of 2002

2.2.1 固着、漂走和外来的秋茄幼苗 人工插植的秋茄胚轴在较短的时间内完成生根固着生长或随水漂走这一过程,在 2002 年 6 月份进行第一次调查测定之后,未再观察到有胚轴漂走。不同样方内秋茄幼苗固着生长的数目不同:样方 1、2 漂走的胚轴个体最少,仅占栽培总体的 12.0%和 8.0%,相应固着个体最多;在样方 3~6 中,胚轴漂走的个体最多,分别占栽培总体的 20.0%、33.3%、34.7%和 33.3%;样方 7、8 内漂走的胚轴数目也较少,占栽培总体的 10.7%和 17.3%。

每年 3~5 月份为漳江口秋茄胚轴大量掉落的时期,因此在第一次调查测定中可以观察到各样方中出现外来并固着生长的当年生秋茄幼苗。不同样方中外来秋茄幼苗的数目也不同:样方 1、2 处于秋茄林下,外来幼苗数目为各样方中最多,分别为 5.3 株/m^2 和 6.7 株/m^2;在样方 3~6 中,基本无外来个体;样方 7 和样方 8 中也有外来的幼苗,分别为 3.3 株/m^2 和 2.3 株/m^2。

2.2.2 秋茄幼苗固着后存活动态 对于那些已经成功固着生长的秋茄幼苗,部分幼苗在随后的生长过程中,受不同因素的影响而死亡。根据观测,死亡的个体可分为损伤性和非损伤性死亡两种类型,前者指秋茄幼苗的枝叶或胚轴被动物取食后而死亡,实验中未发现潮水、大型漂浮物体对样方内幼苗产生的机械损害;后者指秋茄幼苗出现萎蔫、枯死,但幼苗的枝叶或胚轴上未发现有噬咬或机械损伤的痕迹。另有部分幼苗枝条被咬后,仍可以从胚轴顶端长出新的枝条,视为定居成活的个体。

栽培 1 年内,不同样方中秋茄幼苗死亡的个体数不同(表 3):在样方 1、2 中,有 1/3 的幼苗死亡;在样方 3~6 中,死亡的个体数最多,为栽培总体的 45.3%~68.0%;样方 7 幼苗的死亡个体数最少,仅为栽培总体的 13.3%;而样方 8 幼苗死亡的个体数达到栽培总体的 48.0%,主要是由于互花米草生长扩张,导致其中两个小样方的幼苗被覆盖而死亡。

各样方中秋茄幼苗的死亡类型也不同:样方 1 的幼苗以非损伤性死亡为主,占死亡总体的 62.0%;到样方 2,损伤性死亡类型的幼苗占死亡总体的百分比为 58.5%,超过了非损伤性死亡的个体数;在样方 3~6 中,损伤性死亡的个体均占死亡总体的绝大部分,为 83.4%~92.5%;样方 7 中,损伤性死亡的个体相对较多,但从死亡的个体数来看,两种死亡类型幼苗的个体数仅分别为 1.3 株和 2.0 株,远小于其它样方的死亡个体数;样方 8 全部为非损伤性死亡类型的幼苗,其原因是受样方边上的互花米草的覆盖致死。

此后,到 2004 年 4 月,除样方 1、2、7 中仍有秋茄幼苗存活外(2a 成活率依次为 18.7%、23.2%和 71.9%),样方 3~6 和样方 8 中秋茄幼苗全部死亡。其中,随着互花米草的进一步扩张,最后 1 个小样方的幼苗也被覆盖致死。而到 2005 年,既栽培 3a 后,仅样方 7 中有秋茄幼苗进一步长成幼树(3a 成活率为 68.2%),样方 1、2 中秋茄幼苗也全部死亡。

2.3 定居成活秋茄幼苗的生长

各样方 1 年龄秋茄幼苗的生长参数见表 4。各项生长参数均以在林外两个样方最大。而在林内 6 个样方中,秋茄纯林下的幼苗的生长好于白骨壤林。根据上述生长参数并结合历次调查的结果,可以得出:定居成活的秋茄幼苗的总体生长状况为林缘空地和林外滩涂最好,秋茄林内次之,白骨壤林内最差。

此外,相关分析表明,各样方中 1 年龄秋茄幼苗的各项生长参数和相对光照强度之间均呈极显著正相关关系(表 5)。表明光照水平影响秋茄幼苗的光合生产,并进而对秋茄幼苗的根、茎、叶的形态生长和生物量均有影响。

表 4 各样方 1 年龄秋茄幼苗生长参数比较

Table 4 Comparison of some growth parameters of one year old *Kandelia candel* seedlings among sampling sites

项目 Item	样方序号 Sampling site number							
	1	2	3	4	5	6	7	8
茎高 Shoot height (cm)	30.2b ±2.7	29.7b ±2.8	20.5a ±2.2	22.9a ±1.5	26.2 ±2.7 ab	24.4ab ±1.9	52.7c ±3.6	55.5c ±4.5
基径 Stem diameter (cm)	0.48b ±0.04	0.44a ±0.01	0.45a ±0.02	0.42a ±0.02	0.44a ±0.03	0.44a ±0.01	0.91c ±0.01	0.87c ±0.04
每株叶片数 Number of leaves per plant	5.6b ±0.2	5.3b ±0.3	4.0a ±0.0	4.0a ±0.0	4.0a ±0.0	4.0a ±0.0	41.1c ±7.0	38.4c ±6.5
每株叶面积 Leaf area per plant (cm^2)	136.0b ±9.8	81.8ab ±15.5	26.6a ±6.8	27.2a ±8.3	35.2a ±6.5	55.1ab ±11.2	590.9c ±33.2	534.9c ±27.9
每株枝条干重 Shoot dry weight per plant (g)	0.95b ±0.13	0.90b ±0.11	0.54a ±0.10	0.54a ±0.12	0.67a ±0.11	0.71a ±0.12	6.97c ±0.79	6.04c ±0.56
每株叶片干重 Leaf dry weight per plant (g)	1.07b ±0.11	0.81 ab ±0.07	0.42a ±0.06	0.48a ±0.05	0.50a ±0.05	0.36a ±0.05	7.74c ±0.69	6.88c ±0.57
每株根干重 Root dry weight per plant (g)	2.15b ±0.29	1.98b ±0.45	1.55a ±0.26	1.57a ±0.31	1.40a ±0.38	1.71 ab ±0.25	12.88c ±1.51	11.20c ±1.05

同一行数据的不同字母表示多重检验结果差异显著,$p < 0.05$ Mean values in the same row having the different letters are significantly different at $p < 0.05$ level; The values were the means of 6~9 seedlings

表 5 各样方 1 年龄秋茄幼苗生长参数与相对光照强度的相关系数

Table 5 Correlation coefficients between growth parameter of one year old *Kandelia candel* seedlings and relative light intensity

项目 Item	茎高 Shoot height	基径 Stem diameter	叶片数 Number of leaves	叶面积 Leaf area	枝条干重 Shoot dry weight	叶片干重 Shoot dry weight	根干重 Shoot dry weight
相关系数 Correlation coefficients	0.954 **	0.975 **	0.981 **	0.959 **	0.970 **	0.971 **	0.971 **

** $p < 0.01$; * $p < 0.05$

3 讨论

3.1 影响秋茄胚轴早期固着生长的因素

潮间带红树林底质淤泥较为松软,即使是插入淤泥中的胚轴,在涨落潮期间也易于被潮水带走。秋茄胚轴通过迅速长根,获得抵抗潮水冲刷的能力而固着定居[3]。

对红树植物早期定植生长的研究表明,底质土壤盐度、海水浸淹时间以及潮水流速是影响红树植物幼苗早期根系生长的主要原因。底质土壤盐度过高、滩涂潮位太低导致的海水浸淹时间太长都使得红树植物幼苗萌根时间推迟,根系生长不良,当幼苗受到潮水冲刷时易于被带走[3,11,21,27,28]。本研究中,秋茄样品的种源一致,但由于各样方生境不同而影响其萌根,导致了秋茄胚轴固着率差异明显。分析其原因,除互花米草盐沼外侧1个小样方处于小潮沟边上,受潮水冲刷明显而有较多秋茄胚轴漂走外,其它各样方内滩涂较为平整,潮水涨落过程中水流的速度比较一致;此外,红树林对水流的滞缓效应使得林内水流漫溢与排泄流速仅为相应白滩流速的1/3～1/4[12]。林外潮水的流速高于林内,但林缘空地秋茄胚轴漂走的个体数反而低于白骨壤林,因此可以首先排除滩涂潮位、潮水流速不同对样地内秋茄幼苗固着生长带来的影响。再次,相关分析表明,各样方的相对光照强度和底质土壤理化性质和胚轴的固着率没有相关关系($p>0.05$),因此排除了光照以及底质土壤差异对秋茄胚轴萌根的影响。

实验中,观测到白骨壤林下从表土伸出的指状呼吸根密度高达(472 ±63)条/m^2,此外在白骨壤林底质土壤中密生白骨壤细根,而秋茄林、林缘空地以及互花米草盐沼外侧滩涂的土壤均未见白骨壤根系分布。调查结果显示在白骨壤林下秋茄胚轴固着率最低,表明白骨壤根系的存在是抑制秋茄胚轴早期根的萌生的关键因素。但其抑制机理有待进一步研究,包括白骨壤的根系是否对秋茄胚轴存在化感作用[22]以及养分竞争等,这对于了解红树林群落的种类构成,种间相互作用规律有重要意义。

3.2 影响秋茄幼苗存活的因素

已经固着生长的秋茄幼苗在生长过程中,处于高盐、周期性潮水浸淹的逆境中,并遭受昆虫和螃蟹等动物的取食,此外林下幼苗光照强度微弱,因此在栽培3a期间均有不同程度的死亡。本研究中,不同样方土壤理化性质以及潮水浸淹时间和秋茄幼苗的死亡数量没有相关关系($p>0.05$),表明土壤以及水文因素不是引起秋茄幼苗死亡的直接原因。

而从不同样方内秋茄幼苗的死亡个体数和死亡类型来看,昆虫和螃蟹等动物的取食是导致林内秋茄幼苗大量死亡的最主要原因,并以白骨壤林内最具代表性,其损伤性死亡的个体占死亡总体的83.4%～92.5%。实验观测和统计结果表明,林内各样方中秋茄幼苗的胚轴上黑色小孔较多,经取样解剖观察,孔内为昆虫将秋茄胚轴髓部蛀空,而林外秋茄胚轴有见黑色小孔,但孔口已愈合,孔内无昆虫。Sousa等[17]研究发现,在林内遮荫条件下,昆虫生长活动频繁而将幼苗胚轴蛀空,导致幼苗死亡。实验中可以明显看到秋茄幼苗胚轴或顶芽被咬的痕迹,同时观察到螃蟹在幼苗上活动取食,而林外样方被破坏的幼苗个体极少。Minchinton[11]、Krauss等[28]的研究表明,螃蟹对红树植物幼苗危害严重,林窗和林内危害不同,林内遮荫条件下红树植物幼苗胚轴和幼枝较幼嫩,易于被取食。这和本研究的结果一致。

光照是影响红树植物更新的主要生态因子,多数红树幼苗随着遮荫的加重,存活率随之下降[4,5,7]。本实验样地林内相对光照强度仅为林外光滩的7.15%～16.01%,其中以秋茄林下最小。研究结果也表明,林内弱光条件对秋茄幼苗的存活有较大影响:在1年栽培期内,林内秋茄幼苗多集中在栽培最初2个月内枯死,其中以秋茄林内枯死的幼苗个体数相对较多;到2004年栽培2a时,白骨壤林下秋茄幼苗全部死亡,秋茄林下秋茄幼苗仅少量存活(2a成活率为18.7%～23.2%),并在2005年栽培3a时秋茄林下样方内的秋茄幼苗也全部死亡。此外,在调查中也发现漳江口秋茄大量自然更新的地方是在林窗、林缘空地以及林外光滩上,在荫蔽的树冠下,秋茄无法更新或更新很少,特别是在秋茄林下,2年生和3年生的秋茄幼苗数量急剧下降。红树植物胚轴携带的能量用以供给幼苗成长,潮间带的水淹、盐度、林下弱光等不良环境因子不仅影响幼苗的能量支出,还对其能量的收入造成影响。不同的立地自然条件以及胚轴自身的情况决定了胚轴存活与死亡之间必然存在一个"能量的阈限",突破了这个阈限,幼苗才有可能真正在潮间带定植和生长[29]。在1～2年生长期内,由于秋茄幼苗胚轴携带的营养成分仍未耗尽,仍有部分幼苗存活,当秋茄幼苗生长到第3年时,原胚轴的营养物质耗尽,维持幼苗生长的营养成分转由叶片光合作用提供,林下低光照条件下幼苗光合生产能力低下,导致幼苗死亡。此外,在林内弱光条件下,幼苗长势比林外差,并且昆虫以及螃蟹等动物活动更加频繁,进一步促进了秋茄幼苗的死亡。

3.3 影响秋茄幼苗生长的因素

红树幼苗在潮间带滩涂上受到诸如盐、水淹、遮荫、生物干扰等不利因素的胁迫。高盐度、过长的淹水时间对红树幼苗光合速率、气孔导度、蒸腾速率、生长以及其它生理过程造成抑制[28~30]。本研究中,样地处于中高潮带,适合秋茄幼苗的生长。从光滩到林内的高程差仅为 0.39 m,且林内滩涂平整,因此从各样方所处的滩涂位置来看,潮水浸淹时间差别不大;从底质土壤的盐度来看,虽然从林内到林外土壤盐度呈下降趋势,但总体上各样方中土壤的盐度范围在 13.86~21.79 之间,在秋茄幼苗合适的生长盐度范围内[1,25]。此外,林外秋茄幼苗的生长状况比林内还好,因此,土壤盐度不是影响秋茄幼苗生长的主要因素;对土壤有机质以及 pH 值的分析结果也表明它们和幼苗的生长之间无相关关系($p>0.05$)。从上述对样地水文及土壤理化因子的分析来看,虽然不同样方的水文和土壤因子存在差别,但不是造成各样方间秋茄幼苗生长差异的主要影响因子。

红树幼苗在林隙及林荫下胚轴的自然分布密度并无太大的区别,但是林隙下的小苗的密度及其生长速度,明显高于林荫[31]。本研究结果表明,各样方中 1 年龄秋茄幼苗的各项生长参数和相对光照强度之间均呈极显著正相关关系。表明光照水平影响秋茄幼苗的光合生产,并进而对秋茄幼苗的根、茎、叶的形态生长和生物量均有影响。莫竹承[32]的研究结果表明充足的光照对木榄(*Bruguiera gymnorrhiza*)、红海榄(*Rhizophora stylosa*)苗期生长十分重要,解除荫蔽条件可明显促进幼苗生长。Smith[5]对大红树(*Rhizophora mangle*)、叶勇[6]对秋茄(*K. candel*)的研究结果也表明红树植物幼苗在林外比在林内生长得更好。

本研究还发现,在林外滩涂上,由于互花米草迅速生长并成团丛状分布,使其存在的滩涂处于完全荫蔽状态,邻近的秋茄幼苗逐渐被埋没,因缺乏光照无法正常生长而全部死亡。红树林和互花米草二者生态位很接近,并且互花米草生命力与竞争力极强,其扩散蔓延的速度远超过林的天然扩散和更新。有必要进行红树植物和互花米草之间相关关系的研究并采取合适的措施以保护该红树林自然保护区。

References:

[1] Lin P. Mangrove Vegetation. Beijing: China Ocean Press, 1984.

[2] Padilla C, Fortes M D, Duarte C M, *et al*. Recruitment, mortality and growth of mangrove (*Rhizophora* sp.) seedlings in Ulugan Bay, Palawan, Philippines. Trees-structure and function, 2004, 18(5):589~595.

[3] Clarke P J, Kerrigan R A, Westphal C J. Dispersal potential and early growth in 14 tropical mangroves: do early life history traits correlate with patterns of adult distribution? Journal of Ecology, 2001, 89(4):648~659.

[4] Clarke P J, Allaway W G. The regeneration niche of grey mangrove (*Avicennia marina*): effects of salinity, light and sediment facters on establishment, growth and survival in the field. Oecologia, 1993, 93:548~556.

[5] Smith S M, Lee D W. Effects of light quantity and quality on early seedling development in the red mangrove, *Rhizophora mangle* L. Bulletin of Marine Science, 1999, 65(3):795~806.

[6] Ye Y, Tan F Y, Lu C Y. Effects of soil texture and light on growth and physiology parameters in *Kandelia candel*. Acta Phytoecologica Sinica, 2001, 25(1):42~49.

[7] Ball M C. Interactive effects of salinity and irradiance on growth: implications for mangrove forest structure along salinity gradients. Trees-structure and function, 2002, 16(2-3):126~139.

[8] Clarke P J, Myerscough P J. The intertidal distribution of the grey mangrove (*Avicennia marina*) in southeastern Australia: the effect of physical conditions, interspecific competition and predation on propagule establishment and survival. Australian Journal of Ecology, 1993, 18:307~315.

[9] Mckee K L. Soil physicochemical patterns and mangrove species distribotion-reciprocal effects? Journal of Ecology, 1993, 81:477~487.

[10] Lan F S, Li R T, Chen P, *et al*. The relationship between mangrove and soils on the beach of Guanxi. Guihaia, 1994, 14(1):54~59.

[11] Minchinton T E. Canopy and substratum heterogeneity influence recruitment of the mangrove *Avicennia marina*. Journal of Ecology, 2001, 89(5):888~902.

[12] Zhang Q M, Yu H B, Chen X S, *et al*. The relationship between mangrove zone on tidal flats and tidal levels. Acta Ecologica Sinica, 1997, 17(3):258~265.

[13] Ball M C. Salinity tolerance in the mangroves *Aegiceras corniculatm* and *Avicennia marina* Ⅰ. Water use in relation to growth, carbon partitioning and salt balance. Australian Journal of Plant Physiology, 1988, 15:447~464.

[14] Robertson A I, Giddins R, Smith T J Ⅲ. Seed predation by insects in tropical mangroves forests: extent and effects on seed viability and the growth of seedlings. Oecologia, 1990, 83:213~219.

[15] Mckee K L. Mangrove species distribution and propagule predation in Belize — an exception to the dominance predation hypothesis. Biotropica, 1995, 27:

334～345.

[16] Farnsworth E J, Ellison A M. Global patterns of pre-dispersal propagule predation in mangrove forests. Biotropica, 1997, 29(3):318～330.

[17] Sousa W P, Kennedy P G, Mitchell B J. Propagule size and predispersal damage by insects affect establishment and early growth of mangrove seedlings. Oecologia, 2003, 135(4):564～575.

[18] Rabinowitz D. Mortality and initial propagule size in mangrove seedlinds in Panama. Journal of Ecology, 1978, 66:45～51.

[19] Lin G, Sternberg LDA S L. Variation in propagule mass and its effect on carbon assimilation and seedling growth of red mangrove (*Rhizophora mangle*) in Florida, USA. Journal of Tropical Ecology, 1995, 11:109～119.

[20] Rabinowitz D. Dispersal properties of mangrove propagules. Biotropica, 1978, 10:47～57.

[21] Mckee KL. Seedling recruitment patterns in a Belizean mangrove forest: effects of early growth ability and physico-chemical factors. Oecologia, 1995, 101:448～460.

[22] Mo Z C, Fan H Q. Allelopathy of *Bruguiera gymnorrhiza* and *Kandelia candel*. Guangxi Sciences, 2001, 8(1): 61～62.

[23] Lin P ed. The Comprehensive Report of Science Investigation on the Natural Reserve of Mangrove Wetland of Zhangjiang Estuary in Fujian. Xiamen: Xiamen University Press. 2001.

[24] Wu Z Y. Vegetation of China. Beijing: Science Press, 1980.

[25] Liao B W, Zheng D Z, Zheng S F, *et al*. A Study on the Afforestation Techniques of *Kandelia candel* Mangrove. Forest Research, 1996, 9(6):586～592.

[26] Lu R K ed. Analysis Methods of Soil Agricultural Chemistry. Beijing: Chinese Agriculture and Technology Press, 1999.

[27] Elster C, Perdomo L, Schnetter M L. Impact of ecological factors on the regeneration of mangroves in the Cienaga Grande de Santa Marta, Colombia. Hydrobiologia, 1999, (413):35～46.

[28] Krauss K W, Allen J A. Factors influencing the regeneration of the mangrove *Bruguiera gymnorrhiza* (L.) Lamk. on a tropical Pacific island. Forest Ecology and Management, 2003, 176(1-3):49～60.

[29] Yan Z Z, Wang W Q, Huang W B. Development of the viviparous hypocotyl of mangrove and its adaption to inter-tidal habitats: A review. Acta Ecologica sinica, 2004, 24(10):2317～2323.

[30] Clarke P J, Kerrigan R A. Do forest gaps influence the population structure and species composition of mangrove stands in northern Australia? Biotropica, 2000, 32(4):642～652.

[31] Sherman R E, Fahey T J, Battles J J. Small-scale disturbance and regeneration dynamics in a neotropical mangrove forest. Journal of Ecology, 2000, 88:165～178.

[32] Mo Z C, Fan H Q, He B Y. Growth feature of seedlings of two mangroves under mother trees of *Bruguiera gymnorrhiza*. Guangxi Sciences, 2001, 8(3):218～222.

参考文献:

[1] 林鹏. 红树林. 北京: 海洋出版社, 1984.

[6] 叶勇, 谭凤仪, 卢昌义. 土壤结构与光照水平对秋茄某些生长和生理参数的影响. 植物生态学报, 2001, 25(1): 42～49.

[10] 蓝福生, 李瑞棠, 陈平, 等. 广西海滩红树林与土壤的关系. 广西植物, 1994, 14(1):54～59.

[12] 张乔民, 于红兵, 陈欣树, 等. 红树林生长带与潮汐水位关系的研究. 生态学报, 1997, 17(3): 258～265.

[22] 莫竹承, 范航清. 木榄和秋茄的种间化感作用研究. 广西科学, 2001, 8(1): 61～62.

[23] 林鹏著. 福建漳江口红树林湿地自然保护区综合科学考察报告. 厦门: 厦门大学出版社, 2001.

[24] 吴征溢. 中国植被. 北京:科学出版社, 1980.

[25] 廖宝文, 郑德璋, 郑松发, 等. 红树植物秋茄造林技术的研究. 林业科学研究, 1996, 9(6):586～592.

[26] 鲁如坤著. 土壤农业化学分析方法. 北京: 中国农业科技出版社, 1999.

[29] 闫中正, 王文卿, 黄伟滨. 红树胎生现象及其对潮间带生境适应性研究进展. 生态学报, 2004, 24(10):2317～2323.

[32] 莫竹承, 范航清, 何斌源. 木榄母树下2种红树植物幼苗生长特征研究. 广西科学, 2001, 8(3):218～222.

深圳湾引种红树植物海桑的幼苗发生和扩散格局的生态响应*

曾雪琴[1,2,3] 陈鹭真[1,3,4] 谭凤仪[3,5] 黄建辉[1] 徐华林[6] 林光辉[1,4]

(1. 中国科学院植物研究所植被与环境变化国家重点实验室，北京 100093；2. 中国科学院研究生院，北京 100049；3. 深圳福田—城大红树林研发中心，深圳 518040；4. 亚热带湿地生态系统研究教育部重点实验室/厦门大学生命科学学院，厦门 361005；5. 香港城市大学生物及化学系，香港；6. 广东内伶仃福田国家级自然保护区管理局，深圳 518040)

摘要： 海桑(*Sonneratia caseolaris*)是我国华南沿海主要红树林造林树种，在深圳湾引种造林15年后，在天然红树林和光滩中出现了大面积的扩散。为了研究深圳湾红树植物海桑的幼苗扩散及其与生态因子的关系，作者采用样线和样方调查法于2006年9月至2007年9月对深圳福田红树林内天然扩散的海桑幼苗的密度、高度和盖度及其相关生态因子(包括种间竞争、群落类型、光照、扩散距离与滩面高程)进行了6次调查。天然红树林和人工海桑林林下海桑幼苗密度在调查初期分别为24.7棵/m^{2}和19.7棵/m^2，到2007年9月林下的一年生海桑幼苗全部死亡，说明林下的弱光生境显著抑制了海桑幼苗的早期生长和自然更新。不同林型下(包括天然白骨壤林和秋茄林、人工海桑林)的海桑幼苗的密度、高度、盖度差异不显著(P>0.05)；而林中空地各指标显著高于林下(P<0.05)。虽然深圳福田红树林滩面高程介于1.12–2.10 m(黄海平均海平面)之间，海桑幼苗自然扩散分布的最适滩面高程是1.40–1.60 m，属于深圳湾红树林的中高潮滩，但幼苗密度与滩面高程之间相关性较小。海桑具有一定的长距离扩散能力，天然白骨壤林和秋茄林下海桑幼苗密度与其扩散距离(距最近母树的距离)之间呈显著负相关。天然白骨壤林和秋茄林下海桑幼苗密度与光照强度相关性不显著(P>0.05)，而人工海桑林林下的海桑幼苗密度与光照强度呈显著正相关，且相关系数逐次增大，说明海桑幼苗的早期生长受到光照强度的影响极为显著。因此，深圳湾引种海桑的繁殖体在天然白骨壤林和秋茄林下的扩散主要受与母树距离的影响，但在海桑人工林下光照强度是影响幼苗分布的最重要生态因子。

关键词： 红树林, *Sonneratia caseolaris*, 幼苗扩散, 光照强度, 滩面高程, 扩散距离

Seedling emergence and dispersal pattern of the introduced *Sonneratia caseolaris* in Shenzhen Bay, China

Xueqin Zeng[1,2,3], Luzhen Chen[1,3,4] , Nora FungYee Tam[3,5], Jianhui Huang[1] , Hualin Xu[6], Guanghui Lin[1, 4]

1 *State Key Laboratory of Vegetation and Environmental Change, Institute of Botany, Chinese Academy of Sciences, Beijing* 100093

2 *Graduate University of Chinese Academy of Sciences, Beijing* 100049

3 *Futian–CityU Mangrove Research and Development Centre, Shenzhen* 518040

4 *Key Laboratory for Subtropical Wetland Ecosystem Research, Ministry of Education / School of Life Sciences, Xiamen University, Xiamen* 361005

5 *Department of Biology and Chemistry, City University of Hong Kong, Hong Kong SAR, China*

6 *Administrative Bureau of Neilingding Futian National Nature Reserve, Shenzhen* 518040

Abstract: Due to its rapid growth, *Sonneratia caseolaris*, a mangrove species indigenous to Hainan, was introduced to Shenzhen Bay, Guangdong for afforestation purpose during the early 1990s. The seedling

* 中国科学院“百人计划”资助项目、国家自然科学基金(30700092)和中国博士后科学基金(2006400529)
原载于生物多样性，2008，16(3)：236－244

emergence, early growth and dispersal pattern of *S. caseolaris* and their responses to environmental factors have not been well studied in the new habitat. In this study, we evaluated the density, height and coverage of *S. caseolaris* seedlings underneath the canopies of various mangrove forests (including both natural *Kandelia candel* and *Avicennia marina* communities and introduced *S. caseolaris* communities) and on the mudflats without canopy, in Futian Mangroves Natural Reserve of Shenzhen Bay from September 2006 to September 2007. Line intercept and square intercept methods were used in the survey. Tidal elevation, light intensity, community types and the distance between the sample squares and the nearest adult *S. caseolaris* were also recorded. The mean densities of *S. caseolaris* seedlings under the canopies of both the introduced and natural mangrove forests decreased from September 2006 (24.7 seedlings per m^2 and 19.7 seedlings per m^2, respectively) to September 2007 (no seedlings survived). No significant differences were found in the seedling density, height or coverage of *S. caseolaris* among different mangrove communities. However, the density, height and coverage of *S. caseolaris* seedlings were significantly higher on the mudflats without canopy than under the mangrove canopies, indicating that higher light intensity in on the mudflats without canopy promoted the dispersal and vertical growth of *S. caseolaris* seedlings. Although the optimal tidal elevation for *S. caseolaris* seedlings in Shenzhen Bay was between 1.40 m and 1.60 m, an area that falls within the mid-to-high intertidal zones, seedling density and tidal elevation were weakly correlated. The seedling density under the native mangrove canopies was negatively related to dispersal distance. However seedling density were positively correlated with light intensity($P<0.05$), and the correlation coefficients for the introduced *S. caseolaris* forest increased through time with successive surveys. In contrast, there was no significant correlation founded between seedling density and light intensity under native mangrove canopies. We concluded that distance to mother tree was the most important factor determining *S. caseolaris* seedling density under native mangrove canopies, whereas the light intensity was the most important environmental factor for controlling seedling dispersal pattern under the canopy of the introduced *S. caseolaris* forest.

Key words: mangroves, *Sonneratia caseolaris*, seedling dispersal, light intensity, elevation, dispersal distance

红树林是处于陆地与海洋过渡带的一类特殊植被类型，是维持海岸生态平衡的特殊生态系统(林鹏, 1997)，同时也是极易被入侵的生态系统之一(黄建辉等, 2003)。海桑(*Sonneratia caseolaris*)是天然分布于我国海南岛的红树植物，属前缘裸滩定居的先锋树种。由于海桑树体高大、生长迅速、结实率高、耐水淹等特点，近年来被广泛用于华南沿海滩涂红树林的生态恢复和人工造林(郑德璋等, 1999)。海桑1994年首次在广东深圳湾引种并取得成功，1999年出现一定程度的自然更新和扩散现象(王伯荪等, 2002; Zan *et al.*, 2003)，2006 年8月，在深圳湾红树林区出现了海桑幼苗的大面积扩散。由于潮水的携带作用，海桑已经从深圳福田的引种区扩散到对岸的香港米埔湿地，并有进一步扩散的趋势①。快速的扩散能力是入侵植物的重要特征(黄建辉等, 2003)，因此，海桑幼苗在深圳湾的自然扩散已经引起有关部门对其入侵潜力的高度重视和担忧。

幼苗更新是植物种生活史上最重要的阶段之一，而幼苗定居对植物的更新具有重要的筛选作用(Clark *et al.*, 1998; Rey & Alcántara, 2000)。海桑作为深圳红树林自然保护区的引入种，其幼苗的天然更新和扩散及其对各类生态因子的响应还不清楚。本研究拟在对海桑幼苗发生和扩散格局进行调查的基础上，分析幼苗的更新和扩散规律及其生态响应，为其自然扩散范围的预测和入侵潜力的评估提供基础资料。

1 研究地点概况

研究地点位于广东内伶仃福田国家级自然保护区的实验区 (22°32′ N，114°03′ E)(图1)。该区属南亚热带季风气候，年平均气温22℃，最冷月平均14.1℃(1月)，极端最低温0.2℃(1月)，极端最高温38.7℃(7月)，年平均降水量1,927 mm，年均相对湿度79%。海水pH值为7.23–8.05，海水盐度平均为15‰。图2给出了调查期间与历史同期深圳的月均

① Yuk CL (2006) *Study on the Germination Conditions of Two Exotic Species Sonneratia caseolaris and Sonneratia apetala*. Undergraduate thesis, Department of Biology and Chemistry, City University of Hong Kong, Hong Kong.

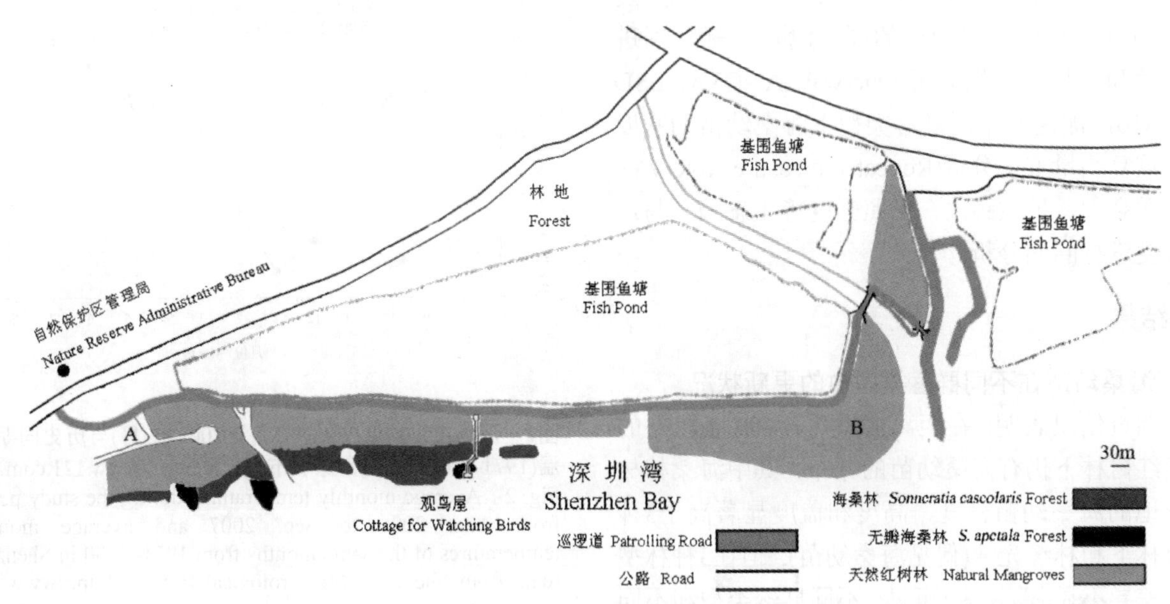

图1　调查区域示意图。调查区域为从A点到B点。天然红树林向陆一侧为秋茄林，向海一侧为白骨壤林。
Fig. 1　Location of the study site. The survey area is from A to B. The landward side of natural mangroves is occupied by *Kandelia candel* forest, and the seaward side is by *Avicennia marina* forest.

温图(1971–2000年) (http://www.121.com.cn)。

调查区域内由陆向海依次分布有天然林(秋茄林和白骨壤林)和人工林(海桑林)，3种林型沿海岸线呈带状分布，林中空地分布在海桑林内的没有冠层的光滩。主要红树植物有本地种白骨壤(*Avicennia marina*)、秋茄(*Kandelia candel*)、桐花树(*Aegiceras corniculatum*)，及引入种海桑和无瓣海桑(*Sonneratia apetala*)。不同红树群落的相对光照强度依次为：海桑人工林12.2%，白骨壤天然林6.64%，秋茄天然林7.18%，林中空地86.44%。

2　研究方法

2.1　样地设置

在保护区的红树林实验区(见图1的A点到B点)随机布置了19条与海岸线垂直的固定样线，从红树林向陆一侧延伸到光滩。样线的长度在30–105 m之间，随红树林林带的宽度变化而变化。其中，天然林中12条，人工海桑林中6条，海桑林林中空地1条。在天然林外侧如果有海桑林，按天然林和海桑林两条样线计算。在样线上每隔5 m 布置一个样点，以样点为中心，设立一个1 m × 1 m的固定样方。总样方数为160个，其中秋茄林46个，白骨壤林50个，人工海桑林50个，林中空地14个。

2.2　样地调查

2006年9月至2007年1月，每隔一个半月调查一次，共调查4次；2007年1月以后，因林下一年生海桑幼苗存活数量很少，故仅在6月和9月各进行1次调查。记录样方所在的林型和样方内低于1 m的一年龄海桑幼苗和本地种苗的密度、高度和盖度。用卷尺测定每种幼苗的高度，用目测法估测幼苗盖度，用GPS定位仪测定天然林下每个样方距最近海桑母树(果期开花结果)的距离。于2006年11月、2007年1月和6月各选取3个天气晴朗的上午，用照度计(TES-1334A)测定每个样方以及实时林外空地上的光照强度。

用连通器法确定每个样方的滩面高程(陈鹭真, 2005)。方法是：两根标杆中间连接一条细小透明塑料管，灌满水，将标杆插入两个相邻样点邻接处. 管内两液面在标杆上的高度差即是两个点的滩面高程差。基准点滩面高程可由香港天文台提供的实时潮汐高度减去测得的实时潮水深度获得。经测量，所有样方的高程范围为1.12–2.10 m(黄海平均海平面，下同)。

2.3　数据分析

应用统计软件SigmaPlot 10.0 (for Windows) 和SPSS 11.0 (for Windows)对数据进行处理。对每次调

查的不同群落下海桑幼苗的密度、高度和盖度，及海桑幼苗与其他红树幼苗的同一测定指标分别进行比较和差异显著性分析(One-way ANOVA, LSD法)；对6次测定的不同群落类型中海桑幼苗的密度进行重复测量方差分析(Repeated measure ANOVA)；对滩面高程、扩散距离、光照强度等生态因子与幼苗密度进行回归分析。

3 结果

3.1 海桑幼苗在不同群落类型中的更新状况

调查结果表明：在距离海桑母树700 m以外的天然红树林下仍有海桑幼苗的分布。其中海桑林林中空地的海桑幼苗密度、高度和盖度显著高于3种红树林下和林缘光滩(偶见海桑幼苗)，但在3种林型之间差异不显著。从表1可见：不同群落类型和不同测定时间的海桑幼苗密度都有极显著差异，且时间和群落类型之间存在显著的交互作用。

如图3所示，3种红树林下海桑幼苗平均密度在2006年9月高达23.6±3.3棵/m^2，其中天然林中平均为24.7±5.7棵/m^2，人工海桑林中为19.7±4.3棵/m^2；而后逐渐减少，到了11月显著下降到4.97±1.28棵/m^2，减少了78.9%；到了2007年1月，已减少到0.70±0.18棵/m^2；至2007年9月全部死亡。海桑林林中空地2006年9月的海桑幼苗平均密度高达69.8±24.0棵/m^2，到2007年1月减少了49.4%，而到2007年9月仍保持为25.6±8.40棵/m^2。林中空地海桑幼苗的平均高度2006年9–12月间从10.0±0.6 cm逐渐增加到26.8±3.9 cm；到了2007年1月却下降到9.38±2.15 cm，叶子所剩无几，并出现了枯顶现象；到2007年9月却又迅速增长到130±16.5 cm，平均盖度高达91±9.2%。

3.2 海桑幼苗与其他红树植物幼苗的种间竞争

前3次调查中，海桑幼苗密度均显著高于其他红树植物(P<0.05)，在前4次调查中，各种林型下其他红树植物的幼苗密度均小于5棵/m^2(图4)。林下不同种红树植物幼苗的平均高度差异也很大，平均盖度均不超过5%。逐步回归分析表明(表2)：海桑幼苗密度与其他红树植物幼苗的密度在2006年9月第一次调查中存在显著的相关性(P<0.05)，但R^2较小(最大为0.054)(表2)，其他测定时间两者相关性不显著。

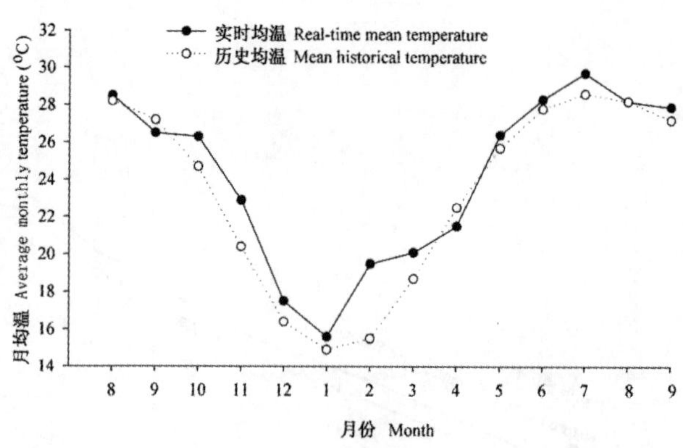

图2 深圳市月均温(2006年8月至2007年9月)与历史同期均温(1971–2000年)(来自深圳市气象局http://www.121.com.cn)

Fig. 2 Average monthly temperatures during the study period from Aug. 2006 to Sep. 2007 and average monthly temperatures of the same months from 1971–2000 in Shenzhen (data from Shenzhen Meteorological Bureau. http://www.121.com.cn)

表1 不同群落类型中海桑幼苗密度的重复测量方差分析结果

Table 1 Repeated measures ANOVA for the seedling density of *Sonneratia caseolaris* in different mangrove communities

变异来源 Sources of variation	自由度 df	F	P
时间 Time	4	15.828	0.000
林型 Forest type	3	26.642	0.000
时间×林型 Time×Forest type	12	3.455	0.000

3.3 滩面高程、扩散距离、光照强度与海桑幼苗分布

深圳湾滩面高程在1.12–2.10 m之间变化，海桑幼苗密度高的区域主要集中在滩面高程1.40–1.60 m之间，属于中、高潮滩，而且在整个调查过程中始终保持着密度上的优势(图5)。除了第一次调查外，天然林中的海桑幼苗密度与滩面高程都达到显著负相关(R^2在0.04–0.08之间，P < 0.05)。

天然红树林下(包括白骨壤林和秋茄林)海桑幼苗扩散距离与幼苗密度呈显著负相关(图6)，特别是在第一次调查中，R^2高达0.30。

在天然林下海桑幼苗密度与光照强度的相关性未达到显著水平(P>0.05)(表3)，而在人工海桑林下的海桑幼苗密度随着林下光照强度的增强呈线性增加，相关程度随着海桑幼苗的生长呈增加趋势(前4次调查中R^2分别为0.160, 0.267, 0.369, 0.364)。

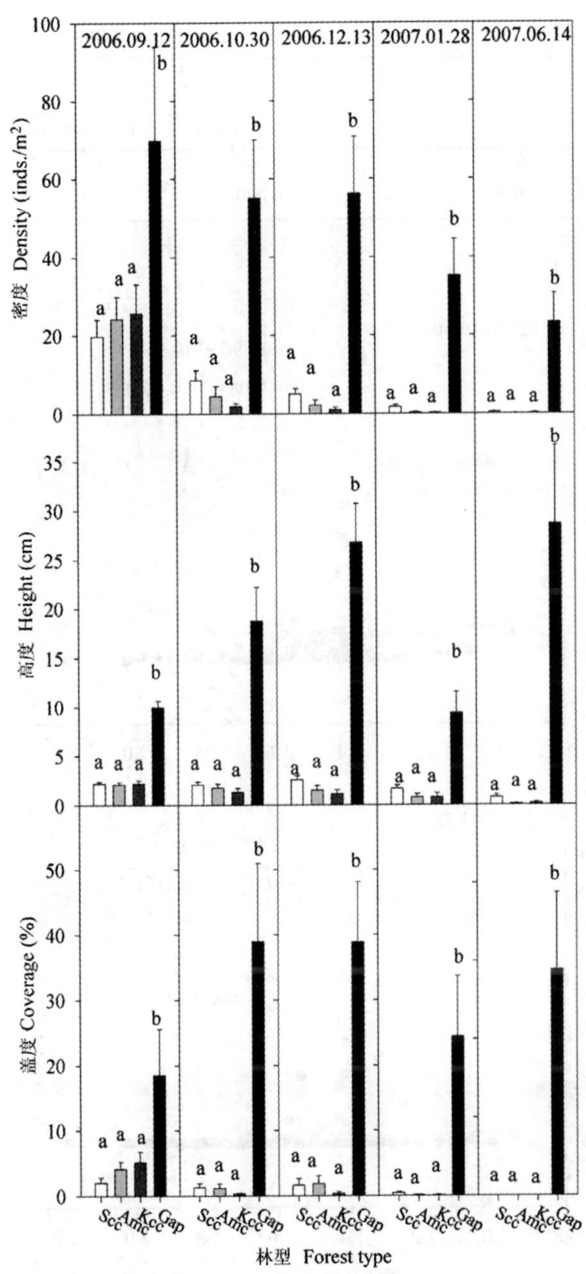

图3　不同群落类型中海桑幼苗的密度、高度和盖度(平均值±标准误)。Scc—海桑林; Amc—白骨壤林; Kcc—秋茄林; Gap—海桑林中空地。不同小写字母代表差异显著(P<0.05)。
Fig. 3　Seedling density, height and coverage of *Sonneratia caseolaris* in different mangrove communities (Means±SE). Scc, *Sonneratia caseolaris* community; Amc, *Avicennia marina* community; Kcc, *Kandelia candel* community; Gap, Forest gap. Different letters indicate significant difference at 0.05 level.

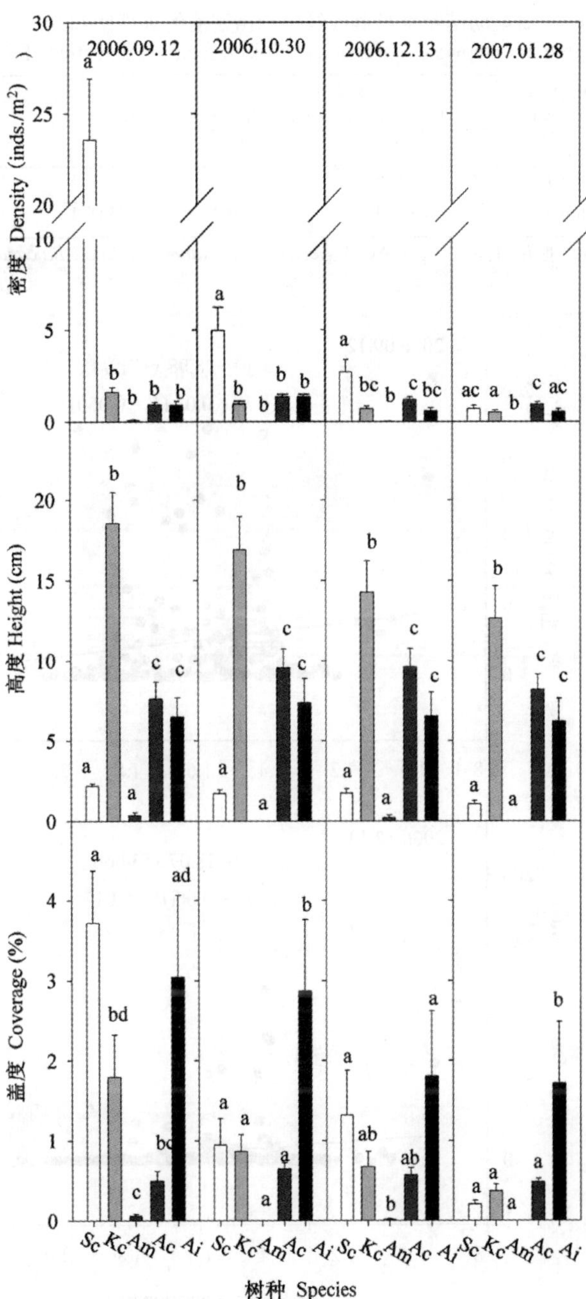

图4　海桑幼苗与其他红树幼苗在不同调查时间的密度、高度和盖度(平均值±标准误)。Sc—海桑; Kc—秋茄; Am—白骨壤; Ac—桐花树; Ai—老鼠勒。不同小写字母代表差异显著(P<0.05)。
Fig. 4　Seedling density, height and coverage of different mangrove species during the first four surveys (Means±SE). Sc, *Sonneratia caseolaris*; Kc, *Kandelia candel*; Am, *Avicennia marina*; Ac, *Aegiceras coniculatum*; Ai, *Acanthus ilicifolius*. Different letters indicate significant difference from each other species at 0.05 level.

4　讨论

4.1　不同生境海桑幼苗的发生和扩散

Herrera等(1994)发现生境类型是影响幼苗定居的主要因素。红树林具有复杂的生境结构, 表现为小尺度生境的多样性(Smith *et al.*, 1994; Sherman *et al.*, 2000)。当繁殖体扩散到安全地点后, 能否成功

表2 海桑幼苗密度与其他红树幼苗密度的逐步回归结果(2006年9月)

Table 2 Stepwise regressions between seedling densities of *Sonneratia caseolaris* and other mangrove species (Sep., 2006)

步数 Step	偏回归系数 Partial regression coefficient		截距 Intercept	R^2	$P>/F/$			
	Ac	*Kc*			*Ac*	*Kc*	截距 Intercept	模型 Model
Step 1	–4.568		29.214	0.028	0.035		0.000	0.035
Step 2	–5.051	2.362	25.974	0.054	0.019	0.040	0.000	0.013

Ac: 桐花树; Kc: 秋茄 Ac, *Aegiceras coniculatum*; Kc, *Kandelia candel*

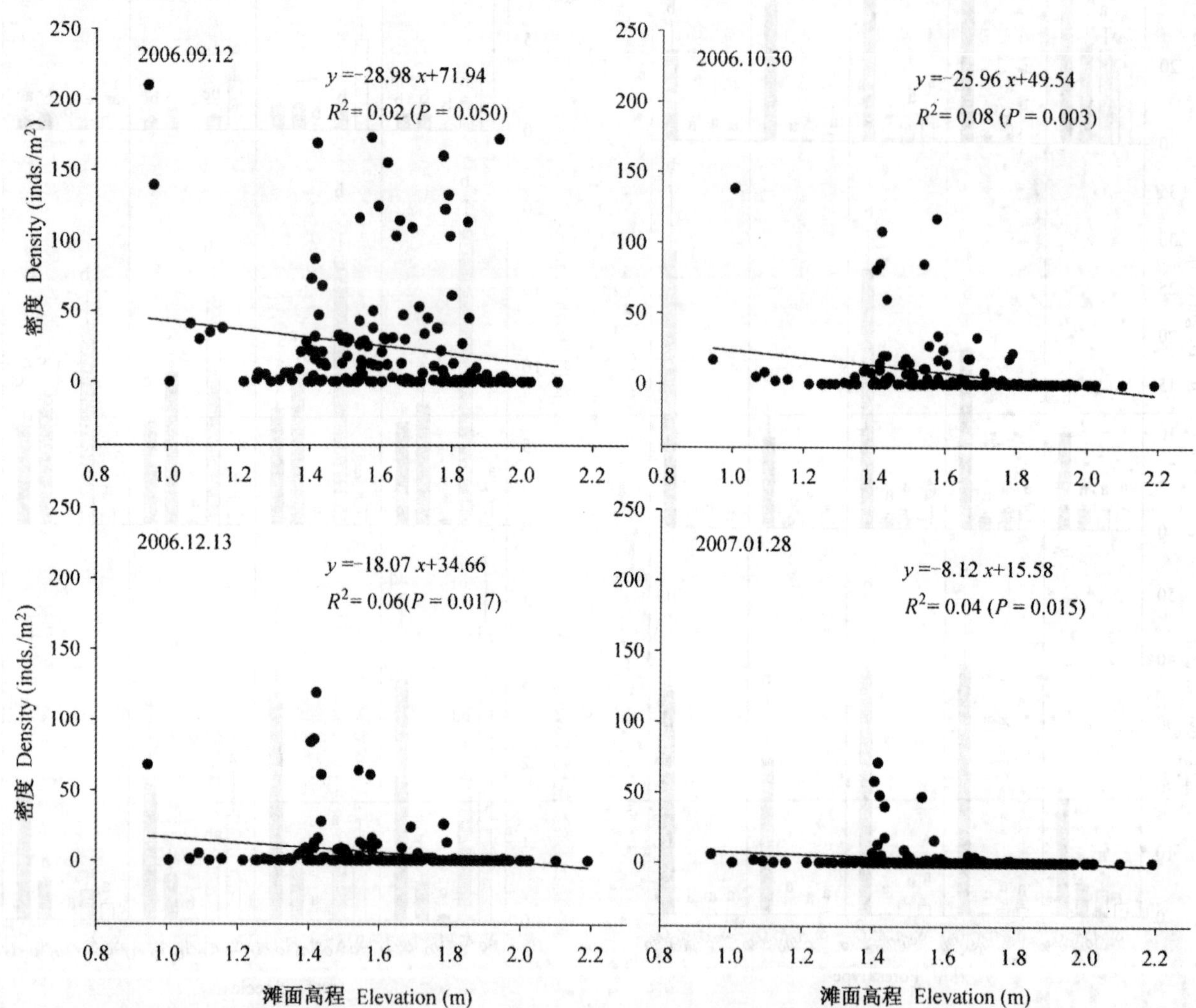

图5 海桑幼苗密度与滩面高程的关系

Fig. 5 The relationship between seedling density of *Sonneratia caseolaris* and tidal elevation

定居在很大程度上取决于小尺度的生境条件(彭闪江等, 2003)。本研究中海桑幼苗的密度变幅很大(0–210棵/m^2), 在样方水平表现出了很大差异, 这跟红树林生态系统的复杂性密切相关。此外, 海桑属繁殖体的扩散是以果实为单位, 果实扩散到安全地点后, 形成小的种子库, 增加了幼苗更新的几率; 而没有果实到达的地方, 幼苗出现的几率小很多(Rabinowitz, 1978)。因此, 海桑幼苗密度因受生境结构和繁殖体机会效应的影响, 在样方尺度上表现出多样性。因此, 海桑繁殖体的扩散特性和红树林生境的复杂性是导致深圳湾海桑幼苗扩散的内因和外因。

4.2 影响海桑幼苗分布格局的主要生态因子

海桑幼苗是阳生苗, 种子的萌发和幼苗的生长都需要较高的光照条件(王伯荪等, 2002)。调查发现: 在一定高程范围内, 林下的光照越强, 海桑的

表3 海桑幼苗密度与相对光照强度的相关分析结果

Table 3 Corelation analysis between the seedlings density of *Sonneratia caseolaris* and relative light intensity

林型 Forest type	测定时间 Measurement time	相关方程 Correlation equation	R^2	P
天然林（白骨壤林+秋茄林） Natural forest	2006.09.12	$y=-18.79x+25.43$	0.001	0.730
	2006.10.30	$y=9.60x+3.30$	0.002	0.635
	2006.12.13	$y=6.28x+1.80$	0.003	0.601
	2007.01.28	$y=4.06x+0.39$	0.005	0.480
人工海桑林 Introduced *S. caseolaris* forest	2006.09.12	$y=78.08x+2.94$	0.160	0.002
	2006.10.30	$y=71.41x-6.33$	0.267	0.000
	2006.12.13	$y=75.62x-10.50$	0.369	0.000
	2007.01.28	$y=46.56x-7.59$	0.364	0.000

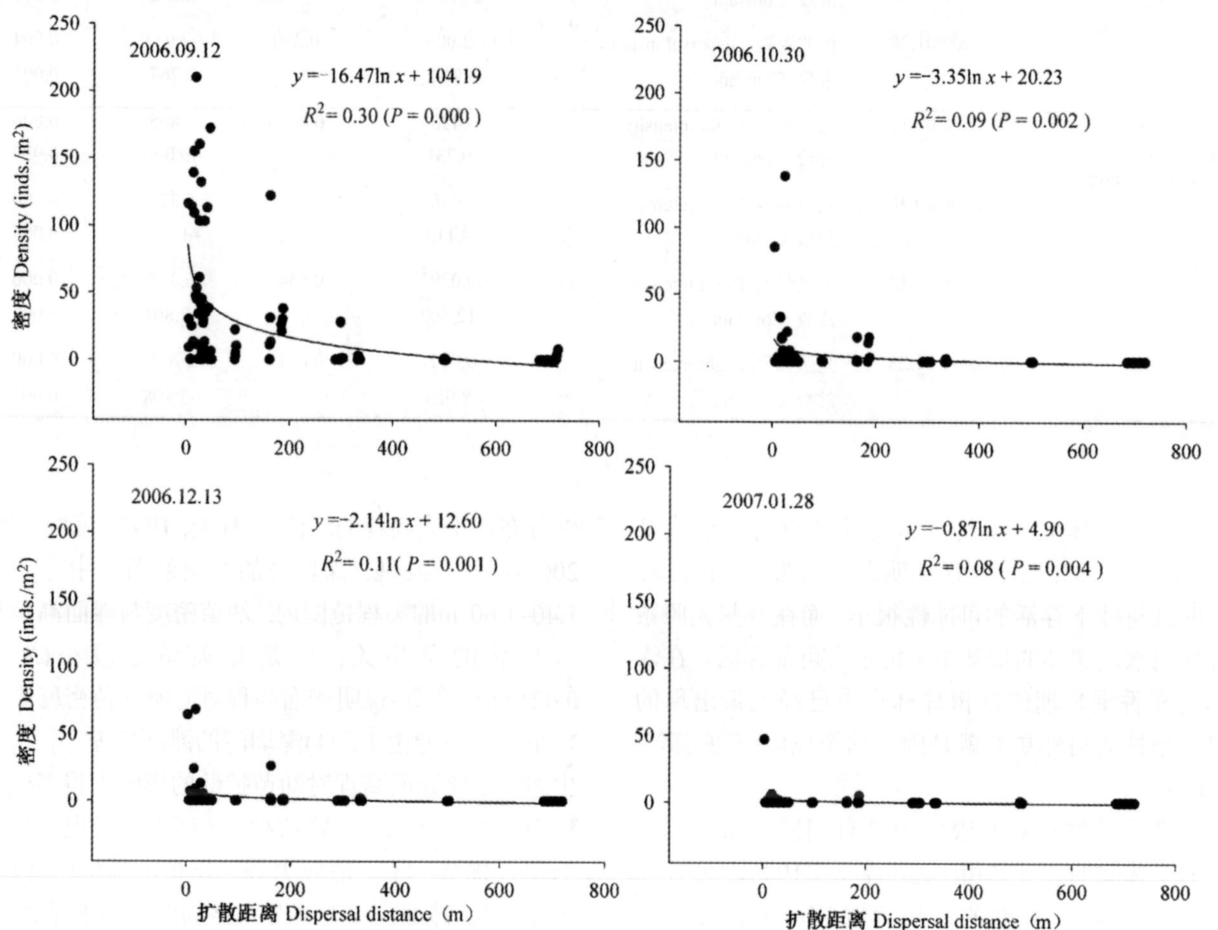

图6 海桑幼苗密度与扩散距离的关系

Fig. 6 The relationship between seedling density of *Sonneratia caseolaris* and dispersal distance

幼苗密度越高。而林中空地的海桑幼苗在密度、高度和盖度上都具有显著优势。因此，光照强度通过影响海桑幼苗的发生和早期生长，从而影响幼苗的分布格局。

调查过程中还发现：2007年1月，林中空地的海桑幼苗受到越冬鸟类取食和低温的影响，生物量大大减少，有些幼苗出现枯顶，甚至死亡。但是，随着气温的升高和候鸟的迁徙，这里的海桑幼苗再次萌发生长，幼苗密度最后维持在25.6棵/m^2。部分海桑幼苗在林中空地能安全越冬并保持较高的生长速

表4　海桑人工林、天然林下海桑幼苗密度与主要生态因子的回归分析结果

Table 4　Multiple regressions between the seedlings densities of *Sonneratia caseolaris* and key ecological factors under introduced *Sonneratia caseolaris* forest and natural forest

林型 Forest type	测定时间 Measurement time	变量 Variables	偏回归系数 Partial regression coefficient	偏相关系数 Partial correlations	t	P
天然林 Natural forest	2006.09.12	扩散距离 Dispersal distance	−37.931	−0.544	−6.518	0.000
		常数 Constant	104.802		8.200	0.000
	2006.10.30	滩面高程 Tidal elevation	−19.785	−0.274	−2.858	0.005
		扩散距离 Dispersal distance	−5.370	−0.214	−2.201	0.030
		常数 Constant	49.093		4.345	0.000
	2006.12.13	扩散距离 Dispersal distance	−3.939	−0.258	−2.689	0.008
		滩面高程 Tidal elevation	−8.248	−0.194	−1.984	0.050
		常数 Constant	24.633		3.631	0.000
	2007.01.28	扩散距离 Dispersal distance	−2.005	−0.280	−2.943	0.004
		常数 Constant	4.904		3.267	0.001
人工海桑林 Introduced *S. caseolaris* forest	2006.09.12	光照强度 Light intensity	0.430	0.460	3.805	0.000
		常数 Constant	−0.731		−0.090	0.929
	2006.10.30	光照强度 Light intensity	0.038	0.587	5.324	0.000
		常数 Constant	−9.113		−1.757	0.085
	2006.12.13	光照强度 Light intensity	0.039	0.654	6.357	0.000
		常数 Constant	−12.502		−2.801	0.007
	2007.01.28	光照强度 Light intensity	0.024	0.671	0.655	0.000
		常数 Constant	−8.983		−3.408	0.001

度；而红树林下海桑幼苗密度随调查时间逐次减少，最后全部死亡。再次证明海桑幼苗在郁闭度较高的红树林下存活的可能性很小，而在一些光照条件好而水流缓和的局部小生境中有可能存活。在深圳湾和香港米埔的红树林林窗中已经大量出现的海桑植株的自然更新就是因为这个原因(王伯荪等, 2002)。

竞争是物种对有限资源具有相同的需求而产生的个体间的相互作用(Begon *et al*., 1986)。植物竞争强度随着其密度的增加而加剧(Gough *et al*., 2002; Wu & Yu, 2004)。调查表明：林下海桑幼苗与其他红树幼苗的密度除了第一次调查外均无显著相关性，说明林下的生存空间对于各种红树植物幼苗的早期生长都是充足的。但是，由于林冠对光的阻挡，大大限制了海桑幼苗的生长。冠层的海桑、白骨壤和秋茄成树与海桑幼苗之间存在着对光资源的竞争。

不同种类的红树植物根据耐淹程度的不同，天然分布在不同高程的滩涂上(林鹏, 1997; 陈鹭真等, 2006)。调查发现：深圳湾的海桑幼苗集中分布在1.40–1.60 m的高程范围内，幼苗密度与滩面高程具有显著的负相关，但是相关系数较小(R^2在0.02–0.08之间)，说明滩面高程对海桑幼苗密度影响较小。这可能由于：(1)深圳湾的滩面落差小，其他生态因子比滩面高程对幼苗扩散的影响大得多；(2)海桑是外缘裸滩的先锋树种，在深圳湾表现出很强的生长优势，环境适应力强，分布范围广(王伯荪等, 2002)，滩面高程对其分布格局的影响不显著。

逃逸假说(Escape Hypothesis)认为：母树附近的种子和幼苗因为容易受到捕食者、寄生者以及病原体等的不利影响，加上与母树存在着对相同资源的竞争，往往具有很高的死亡率；远离母树的种子的出苗成功率较高(Howe & Smallwood, 1982)。天然红树林中的海桑幼苗扩散受到与母树距离的影响显著：离母树越远，幼苗的密度越小(图6)，但在人工海桑林下，海桑幼苗的平均密度没有因为种子供应上的优势而显著高于天然林下(图3)。逃逸假说在一定程度上支持了该结果。

光照强度、滩面高程和扩散距离等各个因子与海桑幼苗密度的多元线性回归分析结果(表4)表明：天然林中，扩散距离是影响海桑幼苗扩散格局的最重要的生态因子；而在海桑人工林中，不存在扩散距离的差异(扩散距离视为0)，光照强度是影响海桑幼苗分布的最重要的生态因子。

潮汐的携带作用为海桑幼苗天然更新和扩散提供了基本条件，但是光照强度和距母树（种子库）距离对于海桑扩散格局的形成具有重要的决定作用，也成为海桑天然扩散的限制因素。

参考文献

Begon M, Harper JL, Townsend CR (1986) *Ecology: Individuals, Populations and Communities.* Blackwell Scientific Publications, Oxford, UK.

Chen LZ (陈鹭真) (2005) *Studies on the Mechanisms of Mangrove Seedlings in Response to Duration of Tidal Immersion* (红树植物幼苗的潮汐淹水胁迫响应机制的研究). PhD dissertation, Xiamen University, Xiamen. (in Chinese with English abstract)

Chen LZ (陈鹭真), Lin P (林鹏), Wang WQ (王文卿) (2006) Mechanisms of mangroves waterlogging resistance. *Acta Ecologica Sinica* (生态学报), **26**, 586–593. (in Chinese with English abstract)

Clark JS, Macklin E, Wood L (1998) Stages and spatial scales of recruitment limitation in southern Appalachian forests. *Ecological Monographs*, **68**, 213–235.

Gough L, Goldberg DE, Hershock C, Pauliukonis N, Petru M (2002) Investigating the community consequences of competition among clonal plants. *Evolutionary Ecology*, **15**, 547–563.

Herrera CM, Jordano P, Lopez-Soria L, Amat JA (1994) Recruitment of a mast-fruiting, bird dispersed tree: bridging frugivore activity and seedling establishment. *Ecological Monographs*, **64**, 315–344.

Howe HF, Smallwood J (1982) Ecology of seed dispersal. *Annual Review of Ecology and Systematics*, **13**, 201–228.

Huang JH (黄建辉), Han XG (韩兴国), Yang QE (杨亲二), Bai YF (白永飞) (2003) Fundamentals of invasive species biology and ecology. *Biodiversity Science* (生物多样性), **11**, 240–247. (in Chinese with English abstract)

Lin P (林鹏) (1997) *Mangrove Ecosystem in China* (中国红树林生态系). Science Press, Beijing. (in Chinese)

Peng SJ (彭闪江), Huang ZL (黄忠良), Peng SL (彭少麟), Xu GL (徐国良) (2003) The processes and mechanisms of the dispersal of fleshy-fruited plants at different spatial scales. *Acta Ecologica Sinica* (生态学报), **23**, 777–786. (in Chinese with English abstract)

Rabinowitz D (1978) Early growth of mangrove seedlings in Panamá, and hypothesis concerning the relationship of dispersal and zonation. *Journal of Biogeography*, **5**, 113–133.

Rey PJ, Alcántara JM (2000) Recruitment dynamics of a fleshy-fruited plant (Oleaeuropaea): connecting patterns of seed dispersal to seedling establishment. *Journal of Ecology*, **88**, 622–633.

Sherman RE, Fahey TJ, Battles JJ (2000) Small-scale disturbance and regeneration dynamics in a neotropical mangrove forest. *Journal of Ecology*, **88**, 165–178.

Smith TJ, Robblee MB, Wanless HR, Doyle TW (1994) Mangroves, hurricanes, and lightning strikes. *BioScience*, **44**, 256–262.

Wang BS (王伯荪), Liao BW (廖宝文), Wang YJ (王勇军), Zan QJ (昝启杰) (2002) *Mangrove Forest Ecosystem and Its Sustainable Development in Shenzhen Bay* (深圳湾红树林生态系统及其持续发展). Science Press, Beijing. (in Chinese)

Wu ZH, Yu D (2004) The effects of competition on growth and biomass allocation in *Nymphoides peltata* (Gmel.) O. Kuntze growing in microcosm. *Hydrobiologia*, **527**, 241–250.

Zan QJ, Wang BS, Wang YJ, Li MG (2003) Ecological assessment on the introduced *Sonneratia caseolaris* and *Sonneratia apetala* at the mangrove forest of Shenzhen Bay, China. *Acta Botanica Sinica* (植物学报), **45**, 544–555.

Zheng DZ (郑德璋), Li Y (李云), Liao BW (廖宝文) (1999) The studies on compatible temperature condition for mangrove species. In: *The Studies on Afforestation and Management Techniques for Main Mangrove Species* (红树林主要树种造林与经营技术研究) (eds Zheng DZ (郑德璋), Liao BW (廖宝文), Zheng SF (郑松发)), pp. 221–228. Science Press, Beijing. (in Chinese)

互花米草混种密度对秋茄幼苗生理生态的影响*

杨 坚 张玲玲 何斌源 黄 旋 林 鹏 郑海雷
(厦门大学生命科学学院,福建 厦门 361005)

摘要: 按照不同密度将互花米草与秋茄进行混种，对比研究胁迫条件下秋茄幼苗生长、光合特性及其渗透调节物质变化规律。结果表明，低密度互花米草促进秋茄的茎长以及各部分生物量，高密度则起抑制作用。随着互花米草密度的增大，秋茄幼苗叶片光合速率、气孔导度、蒸腾速率、水分利用率和蛋白含量均下降；相反，胞间 CO_2 浓度、可溶性糖、淀粉、脯氨酸含量却上升，这些变化有利于对抗互花米草带来的不利影响。

关键词: 互花米草；秋茄；光合特性；渗透调节物质

中图分类号: Q945.78 **文献标识码**: A **文章编号**: 1001-389X(2007)02-0176-04

Effects of mixed-culture densities of *Spartina alterniflora* on eco-physiological characteristics of *Kandelia candel* seedling

YANG Jian, ZHANG Ling-ling, HE Bin-yuan, HUANG Xuan, LIN Peng, ZHENG Hai-lei
(*College of Life Sciences, Xiamen University, Xiamen, Fujian* 361005, *China*)

Abstract: *Kandelia candel* seedlings were mix-cultured with *Spartina alterniflora* of different densities for 120 days, and their growth, photosynthetic characteristics as well as osmotic adjustment law were studied. The results showed that the stem height and biomass of *K. candel* seedlingswere enhanced with low density *S. alterniflora* and inhabited with high density ones. With the density of *S. alterniflora* going up, *Pn*, *Gs*, *Tr*, *WUE* and protein content in *K. candel* seedlings decreased respectively. On the contrary, the content of sugar, starch, proline increased with the density of *S. alterniflora* increasing. These changes in the physiological properties helped *K. candel* seedlings to counteract the adverse effects from *S. alterniflora*.

Key words: *Spartina alterniflora*; *Kandelia candel*; photosynthetic characteristics; osmotic adjustment

互花米草 (*Spartina altemiflora*)是一种原产于美国东海岸滩涂的草本盐沼植物[1]，1979年被初次引进中国，次年 10月在福建沿海等地试种成功，之后陆续扩种和扩散到浙江、江苏、上海、广东和山东等地[2]。由于它具有耐盐、耐潮汐淹没、繁殖力强、根系发达等特点，被认为是保滩护堤、促淤造陆的最佳植物[3]。然而由于其特殊的生物学特性，其生长的速度远超过人们的控制能力，致使大片适宜养殖的滩涂底质被侵占固化，而且使海水营养盐含量下降，浮游生物减少，原有生态环境被破坏。

红树林是国际《湿地公约》、《生物多样性公约》、《联合国海洋法公约》、《中国湿地行动计划》和《中华人民共和国海洋环境保护法》等的重要保护对象之一[2]，它的高生产力、高归还率和高分解率功能给近海海洋生态系统提供强大的物质基础，在防止海岸侵蚀、减缓海平面上升、减轻污染、海洋药物研发、科研教育、生态旅游等方面发挥重要作用[4]。但近年来，互花米草侵占裸露滩涂呈蔓延趋势，而且不同程度地进入原本生长着红树林的林地滩涂，红树林的生存受到威胁[2,5]。文中将互花米草与秋茄 (*Kandelia candel*)以不同的密度混种，对比研究秋茄幼苗的生长、光合特性和渗透调节物质的变化规律，探索互花米草胁迫下红树植物反应的生理生态机制，分析互花米草入侵模式，为对防治互花米草的入侵及设计切实可行的生物替代方案提供科学依据。

* 国家自然科学基金资助项目(30670317,30271065)；福建省自然科学基金资助项目(D0210001)
原载于福建科学院学报,2007,27(2):176—179

1 材料与方法

1.1 实验材料与处理设置

2006年 4月于福建厦门海沧青礁村海滩采集互花米草幼苗和秋茄成熟胚轴，选取大小一致者 (秋茄胚轴选取平均长度为 (20.4 ±1.8) cm，互花米草选取株高为 15 - 25 cm，按照秋茄 互花米草 =1 3、1 1和 3 1的密度将两者混种于沙盆 (口径 35 cm，高 15 cm)中。以单独种植秋茄为对照。每个处理 3个重复，于自然光下培养。幼苗用盐度为 15‰海水培养，每天用自来水补充蒸发水分，15 d更换 1次。

1.2 指标测定与数据分析方法

处理 120 d后随机选取 6株用于幼苗生长数量特征和叶片光合特性测定。采用 CIRAS-1型便携式光合作用测定系统 (英国 PP Systems)测定单片叶片的气孔导度 (*Gs*)、蒸腾速率 (*Tr*)、光合速率 (*Pn*)、胞间 CO_2 浓度 (*Ci*)等光合特性参数。水分利用率 (water use efficiency, WUE)计算公式为：$WUE = Pn/Tr$。幼苗生长数量特征测定后将幼苗分解成胚轴、根、茎和叶 4部分，80 ℃烘干测定生物量。又随机选取秋茄幼苗的第 2、3对成熟叶及尚未木质化的侧根及细根用于相关生理指标测定。用考马斯亮兰 G250法测定可溶性蛋白质[6]。用磺基水杨酸法测定游离脯氨酸含量[6]。用茚三酮比色法测定游离氨基酸含量[7]。用蒽酮比色法测定可溶性糖及淀粉含量[7]。数据用 SPSS统计软件进行单因素方差分析及相关分析，并根据多重比较的结果进行差异标记。

2 结果与分析

2.1 混种密度对秋茄幼苗生长的影响

与米草混种明显地影响秋茄幼苗的形态生长和生物量积累 (表 1)。单因素方差分析显示各处理下秋茄茎长、叶重、茎重和根重在 95%水平上均有显著差异，表明秋茄幼苗的生长受互花米草密度的影响。高密度 (Ⅳ组)下的互花米草显著抑制了秋茄的生长，茎长、叶重、茎重和根重与对照组相比均显著下降，茎长仅为对照组的 55.9%，叶重、茎重和根重仅为对照组的 60%、73.6%和 84.1%，随着密度的降低，抑制作用减弱、消失，甚至转为促进作用。低密度 (Ⅱ组)下的互花米草影响秋茄的茎长、叶重、茎重和根重与对照组相比存在极显著差异 ($P < 0.01$)，与对照组相比分别提高了 55.2%、95.2%、102.8%和 29.2%。

表 1 混种密度对秋茄幼苗茎长和生物量的影响

Table 1 Effects of mixed-culture densities on stem height and biomass of *K. candel* seedlings

组号	处 理	茎长 /cm	叶重 /g	茎重 /g	根重 /g
Ⅰ	单独种植	9.04 ±2.52b	1.25 ±0.29a	0.72 ±0.10a	2.77 ±0.53ab
Ⅱ	秋茄 互花米草 =3 1	14.03 ±4.52c	2.44 ±0.76b	1.46 ±0.21b	3.58 ±0.98b
Ⅲ	秋茄 互花米草 =1 1	11.97 ±45bc	2.41 ±0.59b	1.31 ±0.10b	3.64 ±0.62b
Ⅳ	秋茄 互花米草 =1 3	5.05 ±1.46a	0.75 ±0.33a	0.53 ±0.07a	2.33 ±0.75a

注：表中的数值为平均值 ±标准误，同列数值后附不同字母者表示差异达 0.05显著水平。

实验中，高密度下的互花米草影响了秋茄的生长，表现在秋茄的茎长和各部分生物量明显下降，可能是互花米草分泌出的次生代谢物质[8-9]，致使秋茄的根系受害，吸收能力减弱，生长受到抑制。

任何化感物质对植物的作用都与浓度有关，低浓度促进、高浓度抑制是普遍的现象[10-12]。在本实验中，低密度互花米草混种时，所释放的次生物质没有抑制秋茄幼苗的生长，有些分泌物却促进了秋茄的合成代谢，增加了物质的积累，提高了各部分的生物量。研究发现，比例 1 1以上生物量的累积无明显差异，说明互花米草的密度可能存在一个阈值，控制在一定范围内对红树林可能是无害的。

2.2 混种密度对秋茄幼苗叶片光合特性的影响

混种对秋茄叶片光合作用有明显的影响 (表 2)。单因素方差分析表明，秋茄幼叶 *Gs*、*Tr*、*Pn*、*Ci*、*WUE*在各处理下均存在显著差异 ($P < 0.05$)。在低密度组，*Pn*有所升高，但随着互花米草密度的增加，*Pn*逐渐减少，第 Ⅳ组的 *Pn*仅为对照组的 40%。*Ci*最高值是 366.57 μmol·mol^{-1}，它的 *Gs*仅为对照组的 61%，远低于其它组。当少量互花米草存在时 (Ⅱ组)，对秋茄幼叶 *WUE*的影响不明显 ($P > 0.05$)，但随着互花米草的增多，*WUE*显著减少，第 Ⅳ组的 *WUE*为 1.79 μmol CO_2 ·mmol^{-1} H_2O，仅为对照组的 38.5%。

表 2 混种密度对秋茄幼叶光合特性的影响

Table 2 Effects of mixed-culture densities on photosynthetic characteristics of *K. candel* seedlings

组号	*Gs* $mmol \cdot m^{-2} \cdot s^{-1}$	*Tr* $mmolH_2O \cdot m^{-2} \cdot s^{-1}$	*Ci* $\mu mol \cdot mol^{-1}$	*Pn* $\mu molCO_2 \cdot m^{-2} \cdot s^{-1}$	*WUE* $\mu molCO_2 \cdot mmol^{-1}H_2O$
Ⅰ	73.00 ±2.38b	0.87 ±0.09a	281.43 ±17.45b	6.16 ±0.44c	4.65 ±0.57c
Ⅱ	140.43 ±9.86c	1.52 ±0.11b	261.14 ±10.02a	7.31 ±0.31d	4.85 ±0.49c
Ⅲ	69.86 ±4.70b	2.49 ±0.14c	344.14 ±7.95c	4.01 ±0.45b	2.47 ±0.13b
Ⅳ	44.43 ±4.58a	1.40 ±0.23b	366.57 ±14.67d	2.46 ±0.53a	1.79 ±0.46a

注：表中的数值为平均值 ±标准误，同列数值后附不同字母者表示差异达 0.05显著水平。

当秋茄和高密度的互花米草混种时，可能是其叶片叶绿素含量下降 (数据另文发表)，叶片气孔收缩，气孔导度降低，从而限制了 CO_2 向叶绿体的输送，导致 *Pn*下降。由于叶片的 *Ci*要始终保持低于环境 CO_2 浓度，当 *Ci*升高时，气孔导度会降低以便适应环境的变化[13]。*WUE*由植物的光合速率和蒸腾速率 2方面决定，即消耗单位重量的水植物所固定的碳水化合物。*WUE*的大小往往可以反映植物对环境适应能力的强弱[14]。

综合分析，秋茄的光合作用受到互花米草的显著影响，使幼苗叶片叶绿素含量下降，*Pn*降低，*Gs*减小，*Tr*降低，*WUE*下降，秋茄的生长受到了显著抑制，这与 2.1中各器官生物量显著减少相符。

2.3 混种密度对秋茄幼苗可溶性糖、淀粉含量的影响

低密度混种 (Ⅱ组)时，秋茄幼苗叶片和根中的可溶性糖的含量最高 (图 1)，分别为对照组的 1.24倍和 1.91倍，随着互花米草密度的增大，秋茄幼苗叶和根中可溶性糖的含量逐渐下降至对照组以下。各处理组间秋茄叶片可溶性糖含量与对照差异显著 $(P<0.05)$，根中可溶性糖与对照差异极显著 $(P<0.01)$，表明秋茄幼苗可溶性糖含量受互花米草的影响。高密度组秋茄叶片淀粉含量低于其它组，根中淀粉含量则无显著差异 $(P>0.05)$ (图 1)。

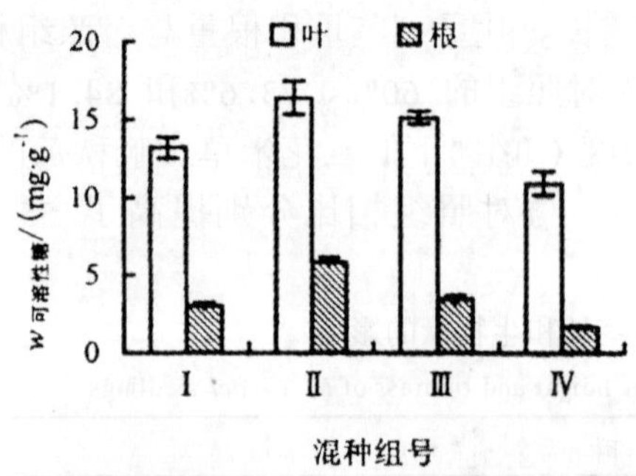

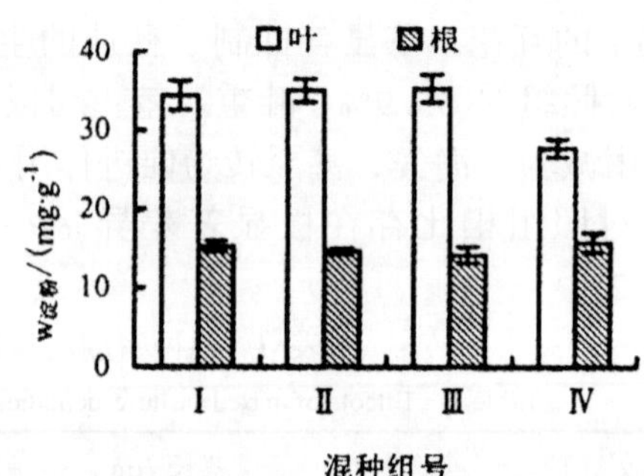

图 1 混种密度对秋茄幼苗可溶性糖、淀粉含量的影响

Figure 1 Effects of mixed-culture densities on sugar and starch content of *K. candel* seedlings

可溶性糖是调节渗透胁迫的小分子物质，在植物对胁迫的适应性调节中，是增加渗透性溶质的重要组成成分。受到互花米草的影响，秋茄在一定程度上通过积累可溶性糖作为渗透调节物质，维持其正常的生理代谢。低密度处理 (Ⅱ组)下，可溶性糖含量升高，一方面是由于光合作用加强，促进了可溶性糖的积累，以提供更多能量促进秋茄的生长；另一方面，也有可能是大分子碳水化合物和蛋白质的合成受到抑制，分解加强，可溶性糖增加。高密度处理 (Ⅳ组)下，秋茄得不到正常的养分供给，通过对淀粉的消耗，从而维持秋茄的生长。

2.4 混种密度对秋茄幼苗可溶蛋白与脯氨酸含量的影响

随着互花米草密度的增加，秋茄幼苗中可溶性蛋白含量显著下降，游离脯氨酸含量逐渐上升 (图 2)，表明二者含量均受互花米草密度的影响。

很多植物叶片中含有大量的酚类物质，它们可以抑制蛋白质合成、改变脂类和有机酸的代谢[15]。本研究的结果也表明，在高密度互花米草混种条件下，秋茄体内蛋白质合成受到抑制，水解加剧，进而导致蛋白含量的下降，氨基酸含量上升，其中脯氨酸的上升最突出，并且随胁迫程度的加重呈上升趋势。

游离脯氨酸含量常作为指示植物遭受水分胁迫程度的指标[16]，作为渗透调节物质，脯氨酸具有水

溶性和水势高的特点，能保持原生质与环境渗透平衡[17]。游离脯氨酸积累还可以降低水分胁迫期间蛋白质水解产生的游离氨的毒害，贮存氮素和碳架，为逆境解除后恢复生长提供呼吸基质和能源[18]。脯氨酸积累是水分胁迫条件下敏感性状的一种表现，秋茄在互花米草的影响下积累了大量的脯氨酸，这是为了对抗胁迫而采取的一种保护性渗透调节反应，通过积累可溶性糖、游离脯氨酸，降低细胞内溶质的渗透势，维持一定的水势，从而维持秋茄细胞的正常生长，对抗高密度互花米草的不利影响。

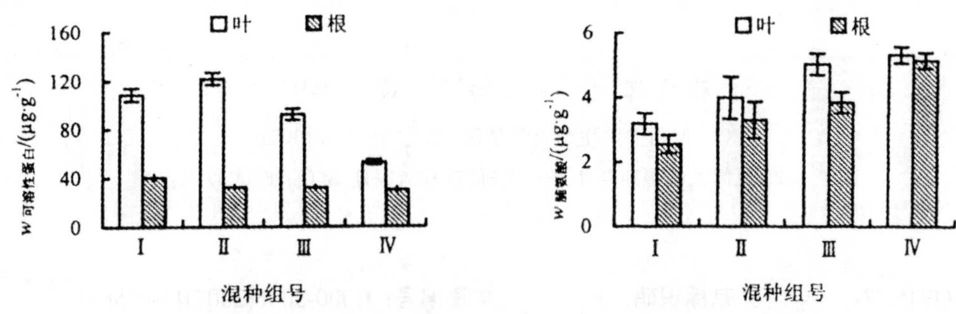

图 2 混种密度秋茄幼苗可溶性蛋白和游离脯氨酸含量的影响

Figure 2 Effects of mixed-culture densities on soluble protein and proline contents of *K. candel* seedlings

3 结论

(1)高密度互花米草(秋 米 =1 3)使秋茄幼苗的茎长、叶重、茎重和根重均减少，抑制秋茄幼苗的生长，与对照相比差异显著 ($P < 0.05$)；低密度互花米草 (秋 米 =3 ∶1)则促进秋茄的生长，使秋茄幼苗茎长、叶重、茎重和根重显著增加。(2)高密度互花米草影响下，秋茄幼叶光合速率降低，气孔导度减小，蒸腾速率降低，水分利用率较低，导致生长受到抑制。(3)秋茄幼苗通过积累可溶性糖和游离脯氨酸，降低细胞内溶质的渗透势，维持一定的膨压，从而维持细胞的正常生长，对抗高密度互花米草的不利影响。(4)在高密度互花米草的影响下，秋茄体内蛋白质水解加剧，进而导致蛋白含量的下降，游离氨基酸含量上升。(5)秋茄幼苗的生长及生理特性的变化可能与互花米草的化感作用有密切关联，关于这些化感物质的种类有待进一步研究。

参 考 文 献:

[1] 钦佩，仲崇信 . 米草的应用研究 [M]. 北京：海洋出版社，1992: 14 - 130.

[2] 杜文琴，马丽娜，刘建，等 . 红树林区内互花米草防除技术研究 [J]. 中国生态农业学报，2006, 14(3): 154 - 156.

[3] 肖强，郑海雷，叶文景，等 . 水淹对互花米草生长及生理的影响 [J]. 生态学杂志，2005, 24(9): 1 025 - 1 028.

[4] 林鹏 . 中国红树林生态系统 [M]. 北京：科学出版社，1997: 297 - 316

[5] 孙书存，朱旭斌，吕超群 . 外来种米草的生态功能评价与控制 [J]. 生态学杂志，2004, 23(3): 93 - 98.

[6] 张志良 . 植物生理实验指导 (第 2版) [M]. 北京：高等教育出版社，1990: 88 - 91.

[7] 邹琦 . 植物生理学实验指导 [M]. 北京：中国农业出版社，2000: 64 - 67, 110 - 115, 173 - 175.

[8] 马永建，李莉，袁宝君，等 . 互花米草成分研究 Ⅰ. GC-MS法研究叶片中脂肪酸 [J]. 中国生化药物杂志，2001, 22(4): 184 - 186.

[9] 马永建，袁宝君，李莉，等 . 互花米草成分研究 Ⅱ. GC-MS法研究挥发性成分 [J]. 中国生化药物杂志，2002, 23(1): 36 - 37.

[10] 林思祖，黄世国，曹光球，等 . 杉木自毒作用的研究 [J]. 应用生态学报，1999, 10(6): 661 - 664.

[11] 李玫，廖宝文，郑松发，等 . 无瓣海桑对乡土红树植物的化感作用 [J]. 林业科学研究，2004, 17(5): 641 - 645.

[12] 莫竹承，范航清 . 木榄和秋茄的种间化感作用研究 [J]. 广西科学，2001, 8(1): 61 - 62.

[13] 王立，杨允菲，孙伟，等 . 两个生态型羊草对 CO_2 浓度倍增的光合生理响应 [J]. 草地学报，2003, 11(1): 52 - 57.

[14] 刘金祥，麦嘉玲，刘家琼 . CO_2 浓度增强对沿阶草光合生理特性的影响 [J]. 中国草地，2004, 26(3): 13 - 23.

[15] 郑丽，冯玉龙 . 入侵植物的生理生态特性对碳积累的影响 [J]. 生态学报，2005, 25(6): 1 430 - 1 438.

[16] 马成仓，高玉葆，蒋福全，等 . 小叶锦鸡儿和狭叶锦鸡儿的生态和水分调节特性比较研究 [J]. 生态学报，2004, 24(7): 1 442 - 1 451.

[17] 马翠兰，刘星辉，胡又厘 . 渗透调节物质和水分状态与琯溪蜜柚抗寒性的关系 [J]. 福建农业大学学报，2000, 29(1): 31 - 34.

[18] 彭志红，彭克勤，胡家金，等 . 渗透胁迫下植物脯氨酸积累的研究进展 [J]. 中国农学通报，2002, 18(4): 80 - 83.

红树植物的盐分平衡机制*

张宜辉[1,2] 王文卿[1,2] 林 鹏[1,2]
(1. 厦门大学生命科学学院,福建 厦门 361005;
2. 厦门大学湿地与生态工程研究中心,福建 厦门 361005)

中图分类号:Q945.78 文献标识码:A 文章编号:1000-3096(2007)11-0086-05

热带、亚热带海岸潮间带的红树林生态系统处于海洋、陆地和大气的动态交界面,作为独特的海陆边缘生态系统在维持海湾河口生态系统的稳定中起着特殊的作用。生长在潮间带高盐环境中的红树植物,经长期的自然选择和进化适应,在生理生化及形态方面形成了一系列适应机制[1~3]。盐生植物体内盐分质量比的调节是植物生存的关键[4],红树植物维持体内盐分平衡的途径包括根系拒盐、叶片泌盐、叶片肉质化、脉内再循环、通过衰老器官的脱落排盐等5种。长期以来一直认为红树植物可分为两类,具盐腺的泌盐红树植物(secreters)和不具盐腺的拒盐红树植物(excluders) [1,3]。老鼠簕属(*Acanthus*)、桐花树属(*Aegiceras*)、阿吉木属(*Aegialitis*)及白骨壤属(*Avicennia*)等属物种的叶片均具盐腺,通过盐腺把过多的盐分排出体外,使叶片得以保持盐分平衡[3];而红树属(*Rhizophora*)、秋茄属(*Kandelia*)、木榄属(*Bruguiera*)及海桑属(*Sonneratia*)等主要是通过木质部的高负压,从含盐基质中分离出淡水,所以称之为拒盐红树植物[1]。不具盐腺的拒盐红树植物通过叶片肉质化来维持体内盐分的平衡,泌盐红树植物由于盐腺泌盐而有效地控制其体内的盐分含量,所以其叶片肉质化程度变化不大[5]。此外,通过衰老器官(尤其是叶片)的脱落排盐是所有红树植物共有的一种排盐方式[3,6~10]。上述就是目前对红树植物维持盐分平衡的主要观点,作者在对红树植物抗盐生理生态进行长期研究的基础上,对红树植物维持盐分平衡的途径进行综述。

1 根系拒盐

红树植物根系拒盐主要是通过根系内皮层中发达的凯氏带起作用[1],它阻止盐分向地上部分的运输。拒盐红树植物如榄李属(*Lumnitzera*)、红树属及海桑属等属根系滞留盐分效率达99%,其木质部液流中的含盐量还不到生长环境基质盐分质量浓度的1%[6]。对红海榄(*Rhizophora stylosa*)不同级别根(分为4级,Ⅰ:d<2 mm,Ⅱ:d为2~5 mm,Ⅲ:d为5~15 mm,Ⅳ:d>15 mm)中盐分分布情况的测定结果表明,红海榄根系中Ⅰ级细根的Na^+、Cl^-含量均低于土壤,这是根内皮层作用的结果。进入Ⅰ级根的盐分马上被Ⅱ、Ⅲ级根滞留,其结果是Ⅱ、Ⅲ级根盐分含量很高,甚至超过叶片,Ⅳ级根的盐分含量已下降了很多[11]。

尽管泌盐红树植物木质部液流中的盐分含量相对较高,达到基质盐分质量浓度的10%,根系对盐分的进入仍具有一定的排斥作用[3]。Waisel 等[12]的研究表明,根系对盐分的排斥作用是泌盐红树植物白骨壤(*Avicennia marina*)最重要的抗盐机制之一,进入根系的盐分有80%被滞留于根系中,叶片盐腺泌盐只能排除进入叶片的盐分的40%或进入根系的盐分的8%;而 Burchett 等[13]和 Scholander[6]的研究表明白骨壤的根系滞留盐分效率分别达 90%和 95%。一般认为

* 国家自然科学基金资助项目(30200031),福建省自然科学基金资助项目(B0410001)
原载于海洋科学,2007,31(11):86-90

泌盐红树植物的盐腺能有效地控制其体内盐分的平衡[3,5,14]，但对白骨壤的研究表明，尽管有盐腺的存在，叶片含盐量还是随基质盐分质量浓度的提高而提高[13,15]。因此对泌盐红树植物来说，仅通过盐腺泌盐并不能维持叶片盐分的平衡。

由此可见，拒盐红树植物和泌盐红树植物在根系排盐效率上有所差别，前者比后者高 10 倍左右，但根系拒盐是它们最重要的排盐机制。

此外，对拒盐红树植物秋茄(*Kandelia candel*)、海莲(*Bruguiera sexangula*)、红海榄以及泌盐红树植物桐花树(*Aegiceras corniculatum*)、白骨壤的研究发现，除个别器官外，秋茄、桐花树、白骨壤、海莲和红海榄植物体所有器官的 Na^+、Cl^-含量均高于土壤，表现为元素富集率大于 1[11]。因此，从这种意义上讲，秋茄、桐花树、白骨壤、海莲和红海榄均需要在体内保持一定的盐分质量比。Medina 和 Francisco[16]对红树属、假红树(*Laguncularia racemosa*)和亮叶白骨壤(*A. germinas*)以及赵可夫等[17]对秋茄和白骨壤渗透调节物质的研究结果与这一观点一致。研究发现，九龙江口秋茄和白骨壤 Na^+、Cl^-、K^+、Ca^{2+}和 Mg^{2+}等无机离子对渗透调节的贡献率约占 88%，其中起主要作用的是 Na^+和 Cl^-[17]。拒盐红树植物如榄李属、红树属及海桑属等的木质部液流中的含盐量还不到基质盐分质量浓度的 1%，但这还是比典型的甜土植物高 10～50 倍[3]。

以上事实说明，一方面，通过根系拒盐保持地上部分较低的盐分质量比是所有红树植物的共同特点。另一方面，在不发生因盐分过多积累而造成伤害的前提下，红树植物尽量多地吸收无机盐(主要是 Na^+和 Cl^-)来进行渗透调节，从能量角度来说，这是最经济的[18,19]。

2 盐腺泌盐

生活在高盐环境中的泌盐红树植物，除了降低根对离子的吸收和阻止离子向地上部的运输外，还利用盐腺这一特殊的分泌结构将已吸收入植物体内的盐分排出体外，以维持体内盐分质量比低于产生生理毒害的浓度。

桐花树、白骨壤和老鼠簕(*Acanthus ilicifolius*)的上、下表皮均有盐腺分布，但前二种植物的盐腺分布于下表皮下陷处，而白骨壤的盐腺突出于表皮之上，而且下表皮上盐腺的数目多于上表皮[20]。对泌盐红树植物叶片盐腺的显微结构的研究表明，不同科的泌盐红树植物盐腺细胞数目和排列方式各不相同。紫金牛科(Myrsinaceae)桐花树的盐腺由 5 个部分组成，从叶肉向叶表面排列的次序是：收集细胞(由若干个细胞组成)、基细胞(仅 1 个扁平细胞)、分泌细胞(由 20 到 42 个细胞组成)、收集室(1 个气室)和盐腺盖(由角质层盖和角质层套构成，并有若干个泌盐孔组成)[21,22]。马鞭草科(Verbenaceae)白骨壤的盐腺有 2 个或 4 个收集细胞和 8 个(有时有 12 个)分泌细胞[23]。

泌盐作用可以去除叶片中由于蒸腾作用而积累的过剩盐分。Atkinson 等[14]证明了红树植物随木质部汁液而达到叶片的氯化物的数量是通过分泌作用来平衡的，因而叶片内的氯化物含量在全天保持相对的数量。泌盐途径一般认为是先将盐腺周围的盐分集中到盐腺基部的收集细胞再到基细胞，然后再运进分泌细胞，再到收集室，然后从壁上的小孔将盐分挤出体外。Ish-Shalom-Gordon 和 Dubinsky 用扫描电镜观察白骨壤的盐腺的超微结构，发现其盐腺的泌盐过程是两种机制控制的：第 1 种是局泌型(merocrine)，即先形成一个囊泡，此囊泡不断扩大直至破裂，盐溶液也随之释放，而后破裂的囊泡解体，并形成新的囊泡；第 2 种是全泌型(holocrine),即植物先在下表皮空间收集待分泌的盐溶液，最后导致表皮破裂，贮存的盐溶液呈小滴状往外流[24]。

基质盐分含量影响泌盐红树植物盐腺的泌盐作用。随着土壤盐度的增加，桐花树叶片下表皮盐腺系统中的分泌细胞平均数和单位面积盐腺数目都有增加，每个盐腺系统分泌细胞由盐度为 3.5 的 19 个上升到盐度为 19.2 的 37 个，相应的下表皮盐腺数目由 7 个/mm^2 上升到 10 个/mm^2 [22]。郑海雷和林鹏[25]对白骨壤和桐花树，Sobrado 和 Greaves[26]对亮叶白骨壤，叶勇等[27]对老鼠簕、桐花树和白骨壤的泌盐特性的研究都发现，叶片泌盐量均随着基质盐度的升高而增加。表明泌盐红树植物幼苗的体内只能容纳一定的盐分，当环境盐度增大，盐分吸收量增加后有能力将相应的盐分排出体外。

3 叶片肉质化

叶片肉质性是拒盐红树植物调节体内盐分平衡的途径之一[3,6]，但林鹏[1]对桐花树、Burchett 等[13]和赵可夫等[17]对白骨壤的研究证实了泌盐红树植物也有肉质性叶片的发育。这说明它们在相同的环境条件下具趋同适应性。叶片肉质化的结果是叶片盐分质量比受叶片年龄的影响不大。Smith 等[5]对直立风车子(*Conocarpus erectus*)的研究表明，肉质化程度高的叶片细胞渗透压大且 NaCl 含量高。也有研究表明，盐胁迫导致的叶片肉质化增大了叶肉细胞表面积与叶片表面积的比值[28~30]，也就是说相对增加了 CO_2 吸收面积，这可能有利于光合碳同化。但是，叶片肉质化只能稀释叶片细胞的盐分，而不能影响叶片盐分含量[7~10]。

肉质性的形成不是细胞分裂的结果，而是细胞体积增大的结果[3,5,31]，因为当叶片发育到最大叶面积一半的时候细胞分裂已停止。这种增大主要是栅栏组织细胞及海绵组织细胞增大的结果，但不同物种有不同的增大方式[28,32]。红树植物直立风车子叶片的肉质化主要是垂直于叶面的细胞长度的增加，且起主要作用的是内层叶肉细胞[5]，假红树属(*Laguncularia*)、榄李属及海桑等属的物种的叶片的肉质化也属这一类；阿吉木属、白骨壤属及红树属等属则主要是内皮层细胞增大的结果；木果楝属(*Xylocarpus*)只有栅栏组织细胞增大，水芫花属(*Pemphis*)则是栅栏组织细胞及海绵组织细胞都增大[3]。

影响盐生植物叶片肉质化程度的两个主要因素是：叶片年龄和基质盐分质量浓度。叶片肉质化程度随叶龄的提高而提高[3,5]，如假红树同一枝条上的老叶比新叶厚 4 倍；基质盐分质量浓度也影响叶片的肉质性程度，一般是叶片的肉质性随基质盐分质量浓度的提高而提高[28,31,33]。但有例外，Burchett 等[13]对白骨壤的研究表明，培养于 50%海水中的白骨壤叶片肉质化程度高于培养于 100%海水中的。

4 脉内再循环

Wignarajah 等[34]发现盐渍条件下菜豆叶片 Na^+可以通过韧皮部运出叶片。对甜菜、玉米和菜豆的研究也表明存在叶片 Na^+、Cl^-的向基性转移，如菜豆顶部的初生叶施加 $^{22}Na^+$ 1 d 后，$^{22}Na^+$几乎全部向基性地转移到根部，且有相当数量进入根外介质[35]。对高粱、玉米及羽扁豆等的研究发现也存在这种情况。80 年代初，提出了一个“脉内再循环”(intravenial recycling)假说[36]。这个假说的主要内容是植物地上部的木质传递细胞可将木质部导管中的 Na^+重新吸收出来，通过韧皮部传递细胞进入筛管，下运到近根部分，其中一部分还可以进入根外介质中。离子从叶片中被重新吸收出来的关键是从木质部到韧皮部的传递，其中传递细胞起了主要作用。其中吸收 Na^+的机理是通过传递细胞中的 K^+与导管中的 Na^+交换进行的。

Waisel 等[12]用放射性同位素 $^{22}Na^+$、$^{36}Cl^-$详细研究了白骨壤叶片的盐分平衡，结果发现进入叶片的盐分只有 40%是通过盐腺排出体外的，而通过韧皮部运输的盐分进仅 27%，且这种运输主要在晚上发生。即使是拒盐红树植物，根系在吸收水分过程中也不能完全排除盐分，盐分必然随蒸腾液流逐渐在叶片中累积。但是，没有证据表明盐分在拒盐红树植物体内有逐步累积的趋势[37]。作者对拒盐红树植物木榄(*Bruguiera gymnorrhiza*)叶片的研究发现，从叶片成熟至开始衰老这段时间里，叶片 Na^+、Cl^-质量比和含量保持不变[7]。木榄叶片具较厚的角质层，通过角质蒸腾或淋溶方式排盐所起的作用不大。因此，必然存在一种将 Na^+、Cl^-重新运出叶片的机制。这也从侧面证明拒盐红树植物体内也存在“脉内再循环”。

5 衰老器官脱落排盐

多年来，一直认为通过衰老器官(尤其是叶片)的脱落排盐是所有红树植物尤其是拒盐红树植物共有的一种排盐方式，因为衰老叶片的盐分质量比远高于成熟叶片[3,6,38]。通过衰老器官尤其是叶片的脱落确实使大量盐分得以离开植物体。

作者对木榄叶片发育及衰老过程中的盐分动态进行了跟踪研究，发现木榄叶片只有在发育过程中累积盐分，此时叶片盐分质量比和含量均升高，自叶片成熟到开始衰老这一段时间里盐分质量比和含量变化不大，在衰老过程中盐分质量比确实升高了很多，落叶 Na^+和 Cl^-质量比分别比成熟叶高 48.8%和 30.6%，含量只分别升高了 6.3%和 4.3%。由此可见，随着叶片的衰老，叶片中盐分质量比升高，但是这一过程并不伴随着盐分向衰老叶片的主动转移，而是由于叶片衰老过程中大量养分转移至植物体的其他器官而造成的一种浓缩效应[7,39]。作者对红树植物秋茄、红海榄的研究也得出了类似结论[8~10]。Tomlinson 也指出，没有证据表明盐分向衰老器官的转移是一种主动行为[3]。因此，通过衰老器官的脱落排盐对维持红树植物的盐分平衡作用有限。

6 结论

Waisel 等[12]通过对白骨壤叶片的盐分平衡的研究后指出，白骨壤通过 3 种不同的途径来适应潮间带高盐环境：(1) 根系拒盐；(2) 高盐环境下维持生理生化代谢活动的正常运行机制；(3) 叶片排盐(这种排盐是通过盐腺泌盐和韧皮部的运输实现的)。许多研究表明，所有红树植物均通过以下一种或多种途径调节体内盐分含量：拒盐、泌盐、肉质化和衰老器官的脱落等[6,38,40]。

作者认为，红树植物主要通过以下几种途径来调节体内盐分平衡：根系拒盐、叶片泌盐、叶片肉质化、脉内再循环和衰老器官的脱落，其中叶片泌盐为泌盐红树植物所特有，衰老器官的脱落的途径作用有限。所有红树植物既是拒盐植物又需要在体内维持一定的盐浓度，在保持细胞渗透调节能力的前提下，尽量减少盐分的吸收是所有红树植物的共性。在不发生因盐分过多积累而造成伤害的前提下，红树植物尽量多地吸收无机盐(主要是 Na 和 Cl)来进行渗透调节，从能量角度来说，这是最经济的。从形态解剖角度把红树植物划分为拒盐红树植物和泌盐红树植物有其合理的成份，但从生理角度来说还有必要对红树植物的盐分平衡机制开展进一步的研究工作。

目前，从细胞水平和分子水平上对红树植物的抗盐机制已开展较为深入的研究，渗透调节物质生物合成相关基因的克隆与转化、盐胁迫的信号传导途径、盐胁迫下离子通道行为等已成为研究的热点。作者从个体水平对红树植物的盐适应机制进行阐述，配合日益发展的生物技术及基因工程手段，将会更进一步了解红树植物抗盐机制的生理学本质，这必将为今后红树植物的引种、驯化以及红树林的生态恢复工程提供更充分的理论依据。

参考文献：

[1] 林 鹏. 红树林[M]. 北京：海洋出版社，1984. 1-104.

[2] Hutchings P, Saenger P. Ecology of mangroves[M]. St Lucia: University of Queensland Press, 1987. 1-388.

[3] Tomlinson P B. The Botany of Mangroves[M]. Cambridge: Cambridge University Press, 1994. 1-419.

[4] Omer L S, Horvath S M, Setaro F. Salt regulation and leaf senescence in aging leaves of *Jaumea carnosa* (Less.) gray (*Asteraceae*), a salt mash species exposed NaCl stress[J]. **American Journal of Botony**, 1983, 70: 363-368.

[5] Smith J A C, Popp M, Luttge U *et al*. Ecophysiology of xerophytic and halophytic vegetation of a coastal alluvial plain in northern Venezuela(Abstract) [J]. **New Phytologist**, 1989, 111: 293-307.

[6] Scholander P F. How mangroves desalinate seawater[J]. **Physiologia Plantarum**, 1968, 21: 251-261.

[7] Wang W, Lin P. Transfer of salt and nutrients in *Bruguiera gymnorrhiza* leaves during development and senescence[J]. **Mangroves and Salt Marshes**, 1999 , 93: 1-7.

[8] 王文卿，林鹏. 红树植物秋茄和红海榄叶片元素含量及季节动态的比较研究[J]. 生态学报，2001，**21**(8): 1 233-1 238.

[9] Lin P, Wang W Q. Changes in the leaf composition, leaf mass and leaf area during leaf senescence in three species of mangroves[J]. **Ecological Engineering**, 2001, **16**(3): 415-424.

[10] Wang W Q, Wang M, Lin P. Seasonal changes in element contents in mangrove element retranslocation during leaf senescence[J]. **Plant and Soil**, 2003, 252: 187-193.

[11] 王文卿，林鹏. 红树植物体内元素分布特点及抗盐机理[J]. 林业科学，2003，**39**(4): 30-36.

[12] Waisel Y, Eshel A, Agami M. Salt balance of leaves of the mangrove *Avicennia marina*[J]. **Physiologia Plantarum**, 1986, 67: 67-72.

[13] Burchett M D, Field C D, Pulkownik A. Salinity, growth and root respiration in the gray mangrove, *Avicennia marina*[J]. **Physiologia Plantarum**, 1984, 60: 113-118.

[14] Atkinson M R, Findlay C P, Hope A B, *et al*. Salt regulations in the mangroves *Rhizophora mucronata* Lamk. and *Aegialitis annulata* R. Br[J]. **Australian Journal of Bilogical Science**, 1967, 20: 589-599.

[15] Downton W J S. Growth and osmotic relations of the mangrove *Avicennia marina* as influenced by salinity[J]. **Australian Journal of Plant Physiology**, 1982, 9: 519-528.

[16] Medina E, Francisco M. Osmolality and $\delta^{13}C$ of leaf tissues of mangrove species from environments of contrasting rainfall and salisity[J]. **Estuarine, Coastal and Shelf Science**, 1997, 45: 337-344.

[17] 赵可夫，冯立田，卢元芳，等. 九龙江口秋茄和白骨壤的渗透调节剂及其贡献[J]. 海洋与湖沼，1999,**30**(1):

58-61.

[18] Flowers T J, Troke P F, Yeo A R. The mechanism of salt tolerance in halophytes[J]. **Annual review of plant physiology**, 1977, 28: 89-121.

[19] Guerrier G. Fluxes of Na^+, K^+ and Cl^- and osmotic adjustment on *Lycopersicon pimpinellifolium* and *L. esculentum* during short and long term exposures to NaCl[J]. **Physiologia Plantarum**, 1996, 97: 583-591.

[20] 林鹏. 中国红树林生态系[M]. 北京：科学出版社，1997，64-66.

[21] Field C D, Hinwood B G, Stevenson I. Structural features of the salt gland of *Aegiceras*[A]. Teas H J. Physiology and management of mangroves[C]. The Hauge: Dr W. Junk Publishers, 1984, 37-42.

[22] 叶庆华，章菽，林鹏. 福建九龙江口桐花树叶片的盐腺系统[J]. 台湾海峡，1988，**7**(3): 264-267.

[23] Drennan P M, Berjak P, Lawton J R, *et al*. Ultrastructure of the salt glands of the mangrove, *Avicennia marina* (Forssk..) Vierh, as indicated by the use of selective membrane staining[J]. **Planta**, 1987, 172: 176-183.

[24] Ish-Shalom-Gordon N, Dubinsky Z. Possible modes of salt secretion in *Avicennia marina* in the Sinai[J]. **Plant Cell Physiology**, 1990, 31: 27-32.

[25] 郑海雷，林鹏. 红树植物白骨壤对盐度的某些生理反应[J]. 厦门大学学报（自然科学版），1997，**36**(1): 135-139.

[26] Sobrado M A, Greaves E D. Leaf secretion composition of the mangrove species *Avicennia germinans* (L.) in relation to salinity: a case study by using total-reflection X-ray fluorescence analysis[J]. **Plant Science**, 2000, 159: 1-5.

[27] 叶勇，卢昌义，胡宏友，等. 三种泌盐红树植物对盐胁迫的耐受性比较[J]. 生态学报，2004，**24**(1): 2 444-2 450.

[28] Longstreth D J, Nobel P S. Salinity effects on leaf anatomy. Consequences for photosynthesis[J]. **Plant Physiology**, 1979, 63: 700-703.

[29] Psaras G K, Rhizopoulou S. Mesophyll structure during leaf development in *Ballota acetabulosa*[J]. **New Phytologist**, 1995, **13**(3): 303-309.

[30] Jafri A Z, Ahmad R. Effect of soil salinity on leaf development, stomatal size and its distribution in cotton (*Gossypium hirsutum* L.)[J]. **Pakistan Journal of Botany**, 1995, **27**(2): 297-303.

[31] Camilleri J C, Ribi R. Leaf thickness of mangroves (*Rhizophora mangle*) growing in different salinities[J]. **Bitrophica**, 1983, 15: 139-141.

[32] Seemann J R, Critchley C. Salinity and nitrogen effects on photosynthesis, ribulose-1,5-bisphosphate carboxylase and metabolite pool size in *Phaseolus vulgaris* L[J]. **Planta**, 1985, 164: 151-162.

[33] Alpha C G, Drake D R, Goldstein G. Morphological and physiological responses of *Scaevola sericea* (Goodeniaceae) seedlings to salt spray and substrate salinity[J]. **American Journal of Botany**, 1996, **83**(1): 86-92.

[34] Wignarajak K, Jennings D H, Handley J F. The effect of salinity on growth of *Phaseolus vulgaris* L. Ⅱ. Effects on internal solute concentration[J]. **Annals of Botany**, 1975, 39: 1 039-1 055.

[35] 马斯纳 H. 曹一平, 陆景陵,等译. 高等植物的矿质营养[M]. 北京: 北京农业大学出版社，1991, 1-327.

[36] 赵可夫，王昭唐. 作物抗性生理[M]. 北京：农业出版社，1990, 1-342.

[37] Field C D. Ions in mangroves[A]. Teas H J. Physiology and management of mangroves[C]. The Hauge: Dr W. Junk Publishers, 1984, 43-48.

[38] Zheng W, Wang W, Lin P. Dynamics of element contents during the development of hypocotyles and leaves of certain mangrove species[J]. **Journal of Experimental Marine Biology and Ecology**, 1999, 233: 247-257.

[39] Cram W J, Torr P G, Rose D A. Salt allocation during leaf development and leaf fall in mangroves[J].**Tree**, 2002, 16: 112-119.

[40] Joshi G V, Jamale B B, Bhosale L J. Ion regulation in mangroves[A]. Walsh G, Snedaker S, Teas H J. Proceedings of International Symposium on Biology and Management of Mangroves[C]. University of Florida Press, 1975, 597-607.

Effect of soil salinity on cold tolerance of mangrove *Kandelia candel* *

YANG Shengchang(杨盛昌) LI Yunbo(李云波) LIN Peng(林 鹏)

(*School of Life Sciences, Xiamen University, Xiamen, Fujian* 361005, *China*)

Abstract Analysis of cold tolerance on mangrove *Kandelia candel* leaf growing in different soil salinity along Jiulong River Estuary in South China showed that the cold tolerance decreased as the increase of soil salinity. The lethal temperatures of *K. candel* leaf were −10.4, −9.9 and −8.6 ℃ in Liaodong, Baijiao and Aotou, respectively. Under 1–2℃ cold stress treatment on detached leaves of *K. candel*, their caloric value gradually decreased, while electrolyte leakage gradually increased. The leaf's caloric value and electrolyte leakage in Aotou with higher soil salinity varied more largely than those in Liaodong with lower soil salinity. In *K. candel* leaf, total water content lowered a little, bound water content rose significantly and free water content dropped significantly with duration of cold stress. At the same time, reduction sugar, soluble sugar and starch content gradually decreased and sucrose content gradually increased. Bound water, free water and sucrose content in *K. candel* leaf from Aotou with higher soil salinity changed more slowly than those from Liaodong with lower soil salinity, but reduction sugar, soluble sugar and starch content in *K. candel* leaf from Aotou had faster variations than those from Liaodong. These data indicated that soil salinity can reduce cold tolerance of *K. candel* leaf by increasing negative effect of salt ions in cell membrane, inhibiting variations of water content, and aggravating consumption of material and energy.

Key words: *Kandelia candel*, salinity, cold tolerance, caloric value

1 INTRODUCTION

As a special woody salt-tolerance plant, mangroves distribute in intertidal zones along tropical and subtropical seashore. Soil salinity and air temperature are the main habitat factors affecting growth and distribution of these plants. Although mangroves have been researched extensively, researches on its cold tolerance are still insufficient (Yang and Lin, 1998). In this work, variations of cold tolerance of *Kandelia candel* plants growing in different salinity habitats along Jiulong River Estuary, Fujian Province, were investigated. At the same time, the effect of cold treatment on the detached leaf' variations of caloric value, membrane permeability, water and carbohydrate content were studied with the aim to elucidate the relationship between physiological processes and cold tolerance of mangrove plants under special habitat, and to provide scientific basis for induction, acclimatization and utilization of mangrove plants.

2 MATERIALS AND METHODS

2.1 Materials

Samples were collected from different *Kandelia candel* communities in Liaodong, Baijiao and Aotou along the southern coastline (about 24° N) of Jiulong River Estuary, Fujian Province, China in January of 1996 and 2000. It was reported that a consistent relationship existed between cold temperation injury to detached leaves and to individuals in fields, and that cold tolerance of detached leaves could reflect the cold tolerance of plant individuals (Sukumaran and Weiser, 1972; Hincha, 1994; Yang and Lin, 1998). So, detached matured leaves including the second and the third pair of leaves from shoot apex to base were collected as research materials in this study. The leaf samples were taken back to laboratory, where about 20 leaves were used for the measurement of cold tolerance, and the

* Supported by the Found for Outstanding Young Teacher, the Ministry of Education, China.

From Chinese Journal of Oceanology and Limnology, 2005, 23(1): 98−103.

rest were wrapped in wet gauze, and put into plastic bag for 1–2℃ cold stress treatment (in refrigerator). Then, some of them were immediately used for the measurement of electrolyte leakage and water content, and the rest were oven-dried at 60℃, grinded into powder, and put into sample bottle for measurement of carbohydrate content and caloric value.

Soil sample collected in 20–40 cm depth below surface *in situ* was crushed and put into sample bottle for the measurement of soil salinity.

2.2 Methods

2.2.1 Cold tolerance

Cold tolerance was measured by electrolyte leakage method of fast frozen leaves (Yang and Lin, 1998). Wash *K. candel* leaf with tap water to remove soil, chemical residue and microbes, and punch discs with 6.5 cm in diameter from leaves. Place five discs in 25 ml glass tube covered with cork, then put them into cool trough for cold treatment at 2℃ increment from −4℃ to −16℃. Measure electrolyte leakage of frozen leaves 1h after cold treatment (Yang and Lin, 1998). A revised logarithmic function can describe the response of the sample's electrolyte leakage to decreasing temperature. The temperature corresponding to the inflection point on curve is taken as the value of cold tolerance of *K. candel* plants.

2.2.2 Electrolyte leakage

Electrolyte leakage was measured based on Yang and Lin (1998), and expressed in percentage of total electrolyte leakage.

2.2.3 Caloric value

The caloric value of *K. candel* leaf was measured by GR 3500 Oxygen Bomb Calorimeter (Lin and Lin, 1991).

2.2.4 Water content

Water content was determined by β-refracting telescope method (Zhang, 1990).

2.2.5 Carbohydrate content

Carbohydrate was determined by Somogyi's method (Yuan and Yang, 1983). Carbohydrate was extracted from leaf sample by 80% alcohol. A part of the extract solution was directly used for measurement of reduction sugar, and the rest for determination of soluble sugar after hydrolysis of 2% HCl (hydrochloric acid). All the residue was used for determination of starch after hydrolysis of α-Amylase. Sucrose was the difference of soluble sugar and reduction sugar.

2.2.6 Soil salinity

Soil salinity was measured by silver nitrate method.

Each measurement mentioned above was repeated 3−4 times.

3 RESULTS

3.1 Relation between the cold tolerance and soil salinity

For the samples taken in January, 2000, variations of cold tolerance of *K. candel* mangrove plants from Liaodong, Baijiao and Aotou along the southern coastline of the Jiulong River Estuary are shown in Fig 1. Soil salinity was 12.55%, 14.96% and 20.21%, respectively for Liaodong, Baijiao and Aotou mangroves; while the cold tolerance of *K. candel* plants was −10.4℃, −9.9 ℃ and −8.6℃, respectively. Student test on the cold tolerance of *K. candel* plant and soil salinity in Liaodong, Baijiao and Aotou revealed significant difference between Aotou and Liaodong, and between Aotou and Baijiao, but no significant difference between Liaodong and Baijiao, showing that the cold tolerance gradually decreased with increase of soil salinity.

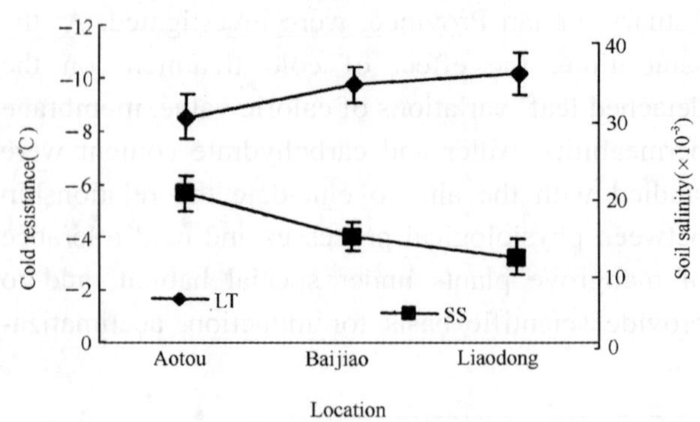

Fig.1 Relation between cold tolerance of *Kandelia candel* leaves and soil salinity at Liaodong,Baijiao and Aotou

3.2 Cold stress on caloric value

As shown in Fig.2, leaf caloric values were 19.41, 19.49 and 19.84 kJ/g, respectively for *K.candel* plants from Liaodong, Baijiao and Aotou. It was obvious that leaf caloric value increased as soil salinity increased. But during cold stress, *K. candel* leaf decreased its caloric values to different extent. Comparison of variations of caloric values of *K. candel* leaf from Liaodong, Baijiao and Aotou revealed that under cold stress, the leaf caloric value had a biggest fall in *K.candel* plants from Aotou, and had a smallest drop in Liaodong *K. candel* plants, showing that leaf caloric value had clear relationship with soil salinity. The higher the soil salinity is, the lower the leaf caloric value.

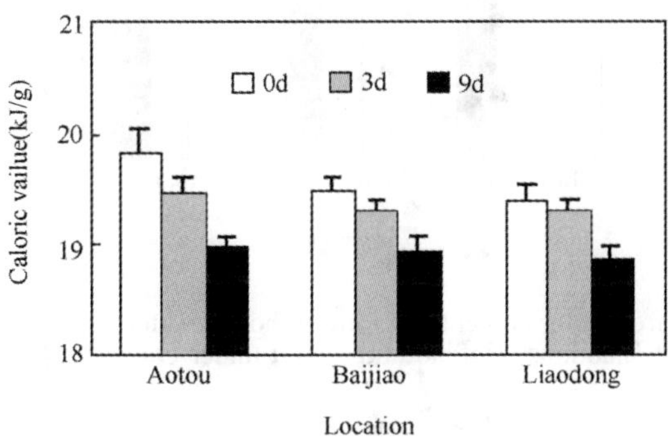

Fig.2 Changes of caloric value of *Kandelia candel* leaves from Liaodong, Baijiao and Aotou under cold stress

3.3 Cold stress on electrolyte leakage

Electrolyte leakage to cold stress from *K. candel* leaf showed no difference among Liaodong, Baijiao and Aotou, and ranged from 24% to 27%. But with duration of cold stress, electrolyte leakage from *K. candel* leaf rose in different degree. In our case, electrolyte leakage from *K. candel* leaves in Aotou with higher soil salinity increased more than that in Liaodong or Baijiao with lower soil salinity. In 18 days after cold stress, electrolyte leakage of leaves from Aotou was 1.96 times that before cold stress; while the values were 1.37 and 1.08 for Baijiao and Liaodong (Fig.3), showing that *K. candel* leaf from high soil salinity had a lower cold toleration than that from low soil salinity. *K. candel* is a salt-excluded plants, where salt accumulated in the plant living in saline soil. As soil salinity increased, more salt distributed in the plant body. However, a large quantity of salt would have negative effect on cell membrane, resulting in chilling injury.

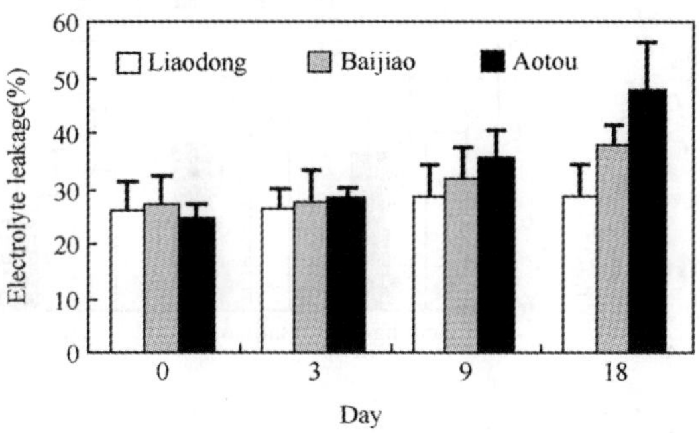

Fig.3 Variations of electrolyte leakage from *Kandelia candel* leaves at Liaodong,Baijiao and Aotou under cold stress

3.4 Cold stress on water content

Under cold stress, total water content decreased only a little in *K. candel* leaf from Liaodong, Baijiao and Aotou. However, bound water content and the ratio of bound water and free water content increased to different extend, with free water content significantly decreased (Fig.4). Comparing the variations of water content in *K. candel* leaf from Liaodong, Baijiao to Aotou under cold stress, it is obvious that the bound water, free water and the ratio of bound water and free water varied most slowly with time-elapsing in Aotou with high soil salinity, and most rapidly in leaves from Liaodong with low soil salinity, so under cold stress, changes of the bound water and free water also had relationship with soil salinity. The higher the soil salinity was, the more slowly the bound water and free water content varied.

3.5 Cold stress on carbohydrate

Before treated in cold, content of various carbohydrate in the leaves are different. Those from higher saline soil contain more carbohydrate. As shown in Fig.5, sugars and starch contents in the leaves decreased and sucrose increased, to some extent after cold treatment. With time elapsing, the sugars and starch contents in the leaves from higher soil salinity decreased faster than those from lower salinity soil. For the leaves of higher salinity mangrove, the sucrose content increased more slowly than those from lower salinity soils did. In other words, with days after the cold treatment, the higher the soil salinity, the faster the decrease of the reduction sugar, soluble sugar and starch content, and the slower the increase of the sucrose content *K. candel* leaf. There existed distinct relationship between variations of carbohydrate content in leaves and soil salinity.

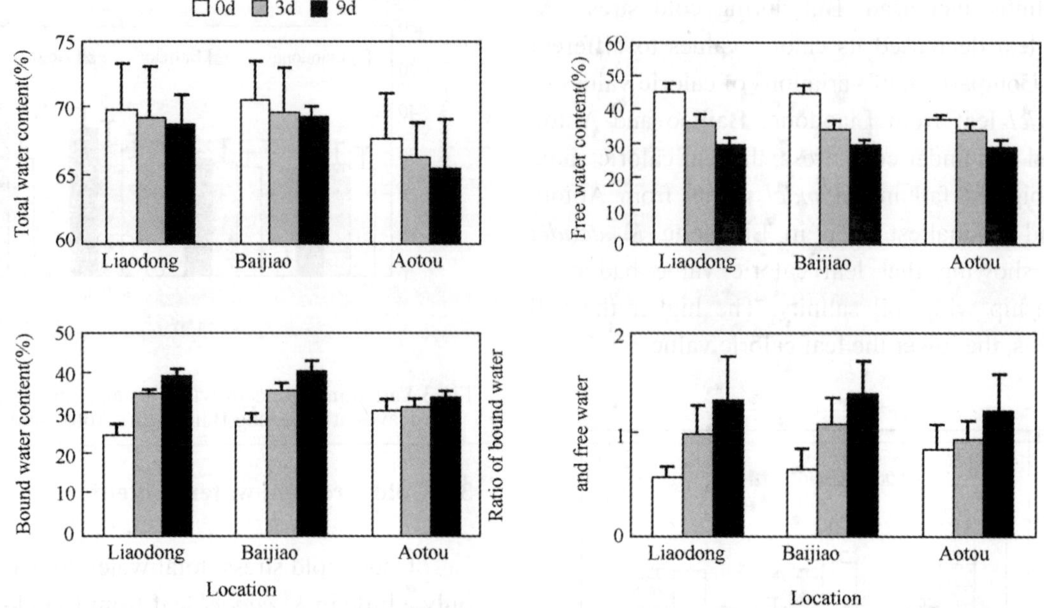

Fig.4 Variations of total water content, bound water content, free water content and ratio of bound water and free water content from *Kandelia candel* leaves at Liaodong, Baijiao and Aotou under cold stress

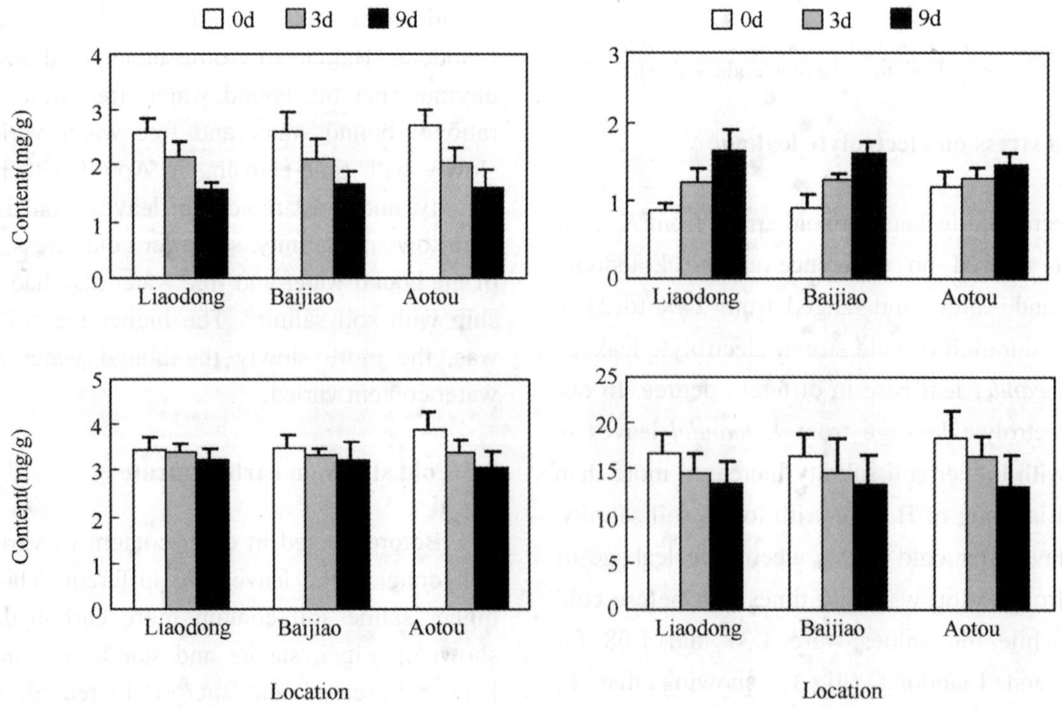

Fig.5 Variations of reduction sugar (RS), sucrose (SU), soluble sugar (SS) and starch (ST) content from *Kandelia candel* leaves at Liaodong, Baijiao and Aotou under cold stress

4 DISCUSSIONS

Limited researches on the effect of soil salinity on cold tolerance of plants yielded different results. Fowler and Hamm (1980) found that wheat and cereal decreased their cold tolerance in saline soil. Suecoff et al. (1976) found that the cold tolerance of woody plant species *Syringa vulgaris* and *Fraxinus permsylvanica* gradually decreased with the increase of soil chlorine but a contrary result occurred in *Spinacia oleracea* plant. Hincha (1994) found that *Spinacia oleracea* leaf would increase its cold tolerance as NaCl concentration increased in culturing media by osmotic adjustment. Syvertsen and Yelenosky (1988) reported that salinity could enhance freezing tolerance of citrus seedlings by modifying growth, water relations and mineral nutrition. Maier and Kappen (1979) found that celluar compartmentalization of salt ions could lessen the effect of salinity, so there was little relation between cold tolerance and soil salinity for halophyte plant species, *Hamimine portulucoides*. Obviously, there is a complicated relation between cold tolerance of plants and soil salinity depending on plant species and ecological conditions.

Mangrove plants grow in saline soil along seacoast. Salt ions could affect the growth and development of mangrove plants. *K. candel* is a salt-excluded plant. We inferred that as soil salinity increasing, salt ions are assembled more in leaf cells of *K. candel*, resulting in chilling injury of cell membrane and reducing cold tolerance of *K. candel* plant. One evidence for above explanation is that electrolyte leakage of *K. candel* leaf in high soil salinity site increased more rapidly than that from low soil salinity site.

Caloric value is a comprehensive reflection of the organic compounds and their components and content in plant materials, and can be considered as effective index for plant growth (Lin and Lin, 1991; Sun *et al*., 1993). In natural condition, as soil salinity increasing, the caloric value of *K. candel* leaf gradually increases too. This probably related to the need from *K. candel* leaf to accumulate more compound matter rich in energy for maintaining energy balance in higher soil salinity. When treated by cold stress, the caloric value of *K. candel* leaf gradually decreased, much more in leaves from high soil salinity than in leaves from low soil salinity; probably because high soil salinity could aggravate the utility of matter and energy in leaves, so the cold tolerance of plant from high soil salinity was lower than that of plant from low soil salinity. The caloric value can be used as an indicator of status of *K. candel* mangrove plant under cold stress.

Loss of water balance is commonly observed in many plant species under cold stress (Wu, 1982). This study revealed that, cold stress caused decrease of free water content, increase of bound water content, only minimal net decrease of total water content in *K. candel* leaf. These changes related with soil salinity, with higher soil salinity resulting in small change of water content. Ordinarily, the water status could reflect metabolism processes of plant cells under stress condition. The higher the free water content, or the lower the bound water content, the more vigorous the cell physiological metabolism, and the weaker the cold stress resistance. On the contrary, the lower the free water content, or the higher the bound water content, the weaker the physiological metabolism, and the stronger the stress resistance (Pan and Dong, 1995). The free water content was high and the bound water content was low in *K. candel* growing in low soil salinity, but due to the rapid drop of free water content and the rapid rise of bound water content under cold stress, the cold tolerance *of K. candel* leaf increased. The free water content was low and the bound water content was high in *K. candel* growing in high soil salinity, but due to the slowly changes of free water content and bound water content under cold stress, so *K. candel* had low cold tolerance at high soil salinity.

Like other plant species (Pan and Dong, 1995), *K. candel* leaf decreased its reduction sugar and starch content, but increased its sucrose content under cold stress. Here, increase of sucrose content was caused by hydrolysis of starch. Fallen amount in reduction sugar content was larger than that of the rise in sucrose content. Soluble sugar content decreased in *K. candel* leaf, too. Therefore, cold stress increased the hydrolysis of carbohydrate, and the use of hydrolysis production such as reduction sugar and soluble sugar. The experiment results also showed that higher soil salinity would led to rapid reduction of sugar and starch, and the sucrose content increased slowly in *K. candel* leaf under cold stress. So a large quantity of salt ions could speed up the use of carbohydrate in *K. candel* under cold stress.

As mentioned above, increasing soil salinity could reduce cold tolerance of *K. candel* mangrove plants by worsening negative effect of salt ions to leaf cell membrane, inhibiting changes of water content, and increasing the exhaustion of materal and energy.

References

Fowler, D. B. and J. W. Hamn, 1980. Crop response to saline soil conditions in the parkland area of Saskatchewan. *Can. J. Soil. Sci.* **60**: 439-449.

Hincha, D. K., 1994. Rapid induction of frost hardiness in spinach seedlings under salt stress. *Planta* **194**: 274-278.

Lin, G. H. and P. Lin, 1991. The change of caloric value of a mangrove species, *Kandelia candel* in China. *Acta Ecologica Sinica* **11**: 44-48. (in Chinese)

Maier, M. and L. Kappen, 1979. Celluar compartmentalization of salt ions and protective agents with respect to freezing tolerance of leaves. *Oecologia* **38**: 303-316.

Nanjing Soil Institute of Chinese Academy of Sciences, 1978. Physical and Chemical Analysis of Soil. Shanghai Science and Technology Publisher: Shanghai, p. 211-213. (in Chinese)

Pan, R. Z. and Y. D. Dong, 1995. Plant Physiology (third edition), High Education Publisher, Beijing, 374. (in Chinese)

Sucoff, E., S. G. Hong and A. Wood, 1976. NaCl and twig dieback along highways and cold hardiness of highway versus garden twigs. *Can. J. Bot.* **54**:2 268-2 274.

Sukumaran, N. P. and C. J. Weiser, 1972. An excised leaflet test for evaluating potato frost tolerance. *Hort Sci.* 7:467-468.

Sun, G. F., Z. M. Zheng, Z. Q. Wang, 1993, Dynamics of calorific values of rice. *Acta Ecologica Sinica* **12**:1-4. (in Chinese)

Syvertsen, J. P. and G. Yelenosky, 1988. Salinity can enhance freezing tolerance of citrus rootstock seedlings by modifying growth, water relations and mineral nutrition. *J. Amer. Soc. Hort. Sci.* **113**:889-893.

Wu, Y. D., 1982. The change of water in the rubber tree on chilling. *Acta Phytophysiologia Sinica* **8**: 17-24. (in Chinese)

Yang, S. C. and P. Lin, 1998. Ecological studies on the resistance and adaptationto cold of some tidal mangrove species in China. *Acta Phytoecologica Sinica* **22**: 60-67. (in Chinese)

Yuan, X. H. and Z. H. Yang, 1983. Experiment for Plant Physiology and Biochemistry. High Education Publisher, Beijing, p. 6-8. (in Chinese)

Zhang, A. L., 1990. Experimental Guide to Plant Physiology (second edition). High Education Publisher, Beijing, 1,6-9. (in Chinese)

不同生境下角果木木材结构变化的适应意义*

邓传远[1,2] 林 鹏[1] 洪志同[2]

(1. 厦门大学生命科学学院,福建 厦门 361005; 2. 福建农林大学园艺学院,福建 福州 350002)

摘要: 研究了中、高潮位生境角果木的木材结构特征。与高潮位生境生长的角果木木材结构特征相比,中潮位生境生长的角果木木材结构特征有利于提高水分输导的安全性和增强抗风浪冲击的能力。

关键词: 角果木;木材结构;潮位;生境

中图分类号: S781.1 **文献标识码**: A **文章编号**: 1002 - 7351(2005)01 - 0004 - 02

Significance of the Adaptivity of *Ceriops tagal* Wood Structure Change under Different Habitats

DENG Chuan-yuan ,LIN Peng ,HONG Zhi-tong

(1. School of Life Sciences, Xiamen University, Xiamen 361005, China;

2. School of Horticulture, Fujian Agriculture and Forestry University, Fuzhou 350002, China)

Abstract: This paper researched the *Ceriops tagal* wood structure characteristics under the middle and high tidemark haditats. The wood structure characteristics of *C. tagal* grew under the middle tidemark habitat advantaged raising the moisture transfusion security and increasing the ability of resisting the wind and wave ,strikes.

Key words: *Ceriops tagal*;wood structure;tidemark;habitat

红树林是生长在热带和亚热带海岸潮间带的森林群落,角果木群落沿中潮区至高潮区呈水平分布带[1]。关于角果木的木材解剖学研究,前人的工作主要集中与其它红树科植物的木材比较解剖学研究上,总结归纳出具有系统生态学意义的木材结构特征[2]。但根据土壤理化因子的动态变化,红树林生境明显可细分为不同的小生境。关于红树植物在不同生境条件下,种内木材结构变动除桐花树外[3],研究很少。用多种观察手段和新近提出的数量特征指标对其次生木质部结构特征进行研究,有重要的理论价值和一定的实践意义。本文将应用光学显微镜、扫描电子显微镜和广泛使用的数量特征指标对中、高潮位生长的角果木次生木质部的解剖特征进行详细研究,旨在探讨适应潮间带生境的角果木木材结构特征。

1 材料和方法

1.1 实验材料

中、高潮位生长的角果木木材取自海南东寨港红树林自然保护区。海南东寨港有自然生长的角果木群落。角果木群落沿中潮区至高潮区呈水平分布带[1]。中潮位生长的角果木的采样位置约在中潮线以上、中高潮线以下的中间地带,其特点是潮水浸没时间较低潮带短。高潮位生长的角果木的采样位置约在中高潮线以上,其特点为潮水淹没的频度小,时间短。实验材料取自株龄在 10 a,株高约 1 m,生长年龄一致的主干,直径在 4 cm 左右。研究植物取自种群中的 5 株标准木,约 40 个样品制成切片。

1.2 制片方法

应用常规制片方法[4]。

1.3 观察与测量

使用光学显微镜、扫描电子显微镜观察。制成的永久切片经激光共聚焦显微镜扫描后,应用 Laser-sharp 软件测量次生木质部数量特征。导管频率、导管比率、射线比率、纤维比率等每项指标测定 30 个以上的值,其余数量特征每项指标测 200 个以上的值。

* 国家自然科学基金"红树植物陆海迁移进化过程的木材结构与能量比较研究"资助项目(49576295)

原载于福建林业科技,2005,32(1):27—29

本文所用术语根据国际木材解剖学家协会制定的多国文字木材解剖学名词汇编[5],对各数量特征的描述和计算依据 Carlquist[6]的专著“Comparative Wood Anatomy”和 Noshiro 等[7]的论文中提出的标准和计算方法。

2 结果

2.1 中、高潮位生长的角果木木材形态结构的差异

通过光学显微镜、扫描电子显微镜观察后发现,中、高潮位生长的角果木木材形态结构的差异主要表现在两方面:①中潮位生长的角果木胶质纤维分布较均匀,数量多(胶质纤维占木材横切面百分率为 2.64 %),范围大;高潮位生长的角果木胶质纤维分布不均匀,数量少(角果木胶质纤维占木材横切面百分率为 1.58 %),范围小。②中潮位生长的角果木的导管较易被胶状物阻塞或较易形成侵填体。其他木材形态结构的差异未观察到。

2.2 中、高潮位生长的角果木木材数量特征的差异

中、高潮位生长的角果木木材数量特征的比较详见表 1、表 2。具体地说,与高潮位生长的角果木相比,中潮位生长的角果木木材结构的数量特征有如下特点:①次生木质部的导管分子长度、管孔弦向直径、管孔面积明显更小;②次生木质部的导管聚合度明显更大;③次生木质部的导管密度明显更大;④木材结构有更小的管孔面积,更厚的纤维壁,更低的导管比率、射线比率,更高的纤维比率;⑤估定有效输导率值更小。

表 1 中潮位和高潮位生长的角果木次生木质部管孔及导管分子数量特征比较

项目	PD/个·mm^{-2}	SPR/%	VCS/个串$^{-1}$	VEL/μm	VWT/μm	TD/μm	TD50/μm	RD/μm	RD50/μm	PA/μm^2	PA50/μm	ESC/μm^2	VP/%	V	M
中潮位	66.44 ±13.92	27.48 ±9.52	1.62 ±1.21	391.55 ±99.17	4.15 ±0.87	30.81 ±8.27	37.26 ±4.46	39.71 ±14.29	51.42 ±6.99	1058.71 ±505.44	1632.57 ±245.89	59.87	6.74 ±3.31	0.464	193.29
高潮位	54.18 ±18.28	40.98 ±11.32	1.27 ±0.49	416.82 ±110.25	5.26 ±1.62	40.70 ±7.71	45.41 ±5.92	52.03 ±11.45	62.25 ±6.15	1690.81 ±566.55	2145.04 ±390.37	148.67	9.16 ±3.07	0.751	294.13
t-检验	**	**	**	*	*	**	**	**	**	**	**		**		

*:PD 为管孔密度,SPR 为单孔率,VCS 为导管聚合度,VEL 为导管分子长度,VWT 为导管壁厚,PA 为管孔平均面积,PA50 为 50 个最大管孔平均面积,ESC 为估定有效输导率,VP 为导管比率,TD 为管孔弦向直径,TD50 为 50 个最大管孔弦向直径,RD 为管孔径向直径,V 为脆度,M 为中性值;**为经 t-检验后,差异极显著($P<0.01$),*为经 t-检验后,差异显著($P<0.05$),NS 为经 T-检验后,无显著差异($P>0.05$)。下同。

表 2 中潮位和高潮位生长的角果木次生木质部射线和纤维数量特征比较

项目	RF/个 mm^{-1}	RH/μm	RW/μm	ARA/μm^2	WRP/%	FL/μm	FWT/μm	FLW/μm	FP/%
中潮位	10.45 ±1.71	598.16 ±259.30	48.90 ±15.81	15608.22 ±10420.56	19.92 ±13.30	931.93 ±184.08	6.35 ±1.46	4.03 ±1.95	72.56 ±37.19
高潮位	12.09 ±1.67	844.52 ±381.51	52.49 ±14.37	31462.81 ±18464.53	34.97 ±20.56	884.27 ±136.54	4.39 ±1.41	4.85 ±2.27	53.39 ±29.46
t-检验	*	**	*	**	**	**	*	NS	**

*:RF 为射线频率,RH 为射线高度,RW 为射线宽度,ARA 为射线平均面积,WRP 为射线比率,FL 为纤维长度,FWT 为纤维壁厚,FLW 为纤维腔径宽,FP 为纤维比率。

3 讨论

与高潮位生长的角果木相比,中潮位生长的角果木木质部导管分子更加“小型化”。生长在中潮位的角果木,由于受海水浸淹的频度更大,其生境的盐离子浓度比生长在高潮位生境的角果木的盐离子浓度高,受水分胁迫的强度更大,木质部导管内有更强的负压,“小型化”导管抗负压,抗栓塞能力更强,有利于输导的安全性[6,8]。

导管聚合度是评估输导安全性的指标[6]。与高潮位生长的角果木相比,中潮位生长的角果木木质部

的导管聚合度更大(表 1) ,导管聚合度增大有利于输导的安全性[6,9] 。

与高潮位生长的角果木相比,中潮位生长的角果木的管孔密度更大(表 1) ,高的管孔密度有助于弥补部分因导管直径减少(小型化导管)而导致输导效率的降低[6,9] 。但总体说来,中潮位生长的角果木更高的管孔密度不能抵消因导管直径减少而导致的输导效率的降低,因为中潮位生长的角果木的估定有效输导率值和中性值小于高潮位生长的角果木的估定有效输导率值和中性值(表 1) 。因此角果木输导安全性的获得是以牺牲输导的有效性为前提的。

与高潮位生境相比,中潮位生境生长的角果木遭受潮水和风浪冲击的强度更大。研究结果表明:与生长在高潮位生境的角果木相比,中潮位生境生长的角果木木材结构有更小的管孔面积,更厚的纤维壁,更低的导管比率和射线比率,更高的纤维比率(表 2) ,胶质纤维分布均匀且数量增加,这些特征有利于中潮位生境生长的角果木提高木材的韧性和强度,增强抗风浪冲击的能力[6] 。

参考文献:

[1]Lin ,P. Mangrove ecosystem in China[M]. Beijing:Science Press,1999. 52 - 68.

[2]林 鹏,林益明,林建辉. 红树植物次生木质部的结构与进化[J]. 海洋学报,1998,20(4):108 - 114.

[3]Sun Q ,Lin P. Wood structure of *Aegiceras corniculatum* and its ecological adaptations to salinities[J]. Hydrobiologia,1997,352:61 - 66.

[4]Miksche J P. Botanical microtechnique cytochemistry[M]. Iowa:The Iowa State University Press,1976. 54 - 129.

[5]IAWA Committee. IAWA list of microscopic feactures for hardwood identification[J]. IAWA Bull NS,1989,10:219 - 332.

[6]Carlquist S. Comparative Wood Anatomy[M]. Berlin:Springer-Verlag,1988.

[7]Noshiro S,Joshi L ,Suzuki M. Ecological wood anatomy of *Alus nepalensis* (Betulaceae) in east Nepel[J].J. Plant Res,1994,107:399 - 408.

[8]Zimmermann MH. Xylem Structure and Ascent of Sap[M]. Berlin:Springer - Verlag,1983.

[9]Tyree MT,Evolution of xylem conduits[J]. IAWA Bull NS. ,1994,15:335 - 360.

显胎生红树植物木榄(*Bruguiera gymnorrhiza*)胎生胚轴发育*

张宜辉[1,2] 王文卿[1,2] 池敏杰[1,2] 林 鹏[1,2]

(1. 厦门大学生命科学学院,福建 厦门 361005;

2. 厦门大学湿地与生态工程研究中心,福建 厦门 361005)

摘要:跟踪测定显胎生红树植物木榄(*Bruguiera gymnorrhiza*)胎生胚轴发育过程中的长度、密度、鲜重、干重、含水量、渗透势、五种主要无机渗透调节离子(Ca^{2+},Mg^{2+},Na^{+},K^{+},Cl^{-})浓度及含量的变化.结果表明:木榄胚轴在发育过程中不断积累有机营养物质以及盐分,并在母树上完成渗透调节、器官形成等生理过程;可以将木榄繁殖体的发育过程划分为果期、生长活跃期和成熟期3个阶段;比较胚轴和生境中土壤、海水的理化指标,可以得出木榄胚轴的密度、渗透势以及盐分水平的变化是以适应母树所处的高盐和周期性海水浸淹生境为目标;探讨了木榄胎生现象的生理生态学意义及适应方式.

关键词:木榄;胎生胚轴;发育;红树林

中图分类号:Q945.78 **文献标识码**:A **文章编号**:0253-4193(2006)02-0121-07

1 引言

热带、亚热带海岸潮间带的红树林生态系统处于海洋、陆地和大气的动态交界面,作为独特的海陆边缘生态系统在维持海湾河口生态系统的稳定中起着特殊的作用.国内外对红树植物适应盐胁迫和海水浸淹胁迫已经进行了大量的研究[1~7],但是与红树林生态恢复相关的许多基本问题还没有得到解决,其中较为关键的一个环节即对红树植物的胎生现象的生理生态学意义缺乏深入的研究.

由于潮间带生境的高度盐渍化以及周期性的海水浸淹,经长期的自然选择和进化适应,红树植物特化出胎生繁殖的方式来适应潮滩生境,即红树植物的果实在成熟后仍留在母树上,种子在母树上的果实内发芽、发育至形成成熟种苗(胚轴)后才脱离母树[8,9].红树植物的胎生现象很早就受到红树林研究者的关注,从1958年金杰里和方亦雄首次从抗盐角度对胎生现象的生理学意义进行研究以来,许多学者进行了大量的研究[10~15],但至今还没有找到比较合理的解释.此外,胎生现象是通过何种方式来适应潮间带生境、胎萌发生的内在机制等问题目前仍然不清楚.因此有必要对胎生现象做综合的生理生态研究,从而对幼苗处理策略、预测幼苗生长和了解生殖特点的演变做全面设计工作[16].

本研究以显胎生红树植物木榄(*Bruguiera gymnorrhiza*)为研究对象,跟踪测定木榄胚轴发育过程中形态及生理指标的变化.通过比较胚轴和母树生境中土壤、海水之间渗透势、盐分水平的关系,力图较为全面地探讨木榄胎生胚轴对高盐、周期性潮水浸淹逆境的适应途径及其胎生现象的生态学意义,并为红树植物生殖生态学的系统研究以及红树林湿地的恢复重建提供可借鉴的科学依据.

2 材料与方法

2.1 样地概况

研究对象的地点位于福建九龙江口龙海市浮宫镇

* 国家自然科学基金资助项目(30200031);福建省自然科学基金资助项目(B0410001)

原载于海洋学报,2006,28(2):121-127

草埔头村附近的红树林引种园(24°24′N,117°55′E).园内木榄为厦门大学红树林科研组 1987 年从海南岛引种,目前生长良好.该地属南亚热带海洋性气候,年均温度 20.9 ℃,极端最低温 - 1.7 ℃.样地处于中潮带至高潮带,2001 年 1 月至 2002 年 12 月实测涨潮时海水的盐度为 4.5～26.0,年均土壤盐度为 15.27 ±2.52.

2.2 材料

于 2002 年 1 月 8 日在引种园内选择年龄、高度及生长状况较一致的木榄 20 株并挂牌标志.该批木榄母树树高(2.88 ±0.13) m,基径(11.1 ±1.8) cm,冠幅 1.5 m ×1.5 m.于每株母树相同高度冠层的东南西北中各个方向挂牌标记 20 个处于相同发育阶段的繁殖体(花刚凋谢而发育为果实,但胚轴尚未露出).从标记日开始,每隔 15 d 于每株母树上各采一个正常生长的繁殖体,同批样品中注意选取大小相当的个体.至 6 月 22 日胚轴成熟,总计采摘 12 批.样品采集后立即带回实验室,从繁殖体中分离出胚轴,经蒸馏水洗涤、纱布擦干后,随机选取其中的 10 个样品,标记并测定每个胚轴的长度、密度、鲜重、干重和含水量,烘干的样品用于测定无机离子的含量;其余 10 个样品用锡箔纸包裹,装入密封袋, - 20 ℃冰冻保存,用于测定渗透势.

2.3 测定与分析方法

胚轴的长度采用标尺测量;密度采用静力称衡法测定[17];烘干法(105 ℃)测定样品的干重和含水量;组织的渗透势用 5520 型蒸汽压渗透压计(美国 WESCOR)测定.1 月 8 日至 2 月 7 日这 3 批样品的胚轴虽已经在果实内萌生,但尚未露出,个体较小而不易测定.因此仅测定 2 月 24 日至 6 月 22 日的 9 批样品;干灰化法测定无机离子的干重浓度[18],其中 Ca^{2+} , Mg^{2+} , Na^{+} , K^{+} 采用原子吸收分光光度法, Cl^{-} 采用 $AgNO_3$ 滴定法测定.并根据胚轴的含水量和干重,由以下公式推导出胚轴的组织液离子浓度[14]:

$$M = RC10^4 / G,$$

式中,M 为组织液中的离子浓度(mmol / dm^3);R 为组分干重和水分含量的比值;C 为离子干重浓度(%);G 为该离子的原子量.

采用下述公式对木榄胚轴进行生长分析:

绝对生长速率 $G = (W_2 - W_1) / (t_2 - t_1)$,

相对生长速率 $RGR = (\ln W_2 - \ln W_1) / (t_2 - t_1)$,

式中,W_1 和 W_2 分别是 t_1 和 t_2 时期的个体生长指标(分别为长度、鲜重和干重).

3 结果

3.1 木榄胚轴发育过程概述

福建九龙江口浮宫引种园里的木榄一年四季均有开花结果,但开花期集中于 8～11 月,胚轴成熟期集中于翌年 6～7 月.木榄为显胎生红树植物,当花凋谢转变为果实后,种子不经休眠而在果实内发芽.刚露出的胚轴呈乳黄色,随着发育进程,颜色逐渐由黄绿色转变成深绿色.胚轴逐渐长长,直径增粗,当胚轴生长到 2.0 cm 时,胚芽原基形成;到 4.0～7.0 cm 时,胚芽由透明逐渐变绿.本实验所测定成熟阶段胚轴的长度为(15.19 ±1.30) cm,直径为(1.57 ±0.11) cm,胚芽长(0.32 ±0.07) cm.

成熟的木榄胚胎由子叶、胚芽、胚轴和胚根 4 个部分组成[19],子叶包被于种皮内,黄色,肥厚,3～4 裂,特化为吸收器官;胚芽包被在子叶内,绿色;胚轴(木榄上胚轴不明显,一般所指的胚轴即下胚轴)伸出果皮之外,深绿色,纺锤型;胚根包被于皮层内,为内生性.从生物学角度来看,一个成熟的胚轴实际上已经是一株完整的幼苗.

3.2 木榄胚轴个体的生长变化

1 月 8 日的第一批样品中,木榄花刚凋谢转变为果实,胚轴尚未露出,将该发育阶段计为 0 d.在发育过程中,木榄胚轴的长度、鲜重和干重都呈持续增长的趋势,到 6 月 22 日(165 d)完全成熟时,其长度、鲜重和干重分别增长了 14.99,193.09 和 188.91 倍(图 1).

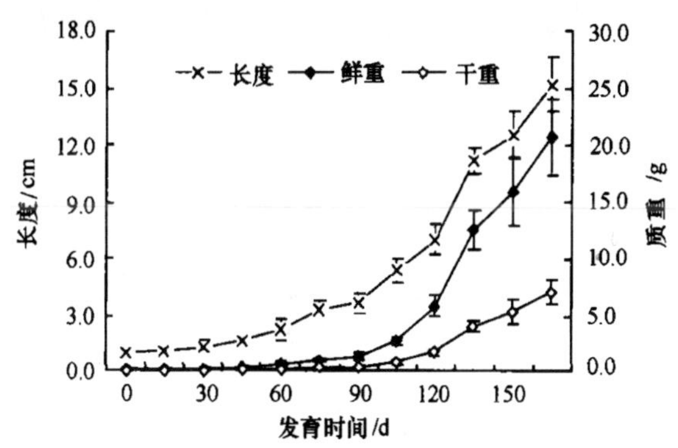

图 1 木榄胚轴发育过程中长度、鲜重和干重的生长变化(数据为 10 个样品的平均值)

计算上述指标每月的绝对生长速率和相对生长速率,可以看出,当胚轴发育至 135 d 时,长度、鲜重和干重的绝对生长速率和相对生长速率都达到最大值(表 1).

表 1 木榄胚轴发育过程中长度、鲜重和干重生长速率的变化

发育时间 /d	绝对生长速率			相对生长速率		
	长度 cm/cm·月$^{-1}$	鲜重 cm/cm·月$^{-1}$	干重 cm/cm·月$^{-1}$	长度 cm/cm·月$^{-1}$	鲜重 cm/cm·月$^{-1}$	干重 cm/cm·月$^{-1}$
15	0.171	0.162	0.052	0.403	0.293	0.230
45	0.614	0.177	0.054	0.466	0.665	0.618
75	1.593	0.589	0.178	0.676	0.963	0.925
105	2.079	1.682	0.469	0.495	1.017	0.952
135	5.798	9.883	3.213	0.737	1.558	1.650
165	4.070	8.071	3.050	0.312	0.498	0.569

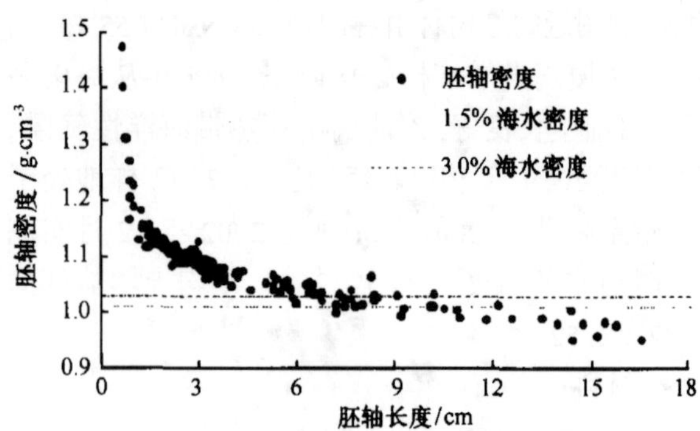

图 2 木榄胚轴发育过程中密度的变化

3.3 木榄胚轴发育过程中若干生理指标的变化

3.3.1 木榄胚轴密度的变化

统计 120 个不同发育阶段的木榄胚轴的密度,可以得出在发育过程中胚轴密度呈下降趋势,其变化范围在 1.472~0.952 g/cm^{3}之间.特别是胚轴长度在 0~2.0 cm 之间时,密度迅速下降.结合发育的时间进程,可以看出发育 0~45 d 胚轴露出的过程中,其密度大幅度下降,此后密度的变化趋于平缓.成熟胚轴的密度小于生境中海水的密度(经测定,25 ℃下,1.5%~3.0%海水的密度为 1.01~1.03 g/cm^3),适于漂浮传播(图 2).

3.3.2 木榄胚轴含水量和渗透势的变化

木榄胚轴发育过程中,含水量的变化呈单峰型:0 d 时的含水量最低,仅为(65.12 ±0.66)%;然后逐渐上升,到 105 d 时达到最高,为(71.01 ±0.71)%;此后随着胚轴的成熟,其含水量下降,到 165 d 时降低为(65.80 ±1.40)%(图 3a).

在发育 45~105 d 期间,胚轴含水量持续上升,水分迅速输入,导致胚轴内渗透调节物质的浓度下降,木榄胚轴的渗透势从 - 2.634 MPa 迅速上升到 - 1.460 MPa;此后直到胚轴成熟的两个月期间,其渗透势较为稳定(- 1.486~ - 1.611 MPa)(图 3b).

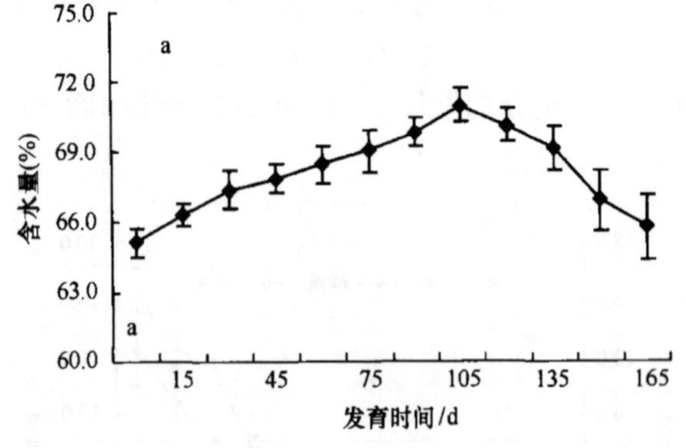

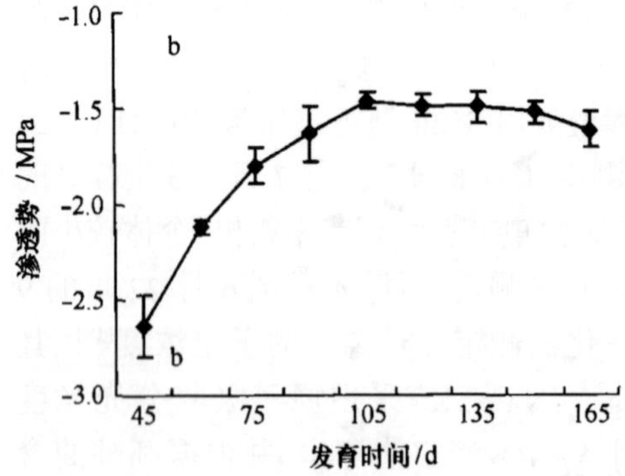

图 3 木榄胚轴发育过程中含水量和渗透势的变化(数据为 10 个样品的平均值)

3.3.3 木榄胚轴中五种主要无机离子浓度及含量动态

木榄胚轴发育 45~135 d 期间,5 种主要无机离子的浓度(mmol/dm^3)都呈降低趋势,此后由于胚轴含水量下降,各离子的浓度略有提高.成熟胚轴(165 d)中 Ca^{2+},Mg^{2+},Na^+,K^+,Cl^- 的浓度比发育 45 d 时分别降低了 85.21%,64.08%,19.37%,8.60%和 52.31%,其中以 Ca^{2+} 的降幅最大(见图 4a).

另一方面,在发育过程中,随着木榄胚轴干重的增长,各离子持续输入,表现为单个胚轴中离子的含量都呈上升的趋势.各离子含量的增长和胚轴生长发育同步,在发育前期,胚轴生长较为缓慢,各离子含量上升也较平缓,到发育 105 d 后,随着胚轴的迅速生长,各离子含量也迅速上升.成熟胚轴中 Ca^{2+},Mg^{2+},Na^+,K^+,Cl^- 的含量分别为发育 45 d 胚轴的 8.12,19.73,44.29,50.20,26.19 倍(见图 4b).

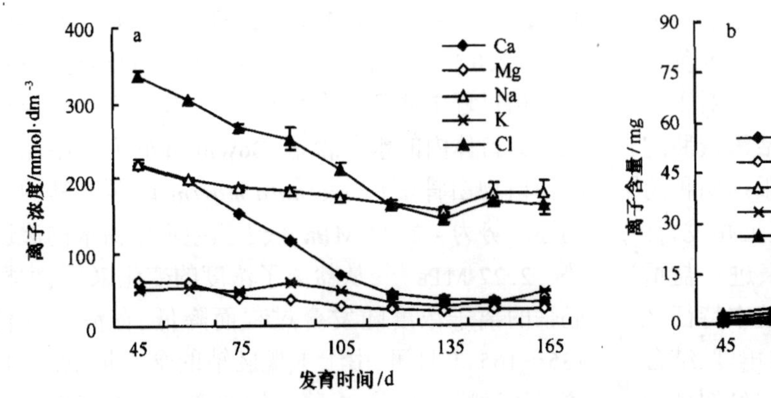

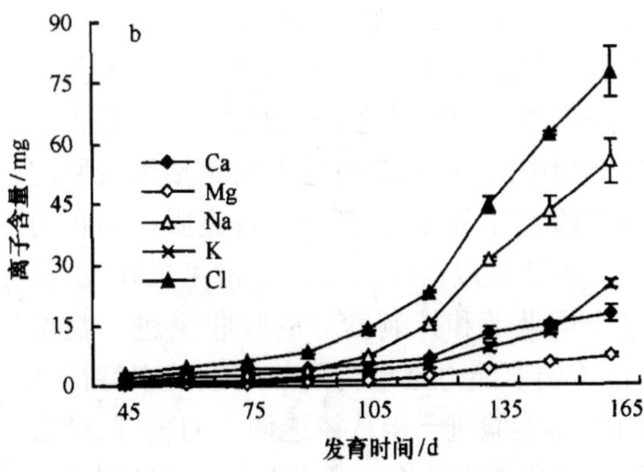

图 4　木榄胚轴发育过程中五种主要无机离子浓度和含量的变化(数据为 10 个样品的平均值)

4　讨论

4.1　木榄胚轴发育阶段的划分

以往对红树植物胎生胚轴的研究取样工作多以胚轴的个体大小来划分其发育等级[10,11,14~16],未能将生理指标变化和生长发育的时间进行对应分析.本研究结果表明木榄胚轴在母树上的生长发育不是呈线性增长,因此不能从胚轴的个体大小来推算它们在母树上生长发育的时间跨度,根据胚轴生长和生理指标的动态,结合发育的时间进程,可以将木榄从花凋谢到最后胚轴成熟这一生长发育过程划分为下述 3 个阶段:

(1) 果期(发育 0～45 d):本阶段木榄的花刚凋谢转变为果实,种子不经休眠,胚轴开始萌动生长,但尚未露出.

(2) 生长活跃期(发育 45～105 d):本阶段在胚轴露出后,持续约两个月.对于胚轴的长度、鲜重、干重、无机离子含量而言,表现为平稳的持续增长;但胚轴内的生理指标是整个生长发育过程中变动最为活跃的阶段,表现为胚轴的含水量持续上升而达到最高、密度和无机离子浓度迅速下降、渗透势迅速上升.从母树向胚轴输入的水分高于有机营养物质以及盐分,是导致这一阶段胚轴内生理指标剧变的主要原因.此外,胚轴顶端胚芽开始形成.本阶段母树的影响仍占主导地位.

(3) 成熟期(发育 105～165 d):本阶段是整个生长发育过程中物质积累最为迅速的阶段,表现为胚轴的长度、鲜重、干重、无机离子含量迅速增长;水分的输入量低于有机营养物质以及盐分,胚轴内含水量转为下降,胚轴密度进一步降低、渗透势略有下降、无机离子浓度略有上升,但后三者的变化都趋于稳定.通过这一阶段的生长,一方面保证了胚轴在掉落前积累足够的营养物质,另一方面,调整了胚轴内渗透势和盐分水平,以适应于母树的生长环境.此外,胚芽由透明变为绿色,绿色的下胚轴可以进行光合、呼吸生理过程[9,13],从生物学角度来看,它已经是一株完整、独立的幼苗.

4.2　木榄胚轴发育过程中渗透势及盐分调节的机理

长期以来认为红树植物的胎生现象是和种苗在母树上进行抗盐锻炼,脱离母树后适应高盐环境密切相关.金杰里认为胎生现象除了有使幼苗迅速抛锚固定的作用外还有另一种生物学的作用,即幼苗在母体植株上发芽时就进行对基质中高浓度盐分的适应,其适应过程是靠从母体植株输送盐分进入幼苗而进行[10].Zheng 等基于对红树植物秋茄(*Kandelia candel*)、木榄(*B. gymnorrhiza*)和海莲(*Bruguiera sexangula*)胎生胚轴发育过程中的离子动态的研究结果,认为它们不是一个盐分累积而获得抗盐锻炼的过程,而是一个低盐化过程,表现出“返祖现象”[12,14].而 Wang 等对发育过程中秋茄(*K. candel*)胚轴主要无机渗透调节物质的测定结果表明秋茄的胎生现象和耐盐性之间不存在直接的关系[15].

本研究中,木榄胚轴发育过程中主要无机渗透调节物质的动态变化为:以单位个体来看,木榄胚轴在发育过程中盐分的含量持续上升,这与金杰里[10]和 Wang 等[15]的结果一致;另一方面,其盐分浓度的变化趋势和 Zheng 等的结果一致[14],表现为低盐化的过程.

此外,在对胎生现象和抗盐性之间关系的研究中,金杰里的研究中仅得出定性结论,其他的研究也多基于对胚轴本身盐分动态变化的探讨.而对于生长在高盐环境中的植物,盐胁迫主要包括渗透胁迫和离子毒害两个方面,植物必须具备克服这两种胁迫的手段才可能适应盐渍生境而正常生长和发育.植物通过渗透调节和控制盐分的吸收来进行盐适应,其中渗透调节能力是植物耐盐的最基本特征之一[20].因此,本实验进一步从渗透调节的角度,结合生境中渗透势和盐离子的浓度水平来探讨红树植物胎生繁殖体和抗盐性之间的关系.

从表2可以看出,木榄胚轴内渗透势和盐离子浓度的变化和生境中土壤以及海水的渗透势和主要盐离子浓度水平密切相关:成熟期(105～165 d)木榄胚轴的渗透势稳定在-1.460～-1.611 MPa之间,低于其生境中1.5%土壤和1.5%海水的渗透势,和2.0%海水相当.这有利于胚轴脱离母体后,在定植过程中从环境中吸收水分,或在海上漂浮过程中保持体内的水分含量.Downton研究了隐胎生红树植物白骨壤(*Avicennia marina*),得出其种子的渗透势为-2.72 MPa,低于其生境中海水的渗透势-2.27 MPa[21];从盐离子浓度的变化来看,木榄胚轴的离子浓度随发育过程而降低,存活在发育135～165 d期间,由于木榄胚轴的含水量下降,其离子浓度略有升高,成熟胚轴的离子浓度和生境中1.5%土壤以及1.0%～1.5%的海水接近.这一结果表明木榄胎生胚轴在发育过程中,可以通过渗透调节和控制盐分的吸收来进行盐适应,最终结果是成熟胚轴的渗透势略低于生境的渗透势水平,离子浓度接近其生境的离子浓度水平.

表2 木榄胚轴和样地土壤以及海水中渗透势、主要盐离子浓度水平的对比

项目	发育时间/d	渗透势/MPa	离子浓度/mmol·dm^{-3}				
			Ca^{2+}	Mg^{2+}	Na^{+}	K^{+}	Cl^{-}
胚轴	45	-2.634	217.10	62.04	219.15	51.40	336.22
	75	-1.803	150.74	38.65	185.17	48.35	265.99
	105	-1.460	69.52	27.19	172.43	47.25	209.67
	135	-1.496	37.27	20.42	154.57	27.11	143.63
	165	-1.611	32.11	22.28	176.70	46.98	160.34
1.5%土壤		-1.168	4.30	19.25	214.84	8.20	238.18
2.5%海水		-1.834	7.19	37.27	328.15	7.15	382.09
2.0%海水		-1.474	5.78	29.96	263.81	5.75	307.19
1.5%海水		-1.111	4.37	22.61	198.84	4.33	231.52
1.0%海水		-0.744	2.93	15.15	133.22	2.90	155.12
0.5%海水		-0.374	1.45	7.61	66.94	1.46	77.93

4.3 木榄胎生现象的生理生态学意义及其对生境的适应途径

已有许多研究讨论了红树植物的胎生现象和它们生长在滨海生境的相关性,包括从适应、形态、生态、生理和进化等方面进行解释[9,22～24].木榄胎生胚轴在其发育过程中,发展了适应于高盐、周期性海水浸淹逆境的生物学特性,包括以下几个方面:

(1)盐分及渗透调节平衡:和许多盐生植物相同[25,26],木榄胎生胚轴在发育过程中进行盐适应的主要途径是通过最大限度利用高盐生境中占优势的离子(Na^{+},Cl^{-})用于渗透调节,并控制胞液中盐离子的水平以减少代谢毒害.从而达到适应生境的盐分及渗透势水平,保证各项生理活动的正常进行;

(2)适应于海流传播的繁殖方式:成熟胚轴的密度小于生境中海水的密度,适于漂浮传播.Tomlinson(1994)观察了柱果木榄(*Bruguiera cylindrica*)的胚轴,其皮层有较多的孔隙,是适于漂浮的一个特征.需要进一步了解不同发育阶段胚轴的解剖结构、物质构成,从而分析密度变化的原因.此外,胚轴皮层富含单宁,胚轴不易腐烂,可以在海水中漂浮数月不死[8];

(3)高盐、周期性海水浸淹逆境的总体策略:胎

生,即种子不经休眠,立即萌发生长.并在母树上完成营养积累、盐分积累、渗透调节平衡、器官形成等生理过程,至掉落时已经成为一个完整的个体.当脱离母体后,可以根据幼苗当时所处的生境条件,采取积极主动的应对策略:如果为掉落扎入土壤,可迅速生根、固着;当遇到涨潮而未能扎根时,则可随水漂浮并推迟萌发,此时由于其体内的盐分、渗透势水平和海水相近,已及单宁等的保护而不至脱水萎蔫或吸水膨胀.从而保证胚轴可以在海水中漂浮较长的时间,保持活力,并向远处传播.

感谢山东农业大学植物科学系高辉远、赵世杰和彭涛老师在渗透势测定实验中提供仪器的使用和指导,山东农业大学植物科学系何宝坤、薛国希同学在实验测定过程中的热心相助!

参考文献:

[1] 范航清,梁士楚. 中国红树林研究与管理 [M]. 北京: 科学出版社. 1995.

[2] 陈桂珠,缪绅裕. 红树林植物秋茄及其湿地系统研究 [M]. 广州: 中山大学出版社, 2000.

[3] HUTCHINGS P, SAENGER P. Ecology of mangroves [M]. St Lucia: University of Queensland Press, 1987.

[4] ROBERTSON A I, ALONGI D M. Tropical mangrove ecosystems [A]. Coastal and Estuarine Studies [M]. Washington DC: American Geophysical Union, 1992.

[5] SMITH S M, SNEDAKER S C. Salinity responses in two populations of viviparous *Rhizophora mangle* L. seedlings [J]. Biotropica, 1995, 27(4):435 —440.

[6] BALL M C. Interactive effects of salinity and irradiance on growth: implications for mangrove forest structure along salinity gradients [J]. Trees, 2002, 16(2-3):126 —139.

[7] CHEVGLu-zhan, WANG Wen-qing, LIN Peng. Influence of water logging time on the growth of *Kandelia candel* seedlings [J]. Acta Oceanologica sinica, 2004, 23(1): 149 —158.

[8] 林 鹏. 红树林 [M]. 北京: 海洋出版社. 1984.

[9] TOMLINSON P B. The botany of mangroves [M]. Cambridge: Cambridge University Press. 1994.

[10] 金杰里,方亦雄. 红树植物胎生现象的生理意义 [J]. 植物学报, 1958, 7(2): 51 —58.

[11] 林 鹏. 中国红树林生态系 [M]. 北京: 科学出版社, 1997.

[12] 郑文教,林 鹏. 红树胚轴和叶片生长发育的元素动态 [J]. 海洋学报, 1997, 19(1): 96 —103.

[13] CHENG Xiao-yang, LIN Peng. A comparison of hypocotyls morphology and seedling growth between normal and albino propagules of *Kandelia candel* (L.) Druce: a reevaluation of the roles of vivipary in mangroves [A]. Marine Biology of the South China Sea[C]. Hong Kong: Hong Kong University Press, 1998, 83 —90.

[14] ZHENG Weng-jiao, WANG Weng-qing, LIN Peng. Dynamics of element contents during the development of hypocotyls and leaves of certain mangroves species [J]. J Exp Mar Biology and Ecol, 1999, 233(2):247 —257.

[15] WANG Weng-qing, KE Lin, TAM NFY, et al. Changes in the main osmotica during the development of *Kandelia Candel* hypocotyl-sand after mature hypocotyls were transplanted in solutions with different salinities [J]. Mar Bio, 2002, 141(6): 1 029 —1 034.

[16] FARNSWORTH E J, FARRANT J M. Reductions in abscisic acid are linked with viviparous reproduction in mangroves [J]. Am J Bot, 1998, 85(6): 760 —769.

[17] 杨述武. 普通物理实验(一、力学及热学部分) [M]. 北京: 高等教育出版社. 2000. 79 —82.

[18] 中国科学院南京土壤研究所. 土壤理化分析 [M]. 上海: 上海科学技术出版社. 1978.

[19] 陈月琴, 蓝崇钰, 黄玉山,等. 秋茄木榄繁殖体的结构及其生态特异性 [J]. 中山大学学报(自然科学版), 1995, 34(4):70 —75.

[20] 刘友良, 汪良驹. 植物对盐胁迫的反应和耐盐性 [A]. 植物生理与分子生物学[M]. 北京: 科学出版社, 1999, 752 —769.

[21] DOWNTON W J S. Growth and osmotic relations of the mangrove *Avicennia marina*, as influenced by salinity [J]. Aust J Plant Physiol, 1982, 9(5):519 —528.

[22] JOSHI G V, PIMPLASKAR M, BHOSALE L J. Physiological studies in germination of mangroves [J]. Bot Mar. 1972, 15(2):91 —95.

[23] JUNCOSA A M. Embryogenesis and seedling developmental morphology of the seedling in *Bruguiera exaristana* Ding Hou (Rhizophoraceae) [J]. Am J Bot, 1984, 71(2): 180 —191.

[24] FARNSWORTH E J. The ecology and physiology of viviparous and recalcitrant seeds [J]. Annu Rev Ecol Syst, 2000, 31:107 —138.

[25] FLOWERS T J, TROKE P F, YEO A R. The mechanism of salt tolerance in halophytes [J]. Ann Rev Plant Physiol, 1977, 28:89 —121.

[26] NIU Xiao-mu, BRESSAN R A, HASEGAWA P A, et al. Ion homeostasis in NaCl stress environments [J]. Plant Physiol, 1995, 109(3): 735 —742.

Development of viviparous hypocotyls of *Bruguiera gymnorrhiza*

Zhang Yi-hui[1,2], Wang Wen-qing[1,2], Chi Min-jie[1,2], Lin Peng[1,2]

(1. *School of Life Sciences, Xiamen University, Xiamen* 361005, *China*; 2. *Research Center for Wetlands and Ecological Engineering, Xiamen University, Xiamen* 361005, *China*)

Abstract: Hypocotyls of *Bruguiera gymnorrhiza* were collected at various stages of development while still on the parent plant. Length, density, fresh weight, dry weight, water content, osmotic potential, ions concentrations and contents (Ca^{2+}, Mg^{2+}, Na^{+}, K^{+}, Cl^{-}) were determined. The result showed that the hypocotyls of *B. gymnorrhiza* undergoed some growth before becoming detached from the parent plant. They accumulated large amounts organic nutriments and ions during the development. The growth of hypocotyls could divided into 3 stages according to the changes of each physiology characteristics: The first 45 d was the fruit stage. The following 60 d were a period of highest activity after the hypocotyl emerged through the fruit wall. In the final 60 d before detachment, the seedlings came into being. The osmotic potential of mature hypocotyls *B. gymnorrhiza* was negative than that of the solution on which - their parent plant vegetated, while the ions concentrations were approaching to those of the solution on which their parent plant vegetating. Thus help the seedlings survive better in the flooding and saline conditions. The density of mature hypocotyls of *B. gymnorrhiza* were lower than seawater, thus the hypocotyls can float and be dispersed by tide.

Key words: *Bruguiera gymnorrhiza*; hypocotyl; development; mangroves

深圳福田几种红树植物繁殖体与不同发育阶段叶片热值研究*

林益明 向 平 林 鹏

(厦门大学生命科学学院,福建 厦门 361005)

摘要: 对深圳福田红树林区的秋茄 (*Kandelia candel*)、木榄 (*Bruguiera gymnorrhiza*)、桐花树 (*Aegiceras corniculatum*)、无瓣海桑 (*Sonneratia apetala*)、海漆 (*Excoecaria agallocha*)、银叶树 (*Heritiera littoralis*)不同发育阶段叶片以及秋茄、木榄、桐花树、无瓣海桑、海漆繁殖体的灰分含量和热值进行研究。结果表明: (1) 不同发育阶段叶片的灰分含量变化趋势没有一定的规律性,6种红树植物中老叶灰分含量均不是最低;植物繁殖体的灰分含量低于成熟叶;(2) 秋茄、无瓣海桑、木榄、桐花树繁殖体的干质量热值和去灰分热值基本上低于不同发育阶段叶片,而海漆繁殖体的干质量热值和去灰分热值高于不同发育阶段叶片;(3) 6种红树植物不同发育阶段叶片的干质量热值与灰分含量具有极显著的线性负相关 ($P<0.01$),不同发育阶段叶片和繁殖体的干质量热值与灰分含量具有显著线性负相关($P<0.05$)。

关键词:繁殖体;不同发育阶段叶片;热值;红树植物;深圳

中图分类号:Q949.7 **文献标识码**:A **文章编号**:1000-3096(2004)02-0043-06

能量是生态学功能研究中的基本概念之一,植物热值是植物含能产品能量水平的一种度量,可反映植物对太阳辐射能的利用状况,也是评价植物营养成分的标志之一。我国对能量生态学研究开始于20世纪80年代初,对于红树植物能量生态学的研究目前主要集中在红树群落的能量贮量、固定量、分布状况[1,2],红树植物叶片热值的季节变化规律[3],繁殖体发育过程的能量变化[4]以及盐胁迫条件下红树植物热值的变化[5],而对红树植物叶片发育及衰老过程中的热值变化还未见报道。Lin[6]对红树植物木榄叶片发育过程中的叶面积、叶重量和营养元素含量的动态变化进行过研究。开展对红树植物叶片发育过程的能量动态研究具有重要的意义。作者对广东深圳福田红树林自然保护区的6种红树植物叶片发育过程的热值和灰分含量的动态变化以及繁殖体的热值进行研究,从能量角度认识红树植物的特性,对红树林的保护、开发利用都具有重要的理论意义。

1 深圳福田红树林自然保护区的自然条件概况

深圳福田红树林自然保护区位于深圳湾的东北部,东经114°05′,北纬22°30′,红树林沿海岸线呈带状分布,长达11 km,林带宽20～400 m不等,该地区与香港米埔红树林保护区相对岸,属南亚热带海洋性季风气候,年平均温度22.4 ℃,1月份平均温度14.1 ℃,7月份平均温度28.1 ℃,年平均降雨量1 950.5 mm,雨量各月分配不均,干湿季交替明显,年平均相对湿度79%。

深圳福田红树林保护区的土壤,淤泥深厚,红树林内枯枝落叶多,林内各种植物生长茂密,物质积累丰富,腐殖质含量较高,一般为3%～5%,表土一般呈弱酸性反应,海水pH值为7.23～8.05,平均为7.66,海水盐度平均为20。

2 材料与方法

2.1 样品采集

于2002年7月从广东深圳福田红树林自然保护区采样。采集的种类有红树科(Rhizophoraceae)的秋茄(*Kandelia candel*)、木榄(*Bruguiera gymnorrhiza*);紫金

* 原载于海洋科学,2004,28(2):43－48.

牛科 (Myrsinaceae) 的桐花树 (*Aegiceras corniculatum*)；海桑科 (Sonneratiaceae) 的无瓣海桑 (*Sonneratia apetala*)；大戟科 (Euphorbiaceae) 的海漆 (*Excoecaria agallocha*)；梧桐科 (Sterculiaceae) 的银叶树 (*Heritiera littoralis*)。选择生长状况较为一致的植株 20 株(银叶树因种类数量有限,故只取 3 株采样),林冠外围随机选取幼叶、成熟叶及老叶(这里指落叶)各 50 片,老叶系轻轻一碰即掉落者(而非落至地面的凋落叶),而成熟叶是指已充分展开且无衰老症状、其下面的叶片已呈衰老态的叶片。除银叶树外,还分别采集 5 种红树植物的繁殖体(胚轴或果实、花)。

2.2　测定方法

样品采集后,经 80 ℃烘干,磨粉处理后过筛贮存备用;另取小样 105 ℃烘干至恒重,求含水量。尔后用热量计法测定其热值含量。仪器采用长沙仪器厂生产的 GR-3500 型微电脑氧弹式热量计。样品热值以干质量热值(每克干物质在完全燃烧条件下所释放的总热量,简称 GCV)和去灰分热值(AFCV)来表示。测定环境是空调控温 20 ℃左右;每份样品多次重复,重复间误差控制在 ±0.2 kJ/g,每次实验前用苯甲酸标定。

灰分含量的测定用干灰化法,即样品在马福炉 550 ℃下灰化 5 h 后测定其灰分含量。之后用以计算样品的去灰分热值,计算方法为:去灰分热值 = 干质量热值/(1 - 灰分含量)。

3　结果与讨论

3.1　灰分含量

灰分是指植物体矿物元素氧化物的总和,不同植物以及不同生长发育时期其含量不同,不同器官的凋落物灰分含量也不同[7]。深圳福田 6 种红树植物不同发育阶段叶片与繁殖体的灰分含量见图 1。可以看出,随着叶片的发育过程 (从幼叶→成熟叶→老叶),秋茄、无瓣海桑的灰分含量是老叶 > 成熟叶 > 幼叶,海漆是成熟叶 > 老叶 > 幼叶,木榄、银叶树是老叶 > 幼叶 > 成熟叶,桐花树是幼叶 > 老叶 > 成熟叶。从本研究看出,不同发育阶段叶片的灰分含量变化趋势没有一定的规律性。

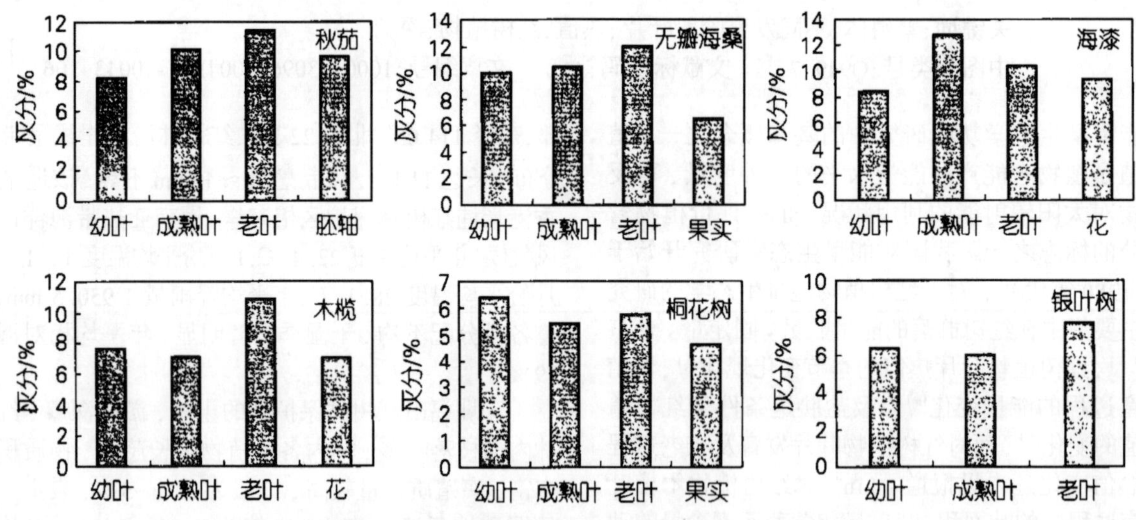

图 1　不同发育阶段叶片与繁殖体的灰分含量

Fig. 1　Ash contents of propagules and leaves at the different development stages

一般认为叶片衰老过程中由于呼吸消耗及碳水化合物、核酸、脂类、蛋白质等降解后小分子物质的外运,使叶片的质量及 N, P, K 等元素的浓度随之下降,老叶的灰分含量低于幼叶、成熟叶。深圳红树林区 6 种红树植物的研究结果表明,秋茄、无瓣海桑、海漆的灰分含量幼叶最低,木榄、银叶树、桐花树的灰分含量成熟叶最低,6 种红树植物中老叶灰分含量均不是最低;老叶灰分含量没有出现预期下降的趋势,说明了叶片可能具有维持自身营养元素平衡的机制;Wang[8] 研究表明,红树植物木榄叶片衰老过程中,叶片中大约 60 %的 N、48 %的 P 和 46 %的 K 转移至多年生的器官和新叶中,而 Ca, Mg 等在叶片衰老过程中含量却增加[8]。对常绿树种,Ca, Mg 表现出负的内吸收率,也就是说它们没有随叶片的衰老而向树体其它部分内吸收,相反却有在老叶中逐步积累的倾向。

无瓣海桑果实、木榄花、桐花树果实的灰分含量均低于不同发育阶段的叶片;而秋茄胚轴、海漆花的灰分含量高于幼叶,低于成熟叶、老叶;比较来看,植物繁殖体的灰分含量低于成熟叶(图 1)。虽然繁殖体对营养元素的需求量大,但同叶一样主要靠蒸腾作用产生动力的从根部向上吸收运输,叶是代谢最活跃的器官,植物根系从土壤中吸收了矿质元素,直接输送到叶中,因此叶的灰分含量较高。

对深圳福田 6 种红树植物不同发育阶段叶片的灰分含量进行比较,桐花树 (5.40%~6.42%)、银叶树 (6.06%~7.72%) 的灰分含量较低, 而秋茄(8.17%~11.31%)、无瓣海桑 (10.10%~12.07%)、海漆(8.46%~12.76%)、木榄(7.14%~10.90 %)的灰分含量较高;种类之间的灰分含量存在差异。一般说来,叶片质地与灰分含量有一定关系,纸质(如白骨壤)、肉质或薄革质(无瓣海桑、海漆)的叶灰分含量高于革质(桐花树)或厚革质(银叶树)。白骨壤的叶被广西沿海人民用作绿肥就是因为其灰分含量高(12.27%),特别是 N、P 含量高,并且质地轻(纸质)易于分解的缘故。从本研究的结果看,无瓣海桑的叶也是做绿肥的好材料。

灰分含量的高低与植物吸收的元素量有关。灰分含量高低可指示植物富集元素的作用,植物各组分对土壤元素的富集量本质上与植物各组分对元素的需求量和土壤中元素的含量及存在形态等有关,而元素的存在形态因不同因素而不同,因此灰分含量与生长的土壤条件有关,不是固定不变的,灰分含量的高低可反映不同植物对矿质元素选择吸收与积累的特点。

3.2 干质量热值

深圳福田 6 种红树植物不同发育阶段叶片与繁殖体的干质量热值见图 2。从图 2 可以看出,除海漆(幼叶 > 成熟叶 > 老叶)外,其余 5 种红树植物成熟叶的干质量热值均高于幼叶、老叶,即,秋茄、银叶树:成熟叶 > 幼叶 > 老叶;无瓣海桑、木榄、桐花树:成熟叶 > 老叶 > 幼叶;成熟叶的生命活动最旺盛,光合能力最强,光合作用积累的有机物最多,而幼叶代谢能力弱、老叶则处于衰退之中,故成熟叶的干质量热值高于幼叶和老叶。

Huges 1971 年在研究英国落叶林植物热值的季

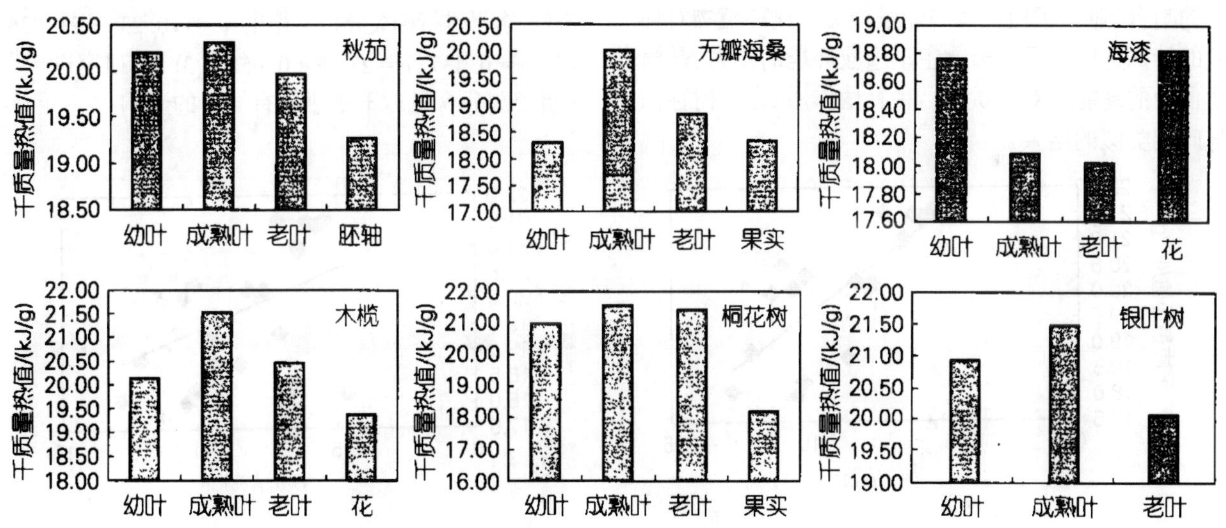

图 2 不同发育阶段叶片与繁殖体的干质量热值

Fig. 2 Gross caloric values of propagules and leaves at the different development stages

节变化时注意到,并称之为“叶脱落时的热值增值(leaf caloric value increment at abscission)”,但未对这种现象加以解释。林光辉和林鹏 1991 年认为这种热值增值主要是由于落叶中含有较多的幼芽和嫩叶,因为幼叶含有较高的热值。作者研究表明,除海漆幼叶 > 成熟叶 > 老叶外,无瓣海桑、木榄、桐花树的幼叶干质量热值都低于成熟叶和老叶。幼叶并不含有较高的干质量热值。干质量热值的高低主要与植物体内所含的内含物有关,并不存在叶脱落时干质量热值增加的现象。

在 6 种红树植物中,秋茄与桐花树不同发育阶段叶片的干质量热值变化平缓,木榄次之,海漆变化较大。说明了不同发育阶段叶片干质量热值的变化趋势因种而异,不是固定不变的。从种间不同发育阶段叶片干质量热值的比较看,无瓣海桑、海漆的干质量热值较低,桐花树、银叶树、秋茄、木榄的干质量热值较高。桐花树不同发育阶段叶片的干质量热值显著高于海漆,经 t 检验,达显著水平($P<0.05$)。

梧桐科的银叶树、紫金牛科的桐花树、红树科的秋茄和木榄不同发育阶段叶片干质量热值均较高,干质量热值在 20.0 kJ/ g 以上 (除了秋茄老叶的干质量热值为 19.96 kJ/ g 外);而海桑科的无瓣海桑、大戟科的海漆干质量热值较低,在 19.0 kJ/ g 以下 (除了无瓣海桑成熟叶的干质量热值为 20.02 kJ/ g 外);这与种类的特性有关。嗜热广布种海漆、嗜热窄布种无瓣海桑干质量热值较低,抗低温广布种秋茄、桐花树干质量热值较高,这与前人研究的抗寒种类干质量热值相对较高的结果一致;嗜热广布种银叶树干质量热值较高可能是叶背密被银白色的小鳞秕的缘故。Howard - Williams 1974 年、Franken 1979 年在研究亚马逊地区的热带雨林时发现有些植物叶的高热值现象。Howard - Williams 指出,亚马逊地区植物叶的高热值现象是在亚马逊地区非常贫瘠的土壤条件下,植物适应环境的结果,在植物叶子中进行了高能化合物的积累。由于土壤贫瘠,养分的有效性较低限制了植物的生长,导致光合作用进入另一渠道生产高能的化合物如蜡、树脂和脂肪,这些化合物保护叶子免受食草动物的啃食,避免了植物体的能量损失。这种植物对环境的特殊适应也反映植物主动适应环境的能力,是植物自身的高能量对物质的一种补偿作用。银叶树存在与此相类似的结果。

不同发育阶段叶片与繁殖体结合起来看 (图 2),秋茄、木榄、桐花树繁殖体的干质量热值低于不同发育阶段叶片(无瓣海桑也基本如此),而海漆繁殖体的干质量热值高于不同发育阶段叶片。通常认为繁殖体比植物体其他组分具有更高的干质量热值,实际上并非如此,作者对福建戴云山自然保护区 31 种植物成熟叶与繁殖体的干质量热值比较后发现,成熟叶与繁殖体的干质量热值相比或高或低,经 t 检验,无显著差异 ($P > 0.05$)①。

3.3 干质量热值与灰分含量相关

植物组分或器官干质量热值的差异主要是受自身组成(所含的营养物质)、结构和功能的影响;其次,还受光照强度、日照长短及土壤类型和植物年龄影响。此外,灰分含量的高低对植物的干质量热值也有一定的影响[9]。

6 种红树植物不同发育阶段叶片的干质量热值与灰分含量具有极显著的线性负相关,相关方程为 $y = -0.3938x + 23.499$,其中 $r = 0.760$,$n = 18$,$P < 0.01$;而将不同发育阶段叶片和繁殖体的干质量热值与灰分含量进行相关分析,得出 $y = -0.23x + 21.726$,其中 $r = 0.437$,$n = 23$,$P < 0.05$(图 3)。本研究进一步证明了灰分含量对干重热值有一定的影响。

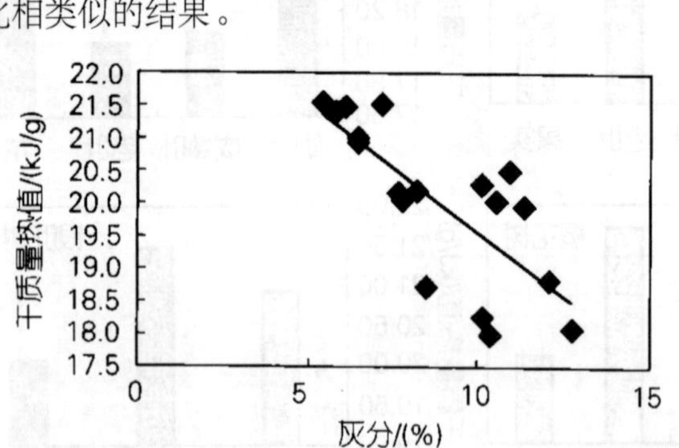

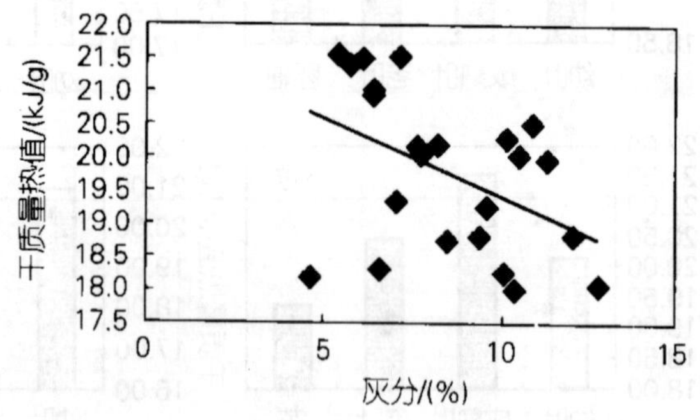

图 3　干质量热值与灰分含量的相关

Fig. 3　Relationships between gross caloric values and ash contents for leaves at the different development stages (left), and for propagules and leaves at the different development stages (right)

左图为不同发育阶段叶片的干质量热值与灰分含量相关,右图为不同发育阶段叶片和繁殖体的干质量热值与灰分含量相关

3.4 去灰分热值

从不同发育阶段叶片的去灰分热值来看(图 4),秋茄、无瓣海桑、木榄、桐花树为成熟叶 > (或≈) 老叶 > 幼叶;海漆、银叶树为成熟叶 > 幼叶 > 老叶。其中秋茄、木榄、桐花树从成熟叶至老叶去灰分热值变化差异不明显 ($P > 0.05$),但总的来看,不同发育阶段叶片的去灰分热值以成熟叶最高。不同发育阶段叶片去灰分热值的变化趋势因种而异,也不是固定不变的。

不同发育阶段叶片与繁殖体结合起来看 (图 4),秋茄、无瓣海桑、木榄、桐花树繁殖体的去灰分热值低于不同发育阶段的叶片,而海漆繁殖体的去灰分热值高于不同发育阶段的叶片。这个结果与干质量热值的结论相同。

4 结论

(1) 不同发育阶段叶片的灰分含量变化趋势没

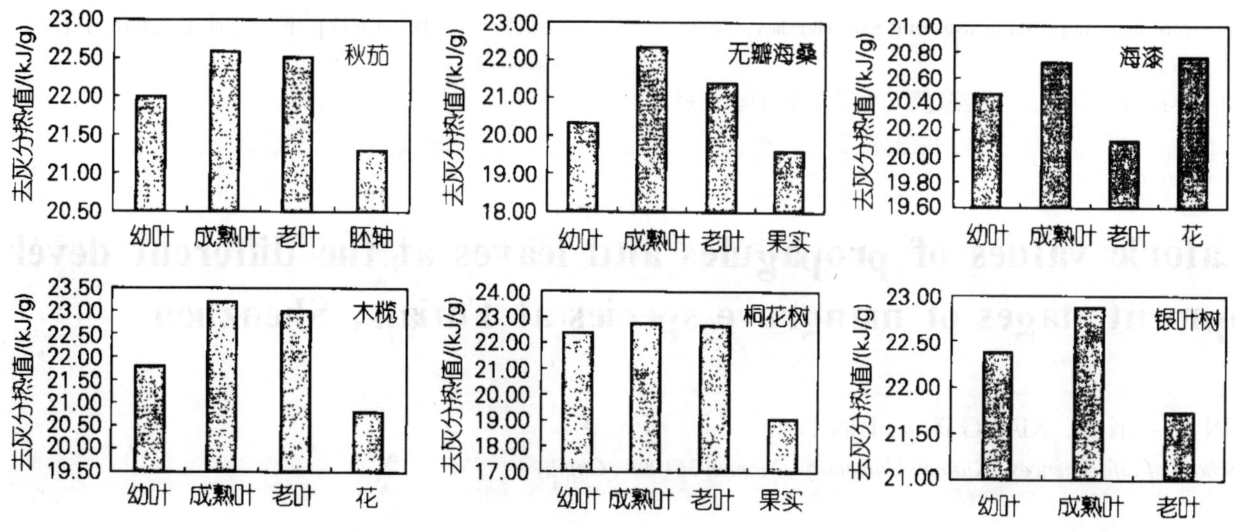

图 4 不同发育阶段叶片与繁殖体的去灰分热值

Fig. 4 Ash free caloric values of propagules and leaves at the different development stages

有一定的规律性;随着叶片的发育过程(从幼叶→成熟叶→老叶),秋茄、无瓣海桑的灰分含量是老叶>成熟叶>幼叶,海漆是成熟叶>老叶>幼叶,木榄、银叶树是老叶>幼叶>成熟叶,桐花树是幼叶>老叶>成熟叶;无瓣海桑果实、木榄花、桐花树果实的灰分含量均低于不同发育阶段的叶片;而秋茄胚轴、海漆花的灰分含量高于幼叶,低于成熟叶、老叶;比较来看,植物繁殖体的灰分含量低于成熟叶。

(2) 不同发育阶段叶片干质量热值的变化趋势因种而异,不是固定不变的;除海漆(幼叶>成熟叶>老叶)外,其余 5 种红树植物成熟叶的干质量热值均高于幼叶、老叶,即,秋茄、银叶树:成熟叶>幼叶>老叶;无瓣海桑、木榄、桐花树:成熟叶>老叶>幼叶;秋茄、木榄、桐花树繁殖体的干质量热值低于不同发育阶段叶片(无瓣海桑也基本如此),而海漆繁殖体的干质量热值高于不同发育阶段叶片。

(3) 6 种红树植物不同发育阶段叶片的干质量热值与灰分含量具有极显著的线性负相关,相关方程为 $y = -0.3938x + 23.499$,其中 $r = 0.760$, $n = 18$, $P < 0.01$;而将不同发育阶段叶片和繁殖体的干重热值与灰分含量进行相关分析,得出 $y = -0.23x + 21.726$,其中 $r = 0.437$, $n = 23$, $P < 0.05$;本研究进一步证明了灰分含量对干重热值有一定的影响。

(4) 从不同发育阶段叶片的去灰分热值来看,秋茄、无瓣海桑、木榄、桐花树为成熟叶>(或≈)老叶>幼叶;海漆、银叶树为成熟叶>幼叶>老叶。其中秋茄、木榄、桐花树从成熟叶至老叶去灰分热值变化差异不明显,但总的来看,不同发育阶段叶片的去灰分热值以成熟叶最高。不同发育阶段叶片去灰分热值的变化趋势因种而异,也不是固定不变的。秋茄、无瓣海桑、木榄、桐花树繁殖体的去灰分热值低于不同发育阶段的叶片,而海漆繁殖体的去灰分热值高于不同发育阶段的叶片。

参考文献:

[1] Lin P. Mangrove Ecosystem In China[M]. Beijing, New York: Science Press, 1999. 182 - 196.

[2] 林益明,林鹏,王通. 几种红树植物木材热值和灰分含量的研究[J]. 应用生态学报, 2000, **11**(2):181 - 184.

[3] 林益明,柯莉娜,王湛昌,等. 深圳福田红树林区 7 种红树植物叶热值的季节变化[J]. 海洋学报, 2002, **24**(3):112 - 118.

[4] 林鹏,吴世军,林益明. 红树植物繁殖体发育过程的能量变化[J]. 海洋科学, 2000, **24**(9):46 - 50.

[5] 林鹏,王文卿. 盐胁迫下红树植物秋茄(*Kandelia candel*)热值变化的研究[J]. 植物生态学报, 1999, **23**(5):466 - 470.

[6] Lin P, Wang W Q. Changes in the leaf composition, leaf mass and leaf area during leaf senescence in three species of mangroves[J]. **Ecological Engineering**, 2001, **16**(3): 415 - 424.

[7] 林益明,杨志伟,李振基. 武夷山常绿林研究[M]. 厦门:厦门大学出版社, 2001.

[8] Wang W Q, Lin P. Transfer of salt and nutrients in *Bruguiera gymnorrhiza* leaves during development and

senescence[J]. **Mangroves and Salt Marshes**, 1999, **3**(1):1- 7.

[9] 林益明, 黎中宝, 陈奕源,等. 福建华安竹园一些竹类植物叶的热值研究[J]. 植物学通报, 2001, 18(3):356- 362.

Caloric values of propagules and leaves at the different development stages of mangrove species at Futian, Shenzhen

LIN Yi - ming, XIANG Ping, LIN Peng

(*School of Life Sciences, Xiamen University, Xiamen* 361005, *China*)

Received: Nov.,20,2002

Key words: propagule; leaves at the different development stages; caloric value; mangrove species; Shenzhen

Abstract: This paper discusses the caloric value and ash content found in the leaves of the mangrove species (*Kandelia candel*, *Bruguiera gymnorrhiza*, *Aegiceras corniculatum*, *Sonneratia apetala*, *Excoecaria agallocha*, *Heritiera littoralis*) and the propagules of *Kandelia candel* during different development stages. The results showed: (1) During different development stages no general trends were found in ash content. Ash content in mature leaves was not the lowest and the content of ash in the propagules were lower than in the mature leaves. (2) The propagules of *Kandelia candel*, *Sonneratia apetala*, *Bruguiera gymnorrhiza* and *Aegiceras corniculatum* had the lowest grass caloric and ash free caloric values than those of any other development stage. This was not the case with Exoecaria agallocha which showed the opposite trend. (3) Gross caloric values were correlated remarkably with ash contents ($P < 0.01$) in the leaves during different development stages also, correlated remarkably with ash contents ($P < 0.05$) for propagules and the leaves during the different development stages.

红树植物桐花树生长发育过程的元素动态与抗盐适应性*

赵 胡 郑文教 孙 娟 林 鹏

(厦门大学生命科学学院,湿地与生态工程研究中心,福建 厦门 361005)

摘要:探讨了泌盐隐胎生红树植物桐花树(*Aegiceras corniculatum*)种苗在母树上的胎生发育、种苗脱离母树后在林地的生长发育、以至成年母树各环节的Cl,Na,K,Ca,Mg及灰分含量动态与抗盐适应性。结果表明,桐花树种苗在母树上的胎生过程是一个低盐环节,也是一个低盐化过程,可以认为这是重演了其祖先的淡生特征。种苗胎生的孕育环境宿存果皮是一个盐分累积提高的高盐环境,这有利于种苗在胎生过程中对盐分的抗性锻炼。胎生种苗脱离母树后,在林地生长发育的初生苗期阶段是一个大量吸收和累积盐分的过程。林地初生小苗的胚轴部位有吸纳累积的大量盐分,这对初生苗期幼苗的抗盐适应具有积极意义。成年母树各部位中,叶片的Cl,Na含量高于幼苗,而其它部位则低于幼苗,母树根系吸收的盐分大量累积于树冠顶部,这有利于盐分从叶片泌盐盐腺排出体外。

关键词:红树林;桐花树(*Aegiceras corniculatum*);生长发育;元素动态;抗盐适应性

中图分类号:Q948.113 **文献标识码**:A **文章编号**:1000-3096(2004)09-0001-05

红树林是热带、亚热带海岸潮间带的木本植物群落,是一项珍贵的生物资源,在海岸河口生态系统中占有重要地位。红树湿地富含Cl,Na,红树植物在长期的演化中对盐分的适应形成了泌盐和拒盐两大植物类群[1,2]。前者具有专司泌盐的盐腺,植物根系吸收盐分并通过盐腺不断排出体外;后者不具备盐腺,主要靠根系的特殊结构来拒挡盐分的大量吸收和随枯枝落叶而排出盐分[2,3]。红树适应生境的另一特性,是不少红树植物具有独特的胎生(显胎生或隐胎生)现象[1,3~5],种子在母树的果实内发芽、发育至形成成熟的胚轴种苗后才脱离母树。作者主要探讨了泌盐隐胎生红树植物桐花树(*Aegiceras corniculatum*)从种苗胚轴在母树上的胎生发育、胎生种苗脱离母树后在林地的生长发育、至成年母树,各生长发育阶段的Cl,Na,K,Ca,Mg及灰分含量变化动态。为揭示红树植物抗盐适应性提供新的科学依据。

1 材料和方法

1.1 实验材料

研究地点位于福建九龙江口红树林自然保护区南岸浮宫镇的海滩红树林(24°24′N,117°55′E)。该红树林湿地受潮汐周期淹浸,林区海水盐度全年波动在18~22,林地土壤为粉泥沉淀物、无结构,表层土壤含盐量13.6,pH7.0。

2002年8~9月于上述红树林区分别采集桐花树母株同一枝条上不同发育阶段的胎生胚轴、林地不同生长发育阶段小苗。其中:母树上胎生胚轴依其大小划分为未成熟(I)和发育成熟(II)两级,并把胎生胚轴与宿存果皮分开。成熟胚轴脱离母树于林地的生长发育,依其长根和叶片萌生与否,划分为尚漂浸于林地未长根叶的胚轴(III)、扎根但未长叶胚轴(IV)、扎根并萌生第一对真叶的初生小苗胚轴(V)共3级,并把初生小苗各组分分开。同时,分别采集林地一年生苗的叶片和茎样品,母树植株根、树干材、树干皮、侧枝、幼枝(当年生枝)和着生胚轴同一枝条的幼叶(第一对叶)、成熟叶(第三对叶)各组分,以及林地土壤样品。所有的样品各分别采集于不少于5个样点,尔后依相应组分混合,各样品取样300~500 g作为分析样品并立刻带回实验室。各植物样品分别用无离子水洗净、烘干、研磨成粉样后贮存待测。生境海水盐度则全年各月测定。

---

* 原载于海洋科学,2004,28(9):1-5

1.2 分析方法

植物样品 Cl 含量采用 Chapman 方法测定[6]。Na, K, Ca, Mg 含量样品干灰化后,采用原子吸收分光光谱仪测定 (WFX-ⅡB 型)。灰分含量采用干灰化法测定 (550 ℃,5h)。生境盐度采用 $AgNO_3$ 滴定法测定[7]。土壤 pH 采用电位法测定 (水土比为 5:1,室温 25 ℃)。以上各测试均设 2~3 个重复。

2 结果

2.1 胎生发育过程中的元素及灰分含量

桐花树种苗在母树上胎生发育过程中,胎生胚轴及孕育包被宿存果皮的 Cl, Na, K, Ca, Mg 及灰分含量变化见表 1。从表 1 可以看出,发育成熟的胎生胚轴 Cl, Na, K, Ca, Mg 及灰分含量均低于发育未成熟胚轴。其中:Cl, Na 含量发育成熟胚轴 (Ⅱ) 分别比发育未成熟胚轴 (Ⅰ) 低 8.3% 和 25.5%, K, Ca, Mg 及灰分含量分别低了 34.0%、16.7%、16.7% 和 17.4%。这一结果表明,桐花树种苗在母树上的胎生发育过程并不累积盐分。

表 1 桐花树种苗在母树上胎生过程中的元素及灰分含量

Tab. 1 Changes in contents of elements and ash in hypocotyles and pericarps during development in the mother tree

组 分	阶 段	元素及灰分含量(%)					
		Cl	Na	K	Ca	Mg	Ash
胎生胚轴	I	0.48	0.55	1.00	0.054	0.030	2.35
	II	0.44	0.41	0.66	0.045	0.025	1.94
宿存果皮	I	1.90	1.09	0.50	0.141	0.198	4.48
	II	1.94	1.20	0.44	0.191	0.211	5.29

从表 1 也可以看出,胎生种苗的孕育包被宿存果皮的元素及灰分含量变化 (除 K 外) 与其内部孕育的胎生种苗不同,发育成熟时的宿存果皮的 Cl, Na, Ca, Mg 及灰分含量均相应高于发育未成熟胚轴的宿存果皮,各元素及灰分含量前者分别比后者高 2.1%, 10.1%, 35.5%, 6.6% 和 18.1%。宿存果皮的 Cl, Na, Ca, Mg 及灰分含量均相应远高于其孕育的胎生胚轴,如胎生发育成熟时 (Ⅱ) 宿存果皮的 Cl 含量是相应胎生胚轴含量的 4.4 倍。这一结果表明:桐花树种苗在母树上的胎生过程中,其孕育包被宿存果皮是一个盐分累积提高的环境。

2.2 胚轴脱离母树后生长发育中的元素及灰分含量

2.2.1 胚轴繁殖体的元素及灰分含量

从表 2 可以看出,胎生胚轴脱离母树后在林地的生长发育中,其 Cl, Na, K, Ca, Mg 及灰分含量均表现为逐步累积提高。特别是当根和叶相继长出 (Ⅴ),胚轴的元素及灰分含量急剧上升,尤其是 Cl 含量。如林地已长根和叶的胚轴 (Ⅴ) 的 Cl, Na, K, Ca, Mg 及灰分含量依次分别是母树上胎生发育成熟胚轴 (Ⅱ) 含量的 13.3, 5.5, 2.3, 8.5, 11.2 和 6.8 倍,其中提高幅度依次为 Cl > Mg > Ca > Na > K > 灰分。这一结果表明,桐花树胚轴种苗脱离母树后,在林地生长发育中是一个明显大量吸收累积盐分的过程。

表 2 桐花树胚轴脱离母树后在林地初生小苗发育中元素及灰分含量

Tab. 2 Changes in contents of elements and ash in hypcotyles during development in the forest

发育阶段	元素及灰分含量(%)					
	Cl	Na	K	Ca	Mg	Ash
II	0.44	0.41	0.66	0.045	0.025	1.94
III	1.66	1.06	1.41	0.112	0.091	4.89
IV	2.46	1.33	1.45	0.141	0.108	6.07
V	5.84	2.25	1.54	0.382	0.281	13.10

2.2.2 林地小苗各组分的元素含量

从表 3 可以看出,林地初生小苗(Ⅳ)各组分的元素含量为苗根 > 胚轴,如苗根的 Cl,Na 含量分别是胚轴含量的 2.7 和 1.6 倍。而具根长叶的初生小苗(Ⅴ)各组分的元素含量则为胚轴 > 苗根 > 苗叶,如胚轴的 Cl,Na 含量分别是苗根含量的 1.2 和 1.5 倍、分别是苗叶含量的 3.8 和 5.4 倍。与初生小苗相比,一年生苗木叶片的元素含量均高于初生小苗叶片,如 Cl,Na 含量一年生苗木叶片分别是初生小苗叶片含量的 1.8 和 2.0 倍。这一结果表明,随着桐花树初生小苗生长发育,苗根的盐分含量降低而叶片的盐分含量则提高,盐分大量累积于胚轴。

表 3 桐花树林地不同发育阶段小苗各组分的元素含量

Tab.3 Features of element contents in different factions of seedlings at various development stages in forest

发育阶段	组 分	元素 含 量(%)				
		Cl	Na	K	Ca	Mg
小苗(Ⅳ)	胚轴	2.46	1.33	1.45	0.14	0.11
	根	6.73	2.06	2.40	0.37	0.50
小苗(Ⅴ)	叶	1.52	0.42	0.76	0.16	0.15
	胚轴	5.84	2.25	1.54	0.38	0.28
	根	4.99	1.53	1.97	0.24	0.39
1 年生苗木	叶	2.71	0.84	1.06	0.22	0.25
	茎	3.47	0.89	0.94	0.25	0.26

2.3 桐花树母树各组分的元素含量

2.3.1 胎生胚轴的元素含量

从表 4 可以看出,桐花树母树植物体各组分中,胎生胚轴的 Cl,Na,K,Ca,Mg 含量普遍远低于其他组分相应元素的含量。如 Cl,Na,胎生胚轴 Cl 含量仅分别是母树同一枝条上幼叶、成熟叶和幼枝含量的 15.0%,8.3%和 20.7%,Na 含量亦仅分别是相应组分含量的 31.8%,23.3%和 64.1%。

2.3.2 林地小苗至成年母树各组分的元素含量

与林地小苗相比:母树根和枝 Cl,Na 含量(3)低于林地小苗相应组分的含量(表 3),而叶片则高于林地小苗叶片的含量。如 Cl 含量,母树的根仅分别是林地初生小苗(Ⅳ)和(Ⅴ)根含量的 29.1%和 39.3%,而母树成熟叶片则是初生小苗(Ⅴ)和一年生小苗叶片含量的 3.6 和 1.9 倍。母树幼枝 Cl,Na 含量仅分别是一年生小苗茎含量的 61.3%,71.9%。

表 4 桐花树母树植株各组分的元素含量

Tab.4 Features of element contents in different factions of *Aegiceras corniculatum* mother tree

组 分	元素含量(%)				
	Cl	Na	K	Ca	Mg
胎生胚轴	0.44	0.41	0.66	0.05	0.03
幼 叶	2.94	1.29	1.27	0.18	0.19
成熟叶	5.32	1.76	1.43	0.34	0.51
幼 枝	2.13	0.64	1.06	0.24	0.44
侧 枝	1.38	0.81	0.69	0.41	0.51
树干皮	1.35	0.64	0.79	0.76	0.53
树干材	0.48	0.39	0.28	0.14	0.10
树 根	1.96	1.12	0.65	0.23	0.41

2.3.3 母树植株各组分的元素含量

从表 4 可以看出,桐花树植物体不同组分的元素含量高低不同。其中:Cl,Na 以叶片、根系和幼枝含量较高,侧枝和树干较低(Na 侧枝高于幼枝)。这表明植株吸收盐分并大量输送至树冠顶部,从而由叶片专司泌盐的盐腺泌出体外。K含量以叶片和幼枝含量较高,树干材含量较低;Ca,Mg 则以叶龄较大的叶片和树干皮含量较高,这与 Ca,Mg 移动性较小,易于在老器官组织中累积有关。

3 讨论

3.1 桐花树胎生种苗盐分含量

从本研究结果看,桐花树胎生种苗不仅 Cl,Na 含量,而且 K,Ca,Mg 含量,均远低于母树各组分的含量,更远低于林地小苗各组分的含量。同时,随着种苗的胎生过程,发育成熟的胚轴各元素含量均低于发育未成熟胚轴的含量。其中 Cl 含量:母树上胎生发育成熟胚轴种苗仅分别只有母树同一枝条上幼叶、成熟叶和幼枝含量的 8%~20%,亦仅为母树侧枝和根含量的 20%~30%,仅分别只有林地初生小苗各组分含量的 7%~30%和一年生小苗叶、茎含量的 12%~15%。这表明,桐花树种苗的胎生过程,在个体生活史中是一个低盐环节。可以认为这是重演了其祖先的淡生特征,也再一次证明了红树植物起源于淡生[1,3,4,8]。

3.2 胎生发育过程种苗的元素动态与抗盐适应性

金杰里等[4]认为红树幼苗在母体上发芽、生长的过程中,就进行了抗盐适应,对盐分适应的过程是靠从母体植株渐渐输送盐分进入胚轴而进行,这一观点也被林鹏所引述[1]。Joshi 等也有类似的看法[5]。而郑文教等则有相反的结论[3]。从本研究结果看:桐花树种苗在母树上的胎生过程,胎生种苗不仅始终远低于母树的盐分含量,同时随着胚轴的发育至成熟,不仅 Cl,Na 含量降低,K,Ca,Mg 及灰分含量也降低。这表明桐花树种苗在母树上的胎生过程,胎生种苗并不是一个盐分含量逐步累积提高的过程,而是一个低盐化的过程。

3.3 胎生发育过程孕育环境的元素动态与抗盐锻炼

桐花树种苗的胎生为"隐胎生",胚轴在胎生发育过程中,始终包被孕育在果实或宿存果皮内。从本研究的结果可以看出:桐花树种苗在胎生过程中,其孕育环境果实或宿存果皮不仅 Cl,Na 而且 Ca,Mg 及灰分含量动态为累积提高,且含量均远高于其内部孕育的胎生胚轴种苗。胚轴成熟时宿存果皮的 Cl,Na,Ca,Mg 及灰分含量是胎生胚轴含量的 4.4,2.9,4.2,8.8 和 2.7 倍。这表明,桐花树种苗在母树上的胎生过程,其发育的孕育环境不仅一个盐分累积提高的环境,而且是一个高盐环境,这对种苗脱离母树后适应高盐生境具有积极意义。可以认为桐花树种苗在胎生发育过程中,对盐分的适应并不是靠胎生胚轴增加盐分的累积而获得,而是在于胎生种苗在富含盐分的果实及宿存果皮的环境中孕育,对盐分的逐步适应而获得抗盐锻炼。这一点与前人的看法不同[1,4,5]。

3.4 种苗脱离母树后生长发育过程的元素动态与抗盐适应性

从本研究结果看,桐花树胎生种苗脱离母树后,在林地生长发育至初生小苗这一阶段,胚轴的 Cl,Na 元素及灰分含量均表现逐步积累提高,特别是当胚轴根和叶相继长出,各元素及灰分急剧提高。这表明桐花树种苗在林地初生小苗生长发育阶段,是一个大量吸收和累积盐分的过程。至具第一对真叶的初生小苗的苗体根和胚轴的 Cl,Na 含量分别是刚脱离母树成熟胎生胚轴含量的 11~13 倍和 4~6 倍。

桐花树属泌盐红树,对盐分的代谢主要是靠叶片的盐腺分泌排除盐分。初生小苗苗体的高盐含量一方面可能与苗叶泌盐系统尚未健全有关,另一方面则可能与初生小苗的抗盐适应性密切相关。初生小苗紧贴于水湿林地与较长时间的海水淹浸,吸水与吸收养分的动力没有如母树有强大的蒸腾流拉力的抽动,吸收水分主要靠代谢性吸收。小苗累积有较高的盐分,有利于组织细胞渗透势的提高,对维持小苗吸水与高盐生境间的渗透平衡与离子平衡有积极意义。对此,可以认为初生苗期小苗相对有较大的盐分累积是非常必须与积极的,是初生苗期小苗适应高盐生境的机制之一。

但盐分在苗体内的过量积累,必将对幼苗造成毒害,初生苗期如何解决盐分过多的累积,值得关注。桐花树在林地初生苗期阶段,原胚轴一直伴随着生于苗体,并随着小苗的发育原胚轴吸水膨胀加粗伸长并逐步老化。这一阶段,胚轴体占有小苗总生物量的 90%以上。从各组分的生物量结合元素含量来计算,初生苗期吸收的盐分总量有 80%以上是贮纳于逐步老化胚轴部位。这随着初生小苗的进一步发育,叶片泌盐机能的逐步完善,过量积累的盐分将被逐步分泌出体外。可以认为,这是桐花树初生苗期抗盐适应性机制,有别于母树,桐花树主要是靠逐步老化的胚轴贮纳大量的盐分,这也是桐花树胎生现象的另一生物学意义,即胎生胚轴在林地初生苗期具有担负贮纳盐分的功能。

参考文献:

[1] 林鹏.红树林[M].北京:海洋出版社,1984. 25-34.

[2] Teas HJ. Siliviculture with saline water [A]. Hollaender A. The Biosaline Concept[C]. New York: Plenum, 1979. 117-161.

[3] Zheng WJ, Wang WQ, Lin P. Dynamics of element contents during the development of hypocotyles and leaves of certain mangrove species[J]. **Journal of Experimental Marine Biology and Ecology**, 1999. 233: 247-257.

[4] 金杰里,方亦雄.红树植物胎生的生理意义[J].植物学报,1958,7(2): 51-58.

[5] Joshi GV, Pimplaskar M, Bhosale LJ. Physiological studies in germination of mangroves[J]. **Botanica Marina**, 1972, 45: 91-95.

[6] Chapman HD, Pratt PF. Chlorine[A]. Chapman HD, Pratt PF. Methods of Analysis for Soil, Plants and Water[C]. California: University of California, 1961. 97-100.

[7] 陈国珍.海水分析化学[M].北京:科学出版社,1965. 17-60.

[8] Dawes CJ. Marine Botany [M]. New York: Wiley, 1981. 521-523.

Dynamics of element levels and adaptation to saline environment during the development in *Aegiceras corniculatum* mangrove

ZHAO Hu, ZHENG Wen-jiao, SUN Juan, LIN Peng

(School of Life Sciences, Research Center for Wetland & Ecological Engineering, Xiamen University, Xiamen 361005, China)

Received: Jun.,9,2003

Key words: mangrove; *Aegiceras corniculatum*; development; element dynamics; salt-resistance

Abstract: The changes in the levels of Cl, Na, K, Ca, Mg and ash in relation to adaptation to saline environment in *Aegiceras corniculatum* mangrove during the development were studied. The viviparous development of seedlings (hypocotyles) on the mother tree was not only a link of lowest salt level in the life but also a lowering of salt, which may be suggested to be a recapitulating sign of the character of the ancestors who lived in fresh-water habits. The fruits or persistent pericarps of breeding seedlings during the viviparous developing were rich in the salt and this may do favorable to adaptation of seedlings to highly saline environment. The development of seedlings in the habitats made an approach to salt absorption and accumulations largely after them drop off from the mother tree. The hypocotyle parties of the young seedlings in the habitats contain a large amount of salts, and this means a great deal to the young seedlings in adapting to saline environment. With the development from seedlings to the mother trees, the levels of Cl and Na in leaves increased. It indicated that the plant roots absorbed salts and transported them to the top of crown and then excluded via salt glands in the leaves.

Litter dynamics and forest structure of the introduced *Sonneratia caseolaris* mangrove forest in Shenzhen, China*

Luzhen Chen[a] Qijie Zan[b,c] Mingguang Li[c] Jinyu Shen[d] Wenbo Liao[c]

(a. *Key Laboratory of the Ministry of Education for Coastal and Wetland Ecosystems, and School of Life Sciences, Xiamen University*, *Xiamen* 361005, *PR China*; b. *Forest Park Administration of Yangtai Mountain*, *Shenzhen* 518048, *PR China*; c. *School of Life Sciences, Sun Yat-Sen University, Guangzhou* 510275, *PR China*; d. *Department of Economics and Management, Beijing Forestry University, Beijing* 100083, *PR China*)

ARTICLE INFO

Article history:
Received 21 May 2009
Accepted 17 August 2009
Available online 26 August 2009

Keywords:
mangroves
Sonneratia caseolaris
litter
introduced species
wetlands
Futian Natural Reserve in Shenzhen of China

ABSTRACT

For the purpose of mangrove restoration in China, *Sonneratia caseolaris* has been introduced and planted in Guangdong Province outside and north of its native habitat, Hainan Province. We monitored the litter fall and forest structure of this *S. caseolaris* forest in Shenzhen City, Guangdong Province, China, from 1996 to 2005. The annual fluctuation in litter fall increased with increases in air temperature from spring to early summer, and reached a maximum in autumn when the fruits matured. The total litter fall was significantly affected by air temperature, day length, and evaporation, rainfall in the previous month and by typhoons. In 1998, the sixth year after cultivation, the total litter production of the mature *S. caseolaris* forest significantly increased. The mean annual total litter production during 1998–2005 was 15.1 t ha^{-1} yr^{-1}, among which, leaves and reproductive materials contributed more than 80% of the total. During the ten years of study, the DBH (diameter at 1.30 m from ground level) and tree height of *S. caseolaris* increased from 5.2 cm to 18.3 cm, and from 4.5 m to 13.4 m, respectively. The litter fall production was strongly correlated with forest structure parameters, such as DBH, tree height, and crown area. The *R* value (the ratio of the maximum total litter fall to the minimum in the same community during the investigation periods) of *S. caseolaris* in the present study was 1.98, indicating a low annual variation of litter fall during these ten years.

1. Introduction

Mangroves, the most productive tropical coastal ecosystems, are facing numerous kinds of destruction on a global scale. During the recent decades, the mangrove area in China has been greatly reduced (Chen et al., 2009). Since the 1990s, the Chinese government has made great efforts to restore mangroves, constructing green windscreen mangrove forests along the coast in China (Liao et al., 2004). *Sonneratia caseolaris*, one of five indigenous *Sonneratia* species in the Hainan Province of China (Ko, 1985), was frequently planted in mangrove afforestation projects. It was introduced to the Futian Mangrove Nature Reserve in Shenzhen, Guangdong Province, in 1993 for afforestation purpose (Wang et al., 2002). This species grows very quickly and forms a dense population after 4–5 years of cultivation (Wang et al., 2002).

Mangrove wetlands are extremely productive ecosystems. Their litter fall provides abundant organic matter, increases microbe biomass (Gee and Somerfield, 1997), and supports a variety of organisms (Teas, 1983). Litter production has been used as a measure of both forest productivity and the litter's contribution to estuarine systems (Lee, 1989; Lin and Lu, 1990; Tam et al., 1998). Studies of seasonal changes (Saenger and Snedaker, 1993; Zhang and Chen, 2003; Hossain et al., 2008) and ecological effects (Wafar et al., 1997; Clough et al., 2000; Arreola-Lizárraga et al., 2004) on mangrove litter production are numerous, and have been carried out in different regions on many types of mangrove forests. Most of this research concerns forests dominated by native mangrove species, except for Cox and Allen (1999), who found out a higher density, biomass and productivity of the non-native *Rhizophora mangle* in Hawaii compared to the same species in its native range.

Sonneratia caseolaris, a non-native, has been established in Shenzhen for more than fifteen years. As a fast growing species, its invasive potential has not yet been fully studied and carefully monitored. High productivity (Zan et al., 2001, 2003), high seed dispersal and germination rates (Wang et al., 2002; Tam, 2007; Zeng et al., 2008), lower leaf photosynthetic regimes than native

* From Estuarine, Coastal and Shelf Science, 2009, 85: 241–246.

mangrove species (Chen et al., 2008) have been reported, which gives it a competitive advantage. Until the present study, the relationship between forest structure and litter fall dynamics of this non-native species was unknown. Some previous studies described that *Sonneratia* species in their native habitats varied in litter fall production (Duke et al., 1981; Sasekumar and Loi, 1983; Duke, 1988). In addition, the native mangrove species, *Kandelia obovata* (formerly *Kandelia candel* (Sheue et al., 2003)) in Shenzhen Bay of China has the total litter production about 10 t ha^{-1} yr^{-1} (Lee, 1989; Tam et al., 1998). Does the productivity of this introduced *S. caseolaris* differ from the native mangroves or of the con-specifics in the native setting?

This study started in 1996; the third year after *Sonneratia caseolaris* was planted in Shenzhen, when the average tree height was about 5 m (Zan et al., 2001). The objectives were (1) to estimate the influence of environmental variables on the litter production of this non-native mangrove population, and (2) to determine the annual production of litter fall during its early growth, the latter of which will further understand the contribution of this non-native mangrove species to the nutrient cycle in the native habitat.

2. Methods

2.1. Study site

The study was carried out in Futian Mangrove Nature Reserve (22°31′N, 114°05′E), located in an estuary of the Zhujiang River in Shenzhen, Guangdong Province, China. The site is characterized by a subtropical monsoonal climate, with 1927 mm of annual precipitation and a rainy season between May and September. The mean annual relative humidity is about 80% and the mean air temperature is 22 °C, with a maximum of 38.7 °C (July) and a minimum of 0.2 °C (January). The coldest month's temperature was 14.1 °C. The tides in Shenzhen Bay are semi-diurnal, with a spring tidal range of about 1.9 m.

In the Futian Nature Reserve in Shenzhen, there are about 75 ha of conserved mangroves distributed along 9 km of the coast. The dominant native species in the Futian mangroves are *Kandelia obovata* and *Avicennia marina*. Non-native *Sonneratia caseolaris*, first introduced to the Futian Nature Reserve in 1993, developed well in the new habitat and formed a dense population covering more than 4 ha (Wang et al., 2002).

2.2. Forest structure

Three transects acted as three replicates of measurement were randomly assigned in the *Sonneratia caseolaris* forest. Each transect was composed of four contiguous sample plots (10 m × 10 m). All trees in the plots were identified and tagged, and the height, DBH (diameter at 1.30 m from ground level) and crown area of each was measured annually in December from 1996 to 2005. The population density was determined by counting the number of individual trees in each 10 m × 10 m plot, and then averaged among the four contiguous plots.

2.3. Litter production

Litter fall was collected in 1 m^2 (1 m × 1 m) baskets with 1.5 mm fiberglass mesh. One basket was placed in each fixed sample plot for forest structure measurement. Litter fall baskets were hung at 2.5 m above the sediment to avoid being dipped in tidal seawater. Litter fall was collected every 10 d from January 1996 through to December 2005. Leaves, branches/twigs, flowers, and fruits were separated, and dried at 80 °C, and the dry weight of each part was weighed and recorded. Monthly litter fall production was calculated based on the accumulated data for each month.

The annual variation of total litter fall was expressed as *R*, which was the ratio of the maximum total litter fall to the minimum total litter fall over ten years (Bray and Gorham, 1964).

2.4. Climatic data

From 1997 to 2005, climatic data, such as air temperature, day length, rainfall, evaporation, and the occurrence of typhoons (1996–2005), were obtained from the website of the Hong Kong Observatory and a weather station of the Shenzhen Meteorological Bureau close to the mangrove reserve.

2.5. Statistical analyses

The mean and standard error (S.E.) of three sample replicates were calculated for monthly and annual litter fall. Annual litter fall of leaves, branches/twigs, flowers and fruits, and forest structure characteristics were analyzed using one-way ANOVA following an LSD (Least Significant Difference) test. The relationships between litter fall products and climatic variables and between each two climatic variables were analyzed using bivariate correlation following the Pearson method. All analyses were carried out using a SPSS13.0 (SPSS Inc., Chicago, IL).

3. Results

3.1. Monthly litter fall and climatic variables

Spring to early summer (April–June) showed the lowest monthly litter fall of *Sonneratia caseolaris* in every year (Fig. 1) as rainfall, evaporation, air temperature, and day length increased. From July to November, monthly litter fall production was high when rainfall was plentiful. An average of four typhoons occurred in Shenzhen each year from 1996 to 2005, but the mean wind speed that month did not show a significant increase (Fig. 1). No significant correlation was found between litter fall and wind speed (Table 1). Rainfall did not significant correlate with litter fall, but did significantly increase the flower, fruit, and total litter fall in the month following (Table 1). Leaf litter fall was strongly correlated to the ratio of rainfall to evaporation. Significantly positive correlations were found between flower and fruit litter fall products and such climatic variables as air temperature, day length, evaporation, and rainfall in the former month (Table 1). Similar results were found in the total litter fall (Table 1). The total litter fall products were lower from December to February than summer during 1996–2001 (Fig. 1). Except for the relationships between day length and wind speed, between day length and rainfall and between wind speed and rainfall in the previous month, the climatic variables were significantly correlated with each other.

3.2. Annual litter fall and forest structure

Except in 1999, the DBH, height, and crown area of *Sonneratia caseolaris* trees increased significantly year after year, and reached their maximums in the last year of investigation, while at the same time the population density decreased (Tables 2 and 3). Similarly, the litter fall of leaves, branches/twigs, flowers, and fruits significantly increased during the ten years of investigation (Fig. 2 and Table 3). In 1998, the litter fall of leaf and reproductive materials increased significantly. From 1998 to 2005, the mean annual total litter fall was 15.1 t ha^{-1} yr^{-1}, among which, leaf and reproductive materials contributing more than 80% of total litter fall; the remaining 20% was mostly twigs and branches.

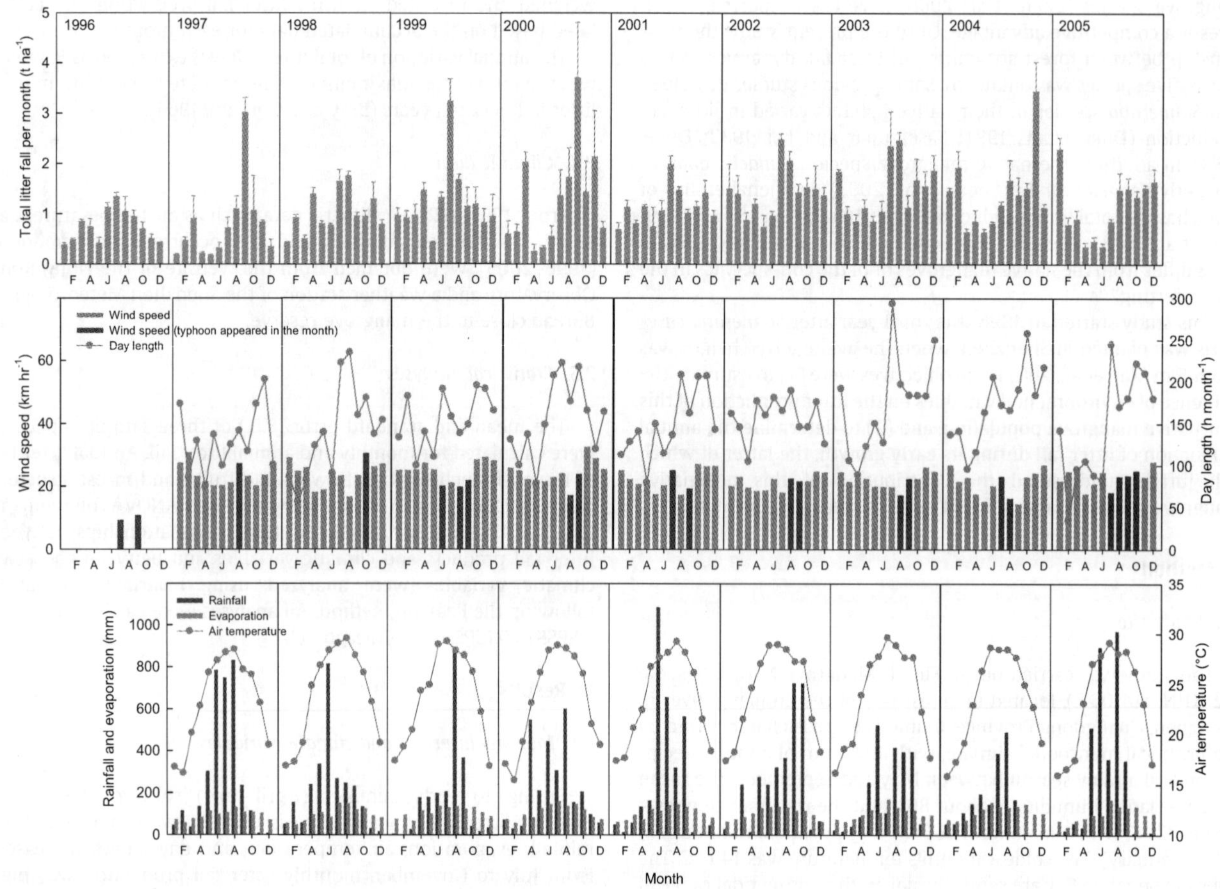

Fig. 1. Total litter fall of non-native *Sonneratia caseolaris* forest and climatic variables during ten annual cycles (1996/1997–2005) in Shenzhen, China.

From 1996 to 2005, Pearson correlations were performed to analyze the relationships between annual litter fall and structural characteristics of *Sonneratia caseolaris* (Table 4). Significant positive correlations were found between branch litter fall, as well as total litter fall and tree DBH, height, and crown area, but were negatively correlated to population density (Table 4). Leaf litter fall was weakly correlated to tree DBH, height, and crown area, but strongly negative to population density (Table 4). Flower and fruit litter fall showed a significant, positive relationship to tree DBH, and strongly negative correlation to population density (Table 4).

The ratio of the maximum to the minimum annual litter fall (*R*) of these ten years was 1.98.

4. Discussion

Seasonal fluctuations have been found in litter fall of several mangrove species, notably of the genera *Avicennia*, *Bruguiera*, *Rhizophora*, and *Sonneratia* (Wium-Andersen and Christensen, 1978; Goulter and Allaway, 1979; Duke et al., 1981; Duke, 1988; Lin and Lu, 1990). Litter fall can also be observed throughout a year with little (Shunula and Whittick, 1999) or marked (Day et al., 1996; Wafar et al., 1997) seasonal variations. In the present study, litter fall took place throughout the year, peaking during the monsoon season. The yearly change of litter fall products increased with an increase in air temperature during spring to early summer (about March–June), and reached a maximum in autumn (Fig. 1). Our site is located in a subtropical monsoonal climate area, in a humid region along the South China Sea. Although some researchers (Saenger and Moverley, 1985) found leaf fall and air temperature to be significantly positively correlated with several mangrove species, the monthly leaf litter fall in the present study was strongly, but negatively, affected by the ratio of precipitation to evaporation, but insignificantly affected by other climatic variables, as shown in Table 1.

In addition to increased temperatures, other environmental factors, such as day length, higher evaporation, or greater rainfall

Table 1

Pearson correlation coefficients of the monthly litter fall of the non-native *Sonneratia caseolaris* forest and climatic variables in Shenzhen from 1997 to 2005 ($n = 108$). Levels of significance are shown as: *0.05, **0.01 and ***0.001.

Litter fall	Air temperature	Day length	Evaporation	Wind speed	Rainfall	Rainfall/evaporation	Rainfall in the previous month
Leaves	−0.017	−0.169	0.077	−0.114	−0.181	−0.212*	0.063
Branches/twigs	0.038	−0.060	−0.041	−0.115	−0.040	−0.071	0.143
Flowers and fruits	0.277**	0.456***	0.388***	−0.015	0.098	−0.002	0.247*
Total	0.199*	0.426***	0.328**	−0.088	−0.027	−0.119	0.236*

Table 2
Structural characteristics (mean ± S.E., $n = 3$) of the non-native *Sonneratia caseolaris* forest in Shenzhen (1996–2005).

Year	DBH (cm)	Tree height (m)	Crown area (m^2)	Population density (individuals per ha)
1996	5.2 ± 0.2	4.5 ± 0.1	7.1 ± 0.3	307 ± 4.4
1997	7.7 ± 0.3	5.5 ± 0.1	6.7 ± 0.4	294 ± 5.0
1998	9.5 ± 0.4	10.7 ± 0.2	6.3 ± 0.7	283 ± 5.2
1999	8.6 ± 0.3	5.8 ± 0.1	9.6 ± 0.3	278 ± 6.0
2000	11.7 ± 0.4	6.7 ± 0.1	11.4 ± 0.5	233 ± 6.3
2001	14.0 ± 0.5	10.2 ± 0.2	13.0 ± 1.2	229 ± 3.0
2002	16.1 ± 0.6	12.2 ± 0.1	26.0 ± 2.4	228 ± 4.6
2003	17.7 ± 0.7	12.9 ± 0.1	26.5 ± 2.1	226 ± 4.4
2004	18.4 ± 0.6	13.2 ± 0.1	31.9 ± 2.4	222 ± 6.0
2005	18.3 ± 0.8	13.4 ± 0.1	32.1 ± 2.3	218 ± 6.6

(which results in more flushing by freshwater), have been proved to be the probably factors affecting litter fall (Ghosh et al., 1990; Chale, 1996; Tam et al., 1998). Similar results were found in the present study (Table 1). The annual total litter fall increased from 1996 to 2003, leveled off, then dropped in 2006 and 2007 (Fig. 2), which may have been the result of abnormal rainfall rates in 2004 (low) and 2005 (much higher) (Fig. 1). In northeastern Australia, *Avicennia* leaf fall (and leaf production) peaked in months when mean daily air temperature was rising to or at 20 °C, while *Sonneratia caseolaris* had maximal leaf fall around 28 °C (Duke, 1988). Similarly, in Shenzhen, high litter fall production was observed during periods from May to October, when the air temperature was higher than 25 °C. A seasonal pattern of litter fall was also observed in mangroves elsewhere (Lin and Lu, 1990; Tam et al., 1998), with the most in summer and least in winter and early spring. The *R* value was used in comparison of the annual variation in litter fall (Bray and Gorham, 1964). It was 1.98 in the present study, indicating little variability of litter fall during these ten years, which may due to the fast growth rate and the early maturation of the *S. caseolaris* forests in Shenzhen. Furthermore, the low *R* value can also be expected that the annual climatic fluctuations are not very severe in these ten years.

Apart from the phenological events, mangrove production cycles appeared to be under the influence of an endogenous rhythm (Duke, 1988). Saplings of *Sonneratia caseolaris* began to reproduce in this new habitat, Shenzhen, at three years (Wang et al., 2002). The major flowering and fruiting season of *S. caseolaris* in Shenzhen is from July to October (Tam, 2007) when the air temperature, day length, and evaporation are at their highest. The trend of total litter fall products followed the changes of flower and fruit litter fall products (Table 1). It was observed that total litter fall decreased in the winter from 1996 to 2001, but remained high after 2002 (Fig. 1), which suggests that high temperature, day length, and low or even no rainfall in winter prolong the flowering and fruiting periods of *S. caseolaris* trees.

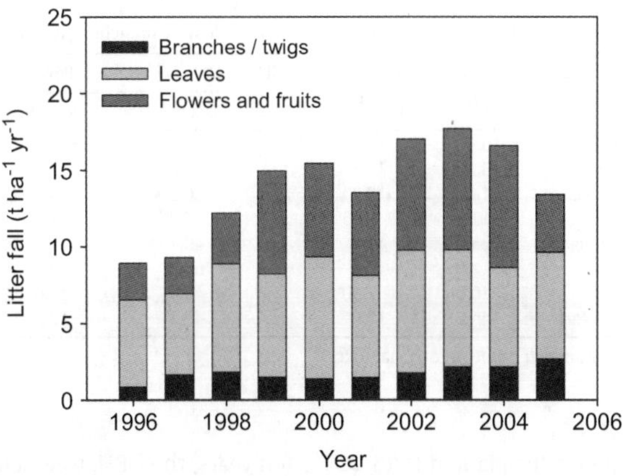

Fig. 2. Components of non-native *Sonneratia caseolaris* litter fall during ten annual cycles (1996–2005) in Shenzhen, China.

Mangroves are located in the coastal areas of the South China Sea, where each summer and early autumn they face typhoons. Over ten years, different times of typhoon have occurred in Shenzhen each year. According to Lu and Lin (1990), litter fall production fluctuated with typhoon occurrence. Though insignificant correlations were found between the wind speed and litter fall, and between the rainfall and litter fall of each component (Table 1), the total litter fall increased after a typhoon hit (Fig. 1). For example, three and four typhoons occurred respectively in July, 2001 and August, 2003, which both significantly increased the total litter fall that month. However, in some case, typhoon did not promote the total litter fall in that month; such as typhoon in September, 2001, as might perhaps be expected that the excessive increase of litter fall production by typhoon in the former months would decrease the litter fall in the following (Lin and Lu, 1990). Furthermore, the influences of typhoons are different due to their different intensities. The damage caused by the typhoon in September, 1999 was severe, which blew down some high trees of *Sonneratia caseolaris* (Chen et al., 2000) and resulted in the decrease in tree height that year (Table 2). However, the litter fall production of September was not the maximum in 1999 (Fig. 1) because two typhoons had hit in former months. In addition, the damage of non-native *S. caseolaris* by typhoon was more serious than native dominant species *Kandelia obovata* (Chen et al., 2000), which further indicated the incompact stem structure of the fast growth tree was weak in resisting the attack of strong wind.

Net primary productivity not only depends on phenological events and abiotic factors, but also on attributes of geographical location and forest structure (Hossain et al., 2008). This ten-year study was started in 1996, the third year after *Sonneratia caseolaris* seedlings had been planted in Shenzhen, and the first year they began to reproduce. The litter fall of leaves and reproductive materials significantly increased in 1998 and remained high reproductive production since then (Fig. 2), which indicated the full

Table 3
Significant values of one-way ANOVA following a Least Significant Difference test (LSD) (*F*-values) for comparisons in annual litter fall and structure characteristics of the non-native *Sonneratia caseolaris* in Shenzhen from 1996 to 2005. Levels of significance are shown as: *0.05, **0.01 and ***0.001.

Items		df	*F* value
Litter fall	Leaves	9, 29	2.413*
	Branches/twigs	9, 29	5.116**
	Flowers and fruits	9, 29	3.877**
	Total	9, 29	5.985***
Forest structure	DBH	9, 492	91.614***
	Height	9, 492	639.814***
	Crown area	9, 492	49.778***
	Population density	9, 492	39.876**

Table 4
Pearson correlation coefficients of the annual litter fall and structure characteristics of the non-native *Sonneratia caseolaris* in Shenzhen from 1996 to 2005 ($n = 10$). Levels of significance are shown as: *0.05, **0.01 and ***0.001.

Litter fall	DBH	Height	Crown area	Population density
Leaves	0.534	0.492	0.382	−0.660*
Branches/twigs	0.817**	0.838**	0.803**	−0.658*
Flowers and fruits	0.651*	0.490	0.583	−0.690*
Total	0.759*	0.634*	0.664*	−0.795**

Table 5
Mean estimates of annual litter fall ($t\ ha^{-1}\ yr^{-1}$) and leaves, branches, flowers and fruits of mangrove species.

Predominant species	Height (m)	Total litter fall ($t\ ha^{-1}\ yr^{-1}$)	Percentage of total litter fall			Locality	Reference
			Leaves (%)	Branches/twigs (%)	Flowers and fruits (%)		
K. obovata[a]	5	11.7	<50	<30	>20	Shenzhen, China	Tam et al. (1998)
K. obovata[a]	3.3	10.3	53.9	6.2	40.7	Hong Kong	Lee (1989)
S. caseolaris	14.0	9.34	60.5	14.5	25	Australia	Duke (1988)
S. alba	12.5	9.05	47.2	11.7	41.1	Australia	Duke (1988)
S. × *gulngai*	25.0	11.7	49.7	19.1	31.2	Australia	Duke (1988)
S. alba	12.5	15.8	67.5	22.5	10.0	Malaysia	Sasekumar and Loi (1983)
S. alba	10.0	7.90	48.8	39.0	21.2	Australia	Duke et al. (1981)
S. caseolaris	4–13	13.9	49.3	12.5	38.2	Shenzhen, China	Present study

[a] It is *Kandelia candel* in original text.

mature of this planted forest. Over ten years, the DBH, tree height and crown area increased except in 1999 when typhoon blew down some trees (Table 2), suggesting the maturation of the forest. Positive correlations between litter fall products and forest structure parameters (Table 4) showed that litter fall of each component and total litter fall significantly increased with the development of the *S. caseolaris* forest.

After *Sonneratia caseolaris* was introduced north of its native habitat of Hainan Province, where the coldest month's temperature was 4.5 °C higher than that in Shenzhen, this species faced and overcame the challenge of low air temperatures in winter in the new habitat (Wang et al., 2002). However, the non-native *Sonneratia* species are more productive than any native mangrove species in Shenzhen. It was reported that the total biomass of 6-year-old *S. caseolaris* population in Shenzhen (Zan et al., 2001) was 40% higher than which of 5-year-old in Hainan (Liao et al., 1990), as was expected that the high N contents in sediment caused by sewage in Shenzhen Bay promoted the growth (Wang et al., 2002). Wang et al. (2002) reported that the averaged tree height and crown area of 7-year-old *S. caseolaris* was respectively 3.15 and 20 times to *Kandelia obovata* saplings in the same age. The litter fall products of *Sonneratia* species in native habitats on a global scare and litter fall of dominant species in Shenzhen, and Hong Kong where is adjacent to Shenzhen, are listed in Table 5. Thought the percentages of leaf and reproduced materials to total litter fall production of non-native *S. caseolaris* in the present study is comparable to the native *K. obovata* in Shenzhen and Hong Kong, the total litter fall production of *S. caseolaris* was higher (Table 5), further indicating less influence from lower winter air temperature. In addition, *S. caseolaris* in the present study had higher litter fall production than trees of the same species in Australia (only 9.34 $t\ ha^{-1}\ yr^{-1}$ compared to more than 13.9 t in Shenzhen) where it is native, also suggesting a good adaptation of this non-native species to new habitat. Comparing to the *Sonneratia* forests on the global scare, *S. caseolaris* in the present study also had higher litter fall production than in Australia, but lower than in Malaysia (Table 5). Similar results were found in the introduced *Rhizophora mangle* L. in Hawaii (Cox and Allen, 1999). Favorable site conditions, lack of competition from other woody plants and predators (Cox and Allen, 1999) would be favorable to this non-native species in Shenzhen.

Exotic tree plantations can cause important changes in diversity and community compositions locally and regionally (Brockerhoff et al., 2001). A large fraction of the organic matter produced by mangrove trees is exported to the coastal ocean, which forms one basis of the food chain (Odum and Heald, 1972). Leaf litter was one potential food source for benthic organisms in mangrove ecosystems (Teas, 1983). The results in present study suggest that the non-native *Sonneratia caseolaris* forest supplied more nutrients and organic matter to the native food web than the dominant native species. However, there is still a lack of basic information on the biodiversity of benthic organisms that are directly correlated to mangrove litter production. Research on the effects on the mangrove biodiversity of planting non-native *Sonneratia* species is also essential in the future.

5. Conclusions

The mean annual litter fall of non-native *Sonneratia casoelaris* in this study was more than 15.1 $t\ ha^{-1}\ yr^{-1}$ after 1997, and leaves and reproductive materials contributed to more than 80% of total litter fall. Owing to heavy rainfall and frequent typhoon events from early summer to autumn, the monsoon season was the peak litter fall season for this planted mangrove forest. The litter fall showed a significant positive correlation to forest structure parameters, such as DBH, tree height, and crown area. This non-native mangrove forest contributes more litter to the native habitat than do the dominant native species. If more widely planted, *S. casoelaris* would greatly contribute to the nutrient cycle and carbon input in this area.

Acknowledgements

The authors thank Lan H.L. in the Administrative Bureau of Neilingding-Futian National Nature Reserve for his help in the fieldwork. This study was partially supported by a National Natural Science Foundation of China grant to L. Chen (30700092), by the Natural Science Foundation of Fujian Province of China grant to L. Chen (2009J05085) and by the technological program of the Shenzhen Bureau of Science Technology and Information of China (2004B-111).

References

Arreola-Lizárraga, J.A., Flores-Verdugo, F.J., Ortega-Rubio, A., 2004. Structure and litter fall of an arid mangrove stand on the Gulf of California, Mexico. Aquatic Botany 79, 137–143.

Bray, J.R., Gorham, E., 1964. Litter production in forest of the world. Advances in Ecological Research 2, 101–157.

Brockerhoff, E.G., Ecroyd, C.E., Langer, E.R., 2001. Biodiversity in New Zealand plantation forests: policy trends, incentives, and the state of our knowledge. New Zealand Journal of Forestry 46, 31–37.

Chale, F.M.M., 1996. Litter production in an *Avicennia germinans* (L.) stearn forest in Guyana, South America. Hydrobiologia 330, 47–53.

Chen, L., Tam, N.F.Y., Huang, J., Zeng, X., Meng, X., Zhong, C., Wong, Y., Lin, G., 2008. Comparison of ecophysiological characteristics between introduced and indigenous mangrove species in China. Estuarine, Coastal and Shelf Science 79, 644–652.

Chen, L., Wang, W., Zhang, Y., Lin, G., 2009. Recent progresses in mangrove conservation, restoration and research in China. Journal of Plant Ecology 2, 45–54.

Chen, Y.J., Zheng, D.Z., Liao, B.W., Zheng, S.F., Zan, Q.J., Song, X.Y., 2000. Researches on typhoon damage to mangroves and preventive measures. Forest Research 13, 524–529.

Clough, B., Tan, D.T., Phuong, D.X., Buu, D.C., 2000. Canopy leaf area index and litter fall in stands of the mangrove *Rhizophora apiculata* of different age in the Mekong Delta, Vietnam. Aquatic Botany 66, 311–320.

Cox, E.F., Allen, J.A., 1999. Stand structure and productivity of the introduced *Rhizophora mangle* in Hawaii. Estuaries 22, 276–284.

Day Jr., J.W., Coronado-Molina, C., Vera-Herrera, F.R., Twilley, R., Rivera Monroy, V.H., Alvarez-Guillen, H., Day, R., Conner, W., 1996. A 7 year record of above-ground net primary production in a southeastern Mexican mangrove forest. Aquatic Botany 55, 39–60.

Duke, N.C., Bunt, J.S., Williams, W.T., 1981. Mangrove litter fall in north-eastern Australia. I. Annual totals by component in selected species. Australian Journal of Botany 29, 547–553.

Duke, N.C., 1988. Phenologies and litter fall of two mangrove trees, *Sonneratia alba* Sm. and *S. caseolaris* (L.) Engl. and their putative hybrid, *S.* × *gulngai* N.C. Duke. Australian Journal of Botany 36, 473–482.

Gee, J.M., Somerfield, P.J., 1997. Do mangrove diversity and leaf litter decay promote meiofaunal diversity? Journal of Experimental Marine Biology and Ecology 218, 13–33.

Ghosh, P.B., Singh, B.N., Chakrabarty, C., Saha, A., Das, R.L., Choudhury, A., 1990. Mangrove litter production in a tidal creek of Lothian Island of Sundarbans, India. Indian Journal of Marine Sciences 19, 292–293.

Goulter, P.F.E., Allaway, W.G., 1979. Litter fall and decomposition in a mangrove stand, *Avicennia marina* (Forsk.) Vierh., in Middle Harbor, Sydney. Australian Journal of Marine and Freshwater Research 30, 541–546.

Hossain, M., Othman, S., Bujang, J.S., Kusnan, M., 2008. Net primary productivity of *Bruguiera parviflora* (Wight & Arn.) dominated mangrove forest at Kuala Selangor, Malaysia. Forest Ecology Management 255, 179–182.

Ko, W.C., 1985. Notes on genus *Sonneratia* (Sonneratiaceae) in S.E Asia. Acta Phytotaxonomica Sinica 23, 311–314 (in Chinese with English abstract).

Lee, S.Y., 1989. Litter production and turnover of the mangrove *Kandelia candel* (L.) druce in a Hong Kong tidal shrimp pond. Estuarine, Coastal and Shelf Science 29, 75–87.

Liao, B.W., Zheng, D.Z., Zheng, S.F., 1990. Studies on the biomass of *Sonneratia caseolaris* stand. Forest Research 3, 47–54 (in Chinese with English abstract).

Liao, B.W., Zheng, S.F., Chen, Y.J., Li, M., Li, Y.D., 2004. Biological characteristics of ecological adaptability for nonindigenous mangroves species *Sonneratia apetala*. Chinese Journal of Ecology 23, 10–15 (in Chinese with English abstract).

Lin, P., Lu, C., 1990. Biomass and productivity of *Bruguiera sexangula* mangrove forest in Hainan Island, China. Journal of Xiamen University (Natural Science) 29, 209–213 (in Chinese with English abstract).

Lu, C., Lin, P., 1990. Studies on litter fall and decomposition of *Bruguiera sexangula* (Lour.) Poir, community on Hainan Island, China. Bulletin of Marine Science 47, 139–148 (in Chinese with English abstract).

Odum, W.E., Heald, E.J., 1972. Trophic analysis of an estuarine mangrove community. Bulletin of Marine Science 22, 671–738.

Saenger, P., Moverley, J., 1985. Vegetative phenology of mangroves along the Queensland coastline. Proceedings of the Ecological Society of Australia 13, 257–265.

Saenger, P., Snedaker, S.C., 1993. Pantropical trends in mangrove above-ground biomass and annual litter fall. Oecologia 96, 293–299.

Sheue, C.R., Liu, H.Y., Yong, J.W.H., 2003. *Kandelia obovata* (Rhizophoraceae), a new mangrove species from Eastern Asia. Taxon 52, 287–294.

Shunula, J.P., Whittick, A., 1999. Aspects of litter production in mangroves from Unguja Isalan, Zanzibar, Tanzania. Estuarine, Coastal and Shelf Science 49, 51–54.

Sasekumar, A., Loi, J.J., 1983. Litter production in three mangrove forest zones in the Malay Peninsula. Aquatic Botany 17, 283–290.

Tam, N.F.Y., 2007. Provision of services for preparation of a management strategy for *Sonneratia* in inner deep bay areas. (Unpublished report).

Tam, N.F.Y., Wong, Y.S., Lan, C.Y., Wang, L.N., 1998. Litter production and decomposition in a subtropical mangrove swamp receiving wastewater. Journal of Experimental Marine Biology and Ecology 226, 1–18.

Teas, H.J., 1983. Biology and Ecology of Mangroves. W. Junk Publishers, The Hague.

Wafar, S., Untawale, A.G., Wafar, M., 1997. Litter fall and energy flux in a mangrove ecosystem. Estuarine, Coastal and Shelf Science 44, 111–124.

Wang, B.S., Liao, B.W., Wang, Y.J., Zan, Q.J., 2002. Mangroves Ecosystem and Sustainable Development in Shenzhen Bay. Science Press, Beijing (in Chinese).

Wium-Andersen, S., Christensen, B., 1978. Seasonal growth of mangrove trees in south Thailand. II. Phenology of *Bruguiera cylindrical*, *Ceriops tagal*, *Lumnitzera littorea*, and *Avicennia marina*. Aquatic Botany 5, 383–390.

Zan, Q.J., Wang, B.S., Wang, Y.J., Li, M.G., 2003. Ecological assessment of the introduced *Sonneratia caseolaris* and *S. apetala* at the mangrove forest of Shenzhen Bay, China. Acta Botanica Sinica 45, 544–551.

Zan, Q.J., Wang, Y.J., Liao, B.W., Zheng, D.Z., 2001. Biomass and net productivity of *Sonneratia apetala*, *S. caseolaris* mangrove man-made forest. Journal of Wuhan Botany Research 19, 391–396.

Zeng, X., Chen, L., Tam, N.F.Y., Huang, J., Xu, H., Lin, G., 2008. Seedling emergence and dispersal pattern of the introduced *Sonneratia caseolaris* in Shenzhen Bay, China. Biodiversity Science 16, 236–244 (in Chinese with English abstract).

Zhang, Q., Chen, Y., 2003. Production and seasonal change pattern of litter fall of *Rhizophora apiculata* in Sanya River mangroves, Hainan Island. Acta Ecologica Sinica 23, 1977–1983 (in Chinese with English abstract).

Effect of different time of salt stress on growth and some physiological processes of *Avicennia marina* seedlings*

Zhongzheng Yan Wenqing Wang Danling Tang

(a. *School of Life Sciences, Xiamen University, Xiamen* 361005 *Fujian, People's Republic of China*; b. *State Kay Laboratory of Marine Environmental Science, Xiamen University, Xiamen* 361005 *Fujian, People's Republic of China*; c. *Key Laboratory of Tropical Marine Environmental Dynamics (LED), South China Sea Institute of Oceanology, Chinese Academic of Science, Guangzhou, People's Republic of China*)

Abstract Growth and physiological characters of *Avicennia marina* seedlings cultured under different levels of salinity were compared at 45 and 100 days after sowing. Based on the growth and physiological responses, the levels of salinity were grouped into two kinds, moderate (5–30‰) and extreme (40 and 50‰ as well as 0‰). Root and shoot length, leaf area, biomass of different organs, and net photosynthesis rate all showed a similar trend: the seedlings grew better at moderate levels of salinity but were adversely affected by extreme levels. Longer exposure (100 days) to salinity markedly enhanced the difference between the effects of the two levels on growth. By 45th day, the cotyledons had withered and fallen off. The concentration of ions (K^+, Na^+, Ca^{2+}, Mg^{2+}, Cl^-) and ash content of the cotyledons were determined before sowing and 45 days later. Ion concentrations and ash content of cotyledons were markedly lower at 45 days—lower than the initial levels—in seedlings irrigated with water at 0‰ salinity level. This suggested that the poor growth of these seedlings at 100 days may be due to lack of ions provided by the cotyledons. The high ion concentrations in the cotyledons grown at moderate salinity levels suggest that these organs may function as ion sinks at this stage, reducing the concentration of ions and consequent toxicity caused by excessive concentrations. Root biomass was higher than shoot biomass 45 days after sowing, whereas after 100 days, shoot biomass was higher. At the early stage of growth (45 days), the rate of photosynthesis at lower levels of salinity (0–30‰) was limited mainly by stomatal closure but at higher levels of salinity (40–50‰), other factors came into play. Later, at 100 days, the causes of reduced photosynthetic rate were other than stomatal closure at both low and high levels of salinity. This indicates that photosynthesis is affected by prolonged exposure to salt stress—including that caused by 0‰ salinity, as shown by poor growth of the seedlings.

Introduction

Although mangroves are a kind of halophyte, their seedlings are sensitive to salt stress (Tomlinson 1986; Lin 1997); a saline substrate affects many aspects of their growth and physiology (Ball and Farquhar 1984a, b; Wang and Lin 1999; Clough 1984; Downton 1982). As is the case with many other halophytes, which grow poorly in a culture medium that lacks sodium chloride (Flower et al. 1977), most studies have found that whereas seedlings grow best at low salinity (25% of sea water), higher salinity (50 or 75% of sea water) or total lack dissolved salts (0‰ salinity, i.e. fresh water) affect growth (Downton 1982; Clough 1984; Wang and Lin 1999). Slow growth in fresh

* From Marine Biology, 2007, 152: 581—587

water is often ascribed to the inability of halophytes to accumulate inorganic ions in quantities sufficient for osmoregulation when the substrate is lacking in sodium chloride (Clough 1984; Jennings 1976; Greenway and Munns 1980; Yeo and Flower 1980). Some authors consider this phenomenon to be the expression of a physiological trait of mangroves that demands salt (Wang and Lin 1999), but few studies have attempted to explain the mechanism.

Seedlings of halophytes can obtain the elements that their growth demands from soil or from the propagules themselves in the case of viviparous mangroves (Tomlinson 1986; Farnsworth 2000). The remarkable propagules accumulate ions and nutrients from the mother tree before being released (Zheng et al. 1999). *Avicennia marina* is a typical salt-secreting cryptoviviparous mangrove, a pioneer species widespread around the world (Tomlinson 1986; Lin 1997) which is regarded as one of the most salt-tolerant species of mangrove (Clough 1984; Tomlinson 1986; Lin 1997). Its cotyledons gain ions and nutrients during propagule development and may act as a sink for ions and nutrients (Wang et al. 2002) that support the seedlings in their early growth in the saline culture medium. But the function of the ions and nutrients accumulated during the propagule's development poses a few questions. After a few days of early growth, the cotyledons of *Avicennia marina* wither and fall off, an event that may mark the watershed between two growth stages. It is known that salt tolerance of a plant is related to the stage of growth (Munns 1986; Zhao and Wang 1990; Wang and Lin 2000). Generally speaking, halophytes are sensitive to salt stress at the seedling stage and the flowering stage but gradually increase their tolerance the longer they are exposed to salt stress (Zhao and Wang 1990; Wang and Lin 2000; Wang et al. 2001). The loss of cotyledons may affect plant growth. We measured some characteristics of growth and physiology of *Avicennia marina* seedlings at different stages of growth, i.e. following shorter or longer duration of exposure to salinity, at different levels of salinity to identify the role cotyledons play in a seedling's early growth.

Materials and methods

Sampling and cultivation

Propagules of *Avicennia marina* were collected from Haimen Island (24°30′N, 117°55′E), Longhai city, Fujian province of China, on 1 October 2003. The collected propagules were more or less uniform in size (2.1 ± 0.3 cm in diameter and weighing 1.8 ± 0.5 g) and free of holes or other signs of damage by insects.

The propagules were raised in a greenhouse in plastic pots (35 cm in diameter and 14 cm deep), each filled with 4 kg of coarse beach sand (particles 2–3 mm in diameter) collected from Xiamen harbor and washed with fresh water before use. The treatments comprised eight levels of salinity, each was replicated three times (0, 5, 10, 15, 20, 30, 40, 50‰). The saline solutions were prepared by mixing fresh water with either sea water (for 0, 5, 10, 15, or 20‰ solutions) or sea salt (for 30, 40, or 50‰ solutions). The salinity level of sea water was 27‰. While the pots were watered with the saline solutions every 15 days throughout the experiment, fresh water was added daily to make up for losses by evaporation or transpiration. The experiment lasted 100 days (October 2003 to January 2004).

Methods of analysis

Six seedlings were collected from each pot on the 45th day and on the 100th day. The length of the stem, mean root length, and leaf area were measured in three seedlings and the other three seedlings were used for determining the biomass: the roots, stems, and leaves were separated, washed with distilled water, and weighed after drying at 75°C for 48 h. The cotyledons, collected on the 45th day, were also dried at 75°C for 48 h and pulverized in a mill until the powder was fine enough to pass through a 1-mm sieve. The powder was stored for analyzing the ion content.

Cl^- was determined by $AgNO_3$ titration. The subsamples (about 0.2 g each) were incinerated at 550°C for 3 h. To avoid the loss of Cl^-, about 0.05 g of calcium oxide was added and mixed with the sample beforehand. The ash was dissolved in distilled water and then filtered, and the filtrate was used to determine Cl^-. Another sample (about 0.1 g) was incinerated at 550°C for 5 h, the ash dissolved in 50 ml hydrochloric acid (1 N), and the solution used in an atomic absorption spectrophotometer (AAnalyst 800, Perkin Elmer Instrument, USA) for determining the concentrations of K^+, Na^+, Ca^{2+}, and Mg^{2+}. Dry powder of the cotyledons was collected on the day the experiment began and on the 45th day to calculate the gross calorific value by using a bomb calorimeter (PARR 1266, USA).

Three seedlings were selected from each pot on 50th and 90th day and nine or more leaves from each treatment were chosen to measure the net photosynthesis rate (P_n, μmol m^2 s^{-1}), stomatal conductance (G_s, μmol m^2 s^{-1}), intercellular concentration of CO_2 (C_i, ppm) and evaporation rate (T_r, μmol m^2 s^{-1}) by the Photosynthesis System (Model CIRAS-1, PPsystem, UK). Photosynthetic photon flux density was maintained at 800 μmol photons m^{-2} s^{-1} on the cuvette surface by a portable light unit. All measurements were carried out between 9 a.m. and 12 noon, when ambient relative humidity was about 80%, the leaf temperature ranged from 29 to 32°C, and ambient CO_2 concentration was about 355 ppm.

Results

Influence of salt stress on growth of *Avicennia marina* seedlings

Root and shoot length, leaf area, biomass of different organs, and net photosynthesis rate all showed a similar response to salinity after short- and long-term exposure to salt stress (Figs. 1, 2, 4). The root: shoot ratio was also unaffected by salinity at day 50 up to 30‰ and then rose above this (Fig. 3). In other words, moderate levels of salinity promoted growth but high levels or total lack of it suppressed growth. The seedlings at 5–20‰ salinity grew fastest. In order to express this difference, we use the *C*-value, which can be calculated by the formula

$$C = \left(\frac{M - T}{T}\right) \times 100\%$$

where *M* is the average value of a parameter (root and shoot length, leaf area, biomass of different organs, or net photosynthesis rate) at low salinities (5–20‰) and *T* is the value at 0‰ salinity. The results, given in Table 1, show that the *C*-values on the 45th day were notably lower than those on the 100th day; root biomass, in particular, showed a value of 21% on the 45th day but of 71% on the 100th day.

The ratio of root biomass to shoot biomass (RB/SB) was also different at the two stages (Fig. 3): on the 45th day, the ratio showed a slight increase with salinity from 0 to 20‰ and a steep rise from 30 to 50‰ whereas on the 100th day, the slight increase was confined to only 0 and 5‰ salinity; as salinity increased further, the value in fact decreased slightly.

Influence of on leaf P_n, T_r, G_s, and C_i of *Avicennia marina* seedlings

The net photosynthesis rate (P_n), stomatal conductance (G_s), intercellular CO_2 concentration (C_i), and transpiration rate (T_r) were all different at the two stages (Fig. 4). On the 90th day, the net photosynthesis rate was markedly depressed at zero salinity in contrast to that at salinity levels from 5 to 20‰ ($P < 0.01$). The*C*-value was 29.8% (Table 1). On the 50th day, leaf P_n was neither depressed (*C*-value was –3.41%) nor differed among salinity levels from 0 to 15‰ ($P > 0.05$) whereas leaf C_i decreased as salinity increased from 0 to 15‰ and increased from 20 to 40‰. On the 100 day, leaf C_i increased with salinity from 0 to 10‰.

Ion contents (K^+, Na^+, Ca^{2+}, Mg^{2+}, Cl^-) of cotyledons initially and on the 45th day

On the 45th day, the contents of sodium, chloride, potassium, and ash at 0‰ salinity, and of potassium at other levels as well, had all declined (Fig. 5) whereas those of calcium and magnesium had increased with salinity levels up to 20‰ and decreased at 40 and 50‰ levels.

Discussion

Effect of salt stress on growth of *Avicennia marina* seedlings

Almost all the studies in which mangrove was grown in a laboratory under controlled conditions confirm that

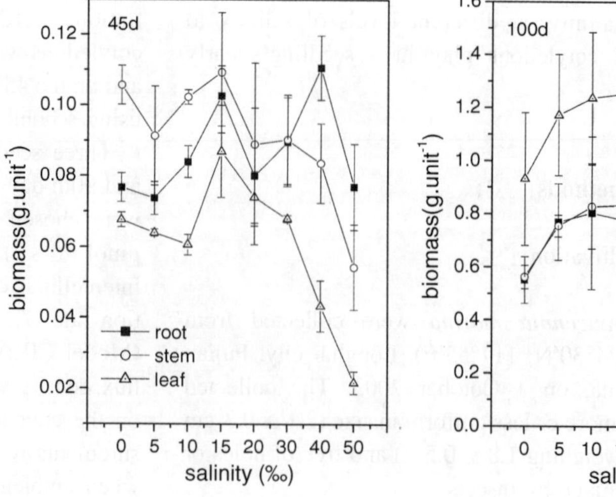

Fig. 1 Biomass of leaf, stem, and root of *Avicennia marina* seedlings on 45th and 100th day at different levels of salinity. Values are means ± standard deviation (n = 6)

Fig. 2 Root and stem length and leaf area of *Avicennia marina* seedlings on 45th and 100th day at different levels of salinity. Values are means ± standard deviation (n = 6)

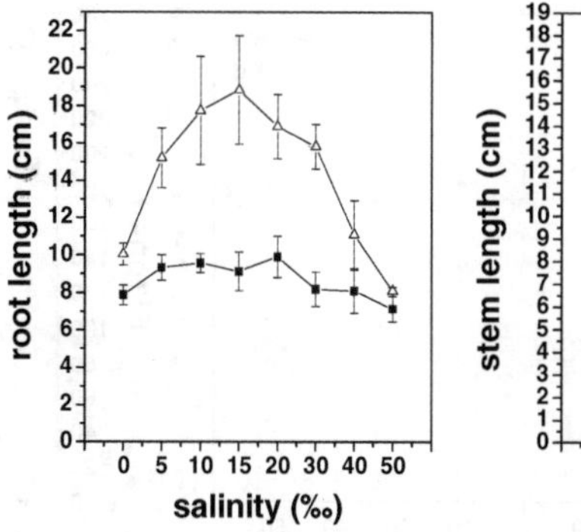

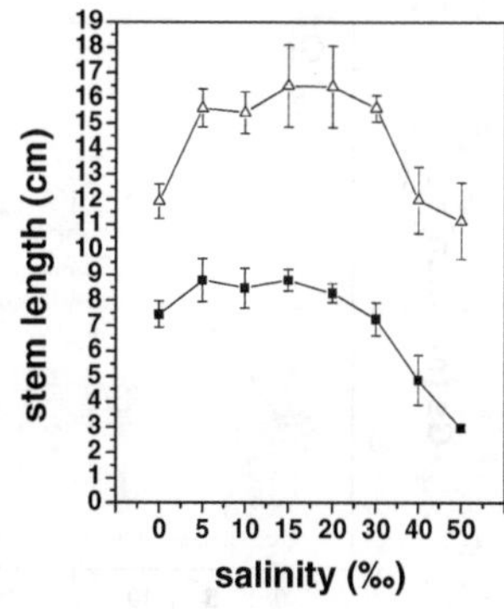

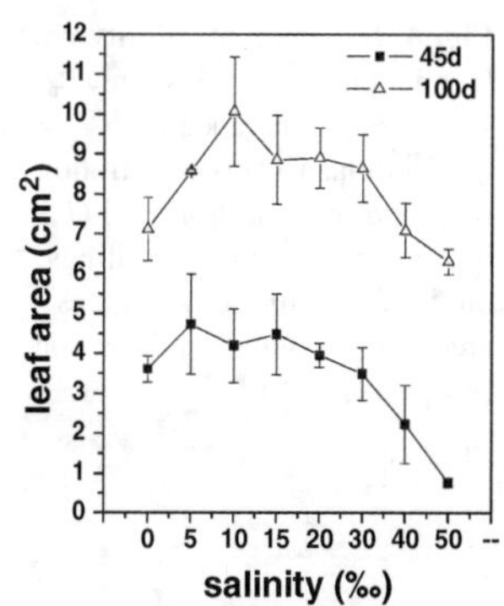

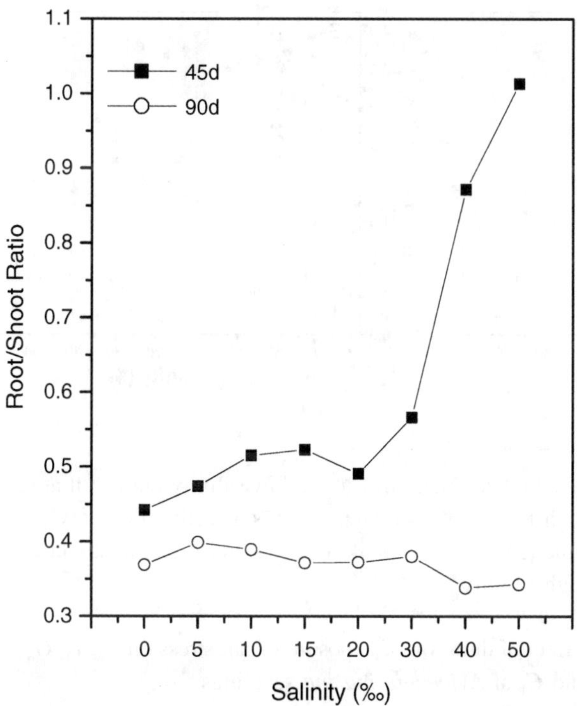

Fig. 3 Root:shoot ratio in *Avicennia marina* seedlings on 45th and 100th day at different levels of salinity

maximum growth does not occur in fresh water (Flower et al. 1977; Downton 1982; Clough 1984; Zheng and Lin 1992; Wang and Lin 1999), and our study was no exception. Moderate levels of salinity always facilitate seedling growth but extreme levels or total lack of it (0, 40, and 50‰) arrest growth. We consider 0‰ salinity as extreme salinity because, in a sense, it *is* a stress as far as the growth of *Avicennia marina* seedlings is concerned. With prolonged exposure to salt stress, growth was poor at the extreme levels (0, 40, and 50‰) and better growth at moderate levels (5–30‰). Downton (1982) found that the seedlings of *Avicennia marina* grew rapidly at 0‰ salinity in the first few weeks but the growth of leaf and bud was affected later and the accumulation of dry matter reduced. Clough (1984) found leaf development inhibited (leaves were smaller and fewer) in *Avicennia marina* and *Rhizophora stylosa* when grown in fresh water, compared with growth 25, 50, and 75% of seawater salinity. Similar results were also obtained by Pezeshki et al. (1990), who found that the total dry weight of *Laguncularia racemosa*, *Rhizophora mangle*, and *Avicennia germinans* increased under conditions of salt stress and salt stress combined with water logging. Although the net carbon assimilation rate (mmol m^{-2} s^{-1}) did not differ between the control and the treatments, the increased leaf area per plant under moderate salinity meant higher net carbon assimilation per plant. This partly explains why some halophytes grow better under moderate salinity than when salinity is lacking altogether.

The possible osmoregulatory function of cotyledons

The poor growth of seedlings exposed for a long time to the stress caused by 0‰ salinity may be explained by the depletion of ions and energy stored in cotyledons during the early growth stage. At 0‰ salinity, ash content and the concentrations of some ions (K^+, Na^+, and Cl^-) had declined by the 45th day (Fig. 5), perhaps because they had been deployed for the growth of other organs: with no supplementary source of ions—the salinity level was 0‰—the supply of ions may have been inadequate to enable maximum growth potential.

On the other hand, ash content and ion concentrations of the cotyledons increased at salinity levels from 5 to 20‰, which may indicate another function of the cotyledons: instead of being a *source* of ions when they are lacking in the culture medium (0‰ salinity), the cotyledons serve as a *store* of ions when, at higher levels of salinity (5–20‰), ions are surplus to requirements. This accumulation of ions in the cotyledons may also reduce the stress caused by high salinity. This high concentration of the ions corresponded with better growth of the seedlings at these levels of salinity. However, at even higher levels salinities, the ion concentration decreased with increasing salinity, and the concentration of Ca^{2+} and Mg^{2+} was lower than the

Fig. 4 Influence of salinity on leaf net photosynthesis rate (P_n), stomatal conductance (G_s), intercellular CO_2 concentration (C_i), and transpiration rate (T_r) in *Avicennia marina* seedlings on 50th and 90th day. Values are means ± standard deviation (n = 3)

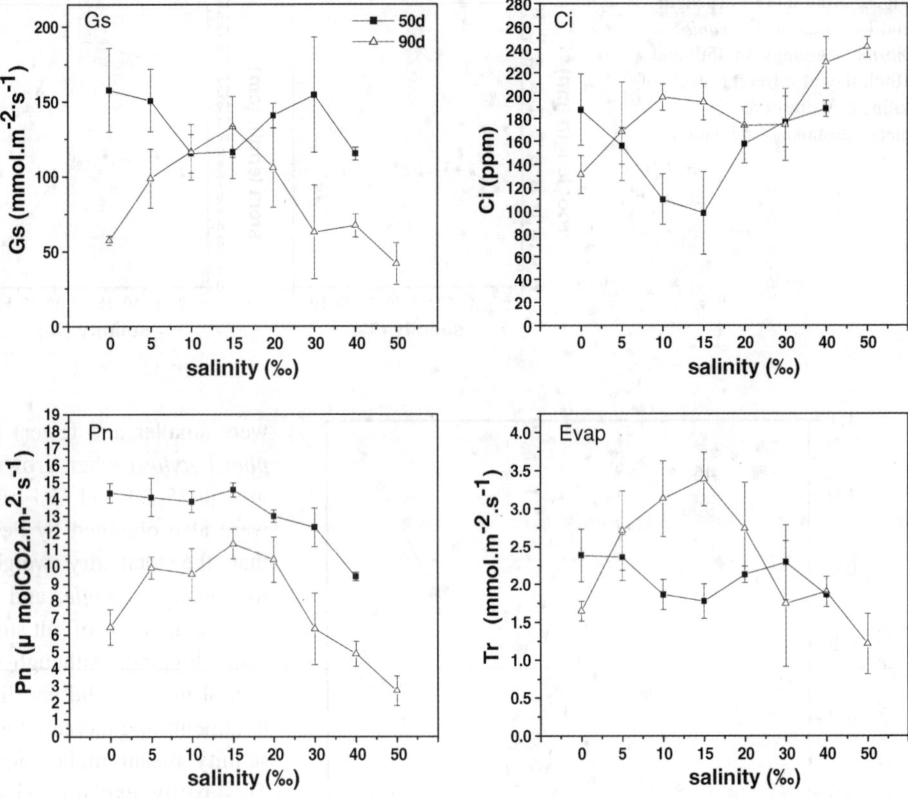

original levels, which suggests that high salinity had adversely affected the sink function. The accumulation of high concentrations of Ca^{2+} and Mg^{2+} at 0‰ salinity was observed also by Clough (1984), who found that root, stem, and leaf tissues of *Avicennia marina* seedlings accumulated markedly more Ca^{2+} and Mg^{2+} at 0‰ salinity than at other levels. Ca^{2+} is often considered an important ion in salt tolerance (Wilson et al. 2000; Bernstein 1975; Cramer et al. 1986). It can be seen from Fig. 5 that cotyledons

Table 1 Growth parameters and rate of photosynthesis in *Avicennia marina* seedlings on 45th and 100th day

Items	*C*-value (%)	
	45 day	100 day
LeafP_n	-3.41	29.80
Leaf area	20.34	27.81
Root length	20.64	71.02
Stem length	15.34	34.01
Root biomass	10.98	41.47
Stem biomass	-7.26	44.97
Leaf biomass	5.54	31.12

C-value can be calculated by the formula $C = \left(\frac{M-T}{T}\right) \times 100\%$ where M is the average value of different growth parameters at moderate levels of salinity (5–30‰) and T is that at a salinity of 0‰

accumulated more of Ca^{2+} at low salinity but lost it at very high levels (30, 40, and 50‰), suggesting that cotyledons function as a Ca^{2+} sink at low salinity and as a source at high salinity.

Effect of duration of exposure to salt stress on P_n, T_r, G_s and C_i of *Avicennia marina* seedlings

Two kinds of factors limit photosynthesis, those related to stomatal function and those unrelated (Farquhar and Sharkey 1982). The intercellular concentration of CO_2 (C_i) is useful in distinguishing between the two limiters (Xu 1997; Guan et al. 1995). If the rate of photosynthesis decreases with decrease in C_i, increased stomatal resistance is implicated; if photosynthetic rate decreases despite increase in C_i, the non-stomatal factors must be working (Xu 1997). Following the above argument, we conclude that after 50 days of moderate (5–20‰) salt stress, photosynthesis proceeds in an atmosphere saturated with CO_2 and is not limited by stomatal conductance. At higher salinity (40‰), the depressed photosynthetic rate is due to inhibition of the photosynthetic apparatus since stomatal conductance decreases only a little but C_i rises substantially. Similarly, after 90 days, it is stomatal conductance that limits photosynthesis at salinity levels up to 20‰, whereas higher salinity damages the photosynthetic apparatus itself,

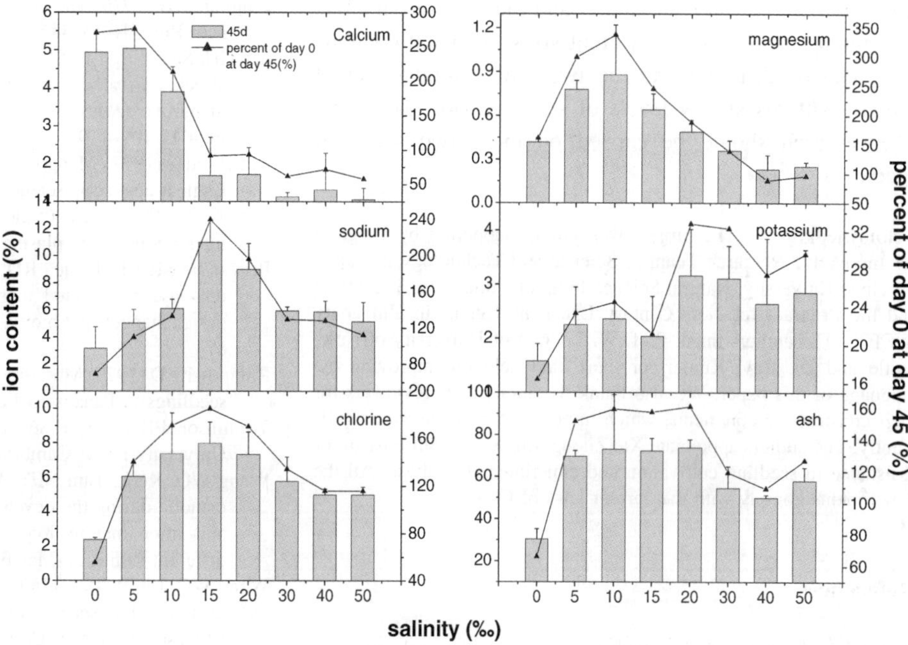

Fig. 5 Concentration of ions (K^+, Na^+, Ca^{2+}, Mg^{2+}, Cl^-) and ash contents of cotyledons on 45th day as percentage of the original values (on day 0) at different levels of salinity. Values are means ± standard deviation (n = 3)

since the rate of photosynthesis falls despite higher C_i, although stomatal conductance is also lower.

Figure 4 shows the U-shaped curve followed by C_i after short time (the first 45 days) duration to stress of varying intensities (different levels of salinity): the values decline at salinity levels from 0 to 15‰ but begin to rise at those from 20 to 40‰, which indicates that the main limiting factors are related to stomatal function at lower levels of salinity and to other functions at higher levels of salinity. After a longer exposure (45–90 days) to salt stress, the shape of the curve is different: C_i values increase with salinity levels from 0 to 10‰ and then decrease slightly up to 30‰, only to rise sharply up to 50‰ This means that following a longer exposure to salt stress, the main limiting factors are other than stomatal functioning at both low salinity levels (0–10‰) and at very high levels (40–50‰). At intermediate salinities (15–30‰), photosynthesis is constrained by stomatal factors: the slight dip in C_i implicates the stomata, which is also consistent with the high photosynthesis rate in seedlings at these levels of salinity. In other words, when salt stress is for a shorter period, seedlings cope first by reducing transpiration (T_r) by reducing stomatal conductance (G_s) and then by 'shipping out' the ions accumulated in toxic proportions by the evaporation. That is, seedlings can adjust their physiological processes to adapt to low levels of salinity in the short term. However, as salinity increases (or is lacking altogether) and the salt stress is prolonged, the seedlings lose this adaptive response and both the rate of photosynthesis and growth of the seedlings are adversely affected, especially at 0, 40, and 50‰ salinity. The limiting factor turns out to be non-stomatal at these salinities and the photosynthesis apparatus of the leaves is damaged.

Conclusion and future prospects

This study focused on the behavior of *Avicennia marina* seedlings in response to the stress occasioned by 0‰ salinity in the short term as well as in the long term. In contrast to the growth at 5–20‰ salinity, growth at 0‰ salinity was particularly affected in the long term. Analysis of ion concentration in the cotyledons showed that cotyledons may play an important role at low salinity levels (especially at 0‰ salinity) during early growth. This is a hypothesis that originates from partial experimentation and demands more direct evidence.

In addition, *Avicennia marina* is a cryptoviviparous species: its propagules are smaller than those of true viviparous species. A review of earlier work showed that the propagule mass (weight or size) in mangrove is closely correlated with their initial growth. For example, the large propagules of *Laguncularia racemosa* represent a stronger ability to endure both waterlogging and dehydration than the small propagules of *Avicennia germinans*. Further, the survival of seedlings also varies with propagule size within species, the percentage survival being greater in seedlings from larger propagules (Rabinowitz 1978). Lin et al. (1995) also found that total leaf area and stem length of the seedlings are positively correlated with initial weight of the mature propagules. Based on these findings, we believe that because the propagules of the truly viviparous *Kandelia candel* and the cryptoviviparous *Avicennia marina* are of different sizes, their ion contents may also be different and the two will exhibit different growth patterns at low levels of salinity. More comparative research should be carried on these two kinds of species.

Acknowledgment The project was jointly supported by Program for Innovative Research Team in Science and Technology in Fujian Province University, Natural Science Fund of China (No. 30200031) and the Program for New Century Excellent Talents in University (NCET). The authors thank Prof. W. J. Cram of University of Newcastle and Dr. Ravi Kumer for their kind help on improving the language of this paper. We also thank the anonymous reviewers for their constructive comments, which improved the manuscript significantly. The authors appreciate Xie Zhong, Lin Xi and Shi Fushan for assistance in seedling cultivation and experimental analysis. All the experiments comply with the current laws of China.

References

Ball MC, Farquhar GD (1984a) Photosynthetic and stomatal responses of two mangrove species, *Aegiceras corniculatum* and *Avicennia marina*, to long term salinity and humidity conditions. Plant Physiol 74:1–6

Ball MC, Farquhar GD (1984b) Photosynthetic and stomata responses of the grey mangrove, *Avicennia marina*, to transient salinity conditions. Plant Physiol 74:7–11

Bernstein L (1975) Effect of salinity and sodicity on plant growth. Ann Rev Phytopathol 13:295–312

Clough BF (1984) Growth and salt balance of the mangrove *Avicennia marina* (Forsk.) Vierh. and *Rhizophora stylosa* Griff. in relation to salinity. Aust J Plant Physiol 11:419–430

Cramer GR, Läuchli A, Epstein E (1986) Effect of NaCl and $CaCl_2$ on ion activities in complex nutrient solutions and root growth of cotton. Plant Physiol 81:792–797

Downton WJS (1982) Growth and osmotic relations of the mangrove *Avicennia marina* as influenced by salinity. Aust J Plant Physiol 9:519–528

Farnsworth EJ (2000) The ecology and physiology of viviparous and recalcitrant seeds. Annu Rev Ecol Syst 31:107–138

Farquhar GD, Sharkey TD (1982) Stomatal conductance and photosynthesis. Annu Rev Plant Physiol 33:317–345

Flowers TJ, Troke PF, Yeo AR (1977) The mechanism of salt tolerance in halophytes. Annu Rev Plant Physiol 28:89–121

Greenway H, Munns R (1980) Mechanisms of salt tolerance in non halophytes. Annu Rev Plant Physiol 31:149–190

Guan YX, Dai JY, Lin Y (1995) The photosynthetic stomatal and nonstomatal limitation of plant leaves under water stress (in Chinese). Plant Physiol Commun 31(4):293–297

Jennings DH (1976) The effects of sodium chloride on higher plants. Biol Rev 51:453–486

Lin GH, Sternberg LDaSL (1995) Variation in propagule mass and its effect on carbon assimilation and seedling growth of red mangrove (*Rhizophora mangle*) in Florida, USA [J]. J Trop Ecol 11:109–119

Lin P (eds) (1997) Mangrove ecosystem in China (in Chinese, with English abstract). Science Press, Beijing

Munns R (1986) Ion activities in solutions in relation to Na^+-Ca^{2+} interactions at the plasma lemma. J Exp Bot 37:320–330

Pezeshki SR, Delanue RD, Patrick WHJ (1990) Differential response of selected mangroves to soil flooding and salinity: gas exchanges and biomass partitioning. Can J For Res 20:869–874

Rabinowitz D (1978) Mortality and initial propagule size in mangrove seedlings in Panama. J Ecol 66:45–51

Tomlinson PB (eds) (1986) The botany of mangroves. Cambridge University Press, Cambridge

Wang WQ, Ke L, Tam NFY, Wong YS (2002) Changes in the main osmotia during the development of *Kandelia candel* hypocotyls and after mature hypocotyls transplanted in solutions with different salinities. Mar Biol 141:1029–1034

Wang WQ, Lin P (1999) Influence of substrate salinity on the growth of mangrove species of *Bruguiera gymnorrhiza* seedling (in Chinese). J Xiamen Univ (Nat Sci) 38(2):273–279

Wang WQ, Lin P (2000) Study on membrane lipid peroxidation of the leaves of *Kandelia candel* seedlings to long-term and short-term salinity (in Chinese). Acta Oceanol Sin 22(3):49–54

Wang WQ, Ye QH, Wang XM, Lin P (2001) Impact of substrate salinity on caloric value, energy accumulation and its distribution in various organs of *Bruguiera gymnorrhiza* seedlings (in Chinese). Chin J Appl Ecol 12(1):8–12

Wilson C, Lesch SM, Grieve CM (2000) Growth stage modulates salinity tolerance of New Zealand Spinach (*Tetragonia tetragonioides* Pall.) and red Orach (*Atriplex hortensis* L.). Annu Bot 85:501–509

Xu DQ (1997) Some problems in stomatal limitation analysis of photosynthesis (in Chinese). Plant Physiol Commun 33(4):241–244

Yeo AR, Flower TJ (1980) Salt tolerance in the halophyte Suaeda maritime L. Dum.: evaluation of the effect of salinity upon growth. J Exp Bot 31:1171–1183

Zhao KF, Wang ST (eds) (1990) Crops resistance physiology. Agriculture Press, Beijing

Zheng WJ, Lin P (1992) Effect of salinity on the growth and some eco-physiological characteristics of mangrove *Bruguiera sexangula* seedlings. J Appl Ecol 3(1):9–14

Zheng WJ, Wang WQ, Lin P (1999) Dynamic of element contents during the development of hypocotyls and leaves of certain mangrove species. J Exp Mar Biol Ecol 233:247–257

Comparison of flooding-tolerance in four mangrove species in a diurnal tidal zone in the Beibu Gulf*

Binyuan He[a,b] Tinghe Lai[b] Hangqing Fan[b] Wenqing Wang[a] Hailei Zheng[a]

(a. *School of Life Sciences, Xiamen University, Xiamen* 361005, *P. R. China*;

b. *Guangxi Mangrove Research Center, Beihai* 536007, *P. R. China*)

Abstract

The flood tolerance of four mangrove species, *Aegiceras corniculatum* (L.) Blanco (AC), *Avicennia marina* (Forsk.) Vierh. (Am), *Bruguiera gymnorrhiza* (L.) Savigny (Bg) and *Rhizophora stylosa* Griff. (Rs) was examined in a field trial conducted from August 2004 until August 2005 in a diurnal tidal zone in Yingluo Bay, Guangxi province, China. In a section of tidal flat, three replicate artificial platforms were constructed for seedling cultivation. Eight different tidal flat elevation (TFE) treatments were created on each platform. After one year of cultivation under the TFE treatments, the survival rate and growth parameters of seedlings were measured. Seedlings of *A. corniculatum* and *A. marina* seedlings survived all treatments. The survival rate of *B. gymnorrhiza* and *R. stylosa* seedlings, however, decreased sharply as the TFE fell; in any treatment, fewer *B. gymnorrhiza* seedlings survived than *R. stylosa* seedlings. Stem elongation in *A. corniculatum* and *A. marina* seedlings was significantly increased by lower TFEs. Lower TFE treatments also increased stem heights in *B. gymnorrhiza* and *R. stylosa* seedlings; however, growth was significantly higher as TFE increased. Leaf number, leaf conservation rate and leaf area per seedling changed relatively little among treatments in *A. corniculatum* and *A. marina* seedlings, while these three indexes in *B. gymnorrhiza* and *R. stylosa* seedlings all decreased dramatically with decreasing TFE. *A. marina* seedlings reached a higher neonatal biomass at lower TFE treatments, whereas *A. corniculatum* seedlings attained a higher biomass under moderate TFEs. In contrast, *B. gymnorrhiza* and *R. stylosa* seedlings accumulated more biomass in the higher TFEs habitats. Biomass partitioning among the components of both *A. corniculatum* and *A. marina* seedlings changed evenly; however, *A. corniculatum* accumulated more biomass in the leaf while *A. marina* accumulated more in the stem. The TFE treatments greatly influenced biomass partitioning in *B. gymnorrhiza* and *R. stylosa* seedlings, with a change from stem to leaf as the TFE increased. Generally speaking, the rank order of tolerance to flooding among these four mangrove species in the diurnal tidal zone was, from most to least tolerant, *A. marina* > *A. corniculatum* > *R. stylosa* > *B. gymnorrhiza*. This conclusion is consistent with the general pattern of natural mangrove species zonation along the Chinese coast in the Beibu Gulf. The critical tidal levels for afforestation with these four mangrove species are proposed.

Keywords: mangrove; flooding tolerance; diurnal tide zone; growth; survival rate; critical tidal level for afforestation

1. Introduction

On December 2004, a fierce tsunami in the Indian Ocean served as a reminder of the well-known but long-ignored importance of mangroves to coastal ecosystems, at the cost of tremendous loss of property and human life. Consequently, the recovery of degraded mangrove forests in Southeastern China and other coastal countries has received increased attention. Rapid economic development along the China coast has however led to increased development in the coastal zone. Intertidal flats have been narrowed, and many of the higher intertidal zones have disappeared, leaving only the lower intertidal zone; this remaining habitat zone is unsuitable for most mangrove species. As a result, mangrove afforestation along the China coast has become especially difficult (Fan and Li, 1997).

Mangrove rehabilitation is dependent on the presence of tidal flats that are suitable for afforestation. Under natural circumstances, zonation of mangrove species is related to their

* From Estuarine, Coastal and Shelf Science, 2007, 74: 254—262

adaptability, especially tolerance to flooding. Although mangroves are often credited with a special adaptability to seawater flooding, they can cope with only limited frequency and duration of flooding; otherwise, growth, anatomical development, gas exchange, materials accumulation, biomass partition, anti-stress enzymes and hormone levels are affected (Skelton and Allaway, 1996; Youssef and Saenger, 1996; Naidoo et al., 1997; Ye et al., 2001, 2003, 2004a,b; Kitaya et al., 2002; Chen et al., 2004, 2005, 2006).

At different localities, certain mangrove species may naturally occupy the intertidal flats at different elevations. Correspondingly, the critical tidal level (CTL) for afforestation of a species varies between localities. For *Kandelia candel* (L.) Druce, Liao et al. (1996) suggested that the CTL should not be more than 22 cm lower than the local mean sea level (MSL) in Shenzhen (22°32′ N, 114°03′ E), and not more than 30 cm lower than MSL in Dongzaigang (19°56′ N, 110°34′ E). Chen et al. (2006) proposed that the CTL for *K. candel* should be higher than 455 cm above the Yellow Sea Datum (YSD) in Xiamen (24°27′ N, 118°02′ E).

Afforestation guidelines based on particular environmental characteristics requires specific information on the anatomical, morphological and physiological responses of mangrove plants to flooding. The coastal zone along Beibu Gulf supports the largest area of mangrove forest of China, and has extensive areas of intertidal flats suitable for afforestation. The tidal regime in Beibu Gulf is diurnal, which is rarely found in other regions of China. Diurnal and semi-diurnal tide zones differ in hydrological, climatic and edaphic properties, and mangrove afforestation guidelines and techniques also differ between these zones. Research on the flooding-tolerance of mangrove plants in diurnal tidal habitats is however lacking for China.

Numerous greenhouse experiments have simulating flooding to quantify impacts on mangrove species. The results obtained in greenhouse experiments may however differ from those in the field trials (Chen et al., 2006). In a field trial that included plantings, Komiyama et al. (1996) found that other environmental factors overshadowed the effect of flooding. To address these potential confounding factors, we constructed experimental platforms to manipulate tidal elevation and flooding stress; seedlings were exposed to similar environmental conditions of light, salinity, substrate properties and nutrient supply.

In this experiment, we tested the flooding-tolerance responses of four mangrove species, i.e., *Aegiceras corniculatum* (L.) Blanco, *Avicennia marina* (Forsk.) Vierh., *Bruguiera gymnorrhiza* (L.) Savigny and *Rhizophora stylosa* Griff. These species are commonly used in afforestation practices in China. Our specific objectives were to: (1) examine their differential strategies in survival and growth under gradients of flooding, (2) determine the ranking order of tolerance among these species, and (3) propose their CTLs in a diurnal tidal zone.

2. Materials and methods

2.1. Descriptions of research area

Our experiment was conducted in Yingluo Bay (Fig. 1), a core zone within the Shankou Mangrove Reserve (center coordinates 109°43′ E, 21°28′ N) of Guangxi, China. Average annual air temperature is 23.4 °C, with average temperature in the coldest month, January, ranging between 14.2 °C and 14.5 °C; minimum recoded temperature in January is 2 °C. Annual rainfall ranges between 1500 mm and 1700 mm, and annual evaporation between 1000 mm and 1400 mm. The annual relative humidity reaches 80%. The tidal regime in Yingluo Bay is diurnal, a condition that prevails in Beibu Gulf. The average tidal amplitude is 2.52 m, with the maximum recorded as 6.25 m. The local MSL elevation is 359 cm YSD. Without riverine freshwater inputs, the average salinity of the seawater was 28.9 (Fan et al., 2005). Mangrove forest covers 80 ha in Yingluo Bay, with 9 mangrove species recorded. *Rhizophora stylosa* dominates most of the local mangrove forest. In this area, *Avicennia marina* and *Aegiceras corniculatum* are pioneer species, *Bruguiera gymnorrhiza* is a late-succession species, and *R. stylosa* and *Kandelia candel* are intermediate-stage species (Wen et al., 2002).

2.2. Method

2.2.1. Description of the experimental platforms

Experimental platforms were constructed on bare intertidal flat, nearly adjacent to the seaward mangrove forest and at an elevation of about 300 cm YSD. Several candidate sites were proposed based on the tide forecast table issued by the State Oceanic Administration of China. Then the precise elevations of the sites were measured (mean error within ± 5 cm) by using a Total Station instrument (Topcon GTS721, Japan). The selected site (109°45′52″ E, 21°29′27″ N) was leveled off at an elevation of 300 cm YSD, covering an area of 200 m^2.

Three replicate woody platforms were established, with a distance of about 1 m between neighboring platforms. Each platform was 6.4 m in width and 8 m in length. Platform slopes faced seaward and paralleled to the direction of the spring tide. The upper planes of the seedlings' substrates were regarded as the elevations of the treatments. Eight TFE treatments were created on each platform in the form of eight 80-cm wide steps constructed using timbers, boards and bricks. The treatments were 320 cm, 330 cm, 340 cm, 350 cm, 360 cm, 370 cm, 380 cm and 390 cm YSD. As 360 cm YSD is close enough to 0 cm to the local MSL in Yingluo Bay, the relative elevations used as treatments then could be transformed as follows: −40 cm, −30 cm, −20 cm, −10 cm, 0 cm, 10 cm, 20 cm, and 30 cm, respectively.

2.2.2. Propagules collection and cultivation

Mature propagules of *Rhizophora stylosa*, *Bruguiera gymnorrhiza*, *Avicennia marina* and *Aegiceras corniculatum*, of similar size and fresh weight, were collected from parent trees in the local mangrove forest in August 2004. The lengths of selected propagules ranged between 28.9 cm and 31.1 cm in *R. stylosa* and between 18.8 cm and 21.0 cm in *B. gymnorrhiza*. The fresh weights ranged between 26.2 g and 28.9 g, 27.9 g and 29.0 g, 2.4 g and 2.6 g, and 1.1 g and 1.2 g in *R. stylosa*, *B. gymnorrhiza*, *A. marina*, and *A. corniculatum*, respectively.

Plastic nursery bags, 15 cm in diameter and 20 cm high, were packed with soil from the local intertidal flat and

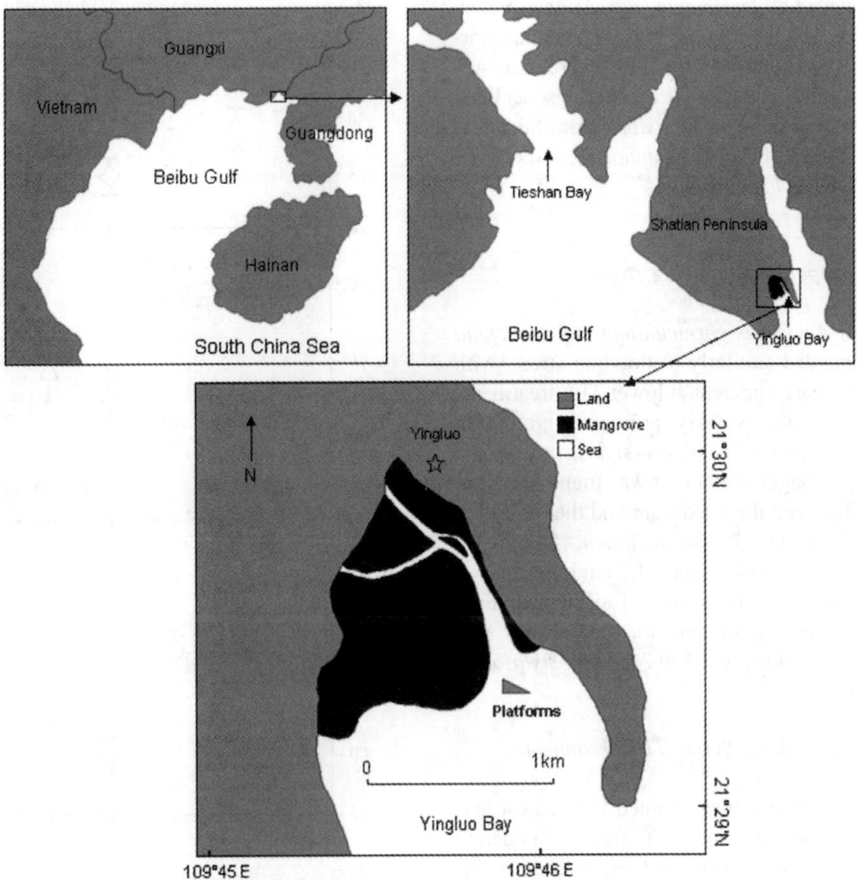

Fig. 1. Map depicting the location of the platforms for flooding tolerance experiment of four mangrove species in the diurnal tidal zone of China.

attached to the platforms before cultivation, making the upper planes of substrates equal to the designated TFEs. Propagules of *Aegiceras corniculatum* and *Avicennia marina* were submerged in seawater until the seed capsules were breached (about 2 days), and were then planted in the prepared nursery bags. Propagules of *Bruguiera gymnorrhiza* and *Rhizophora stylosa* were planted directly, with one third of their lengths in the soil. One propagule was planted in each nursery bag. Fifty propagules of each species were allocated to one treatment on each platform, so a total of 1200 propagules of each species were included in the experiment.

The soil used had the following characteristics: pH value of 4.5; organic matter, 1.6%; total nitrogen, 0.014%; total phosphorus, 0.021%; available phosphorus, 9.34×10^{-6}; total salt, 12.5‰; and clay content, 5.3%.

He (2002) reported that in Yingluo Bay, April, October and November were the most common months for attachment of fouling fauna to mangrove plants. In view of this, the seedlings were sprayed every 3 days with diluted low-toxic pesticide during these three months, and every 14 days in other months.

2.2.3. *Measurement of growth characteristics indexes and data analysis*

In August 2005, the survival rate in each treatment was counted in situ. Twenty seedlings in each treatment were randomly harvested and separated into leaves, stems, roots and remaining propagules (*Avicennia marina* seedlings were separated into leaves, stems and roots). Growth parameters, including stem height, leaf scar, leaf number and leaf area per seedling, were measured. Leaf conservation rate was calculated as: Leaf conservation rate (%) = Leaf numbers/ (scars + leaf numbers) × 100%

Seedling components were oven-dried to constant weight at 80 °C, and then weighed.

Statistical analysis was done in SPSS software without data transformation. Mean and standard deviation (SD) values of 20 replicates in each treatment were calculated. As most of the measured growth parameters among four mangrove species differed greatly, the significance test and correlation regression were conducted within eight treatments of the same species only. The flooding effects on each species were analyzed by one-way ANOVA using a *t*-test.

3. Results

3.1. *Survival status of 1-year-old mangrove seedlings under different TFE treatments*

All propagules germinated and formed a first pair of leaves within one month. Seedlings of *Aegiceras corniculatum* and

Avicennia marina survived all treatments (Fig. 2). In contrast, *Bruguiera gymnorrhiza* and *Rhizophora stylosa* seedling mortality occurred in all treatments. As the TFE decreased, survival rates in 1-year-old *B. gymnorrhiza* seedlings declined from 73.6% to 35.0%, and those in *R. stylosa* from 88.9% to 40.0%. Within a treatment, fewer *B. gymnorrhiza* seedlings survived compared to *R. stylosa* seedlings.

3.2. Stem elongation response to flooding stress

Stem elongation of *Aegiceras corniculatum* and *Avicennia marina* seedlings responded similarly to flooding stress, with a significant growth increase apparent at lower TFE treatments (Fig. 3). A lower TFE also weakly promoted growth in *Bruguiera gymnorrhiza* and *Rhizophora stylosa* seedlings; however, significantly stronger promotion was found in higher habitats. Differences between the maximum and the minimum elongation were 1.3 cm in *A. corniculatum*, 7.1 cm in *A. marina*, 9.7 cm in *B. gymnorrhiza* and 14.7 cm in *R. stylosa*, respectively, and the ratios of the maximum to the minimum elongation were 132.6% in *A. corniculatum*, 123.8% in *A. marina*, 135.8% in *B. gymnorrhiza* and 139.2% in *R. stylosa*.

3.3. Foliar development under different TFE treatments

Aegiceras corniculatum seedlings attained the highest leaf number, leaf conservation rate and leaf area in moderate TFE habitats (Fig. 4). *Aegiceras corniculatum* seedlings maintained a higher leaf conservation rate, with a range of 56.5% to 70.8% among treatments.

The leaf number in *Avicennia marina* changed irregularly across gradients of flooding stress. Leaf conservation rate in *A. marina* showed little response, as it fluctuated only between 32.8% and 39.0%; and leaf areas in *A. marina* were greater under lower TFE treatments (Fig. 4).

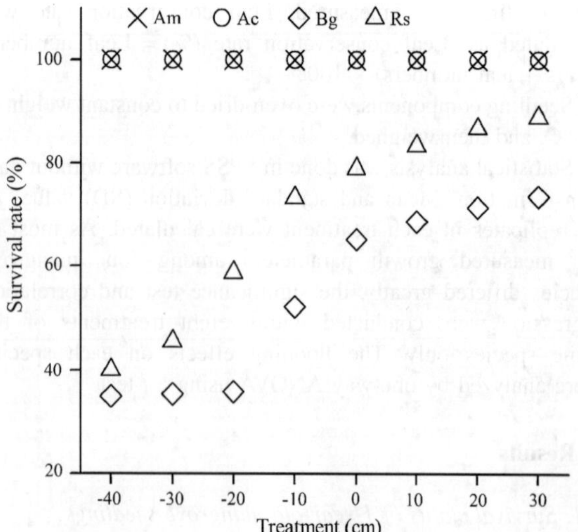

Fig. 2. Difference of survival rates of 1-year-old seedlings under various treatments. Ac, *Aegiceras corniculatum*; Am, *Avicennia marina*; Bg, *Bruguiera gymnorrhiza*; Rs, *Rhizophora stylosa*.

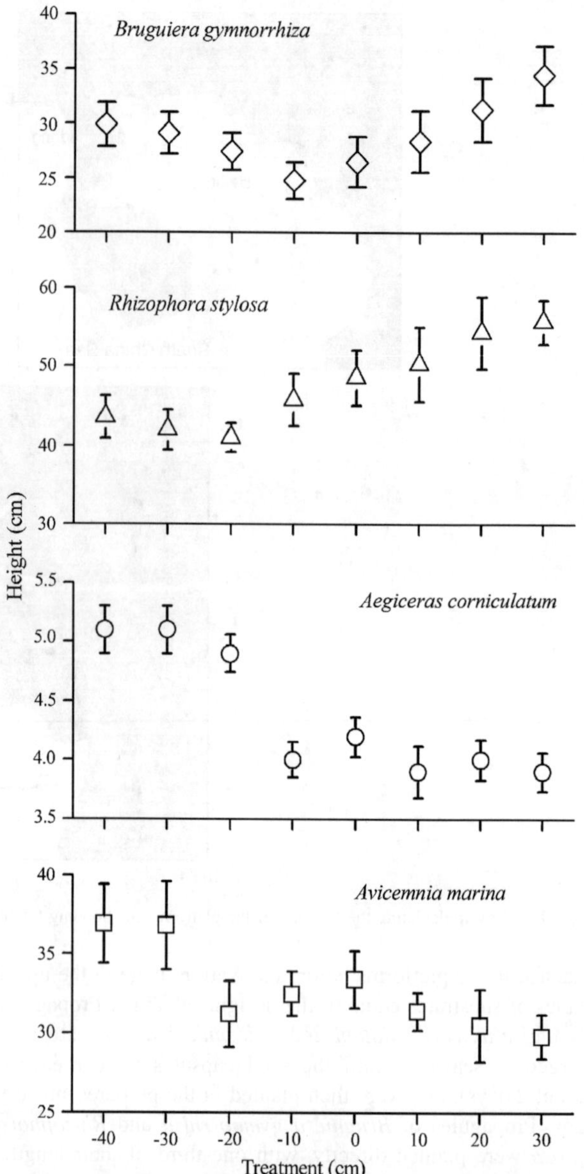

Fig. 3. Stem heights (±standard error) of 1-year-old seedlings under different treatments.

In contrast, leaf number, leaf conservation rate and leaf area in both *Bruguiera gymnorrhiza* and *Rhizophora stylosa* increased dramatically with increasing TFE (Fig. 4). The maximum foliar indexes in both species occurred under the highest TFE treatment in this experiment. And under most circumstances, significant differences existed ($p < 0.05$) between the neighboring treatments, especially in leaf area. Intensive correlations prevailed among these three indexes ($p < 0.001$). A lower conservation rate in leaves, namely earlier defoliation, led to a reduction in leaf number which, in turn, brought about a narrower leaf area which reduced the growing time. As a result, these three indexes responded congruously to flooding stress.

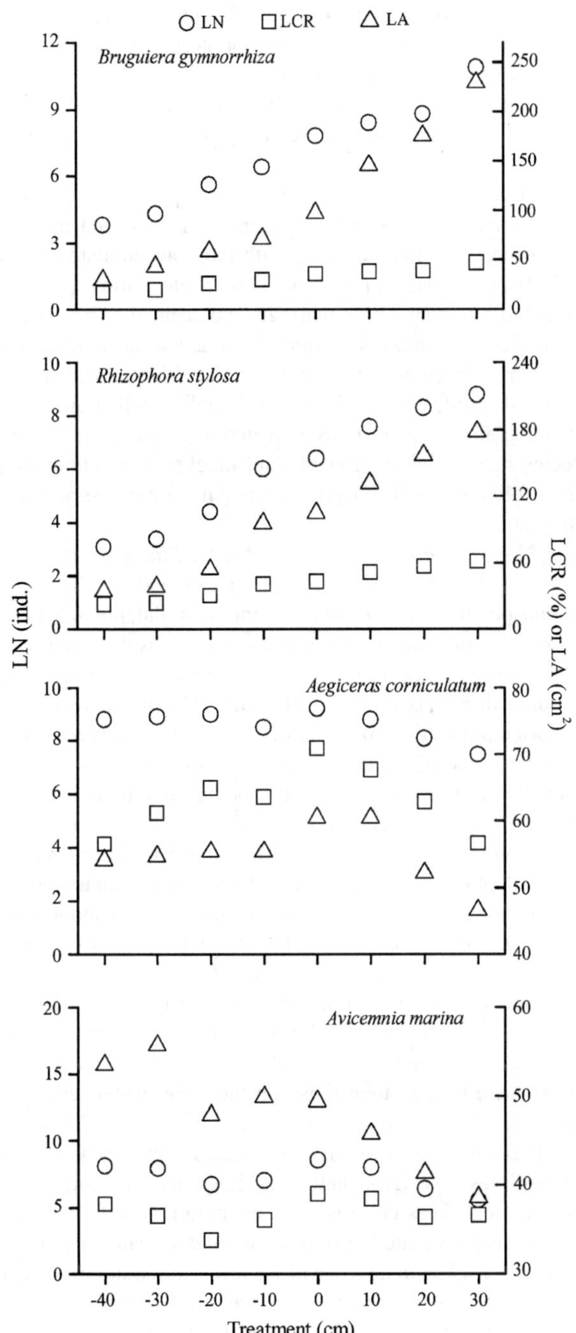

Fig. 4. Leaf numbers (LN), leaf area (LA) per seedling and leaf conservation rates (LCR) of 1-year-old seedlings under different treatments.

In lower TFE habitats, these three foliar indexes in *Aegiceras corniculatum* and *Avicennia marina* seedlings were higher than those in *Bruguiera gymnorrhiza* and *Rhizophora stylosa* seedlings; but in higher habitats, there was the opposite tendency. These results indicate that, concerning leaf growth and conservation, lower TFE habitats favored *A. corniculatum* and *A. marina* seedlings while higher ones favored *B. gymnorrhiza* and *R. stylosa* seedlings.

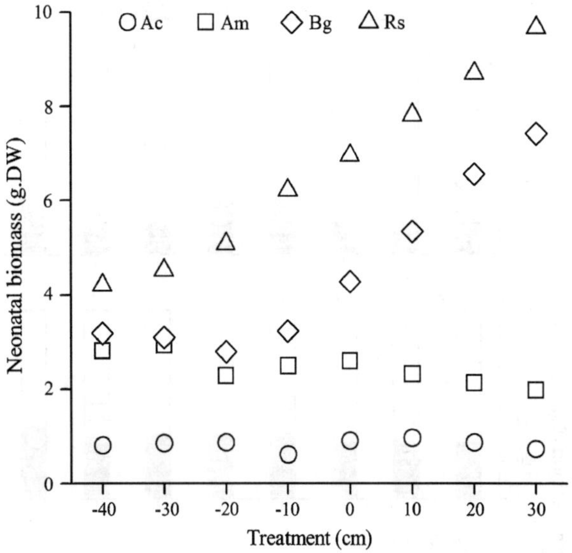

Fig. 5. Neonatal biomasses of 1-year-old seedlings under different treatments. Ac, *Aegiceras corniculatum*; Am, *Avicennia marina*; Bg, *Bruguiera gymnorrhiza*; Rs, *Rhizophora stylosa*.

3.4. Biomass partition among organs at different TFEs

The neonatal biomass in *Avicennia marina* seedling accumulated more at lower TFEs than at higher ones (Fig. 5). *Aegiceras corniculatum* seedlings acquired a higher biomass in moderate TFE habitats. Ratios of the maximum to the minimum biomass accumulation were 147.5% in *A. marina* and 158.1% in *A. corniculatum*, respectively.

Biomass accumulation was weakly promoted in *Bruguiera gymnorrhiza* seedlings by lower TFEs, but on the whole, *B. gymnorrhiza* seedlings attained more biomass at higher TFEs (Fig. 5). And biomass in *Rhizophora stylosa* seedlings exhibited a linear increase with the increasing TFE. Great differences existed among the results for different treatments ($p < 0.05$ or less), and the ratios of the maximum to minimum biomass accumulation were 230.2% in *R. stylosa* and 266.6% in *B. gymnorrhiza*, respectively.

In *Aegiceras corniculatum* seedlings, the biomass partition in the eight treatments descended in order from leaf to root to stem (Fig. 6). Leaf dominated the biomass of *A. corniculatum* seedlings, although its proportion decreased slowly from 53.4% to 42.3% as the TFE increased. In *Avicennia marina* seedlings, however, the stem occupied the largest proportion, ranging from 54.6% to 59.3% among treatments; the root biomass was next, and the leaf biomass ranked third.

The proportions of partition among neonatal organs in *Rhizophora stylosa* seedling changed greatly (Fig. 6). As the TFE went up, root proportion increased smoothly; the stem proportion decreased dramatically from 69.6% to 39.8%; while the leaf proportion soared from 9.5% to 35.4%. It seemed to be a trend that the allocation of neonatal biomass was moved from stem to leaf.

Partition of neonatal biomass in *Bruguiera gymnorrhiza* varied much more strongly. As the TFE increased, the root

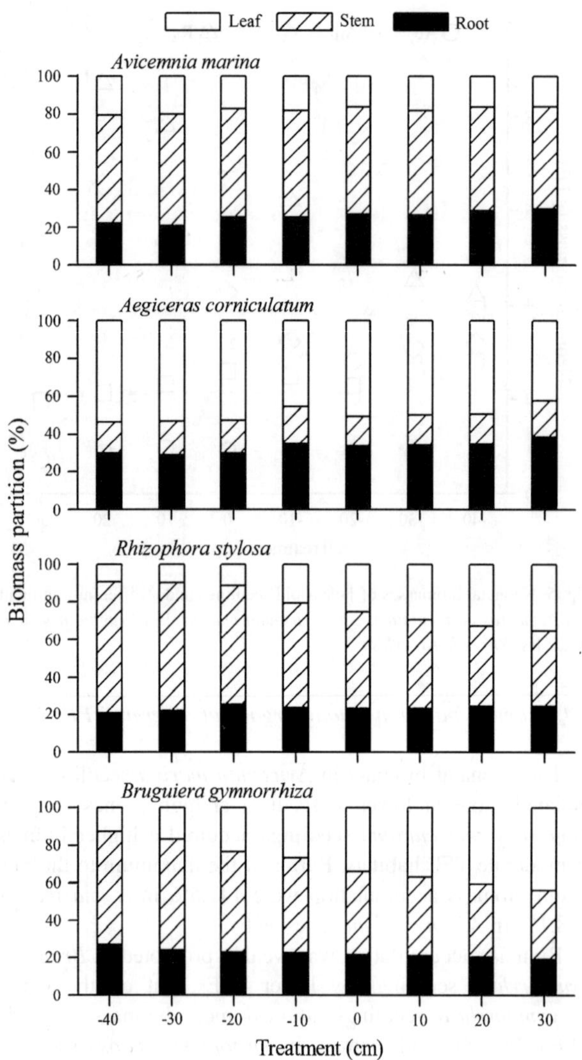

Fig. 6. Variation of the proportions (%) of stem, root and leaf in neonatal biomass at different treatments.

proportion decreased from 27.0% to 19.4%, stem proportion dropped steeply from 60.1% to 36.6%; and the leaf proportion rose rapidly from 12.9% to 44.0%. The allocation of biomass among neonatal organs had already moved from stem to leaf.

In *Bruguiera gymnorrhiza* and *Rhizophora stylosa* seedlings, it was obvious that flooding stressed the leaf more than other organs. Longer duration of submergence under the lower TFE treatments led to earlier defoliation and less leaf area available for photosynthesis which, in turn, led to inadequate production and lower biomass accumulation.

4. Discussion

4.1. Growth responses of mangrove species to flooding

Examining the effect of flooding on the survival and growth of mangrove seedlings is important for successful mangrove afforestation in the intertidal zone. This information will hopefully be helpful for selecting appropriate species and sites in mangrove rehabilitation projects.

In our experiment, the growth responses to gradients of flooding differed among these four species. They exhibited different strategies to adapt to the stressing habitats. In *Aegiceras corniculatum*, stem elongation was promoted at lower TFEs; leaves grew better in moderate habitats; and biomass proportions changed narrowly, with more accumulated in the leaf. *Avicennia marina* seedling's stem elongation was also promoted by lower TFEs, while its leaf seemed to be insensitive to flooding stress, and more biomass was accumulated in the stem. *Bruguiera gymnorrhiza* and *Rhizophora stylosa* exhibited similar trends in survival and growth traits, with intense gradients of responses to flooding. Seedlings of both species had a higher survival rate, quicker stem elongation, larger foliar area and conservation, and more biomass accumulation at higher TFEs.

In North American mangrove forests, *Rhizophora mangle* L. is a widespread and dominant species and usually forms a monospecific forest on the lower intertidal flat, playing a crucial role in constructing the mangrove forest as *Aegiceras corniculatum* and *Avicennia marina* do in Guangxi coast of China. Flooding durations between 50% and 75% of the year often supported populations of *R. mangle*, while *Laguncularia racemosa* (L.) Gaertn. f. and *Avicennia germinans* (L.) L. were generally restricted to areas with flooding durations less than 50% of the year (Krauss et al., 2006).

As a pioneer species, it seems likely that *Rhizophora mangle* would adapt well to the intertidal habitat in North American, repeating the performance of *Aegiceras corniculatum* and *Avicennia marina* in the Beibu Gulf. Ellison and Farnsworth (1993) reported that the surviving 1-year-old *R. mangle* seedlings on the lowest low water tidal flat grew quicker in terms of height, diameter, leaf production and biomass than those on the mean water flat, while survival of seedlings on the lowest low water flat was a little less than those on the mean water level flat, and no seedling survived on the highest high water flat. However, Pezeshki et al. (1989) recorded that leaf growth of *R. mangle* seedlings was significantly inhibited under flooding stress, and total leaf areas decreased. Simulating a semidiurnal high tide that approximated current conditions (mean water treatment, MW), a 16 cm increase in sea level (low water treatment, LW) and a 16 cm decrease in sea level (high water treatment), Ellison and Farnsworth (1997) found out that although *R. mangle* seedlings under LW grew faster during the first year, after 2.5 years, the saplings under MW treatment grew better, with a 10–20% increase in all measured growth parameters compared with those under LW. In longer experiments, *R. mangle* favored the moderate intertidal flat, indicating although it is a pioneer species of mangrove community; its occupation of the lower intertidal flat seems to be rather an ecological adaptation than a physiological one.

In contrast, *Rhizophora mangle* was insensitive to hydroperiod changes in some greenhouse experiments. Cardona-Olarte et al. (2006) documented that, except for the leaf area ratio at low salinity, 1-year-old *R. mangle* seedlings exhibited few changes in plant traits with different hydroperiod treatments (permanent and tidal flooded). Krauss et al. (2006) reported that heights of 2a *R. mangle* seedlings were not significantly different among hydroperiod treatments (unflooded, intermittently and continuously flooded).

Frequently other mangrove species such as *Laguncularia racemosa* and *Avicennia germinans* replaced *Rhizophora mangle* in landward forest belt (Ellison and Farnsworth, 1993). Under three treatments, i.e. flooding according to tidal cycle (FT), no flooding (NF), and continuous flooding (F), stem growth of established *L. racemosa* seedlings differed significantly among treatments, following the gradient: FT > NF > F; while the height increment of *A. germinans* seedlings did not differ significantly among treatments (Delgado et al., 2001). Similarly, Krauss et al. (2006) reported that flooding treatments promoted higher increments of height, diameter and total biomass compared to the unflooded treatment in *L. racemosa*, while *A. germinans* was relatively insensitive. In a field trial (Ellison and Farnsworth, 1993), 47% of *A. germinans* seedlings survived 1 year at MW, while no individual survived on tidal flats at either highest high water or lowest low water. It seems that *A. germinans* is less adaptable to flooding than *L. racemosa*. As to leaf growth, both *A. germinans* and *L. racemosa* seedlings were all significantly inhibited under flooding stress, and total leaf areas declined.

In Chinese mangrove forests, the role of *Kandelia candel* is complicated; it forms either a monospecific stand or being pushed out to the fringes. However, *K. candel* is not the dominant species in tropical mangroves, even along the Guangxi coast. The flooding-tolerance of *K. candel* has been fully studied in China, but rarely in other countries. Liao et al. (1996) planted *K. candel* seedlings on three plots below the local MSL in Shenzhen, and found the 3a survival rate and total biomass per seedling decreased rapidly as the TFE declined, with maximal survival (62.2%) recorded at the highest treatment (near the local MSL). In both the mesocosm experiment and field trial conducted in Hong Kong (Ye et al., 2004a), simulating a water level rise of 30 cm stimulated the *K. candel* propagule germination and early increment of stem height within 4 months. In a greenhouse experiment simulating the semidiurnal tide of Xiamen (Chen et al., 2004), *K. candel* seedlings under the treatment of 2 h submergence per tidal cycle attained the highest biomass, had the lowest mass loss of propagules and enlarged the photosynthetic area by significantly increasing the area per leaf after 70 days of cultivation. Field data from intertidal flats of Xiamen (Chen et al., 2006) showed that at 162 cm above the zero tidal level of Huang Ocean, *K. candel* seedlings had the best growth and the highest photosynthetic assimilation, with a survival rate of 90%. Given the larger range of tidal flooding, it seemed that the moderate and/or less degree of flooding would be more favorable for the *K. candel* seedlings, either in greenhouse experiments or on intertidal flats. In a tank experiment simulating a semidiurnal tide, *Avicennia marina* responded similarly to *K. candel* (Hovenden et al., 1995): the maximal biomass per seedling was obtained at intermediate-length flooding periods (especially 3.75–4 h per tide); those seedlings that had been submerged the longest (8.5 h per tide) had the least root tissue.

Generally, genera *Bruguiera* and *Xylocarpus* are located at the landward stands, even those which are occasionally inundated by exceptional or equinoctial tides. The greater area and thickness of the full-grown leaf in *Bruguiera gymnorrhiza* plants responding to periodic submergence may help in overcoming the limitations imposed on photosynthesis (Misra et al., 1984). But the relative growth rate of *B. gymnorrhiza* seedlings decreased significantly with the duration of waterlogging (Ye et al., 2003). Kitaya et al. (2002) documented that no 1-year-old *Bruguiera cylindrica* (L.) Blume plants survived on the flats with flooding depths of higher than 190 cm at the highest high tide. Allen et al. (2003) simulated a semidiurnal tide with 2 h of flood cycle and 4 cm of flood height above the soil surface, and found that the *Xylocarpus granatum* J.Koenig seedlings under this simulated flood treatment showed few signs of stress in stem height and diameter compared to other treatments with no change in water level. It seemed that the flood treatment was insufficient to induce any growth variation for *X. granatum* seedlings. In a field trial relating to a flooding amplitude of 90 cm to 270 cm (at the highest high tide), only 20% of *X. granatum* plants survived even on the highest intertidal flat, and no plant survived on the flats with flooding depths of higher than 170 cm, while the remaining plants exhibited lower stem elongation in the second half-year than in the first half-year (Kitaya et al., 2002).

4.2. Comparison of tolerance to flooding among species

Comparison of tolerance to flooding is based mainly on the survival and growth of seedlings. Flooding tolerance for these species studied decreased in the following order: *Avicennia marina* > *Aegiceras corniculatum* > *Rhizophora stylosa* > *Bruguiera gymnorrhiza*. As a general tendency, *A. marina* was adapted to broader habitats and grew well even in the low intertidal zone; *A. corniculatum* grew better in the zone a few cm higher than *A. marina*; a higher TFE was the favored habitat for *R. stylosa*, although a few *R. stylosa* seedlings survived at a lower TFE; and *B. gymnorrhiza* was adapted to the high intertidal zone only. These conclusions accorded with the suggested rule in some studies of mangrove species zonation in Guangxi, China (Lin, 1999; Fan, 2002; Wen et al., 2002). In Phang Nga province (8°20′ N, 98°25′ E), southern Thailand, the flooding tolerance of seven mangrove species was ranked in descending order as follows: *Rhizophora mucronata* Lam., *Sonneratia alba* J. Sm., *Rhizophora apiculata* Blume, *Avicennia officinalis* L., *Ceriops tagal* (Perr.) C.B. Rob., *Bruguiera cylindrica* and *Xylocarpus granatum* (Kitaya et al., 2002). In a Hong Kong study, *B. gymnorrhiza* was found to be less flood-tolerant than *Kandelia candel* (Ye et al., 2004a).

4.3. The local MSL and the distribution of mangrove plants

The critical tidal level for afforestation was regarded as one of the most important indexes in the guideline for selecting plantable intertidal flats (Lin, 2003: Zheng et al., 2003; Liao et al., 2005). Zhang et al. (1997, 2001) maintained that the local MSL determines the lowest elevation of the natural distribution of mangrove forest and so it could be regarded as the critical tidal level for mangrove afforestation. In contrast, Mo (2002) found that the seaward fringe of mangrove forest overlapped approximately with the local MSL, but it could be higher or lower in particular areas. Fan (2000) found that, mangrove forest developed into big patch of forest even when it occupied flats below the local MSLs along some well-shaded coasts of bays and lagoons.

According to our surveys, Yingluo Bay supports a mangrove forest of 80 ha, of which about 50% is located below the local MSL. In Beilun Estuary Mangrove Reserve of Guangxi, China, three localities, i.e. Jiaodong (108°11′14″ E, 21°35′52″ N), Shijiao (108°14′0.97″ E, 21°36′53″ N) and Malanji (108°14′49″ E, 21°36′27″ N), have 42.5, 28.0 and 65.0 ha of mangrove forest growing below the local MSL respectively, occupying 25.0%, 28.7% and 60.9% of their individual patch areas in each case. In this experiment, all the *Aegiceras corniculatum* and *Avicennia marina* seedlings survived at the local MSL, and more than half of *Bruguiera gymnorrhiza* and *Rhizophora stylosa* seedlings survived as well. Our results are consistent with the conclusion that mangrove forests can be located below the local MSL.

Furthermore, it is reasonable to deduce that, after cautious species selection and careful fostering, planting mangrove below the local MSL may be practicable. If some additional tidal-flat-elevating methods were applied (Riley and Kent, 1999; Kent and Lin, 1999), the possibility of successful afforestation would be enhanced greatly. Provided all these conditions applied, the erosion damage to the coastlines and dykes which lack protection from mangrove forest should consequently be decreased.

4.4. Proposal of the critical tidal levels for these four mangrove species

In China, mangrove afforestation has been practiced for more than 100 years, and a great deal of nationwide practical experience and relevant research has been summarized and reviewed (Lin, 2003: Zheng et al., 2003; Liao et al., 2005). However, the successful afforestation activities are still rarely reported. One of the most important reasons is that neither nationwide nor regional technological guidelines for mangrove afforestation have yet been developed.

In order to afforest successfully, not only the ranking order of flooding-tolerance is needed, but also the CTL for particular species. By correlating the growth parameters with the precise tidal flat elevation, some CTLs for *Kandelia candel* afforestation were further proposed, with fairly large differences among localities (Liao et al., 1996; Chen et al., 2006).

Chen et al. (2006) suggested the definition of the CTL for *Kandelia candel* in Xiamen should be where the seedlings attain a survival rate of above 65% and heights of above 35 cm within 6 to 12 months from colonization. Based on our experimental results, the CTL for planting these four mangrove species in Beibu Gulf are discussed. The proposed CTLs (PCTLs) for these species in diurnal tide zones are listed in Table 1. The survival rate was regarded as the most important index; and the height, leaf area and biomass were also taken into account. The PCTL for *Avicennia marina* was located at 29 cm under local MSL, while those for *Aegiceras corniculatum* and *Rhizophora stylosa* were near the local MSL, and that for *Bruguiera gymnorrhiza* was 21 cm above local MSL. We hope that our proposals will afford some helpful information for further mangrove afforestation planning in diurnal tide zones.

Table 1
Proposed critical tidal levels (PCTL) for *Avicennia marina* (Am), *Aegiceras corniculatum* (Ac), *Rhizophora stylosa* (Rs) and *Bruguiera gymnorrhiza* (Bg) in diurnal tide zone

Species	Am	Ac	Rs	Bg
PCTL in this paper (cm above Yellow Sea Datum)	330	360	360	380
PCTL contrast to local MSL (cm)	−29	+1	+1	+21
Proposed survival rates at 1a (%)	95	95	75	70
Actual survival rates under PCTL (%)	100	100	78.9	71.5
Ratio of height under PCTL to maximum (%)	100	81.8	87.2	90.6
Ratio of leaf area under PCTL to maximum (%)	100	99.9	80.3	76.6
Ratio of biomass under PCTL to maximum (%)	100	91.9	82.4	88.5

5. Conclusions

In our experiment, these four mangrove species exhibited different growth strategies to respond to flooding stress. Being the pioneer species in mangrove communities, *Avicennia marina* and *Aegiceras corniculatum* were adapted to broader habitats than *Rhizophora stylosa* and *Bruguiera gymnorrhiza*, although stem elongation and biomass accumulation in *A. marina* and *A. corniculatum* were inferior to *R. stylosa* and *B. gymnorrhiza*. In *R. stylosa* and *B. gymnorrhiza* seedlings, flooding stress was greater for leaves than other organs. In general, the ranking order of tolerances was determined as *A. marina* > *A. corniculatum* > *R. stylosa* > *B. gymnorrhiza*. These results can illuminate the natural rule of mangrove species zonation in Guangxi, China in terms of growth adaptability. And it is more important that, supported by this information, we can propose the critical tidal levels for planting these four mangrove species in diurnal tide zones.

Acknowledgements

This research work was supported by grants from Guangxi Science Foundation of P.R. China (Grants No. 0447007 and No. 0640014), and also by the Program for New Century Excellent Talents in University of P.R. China. The authors thank the staff of Chinese National Shankou Mangrove Reserve for long-term assistance in field work.

References

Allen, J.A., Krauss, K.W., Hauff, R.D., 2003. Factors limiting the intertidal distribution of the mangrove species *Xylocarpus granatum*. Oecologia 135, 110–121.

Cardona-Olarte, P., Twilley, R.R., Krauss, K.W., Rivera-Monroy, V., 2006. Responses of neotropical mangrove seedlings grown in monoculture and mixed culture under treatments of hydroperiod and salinity. Hydrobiologia 569, 325–341.

Chen, L.Z., Wang, W.Q., Lin, P., 2004. Influence of waterlogging time on the growth of *Kandelia candel* seedlings. Acta Oceanologica Sinica 23, 149–158.

Chen, L.Z., Wang, W.Q., Lin, P., 2005. Photosynthetic and physiological responses on *Kandelia candel* L. Druce seedlings to duration of tidal immersion in artificial seawater. Environmental and Experimental Botany 54, 256–266.

Chen, L.Z., Yang, Z.W., Wang, W.Q., Lin, P., 2006. Critical level for planting *Kandelia candel*seedlings in Xiamen. Chinese Journal of Applied Ecology 17, 177–181 (in Chinese with English abstract).

Delgado, P., Hensel, P.F., Jiménez, J.A., Day, J.W., 2001. The importance of propagule establishment and physical factors in mangrove distributional patterns in a Costa Rican estuary. Aquatic Botany 71, 157–178.

Ellison, A.M., Farnsworth, E.J., 1993. Seedling survivorship, growth, and response to disturbance in Belizean mangal. American Journal of Botany 80, 1137–1145.

Ellison, A.M., Farnsworth, E.J., 1997. Simulated sea level change alters anatomy, physiology, growth, and reproduction of red mangrove (*Rhizophora mangle* L.). Oecologia 112, 435–446.

Fan, H.Q., 2000. Coastal Guarder—mangrove. Guangxi Science and Technology Press, Nanning (in Chinese).

Fan, H.Q., Li, G.Z., 1997. Effect of sea dike on the quantity, community characteristics and restoration of mangrove forest along Guangxi coast. Chinese journal of applied Ecology 8, 240–244 (in Chinese with English abstract).

Fan, H.Q., Chen, G.H., He, B.Y., Mo, Z.C., 2005. Mangrove coastal wetland in Shankou and management. Ocean Press, Beijing, 126 pp. (in Chinese).

He, B.Y., 2002. Research on the ecology of the fouling fauna community in mangrove. Guangxi Science 9, 133–137 (in Chinese with English abstract).

Hovenden, M.J., Curran, M., Cole, M.A., Goulter, P.F.E., Skelton, N.J., Allaway, W.G., 1995. Ventilation and respiration in roots of one-year old seedlings of grey mangrove *Avicennia marina* (Forsk.) Vierh. Hydrobiologia 295, 23–29.

Kent, C.P.S., Lin, J., 1999. A comparison of Riley encased methodology and traditional techniques for planting red mangroves (*Rhizophora mangle*). Mangroves & Salt marshes 3, 215–225.

Kitaya, Y., Jintana, V., Piriyayotha, S., Jaijing, D., Yabuki, K., Izutani, S., Nishimiya, A., Iwasaki, M., 2002. Early growth of seven mangrove species planted at different elevations in a Thai estuary. Trees 16, 150–154.

Komiyama, A., Santiean, T., Higo, M., Patanaponpaiboon, P., Kongsangchai, J., Ogino, K., 1996. Microtopography, soil hardness and survival of mangrove (*Rhizophora apiculata* BL.) seedlings planted in an abandoned tin-mining area. Forest Ecology and Management 81, 243–248.

Krauss, K.W., Doyle, T.W., Twilley, R.R., Rivera-Monroy, V.H., Sullivan, J.K., 2006. Evaluating the relative contributions of hydroperiod and soil fertility on growth of south Florida mangroves. Hydrobiologia 569, 311–324.

Liao, B.W., Zheng, D.Z., Zheng, S.F., Li, Y., Chen, X.R., Chen, Z.T., 1996. A study on the afforestation techniques of *Kandelia candel* mangrove. Forest Research (China) 9, 586–592 (in Chinese with English abstract).

Liao, B.W., Zheng, S.F., Chen, Y.J., Li, M., 2005. Advance in researches on rehabilitation technique of mangrove wetlands. Ecologic science (China) 24, 61–65 (in Chinese with English abstract).

Lin, P., 1999. Mangrove Ecosystem in China. Science Press, Beijing & New York, 271 pp.

Lin, P., 2003. The characteristics of mangrove wetlands and some ecological engineering questions in China. Engineering Science (China) 5, 33–38 (in Chinese with English abstract).

Misra, S., Choudhurya, A., Ghosh, A., 1984. The role of hydrophobic substances in leaves in adaptation of plants to periodic submersion by tidal water in a mangrove ecosystem. Journal of Ecology 72, 621–625.

Mo, Z.C., 2002. The preliminary study on the mangrove habitat conditions in Guangxi. Guangxi Forestry Science (China) 31, 122–127 (in Chinese with English abstract).

Naidoo, G., Rogalla, H., von Willert, D.J., 1997. Gas exchange responses of a mangrove species, *Avicennia marina*, to waterlogged and drained conditions. Hydrobiologia 352, 39–47.

Pezeshki, S.R., Delaune, R.D., Patrick, W.H.J., 1989. Differential response of selected mangroves to soil flooding and salinity: gas exchanges and biomass partitioning. Canadian Journal of Forest Research 20, 869–874.

Riley, R.W.J., Kent, C.P.S., 1999. Riley encased methodology: principles and processes of mangrove habitat creation and restoration. Mangroves & Salt Marshes 3, 207–213.

Skelton, N.J., Allaway, W.G., 1996. Oxygen and pressure changes measured in situ during flooding in roots of the grey mangrove *Avicennia marina* (Forsk.) Vierh. Aquatic Botany 54, 165–175.

Wen, Y.G., Liu, S.R., Yuan, C.A., 2002. The population distribution of mangrove at Yingluogang of Guangxi, China. Acta Ecologica Sinica 22, 1160–1165 (in Chinese with English abstract).

Ye, Y., Lu, C.Y., Tan, F.Y., 2001. Studies on differences in growth and physiological responses to waterlogging between *Bruguiera gymnorrhiza* and *Kandelia candel*. Acta Ecologica Sinica 21, 1654–1661 (in Chinese with English abstract).

Ye, Y., Tam, N.F.Y., Wong, Y.S., Lu, C.Y., 2003. Growth and physiological responses of two mangrove species (*Bruguiera gymnorrhiza* and *Kandelia candel*) to waterlogging. Environmental and Experimental Botany 49, 209–221.

Ye, Y., Lu, C.Y., Zheng, F.Z., Tam, N.F.Y., 2004a. Effects of simulated sea level rise on the mangrove *Kandelia candel*. Acta Ecologica Sinica 24, 2238–2244 (in Chinese with English abstract).

Ye, Y., Tam, N.F.Y., Wong, Y.S., Lu, C.Y., 2004b. Does sea level rise influence propagule establishment, early growth and physiology of *Kandelia candel* and *Bruguiera gymnorrhiza*? Journal of Experimental Marine Biology and Ecology 306, 197–215.

Youssef, T., Saenger, P., 1996. Anatomical adaptive strategies to flooding and rhizophere oxidation in mangrove seedlings. Australian Journal of Botany 44, 297–313.

Zhang, Q.M., Yu, H.B., Chen, X.S., Zheng, D.Z., 1997. The relationship between mangrove zone on tidal flats and tidal levels. Acta Ecologica Sinica 17, 258–265 (in Chinese with English abstract).

Zhang, Q.M., Su, S.Z., Zhang, Y.C., Yu, H.B., Sun, Z.X., Wen, X.S., 2001. Marine environmental indexes related to mangrove growth. Acta Ecologica Sinica 21, 1427–1437 (in Chinese with English abstract).

Zheng, D.Z., Li, M., Zheng, S.F., Liao, B.W., Chen, Y.J., 2003. Headway of study on mangrove recovery and development in China. Guangdong Forestry Science (China) 19, 10–14 (in Chinese with English abstract).

真红树和半红树植物体内盐分分布的比较研究*

蒋巧兰　牟美蓉　王文卿　赵杭平

(厦门大学生命科学学院,福建 厦门 361005)

摘要: 研究了典型的真红树植物(红树 *Rhizophora apiculata*、木榄 *Bruguiera gymnorhiza*、海桑 *Sonneratia caeseolaris*、桐花树 *Aegiceras corniculatum*、白骨壤 *Avicennia marina*)、半红树植物(黄槿 *Hibiscus tiliaceus*、水黄皮 *Pongamia pinnata*)和目前还存在争议的物种银叶树(*Heritiera littoralis*)体内各器官的 K、Na、Cl 元素分布特征. 结果表明:(1)真红树植物和半红树植物体内盐分分布存在明显的区别,真红树植物体内各器官的 Na、Cl 含量明显比半红树高,叶片的 K/Na 值比半红树低;(2)真红树植物(除红树和木榄)和半红树植物的 K:Na 离子吸收选择性(USR)没有差异,但半红树植物的 K:Na 离子运输选择性(TSR)明显比真红树高;(3)真红树植物根系 Cl 含量随径级增加有先上升后下降趋势,与半红树植物不同. 从盐分分布特征来看,银叶树最好归为半红树植物.

关键词: 真红树植物;半红树植物;盐分

中图分类号:Q 945　　**文献标识码**:A　　**文章编号**:0438-0479(2006)06-0867-06

红树林常分布于热带、亚热带海岸潮间带,主要由一些盐生的木本植物组成,通常分为真红树植物和半红树植物. 早在 1905 年,Tansley 和 Fritsch 在《Sketches of vegetation at home and abroad 1. The flora of the Ceylon littoral》一书中就提出了半红树植物的概念[1]. Saenger 将红树林生境专有的红树植物称为真红树,而将既可在红树林生境出现也可在中高潮带以上生境出现的称为"非专有红树"、"后红树"、"半红树"等[2]. Tomlison 为真红树植物的鉴别做了具体标准[3];林鹏和傅勤[4]提出了非常明确的"真红树"和"半红树"的概念及鉴别标准. 但这些标准主要是从植物的一些外部形态特征及生境潮汐特点上进行划分,未从内部生理水平加以比较. 加之不同学者对这些标准的理解不同,导致对中国及国内各省区的红树植物种类组成报道有所不同,一些经常存在争议的物种如:银叶树(*Heritiera littoralis*)、卤蕨(*Acrostichum aureum*)、海漆(*Excoecaria agallocha*)等[5~9]. 但对于一个国家或地区而言,了解红树植物种类组成,是研究红树林结构和功能的前提,也可以加强我们对红树林的保护和管理[8,9].

在红树植物抗盐机理方面,多数研究集中在光合、呼吸、生长、基因调控,从盐分的吸收、转移和分配角度研究报道不多,且集中于泌盐红树植物白骨壤属的植物,对真红树和半红树之间的比较研究较少[10]. 而研究植物的抗盐性必须跟踪盐分的吸收、运输和分配,盐生植物体内盐分浓度的调节是植物生存的关键[11].

聚类分析方法被大量应用于系统分类学和群落调查,但利用聚类方法研究植物抗性的报道不多. 任文伟等[12,13]运用聚类分析方法对不同种源羊草(*Leymus chinensis*)的抗旱性进行了研究.

本研究着重于比较真红树和半红树在体内各器官盐分分布上的区别,从而从内部离子分布上为真红树和半红树划分提供鉴别依据,并对它们的抗盐机理进行深入了解.

1 材料与方法

1.1 样品采集

样品于 2005 年 7 月和 11 月采自海南东寨港红树林自然保护区(19°51′N,110°24′E)和文昌清澜港红树林自然保护区(19°22′~19°35′N,110°40′~110°48′E).

在样地内选择红树 *Rhizophora apiculata*、木榄 *Bruguiera gymnorhiza*、海桑 *Sonneratia caeseolaris*、桐花树 *Aegiceras corniculatum*、白骨壤 *Avicennia marina*、黄槿 *Hibiscus tiliaceus*、水黄皮 *Pongamia pinnata*、银叶树 *Heritiera littoralis* 标准木各 3 棵,分别采集成熟叶、幼枝、多年生枝、树干材、树干皮、地下根(细分成 4 级, Ⅰ:$d<2$ mm, Ⅱ:d 2~5 mm, Ⅲ:d 5~15 mm, Ⅳ:$d>15$ mm). 其中叶片和幼枝取自树冠外围,所有植物样品用纯净水清洗干净、称重后, 80 ℃

* 国家自然科学基金(30200031)资助
原载于厦门大学学报(自然科学版),2006,45(6):867-872

烘干,磨粉,贮存备用.采集植物样品的同时采集 20～30 cm 层土壤样品,风干磨碎过 1 mm 筛备用.

1.2 实验方法

植物和土壤样中的 Cl 含量用 $AgNO_3$ 滴定法测定;土壤样品和植物样品分别经微波消解和马福炉灰化后,用 ICP-Mass(ELAN DRC-e,Axial Field Technology)测定 K、Na 含量.各元素含量均以干样质量计,单位为 mg · g^{-1}.

植物对 K^+、Na^+ 吸收和运输的选择性[14]用 K:Na 离子吸收选择性和 K:Na 离子运输选择性表示.

K:Na 离子吸收选择性(K:Na USR) =

{根系[K^+]/[Na^+]}/{土壤[K^+]/[Na^+]},

K:Na 离子运输选择性(K:Na TSR) =

{叶片[K^+]/[Na^+]}/{根系[K^+]/[Na^+]}.

2 结 果

2.1 真红树植物与半红树植物的根系元素含量

从图 1 可看出,红树植物(包括半红树植物)根各个径级的 Cl 元素含量都要比 K 和 Na 高.真红树植物根系不同径级的 Cl 含量差别较大,随着根直径的增加,红树 Ⅲ级根的 Cl 含量较 Ⅰ级根明显升高,在 Ⅳ级根中的含量又明显下降;木榄 Ⅲ级根的 Cl 含量与 Ⅱ级根相比明显升高,在 Ⅳ级根中又明显下降;海桑和桐花树 Cl 含量在 Ⅱ级根中明显升高,而后在 Ⅲ级根中的含量又明显下降;白骨壤的各径级根系 Cl 含量有先上升后下降的趋势,但没有构成显著性差异.总之,真红树植物(除白骨壤)Cl 含量在 Ⅱ级或 Ⅲ级根中明显升高($p<0.05$),而后在 Ⅳ级根中又明显下降($p<0.05$).

半红树植物各级根系的 Cl 含量没有明显差异,即无先上升后下降的趋势.银叶树的根系 Cl 含量趋势与半红树相似,无先升高后下降的趋势.

2.2 真红树植物与半红树植物地上各器官元素含量

从图 2 可以看出,红树(除树干材)和木榄地上部分各器官的元素含量基本一致:Cl > Na > K;海桑幼枝、多年生枝、树干材元素含量顺序为:Cl > K > Na,叶片、树皮的顺序:Cl > Na > K;桐花树地上部分各器官的元素含量为:Cl > Na > K;白骨壤的叶片元素含量顺序:Cl > Na > K,幼枝、树皮:Cl > K > Na,多年生枝:K > Cl > Na,树干材:K > Na > Cl;半红树植物黄槿和水黄皮的地上部分器官元素含量一致,其幼枝、多年生枝、树皮、叶片的元素顺序都为:K > Cl > Na,树干材:K > Na > Cl;而银叶树幼枝、多年生枝、叶片,树干材都为:K > Cl > Na,树皮:Cl > K > Na.

总的来说,真红树植物体内的 Na、Cl 含量明显要比半红树植物高(表 1),且真红树地上各器官的 Cl 元素含量要比 K 元素高.相比之下,半红树植物黄槿、水黄皮的 Na、Cl 含量明显较低(表 1),地上各器官的 K 要比 Na、Cl 高.争议物种银叶树 Na 和 Cl 含量较低,地上各器官的元素顺序也与半红树植物较接近.

此外,对红树植物不同器官元素含量之间的关系进行分析,发现真红树植物和半红树植物各器官的 Na 和 Cl 含量之间均存在正相关关系,且均达到极显著水

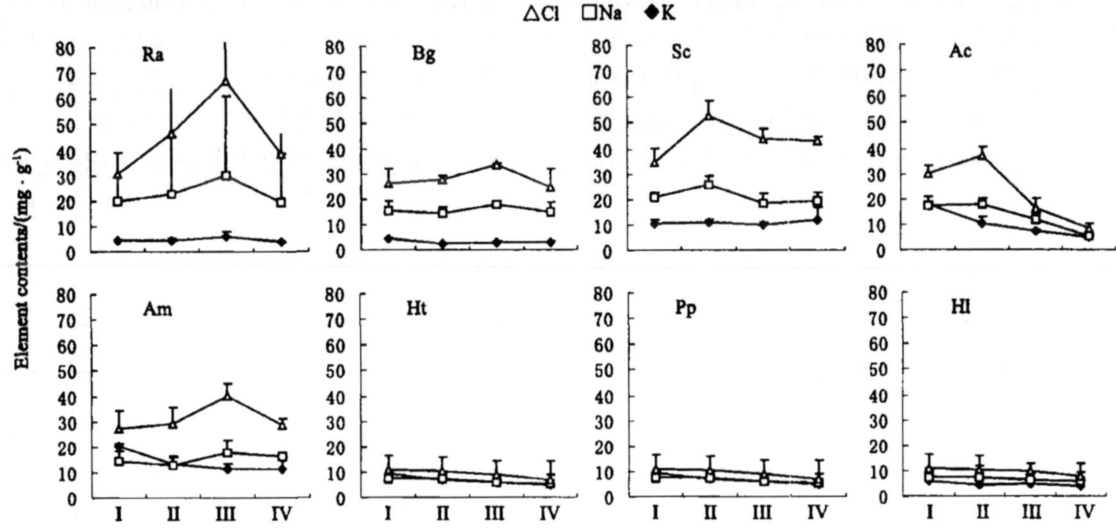

图 1 真红树植物和半红树植物不同径级根系 K、Na、Cl 含量

Ra:红树;Bg:木榄;Sc:海桑;Ac:桐花树;Am:白骨壤;Ht:黄槿;Pp:水黄皮;Hl:银叶树

Ⅰ:$d<2$ mm,Ⅱ:d 2～5 mm,Ⅲ:d 5～15 mm,Ⅳ:$d>15$ mm

Fig.1 Element contents in roots of different diameter in true mangrove plants and mangrove associates

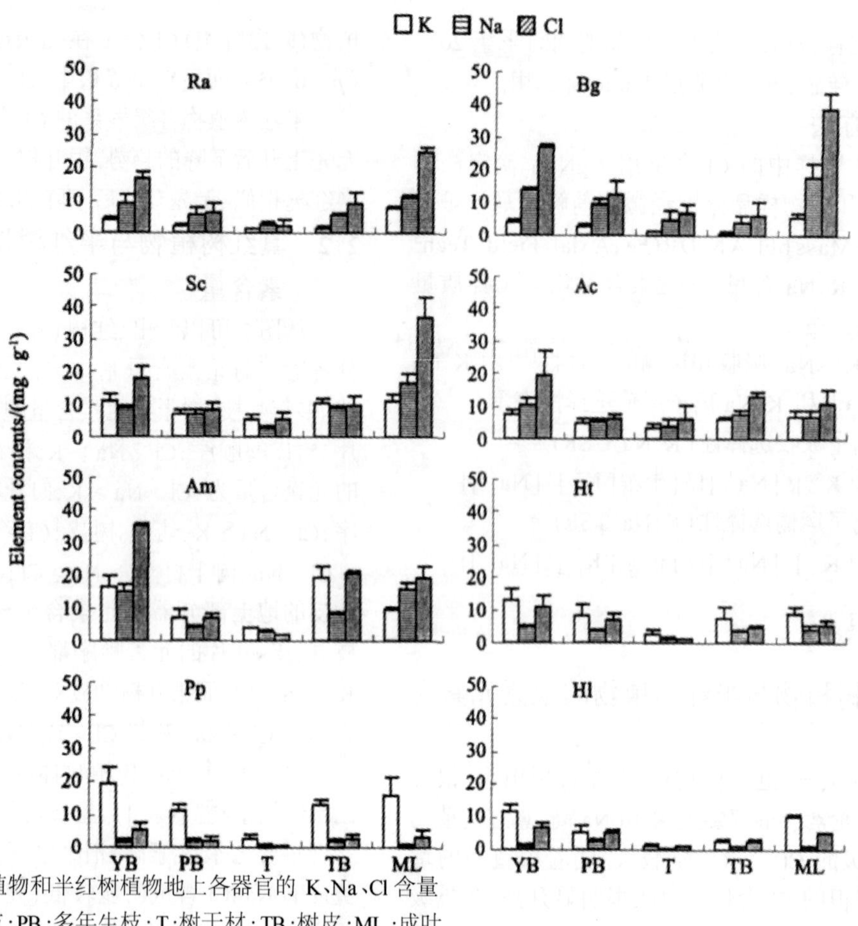

图2 真红树植物和半红树植物地上各器官的 K、Na、Cl 含量

YB:幼枝;PB:多年生枝;T:树干材;TB:树皮;ML:成叶

Fig. 2 Element contents in various over-ground organs of true mangrove plants and mangrove associates

平(表2).真红树植物的 K和 Na 含量之间也都存在正相关关系,除白骨壤和红树外,其余均达到显著水平,桐花树达到极显著水平;而半红树植物的 K和 Na 之间存在负相关.真红树的 Cl 和 K含量之间正相关,但只有木榄和桐花树达到极显著水平,海桑达到显著水平;半红树植物水黄皮的 Cl 和 K含量负相关,黄槿为正相关.银叶树的 Cl 和 K、K和 Na 含量之间都为负相关.

2.3 叶片 K/ Na

从统计结果(表1)可以看到,真红树植物叶片的 K/ Na 值在 0.34～0.98 之间,相比而言,半红树植物叶片 K/ Na 值比较高,黄槿为 4.62,水黄皮为 16.30.争议物种银叶树叶片的 K/ Na 值为 8.61,说明银叶树在叶片的离子关系上接近半红树植物.

2.4 K:Na 离子吸收选择性与运输选择性

不同直径的根系 K:Na 离子吸收选择性见表3.

表1 真红树和半红树植物体内 Na、Cl、K元素含量的统计结果

Tab. 1 Statistics of element contents in true mangrove plants and mangrove associates

类别	树种	Na/ (mg · g⁻¹)	Cl/ (mg · g⁻¹)	K/ (mg · g⁻¹)	叶片 K/ Na
真红树	红树 *Rhizophora apiculata*	14.12 ±0.44	26.86 ±3.35	3.97 ±0.15	0.69 ±0.06
	木榄 *Bruguiera gymnorhiza*	13.10 ±1.15	22.87 ±1.35	3.09 ±0.35	0.34 ±0.12
	海桑 *Sonneratia caeseolaria*	14.63 ±0.83	28.07 ±1.22	9.34 ±0.26	0.69 ±0.25
	桐花树 *Aegiceras corniculatum*	9.66 ±0.58	16.59 ±2.07	7.67 ±0.14	0.98 ±0.30
	白骨壤 *Avicennia marina*	12.02 ±1.27	23.34 ±2.64	12.66 ±1.16	0.63 ±0.11
半红树	黄槿 *Hibiscus tiliaceus*	5.01 ±1.82	7.55 ±1.95	6.71 ±1.64	4.62 ±2.38
	水黄皮 *Pogamia pinnata*	3.73 ±1.18	5.81 ±2.38	10.15 ±0.53	16.30 ±6.95
	银叶树 *Heritiera littoralis*	3.14 ±1.06	7.14 ±2.22	4.97 ±0.12	8.61 ±1.63

表 2 同一植物不同器官元素含量之间直线回归相关关系①

Tab. 2 Linear regressive coefficients between the contents of two elements in various organs of the same mangroves

类别	树 种	Na-Cl	K-Na	Cl-K
真红树	红树 *Rhizophora apiculata*	0.9852**	0.6041	0.6451
	木榄 *Bruguiera gymnorhiza*	0.9642**	0.7977*	0.8197**
	海桑 *Sonneratia caeseolaria*	0.9742**	0.7504*	0.7644*
	桐花树 *Aegiceras corniculatum*	0.9696**	0.8752**	0.8177**
	白骨壤 *Avicennia marina*	0.9008**	0.4556	0.5996
半红树	黄槿 *Hibiscus tiliaceus*	0.8456**	-0.2042	0.3018
	水黄皮 *Pogamia pinnata*	0.9376**	-0.4125	-0.1624
	银叶树 *Heritiera littoralis*	0.9362**	-0.3203	-0.0036

①**: $p<0.01$; *: $p<0.05$.

表 3 真红树和半红树 K:Na 离子吸收选择性及运输选择性

Tab. 3 The K:Na uptake selectivity ratios and transport selectivity ratios of true mangrove plants and mangrove associates

类别	树 种	K:Na 吸收选择性				K:Na 运输选择性
		<2 mm	2～5 mm	5～15 mm	>15 mm	
真红树	红树 *Rhizophora apiculata*	0.82 ±0.11	0.68 ±0.10	0.69 ±0.08	0.74 ±0.18	3.39 ±0.37
	木榄 *Bruguiera gymnorhiza*	1.11 ±0.35	0.57 ±0.08	0.64 ±0.07	0.66 ±0.03	1.72 ±0.50
	海桑 *Sonneratia caeseolaria*	1.45 ±0.12	1.23 ±0.18	1.43 ±0.22	1.73 ±0.44	1.46 ±0.72
	桐花树 *Aegiceras corniculatum*	2.93 ±0.48	1.71 ±0.23	1.81 ±0.40	2.81 ±0.29	1.28 ±0.46
	白骨壤 *Avicennia marina*	4.17 ±1.07	3.07 ±0.74	1.85 ±0.37	2.03 ±0.80	0.66 ±0.19
半红树	黄槿 *Hibiscus tiliaceus*	1.36 ±0.55	1.21 ±0.25	1.63 ±0.37	1.30 ±0.19	7.04 ±3.45
	水黄皮 *Pogamia pinnata*	2.55 ±0.67	1.93 ±0.81	2.30 ±0.88	4.13 ±2.82	14.36 ±7.58
	银叶树 *Heritiera littoralis*	1.98 ±0.82	2.71 ±1.38	2.89 ±1.98	3.12 ±1.73	13.08 ±5.23

注: <2 mm,2～5 mm,5～15 mm,>15 mm 分别代表不同直径的根系.

实验结果表明,真红树植物中的白骨壤($d<2$ mm 和 d 2～5 mm)较高,而拒盐红树植物红树、木榄较其它的树种偏低;半红树植物黄槿相对于水黄皮和银叶树偏低.但总体来说,真红树和半红树的吸收选择性没有存在明显差异.

真红树植物的 K:Na 运输选择性较低(0.66～3.39).而半红树植物黄槿和水黄皮的 K:Na 运输选择性,在 7.04～13.08 之间.说明半红树植物对 K^+ 具有较高的运输选择性.争议物种银叶树 K/Na 吸收运输率为 13.08,与半红树植物一致.

2.5 聚类分析

运用 SPSS10.0 软件,按欧氏距离法对红树植物各器官的 K、Na、Cl 含量进行分层聚类分析,结果如图 3.可以把红树植物分为 3 类,银叶树、水黄皮、黄槿为一类;红树和木榄一类;海桑、白骨壤、桐花树为一类.

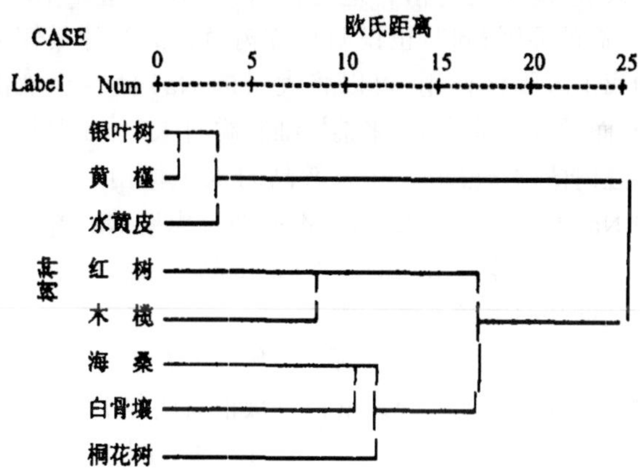

图 3 红树植物各器官 K、Na、Cl 含量聚类树状图

Fig. 3 The map of group agglomerative classification of K,Na,Cl contents

3 讨 论

红树植物由于其生境的高含盐量,面临着如何维持其组织、细胞膨压稳定性问题.这主要通过在液泡中积累大量的无机离子来完成.因此红树植物其体内的含盐量相对于陆生植物要高些[15].从我们的研究结果

看出,半红树植物虽然体内含盐量也高,但其体内各器官的 Na^+、Cl^- 含量明显比真红树植物低. 原因是真红树植物主要分布于潮间带,生境的含盐量要比半红树植物高得多,故其主要靠积累 Na^+、Cl^- 来进行渗透调节,从而使自身的水势比外界土壤更低,而半红树可能很大程度上依赖 K^+ 和糖类等渗透调节物质[16]. 并且半红树植物叶片中的 K/Na 值比真红树高,因为对某些盐生植物来说,地上器官保持较高的 K/Na 比,比单纯的减少 Na^+ 含量对其提高耐盐能力更重要[17].

经统计发现,真红树和半红树植物体内各器官的 Na 和 Cl 之间都存在极显著的正线性关系,说明两者进入植物体后的行为相同;真红树的 K 和 Na、K 和 Cl 之间也存在正相关关系,但在半红树植物体内却不呈正相关,在某些种内为明显负相关(黄槿和银叶树). 说明 K 和 Na、K 和 Cl 在真红树植物体内是协同关系,在半红树植物体内却是拮抗关系,也反映了半红树植物对 Na 和 Cl 有一定的排斥作用.

K 对于植物体来说是一种重要的营养元素,光合作用及细胞内一些重要的代谢反应都需要 K 的存在. 从真红树和半红树植物的 K:Na 离子吸收选择性可看出,真红树植物中白骨壤 K:Na 离子吸收选择性较高,而拒盐红树植物红树和木榄的值较低. 这与之前相关文献报道白骨壤对 K 有较强的选择吸收性相符[18,19].

从总体上讲,真红树植物和半红树植物的根系在 K:Na 离子吸收选择性上没有存在明显差异. 但从 K:Na 离子运输选择性看,半红树植物却比较高,说明进入根系的部分 Na^+,从根系共质体进入木质部时受到卸载,也是半红树植物地上器官的 K^+ 含量比 Na^+、Cl^- 高的原因;同时也说明真红树植物的叶片具较强的离子区隔化作用,可以将大量的 Na^+ 储存于液泡中,而通过少量的 K^+ 来维持细胞质的 K^+/Na^+[20]. 而半红树植物虽然叶片中含较高的 K^+,但由于植物体内 Na^+ 和 Cl^- 含量太低,即不能通过积累 Na^+ 和 Cl^- 来降低植物体各组织的渗透势,限制了半红树植物只能在低盐的生境下生长[10].

从不同径级根系中的 Cl^- 含量可看出,真红树植物 Ⅱ级或 Ⅲ级根的含氯量较 Ⅰ级根有明显的上升,说明盐分在此被滞留,导致了 Ⅱ级或 Ⅲ级根的盐分含量很高,甚至超过叶片(木榄除外),在 Ⅳ级根的含量又有所下降. 王文卿和林鹏[15]对红海榄不同径级根系的研究结果也发现 Na、Cl 含量在不同径级根系中的分布有类似趋势. 半红树植物的根系含氯量却无此特征,可能是半红树植物的根系对盐分的外泄能力较强. 对这其中存在的抗盐机理还有待进一步研究.

争议物种银叶树被许多学者认为是嗜好淡水的半红树植物[3,21,22],但也有些学者将其归为真红树[23~25]. 从我们的研究结果看出,银叶树在体内 Na^+、Cl^- 含量,叶片 K/Na,不同径级根系 Cl^- 的变化趋势都与半红树植物相似. Paliyavuth[26] 等对白骨壤、木榄、木果楝(*Xylocarpus granatum*)、银叶树的研究结果也表明,银叶树叶片的钠、氯含量相对于其它几个种明显偏低,并认为是一些有机物质在维持渗透调节及水分平衡上起着重要作用.

聚类分析结果表明,8 种红树植物可分为三类:银叶树、黄槿和水黄皮为一类,即半红树植物;红树和木榄归为一类;而海桑和桐花树、白骨壤归为一类. 把泌盐红树植物与拒盐红树植物分开. 说明红树和木榄在体内盐分分布上相似. 而海桑虽然为非胎生红树植物,但其耐盐能力却较强,野外环境下有时与桐花树、白骨壤分布在同一潮带,导致其在盐分分布特点上与泌盐红树植物接近.

本实验结果表明,对真红树和半红树植物体内盐分分布特征进行对比研究,探讨它们在抗盐机理上的区别,可以从内部生理水平上区别真红树和半红树,并对一些争议物种做出初步判断,对加强红树林的管理和保护提供帮助.

参考文献:

[1] Tansley A G, Fritsch F E. Sketches of vegetation at home and abroad (Ⅰ):the flora of the ceylon littoral[J]. New Phytol. ,1905,4:1 - 17,27 - 55.

[2] Sanger P, Hegerl E J, Davie J D S. Global status of mangrove ecosystems [J]. The Environmentalist, 1983, 3 (Suppl.):1 - 88.

[3] Tomlinson P B. The Botany of Mangroves [M]. Cambridge:Cambridge University Press,1986.

[4] 林鹏,傅勤. 中国红树林的环境生态及经济利用[M]. 北京:高等教育出版社,1997.

[5] 林鹏. 中国红树林种类及其分布[J]. 林业科学,1987,23:481 - 490.

[6] 林鹏. 中国红树植物种类分布和林相类型[M]//李振基. 环境与生态论丛. 厦门:厦门大学出版社,1993:74 - 79.

[7] 王瑞江,陈忠毅,黄向旭. 国产红树林植物的染色体计数[J]. 热带亚热带植物学报,1998,6(1):40 - 46.

[8] 王伯荪,梁士楚,张炜银,等. 世界红树植物区系[J]. 植物学报,2003,45(6):644 - 653.

[9] 赵亚,郭跃伟. 真红树林植物化学成分及生物活性研究概况[J]. 中国天然药物,2004,2(3):135 - 140.

[10] Naidoo G, Tuffers A V, von Willert D J. Changes in gas exchange and chlorophyll fluorescence characteristics of two mangroves and a mangrove associate in response to salinity in the natural environment[J]. Trees. ,2002,16:

140 - 146.

[11] Omer L S, Horvath S M, Setaro F. Salt regulation and leaf senescence in aging leaves of *Jaumea carnosa* (Less.) gray (Asteraceae), a salt mash species exposed NaCl stress[J]. American Joural of Botany., 1983, 70 (3):363 - 368.

[12] 任文伟,钱吉,马骏,等.不同地理种群羊草在聚乙二醇胁迫下含水量和游离脯氨酸含量的比较[J].生态学报, 2000, 2(20):349 - 352.

[13] 任文伟,罗岫泉,郑师章.不同种源羊草的 SOD、POD 的活性及丙二醛含量的比较[J].植物生态学报, 1997, 21 (1):77 - 82.

[14] 赵可夫.植物抗盐生理[M].北京:科学出版社, 1993:58 - 60, 180 - 184.

[15] 王文卿,林鹏.红树植物体内元素分布特点与抗盐机理[J].林业科学, 2003, 39(4):30 - 36.

[16] Popp M. Salt resistance in herbaceous halophytes and mangroves[J]. Progress in Botany., 1995, 56:416 - 429.

[17] Maathuis F I M, Amtmann A. K^+ nutrition and Na^+ toxicity: the basis of cellular K^+/Na^+ ratios[J]. Ann. Bot., 1999, 84:123 - 133.

[18] Ball M C, Farquhar GD. Photosynthetic and stomatal responses of two mangrove species *Aegiceras corniculatum* and *Avicennia marina*, to long term salinity and humidity conditions[J]. Plant Physiol., 1984, 74:1 - 6.

[19] Medina E, Francisco M. Osmolality and $\delta^{13}C$ of leaf tissues of mangrove species from environments of contrasting rainfall and salinity[J]. Estuarine, Coastal and Shelf Science, 1997, 45:337 - 344.

[20] Tester M, Davenport R. Na^+ tolerance and Na^+ transport in higher plants[J]. Ann. Bot., 2003, 91:503 - 527.

[21] Kartawinata K, Adisoemarto S, Soemodihardjo S, et al. Status Pengetahuan Hutan Bakau di Indonesia[M]//Soemodihardjo S, Nontji A, Djamali A. Proceedings of Seminar Ecosystem Hutan Mangrove. Jakarta: LIPI, 1979:21 - 39.

[22] Mukherjee A K, Acharya L K, Mattagajasingh I, et al. Molecular characterization of three *Heritiera* species using AFLP markers[J]. Biol. Plant., 2003, 47(3):445 - 448.

[23] Santisuk T. Taxonomy of the terrestrial trees and shrubs in the mangrove formation in Thailand[M]//The First UNDP/UNESCO Regional Training Course on Introduction to Mangrove Ecosystem. Bangkok: National Research Council, 1983.

[24] Das A B, Basak U C, Das P. Karyotype diversity in three species of *Heritiera*, a common mangrove tree on the Orissa Coast[J]. Cytobios., 1994, 80:71 - 78.

[25] Parani M, Lakshmi M, Senthilkumar P, et al. Molecular phylogeny of mangroves V. Analysis of genome relationships in mangrove species using RAPD and RFLP markers[J]. Theor. Appl. Genet., 1998, 97:617 - 625.

[26] Paliyavuth C, Clough B, Patanaponpaiboon P. Salt uptake and shoot water relations in mangroves[J]. Aquat. Bot., 2004, 78(4):349 - 360.

The Salt Distribution Characteristics in True Mangrove Plants and Mangrove Associates

JIANG Qiao-lan, MU Mei-rong, WANG Wen-qing, ZHAO Han-ping

(School of Life Sciences, Xiamen University, Xiamen 361005, China)

Abstract: The distribution of K, Na, Cl in different organs of unanimously true mangrove plants (*Rhizophora apiculata, Bruguiera gymnorhiza, Sonneratia caeseolaria, Aegiceras corniculatum, Avicennia marina*), mangrove associates (*Hibiscus tiliaceus, Pongamia pinnata*) and the still controversial species *Heritiera littoralis* were studied in this paper. The results indicated: (1) Salt distribution between true mangroves and mangrove associates were significantly different. The Na and Cl contents in true mangroves were significantly higher than that of mangrove associates, but the K/Na ratios in leaves were lower. (2) The K:Na uptake selectivity ratios were not different between true mangrove plants (except *R. apiculata* and *B. gymnorhiza*) and mangrove associates. The K:Na transport selectivity ratios were significantly higher in mangrove associates than that of true mangrove plants. (3) In the true mangrove plants, the root Cl contents increased but later decreased with the increase of diameter, the mangrove associates hadn't. *H. littoralis* is better classified as mangrove associate from the characteristics of salt distribution.

Key words: true mangrove plants; mangrove associates; salt

能量标签技术及其在红树林生态系统能流研究中的应用*

黄振远 王 瑁 王文卿

(厦门大学生命科学学院,福建 厦门 361005)

摘要:传统上认为红树林输出的有机质产生巨大的能流,支持了巨大的河口和近岸水域生态系统的次级生产。但能量标签技术的研究结果却显示红树林输出的有机质的作用并没有如此巨大。用红树碎屑难消化特性来解释此现象,此外数学模型模拟分析发现潮汐的稀释作用也可以解释这种现象。但这两者都不能解释,在其他初级生产者稀少时,红树材输出的有机质可以被大量利用的现象。在有红树林的河口和近海岸水域生态系统中,藻类等非红树初级生产者具有比红树植物更高的初级生产力,而且更容易被动物获得和消化。可以认为是藻类等巨大初级生产力的竞争作用导致红树初级生产在消费者组织中很难被发现,如此上面提到的难题就能得到很好的解决。此外能量标签技术检测出的是红树的初级生产在消费者组织中的相对比率,不是绝对数量值,从此角度看,能量标签技术的结果与传统观点不是矛盾而是互相补充的关系。由此推测红树的初级生产应该还是被消费者所利用,只是它们在消费者初级营养来源组成中占的比例并不大,但其绝对数量并不少。这与传统观点认为的红树的初级生产被大量利用,支撑了具有巨大的次级生产稍有不同。此外,能量标签技术在红树林生态系统中的适用性尚未检验;计算食物组成的数学工具不是很完善;实验设计上考虑的不够全面;对定量研究有一定的影响。

关键词:稳定性同位素;^{13}C能量输出;消费者;初级生产

文章编号:1000-0933(2007)03-1206-11 **中图分类号**:Q948 **文献标识码**:A

Energy signature technology and its application on energy flux in mangrove ecosystems: a review and outlook

HUANG Zhen-Yuan, WANG Mao , WANG Wen-Qing

School of Life Sciences, Xiamen University, Xiamen 361005, China

Acta Ecologica Sinica, **2007, 27(3): 1206~1216.**

Abstract: A central view of tropical estuarine ecology is that export of organic matter from mangroves represents a major energy pathway and support much of the secondary production of estuaries and nearshore waters. Nevertheless, recent results, especially those obtained using energy signature technology, contradict this paradigm. Overall, most studies found a limited role of mangrove detritus in estuarine food webs. Usually only animals collected inside mangrove swamps or in mangrove - lined waterways have depleted carbon isotopic signatures characteristic of mangrove detritus. These findings are explicable by the difficulty in assimilation of mangrove carbon and the dilution effect of the tides. These factors, however, cannot explain the heavy use of mangroves in estuarine habitats when other primary production sources are scarce. Algae and other primary production sources (e. g. seagrass) have much higher productivity than mangrove in estuaries and nearshore

* 国家自然科学基金资助项目(40376025)

原载于生态学报,2007,27(3):1206－1216

waters, and are easy to obtain and digest So these alternative carbon sources and the complicated food web structure mask the nutrition role of mangroves If this hypothesis is true then it could offer an explanation for the observation that mangrove litter is heavily used by consumers when other primary production sources are scarce, a phenomenon that cannot be explained by the difficulty in assimilation or tidal dilution Energy signature technology addresses the relative contribution of mangroves to the consumer's carbon sources, not the biomass in numbers Energy signature technology addresses the relative proportion of the assimilated carbon from mangrove to all the consumer carbon, not the biomass in numbers The traditional view is that mangrove primary production supports large consumer biomass but not the relative proportion With that in mind energy signature technology is not inconsistent with the traditional view but is complementary We hypothesize that all of mangrove litter production is used by aquatic consumers The mode of the use is, however, not in terms of dominance in the consumer(s tissue but small contributions to many individuals One blemish of past isotopic studies is that none has directly tested the applicability of the approach to food web links between consumer and mangrove Another blemish is that what isotopic data indicated was the proportion of the assimilated carbon from mangrove to all the consumer carbon, not its importance For example algae could get into food webs directly through consumption by all kinds of fishes but mangrove litter must be consumed by crabs first and then perhaps by fish So in fish tissue there will be much less carbon from mangrove than from algae if the initial primary production in terms of carbon mass of mangrove and algae is equal Because the carbon mass of mangrove is much reduced after crab digestion and absorption Digestion and absorption will affect the importance of different carbon sources too, but this issue has been ignored in past studies The study of energy flux in mangrove ecosystems only focused on particulate organic matter but has ignored dissolved organic carbon (DOC), which is the dominant form of carbon exported from mangroves N and P export is also important It is apparent that mangrove litter is not the main source of primary production fuelling the food chains in estuaries and nearshore waters Isotope signature technology is not yet a perfect tool for studying these trophic linkages, but the imperfection is in the details

Key Words: stable isotope; ^{13}C ; energy export; mangrove; consumer; primary production

红树林是热带亚热带海岸潮间带的木本植物群落,是海湾河口生态系统重要的第一生产者,也是世界"四大最丰富生物多样性海洋生态系统"之一,它具有高生产、高归还和高分解三高特点[1]。红树林作为沿海地区重要的湿地,是许多珍稀动物的栖息地,也是生物多样性保护的热点地区[1];同时它以其巨大的生物量和生产力对其附近的河口近岸水域的次级生产有着巨大影响[2]。红树林区的渔业产值也比其他海洋生态系统的高,例如澳大利亚北部和美国佛罗里达红树林区的鱼类生物量分别比其邻近的海草场高 4～10倍和 35倍[3],在澳大利亚红树林中年捕捞生物量高达 5840kg/hm^2,市场价值约为 5330美元[4]。研究显示红树林区的渔业产值很高[3~9],这是红树林被保护的重要原因之一[10]。

能量标签技术即稳定性同位素技术在红树林生态系统中的应用,在研究动物的食物来源,食物链、食物网、群落结构等方面与传统方法相比具有很大的优势[11,12],特别是在研究小型甚至是微型动物例如蚯蚓等食物来源方面[13~16]。国外的学者应用此技术解决了许多传统方法不能解决的难题,取得了一定的成果[11]。例如发现在红树林生态系统中,动物组织的 C主要来源不是红树的初级生产[17,18],从而引起研究者的重视。然而学者们从不同角度研究后得出的结果不一致,存在争议[14]。

与国外大量的研究相比,能量标签技术在我国生态学和环境科学研究中的应用才刚刚起步[11],国内对红树林与林区水生生物关系的研究还主要集中于鱼类和底栖无脊椎动物的生物多样性调查,季节变化和经济鱼类产量的研究[19~22];而在应用能量标签技术研究红树林与河口近海岸水域中消费者的关系方面还是空白。因此本文对此问题作一简单的总结并对某些问题提出一些自己看法,促进对红树林的保护与研究。

1 能量标签技术简介

1.1 稳定同位素的特点

元素一般都有几种同位素,例如 C有^{14}C、^{13}C、^{12}C,N有^{15}N、^{14}N;其中^{14}C为放射性同位素,^{13}C、^{15}N为稳定

性同位素。稳定性同位素在大气中的含量在一定程度上是恒定的,当自养生物进行光合作用时,由于光合途径和同位素热力学性质的不同[23]从而导致光合作用的底物与产物中的^{13}C的含量不同,这就是稳定性同位素分馏。动物在同化食物中的C时基本不发生分馏,因而可以用它示踪动物食物的来源。N在光合过程中不发生分馏但在动物的新陈代谢过程中由于酶对^{15}N的偏好[24]导致动物组织中的^{15}N比其食物富集(一般是3‰左右)因而可以利用用动物组织中^{15}N的富集程度推测动物所处的营养级[11,12]。

1.2 稳定同位素的测定方法

稳定同位素的测量一般要经过以下几个过程:(1)预处理,(2)气化,(3)干扰物去除,(4)质谱分析[12]。

由于稳定同位素在物质中含量很低,通常采用相对法进行量度,以国际通用的"δ"标记法表示,用公式可以表示为:

$$\delta X = [(R_{sam} - R_{std}) / R_{std}] \times 1000$$

式中,R_{sam}是样品中元素的重轻同位素丰度之比如($^{13}C_{sam}/^{12}C_{sam}$);$R_{std}$是国际通用标准物的重轻同位素丰度之比(如$^{13}C_{std}/^{12}C_{std}$)[25]。

1.3 同位素质量平衡方程

计算动物食物组成时的数学工具比较多,主要有质量守恒模型和欧几里得距离模型[26]。比较准确且在研究红树林能量流中使用较多的是同位素质量平衡方程[26,27]。李忠义等详细介绍了各种计算方法[28]。同位素质量平衡方程如下:

$$\delta^{13}C'_i = \sum_{j=1}^{n}[f'_{ij}(\delta^{13}C_j + \Delta'_C)]$$

$$\delta^{15}N'_i = \sum_{j=1}^{n}[f'_{ij}(\delta^{15}N_j + \Delta'_N)]$$

$$\sum_{j=1}^{n}f'_{ij} = 1$$

式中,$\delta^{13}C'i$和$\delta^{15}N'i$分别是消费者C和N同位素组成;$\delta^{13}C_j$和$\delta^{15}N_j$分别为食物C和N同位素组成;$\Delta'C$和$\Delta'N$分别为C和N同位素分馏值;f'_{ij}是不同食物在整体食物中所占比例;n代表消费者全部食物种类

1.4 稳定同位素技术在红树林生态系统中的运用——能量标签技术

早在1979年就有人提出:消费者的C组成决定于其食物的C组成。不同的生产者有不同的δ值,但生产者和消费者在δ值统一的基础上存在分馏[29,30]。Twilley将稳定C同位素技术应用于红树林生态系统提出能量标记假说发现红树林输出碎屑的数量和周期性变化取决于红树林的环境特性,Mancera和Twilley用稳定性同位素^{13}C的比率验证了能量标签假说[31],认为应用能量标签技术研究红树林的能流非常合适。

2 红树林与其附近水域中消费者的关系

2.1 传统的研究方法及其缺陷

红树林是热带海岸潮间带最典型最重要的生物群落,有着巨大的面积和生物量,但人们对红树林的了解却一直不够。直到1968年Odum在第二次海洋大会上发表演讲认为:海岸湿地输出的巨大有机质产生巨大的能流物流,支持了河口和近岸水域生态系统中巨大的次级生产[2]。支持Odum观点的研究结果主要有以下3个:(1)消费者消化道内含物与栖息地两方面分析的结果显示红树林支持了很高的次级生产[32,33];(2)许多研究发现红树林与其附近的多数近海鱼类有很大的相关性[3~8];(3)物质平衡计算方面显示红树林是其周围水域次级生产的主要支持者[34]。由于以上几个方面证据的支持Odum的观点影响很大[31]。

但消化道内含物分析、统计相关性分析、物质平衡计算,这3种方法都是间接的推断红树林与消费者的营养关系,不够直接精确。另外这些方法本身还存在一定的缺陷,例如消化道内含物分析,通常发现的只是动物摄食的大型的和不容易消化的食物,小型或微型食物还有很容易被消化的食物则通常被忽略。统计相关分析法则只能说明二者具有很大的关系,不能说明具体是什么关系。物质平衡计算有一定说服力但不够精确。而能量标签技术则通过检测消费者组织中的C的组成来揭示消费者的营养来源,比传统方法更直接更有说服

力。因此获得广泛的应用[11,12]。

2.2 能量标签技术在研究消费者的营养来源中的应用及结果

2.2.1 红树林对消费者的营养贡献

传统认为:红树林能比其他栖息地为鱼类提供更丰富的食物[7]。总体上红树林对林区食物网的贡献有以下几个方面:(1)红树凋落物是碎屑食物网的基础;(2)红树林林区较其他地区丰富的无脊椎动物的幼虫是鱼类的重要食物;(3)红树林林内和滩涂中的沉积有机质含量较高,营养丰富,是某些动物的重要食物;(4)红树根系的分泌物有利于藻类等浮游生物的生长。这些使红树林生态系统的食物网高度发达且有效率[35](图1)。

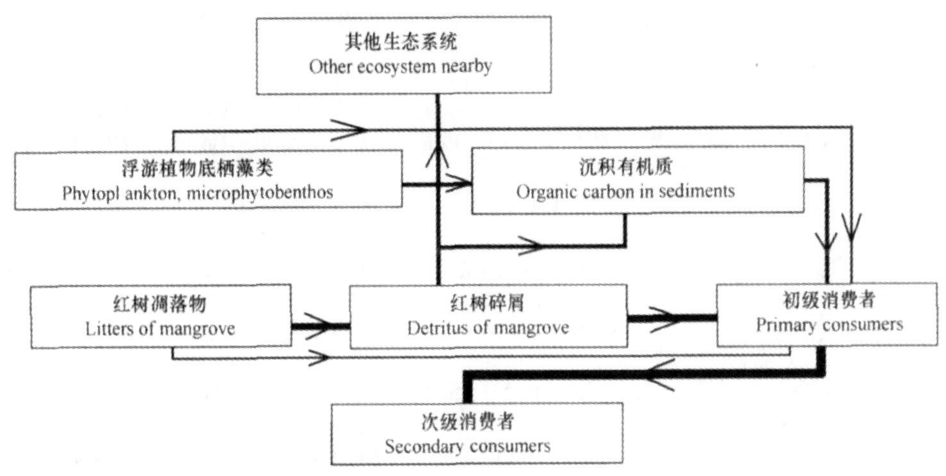

图 1 传统上的红树林生态系统能流 (线条粗细代表能流的大小)

Fig 1 Energy output flux of mangrove ecosystem in tradition (thread show the scale of energy output flux)

普遍接受的红树林与鱼类高的相关性的主要原因之一,是红树林为鱼类提供了优良的摄食场[7]。但能量标签技术却研究发现,红树林为鱼类提供优良摄食场的观点与Odum的物质输出的理论一样,都是非常片面的[36]。例如Melville和Connolly发现,澳大利亚昆士兰的Moreton湾红树林边缘泥滩上分布的鱼类的营养最多有30%来自^{13}C含量低的初级生产者(红树、盐沼肉质植物)[37],红树对鱼类的营养作用则更低,具体的需进一步通过另一种稳定同位素,例如^{34}S在鱼类和红树以及中的分布得到[38]。在印度kakinada湾,潮下带底栖无脊椎动物群具有明显的选择浮游植物和底栖微藻为食物的习性[39,40]。Bouillon在研究了Wildlife Sanctuary的沉积物、初级生产者(红树、MPB、浮游生物等)、22种潮间带无脊椎动物的δ^{13}C和δ^{15}N值后发现,大多数潮间带无脊椎动物是以本地和周围水域中的藻类为主要营养来源[41]。

Melville和Connolly分析了所有可能的生产者(红树、藻类、大型海藻附生植物、底栖微藻、海草、盐沼植物)与3种主要的经济鱼类^{13}C组成的空间变化的相关性后发现,红树林与河口海岸生态系统的营养相关性很低,消费者主要的营养来源只能是其他生产者[42]。运用能量标签技术对红树林林内及其周围的水域中的DOC、POC进行追踪研究,也发现红树林对周围生态系统的作用被高估了[43]。Lee综合了现有的能量标签技术的结果后,认为支持红树林的初级生产支撑了巨大的近海岸水域次级生产的证据并不充分[36](表1)。虽然Thimdee等在泰国的Khung Krabaen湾的研究发现红树林支持了红树林内生活的部分动物的次级生产,但到目前为止绝大部分的研究结果显示:红树林输出的C仅仅被大量发现于红树林内部极小区域内的不多的物种[41~60]。

所以,来自红树林的C并不是河口海岸生态系统水域中消费者组织中C的主要组成部分,而林区及附近水域的浮游植物,滩涂上的底栖藻类,还有来自附近其他的生态系统如盐沼、海草场等的有机质中的C,才是水域中消费者组织中C的主要组成部分,但具体到某个地区的红树林,这几种C来源的相对重要性会有所不同。具体食物网应修正为如下图2。

表 1 关于应用稳定同位素示踪技术研究红树林与近海岸营养联系结果的汇总[36]

Table 1 A summary of the results of energy signature technology studies investigating the mangrove - offshore tropic connection[36]

游泳动物 Nekton species	示踪同位素 Isotope tracers	游泳动物 δ值 Signature of nekton(%)	红树植物 δ值 Signature of mangroves(‰)	红树-消费者 δ值之差 Difference(‰) (consumer-mangrove)
对虾属 *Penaeus* spp.		δ^{13}C: - 14.9～17.9		δ^{13}C: 6.6～13.6
新对虾属 *Metapenaeus* spp.	^{13}C		δ^{13}C: - 24.5～- 28.8	
细巧仿对虾 *Parapeneopsis* spp.		δ^{13}C: - 13.1～- 18.6		δ^{13}C: 5.9～15.4
对虾属 *Penaeus* spp.	^{15}N, ^{13}C	δ^{13}C: - 15.3～- 17.8	δ^{13}C: - 26.7～- 29.8	δ^{13}C: 8.9～14.5
细巧仿对虾 *Parapeneopsis* spp.	^{34}S	δ^{15}N: 11.1～12.1 δ^{34}S: - 10.1～- 12.4	δ^{15}N: 2.2～6.3 δ^{34}S: - 1.4～- 5.1	δ^{15}N: 4.8～9.9 δ^{34}S: - 5.0～- 13.8
对虾属 *Penaeus* spp.	^{15}N, ^{13}C	δ^{13}C: - 15.5～- 19.6	δ^{13}C: - 26.9～- 30.0	δ^{13}C: 7.3～14.5
新对虾属 *Metapenaeus* spp.		δ^{15}N: 8.4～11.3	δ^{15}N: 6.3～8.0	δ^{15}N: - 0.4～5.0
对虾属 *Penaeus* spp.	^{15}N, ^{13}C	δ^{13}C: - 14～- 17	δ^{13}C: - 27.0～- 28.8	δ^{13}C: 10～15
新对虾属 *Metapenaeus* spp.		δ^{15}N: 5.210.0	δ^{15}N: 1.3～3.7	δ^{15}N: - 1.5～8.7
细巧仿对虾 *Parapeneopsis* spp.	^{15}N, ^{13}C	δ^{13}C: - 18.2～- 20.2	δ^{13}C: - 24.4～- 28.1	δ^{13}C: 4.2～9.9
新对虾属 *Metapenaeus* spp.		δ^{15}N: 10.0～12.6	δ^{15}N: 4.0～12.6	δ^{15}N: 40～8.6
对虾属 *Penaeus* spp.	^{15}N, ^{13}C	δ^{13}C: - 17.0～- 25.1	δ^{13}C: - 28.7～- 26.7	δ^{13}C: 1.6～11.7
新对虾属 *Metapenaeus* spp.		δ^{15}N: 8.4～11.3	δ^{15}N: 4.4～6.3	δ^{15}N: 2.1～6.9
十足目 Decapod larve	^{15}N, ^{13}C	δ^{13}C: - 17～- 23 δ^{15}N: 6.2～7.7	δ^{13}C: - 25.5～- 26.8 δ^{15}N: 4.4	δ^{13}C: 2.5～9.8 δ^{15}N: 1.8～3.3
平均值 Mean value		δ^{13}C: - 15.6～- 19.9		δ^{13}C: 5.9 ±3.0～13.1 ±2.3

2.2.2 消费者不以红树林碎屑为主要食物的原因

在热带河口海岸,红树林有非常广阔的面积与很高的净初级生产力[61]。早期就是用其生产力和物质守恒律[34]去论证它们对生物网的重要贡献,然而新的实验结果却显示红树林的作用并没有如此巨大。如此就存在一个问题:红树林如此巨大的生产力为何不为消费者所利用,它的最终去向又是哪里?

Rodell等发现:红树碎屑很难被消费者所同化,原因是碎屑的不溶解性和高的 C/N比[17]。理论模拟也发现,潮汐的稀释作用会使得红树林输出的有机质对周围的生态系统的影响不明显[62]。实地研究也有相同的结果,例如 Chong等发现在上游的红树林地区,对虾组织中的 C有高达 84%来自红树植物,但在下游随着浮游植物和潮汐影响的增加,对虾来自红树植物的 C下降到 16%～24%[56]。Rodelli等发现在潮沟中消费者同化的 C中 65%是来自红树植物,但比例随着潮沟与红树林距离的增加逐渐下降[17]。虽然以上这两种解释各自能解释一部分实验结果,但都不能解释当其他食物来源稀少时,来自红树的 C可以占消费者组织中 C的 30%～50%[10]。而且用高 C/N比和难溶解性来解释红树碎屑不能被同化,也不是很确切。因为红树凋落物在凋落的初期 C/N比碎屑的 C/N还要高,只是在随后的分解的过程中,由于糖等易溶物质的快速丢失而使碎屑的 C/N降低,同时难溶解性增加营养价值减少[63]。此外红树的净初级生产只有大约 30%是以凋落物的形式输出的[36]。

红树林区的浮游植物虽然数量不多但世代更新快,在单位时间上输出的生物量相当巨大[1]。Wafar等估

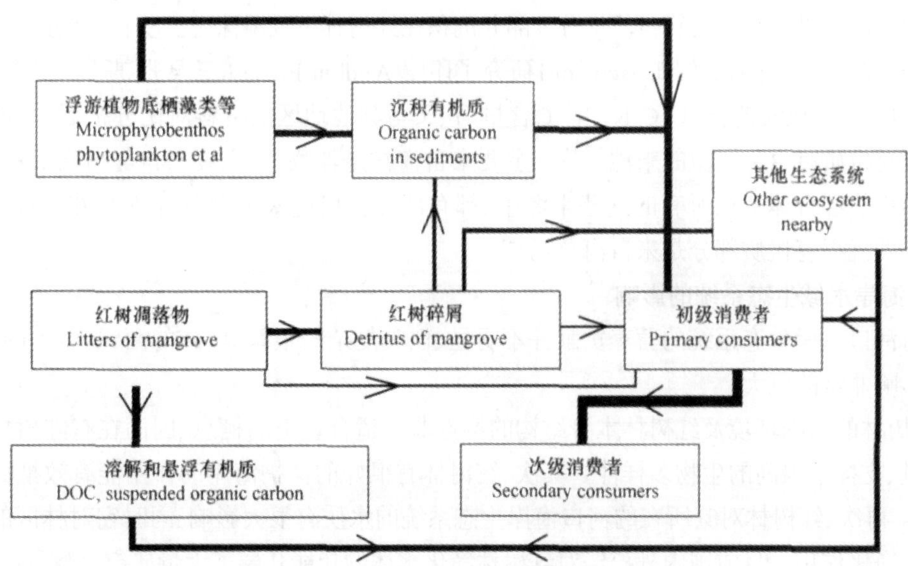

图 2 新的红树林生态系统能流 (线条粗细代表能流的大小)

Fig 2 New energy output flux of mangrove ecosystem (thread show the scale of energy output flux)

计在 western Indian estuary所有潜在的红树林输出的 C仅仅是浮游植物输出 C的 37%,而 N和 P等则降到 3% ~4%[64];Li和 Lee估计在 Deep Bay红树林仅仅贡献了全部 C库的 1. 8%[65];Dehairs等使用凋落物测量法,测得浮游植物的平均重量高达 1. 2g C/(m^2·d)[66],而在热带红树仅为 5~10Mg C/(hm^2·a)[36]。所有这些结果都显示浮游植物有巨大的生产力,可以为水域中的次级生产提供有力的营养支撑。

此外,虽然红树林内有限的光线和可溶性单宁的作用,限制了林内底栖微藻 (microphytobenthos MPB)的生产力[67,68],但泥滩内有大量的底栖微藻 (MPB),他们被无脊椎动物所摄食,底栖微藻 (MPB)也是其他潮间带生态系统例如盐沼的重要 C来源[69~71],有学者建议对它们在红树林生态系统中的作用作进一步研究[72,73]。

稳定同位素分析显示:红树林附近的海草、海草附生植物、盐沼肉质植物等与藻类和浮游生物一样生产力很高[51,74]。同时发现红树林附近的海草场才是鱼类的主要摄食场所[51],只在雨季时红树林的林缘潮沟的狭小区域,才是鱼虾等的主要摄食区域[74]。

因此笔者认为:在有红树林的河口海岸生态系统中,红树林的初级生产不是消费者营养最初来源的主要原因是,由于红树林生态系统中具有更加巨大生产力并且容易被采食和消化的藻类等其他初级生产者的竞争,掩盖了红树林初级生产的作用。潮汐的稀释和碎屑的难溶特性也有一定的作用。理由如下: (1)其他初级生产者稀少的情况下红树的初级生产会被大量利用[10]; (2)藻类等其他初级生产者巨大的更容易消化的初级生产力的存在[65,66],会对红树初级生产的作用产生竞争性遮盖; (3)能量标签技术检测出的,是消费者不同营养来源的比例,是相对值而不是数量上的绝对值。据此评价红树初级生产的重要性,会低估其重要性。因此,能量标签技术的结果与传统的认为红树的初级生产支撑了巨大的海岸水域次级生产的观点,不是冲突而是互相补充的关系。 (4)传统的统计相关性分析、物质平衡计算和胃含物分析的方法虽然不够精确但没有错误。

据此可以推测红树的初级生产,应该最终还是被鱼类等水生生物所利用,只是利用的方式与传统的认识不同,存在两种情况:一在小范围少数物种中被大量利用,例如红树林中存在大量以红树叶为主食的蟹类;二多数物种的单个生物体对红树碎屑的利用不多,个体数却是很多,因此红树输出有机物的影响范围要比传统上认为的大得多。当然这还需要实验检验。

2.2.3 红树林输出的有机质的最终去向

Lee发现:红树林生产力的输出主要是以溶解有机碳 (DOC)的形式[61],碎屑并不是主要的有机质输出。

而且碎屑的最大的一部分进入细菌食物链[64];另一部分沉积在红树林下或滩涂上,还有一部分由潮汐转运到其他地区沉积下来,或进入食物网。例如 Bouillon1研究了印度 Andhra Pradesh三角洲、斯里兰卡西南部 Galle and Pambala等地的沉积有机质的含量、C/N、和 ^{13}C含量后,又与其他地区的资料对比分析后,发现红树林内以及泥滩上的沉积有机质有 3种类型的来源[75]:一主要来自海洋悬浮物;二主要来自红树凋落物;三是海洋悬浮物与红树凋落物的混合体。在所有的这些来源中,都有部分直接进入了消费者体内 (图 2)。Hemminga等也发现红树林下的沉积有机质部分是来自海草场[43]。

3 红树林对河口海岸水域生态系统的影响

虽然红树林对河口海岸生态系统的营养贡献并不明显,但其复杂的根系、广阔的面积、巨大的生产力,对周围生态系统的影响非常的巨大。

首先,由于红树林的周围环境及红树林本身结构的特点非常适合鱼类的栖息,因而在有红树林的地区富集了巨大的生物量,支撑了很高的生物多样性。其次,红树林有很好的保护海岸的作用能有效抵御四十年一遇的台风侵袭[76]。再次,红树林对其所在的河口海岸生态系统的水质有很大影响,同时它对林区的养殖鱼塘的废水也有明显的净化作用。目前国内外已经对红树林净化水体的功能开展了大量研究,已经成为热点[77]。最后,红树林向周围水域环境中的 C输出,很大程度上依赖于生态系统的地形和潮汐特征[31]。虽然红树林输出的 C对周围生态系统的影响范围很小,一般只能影响到红树林周围几米的范围[78]。但是红树林红树林可以为生活在其周围生态系统中的鱼类在某些时候提供栖息地和避难所[37],能明显的提高其附近珊瑚礁等生态系统的鱼类多样性[79],对海草场的鱼类也有相似的影响。

国外学者普遍接受的红树林与林区鱼类高相关性观点的原因主要有 3点[35]:一红树林本身的结构多相性对它们有特殊的吸引力;二与其他栖息场所相比,红树林复杂的结构可以降低幼鱼的被捕食率;三红树林区鱼类的食物非常丰富。红树林被保护主要是由于它是许多经济鱼类和无脊椎动物的重要栖息地以及林区生物多样性高的特点,在美国佛罗里达和澳大利亚的昆士兰,红树林被保护就是这种情况;而在某些发达国家或地区红树林被保护却是因为其可作为旅游点[63]。

总之,红树林的初级生产并不是鱼类的主要营养来源。结构的多相性、林区丰富的食物来源以及其所在区域的环境特点,是红树林与林区鱼类以及其周围的其他生态系统的密切关系的主要原因。

4 问题与讨论

红树林生态系统一般处于海陆交错带的河口区,海洋、陆地、河流对它都有很大的影响;同时受到气候、地形、潮汐、人类活动等因素的影响;此外各个地区的红树林系统间存在着很大的差异;能量来源和流动途径复杂。因此,红树林与林区及周围环境中的其他生物的关系极其错综复杂,再加上研究方法的多样性,而各种方法又不是很完美,因而研究中存在许多问题。

4.1 稳定性同位素技术本身有缺陷,可能没有真实的测出红树林初级生产与消费者间的关系。

(1)计算动物食物来源和不同来源食物所占的比例时,必须与消化道内含物分析相结合,并且各食物的同位素组成必须有明显差异[11]。在研究的生物有多种食物来源的时候 (多于 4种),对生物主要食物来源的分析就可能存在困难,需要一定的策略或多种稳定同位素进行复合分析[37,38],在某些情况下则可能没办法得出结果。

(2)目前仅利用同位素物质平衡方程等工具研究了不同来源的 C在动物组织中的比例,没有考虑不同来源的各种初级生产的相对重要性。消费者特别是高级消费者组织中的 C是经过生物链的传递而来,而动物食物组成通常不单一,并且相邻营养级间生物量差别巨大。因此一旦动物的不同类食物的营养级不一致,同位素物质平衡方程计算出的结果就不能反映不同初级生产者的相对重要性。例如,红树凋落物必须经过蟹类的处理后,才能被鱼类所利用[36],藻类等则可直接被鱼类所采食。如此,当有相同量的红树凋落物和藻类同时进入食物链时,由于红树凋落物要比藻类多经过一个营养级,最后进入鱼类组织的藻类的初级生产就会比红树的多 10倍 (相邻营养级间的生物量大约是 10倍的关系)左右。同时目前所有的研究都没有整合食物的吸收转化效率，很多食物特别是富含纤维的食物的吸收转化率很底[80]，而红树恰好就是此类食物。在此特殊条件下，能量标签技术的结果就会有很大的误差甚至是错误的，但幸好此情况并不多见。

(3)红树叶腐烂过程中只发生很小的同位素分馏[45,60]，但是在进一步的同化过程中有没有分馏却没有实验验证。虽然"消费者的 C组成决定于其食物的 C组成"的 C稳定同位素规则在盐沼生物中适用，但是在红树林生物中却适用得不是很好[29]。实验室和自然条件下都发现真菌可以使^{13}C富集 6‰，某些红树林内的蟹类在同化过程中也可以使^{13}C富集 3‰[32]。所以，在真菌(真菌是食碎屑食物网的重要中间环节)和红树林蟹类有重要作用的红树林区，C稳定同位素规则就可能不能完全适用。对陆生蚯蚓(多毛类)的研究也显示它们在同化 C时可以使^{13}C富集 3‰～4‰[13～16]，而在红树林区的多毛类也是红树凋落物的重要消费者。

但是，如果这种分馏作用在红树林的多毛类中也真实的存在，那么红树林与它们就有 100 %的营养相关性[10] (消费者组织的^{13}C大约是 -21‰～-24‰，红树大约是 -27‰)。然而这成立的不能性不大，因为这与红树林周围水域中存在有其他具有巨大生产力而且容易被获得和消化的初级生产者(浮游植物、藻类等)相矛盾[1]，而且林区动物的食性也不可能如此单一到完全以红树输出的有机质为食。

(4)通常鱼类只在红树林内度过其部分生活史，而能量标签技术检测出的却是生物长期的食物来源状况。另外"同位素印迹"现象[80,81]的存在也会增加实验误差。

需要强调的是，虽然能量标签技术存在以上细节问题，但基本上可以确定红树的初级生产不是鱼类等最初营养的主要来源[36]。此外，能量标签技术检测出的是消费者不同营养来源在比例上的相对值，而不是数量上的绝对值。从这个角度看，能量标签技术的结果与传统认为红树的初级生产支撑了巨大的海岸水域次级生产观点，不是对立而是互相补充的关系。

4.2 物质平衡计算和能量标签技术研究都显示：红树林输出的 N、P在河口海岸的营养循环中有重要作用，因而存在红树林向消费者提供 N、P而不是 C的可能[82～85]。

4.3 红树凋落物的大部分，并不是以前学者们认为的被埋在了泥滩里[86]，也不是快速的被消费者采食，而是随潮水向潮沟、林外滩涂、近海转移。凋落物的分解也不是在泥下发生。红树树下和泥滩上的沉积有机质不是原生性的由红树凋落物直接在泥滩内分解形成，而是有相当一部分来自其他生态系统的次生性沉积[75]，至于比例的大小要由具体的红树林所在的环境特点所决定。

5 总结与展望

综上所述，鱼类等水生生物体内 C的主要来源不是红树，而是浮游植物、底栖藻类、海草等其他初级生产者。红树林生态系统中藻类等容易被获得和消化而且具有巨大生产力的初级生产者的竞争性掩盖作用，是红树林区鱼类不以红树初级生产为主要食物的主要原因。能量标签技术检测出的，是消费者不同营养来源的相对比例，而不是数量上的绝对值。因此，它与传统的认为红树的初级生产支撑了巨大的海岸水域次级生产观点不是冲突而是补充关系。推测红树初级生产应该最终还是被鱼类等水生生物所完全利用。今后要作的工作主要有以下几个方面：

(1)完善能量标签技术。

(2)开展红树林食物网的细节研究，追踪红树林输出的溶解有机物的最终去向。

(3)寻找新的可以检测红树林初级生产营养作用的研究方法，以及其他实验结果以检验上述的观点。

References:

[1] Lin P. Mangrove Ecosystem in China. Beijing, New York: Science Press, 1999.

[2] Odum E P. A research challenge: evaluating the productivity of coastal and estuarine water. University of Rhode Island: Proceedings of the Second Sea Grant Conference, 1968. 63-64.

[3] Ronnback P. The ecological basis for economic value of seafood production supported by mangrove ecosystems. Ecological Economics, 1999, 29 (2): 235-252.

[4] Morton R M, Community structure, density and standing crop of fishes in a subtropical Australian mangrove area. Marine Biology, 1990, 105(3): 385-394.

[5] Nickerson D J. Trade-offs of mangrove area development in the Philippines Ecological Economics, 1998, 28(2): 279-298.

[6] Laegdsgaard P, Johnon C R. Mangrove habitats as nurseries: unique assemblages of juvenile fish in subtropical mangrove in eastern Australia Marine Ecology Progress Series, 1995, 126(1-3): 67-81.

[7] Laegdsgaard P, Johnson C R. Why do juvenile fish utilize mangrove habitats? Journal of Experimental Marine and Ecology, 2001, 257(2): 229-253.

[8] Roberston A I, Duck N C. Mangrove-fish communities in tropical Queensland Australia: spatial and temporal patterns in densities, biomass and community structure. Marine Biology, 1990, 104(3): 369-379.

[9] Roberston A I, Duck N C. Recruiment growth and residence time of fish in a tropical Australian Mangrove Estuarine Coastal and Shelf Science, 1990, 31(5): 725-745.

[10] Fry B, Ewel K C. Using stable isotope in mangrove fisheries research a review and outlook Isotopes in Environmental and Health Studies, 2003, 39(3): 191-196.

[11] Wang J Z, Lin G H, Huang J H, *et al* The application of isotope in the study of relationships between animals and plants in the land ecosystem. Chinese Science Bulletin, 2004, 49(21): 2141-2149.

[12] Cai D L, Zhang J H, Zhang J. Applications of stable carbon and nitrogen isotope methods in ecological studies Journal of Ocean University of Qingdao, 2002, 32(2): 287-295.

[13] Hendrix P E, Callaham M A, Jr Lachicht S T, *et al* Stable isotope studies of resource utilization by nearctic earthworms (*Diplocardia, Oligochaeta*) in subtropical savanna and forest ecosystems Pedobiologia, 1999, 43(9): 818-823.

[14] Neilson R, Boag B, Smith M. Earthworm $\delta^{13}C$ and $\delta^{15}N$ analyses suggest that putative functional classifications of earthworms are site-specific and may also indicate habitat diversity. Soil Biology and Biochemistry, 2000, 32(8-9): 1053-1061.

[15] Spain A V, Saffigna P G, Wood A W. Tissue carbon sources for *Pontoscolex corethrurus* (Oligochaeta: Glossoscolecidae) in a sugarcane ecosystem. Soil Biology and Biochemistry, 1990, 22: 703-706.

[16] Spain A, Le Feuvre R. Stable C and N isotope values of selected components of a tropical Australian sugarcane ecosystem. Biology and Fertility of Soils, 1997, 24(1): 118-122.

[17] Rodelli M R, Gearing J N, Gearing P J, *et al* Stable isotope ratio as a tracer of mangrove carbon in Malaysian ecosystems Oecologia, 1984, 61(3): 326-333.

[18] Stoner A W, Zimmerman R J. Food pathways associated with penaeid shrimps in a mangrove-fringed estuary. Fisheries Bulletin, 1988, 86(3): 543-551.

[19] Fan H Q, He B Y, Wei S Q. Influences of traditional fishing on the mangrove fisheries and management countermeasures in Yingluo Bay, Guangxi Province. Chinese Biodiversity, 1996, 4(3): 167-174.

[20] Fan H Q, Wei S Q, He B Y, *et al* The seasonal dynamics of nekton assemblages in mangrove-fringed tidal waters of Yingluo Bay, Guangxi Guangxi Sciences, 1998, 5(1): 45-50.

[21] He B Y, Fan H Q, Mo Z C. Study on species diversity of fishes in mangrove area of Yingluo Bay, Guangxi Province. Journal of Tropical Oceanography, 2001, 20(4): 74-79.

[22] Lin P. The Synthetical Scientific Investigation of Zhangjiangkou Mangrove Wetland Natural Reserve in Fujian Xiamen: Xiamen Unversity Press 2001. 68-103.

[23] Nier A O. A mass spectrometer for isotope and gas analysis Review of Scientific Instruments, 1947, 18: 398-411.

[24] Davidson D W, Cook S C, Snelling R R, *et al* Explaining the abundance of ants in lowland tropical rainforest canopies Science, 2003, 300(5621): 969-972.

[25] Zheng S H, Zheng S C, Mo Z C. Isotope analysis of the geochemistry. Beijing: Beijing University Press 1986. 309.

[26] Phillip s D L. Mixing models in analyses of diet using multiple stableiso tope: a critique Oecologia, 2001, 127 (2) : 166~170.

[27] Saito L, Johnson B M, Bartholow J, *et al* Assessing ecosystem effects of reservoir operations using food web-energy transfer and water quality models Ecosystems, 2001, 4: 105-125.

[28] Li ZY, Jin X S, Zhuang ZM, *et al* Appl ica t ion s of stable isotope techniques in aquatic ecological studies Acta Ecologica Sinica, 2005, 25(11): 3052-3060.

[29] Haines E B, Montague C L. Food sources of estuarine invertebrates analyzed using $^{13}C/^{12}C$ ratios Ecology, 1979, 60(1): 48-56.

[30] Fry B, Sherr E B. ^{13}C measurements as indicators of carbon flow in marine food webs Contribution in Marine Science, 1984, 27(1): 15-47.

[31] Mancera J E, Twilley R R. Testing the energy signature hypothesis of mangroves using carbon isotope ratio analysis 2003 FCE LTER All Scientists Meeting, 2003.

[32] Odum W E, Heald E J. Trophic analyses of an estuarine mangrove community. Bulletin of Marine Science, 1972, 22(3): 671-738.

[33] Odum W E, Heald E J. The detritus-based food web of an estuarine mangrove community. In: Wiley M. ed Estuarine Research I New York: Academic Press, 1975. 265-286.

[34] Twilley R R. Coupling of mangroves to the productivity of estuarine and coastal waters In: Jansson B. O. ed Coastal offshore Ecosystem Interactions Berlin: Springer-Verlag 1988. 155-180.

[35] Shi F S, Wang M, Wang W Q, *et al* A review of the relationships between mangroves and fishes Marine Sciences, 2005, 29(5): 54-59.

[36] Lee S Y. Exchange of Organic Matter and Nutrients Between Mangroves and Estuaries: Myths, Methodological Issues and Missing Links International Journal of Ecology and Environmental Sciences, 2005, 31(3): 163-176.

[37] Melville A J, Connolly R M. Food webs supporting fish over subtropical mudflats are based on transported organic matter not in situ microalgae Marine Biology, 2005. 148(2): 363-371.

[38] Connolly R M, Guest M A, Melville A J, *et al* Sulfur stable isotopes separate producers in marine food-web analysis Oecologia, 2004, 138(2): 161-167.

[39] Bouillon S, Ramen A V, Dauby P, *et al* Carbon and nitrogen stable isotope rations of subtidal benthic invertebrates in an estuarine mangrove ecosystem (Andhra Pradesh, India). Estuarine Coastal and Shelf Science, 2002, 54(5): 901-913.

[40] Bouillon S, Koedam N, Baeyens W, *et al* Selectivity of subtidal benthic invertebrate communities for local microalgal production in an estuarine mangrove ecosystem during the post-monson period Journal of Sea Research, 2004, 51(2): 133-144.

[41] Bouillon S, Koedam N, Raman A V, *et al* Primary producers sustaining macro-invertebrate communities in intertidal mangrove forests Oecologia, 2002, 130(3): 441-448.

[42] Melville A J, Connolly R M. spatial analysis of stable isotope data to determine primary sources of nutrition for fish Oecologia, 2003, 136(4): 499-507.

[43] Hemminga M A, Slim F J, Kazungu J, *et al* Carbon outwelling from a mangrove forest with adjacent seagrass beds and coral reefs (Gazi Bay, Kenya). Marine Ecology Progress Series, 1994, 106(3): 291-301.

[44] Fry B. $^{13}C/^{12}C$ ratios and the trophic importance of algae in Florida *Syringodium filiforme* seagrass meadows Marine Biology, 1984, 79(1): 11-19.

[45] Zieman J C, Macko S A, Mills A L. Role of seagrasses and mangroves in estuarine food webs: temporal and spatial changes in stable isotope composition and amino acid content during decomposition Bulletin of Marine Science, 1984, 35(3): 380-392.

[46] Harriga P, Ziema J C, Macko S A. The base of nutritional support for the gray snapper (*Lutjanus griseus*): an evaluation based on combined stomach content and stable isotope analysis Bulletin of Marine Science, 1989, 44(1): 65-77.

[47] Fleming M, Li G, Sternberg L, *et al* Influence of mangrove detritus in an estuarine system. Bulletin of Marine Science, 1990, 47(3): 663-669.

[48] Newell R I E, Marshall N, Sasekumar A, *et al* Relative importance of benthic microalgae phytoplankton and mangroves as sources of nutrition for penaeid prawns and other coastal invertebrates from Malaysia Marine Biology, 1995, 123(3): 595-606.

[49] Primavera J H. Stable carbon and nitrogen isotope ratios of penaeid juveniles and primary producers in a riverine mangrove in Guimaras Philippines Bulletin of Marine Science, 1996, 58(2): 675-683.

[50] Dittel A I, Epifanio C E, Cifuentes L A, *et al* Carbon and nitrogen sources for shrimp postlarvae fed natural diets from a tropical mangrove system. Estuarine and Coastal Marine Science, 1997, 45(5): 629-637.

[51] Marguillier S, van der Velde G, Dehairs F, *et al* Tropical relationship in an interlinked mangrove-seagrass ecosystem as traced by $\delta^{13}C$ and $\delta^{15}N$. Marine Ecology Progress Series, 1997, 151(1-3): 113-121.

[52] France R. Estimating assimilation of mangrove detritus by fiddler crabs in Laguna Joyuda, Puerto Rico, using dual stable isotopes Journal of Tropical Ecology, 1998, 14(4): 413-425.

[53] Hayas S, Ichikawa T, Tanaka K. Preliminary report on stable isotope ratio analysis for samples from Matang mangrove brackish water ecosystems Japan Agricultural Research Quarterly, 1999, 33(3): 215-221.

[54] Bouillon S, Chandra Mohan P, Sreenivas N, *et al* Sources of suspended organic matter and selective feeding by zooplankton in an estuarine mangrove ecosystem as traced by stable isotopes Marine Ecology Progress Series, 2000, 208: 79-92.

[55] Lee S Y. Carbon dynamics of Deep Bay, eastern Pearl River estuary, China, II: trophic relationship based on carbon- and nitrogen-stable isotopes Marine Ecology Progress Series, 2000, 205: 1-10.

[56] Chong V C, Low C B, Ichikawa T. Contribution of mangrove detritus to juvenile prawn nutrition: a dual stable isotope study in a Malaysian mangrove forest Marine Biology, 2001, 138(1): 77-86.

[57] Thimdee W, Deein G, Sangrungruang C, *et al* Stable carbon and nitrogen isotopes of mangrove crabs and their food sources in a mangrove fringed estuary in Thailand Benthos Research, 2001, 56(1): 73-80.

[58] Hsieh H L, Chen C P, Chen Y G, *et al* Diversity of benthic organic matter flows through polychaetes and crabs in a mangrove estuary: $\delta^{13}C$ and $\delta^{34}S$ signals Marine Ecology Progress Series, 2002, 227: 145-155.

[59] Schwamborn R, Ekau W, Voss M, *et al* How important are mangroves as a carbon source for decapod crustacean larvae in a tropical estuary? Marine Ecology Progress Series, 2002, 229, 195-205.

[60] Fry B, Smith T J III Stable isotope studies of red mangroves and filter feeders from the Shark River estuary, Florida Bulletin of Marine Science, 2002, 70(3): 871-890.

[61] Lee S Y. Mangrove outwelling-a review. Hydrobiologia, 1995, 295(1-3): 203-212.

[62] Odum W E. Dual gradient concept of detritus transport and processing in estuaries Bulletin of Marine Science, 1984, 35(3): 510-521.

[63] Farnsworth E J, Elliso A M. The global conservation status of mangroves Ambio, 1997, 26(6): 328-334.

[64] Wafar S, Untawale A G, Wafar M. Litter fall and energy flux in a mangrove ecosystem. Estuarine, Coastal and Shelf Science, 1997, 44(1): 111-124.

[65] Li M S, Lee S Y. The particulate organic matter dynamics of Pearl Bay, eastern Pearl River estuary, China I Implications for waterfowl conservation Marine Ecology Progress Series, 1998, 172: 73-87.

[66] Dehairs F, Rao R G, Chandra Mohan P, *et al* Tracing mangrove carbon in suspended matter and aquatic fauna of the Gautami-Godavari Delta, Bay of Bengal (India). Hydrobiologia, 2000, 431(2-3): 225-241.

[67] Twilley R R, Poro M, Garcia V H, *et al* Litter dynamics in riverine mangrove forests in the Guayas River Estuary, Ecuador Oecologia, 1997, 111(1): 109-122.

[68] Lee S Y. Ecological role of grapsid crabs in mangrove ecosystems: a review. Marine and Freshwater Research, 1998, 49(4): 335-343.

[69] Sullivan M J, Moncreiff C A. Edaphic algae are an important component of salt marsh food webs: evidence from multiple stable isotope analysis Marine Ecology Progress Series, 1990, 62: 149-159.

[70] Currin C A, Newell S Y, Paerl H W. The role of standing dead *Spartina alterniflora* and benthic microalgae in salt marsh food webs: considerations based on multiple stable isotope analysis Marine Ecology Progress Series, 1995, 121: 99-116.

[71] Page H M. Importance of vascular plant and algal production to macro-invertebrate consumers in a southern California salt marsh Estuarine, Coastal and Shelf Science, 1997, 45(6): 823-834.

[72] Micheli F. Feeding ecology of mangrove crabs in North Eastern Australia: mangrove litter consumption by *Sesarma messa* and *Sesarma smithii* Journal of Experimental Marine Biology and Ecology, 1993, 171(2): 165-186.

[73] Dennison W C, Abal E G Moreton Bay study: A scientific basis for the healthy waterways campaign In: South East Queensland Regional Water Quality Management Strategy. Brisbane: 1999.

[74] Loneragan N R, Bunn S E, Kellaway D M. Are mangroves and seagrass sources of organic carbon for penaeid prawns in a tropical Australian estuary? A multiple stable-isotope study. Marine Biology, 1998, 30(2): 289-300.

[75] Bouillonl S, Dahdouh-Guebas F, Rao A V V S, *et al* Sources of organic carbon in mangrove sediments: variability and possible ecological implications Hydrobiologia, 2003, 495(1): 33-39.

[76] Han W D, Gao X M, Lu C Y, *et al* The ecological values of mangrove ecosystems in China Ecologic Science, 2000, 19(1): 40-46.

[77] Costanza R, d'Arge R, de Groot R, *et al* The value of the world(s ecosystem services and natural capital Ecological Economics, 1997, 387(1): 253-260.

[78] Guest MA, Connolly RM. Fine-scale movement and assimilation of carbon in saltmarsh and mangrove habitat by resident animals Aquatic Ecology, 2004, 38(4): 599-609.

[79] Mumby P J, Edwards A J, J E Arias-Gonz lez *et al* Mangroves enhance the biomass of coral reef fish communities in the Caribbean Nature, 2004, 427 (6974): 533-536.

[80] Schwarcz H P. Some theoretical aspects of isotope paleodiet studies Journal of Archaeology Science, 1991, 18(3): 261～275.

[81] Gannes L Z, Martínez del Rio C, Koch P. Natural abundance variations in stable isotopes and their potential uses in animal physiological ecology. Comparative Biochemistry and Physiology, 1998, 199(3): 725-737.

[82] Torgersen T, Chivas A R. Terrestrial organic carbon in marine sediment a preliminary balance for a mangrove environment derived from ^{13}C. Chemical Geology, 1985, 52: 379-390.

[83] Twilley R R. The exchange of organic carbon in basin mangrove forests in a southwest Florida estuary. Estuarine and Coastal Marine Science, 1985, 20(5): 543-557.

[84] Cifuentes L A, Coffin R B, Solorzano L, *et al* Isotopic and elemental variations of carbon and nitrogen in a mangrove estuary. Estuarine and Coastal Marine Science, 1996, 433(6): 781-800.

[85] Dittmar T, Lara R J. Do mangroves rather than rivers provide nutrients to coastal environments south of the Amazon River? Evidence from long-term flux measurements Marine Ecology Progress Series, 2001, 213: 67-77.

[86] Holguin G, Vazquez P, Bashan Y. The role of sediment microorganisms in the productivity, conservation, and rehabilitation of mangrove ecosystems: an overview. Biology and Fertilization Soils, 2001, 33(4): 265-278.

参考文献:

[11] 王建柱, 林光辉, 黄建辉, 等. 稳定同位素在陆地生态系统动-植物相互关系研究中的应用. 科学通报, 2004, 49(21): 2141～2149.

[12] 蔡德陵, 张淑芳, 张经. 稳定碳、氮同位素在生态系统研究中的应用. 青岛海洋大学学报, 2002, 32(2): 287～295.

[19] 范航清, 何斌源, 韦受庆. 传统渔业活动对广西英罗港红树林区渔业资源的影响与管理对策. 生物多样性, 1996, 4: 167～174.

[20] 范航清, 韦受庆, 何斌源, 等. 英罗港林缘潮水中游泳动物的季节动态. 广西科学, 1998, 5(1): 45～50.

[21] 何斌源, 范航清, 莫竹承. 广西英罗港红树林区鱼类多样性研究. 热带海洋学报, 2001, 4: 74～79.

[22] 林鹏. 福建漳江口红树林湿地自然保护区综合科学考察报告. 厦门:厦门大学出版社, 2001. 68～103.

[25] 郑淑惠, 郑斯成, 莫志超. 稳定同位素地球化学分析. 北京: 北京大学出版社, 1986. 309.

[35] 施富山, 王瑁, 王文卿, 等. 红树林与鱼类关系的研究进展. 海洋科学, 2005, 29(5): 54～59.

[76] 韩维栋, 高秀梅, 卢昌义, 等. 中国红树林生态系统生态价值评估. 生态科学, 2000, 19(1): 40～46.

梯度淹水胁迫下全日潮海区秋茄幼苗的生长和生理反应*

何斌源[1,2] 赖廷和[1] 王文卿[2] 陈剑锋[1] 邱广龙[1]

(1. 广西红树林研究中心,广西 北海 536007;2. 厦门大学生命科学学院,福建 厦门 361005)

摘 要:以秋茄(*Kandelia candel* L. Druce)幼苗为材料,在全日潮海区广西英罗湾开展为期 5 个月的野外梯度淹水胁迫实验。3 座平台作为重复,每座设置 8 个梯度,相邻高程组间的高度相差 10 cm。实验 5 个月后测定幼苗的生长状况和一些生理指标。结果表明:秋茄幼苗的生长高度、节数、叶数和大根数这 4 个外部形态特征指标的最大值均出现在 360 cm 高程组。生长高度、节数、叶数的曲线较为一致,幼苗大根数的略有不同,但总体上可看出太低的高程均不利于这 4 个指标的增长。叶面积对淹水胁迫的反应与叶数一致,但高程组之间的差异更大。叶绿素含量则表现长时间淹水胁迫促进含量增高。在任一高程组,秋茄幼苗各新生器官生物量分配均表现为:茎 > 根 > 叶。各新生器官及全株生物量的最大值出现在 360 cm 高程组,它们的干重曲线表现比较一致。幼苗根系中的活性氧清除酶类活性均高于叶片。根系中 SOD 酶和 POD 酶的分布规律较为一致,都以 360 cm 高程组的活性最高。叶片中 SOD 酶和 POD 酶的变化则相反,叶片 SOD 酶随高程增大而降低,POD 酶则随高程增大而增大。根系中活性氧清除酶类活性的水平与幼苗全株生物量的关系均呈显著相关,叶片中的则无此相关性。360 cm 及以上高程组幼苗全部存活,较低高程组有部分死亡。综合本文实验结果,可初步判断北部湾沿海秋茄造林的滩涂高程不宜低于当地平均海平面。

关键词:秋茄幼苗;全日潮;海面高程;淹水胁迫;生长反应

中图分类号:Q178.53 **文献标识码**:A **文章编号**:1001-6932(2007)02-0042-0008

历史上我国东南沿海曾分布着大面积茂盛的红树林,然而不合理的开发活动已造成大量原生红树林消失,残存的红树林普遍出现次生化,从而导致系统功能的退化、生态服务价值总量锐减,危及近海渔业和养殖业[1]。保护和恢复红树林已成为我国东南沿海海岸生态安全建设的重要内容之一。

红树林生态恢复工程的关键是宜林地的选择标准[2]。自然环境中不同的红树种类在滩涂上呈带状分布,因此对应不同的红树种类,宜林地应是多样化和梯度化的。宜林地滩涂高程范围尤其是最低高程线,很大程度上取决于红树植物在淹水环境中生理能力的强弱。关于淹水胁迫下红树植物在生长、营养发育、生物量分配、相关酶系、激素水平等方面的生理生态反应,目前国外已有较多的研究[3-12],国内也开展了部分工作[13-17]。陈鹭真等[17]指出以往研究的淹水胁迫方式是土壤表面渍水,并且该文采用了幼苗被周期性淹没的处理方式。由于全日潮潮汐类型少见,上述文献集中研究了红树在半日潮海区一些潮汐特征下的淹水胁迫的生理生态反映,且大多采用室内模拟试验。

我国红树林的造林技术有很强的适地性,一地成功的造林经验往往难于直接应用到另一地,因此造林理论和技术必然是多元化和区域化的。北部湾的广东、广西海岸是我国红树林主要分布区,也是主要的红树林造林区。北部湾的潮汐类型是世界少见的全日潮,水文、光照条件和沉积物等因素不同于半日潮海区。在全日潮海区,红树受浸淹频度相对较低而单个潮日内连续受浸淹时间较长,而在半日潮海区则情况相反。为了科学精细地确定全日潮海区淹水条件下红树林的宜林地临界线,为我国全日潮海区红树植物秋茄宜林地划分和造林成功率提高提供科学依据,本文选择中国红树植物中分布最广的秋茄作为研究材料,在野外自然状态下开展了围绕平均海面线共 8 个海面高程梯度的幼苗淹水实验,对比研究幼苗在生长形态、叶片和根系中抗逆境的活性氧清除酶类(SOD 和 POD)活性等方面的生理、生态差异。

* 广西科学基金项目(桂科自 0447007)

原载于海洋通报,2007,26(2):42-49

1 样地概况

本文实验基地英罗湾(109° 43′ E,21° 28′ N)是广西山口国家级红树林自然保护区的核心区,位于广西东海岸。这里属南亚热带海洋性性季风气候,年平均气温 23.4 ℃,1 月均温 14.2 ℃~14.5 ℃,极端最低气温 2 ℃。年降水量 1 500~1 700 mm,蒸发量 1 000~1 400 mm,年均相对湿度为 80 %。英罗湾海区的潮汐类型属非规则全日潮,全日潮占一年中约 60 % 的天数。多年平均潮差为 2.52 m,最大潮差 6.25 m。多年平均海水盐度为 28.9 [18]。湾内生长的红树林面积 80 hm^2,绝大部分为红海榄(*Rhizophora stylosa*)种群,其他的红树种类有桐花树(*Aegiceras corniculatum*)、白骨壤(*Avicennia marina*)、秋茄和木榄(*Bruguiera gymnorrhiza*)等。

2 材料与方法

实验时间为 2005 年 4-9 月,历时 5 个月;实验材料为在当地采集的秋茄幼苗:筛选比较一致的胚轴用于实验,选取胚轴鲜重范围为 11~12 g,长度为 22~24 cm;实验场地为在红树林外缘平坦光滩上搭建的 3 个各有 8 级梯级的实验平台,相邻梯级间高度差为 10 cm。采种后当天将胚轴直接插播在已装好淤泥的育苗袋中,摆放到 3 个野外实验平台上,每种每层摆放 30 袋。根据多次实地测验,从平台的最底层至最高层,摆放幼苗后 8 个梯级基质表面的高程分别相当于 320 cm, 330 cm, 340 cm, 350 cm, 360 cm, 370 cm, 380 cm 和 390 cm(英罗湾当地潮高基准面在平均海面下 359 cm,即为当地平均海面高程)。试验平台的高程确定方法如下:首先通过多次对照国家海洋局潮汐表而粗定建造地址(高程大约为 300 cm),接着采用全站仪法测定较精确的滩涂高程(平均误差在 ±5 cm 以内),然后用木板和石块铺垫使之达到目的高程。

实验结束后将全部幼苗带回实验室,测定幼苗外部形态指标、各器官生物量和一些生理指标。叶面积测定采用剪纸称重法。叶片叶绿素含量、叶片和根系中的活性氧清除酶类活性测定取样时,叶片取完全展开的第 2 对真叶,根系则随机取样,每一平台的幼苗重复取样 3 次。叶片叶绿素含量测定采用混合液提取法[19],提取时间为 48 h。参照赵世杰等(2002)的方法[20],超氧物歧化酶 SOD 采用氮蓝四唑法(提取酶的磷酸缓冲液中加 1 % 的聚乙烯吡咯烷酮 PVP)。过氧化物酶 POD 活性测定采用比色法[21]。

3 结果与分析

3.1 不同高度梯度淹水胁迫下秋茄幼苗的生长反应

3.1.1 茎的生长反应　360 cm 高程组是所有高程组中生长最快的(图 1),370 cm 高程组次之,比前者生长高度仅低 1.6 %;360 cm 高程组幼苗茎的生长高度比其余的处理组高出 10 % 以上,其中比最低的 340 cm 高程组显著地高出了 32.1 %。显然 360 cm 海面高程滩涂最适合于秋茄茎的生长。由图 1 可看出,在 320~340 cm 低海面高程上对茎的高度增长有一定促进作用,但生长情况劣于 350 ~390 cm 的 5 个高程组,则表明低海面高程下对茎生长的促进作用只具有相对意义,总体上太低的海面高程不利于秋切幼苗茎的生长。

不同海面高程组幼苗茎节数的生长规律与生长高度的情况相似(图 1),总体上也是太低的海面高程不利于秋茄幼苗节数的增长。320~340 cm 高程组的平均节数为 6.1 节, 350~390 cm 高程组的平均值为 6.7 节。360 cm 高程组是所有高程组中节数最多的,370 cm 高程组次之,前者生长高度仅比后者的高出 3.0 %。340 cm 高程组幼苗节数最少,但与最大值的差别不如在生长高度上的那么大,360 cm 高程组的节数比 340 cm 高程组多出 22.0 %。

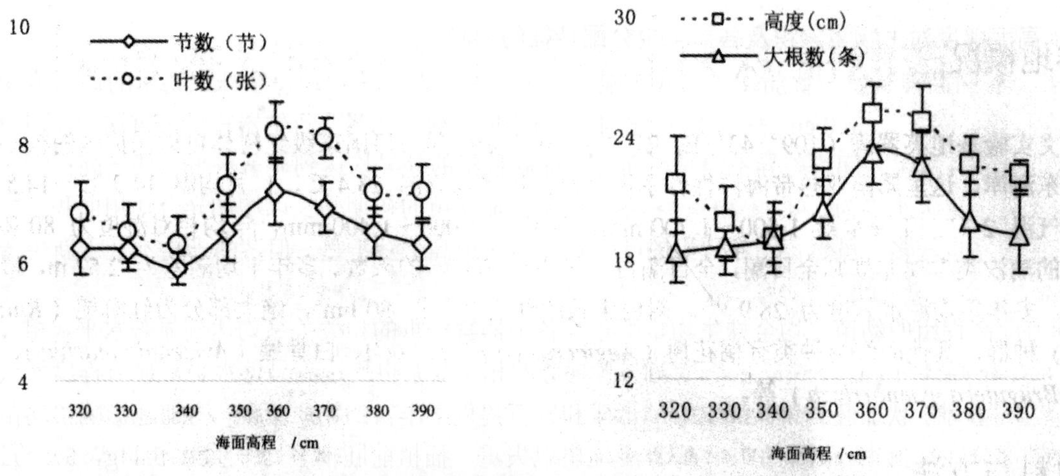

图 1 梯度海面高程的滩涂上秋茄幼苗的生长形态

Fig.1 Growth state of *K. candel* seedlings on the artificial tidal flats with gradients of sea level altitudes

3.1.2 叶数和叶面积的反应 8 个高程组中，360 cm 高程组秋茄幼苗的叶数最多（图 1），370 cm 处理组次之，前者叶数仅多出后者 1.2 %。360 cm 高程组显著地高出叶数是最少的 340 cm 高程组 30.2 %。在 320～340 cm 低海面高程组，长时间淹水对促进叶数的增长起一定作用，但这 3 个高程组的平均叶数为 6.5 张，最大值为 6.8 张，而 350～390 cm 这 5 个组平均叶数为 7.5 张，最小值 7.2 张，说明太低的海面高程不利于秋茄幼苗叶数的生长和保存。

360 cm 高程组幼苗的平均每叶叶面积仅高出 370 cm 高程组的 1.1 %（图 2）。这两组与其余的高程组的差异均达到显著水平，其中 360 cm 高程组比 320 cm 高程组显著地高出了 72.5 %。350～390 cm 等 5 个高程组的平均每叶叶面积均大于 320～340 cm 等 3 个高程组，可见过低的海面高程不利于秋茄幼苗叶面积发展。

3.1.3 大根数量的反应 秋茄幼苗的大根数曲线与其他外部形态特征值略有差异，表现为单峰分布（图 1）。大根数在 360 cm 高程组出现最大值，高于或低于此高程大根数量对淹水胁迫的表现均为下降的反应趋势。370 cm 高程组仅比 360 cm 组低 3.1 %，390 cm 高程组的大根数也高于 320 cm 组，所以总体趋势是较低高程对秋茄幼苗的大根发育具有更大的抑制作用。

3.3 淹水高度梯度对叶绿素含量的影响

320～340 cm 这 3 个高程组秋茄幼苗叶片的叶绿素 a、b 和总含量较相近，且明显高于 350 至 390 cm 等 5 个处理组（图 3）。低海面高程显著促进幼苗叶片叶绿素含量提高，较大海面高程对之则起抑制作用或没有促进作用。这表明在一定条件下，秋茄幼苗可通过提高光合作用来抵抗淹水胁迫的影响。

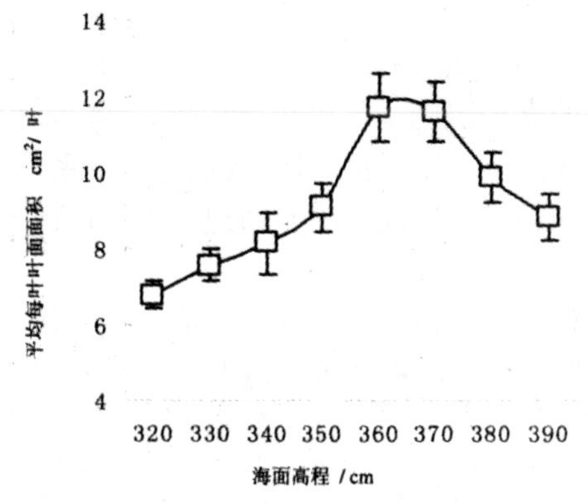

图 2 不同高程滩涂上秋茄幼苗的平均每叶叶面积

Fig.2 Average leaf areas of *K. candel* seedlings on the artificial tidal flats with gradients of sea level altitudes

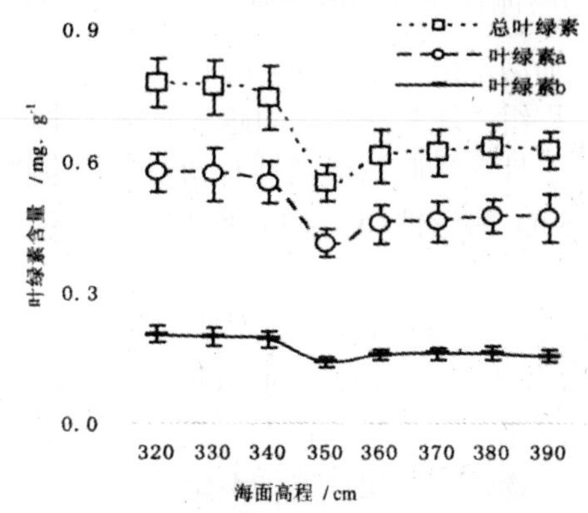

图 3 不同高程滩涂上秋茄幼苗的叶绿素含量

Fig.3 Contents of Chl-a, Chl-b and total Chl in mature leaves of *K. candel* seedlings on the tidal flats with gradients of sea level altitudes

3.4　淹水高度梯度对生物量累积及各器官中分配特征的影响

在 8 个海面高程组中，秋茄幼苗的叶、根系、茎、新生器官总干重及全株生物量的最大值均出现在 360 cm 高程组，次之为 370 cm 高程组（图 4）。对于叶、根系、茎及新生器官总干重这 4 个指标，360 cm 与 370 cm 高程组之间均差异不显著，而与其他高程组的差异均达到显著水平。任一高程组的秋茄幼苗的新生器官干重大小均为：叶 > 根系 > 茎，可见秋茄幼苗的初期生长以增加光合作用面积为主，用于固定自身的根系也较多。虽然每一高程组的叶干重均显著地高于茎干重，但茎的干重不低于新生器官总干重的 25 %。

不同海面高程组的幼苗平均全株生物量干重与各个器官干重的相关关系均达到显著水平，相关系数大小为：$r_{全株—茎}>r_{全株—根}>r_{全株—叶}>r_{全株—胚轴}$。幼苗与各部分的相关关系也均达到显著水平，显著程度依次为：根 > 叶 > 茎 > 胚轴。秋茄胚轴为新生器官的形成和发育提供了丰富的物质基础，密切地影响了幼苗生长进程。原胚轴部分在幼苗生长过程中均不同程度地出现失重，插植前胚轴平均干重为 5.44g，5 个月后这一部分仅剩 3.10～3.82 g，320 cm, 330 cm 和 390 cm 这 3 个高程组胚轴失重最严重，340～380 cm 等 5 个高程组的失重程度则较为接近。

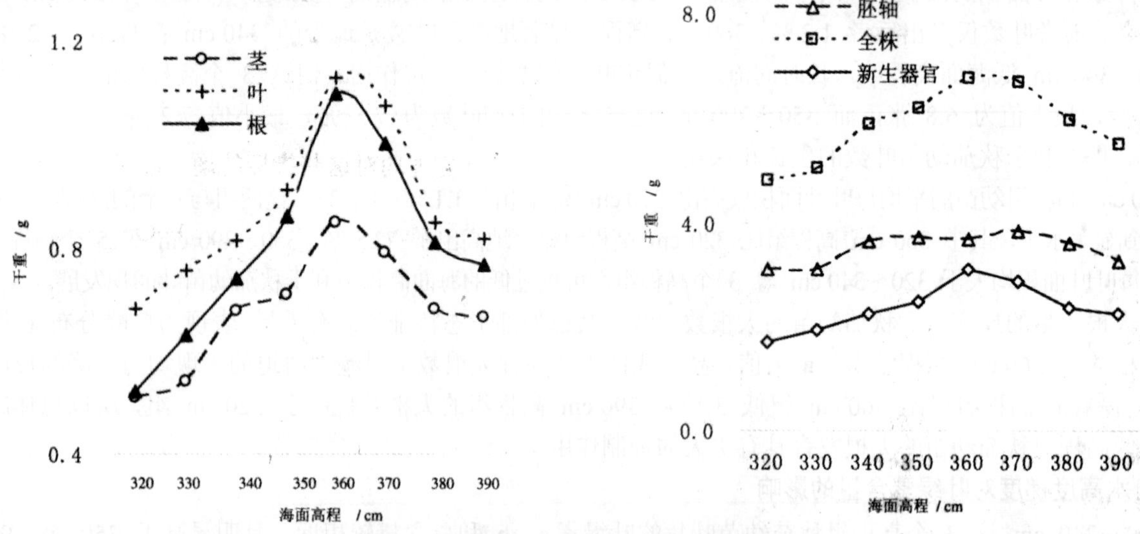

图 4　不同高程滩涂上秋茄幼苗的生物量分配

Fig.4　Biomass proportions of *K. candel* seedlings on the artificial tidal flats with gradients of sea level altitudes

3.5　与耐淹性相关的活性氧清除酶系对淹水高度梯度的反应

秋茄幼苗根系中 SOD 酶和 POD 酶活性的最大值均出现在 360 cm 海面高程组（图 5），大于此高程的高程组表现为淹水时间越长而酶活性越高，而低于此高程的高程组则呈现出相反趋势。同时，根系中的两种活性氧清除酶类活性与幼苗全株生物量的关系均呈显著相关。相同高程组的秋茄幼苗根系中 SOD 酶和 POD 酶活性均大于叶片中的，显然是由于根系更加频繁和长时间地面临淹水胁迫，更多地诱导这些酶的活性上升。

叶片中 SOD 酶活性大体上表现为较低的高程对 SOD 酶活性有抑制作用，其最大值出现在最大高程 390 cm 高程组。叶片中的 POD 酶则相反，总的表现为酶活性随海面高程增大而减弱，即长时间的淹水对 POD 酶活性有促进作用，其最小值出现在 390 cm 最大高程组。叶片中 SOD 酶和 POD 酶活性与幼苗全株生物量之间相关关系不显著。

3.6　淹水胁迫下幼苗存活率

在 5 个月的淹水实验期间，360 cm 及其以上高程组的幼苗存活率均为 100 %（图 6），其下从 350 cm 高程组开始出现死亡，存活率随海面高程降低而降低，最低存活率为 83 %。

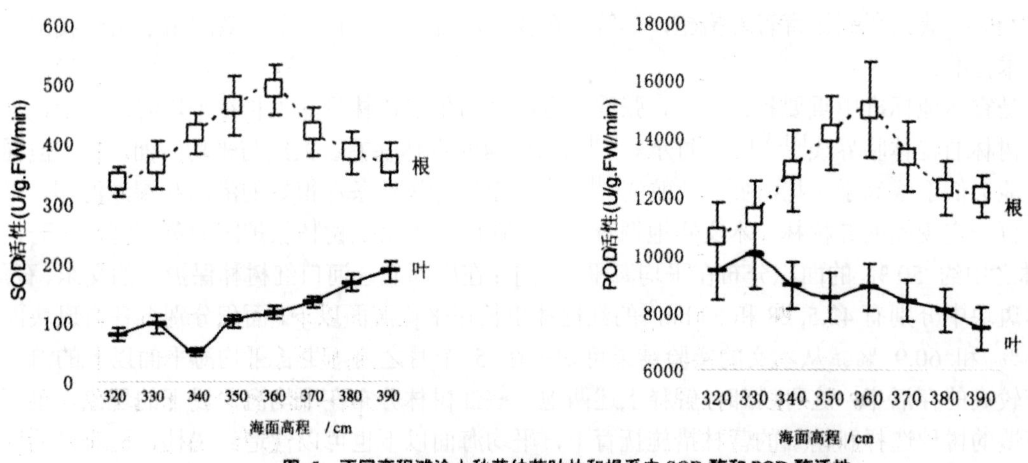

图 5 不同高程滩涂上秋茄幼苗叶片和根系中 SOD 酶和 POD 酶活性

Fig.5 Activities of SOD and POD in mature leaves and roots of *K. candel* seedlings on the tidal flats with gradients of sea level altitudes

4 讨 论

4.1 秋茄抗淹生理生态能力与现实生态位的差异

从幼苗的外部形态特征指标、生物量分配来看，太低的高程处理均对这些指标的增长起了抑制作用。在生理指标上，低海面高程处理使叶绿素含量提高，有利于加强光合作用抵抗淹水胁迫；秋茄幼苗根系中活性氧清除酶类活性大于叶片，而且根系中的这两种酶与幼苗全株生物量的关系均呈显著相关。根系活性氧清除酶类起着清除植物体内的氧自由基、防止脂膜过氧化的作用，保障细胞正常功能，促进干物质的累积，是秋茄幼苗抵抗淹水胁迫关键因素。

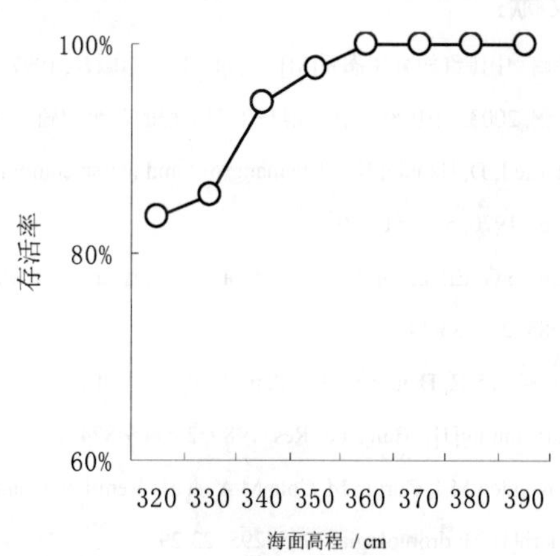

图 6 不同高程滩涂上秋茄幼苗存活率

Fig.6 Survival rates of *K. candel* seedlings on the tidal flats with gradients of sea level altitudes

本文实验设置的最低高程组的海面高程为 320 cm，尽管低于当地平均海面线约 40 cm，但供试的秋茄幼苗仍有较高的存活率。在广西英罗湾观察到，虽然在向海林带外缘滩涂上分布着一些秋茄成树，有一定的种源补充，但与本研究最低处理组平行的滩涂上却罕见红树幼苗生长。在自然光滩上，还存在着一些不利于秋茄胚轴着生的复杂的胁迫因素，如沉积物性质的影响、潮汐冲刷、植物种间竞争、动物啃咬、人为破坏，以及“光眠现象”抑制秋茄根的萌发等等，在这些复杂的因素胁迫下秋茄幼苗即使发挥了其最大的抗淹生理能力，也可能难以完全占据该生态位，这反映植物的抗淹生理生态能力与现实生态位之间存在一定差异。本实验实际上已经给秋茄提供了去除“光眠现象”抑制作用、稳定的沉积物着生基质、减少或缺失种间竞争、以及精细管理和抚育等这些对秋茄胚轴萌发和幼苗生长起着至关重要的生长环境。在人类的干预或抚育下，使植物的最大生理能力得以实现，这也是人工育苗和造林所希望达到的目的。

4.2 探讨秋茄的宜林临界线

秋茄是我国红树林造林最重要的树种之一。它分布广，适应性强，在我国红树植物耐寒性等级序列中被列为最耐寒的第 I 级[1]。在我国红树林引种北界的浙江省，秋茄甚至成了当地红树林造林惟一的选择。我国的红树林造林历史长远、实践经验丰富，秋茄造林理论和经验有较多的总结报道[22-30]。但我国的红树林造林成功率低是一个普遍的现象，其原因是多方面的，如：宜林地标准模糊、不了解红树林的海洋属性而照搬陆地造林经验、缺乏后期抚育管理措施，等等。在我国，目前尚没有一个有效的用于指导我国红树林造林行动的技术标准。

宜林临界线是宜林地标准中重要指标之一，张乔民等研究指出红树林只能生长在平均海面之上，平均海面线可作为红树林的宜林临界线[31, 32]。莫竹承等[33]指出广西的宜林临界线大约与平均海面重叠，但同时指出个别地方可能会低于或高于平均海面。范航清[34]认为在个别避风浪条件很好的港湾和潟湖海岸，平均海面以下也可发育一定规模的红树林。本文实地调查发现：在广西山口红树林保护区的英罗湾，在一片约 80 hm^2 的红树林之中约 50 % 的面积分布在平均海平面之下；在广西北仑河口红树林保护区的交东、石角、马兰基的红树林斑块中分别有 42.5, 28 和 65 hm^2 的红树林生长在平均海面以下，面积分别占各自斑块面积的 25 %，28.7 % 和 60.9 %。从本文的实验结果可知：在 5 个月之内，生长平均海平面以下的滩涂上的秋茄幼苗仍有较高的存活率，这可以部分解释上述两地一些红树林分布在平均海面以下的现象，也可以初步说明：在谨慎的树种选择和精细的营林措施抚育下，平均海面以下也可以营造红树林。这对目前许多处于潮间带偏下海滩和堤前无红树林保护的海堤的抗风浪治理，无疑可提供借鉴作用。

对于秋茄这一种红树植物，廖宝文等[25]研究指出在深圳赤湾用秋茄造林的潮滩高程不低于当地平均海面 22 cm；而在海南东寨港，造林潮滩的高程不低于当地平均海面 30 cm。表明因潮汐类型、水热条件、土壤条件的差异，同一种红树植物在不同地区可有不同的宜林临界线。考虑到实验时间越长，将有更多的植株死亡，结合秋茄自身的抗淹生理耐受能力，可初步判断：在全日潮海区，秋茄的宜林地滩涂高程不应低于当地平均海平面。

参考文献：

[1] 林鹏. 中国红树林生态系[M]. 北京: 科学出版社, 1997.

[2] 林鹏. 2003. 中国红树林湿地与生态工程的几个问题. 中国工程科学. 5(6): 33-38

[3] Clarke L D, Hannon N J. The mangrove and marsh communities of the Sydney district: III. Plant growth in relation to salinity and waterlogging [J]. J Ecol, 1970, 58: 351-369.

[4] Naidoo G. Effects of waterlogging and salinity on plant-water relations and on the accumulation of solutes in three mangrove species [J]. Aquat. Bot.，1985, 22: 133-143.

[5] Pezeshki S R, Delaune R D，Patrick W H Jr. Differential response of selected mangroves to soil flooding and salinity：gas exchanges and biomass partitioning[J]. Can J For Res, 1989, 20: 869-874.

[6] Hovenden M J, Curran M, Cole M A, et a1. Ventilation and respiration in roots of one-year old seedlings of grey mangrove Avicennia marina (Forsk.) Vierh[J]. Hydrobiologia, 1995, 295: 23-29.

[7] Ellison A M, Farnsworth E J. Simulated sea level change alters anatomy，physiology，growth，and reproduction of red mangrove (Rhizophora mangle L)[J]. Oceanographic Literature Review, 1998, 45(6): 1003-1004.

[8] Misra S, Choudhurya A, Ghosh A. The role of hydrophobic substances in leaves in adaptation of plants to periodic submersion by tidal water in a mangrove ecosystem[J]. J Ecol, 1984, 72(2): 621-625.

[9] Skelton N J, Allaway W G. Oxygen and pressure changes measured in situ during flooding in roots of the grey mangrove Avicennia marina (Forsk.) Vierh[J]. Aquat Bot, 1996, 54(2，3): 165-175.

[10] Youssef T, Saenger P. Anatomical adaptive strategies to flooding and rhizophere oxidation in mangrove seedlings [J]. Aust J Bot，1996，44：297- 313.

[11] Koch M S, Snedaker S C. Factors influencing Rhizophora mangle L seedling development in everglades carbonate soils[J]. Aquat Bot，1997，59(1，2): 87-98.

[12] Pezeshki S R, Delaune R D, Meeder J F. Carbon assimilation and biomass partitioning in Avicennia germinans and Rhizophora mangle seedlings in response to soil redox conditions[J]. Environ Exp Bot，1997，37(2-3)：161-171.

[13] 叶勇, 卢昌义, 谭凤仪. 木榄和秋茄对水渍的生长与生理反应的比较研究 [J]. 生态学报, 2001, 21(10): 1 654-1 661.

[14] Ye Y et al. Growth and physiological responses of two mangrove species (Bruguiera gymnorrhiza and Kandelia candel) to waterlogging. Environmental and Experimental Botany, 2003, (49): 209-/221.

[15] Ye Y et al. Does sea level rise influence propagule establishment, early growth and physiology of *Kandelia candel* and *Bruguiera gymnorrhiza*? Journal of Experimental Marine Biology and Ecology. 2004, 306(2): 197-215。

[16] Chen L Z, Wang W Q, Lin P. Photosynthetic and physiological responses on *Kandelia candel* L. Druce seedlings to duration of tidal immersion in artificial seawater [J]. Environmental and Experimental Botany, 2005, 54: 256-266.

[17] 陈鹭真, 王文卿, 林鹏. 潮汐淹水时间对秋茄幼苗生长的影响[J]. 海洋学报, 2005, 27(2): 141-147.

[18] 范航清, 陈光华, 何斌源, 等. 山口红树林滨海湿地与管理[M]. 北京: 海洋出版社, 2005.

[19] 何斌源, 梁士楚, 凌俊文. 红树植物叶片叶绿素提取方法比较及其活体测定 [J]. 广西科学院学报, 1993, 9(2): 77-81.

[20] 赵世杰, 等主编. 植物生理学实验指导[M]. 北京: 中国农业科学技术出版社, 2002.

[21] 张志良, 瞿伟菁. 植物生理学实验指导[M]. 北京: 高等教育出版社, 2004.

[22] 汪惟礼. 谈谈秋茄引种栽植技术 [J]. 浙江林业科技, 1988, 8(3): 42-44.

[23] 卢昌义, 林鹏. 秋茄红树林的造林技术及其生态学原理[J]. 厦门大学学报 (自然科学版), 1990, 29(6): 694-698.

[24] 刘治平. 秋茄和木榄的海上育苗研究[J]. 生态科学, 1991, (1): 72-75.

[25] 廖宝文, 郑德璋, 郑松发, 等. 红树植物秋茄造林技术的研究[J]. 林业科学研究, 1996, 9(6): 586-592.

[26] 李建清, 徐何方, 李克思, 等。秋茄红树林北移引种造林技术[J]. 浙江亚热带作物通讯, 2000,23(1): 6-8.

[27] 李巧姿, 沿海滩涂生态因子对秋茄生长的影响分析[J]. 福建林业科技, 2000, 27(4): 31-34.

[28] 李建清, 徐何方, 李克思, 等. 温州沿海海涂秋茄红树林引种造林及开发前景[J]. 华东森林经理, 2001, 15(3): 24-25.

[29] 昝启杰, 王勇军, 廖宝文, 等. 秋茄种源引种深圳湾后幼苗生理生态研究[J]. 生态学报，2001, 21(10): 1 662-1 669.

[30] 林光平. 秋茄红树林造林技术[J]. 林业实用技术, 2005, (1): 17-18

[31] 张乔民, 于红兵, 陈欣树, 等. 红树林生长带与潮汐水位关系的研究[J], 生态学报, 1997, 17(3): 258-265.

[32] 张乔民, 等. 红树林宜林海洋环境指标研究 [J]. 生态学报, 2001, 21(9): 1427-1437.

[33] 莫竹承. 广西红树林立地条件研究初报 [J]. 广西林业科学, 2002, 31(3): 122-127.

[34] 范航清. 红树林—海岸环保卫士[M]. 南宁: 广西科学技术出版社.

Growth and Physiological Response of *Kandelia candel* L. Druce Seedlings to Gradients of Waterlogging Stress in the Diurnal Sea Area

HE Binyuan [1,2], LAI Tinghe[1], WANG Wenqing[2], CHEN Jianfeng[1], QIU Guanglong[1]

(1. Guangxi Mangrove Research Center, Beihai 536007, Guangxi, China; 2. School of Life Science, Xiamen University, Xiamen 361005, Fujian, China)

Abstract: A field waterlogging stress experiment on the *Kandelia candel* L. Druce seedlings were conducted in the Yingluo Bay (109° 43′ E, 21° 28′ N) of Guangxi, China, where the tide type was a diurnal one. Three wooden platforms as three replicates were built for the seedlings cultivation and eight gradients of sea level altitudes (abbreviated as SLA) on each platform were provided, with a difference of 10 cm in height between neighboring treatments. After five months of cultivation under gradients of SLA , the growth states of the seedlings and some physiological responses in relation to the waterlogging tolerant capacity were measured. The results showed that all the four maximums of height, node number, leaf number and cable root number occurred under the 360cm SLA group. As a whole, too low SLA hampered the development of the height, node, leaf and cable root. The leaf areas followed the same rule as the leaf numbers', while the differences in of the former between SLA groups were more significant than in the latter. The chlorophyll contents of lower SLA groups were higher than the higher ones, indicating that the longer waterlogging had a positive promotion to the increase of chlorophyll contents. In each SLA group, the order of biomass composition in different neonatal organs was: leaf >root > stem for *K. candel* seedling. All the maximums of neonatal organs and total biomass of the seedling occurred in the 360cm SLA group, and their biomass curves were very similar. The activities of reactive-oxygen-processed enzymes in root were higher than those in leaf in the same gruop. In root, the SOD followed the same rule as POD, and both of their maximums occurred in the 360 cm SLA group. In leaf, there showed an opposite trend, with the SLA increased, the SOD activity decreased while the POD increased. There existed significant correlation between the total biomasses of the seedling and the activities of reactive-oxygen-processed enzymes in root, but not in leaf. Seedlings under the 360 cm-and-above SLA all survived through the duration of experiment, however, some seedlings under a lower SLA died. Comprehensively considering, a preliminary proposal can be presented that the favorable SLA of tidal flats for *K. candel* afforestation forestation along the Beibu Gulf seacoast should not be lower than the local mean sea level.

Keywords: *Kandelia candel* L. Druce seedling; diurnal tide; sea level altitudes; waterlogging stress; growth response

真红树和半红树植物叶片氯含量及叶性状的比较*

牟美蓉 蒋巧兰 王文卿

(厦门大学生命科学学院,厦门 361005)

摘 要 依据红树植物在潮间带的分布,将其分为真红树植物和半红树植物两大类。但对一些过渡地带种类的归属问题一直存在争议。该研究选取国内大部分红树植物,比较其成熟叶片中的 Cl 含量、肉质化程度、比叶面积(*SLA*)、单位重量叶氮含量(N_{mass})和单位面积叶氮含量(N_{area}),并对争议树种重新进行界定。结果表明:1)真红树植物叶片中 Cl 含量和肉质化程度远高于半红树植物;2)真红树植物具有低 *SLA* 和高 N_{area}的特点,除水芫花(*Pemphis acidula*)外半红树植物具有高 *SLA* 和低 N_{area}的特点。3)争议的 7 种红树植物中,银叶树(*Heritiera littoralis*)、海漆(*Excoecaria agallocha*)、卤蕨(*Acrostichum aureum*)和尖叶卤蕨(*Acrostichum speciosum*)归为半红树植物更合适;老鼠簕(*Acanthus ilicifolius*)和小花老鼠簕(*Acanthus ebrecteatus*)归为真红树植物。木果楝(*Xylocarpus granatum*)有待进一步研究。

关键词 真红树植物 半红树植物 界定 叶片氯含量 叶性状

COMPARISONS OF LEAF CHLORIDE CONTENT AND LEAF TRAITS BETWEEN TRUE MANGROVE PLANTS AND SEMI-MANGROVE PLANTS

MU Mei-Rong, JIANG Qiao-Lan, and WANG Wen-Qing

School of Life Sciences, Xiamen University, Xiamen 361005, China

Abstract Aims Mangrove plants are usually categorized as true mangrove plants and semi-mangrove plants on the basis of their distribution in inter-tidal regions. However, the identification of some fringe mangrove species found mainly on the landward transitional zones is controversial. Specific leaf area (*SLA*, leaf area per unit dry mass) and mass- and area-based leaf nitrogen concentrations (N_{mass} and N_{area}) are important leaf traits for plants, but relevant comparative research on true and semi-mangrove plants is unavailable. Our objective was to determine differences between the two groups and to classify the controversial species according to their leaf traits. Ultimately, this will assist in the management, protection and utilization of mangrove forest.

Methods Three individuals in similar growth sites were chosen for each species from Hainan Island. Fully expanded mature leaves were sampled from the upper canopy of all plants. Succulence (water content per unit leaf area), *SLA*, N_{mass} and N_{area} of mature leaves were studied for 33 species, representing all but three of the mangrove species in China.

Important findings True mangrove plants accumulated more Cl and water per unit leaf area than semi-mangrove plants, except for *Pemphis acidula*, *Hernandia sonora* and *Clerodendrum inerme*. Cl and water content per unit leaf area of true mangrove plants were generally >2.5 mg·cm^{-2} and >2.4 g·dm^{-2}, respectively. Cl concentrations were positively related to succulence for all mangrove species. True mangrove plants had low *SLA* (<100 cm^2·g^{-1}) and high N_{area}; however, semi-mangrove plants had high *SLA* (mean of 160.4 cm^2·g^{-1}). *Pemphis acidula* had much lower *SLA* than other semi-mangrove species. Our study suggested that there are significant differences between true mangrove plants and semi-mangrove plants in leaf Cl concentration, succulence, *SLA*, N_{mass} and N_{area}. *Heritiera littoralis*, *Excoecaria agallocha*, *Acrostichum aureum* and *Acrostichum speciosum* are better classified as semi-mangrove plants, while *Acanthus ilicifolius* and *Acanthus ebrecteatus* are classified as true mangrove plants, and *Xylocarpus granatum* needs further research.

Key words true mangrove plants, semi-mangrove plants, identification, leaf chloride conten, leaf traits

红树植物是生长于受潮汐影响的热带、亚热带海岸潮间带,具有特殊形态结构和生理适应的木本植物,通常分为真红树植物和半红树植物(王伯荪等,2003)。目前已经有非常明确的真红树植物和半红树植物划分标准(林鹏和傅勤,1995)。但一直以来人们对红树植物的划分都是在野外调查的基础

* 厦门大学新世纪优秀人才支持计划"资助课题和国家自然科学基金项目(30200031)

原载于植物生态学报,2007,31(3):497—504

上,根据其在潮间带的分布格局进行的,而没有从物种的形态结构和内部生理水平加以说明。由于野外生境的复杂性以及人们对划分标准理解的不同,导致对许多过渡地带的红树植物种类的界定产生了争议。真红树和半红树物种数统计数据的不一致,将不利于红树林的保护、管理与开发利用(王伯荪等,2003;赵亚和郭跃伟,2004)。

红树植物在潮间带的分布主要取决于对高盐、潮汐、贫瘠和生理干旱环境的适应能力,其中耐盐和耐水淹能力最为关键。但对真红树和半红树植物而言,耐水淹能力不是主要因素。红树林生境所有阴离子中,Cl^- 是与抗盐机制相关的最重要和最关键的一种元素(Joshi *et al*., 1972),它与 Na^+ 一起成为无机渗透调节最主要的贡献者(赵可夫等,1999)。叶片肉质化是红树植物调节体内盐分平衡的途径之一(Scholander, 1968; Tomlinson, 1999)。此外,植物的耐盐性还与水分利用效率相关,白骨壤(*Avicennia marina*)比桐花树(*Aegiceras corniculatum*)更耐盐的主要原因是前者具有更高的水分利用效率(Ball, 1988; Khan & Aziz, 2001)。一般来说,水分利用效率与单位面积叶氮含量成正相关,与比叶面积成负相关(Wilson *et al*., 1999; Poorter & de Jong, 1999)。

目前对红树植物生理生态特征的研究多集中在真红树植物,对半红树植物的相关研究非常少(Naidoo *et al*., 2002),而对真红树植物和半红树植物生理生态特征的比较研究就更少,而从叶性特征上来对真红树植物和半红树植物进行的比较研究几乎没有。相关研究的不足导致真红树植物和半红树植物界定的困难。本研究将通过对国内红树植物叶片盐分(主要是 Cl 元素)及叶性状(肉质化程度、比叶面积和叶氮含量)的研究来对真红树和半红树植物进行比较,并试图对一些有争议的物种:银叶树(*Heritiera littoralis*)、海漆(*Excoecaria agallocha*)、木果楝(*Xylocarpus granatum*)、卤蕨(*Acrostichum aureum*)、尖叶卤蕨(*Acrostichum speciosum*)、老鼠簕(*Acanthus ilicifolius*)和小花老鼠簕(*Acanthu ebrecteatus*)(Wang *et al*., 2003a)进行重新界定,为科学地保护、管理和利用红树林提供依据。

1 材料和方法

研究地点位于海南东寨港红树林自然保护区(19°51′N, 110°24′E)和文昌清澜港红树林自然保护区(19°22′~19°35′N, 110°40′~110°48′E)。各物种取样点、分布潮滩和土壤含盐量见表1。

对国内大部分红树植物进行采样,共33种(水椰(*Nypa fruticans*)和海南海桑(*Sonneratia hainansis*)因为个体较稀少未能进行采样)。其中,真红树16种,半红树10种,争议物种7种。在样地内选择标准木3棵,分别采集树冠外围完全展开且保持完整的成熟叶片20~50 g,回室内经清洗干净,测定叶面积。105 ℃杀青10 min,80 ℃烘干、磨粉,贮存备用。水芫花(*Pemphis acidula*)、莲叶桐(*Hernandia sonora*)、海滨猫尾木(*Dolichandrone spathaceae*)、海芒果(*Cerbera manghas*)、老鼠簕和小花老鼠簕因为样地内个体较少未重复采样。

植物 Cl 用 $AgNO_3$ 滴定法测定。称重法测定叶片饱和含水量,剪纸恒重法测定叶片面积,肉质化程度=饱和水分含量(g)/表面积(dm^2)。以每种植物的30枚叶片面积除以对应叶片的干重得到该种植物的比叶面积(*SLA*)(李玉霖等,2005)。植物样品经 H_2SO_4-H_2O_2 消化后N含量采用纳氏比色法测定(华南热带作物研究院,1974),所测值为单位重量叶氮含量(N_{mass});而单位面积叶氮含量(N_{area}) = N_{mass}/*SLA*。

2 结果

2.1 叶片中 Cl 元素含量和肉质化

2.1.1 Cl 元素含量

所测33种红树植物叶片 Cl 含量总体变异系数较大(78.7%)(表1),反映了红树林生境中植物对 Cl 的积累存在较大的差异性。其中,半红树的变异系数高达172.2%,真红树变异相对小一些,为32.9%。

真红树植物,尤其是大部分拒盐红树植物(红海榄(*Rhizophora stylosa*)、正红树(*R. apiculata*)、木榄(*Bruguiera gymnorhiza*)、角果木(*Ceriops tagal*)、瓶花木(*Scyphiphora hydrophyllacea*)、榄李(*Lumnitzera racemosa*)、红榄李(*L. littorea*)、海桑(*Sonneratia caseolaris*)、拟海桑(*S. ×gulngai*)、杯萼海桑(*S. alba*)和卵叶海桑(*S. ovata*))叶片中 Cl 含量较高,秋茄(*Kandelia obovata*)和泌盐红树植物(白骨壤和桐花树)叶片 Cl 含量略低。白骨壤和桐花树由于叶片具有盐腺能够泌盐,所以可以维持叶片较低的盐分浓度。拒盐红树植物秋茄是北半球最耐寒的拒盐红树植物,其叶片内较低的 Cl 含量也可能表明其有机渗透调节物质较其它真红树植物多,无机渗透调节物质相对来说较少,不过还有待研究。但比起大多数半红树植物来说,真红树植物叶片中的 Cl 含量相对

表1 国内33种红树植物单位叶面积Cl含量(mg·cm^{-2})、肉质化程度(单位叶面积含水量,g·dm^{-2})和各树种取样地、分布潮滩及生境土壤Cl含量(mg·g^{-1})

Table 1 Cl content per unit area (mg·cm^{-2}) and succulence (g·dm^{-2}) of mature leaves for 33 mangrove species and their sites, tides and Cl content (mg·g^{-1}) of soil for each species in China

类别 Groups	物种名 Plant species	单位叶面积Cl含量 Cl content per unit leaf area	肉质化程度 Succulence	取样地及分布潮滩 Site and tide	土壤Cl含量 Cl content of soil
真红树植物 True mangrove plants	1. 红海榄 *Rhizophora stylosa*	0.65 ±0.04	5.03 ±0.02	东寨港(中潮滩) Dongzhai Harbour (Meso-tide bank)	8.36
	2. 正红树 *R. apiculata*	0.57 ±0.06	4.68 ±0.27	清澜港(中高潮滩) Qianglan Harbour (Meso- and high-tide bank)	8.23
	3. 木榄 *Bruguiera gymnorhiza*	0.57 ±0.12	3.48 ±0.71	东寨港(中高潮滩) Dongzhai Harbour (Meso- and high-tide bank)	10.04
	4. 海莲 *B. sexangula*	0.31 ±0.02	2.41 ±0.30	东寨港(中高潮滩) Dongzhai Harbour (Meso- and high-tide bank)	15.56
	5. 尖瓣海莲 *B. sexangula* var. *rhynchopetala*	0.29 ±0.02	2.53 ±0.27	东寨港(中高潮滩) Dongzhai Harbour (Meso- and high-tide bank)	13.04
	6. 秋茄 *Kandelia obovata*	0.26 ±0.03	3.43 ±0.08	东寨港(中潮滩) Dongzhai Harbour (Meso-tide bank)	6.6
	7. 角果木 *Ceriops tagal*	0.47 ±0.16	4.39 ±1.37	东寨港(高潮滩) Dongzhai Harbour (High-tide bank)	8.82
	8. 瓶花木 *Scyphiphora hydrophyllacea*	0.66 ±0.04	4.58 ±0.19	清澜港(中高潮滩) Qianglan Harbour (Meso- and high-tide bank)	9.20
	9. 榄李 *Lumnitzera racemosa*	0.62 ±0.10	4.82 ±0.85	清澜港(中高潮滩) Qianglan Harbour (Meso- and high-tide bank)	9.20
	10. 红榄李 *L. littorea*	0.63 ±0.05	4.73 ±0.29	铁炉港(高潮滩) Tielu Harbour (High-tide bank)	5.34
	11. 海桑 *Sonneratia caeseolaris*	0.44 ±0.04	4.65 ±0.36	东寨港(中低潮滩) Dongzhai Harbour (Meso- and low-tide Bank)	5.81
	12. 拟海桑 *S.* × *gulngai*	0.35 ±0.10	3.59 ±0.39	东寨港(中低潮滩) Dongzhai Harbour (Meso- and low-tide Bank)	6.75
	13. 杯萼海桑 *S. alba*	0.43 ±0.14	4.15 ±0.58	东寨港(中低潮滩) Dongzhai Harbour (Meso- and low-tide Bank)	6.32
	14. 卵叶海桑 *S. ovata*	0.44 ±0.01	4.28 ±0.07	东寨港(高潮滩) Dongzhai Harbour (High-tide Bank)	5.11
	15. 白骨壤 *Avicennia marina*	0.28 ±0.04	2.82 ±0.19	东寨港(中低潮滩) Dongzhai Harbour (Meso- and low-tide Bank)	7.07
	16. 桐花树 *Aegiceras corniculatum*	0.25 ±0.03	2.81 ±0.05	东寨港(中低潮滩) Dongzhai Harbour (Meso- and low-tide Bank)	8.15
	平均值 Average	0.45 ±0.15	3.90 ±0.89		8.35 ±2.78
半红树植物 Semi-mangrove plants	17. 黄槿 *Hibiscus tiliaceus*	0.04 ±0.00	1.70 ±0.18	清澜港(高潮滩) Qinglan Harbour (High-tide bank)	0.59
	18. 水黄皮 *Pongamia pinnata*	0.02 ±0.00	0.81 ±0.11	清澜港(高潮滩) Qinglan Harbour (High-tide bank)	1.02
	19. 杨叶肖槿 *Thespesia populnea*	0.09 ±0.01	1.59 ±0.02	东寨港(高潮滩) Dongzhai Harbour (High-tide bank)	1.57
	20. 海滨猫尾木 *Dolichandrone spathaceae*	0.03	1.38	清澜港(特大高潮滩) Qinglan Harbour (Megalo-tide bank)	0.50
	21. 阔苞菊 *Pluchea indica*	0.09 ±0.01	2.26 ±0.20	东寨港(特大高潮滩) Dongzhai Harbour (Megalo-tide bank)	2.00
	22. 玉蕊 *Barringtonia racemosa*	0.06 ±0.01	2.65 ±0.13	清澜港(高潮滩) Qinglan Harbour (High-tide bank)	1.57
	23. 海芒果 *Cerbera manghas*	0.036	3.23	清澜港(高潮滩) Qinglan Harbour (High-tide bank)	0.58
	24. 莲叶桐 *Hernandia sonora*	0.26	3.58	清澜港(高潮滩) Qinglan Harbour (High-tide bank)	-

表1 (续) Table 1 (continued)

类别 Groups	物种名 Plant species	单位叶面积Cl含量 Cl content per unit leaf area	肉质化程度 Succulence	取样地及分布潮滩 Site and tide	土壤Cl含量 Cl content of soil
	25. 许树 *Clerodendrum inerme*	0.39 ±0.12	3.75 ±0.93	东寨港(高潮滩) Qinglan Harbour (High-tide bank)	2.98
	26. 水芫花 *Pemphis acidula*	1.32	6.99	清澜港(高潮滩) Qinglan Harbour (High-tide bank)	-
	平均值 Average	0.23 ±0.40	2.79 ±1.77		1.35 ±0.86
争议物种 Controversial species	27. 银叶树 *Heritiera littoralis*	0.06 ±0.00	1.88 ±0.19	清澜港(高潮滩) Qinglan Harbour (High-tide bank)	5.19
	28. 卤蕨 *Acrostichum aureum*	0.15 ±0.03	2.23 ±0.12	清澜港(高潮滩) Qinglan Harbour (High-tide bank)	2.97
	29. 尖叶卤蕨 *A. speciosum*	0.15 ±0.02	2.26 ±0.10	清澜港(高潮滩) Qinglan Harbour (High-tide bank)	1.77
	30. 海漆 *Excoecaria agallocha*	0.15 ±0.06	2.25 ±0.38	清澜港(高潮滩) Qinglan Harbour (High-tide bank)	2.82
	31. 木果楝 *Xylocarpus granatum*	0.29 ±0.03	2.72 ±0.12	清澜港(高潮滩) Qinglan Harbour (High-tide bank)	3.40
	32. 老鼠簕 *Acanthus ilicifolius*	0.44	4.47	东寨港(高潮滩) Dongzhai Harbour (High-tide bank)	7.58
	33. 小花老鼠簕 *A. ebrecteatus*	0.42	4.58	东寨港(高潮滩) Dongzhai Harbour (High-tide bank)	4.70
	样品总体平均值 Average in total	0.34 ±0.27	3.36 ±1.33		6.02 ±4.08

来说仍然较高,平均值为0.45 mg·cm^{-2}。

半红树植物叶片中Cl含量普遍偏低,大部分仅为真红树的1/10左右。莲叶桐、许树(*Clerodendrum inerme*)和水芫花叶片Cl含量较高,特别是水芫花Cl含量高达1.32 mg·cm^{2},远远超过所有真红树种类。

争议的7种红树植物中,除老鼠簕和小花老鼠簕叶片Cl含量接近真红树平均值外,其余均介于真红树和半红树植物之间,但更接近后者。其中银叶树叶片Cl含量非常低,只有0.06 mg·cm^{-2},跟大多数半红树差不多。

2.1.2 叶片肉质化程度比较

红树植物叶片肉质化程度相对Cl含量来说,变异系数没那么高,为39.8%(表1)。同样,半红树植物变异系数远大于真红树植物,分别为63.4%和22.8%。

真红树中除了海莲(*Bruguiera sexangula*)、尖瓣海莲(*B. sexangula* var. *rhynchopetala*)、白骨壤和桐花树肉质化程度偏低外,其余都非常高,单位叶面积含水量平均值为3.9 g·dm^{-2}。半红树植物肉质化程度普遍偏低。除许树、莲叶桐和水芫花肉质化程度较高外,大多数半红树植物单位叶面积含水量平均值仅为1.95 g·dm^{-2}。

争议物种中,只有老鼠簕和小花老鼠簕肉质化程度较高,达到了真红树的平均值3.9 g·dm^{-2}。

值得注意的是半红树植物许树、莲叶桐和水芫花,其叶片中Cl元素含量和肉质化程度都远高于其它半红树植物。特别是水芫花,该种植物生境非常特殊(迎风石块中),不仅土壤中盐分可以通过蒸腾流到达叶片,海水中盐分还可以通过浪花飞溅进入,从而造成水芫花叶片中Cl含量比真红树植物还高。莲叶桐和许树叶片Cl含量为何较高需进一步研究。

2.1.3 叶片肉质化程度和Cl含量的关系

图1给出了红树林生境中真红树植物和半红树植物及全部植物叶片Cl含量和肉质化程度的散点分布。从图上看出,Cl含量和肉质化程度呈正相关。不管是真红树植物还是半红树植物,所有植物叶片内的Cl含量与其肉质化程度显著相关($p < 0.01$)。

2.2 比叶面积和叶氮含量

2.2.1 比叶面积(*SLA*)

所研究的的33种红树植物的*SLA*变异系数仍然较大(44.5%)(图2),但两类红树植物内部变异减小(真红树植物22.7%,半红树25.9%)。其中真红树角果木和半红树海芒果分别具有最低的*SLA*(42.3 cm^{2}·g^{-1})和最高的*SLA*(222.1 cm^{2}·g^{-1}),其差值高达4.2倍。说明真红树和半红树植物存在较明显的生理差异,对盐渍生境有着不同的适应策略。

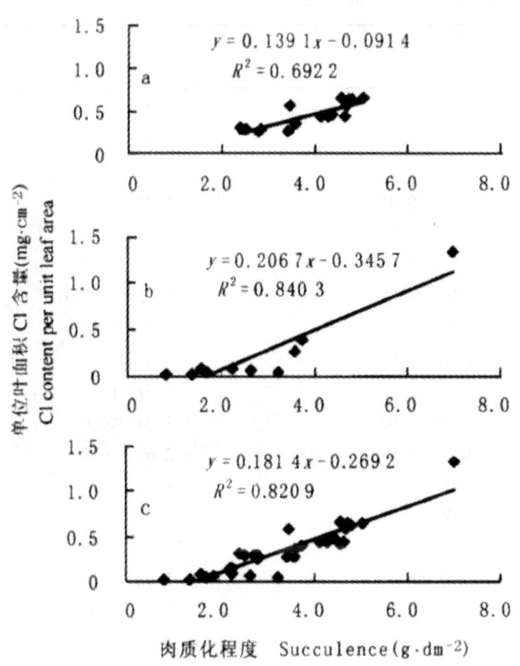

图 1 红树植物叶片 Cl 含量与肉质化程度的关系

Fig. 1 The relationship between leaf Cl content and succulence for mangroves in China

a:真红树 True mangrove plants b:半红树 Semi-mangrove plants c:全部植物 All species

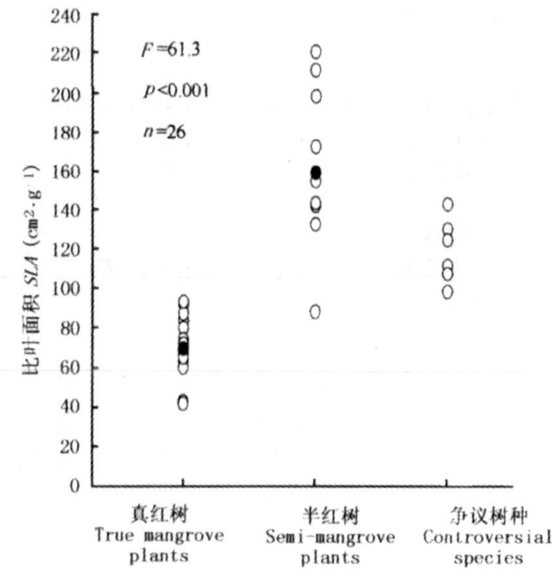

图 2 国内 33 种红树植物比叶面积($cm^2\ g^{-1}$)

Fig. 2 The specific leaf area (SLA) for 33 mangrove species in China

○:每种植物比叶面积的平均值 Represent mean SLA for each species ●:两类红树植物的平均值 Represent the average values for two groups of mangroves F 值和 p 值为两类红树植物的方差分析结果 F-value and p-value, and sample number (n) are given for One-Way ANOVAs

真红树一般具有较低的 SLA,平均值为 70.6 $cm^2\ g^{-1}$,且最高值未超过 100.0 $cm^2\ g^{-1}$。而半红树植物平均值都高达 160.4 $cm^2\ g^{-1}$,其中只有水芫花的 SLA 值略低于 100.0 $cm^2\ g^{-1}$。

争议物种中,银叶树和老鼠簕具有较小的 SLA,分别为 71.5 $cm^2\ g^{-1}$和 99.5 $cm^2\ g^{-1}$。其余几种植物 SLA 在 108.0~144.0 $cm^2\ g^{-1}$之间。

2.2.2 叶氮含量

真红树植物一般单位重量叶氮含量(N_{mass})较低,平均值为 16.34 $mg\ g^{-1}$(表 2)。但杯萼海桑、海桑、拟海桑和白骨壤具有较高的 N_{mass}值。较高的 N_{mass}表明叶片光合能力较强(Körner, 1989),这与野外观察到杯萼海桑、海桑、拟海桑在潮滩生长较茂盛相一致。白骨壤虽然 N_{mass}也较高,但由于将大部分光合产物投资于地下部分来维持较高的水分利用效率以此增加耐盐性,从而牺牲了地上部分的生长(Ball, 1988)。半红树植物 N_{mass}都较高,平均值高达 26.28 $mg\ g^{-1}$。

从单位面积叶氮含量(N_{area})来看,真红树植物除了瓶花木、榄李和红榄李较低外,其余均较高,整体平均值达 2.35 $g\ m^{-2}$。瓶花木、榄李和红榄李多分布在高潮带,耐盐性不强,其相对较低的 N_{area}也属正常。而半红树植物 N_{area}普遍偏低,平均值为 1.51 $g\ m^{-2}$。

所有争议物种的 N_{mass}和 N_{area}介于真红树和半红树之间。不过银叶树和老鼠簕却有较高的 N_{area}值,分别为 2.36 和 2.21 $g\ m^{-2}$。此外,所测 33 种植物的 N_{mass}和 N_{area}变异系数也不低,分别为 36.8%和 36.1%。

综合来看,真红树植物具有较低的 SLA 和较高的 N_{area},表明其具有较高的水分利用效率,耐盐性较强。而半红树植物除水芫花 SLA 略为偏低外,其余均较高。此外,半红树植物的 N_{area}普遍偏低。

2.2.3 SLA 与 N_{area}的关系

图 3 表明,SLA 和 N_{area}呈负相关,SLA 较大的植物,其 N_{area}较小。但两类植物各自的 SLA 和 N_{area}都不形成显著相关($p > 0.05$),尤其是真红树植物。

3 讨 论

自然条件下,红树林生境是一种寡养分生境,土壤养分尤其是 N 严重不足,红树林内 N 含量远低于临近光滩(Tomlinson, 1999; Wang *et al.*, 2003b)。红树植物虽然受潮水浸淹,但由于土壤含盐量高,土壤渗透压高,植物吸水困难,故而呈现生理性干旱的

表2 国内33种红树植物单位重量叶氮含量(N_{mass})和单位面积叶氮含量(N_{area})

Table 2 Measurements of N_{mass}, N_{area} for 33 mangroves species in China

类别 Groups	物种 Plant species	单位重量叶氮含量 N_{mass}(mg g^{-1})	单位面积叶氮含量 N_{area}(g m^{-2})
真红树植物 True mangrove plants	1. 红海榄 *Rhizophora stylosa*	14.22 ±1.35	2.83 ±0.03
	2. 正红树 *Rhizophora apiculata*	12.13 ±2.30	2.77 ±0.04
	3. 木榄 *Bruguiera gymnorhiza*	15.11 ±2.04	2.17 ±0.00
	4. 海莲 *B. sexangula*	13.19 ±0.44	1.98 ±0.04
	5. 尖瓣海莲 *B. sexangula* var. *rhynchopetala*	13.92 ±1.17	1.66 ±0.02
	6. 秋茄 *Kandelia obovata*	18.96 ±1.58	2.96 ±0.03
	7. 角果木 *Ceriops tagal*	8.56 ±0.93	2.12 ±0.05
	8. 瓶花木 *Scyphiphora hydrophyllacea*	10.06 ±1.43	1.24 ±0.02
	9. 榄李 *Lumnitzera racemosa*	14.03 ±1.67	1.51 ±0.01
	10. 红榄李 *L. littorea*	9.56 ±0.84	1.57 ±0.01
	11. 海桑 *Sonneratia caeseolaris*	26.66 ±0.92	3.04 ±0.04
	12. 拟海桑 *S.* × *gulngai*	24.99 ±2.89	2.69 ±0.06
	13. 杯萼海桑 *S. alba*	30.89 ±1.53	4.10 ±0.03
	14. 卵叶海桑 *S. ovata*	13.00 ±0.97	1.73 ±0.01
	15. 白骨壤 *Avicennia marina*	22.86 ±1.37	3.14 ±0.02
	16. 桐花树 *Aegiceras corniculatum*	13.35 ±0.51	2.02 ±0.01
	平均值 Average	16.34 ±6.60	2.35 ±0.77
半红树植物 Semi-mangrove plants	17. 黄槿 *Hibiscus tiliaceus*	25.95 ±1.74	2.02 ±0.01
	18. 水黄皮 *Pongamia pinnata*	36.07 ±6.41	1.81 ±0.03
	19. 杨叶肖槿 *Thespesia populnea*	15.93 ±1.81	1.11 ±0.00
	20. 海滨猫尾木 *Dolichandrone spathaceae*	29.02	1.36
	21. 阔苞菊 *Pluchea indica*	30.22 ±3.41	1.74 ±0.01
	22. 玉蕊 *Barringtonia racemosa*	23.69 ±1.18	1.64 ±0.01
	23. 海芒果 *Cerbera manghas*	36.46	1.6
	24. 莲叶桐 *Hernandia sonora*	24.94	1.86
	25. 许树 *Clerodendrum inerme*	24.58 ±2.81	1.60 ±0.03
	26. 水芫花 *Pemphis acidula*	15.93	1.78
	平均值 Average	26.28 ±7.06	1.51 ±0.54
争议树种 Controversial species	27. 银叶树 *Heritiera littoralis*	16.68 ±0.62	2.36 ±0.04
	28. 卤蕨 *Acrostichum aureum*	16.14 ±1.32	1.52 ±0.05
	29. 尖叶卤蕨 *A. speciosum*	22.24 ±2.03	1.55 ±0.01
	30. 海漆 *Excoecaria agallocha*	23.14 ±2.06	1.80 ±0.03
	31. 木果楝 *Xylocarpus granatum*	22.82 ±1.43	1.82 ±0.02
	32. 老鼠簕 *Acanthus ilicifolius*	21.96	2.21
	33. 小花老鼠簕 *A. ebrecteatus*	18.93	1.74
	样品总体平均值 Average in total	20.19 ±7.42	1.99 ±0.72

N_{mass}: Mass-based leaf nitrogen concentration N_{area}: Area-based leaf nitrogen concentration

特征(Levitt, 1980)。叶氮含量、比叶面积和叶寿命等叶性因子都是植物适应环境所表现出的重要结构参数(Körner, 1991),并通过其相互作用而影响叶的功能性状,如光合、呼吸等,从而影响植物的生长与分布格局。

本研究表明,真红树植物叶片 Cl 含量和肉质化程度远高于大多数半红树植物,并且具有低 *SLA* 和高 N_{area}的特点。这说明真红树植物具有积累大量无机离子来进行渗透调节降低水势的能力;并且将较多的氮投资于保护构造上(防止失水过多等)和增加叶肉细胞密度上(Bazzaz, 1997),水分利用效率高,从而提高其耐盐性来适应高盐生境。半红树植物莲叶桐、许树虽然叶片 Cl 含量和肉质化程度较高,但它们的高 *SLA* 和低 N_{area}特征都与其它半红树植物相类似。水芫花为了适应极度高盐,贫瘠和风浪袭击的恶劣生境,在 Cl 含量和叶性状方面表现出了特殊性。

根据以上我们对一些真红树植物和半红树植物叶片中 Cl 含量和叶性状的比较研究,并参照林鹏和傅勤(1995)提出的真红树植物和半红树植物的鉴别标准,对 7 种争议树种的界定如下:

银叶树:银叶树是目前人们争议较多的物种。

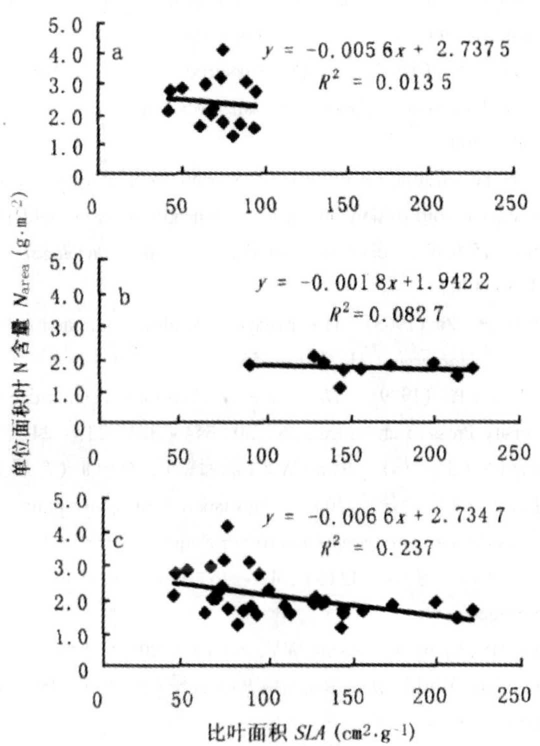

图3 红树植物单位面积叶N含量与比叶面积的关系

Fig. 3 The relationship between area-based leaf nitrogen content (N_{area}) and specific leaf area (SLA) for mangroves

a、b、c: 见图1 See Fig. 1

由于其多分布在红树林林缘,不耐高盐,许多学者将其归为半红树(林鹏, 1987; Tomlinson, 1999; Wang *et al*., 2003a)。Mukherjee 等(2003)通过分子标记也认为将银叶树归为半红树植物比较合适。但由于银叶树可以形成纯植丛,且相对其它半红树植物较耐盐,在海岸防护林体系中作用非同小可,国内不少人将其归为真红树植物(林鹏, 1993; 范航清, 2000)。就目前研究来看,银叶树叶片 Cl 含量和肉质化程度相当低;且野外观察其生境具有两栖性,符合半红树特征,故应归为半红树植物。其低的 SLA 和较高的 N_{area},说明该树种水分利用效率高,较耐盐。同时,低的 Cl 含量和 SLA 表明银叶树是以有机渗透调节为主,渗透调节物质可能是一些分子量较大的有机物(如柠檬酸等)(Paliyavuth, 2004)。

海漆:该种植物对红树林生境没有明显的适应特征,且可以在海拔 400 m 的开阔地带生长,不少国内外学者认为海漆是半红树植物(林鹏, 1987; Tomlinson, 1999; Wang *et al*., 2003a)。但国内也有很多人认为海漆属于真红树植物(林鹏, 1993; 郑德璋等, 1995)。Moorthy 和 Kathiresan(1997)发现海漆与白骨壤、木榄和角果木在光合、色素组成等方面非常接近,认为海漆是真红树植物。但本研究表明,海漆无论是从叶片 Cl 含量、肉质化程度,还是从叶 N 含量和 SLA 来看都应归为半红树。而且据野外调查,海漆大多分布在高潮带偏上的位置,有的甚至生长在潮水完全不可淹及的地方。

卤蕨和尖叶卤蕨:人们对该属植物的界定争议也很大。卤蕨属植物多分布在红树林群落内缘或高潮带滩涂,因而很多人赞同将其归为半红树植物(林鹏, 1987; Tomlinson, 1999)。但后来林鹏(1993)和范航清(2000)把它们归为真红树植物。而 Wang 等(2003a)和王伯荪等(2003)认为它们不属于木本植物,所以既不是真红树植物也不是半红树植物,而是红树林伴生植物。但就目前研究而言,卤蕨和尖叶卤蕨应归为半红树植物。原因在于该类植物叶片 Cl 含量和肉质化程度较低,不同于真红树植物;其叶氮含量和 SLA 与半红树植物更接近。

木果楝:林鹏(1987)将其定为半红树植物,但后来很多人认为木果楝属于真红树植物(林鹏, 1993; Wang *et al*., 2003a)。Tomlinson(1999)对该红树植物的归属问题一直未作定论。本研究中,木果楝叶片 Cl 含量、肉质化程度和叶 N 含量介于半红树和真红树之间,SLA 略微偏大。木果楝究竟属于真红树植物还是半红树植物还有待进一步研究。

老鼠簕和小花老鼠簕:该属植物是灌木和亚灌木,一般生长于红树林林缘或与半红树植物生长在一起,有时可以生长在几乎不受潮水影响的低盐的河岸,因此被认为是典型的半红树植物(Tomlinson, 1999; Wang *et al*., 2003a;王伯荪等,2003)。但国内大部分人还是将老鼠簕属植物归为真红树植物(Lin, 1987, 1993; 郑德璋等, 1995; 范航清, 2000)。通过研究,发现老鼠簕和小花老鼠簕叶片 Cl 含量和肉质化程度高于真红树平均值,叶 N 含量和 SLA 都比较接近真红树。所以我们赞成将它们归为真红树植物。

参 考 文 献

Ball MC (1988). Salinity Tolerance in the Mangroves *Aegiceras corniculatum* and *Avicennia marina*. I. Water use in relation to growth, carbon partitioning, and salt balance. *Australian Journal of Plant Physiology*, 15, 447 - 464.

Bazzaz FA (1997). Allocation of resources in plants: state of the science and critical questions. In: Bazzaz FA, Gracedds J eds. *Plant Resource Allocation*. Academic Press, New York, 1 - 37.

Fan HQ (范航清) (2000). *Mangrove —Safeguard of Environment Protection* (海岸环护卫士 ——红树林). Guangxi Science and

Technology Press, Nanning, 18 - 23. (in Chinese)

Institute of Tropic Crop of South China (华南热带作物研究院) (1974). Measurement for nitrogen of latex by Nessler's reagent colorimetric method. *Communication for Tropical Crop* (热作科技通讯), (5), 12 - 13. (in Chinese)

Joshi GV, Pimplaskar M, Bhosale LJ (1972). Physiological studies in germination of mangroves. *Botanica Marina*, 45, 91 - 95.

Khan MA, Aziz I (2001). Salinity tolerance in some mangrove species from Pakistan. *Wetlands Ecology and Management*, 9, 219 - 223.

Körner C (1989). The nutrient use efficiency and fertility in forest ecosystems. *Oecologia*, 81, 379 - 391.

Körner C (1991). Some overlooked plant characteristics as determinants of plant growth: a reconsideration. *Functional Ecology*, 5, 162 - 173.

Levitt J (1980). *Responses of Plant to Environmental Stress* 2nd edn. Academic Press, New York, 365 - 488.

Li YL (李玉霖), Cui JY (崔建垣), Su YZ (苏永中) (2005). Specific leaf area and leaf dry matter content of some plants in different dune habitats. *Acta Ecologica Sinica* (生态学报), 25, 304 - 311. (in Chinese with English abstract)

Lin P (林鹏) (1987). Distribution of mangrove species. *Scientia Silvae Sinicae* (林业科学), 23, 481 - 490. (in Chinese with English abstract)

Lin P (林鹏) (1993). Species distribution and type of forest aspect of mangrove in China. In: Li ZJ (李振基) ed. *Issue of Environment and Ecology* (环境与生态论丛). Xiamen University Press, Xiamen, 74 - 79. (in Chinese with English abstract)

Lin P(林鹏), Fu Q(傅勤) (1995). *Environmental Ecology and Economic Utilization of Mangroves in China* (中国红树林环境生态及经济利用). Higher Education Press, Beijing, 23 - 31. (in Chinese)

Moorthy P, Kathiresan K (1997). Photosynthetic pigments in tropical mangroves: impacts of seasonal flux of UV-B radiation and other environmental attributes. *Botanica Marina*, 40, 341 - 349.

Mukherjee AK, Acharya LK, Mattagajasingh I, Panda PC, Mohapatra T, Das P (2003). Molecular characterization of three *Heritiera* species using AFLP markers. *Biologia Plantarum*, 47, 445 - 448.

Naidoo G, Tuffers AV, von Willert DJ (2002). Changes in gas exchange and chlorophyll fluorescence characteristics of two mangroves and a mangrove associate in response to salinity in the natural environment. *Trees*, 16, 140 - 146.

Paliyavuth C, Clough B, Patanaponpaiboon P (2004). Salt uptake and shoot water relations in mangroves. *Aquatic Botany*, 78, 349 - 360.

Poorter H, de Jong JR (1999). A comparison of specific leaf area, chemical composition and leaf construction costs of field plants from 15 habitats differing in productivity. *New Phytologist*, 143, 163 - 176.

Scholander PF (1968). How mangrove desalinate seawater. *Physiologia Plantarum*, 21, 251 - 256.

Tomlinson PB (1999). *The Botany of Mangroves*. Cambridge University Press, Cambridge, 26 - 30, 374 - 381, 237 - 242.

Wang BS (王伯荪), Zhang WY (张炜银), Zan QJ (昝启杰), Liang SC(梁士楚) (2003). Annotation of mangrove plant. *Acta Scientiarum Naturalium Universitatis Sunyatseni* (中山大学学报(自然科学版)), 42(3), 42 - 46. (in Chinese with English abstract)

Wang BS, Liang SC, Zhang WY, Zan QJ (2003a). Mangrove Flora of the World. *Acta Botanica Sinica* (植物学报), 45, 644 - 653.

Wang WQ, Wang M, Lin P (2003b). Seasonal changes in element levels in mangrove leaves and element retranslocation during leaf senescence. *Plant and Soil*, 252, 187 - 193.

Wilson PJ, Thompson K, Hodgson J (1999). Specific leaf area and leaf dry matter content as alternative predictors of plant strategies. *New Phytologist*, 143, 155 - 162.

Zhao Y (赵亚), Guo YW (郭跃伟) (2004). The proceeding of chemical constituents and pharmacological activities of mangrove. *Chinese Journal of Natural Medicines* (中国天然药物), 2, 135 - 140. (in Chinese with English abstract)

Zhao KF (赵可夫), Feng LT (冯立田), Lu YF (卢元芳), Fan H (范海) (1999). The osmotica and their contributions to the osmotic adjustment in two mangrove species—*Avicennia marina* (Forsk) Vierh and *Kandelia candel* (L.) Druce growing in the Fujian Jiulongjia River Estuary of China. *Oceanologia et Limonolgia Sinica* (海洋与湖沼), 17, 58 - 61.

Zheng DZ (郑德璋), Zheng SF (郑松发), Liao BW (廖宝文) (1995). *The Dynamic Studies on Mangrove of Qinglan Harbour in Hainan Island* (海南岛清澜港红树林发展动态研究). Guangdong Science and Technology Press, Guangzhou, 133 - 137. (in Chinese)

红树植物桐花树大小孢子发生及雌雄配子体发育的观察*

游学明　田惠桥　杨盛昌

（厦门大学生命科学学院，植物基因与基因技术研究所，福建 厦门 361005）

摘要：用常规石蜡切片法和显微镜观察，对桐花树的胚胎发育早期进行研究，得到以下结果：1）花药由 4 个花粉囊组成，药壁包括表皮、药室内壁、中层及绒毡层，绒毡层为腺质绒毡层；2）小孢子母细胞减数分裂为同时型，四分体排列方式为四面体型；3）成熟花粉二细胞，有 4 个萌发孔；4）特立中央胎座，薄珠心，大孢子母细胞减数分裂形成 T 形排列的 4 个大孢子，合点端的大孢子具功能，成熟胚囊为七胞八核结构，胚囊发育类型为蓼型；5）花药壁中含单宁细胞团，发育过程中子房内的胎座出现类似于盐腺的结构。

关键词：桐花树；大孢子发生；小孢子发生；雌配子体；雄配子体

中图分类号：Q 945.41　**文献标识码**：A　**文章编号**：0438-0479(2005)05-0718-05

胎生现象是红树植物最显著的特征之一，很早就受到研究者的注意，并开展了相关研究。其中，侯宽昭很早就详细描述了红树植物的胎生繁殖体[1]；Juncosa 对红树科的红海榄，大红树进行了发育形态学研究，阐明了红海榄与大红树从受精卵到成熟胚轴的发育过程[2~5]；陈月琴等研究了秋茄，木榄的繁殖体的结构及其生态特异性，表明了它们的胚轴在胚胎发育过程中，出现了一些与外界相适应的结构变化[6]；Tomlinson 研究了红树科植物繁殖体从掉落到定植过程中此生木质部解剖结构的变化[7]；Ma 等对 5 种红树植物进行了花的分类学和系统发生学研究，表明了花各部分形态发育与分类的关系[8]。但是对红树植物的胚胎发育早期（即雌雄配子体发育）方面的工作还未见报道。

胎生有显胎生与隐胎生二种形式。其中显胎生的红树植物有秋茄（*Kandelia candel*），木榄（*Bruguiera gymnorrhiza*），红海榄（*Rhizophora. stylosa*），角果木（*Ceriops tagal*）等。隐胎生的有桐花树（*Aegiceras corniculatum*），白骨壤（*Avicennia marina*）等。桐花树属紫金牛科，为红树植物隐胎生的代表，广布于亚洲至大洋洲热带海岸。在我国广东、福建、广西、海南都有分布[9]。胚胎学研究是植物系统分类和进化的重要依据之一。本文以桐花树为材料，研究其大小孢子的发生及雌雄配子体的发育，对胎生现象及红树植物的系统分类及进化研究有重要意义。

1　材料和方法

1.1　植物材料

实验所用的材料取自福建省龙海市的红树林自然保护区内，地点为浮宫镇草埔头村。于 2003 年与 2004 年每年 3 月初开始，以 3 d 为间隔，从幼小花序开始固定，固定液为卡诺固定液（95 %V（乙醇）∶V（冰醋酸）= 3∶1），固定 24 h 后转至 70 %乙醇 4 ℃冰箱保存。

1.2　方　法

爱氏苏木精整体染色后，常规石蜡包埋，切片，切片厚度 7～9 μm，Leica 显微镜观察并摄影。

2　结　果

2.1　小孢子发生与雄配子体发育

桐花树每朵花具雄蕊 5 枚，雌蕊 1 枚。成熟雄蕊长约 4 mm，每个花药有药囊 4 个。从雄蕊原基横切面上，可见 4 个角隅处分化出单个胞原细胞，胞原细胞平周分裂形成初生壁细胞和初生造胞细胞，初生造胞细胞重复有丝分裂产生次生造胞细胞，次生造胞细胞进一步发育为小孢子母细胞。小孢子母细胞体积较大，细胞质浓厚（图版 1-1～3）。随后小孢子母细胞开始第 1 次减数分裂，产生 2 个核（图版 1-4），再经第 2 次减数分裂，形成 4 个核（图版 1-5）。之后胞质分裂形成四面体形的四分孢子，刚形成的四分孢子，被胼胝质壁所包围（图版 1-6），因此桐花树小孢子母细胞的减数分裂方式为同时型。随着胼胝质溶解，4 个小孢子释放出来，彼此分开。刚释放出来的小孢子细胞质浓厚，细胞

* 原载于厦门大学学报（自然科学版），2005，44(5)：718－722

核位于中央,细胞壁加厚(图版 1-7). 细胞质里的小液泡逐渐融合成中央大液泡,细胞核被推向花粉壁一侧,形成单核靠边期的小孢子(图版 1-8). 小孢子核在贴近花粉壁的位置进行不均等的有丝分裂,分裂结果形成两个子核:营养核与生殖核(图版 1-9). 接着发生不等的胞质分裂,形成大小不等的两个细胞,大的为营养细胞,占据花粉的绝大部分体积,细胞核为圆球形;小的为生殖细胞,呈凸透镜状,紧贴花粉壁,只有少量细胞质而无液泡,细胞核椭圆形. 生殖细胞与营养细胞之间存在着明显的界限(图版 1-10). 此后,生殖细胞逐渐移向营养细胞的胞质里,形状变为长纺锤形,浸没在营养细胞中(图版 1-11). 成熟花粉粒有 4 个萌发孔,二胞型.

2.2 花药壁的发育

初生壁细胞经过一次平周分裂和多次垂周分裂形成两层细胞,其外层细胞又经一次平周分裂和多次垂周分裂,最后形成连同表皮在内的 4 层细胞. 小孢子母细胞阶段,花药壁基本分化完成,自外向内依次为表皮层 1 层、药室内壁 1 层、中层 1～2 层和绒毡层 1 层. 小孢子母细胞进入减数分裂时,绒毡层细胞径向伸长,体积明显增大,细胞内进行核分裂,形成双核细胞. 此时中层细胞被挤压成扁平形. 单核小孢子时期,绒毡层细胞在原位开始出现解体的迹象,其绒毡层为分泌型. 此时药室内壁细胞径向延长. 二细胞花粉时期,表皮细胞为扁平状,药室内壁径向延长并出现纤维加厚,被挤压的中层细胞恢复,仍可见绒毡层细胞的残体. 之后药室内壁的纤维加厚结束,整个药室为表皮层及加厚的药室内壁和中层包围,两个相邻药室之间的药隔进而破坏,两个药室间连通. 成熟花药的开裂方式为纵裂.

另外,在花药壁上,间隔分布有单宁细胞团构成的腔室(图版 1-12).

2.3 大孢子发生与雌配子体发育

桐花树每朵花有子房 1 个. 胎座为特立中央胎座,胚珠多数,薄珠心,彼此紧密排列成圆球形. 大孢子母细胞体积较大,细胞质浓厚,具显著的细胞核,与周围珠心细胞有明显区别(图版 2-1). 大孢子母细胞进行减数分裂,形成二分体(图版 2-2),二分体珠孔端的大孢子垂直分裂,合点端大孢子横向分裂(图版 2-3),形成 T 形排列的四分体(2-4). 四分体中合点端的大孢子为功能孢子,其它 3 个退化. 功能孢子体积增大,液泡化,此时形成单核胚囊(图版 2-5). 单核胚囊经第 1 次有丝分裂形成二核胚囊,(图版 2-6),再经第 2 次有丝分裂形成四核胚囊(图版 2-7). 由四核胚囊继续发育,形成七胞八核胚囊(图版 2-8). 随后,合点端先分化出 3 个反足细胞并随之退化,下极核与上极核汇合,珠孔端分化出卵器(图版 2-9). 桐花树胚囊的发育类型为蓼型.

2.4 子房中的其它结构的发育

在雌配子体发育过程中,发现子房内子房壁与胚囊之间的胎座上分布有分泌结构,它们从大孢子母细胞时期到形成成熟卵细胞这段时期都存在. 在解剖镜下可见其环绕胎座都有分布. 图版 2-10 为子房的一个纵切面,从中可见其着生位置,图版 2-11 为该器官的纵切面. 图版 2-10 为其横切面. 它们头部都由 10～12 个分泌细胞组成. 与子房胎座相连的收集细胞一个. 柄细胞有一个或多个,头尾相连,因此各个分泌结构长短不一. 柄细胞内可见有大小不一的液泡(图版 2-11),从其解剖结构特点来看,这与桐花树叶片表面的盐腺结构类似.

3 讨 论

1) 本文对红树植物桐花树的胚胎发育早期进行了研究. 结果表明,桐花树胚胎早期的发育特点为:a) 小孢子母细胞的减数分裂方式为同时型,小孢子四分体为四面体型,花药绒毡层为腺质绒毡层,成熟花粉粒有 4 个萌发孔. b) 在大孢子母细胞形成四分体过程中,其大孢子母细胞分裂形成的大孢子四分体为 T 型排列. 成熟胚囊为七胞八核结构. 胚囊发育类型为蓼型.

2) 大多数植物形成的大孢子四分体都为线形排列,另外还有等二列排列的,成 T 形排列的较少,但也有少量报道,如在羊草中的 4 个大孢子就是 T 形排列的[10]. 植物花药中层细胞一般含淀粉或其它储藏物,在小孢子发育过程中储藏物被吸收,中层细胞趋于解体,在植物成熟花药中一般不存在中层,但在桐花树花药中,最外层的中层细胞一直保持到花药成熟,而且象药室内壁细胞一样发生纤维状加厚. 这种中层细胞不退化的例子在百合花药中也存在[11].

3) 在桐花树花药及子房壁中,观察到单宁细胞团存在. 单宁细胞多分布于植物枝茎和叶片等处,单宁的存在有助于红树植物抵抗海水腐蚀[9]. 桐花树花药及子房壁中存在单宁,说明红树植物的繁殖器官在进化中也发展并形成了与盐生环境相适应的结构特征.

4) 在子房中观察到的分泌结构,与 Tomlinson[12] 所述桐花树叶片表面的盐腺结构类似. 金杰里认为,红树植物胎生的生物学意义是,幼苗在母体植株上发芽生长的过程中进行对高浓度盐分的适应作用. 在含盐量不多的种子在母体发芽时,由于生长的关系开始从植株流入有机与无机的物质,这样幼苗也得到了盐分,逐渐对盐分适应而获得高的抗盐性.在幼苗中含盐量是逐渐增加的[13].郑文教根据红树植物胎生胚轴发育过程中的元素和灰分动态的研究结果,认为胎生红树植物胚轴发育过程不是一个盐分积累而获得抗盐锻炼的过程,而是一个低盐化过程,表现出"返祖现象"[14,15].张宜辉根据无机离子变化和分布特点,发现木榄及桐花树繁殖体在胚轴发育早期花萼中的离子浓度比胚轴高,胚轴中的离子浓度是逐步增加的,因而认为胎生应是一个盐分积累的过程[16].本实验中在桐花树子房内部观察到分泌结构存在,表明桐花树可以通过特殊的分泌结构调节胚胎发育早期(雌雄配子体发育至种子形成过程)胚胎中的盐分含量,避免了盐离子对植物胚胎产生毒害.这也可能是造成胚轴发育早期低盐含量的原因之一.

参考文献:

[1] 侯宽昭,何椿年.中国的红树林[J].生物学通报,1953,(10):365 - 369.

[2] Juncosa A M. Embryo and Seeding Development in the Rhizophoraceae[D]. Durham,North Carolina:Duke University,1982.

[3] Juncosa A M. Developmental morphology of the embryo and seeding of *Rhizophora mangle* L.(Rhizophoraceae) [J].Am.J.Bot.,1982,69:1 599 - 1 611.

[4] Juncosa A M. Embryogenesis and seedling development in *Cassipourea elliptica (Sw.)* Poir.(Rhizophoraceae)[J]. Am.J.Bot.,1984,71:170 - 179.

[5] Juncosa A M. Embryogenesis and seedling development morphology of the seedling in *Bruguiera exaristata* Ding Hou (Rhizophoraceae)[J]. Am.J.Bot.,1984,71:180 - 191.

[6] 陈月琴,蓝崇玉,黄玉山,等.秋茄、木榄繁殖体的结构及其生态特异性[J].中山大学学报(自然科学版),1995,34(4):70 - 75.

[7] Tomlinson P B,Cox P A. Systematic and functional anatomy of seedlings in mangrove Phizophoraceae:vivipary explained[J].Bot.J.Linn.Soc.,2000,134:215 - 231.

[8] Ma O S W,Saunders R M K. Comparative floral ontogeny of Maesa (Maesaceae),Aegiceras (Myrsinaceae) and Embelia (Myrsinaceae):taxonomic and phylogenetic implications[J]. Plant Syst. Evol.,2003,243:39 - 58.

[9] 林鹏.红树林[M].北京:海洋出版社,1984.

[10] 卫星,申家恒.羊草大小孢子的发生与雌雄配子体的发育的观察[J].西北植物学报,2003,23(12):2 058 - 2 066.

[11] 胡适宜.被子植物胚胎学[M].北京:人民教育出版社,1982.

[12] Tomlinson P B. The Botony of Mangroves[M]. New York,USA:Cambridge University Press,1994.

[13] 金杰里,方亦雄.红树植物胎生的生理意义[J].植物学报,1958,7(2):51 - 58.

[14] 郑文教,林鹏.红树胎生胚轴发育和叶片发育的Cl量动态[J].厦门大学学报(自然科学版),1992,31(5):537 - 542.

[15] Zheng W J,Wang W Q,Lin P. Dynamics of element contents during the development of hypocotyls and leaves of certain mangrove species[J]. J. Exp. Mar. Biology and Ecol.,1999,233:247 - 257.

[16] 张宜辉.几种红树植物繁殖体发育和幼苗成长过程的生理生态学研究[D].厦门:厦门大学,2003.

Mega and Microsporogenesis and Development of Female and Male Gamepothytes in *Aegiceras corniculatum*

YOU Xue-ming,TIAN Hui-qiao,Yang Sheng-chang

(School of Life Sciences,Institute of plant gene and gene technology,Xiamen University,Xiamen 361005,China)

Abstract: Viviparity is the distinct characteristic of some mangrove plants. Most researches focus on its physiology and ecology, but little is known on the early development stage of the embryo. In this paper,the sporogenesis and the development of female and male gamepothytes in *Aegiceras corniculatum* were studied by the means of paraffin-embed and microscope obesrvation. The results show that:(1) the wall of anther consists of four layers:epidermis,endothecium,middlelayers and tapetum from out to inside,and tapetum cell was secretory type;(2) cytokinesis of microsporocyte meiotic division was simultaneous type,and tetrasporophyte was tetrahedral;(3) mature pollen grain was 2-cell type,and contained four germ pores;(4) the ovule was double integuments,tenuinucellate,and megasporocyte divides into four T-type-allied megaspores by meiotic division. The megaspore near chalazae was the functional one. Embryo sac was Polygonum type. (5) there were groups of tannin-cell in the wall of anther,and secretory structures like the salt gland in the ovarys,showing that mangrove plants have developed some special systems in the reproduce apparatus to adapt themselve to the salt habitat.

Key words: *Aegiceras corniculatum*;megaspore microspore;male gametophyte;female gametophyte

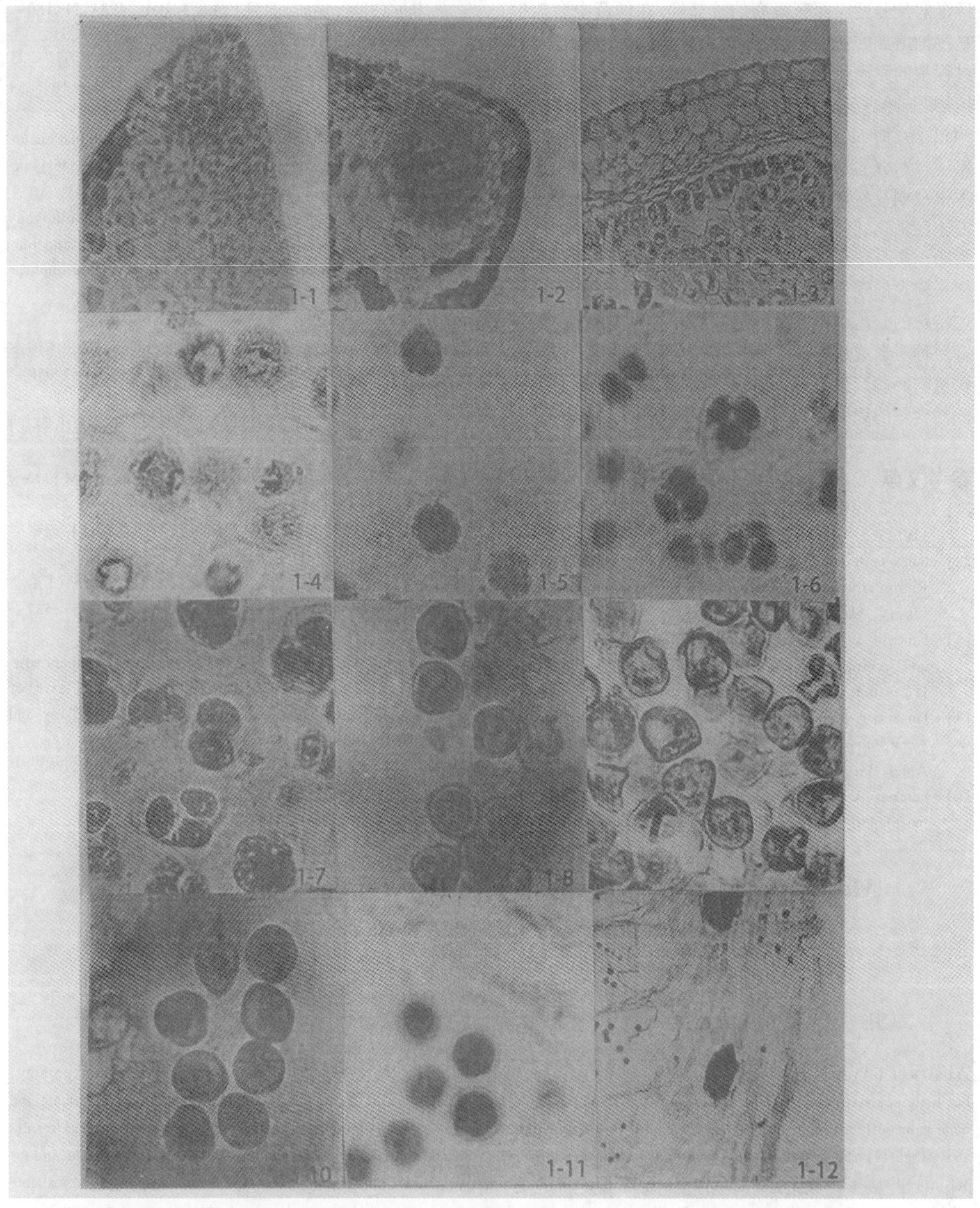

图版 1　桐花树大、小孢子发生与雌、雄配子体发育的观察

1-1～3 小孢子母细胞的发育过程

1-4 第 1 次减数分裂后期(×1 000)　1-5 第 2 次减数分裂(×1 000)　1-6 形成的 4 分体(×1 000)

1-7 刚释放出来的小孢子(×1 000)　1-8 小孢子单核靠边期(×1 000)　1-9 小孢子有丝分裂末期(×1 000)

1-10 形成的营养细胞与生殖细胞(×1 000)　1-11 成熟 2 胞花粉(×1 000)　1-12 花药壁上的单宁细胞腔室(×100)

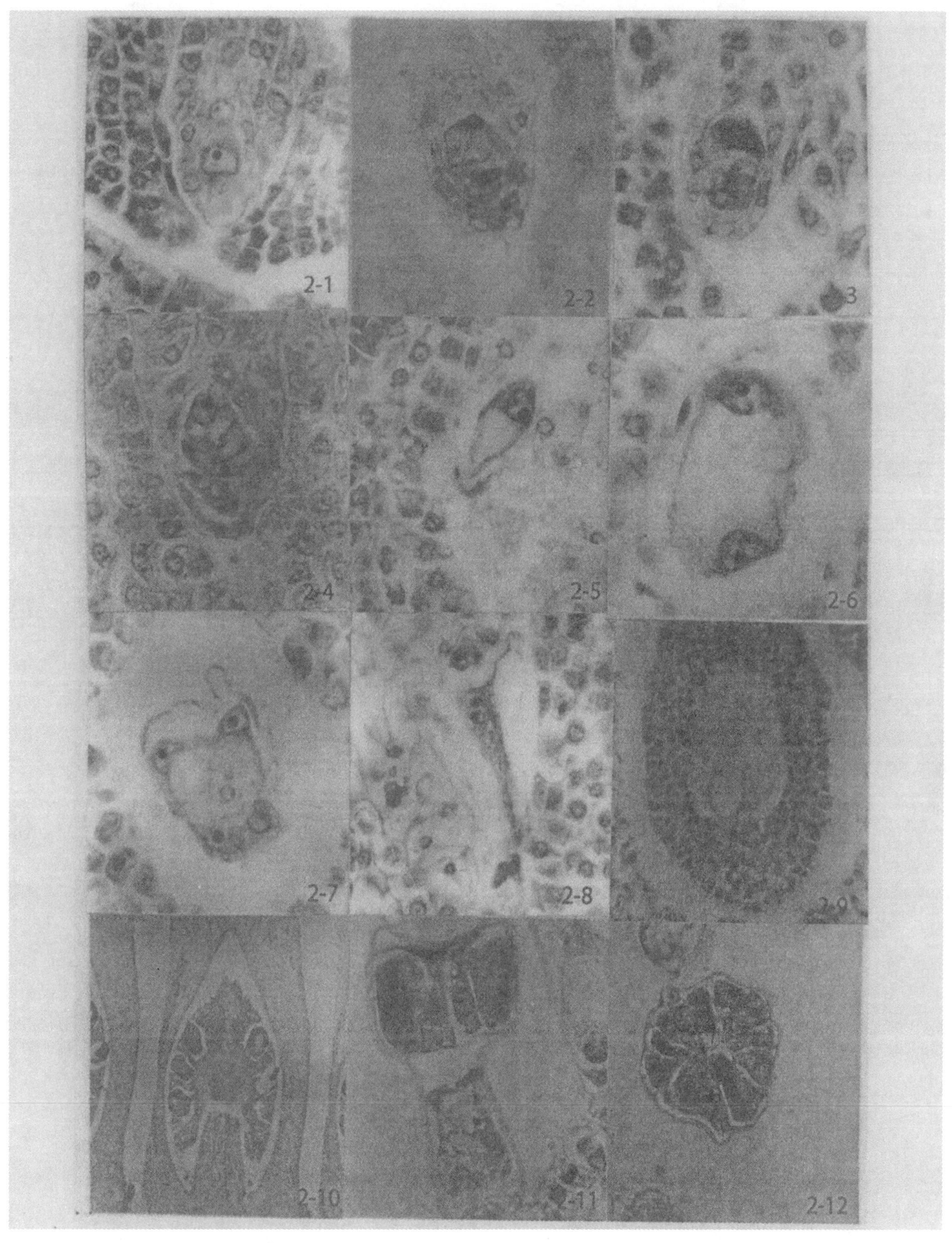

图版 2 桐花树大、小孢子发生与雌、雄配子体发育的观察

2-1 大孢子母细胞(×1 000) 2-2 2 分体(×1 000) 2-3 示合点端的大孢子横向分裂(×1 000)

2-4 T 形排列的 4 分体(×1 000) 2-5 单核胚囊(×1 000) 2-6 2 核胚囊(×1 000)

2-7 4 核胚囊(×1 000) 2-8 7 胞 8 核时期胚囊(还有两核在相邻切面上)(×1 000)

2-9 成熟卵细胞与极核(×1 000) 2-10 示子房中分布的分泌腺的位置(×40)

2-11 分泌腺的横切面(×1 000) 2-12 分泌腺的纵切面(×1 000)

二 红树林生物多样性

PART II MANGROVE BIODIVERSITY

红树林考氏白盾蚧的初步研究*

张飞萍[1] 杨志伟[2] 江宝福[1] 林 鹏[2]

（1. 福建农林大学林学院，福建 福州 350002；2. 厦门大学生命科学学院，福建 厦门 361005）

摘要：首次发现陆生考氏白盾蚧严重为害潮间带红树植物秋茄。通过野外定位调查和室内饲养观察，研究其危害特征、种群分布、生殖力和自然死亡率。结果表明，考氏白盾蚧主要分布在秋茄叶片主脉两侧，叶背虫口数显著多于叶面（$P<0.01$），单叶虫口数多为 1 - 50头，平均 25.65头，最高 418头；秋茄植株不同垂直空间层次的虫口密度无显著差异，但聚集度随垂直空间下降而增大；种群平均产卵量每雌 52.22粒，自然死亡率 69.35%，寄生蜂寄生是最主要的死亡因子，寄生率达 34.23%；与几种陆生寄主植物相比，红树林考氏白盾蚧的种群密度、生殖力、叶最高虫口数均较高，该虫对潮间带生境及寄主秋茄具有适应性。

关键词：红树林；秋茄；考氏白盾蚧

中图分类号：S763.3 **文献标识码**：A **文章编号**：1001 - 389X(2008)03 - 0220 - 05

Preliminary studies on the scale insect pest of *Pseudau lacaspis cockerelli* in mangrove

ZHANG Fei-ping[1], YANG Zhi-wei[2], JIANG Bao-fu[1], LIN Peng[2]

(1. College of Forestry, Fujian Agriculture and Forestry University, Fuzhou, Fujian 350002, China;
2. College of Life Science, Xiamen University, Xiamen, Fujian 361005, China)

Abstract: The scale insect, *Pseudaulacaspis cockerelli* is an important pest for many kinds of plants for a long time in the land. In 2006, it was found firstly attacking the mangrove plant of *Kandelia obovata* in the tideland in Xiamen City, and had led to badly damages. Based on the field surveys and lab feeding, this paper dealed with its damage characteristic, population distribution, fecundity and natural mortality etc. The results showed that the scale insect mainly distributed on both sides of the midvien of *K. obovata* leaves, but the individual on the abaxial surface was significantly more than those on the adaxial surface ($P<0.01$). The individual number of one leaf mostly fluctuated from 1 to 50, and the average and tiptop numbers were 25.65 and 418, respectively. The pest population density had no remarkable discrepancies among different vertical spatial layer of *K. obovata*, however, its aggregating degree increased with falling of the layer. The population had an avarage fecundity of 52.22 each female and a high natural mortality of 69.53%. The main natural death factor were two species of parasites (Hymenoptera: Chalcididae), and their total parasitical ratio reached 34.23%. In the mass, the pest population density, fecundity and tiptop individual number of a leaf in the mongrove were all more than those in serval other host plants on land, which implied that this scale insect had a high adaptability to the habitat of tideland and the plant of *K. obovata*.

Key words: mongrove; *Kandelia obovata*; *Pseudaulacaspis cockerelli*

在气候变暖和海平面逐渐上升的背景下，红树林因具有防浪护堤、促淤造陆和净化海水等诸多功能而倍受关注。长期以来，由于人类自身认识的不足和过度追求经济发展，全球红树林面积锐减。保护现存的红树林资源、修复受损的红树林生态系统和人工营造新的红树林已成为世界生态环境保护工程的热点[1-7]。然而，人工进行单一红树植物高密度造林引发了病虫暴发危害等新问题，成为红树林保护工程的重要障碍[8-11]。

对于大多数昆虫而言，红树林生长的潮间带生境相对恶劣，且许多红树植物富含抗虫成分，因而能够暴发成灾的害虫极少，相关的研究也不多见[1,8-9,11-12]。考氏白盾蚧［*Pseudaulacaspis cockerelli* (Cooley)］是近些年在厦门市海沧区首次发现的一种红树林新害虫，主要为害重要造林树种秋茄（*Kandelia obovata* Sheue, Liu & Yong），以极高的种群密度在叶和嫩梢上刺吸危害，引起落叶和生长不良。该虫同时也是多

* 原载于福建林学院学报，2008，28(3)：220－224

种陆生园林植物和果树的重要害虫,前人针对其陆生种群的生物学、生态学等开展了系列研究,并构建了有效的综合防治技术[13-16]。然而,缺乏有关红树林考氏白盾蚧的研究资料,鉴于红树植物的重要性及其所在生境的特殊性,文中描述了该虫的危害特征,分析了其种群密度及分布规律,并以陆生寄主植物为对照,初步探讨了其种群生殖和自然死亡情况,以期为深入揭示考氏白盾蚧对潮间带逆境的适应性和建立有效的控制技术提供基础依据。

1 材料与方法

1.1 试验地概况

研究地位于福建省厦门市海沧区,红树林面积约 0.53 hm^2, 1998年人工造林, 1999年补植部分秋茄。林内以秋茄为主,间种少量桐花树 (*Aegiceras corniculatum*)、红海榄 (*Rhizophora stylosa*)、白骨榄 (*Avicennia marina*)和木榄 (*Bruguiera gymnorrhiza*)。秋茄造林时系胚轴引种,种源取自福建省漳州市九龙江红树林自然保护区,林分密度 52 890株 · hm^{-2},平均株高 2.08 m,海水涨潮时可淹没所有红树植物。该林分近岸一侧滩涂层厚 20.06 cm,远岸一侧 49.87 cm,长期实行封闭管理,无垦覆和施肥等抚育措施。试验地周边陆地植被以行道树和禾本科杂草为主。含笑 (*Michelia figo*)和白兰花 (*Michelia celba*)等陆生植物考氏白盾蚧的调查取样在厦门和福州两地市区进行。

1.2 红树林考氏白盾蚧种群密度调查

根据海水涨落潮高度,将秋茄植株沿垂直高度划分为高潮区 (小潮的高潮线与大潮高潮线之间)、中潮区 (小潮的高潮线与大潮低潮线之间)和低潮区 (小潮的低潮线与大潮的低潮线之间),并进一步按等分法将高、中潮区分别划分为上、中、下共 6个垂直空间层次 (低潮区无叶片,不调查),从高到低,各空间层次单日受海水浸泡的时间逐渐减少。抽样调查采用棋盘式方法确定 30样株,对每样株的各空间层次均按东、西、南、北、中 5个方位各随机抽取 2叶,每株共 60叶,按样株 ×空间层次袋装后,带回室内镜检,统计各样株、各层次、各叶考氏白盾蚧的数量。为保证抽取样叶的可比性,每次取样均抽取顶梢下第 1、2轮叶。取样时间为 2006、2007年 4 - 5月虫口高峰期。

1.3 红树林考氏白盾蚧的空间格局

基础数据来源于 1.2,选用平均拥挤度 (M^*)、丛生指标 (I)、聚块性指标 (M^*/M)、Cassie指标 (Ca)、扩散系数 (C)和负二项分布 K值等指标测定。

1.4 不同寄主植物考氏白盾蚧叶面、叶背虫口数量的比较

随机抽取白兰花、含笑、米兰 (*Aglaia odorata*)和夹竹桃 (*Nerium indicum*) 4种受考氏白盾蚧为害的植物顶梢下第 1、2轮叶各 50 - 100片,分别统计各叶叶面、叶背的虫口数量,结合 1.2中的秋茄林调查数据,一并比较分析考氏白盾蚧的叶背、叶面分布特征。

1.5 不同寄主植物考氏白盾蚧的生殖力

随机抽取秋茄、白兰花、含笑 3种植物上的考氏白盾蚧,选择雌成虫进行解剖,统计其孕卵和卵壳数量。每种寄主植物均解剖约 50头雌成虫,分别统计其最大产卵量、最小产卵量、平均产卵量和方差。取样时间同 1.2。

1.6 不同寄主植物考氏白盾蚧种群自然死亡率

野外调查表明,考氏白盾蚧种群的自然死亡因子主要有寄生、捕食和其他不明原因。随机抽取秋茄、含笑、白兰花 3种受考氏白盾蚧危害的植物叶片各约 40 - 100片 (保证检查的虫口数大于 300),对叶片上的所有蚧虫均挑去蚧壳,判断其存活情况,并逐个查明死亡原因。其中被寄生虫口以蚧壳有寄生蜂羽化孔,和解剖后体内有寄生蜂幼虫或蛹为判别标准;被捕食虫口以虫体残缺或蚧壳内有捕食性节肢动物为判别标准;其余如虫体干瘪、变色等为其他不明原因死亡;存活虫口以虫体饱满、有光泽为判别标准。另取部分秋茄带虫叶片置于指型管中,用湿棉花塞住管口,逐日观察并收集羽化的寄生蜂或捕食性天敌。取样时间同 1.2。

2 结果与分析

2.1 红树林考氏白盾蚧的鉴定、寄主、形态与危害状

红树林考氏白盾蚧种名由北京林业大学蚧虫分类专家武三安先生确认。野外调查表明:该虫主要寄生于秋茄叶的正、反面,少数寄生于嫩梢;在另一红树植物桐花树上偶见该虫个体,但数量极少,不形成种群;形态学上与寄生于含笑、白兰花、米兰和夹竹桃等陆生植物的个体无明显差异,但秋茄种群中雌成虫和2龄若虫的蚧壳稍厚,色更白;该虫在秋茄叶片上主要分布于主脉两侧,受害部位出现黄色褪绿斑,易脱落,新叶受害稍卷曲、叶小、生长不良。红树林考氏白盾蚧的危害状见图 1。

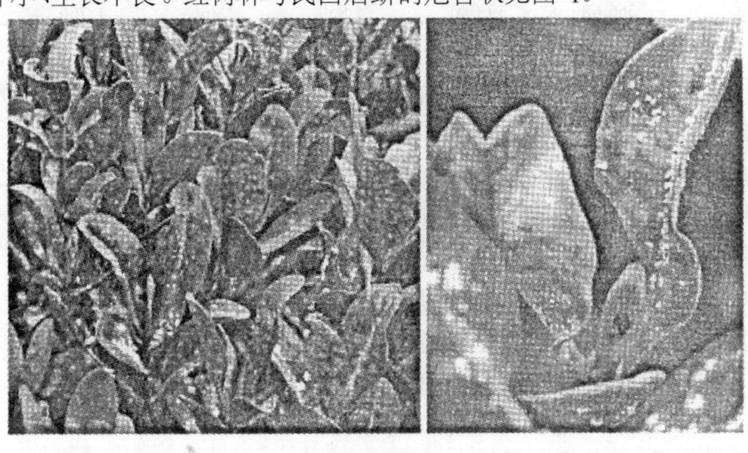

图 1 考氏白盾蚧危害状

Figure 1 Damage symptoms of *P. cockerelli* in the mangrove

2.2 红树林考氏白盾蚧的种群密度

从高、中潮区上、中、下 6层次的虫口密度(表 1)可以看出,该秋茄林各空间层次株受害率均为 100%,叶受害率基本大于 70%;各层次考氏白盾蚧平均虫口数为 14.13 - 35.20头 ·叶$^{-1}$,不同层次间无显著差异。叶最高虫口数以高潮区低层最高,达 418头 ·叶$^{-1}$。按单叶虫口数 0、1 - 10、11 - 50、51 - 100、101头以上 5个密度梯度将样本分类,统计各密度梯度的叶比率。结果表明:各空间层次的叶虫口数多为 1 - 50头,其叶比率均超过 60%,100头 ·叶$^{-1}$以上叶比率则低,均小于 9%。秋茄林各空间层次间的虫口密度无显著差异,这表明在高、低潮转变期间内,海水浸泡时间的长短对该虫种群密度无显著影响。

表 1 红树林考氏白盾蚧的种群密度

Table 1 Population density of *P. cockerelli* in the mangrove

潮区	层次	叶受害率 %	株受害率 %	不同虫口密度叶比率 /%					最高虫口数 头 ·叶$^{-1}$	平均虫口数 头 ·叶$^{-1}$
				0	1 - 10	11 - 50	51 - 100	≥101		
高潮区	高	82.50	100	17.50	32.08	32.92	12.08	5.42	313	25.15 a
	中	87.50	100	12.50	31.67	33.75	15.42	6.67	296	33.03 a
	低	76.25	100	23.75	39.58	25.83	8.33	2.50	188	16.13 a
中潮区	高	77.92	100	22.08	27.08	32.08	11.25	7.50	247	29.60 a
	中	90.83	100	9.17	34.17	33.75	14.58	8.33	418	35.20 a
	低	68.33	100	31.67	32.50	28.33	5.83	1.67	281	14.13 a

2.3 红树林考氏白盾蚧的空间格局

不同空间层次考氏白盾蚧种群聚集度指标见表 2。从表 2可以看出,各层次考氏白盾蚧秋茄种群的 M^*/M、I均大于 1, Ca值均大于 0, K值均小于 8,且大多数均小于 1,说明不同空间层次的考氏白盾蚧均呈聚集分布。由于聚集度指标 K值与虫口密度无关,适合于相同样方之间的比较,进一步以 K值为指标,比较考氏白盾蚧不同空间层次的聚集程度。从表 2可以看出,自上而下, K值逐渐变小,说明随着空间层次的下降,考氏白盾蚧的聚集程度逐渐升高。显然,随着垂直空间下降,植株受海水浸泡的时间越长,而种群的高度聚集有利于其度过这一逆境。考氏白盾蚧聚集程度的空间层次变化,在一定程度上体现了其对海陆交错带恶劣生境的适应。

表 2 红树林考氏白盾蚧的种群聚集度指标

Table 2 Aggregation index of *P. cockerelli* in the mangrove

潮区	层次	S^2	M^*	I	M^*/M	Ca	C	K
高潮区	高	635.41	49.415	24.265	1.965	0.965	25.265	1.036
	中	1 507.04	77.656	44.626	2.351	1.351	45.626	0.740
	低	397.59	39.779	23.649	2.466	1.466	24.649	0.682
中潮区	高	1 319.29	73.171	43.571	2.472	1.472	44.571	0.679
	中	1 939.29	89.293	54.093	2.537	1.537	55.093	0.651
	低	366.68	39.080	24.950	2.766	1.766	25.950	0.566

2.4 不同寄主植物考氏白盾蚧叶面、叶背虫口数的比较

从秋茄、白兰花、含笑、米兰、夹竹桃 5种植物受害叶叶面、叶背虫口数的统计结果 (表 3)可以看出,各寄主植物叶面、叶背的平均虫口数均具有极显著差异,说明考氏白盾蚧对寄生部位具有明显的选择性。但是,秋茄以叶背的虫口数显著较多,其他植物则均以叶面显著较多,这一差异可能与潮间带光照强度大、风速高以及秋茄叶片结构的特异性等有关。从叶最大虫口数看,以秋茄和白兰花最多,平均虫口数也明显较高,而秋茄的单叶面积明显小于白兰花,可见考氏白盾蚧在秋茄林具有更为繁荣的种群。

表 3 不同寄主条件下考氏白盾蚧的叶面、叶背虫口数量

Table 3 Individuals of *P. cockerelli* on the adaxial and abaxial surface of leaf in different host plants

寄主植物	寄生部位	最大虫口数/头	平均虫口数 头·叶$^{-1}$	t值
秋茄	叶面	201	11.32	4.691 8**
	叶背	217	14.33	
白兰花	叶面	261	52.90	5.874 9**
	叶背	117	15.52	
含笑	叶面	22	5.63	5.928 8**
	叶背	16	2.07	
米兰	叶面	8	2.16	8.188 9**
	叶背	4	0.25	
夹竹桃	叶面	80	15.65	6.423 2
	叶背	6	1.22	

2.5 不同寄主植物考氏白盾蚧的生殖力

秋茄、白兰花、含笑 3种植物上考氏白盾蚧雌成虫产卵数见表 4。从表 4可以看出,该虫在秋茄、含笑上的最大产卵量、最小产卵量和平均产卵量均显著大于白兰花,而秋茄和含笑之间无显著差异。这说明寄主植物对考氏白盾蚧的种群生殖力具有显著影响。取食秋茄的考氏白盾蚧具有繁荣的种群与其较高的生殖力密切相关。

表 4 不同寄主条件下考氏白盾蚧的生殖力

Table 4 Fecundity of *P. cockerelli* in different host plants

寄主植物	最大产卵量 粒·雌$^{-1}$	最小产卵量 粒·雌$^{-1}$	平均产卵量 粒·雌$^{-1}$	方差
秋茄	135	13	59.22 a	744.137 3
含笑	135	17	62.48 a	971.431 0
白兰花	85	4	36.86 b	337.400 7

2.6 不同寄主植物考氏白盾蚧的自然死亡率

从不同寄主植物考氏白盾蚧的自然死亡和存活情况 (表 5)可以看出,与含笑和白兰花相比,考氏白盾蚧秋茄种群的存活率最低,仅 30.65%。分析其原因,主要是寄生率和其他原因死亡率明显较高。

表 5 不同寄主条件下考氏白盾蚧种群死亡率

Table 5 Mortality of *P. cockerelli* population in different host plants %

寄主植物	寄生率	捕食率	其他死亡率	存活率
秋茄	34.23	7.37	27.75	30.65
含笑	17.53	24.03	19.48	38.96
白兰花	25.42	11.03	13.73	49.82

3 讨论

我国的红树林由于现存面积小、造林存活率低和生态功能强大等而成为珍贵的森林资源,建立红树林自然保护区和人工造林是当前保护和修复红树林生态系统的主要措施[1-2]。然而,人工造林中由于可供选择的树种极少,投入大、存活率低,现有人造红树林基本上均为密度极高的纯林。可以预见,随着人工造林面积的增多,病虫暴发危害将严重制约红树林的发展,但这一问题尚未得到足够重视。

首次发现考氏白盾蚧对秋茄造成严重危害,由于该试验林造林时间较短,且潮间带生境特殊,有必要充分探讨其考氏白盾蚧的来源。该林分系胚轴造林,野外调查未发现考氏白盾蚧寄生胚轴,对种源地的实地考察也未见该虫危害,而在试验地周边园林植物上发现考氏白盾蚧,因此判断红树林中的考氏白盾蚧为陆地迁移所致。文中考氏白盾蚧种群密度在垂直空间层次上无显著差异 (表 1),种群聚集度随着空间层次的下降而增大 (表 2),与陆生植物相比具有较高的繁殖力 (表 4)等说明,长期生长于陆地的考氏白盾蚧

迁移到潮间带,在频繁地经受海水浸泡后仍能够形成繁荣的种群,可见该虫对潮间带生境和寄主秋茄具有高度的适应能力。一些研究认为,海洋和潮间带由于恶劣的生境条件而极少昆虫涉足[12],且蚧类昆虫主要营固定取食,不具备主动避难功能,该类昆虫在这一生境中长期生存的可能性较小,因此,考氏白盾蚧的这一适应能力值得关注。

自然条件下红树林考氏白盾蚧种群具有极高的死亡率,除了寄生这一重要的生物致死因子外,其它原因死亡也是一个重要的因子,其引起的死亡率为 27.75%,明显大于其他 2种陆生植物上的种群,这无疑是潮间带逆境胁迫带来的结果。然而,考氏白盾蚧凭借其较高的种群生殖力适应策略,仍然在这一逆境中维持着繁荣的种群。还值得一提的是,由于潮间带周期性的海水浸泡及其强大的冲击力,难以对红树林考氏白盾蚧实施化学防治,而本文发现该虫自然种群最主要的致死因子为 2种寄生蜂 (隶属于小蜂总科,种名待定)的寄生作用,致死率高达 34.23%,因此,保护利用寄生蜂实施生物防治应是该虫综合治理研究的重点内容。

参考文献

[1] 林鹏,傅勤. 中国红树林环境生态及经济利用 [M]. 北京:高等教育出版社, 1994: 1 - 3, 61 - 71.

[2] 王文卿,王瑁. 中国红树林 [M]. 北京:科学出版社, 2007: 63 - 68, 95 - 105, 143 - 159, 161 - 167.

[3] Lugo A C, Snedaker S C. The ecology of mangroves[J]. Annual Review of Ecology and Systematics, 1974, 5: 39 - 64.

[4] Othman M A. Value of mangroves in coastal protection[J]. Hydrobiologia, 1994, 285: 277 - 282.

[5] Ong J E. The ecology of mangrove conservation and management[J]. Hydrobiologia, 1995, 295: 343 - 351.

[6] Cabrera M A, Seijo J C, Euan J. Economic values of ecological services from a mangrove ecosystem [J]. International Newsletter of Coastal Management, 1998, 33: 1 - 2.

[7] Bandaranayake W M. Traditional and medicinal uses of mangroves[J]. Mangroves and Salt Marshes, 1998, 2: 133 - 148.

[8] 贾凤龙,陈海东,王勇军,等. 深圳福田红树林害虫及其发生原因. 中山大学学报:自然科学版, 2001, 40(3): 88 - 91.

[9] 范航清,邱广龙. 中国北部湾白骨壤红树林的虫害与研究对策 [J]. 广西植物, 2004, 24(6): 558 - 562.

[10] 伍荔霞. 关注红树林湿地 [N]. 广西日报, 2006 - 12 - 04(10).

[11] Ozaki K, Kitamura S, Subiandoro E. Life history of *Aulacaspis marina* Takagi and Williams (Hom., Coccoidea), a new pest of mangrove plantations in Indonesia, and its damage to mangrove seedlings[J]. Journal of Applied Entomology, 1999, 123(5): 281.

[12] 张小斌,陈学新,程家安. 为何海洋中的昆虫种类如此稀少 [J]. 昆虫知识, 2005, 42(4): 471 - 475.

[13] 胡兴平. 考氏白盾蚧形态研究 [J]. 山东农业大学学报, 1991, 22(3): 221 - 226.

[14] 胡兴平,周朝华. 观赏植物上考氏白盾蚧生物学与防治 [J]. 山东农业大学学报, 1993, 24(1): 99 - 101.

[15] 林克明,简翠馨,凌远方. 考氏白盾蚧的生物学特性及防治 [J]. 昆虫知识, 1994, 31(2): 91 - 94.

[16] 罗佳,葛有茂. 考氏白盾蚧生物学与天敌初步研究 [J]. 福建农业大学学报, 1997, 26(2): 194 - 199.

厦门东屿红树林湿地鸟类资源及其分布*

林清贤 陈小麟 林 鹏

(厦门大学生命科学学院,福建 厦门 361005)

摘要: 根据2001年1月～2003年3月的调查结果表明,厦门东屿红树林湿地鸟类共有鸟类8目27科97种,其中冬候鸟38种、留鸟32种、旅鸟22种、夏候鸟5种;古北种鸟类57种,东洋种25种,广布种15种;厦门东屿红树林湿地鸟类以水鸟为主,有55种,占56.7%;优势种类为环颈鸻 *Charadrius alexandrinus*、铁嘴沙鸻 *Charadrius leschenaultii*、黑腹滨鹬 *Calidris alpina*、白鹭 *Egretta garzetta*、苍鹭 *Ardea cinerea* 和麻雀 *Passer montanus*. 文中还对红树林周边不同生境中鸟类的分布情况进行分析.

关键词: 鸟类;红树林;东屿;厦门

中图分类号: Q 958.2 **文献标识码**:A **文章编号**:0438-0479(2005)Sup-0037-06

红树林湿地具有高光合率、高呼吸率和高归还率的三高特点[1]. 因此红树林生长的潮滩往往含有大量的有机残物,为生活在红树林滩涂的底栖动物提供了大量的能量来源. 底栖动物的增加又为生活在红树林的鸟类鱼类等脊椎动物提供了丰富的饵料来源. 由于有了充足的食物资源,红树林湿地的鸟类多样性较高,在国际性水鸟保护协定中红树林湿地的作用是不可或缺的一环,对红树林湿地鸟类研究具有重要的意义.

在我国红树林湿地鸟类研究方面,香港米埔、台湾、深圳福田、海南东寨港和广西山口红树林保护区都做了不少研究[2~10],在福建,宋晓军对云霄竹塔、龙海浮宫、厦门东屿和福鼎鲎屿4个红树林区进行了鸟类调查,初步了解了福建红树林区鸟类的种类和区系组成[11],本文在宋晓军调查基础上进一步对厦门东屿红树林湿地鸟类资源及其分布进行了调查和研究.

1 自然概况与研究方法

1.1 自然概况

厦门东屿红树林湿地位于厦门市海沧区东屿村,地理坐标为东经118°03′,北纬24°31′,年均温21.1 ℃,年降雨量1 036 mm.. 红树林分布在中高潮带,主要树种为白骨壤,间有零星的秋茄,高度为1～2 m,呈斑块状分布,面积原有67 hm^2,在1997年剩20 hm^2 左右,到2000年仅存6 hm^2 左右,2001年开始的海沧滨海大道建设中,又破坏了大量的红树林,现在剩下的红树林面积已不足2 hm^2,红树林斑块生长密集,难以通行,斑块间具有较大空地,有些斑块被基围鱼塘所包围,在大潮时,所有红树林林冠都可被潮水淹没. 红树林带外为泥沙质滩涂,宽度150～250 m,中潮区部分养殖缢蛏,低潮区有牡蛎养殖. 南侧堤岸上有稀疏的木麻黄防护带,堤后为一条水渠及虾池,在2001年,海沧滨海大道建设从滩涂中间开出约110 m宽的公路,不仅破坏了不少红树林,还使部分红树林与滩涂隔离;堤岸北侧为1997年围海造陆形成的荒废裸地. 在裸地后为村庄、鱼塘及荒废空地.

1.2 研究方法

2001年1月～2003年3月对厦门东屿红树林湿地鸟类进行为期2年3个月的调查,在候鸟迁徙季节每月(3～5月和9～11月)调查2～4次,其它月份1～2次.

调查采用路线调查和高位定点及直线统计相结合方法[11],对红树林带、堤岸、农田、鱼塘、荒废裸地等生境带采用路线调查法进行统计,调查行进速度约1 km/h,调查宽度约50 m,对在泥滩和海面觅食、栖息的鸟类采用高位定点观察和直接进入泥滩观察的方法统计绝对数量. 调查时记录时间、地点、鸟类的种类、数量、栖息的生境类型、表现行为(觅食、休息、繁殖)等.

鸟类的优势度等级根据频度指数(*RB*)判定[12]:

$$RB=\frac{d}{D}\times\frac{N}{D}$$

式中:*RB* 为频率指数;*d* 为遇见该种鸟类的天数;*N* 为遇见该种鸟类的总数量;*D* 为调查总天数.

RB 在10以上为优势种(++++),5～10为常见种(+++),1～5为少见种(++),1以下为偶见种(+).

---

* 国家自然科学基金(40276028)资助

原载于厦门大学学报(自然科学版),2005,44(Sup.):37－42

表1 厦门东屿红树林区鸟类种类组成

Tab.1 Checklist of bird species in mangrove areas of Dongyu Islet ,Xiamen

中文名	学名	区系型	生态类群	居留型	保护级别	优势等级
I 雁形目	ANSERIFORMES					
1.鸭科	Anatidae					
赤颈鸭	*Anas penelope*	p	sw	w	C-J	+
绿翅鸭	*Anas crecca*	p	sw	w	C-J	+ +
II 戴胜目	UPUPIFORMES					
2.戴胜科	Upupidae					
戴胜	*Upupa epops*	w	c	r		+
III 佛法僧目	CORACIFORMES					
3.翠鸟科	Alcedinidae					
普通翠鸟	*Alcedo atthis*	w	c	r		+ +
白胸翡翠	*Halcyon smyrnensis*	o	c	r		+
蓝翡翠	*Halcyon pileata*	o	c	r		+
4.鱼狗科	Cerylidae					
斑鱼狗	*Ceryle rudis*	o	c	r		+
IV 雨燕目	APODIFORMES					
5.雨燕科	Apodidae					
小白腰雨燕	*Apus affinis*	o	c	r	C-J	+
V 鸽形目	COLUMBIFORMES					
6.鸠鸽科	Columbidae					
珠颈斑鸠	*Streptopelia chinensis*	p	l	r		+
VI 鹤形目	GRUIFORMES					
7.秧鸡科	Rallidae					
白胸苦恶鸟	*Amaurornis phoenicurus*	o	w	r		+
黑水鸡	*Gallinula chloropus*	o	w	r	C-J	+ + +
VII 鹳形目	CICONIIFORMES					
8.丘鹬科	Scolopacidae					
小杓鹬	*Numenius minutus*	p	w	t	Ⅱ,C-A	+
中杓鹬	*Numenius phaeopus*	p	w	t	C-A,C-J	+ +
白腰杓鹬	*Numenius arquata*	p	w	w	C-A,C-J	+ +
大杓鹬	*Numenius madagascariensis*	p	w	t	C-A,C-J	+
红脚鹬	*Tringa totanus*	p	w	t	C-A,C-J	+ + +
泽鹬	*Tringa stagnatilis*	p	w	w	C-A,C-J	+
青脚鹬	*Tringa nebularia*	p	w	w	C-A,C-J	+ + +
小青脚鹬	*Tringa guttifer*	p	w	t	Ⅱ,C-J	+
白腰草鹬	*Tringa ochropus*	p	w	w	C-J	+
林鹬	*Tringa glareola*	p	w	t	C-A,C-J	+
翘嘴鹬	*Xenus cinerea*	p	w	t	C-A,C-J	+ +
矶鹬	*Actitis hypoleucos*	p	w	w	C-A,C-J	+ +
灰尾[漂]鹬	*Heteroscelus brevipes*	p	w	t	C-A,C-J	+ +
翻石鹬	*Arenaria interpres*	p	w	t	C-A,C-J	+
半蹼鹬	*Limnodromus semipalmatus*	p	w	t	C-A	+
大滨鹬	*Calidris tenuirostris*	p	w	t	C-A,C-J	+
红颈滨鹬	*Calidris ruficollis*	p	w	t	C-A,C-J	+ +
青脚滨鹬	*Calidris temminckii*	p	w	w	C-J	+
长趾滨鹬	*Calidris subminuta*	p	w	w	C-A,C-J	+
尖尾滨鹬*	*Calidris acuminata*	p	w	t	C-A,C-J	+ +
黑腹滨鹬	*Calidris alpina*	p	w	w	C-A,C-J	+ + + +
弯嘴滨鹬	*Calidris ferruginea*	p	w	t	C-A,C-J	+ +

续表 1

中文名	学名	区系型	生态类群	居留型	保护级别	优势等级
9. 鸻科	Charadriidae					
黑翅长脚鹬	*Himantopus himantopus*	p	w	t	C-J	+
金斑鸻	*Pluvialis fulva*	p	w	t	C-A, C-J	+
灰斑鸻	*Pluvialis squatarola*	p	w	w	C-A, C-J	+ +
剑鸻	*Charadrius hiaticula*	p	w	w	C-A	+
金眶鸻	*Charadrius dubius*	p	w	r	C-A	+ +
环颈鸻	*Charadrius alexandrinus*	p	w	w		+ + + +
蒙古沙鸻	*Charadrius mongolus*	p	w	t	C-A, C-J	+
铁嘴沙鸻	*Charadrius leschenaultii*	p	w	t	C-A, C-J	+ + + +
灰头麦鸡	*Vanellus cinereus*	p	w	t		+
10. 燕鸻科	Glareolidae					
普通燕鸻	*Glareola maldivarum*	p	w	t	C-A, C-J	+
11. 鸥科	Laridae					
黑尾鸥	*Larus crassirostris*	p	sw	w		+
海鸥	*Larus canus*	p	sw	w	C-J	+
红嘴鸥	*Larus ridibundus*	p	sw	w	C-J	+ +
黑嘴鸥	*Larus saundersi*	p	sw	w		+
鸥嘴噪鸥	*Gelochelidon nilotica*	p	sw	w		+
白额燕鸥	*Sterna albifrons*	w	sw	s	C-A, C-J	+ + +
须浮鸥	*Chlidonias hybrida*	p	sw	w		+
白翅浮鸥	*Chlidonias leucoptera*	p	sw	t	C-A	+
12. 鹰科	Accipitridae					
鹗	*Pandion haliaetus*	w	r	w	Ⅱ	+
黑耳鸢	*Milvus lineatus*	w	r	r	Ⅱ	+
13. 隼科	Falconidae					
红隼	*Falco tinnunculus*	w	r	w	Ⅱ	+
14. 䴙䴘科	Podicipedidae					
小䴙䴘	*Tachybapus ruficollis*	w	sw	r		+
15. 鸬鹚科	Phalacrocoracidae					
[普通]鸬鹚	*Phalacrocorax carbo*	p	sw	w		+
16. 鹭科	Ardeidae					
白鹭	*Egretta garzetta*	w	w	r		+ + + +
苍鹭	*Ardea cinerea*	p	w	w		+ + + +
大白鹭	*Casmerodius albus*	w	w	w	C-A, C-J	+ +
中白鹭	*Mesophoyx intermedia*	w	w	w	C-J	+
牛背鹭	*Bubulcus ibis*	o	w	s	C-A, C-J	+
池鹭	*Ardeola bacchus*	o	w	r		+ + +
绿鹭	*Butorides striatus*	o	w	s	C-J	+
夜鹭	*Nycticorax nycticorax*	o	w	r	C-J	
黄苇鳽	*Ixobrychus sinensis*	o	w	s	C-A, C-J	
VIII 雀形目	PASSERIFORMES					
17. 伯劳科	Laniidae					
棕背伯劳	*Lanius schach*	o	si	r		+ +
18. 鹟科	Museicapidae					
蓝矶鸫	*Monticola solitarius*	o	si	r		+
乌鸫	*Turdus merula*	o	si	r		+
鹊鸲	*Copsychus saularis*	o	si	r		+
北红尾鸲	*Phoenicurus auroreus*	p	si	w	C-J	+
黑喉石䳭	*Saxicola torquata*	p	si	w	C-J	+

续表 1

中文名	学名	区系型	生态类群	居留型	保护级别	优势等级
19. 椋鸟科	Sturnidae					
黑领椋鸟	*Sturnus nigricollis*	o	si	r		+
八哥	*Acridotheres cristatellus*	o	si	r		+
20. 燕科	Hirunxinidae					
家燕	*Hirundo rustica*	w	si	s	C-A, C-J	+ + +
金腰燕	*Hirundo daurica*	w	si	r	C-J	+
21. 鹎科	Pycnonotidae					
白头鹎	*Pycnonotus sinensis*	o	si	r		+ +
白喉红臀鹎	*Pycnonotus aurigaster*	o	si	r		+
22. 扇尾莺科	Cisticolidae					
棕扇尾莺	*Cisticola juncidis*	o	si	r		+
黄腹鹪莺	*Prinia flaviventris*	o	si	r		+
褐头鹪莺	*Prinia inornata*	o	si	r		+
23. 绣眼鸟科	Zosteropidae					
暗绿绣眼鸟	*Zosterops japonicus*	o	si	r		+
24. 莺科	Sylviidae					
黑眉苇莺	*Acrocephalus bistrigiceps*	p	si	w	C-J	+
大苇莺	*Acrocephalus arundinaceus*	p	si	t	C-A, C-J	+
褐柳莺	*Phylloscopus fuscatus*	p	si	w		+
25. 百灵科	Alaudidae					
小云雀	*Alauda gulgula*	o	si	w		+
26. 麻雀科	Passeridae					
[树]麻雀	*Passer montanus*	w	si	r		+ + + +
白鹡鸰	*Motacilla alba*	w	si	r	C-A, C-J	+ + +
黄鹡鸰	*Motacilla flava*	p	si	w	C-A, C-J	+
灰鹡鸰	*Motacilla cinerea*	p	si	w	C-A	+
理氏鹨	*Anthus richardi*	p	si	w	C-J	+
树鹨	*Anthus hodgsoni*	p	si	w	C-J	+
红喉鹨	*Anthus cervinus*	p	si	w	C-J	+
斑文鸟	*Lonchura punctulata*	o	si	r		+ +
27. 燕雀科	Fringillinae					
金翅[雀]	*Carduelis sinica*	w	si	r		+
黄胸鹀	*Emberiza aureola*	p	si	w	C-J	+
灰头鹀	*Emberiza spodocephala*	p	si	w	C-J	+
红颈苇鹀	*Emberiza yessoensis*	p	si	w		+

注:区系:w:广布种(widespread),p:古北界(palaearctic),o:东洋界(oriental);生态类群:sw:游禽(swimmer),w:涉禽(wader),r:猛禽(raptor),l:陆禽(land-bird),c:攀禽(climbing-bird),si:鸣禽(singing-bird);居留型:r:留鸟(resident),w:冬候鸟(wintering),t:旅鸟(traveling),s:夏候鸟(summering);保护级别:C-A:中澳联合保护候鸟,C-J:中日联合保护候鸟,Ⅱ:国家Ⅱ级重点保护鸟类;优势等级:+:偶见种,++:少见种,+++:常见种,++++:优势种.

2 结果分析

2.1 种类组成

厦门东屿红树林湿地共发现鸟类 97 种,分属 8 目 27 科(见表 1). 东屿红树林湿地鸟类的居留型组成主要以冬候鸟和留鸟为主,分别有 38 种和 32 种,另有旅鸟 22 种和夏候鸟 5 种;鸟类生态类群主要以涉禽和鸣禽为主,分别为 43 种和 32 种,其它生态类群有游禽 12 种、猛禽 3 种、陆禽 1 种、攀禽 6 种,其中水鸟(涉禽和游禽)有 55 种,占 56.7%;在区系组成上,古北界鸟类 57 种,东洋界鸟类 25 种,广布种 15 种;厦门东屿红树林湿地鸟类的优势种以涉禽为主,有环颈鸻、铁嘴沙鸻、黑腹滨鹬、白鹭和苍鹭,非涉禽类的只有麻雀,常见鸟有 7 种,少见鸟有 17 种,偶见鸟 67 种.

厦门东屿红树林湿地存在着不少珍稀濒危鸟类,

表 2 厦门东屿红树林区各生境带的鸟类组成

Tab.2 The components of bird species in each habitat at Dongyu mangrove area

类群	海面	滩涂	红树林	堤岸带	荒废裸地	塘区	空中
游禽	8	8	3	0	1	1	0
涉禽	0	26	21	5	27	11	0
猛禽	0	1	2	0	2	2	3
陆禽	0	0	0	0	0	1	0
攀禽	0	3	5	1	1	3	2
鸣禽	0	2	13	7	20	26	2
总计	8	40	44	13	51	44	7

其中国家 Ⅱ级重点保护鸟类有黑耳鸢、鹗、红隼、小杓鹬、小青脚鹬 5 种,列入《中华人民共和国政府和澳大利亚政府保护候鸟及其栖息环境协定》的有 36 种,占总数 81 种的 44.4 %;列入《中华人民共和国政府和日本政府保护候鸟及其栖息环境协定》的有 52 种,占总数 227 种的 22.9 %(见表 1). 另外列入其它保护的种类有:

(1) 列入濒危野生动植物种国际贸易公约(CITES,1995)附录 Ⅰ的有小青脚鹬,小杓鹬,附录 Ⅱ的有鹗、黑耳鸢、红隼;

(2) 列入世界自然保护联盟(IUCN)濒危等级的有小青脚鹬、黑嘴鸥;

(3) 列入中国濒危动物红皮书稀有等级的有半蹼鹬、鹗,未定等级小青脚鹬.

2.2 厦门东屿红树林区各生境带的鸟类种类组成

根据生境差别将厦门东屿红树林湿地分为海面、滩涂、红树林带、堤岸带、荒废裸地、塘区和空中 7 个生境带,海面种类都为游禽,由鸥科、鸭科和鸬鹚科组成. 滩涂种类主要由涉禽的鸻鹬类和鹭科鸟类组成,大部分游禽也经常在滩涂上觅食和休息,部分猛禽和攀禽经常停在滩涂上的竹竿上休息,鸣禽的白鹡鸰和家燕经常出现在滩涂上. 红树林带主要为涉禽和鸣禽,分别占 44.7 %和 29.5 %,涉禽中的鸻鹬类主要出现在红树林斑块中的光滩上,鹭类在红树林中和光滩上都有,一些其它生态类群种类经常在红树林带中出现. 堤岸带主要为涉禽和鸣禽,分别占 38.4 %和 53.8 %,其它有攀禽普通翠鸟. 荒废裸地的主要种类为涉禽和鸣禽,分别占 52.9 %和 39.2 %,大部分涉禽特别是鸻鹬类在退潮时在滩涂上觅食,涨潮时在荒废裸地上休息,另有游禽白额燕鸥,为该地繁殖鸟类,部分猛禽和攀禽也经常出现在该生境带. 塘区主要为鸣禽和涉禽,分别占 59.1 %和 25.0 %,另有游禽红嘴鸥;攀禽普通翠鸟、斑鱼狗和白胸翡翠;猛禽黑耳鸢和红隼;陆禽珠颈斑鸠. 空中种类由猛禽黑耳鸢、红隼和鹗,攀禽斑鱼狗和小白腰雨燕及鸣禽中的家燕和金腰燕组成. 厦门东屿红树林湿地各生境带中各生境带鸟类物种数量的排列顺序为荒废裸地 > 红树林带 = 塘区 > 滩涂 > 堤岸带 > 海面 > 空中(见表 2).

3 讨 论

厦门东屿红树林湿地是厦门最重要的冬候鸟觅食地和栖息地之一,大量的鸻鹬类在此越冬,不少国际或国内的珍稀濒危鸟类也在此出现. 但近年来由于围海造陆及滨海大道的建设,红树林和滩涂面积大量减少,人为干扰严重,导致了当地鸟类资源急剧减少. 与 1996~1997 年调查结果[12]相比较,厦门东屿红树林湿地的优势种类有明显的变化,如鹭科鸟类的池鹭和大白鹭分别由优势种变为常见种和少见种,鸻鹬类的红脚鹬和青脚鹬由优势种变为常见种,鸥类的红嘴鸥由优势种变为少见种,鸭类的绿翅鸭由优势种变为偶见种,而麻雀由原来的偶见种变为优势种,说明了厦门东屿红树林湿地正在逐步向人工环境转变. 从数量上看,在 1996~1997 年间红嘴鸥的越冬数量达到 2 000 多只,2001~2003 年只能见到几只或几十只,绿翅鸭在 1996~1997 年 1 月间累计调查到 1 361 只,单次调查最大数量为 431 只,而在 2001~2003 年 3 月共调查到 68 只,单次调查最大数量为 30 只. 从种类上来看,环境的变化还导致了国家 Ⅱ级重点保护动物黑脸琵鹭和海鸬鹚等珍稀濒危鸟类在厦门消失. 由两次调查结果比较中可以看出,厦门东屿红树林湿地的水鸟资源已经受到严重的破坏. 夏季在厦门东屿红树林湿地仍能见到不少冬候鸟度夏,如青脚鹬、红脚鹬、灰尾鹬、白腰杓鹬和铁嘴沙鸻等,这些可能是些尚未性成熟或老、弱、病的个体,有待进一步研究.

厦门东屿红树林区边缘的荒废裸地是近年来新形成的生境,它是许多在东屿红树林区及其附近滩涂觅食的鸟类,特别是鸻鹬类和鹭类等涉禽在涨潮时重要休息地，因此涨潮时该荒废裸地鸟类数量比较多，冬季最多时有 2 000 多只的鸻鹬类和鹭科鸟类. 东屿红树林区边缘的荒废裸地上的鸟类物种也相当丰富，是红树林区鸟种最多的生境带. 因而对于红树林区涉禽类的保护，不仅要考虑到它们的滩涂觅食生境，还要注意保护其休息生境.

参考文献：

[1] 林鹏. 中国红树林生态系[M]. 北京：科学出版社，1997. 1 - 10，157 - 177.

[2] 邓巨燮，关贯勋，徐利生. 深圳市福田红树林鸟类保护区的鸟类及无脊椎动物调查报告[J]. 生态科学，1986，1：44 - 50.

[3] Earles S. Birds of Mai Po Marshes[J]. Birds International，1990，2(3)：11 - 21.

[4] 王勇军，刘治平，陈相如. 深圳福田红树林冬季鸟类调查[J]. 生态科学，1993，2：74 - 84.

[5] 王勇军，林鹏，宋晓军. 深圳湾福田红树林湿地水鸟的周年动态[J]. 厦门大学学报(自然科学版)，1998，38(1)：122 - 129.

[6] 宋晓军，林鹏，苏文拔. 我国红树林区的动物多样性和持续利用. 中国湿地研究和保护[M]. 上海：华东师范大学出版社，1998. 93 - 101.

[7] 王勇军，昝启杰，林鹏. 深圳市福田红树林陆鸟类变迁及保护[J]. 厦门大学学报(自然科学版)，1999，38(1)：137 - 144.

[8] 王勇军，陈桂株. 深圳福田红树林湿地鸟类研究 I、II、III. 中国湿地研究和保护[M]. 上海：华东师范大学出版社，1998. 179 - 195.

[9] 周放，房慧伶，张红星. 山口红树林鸟类多样性初步研究[J]. 广西科学，2000，7(2)：154 - 157.

[10] 常弘，毕肖峰，陈桂珠，等. 海南岛东寨港国家级自然保护区鸟类组成和区系的研究[J]. 生态科学，1999，2：53 - 61.

[11] 宋晓军，林鹏. 福建红树林湿地鸟类区系研究[J]. 生态学杂志，2002，21(6)：5 - 10.

[12] 盛和林，王岐山. 脊椎动物野外实习指导[M]. 北京：高等教育出版社，1987. 408 - 10.

Bird Resources and its Distribution in Mangroves at Dongyu Islet, Xiamen, China

LIN Qing-xian, CHEN Xiao-lin, LIN Peng

(School of Life Sciences, Xiamen University, Xiamen 361005, China)

Abstract: Census of birds was made from January in 2001 to March in 2002 at the mangrove areas in Dongyu Islet, Xiamen, China. 97 species belonging to 27 families in 8 orders were recorded, in which there were 38 species of winter visiting birds, 22 species of traveling birds, 5 species of summer visiting birds and 32 species of resident birds. 55 species of avian fauna belong to the Palaearctic Realm, 25 species belong to Oriental Realm, and 15 species belong to Widespread. There were 55 species of waterbirds (including swimming birds and waders), which were the main composition of bird at the mangrove areas in Dongyu, Xiamen, China. The most abundant species were *Charadrius alexandrinus*, *Charadrius leschenaultia*, *Calidris alpine*, *Egretta garzetta*, *Ardea cinerea* and *Passer montanus*. The distributions of birds in different habitats were analyzed in paper also.

Key words: bird; mangroves; Dongyu; Xiamen

厦门东屿红树林区环境变迁对鸟类的影响*

林清贤　陈小麟　林　鹏

(厦门大学生命科学学院,福建 厦门 361005)

摘要:比较厦门东屿红树林区两次鸟类调查数据(2002 年 1 月～2003 年 1 月与 1996 年 1 月～1997 年 1 月),分析当地生境的变化对鸟类物种组成及主要鸟类类群数量的影响.与 1996 年相比,2002 年的鹭类、鸥类和鸭类年累计数量分别从 2 444、3 023 和 873 只减少到 352、73 和 10 只,而鸻鹬类从 3 088 只增加到 5 478 只.滩涂、红树林面积减少及人为活动增加是鹭类、鸥类和鸭类数量减少的主要原因,而鸻鹬类数量增加与新增荒废裸地生境为其提供理想休息场所关系密切.

关键词:厦门东屿;环境变迁;鸟类

中图分类号:Q 958.12　　**文献标识码**:A　　**文章编号**:0438-0479(2007)01-0104-05

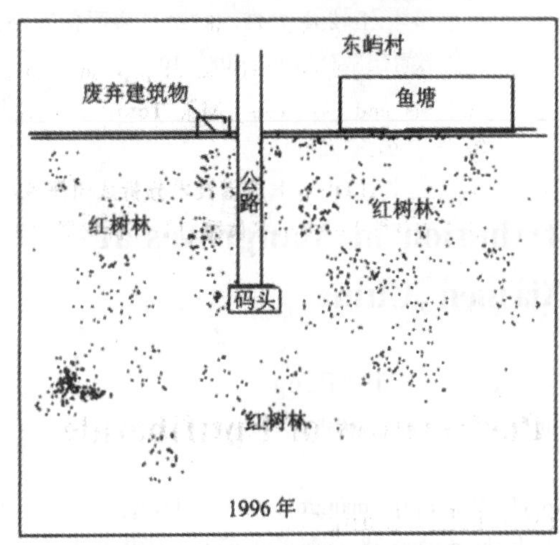

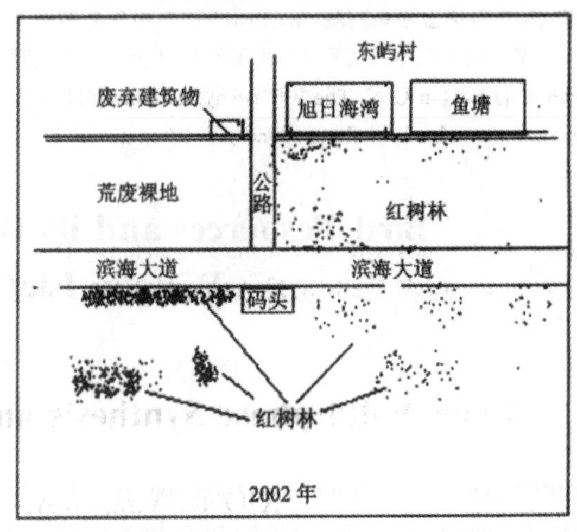

图 1　1996 和 2002 年厦门东屿红树林区生境变迁示意图

Fig. 1　Sketch map of habitat variance in Dongyu mangrove area in 1996 and 2002

厦门东屿红树林区近年来由于经济建设原因,环境破坏比较严重,对当地鸟类有较大影响.通过 2002 年 1 月～2003 年 1 月的调查和宋晓军在 1996 年 1 月～1997 年 1 月的调查数据进行比较,分析生境变化对该地鸟类种类、数量及物种多样性的影响.

1　样地与调查方法

1.1　样地概况

厦门东屿红树林区位于厦门市海沧区东屿村,地理坐标为东经 118°03′,北纬 24°31′,年均温 21.1℃,年降雨量 1 036 mm.红树林分布在中高潮带,主要树种为白骨壤,间有零星的秋茄,高度为 1～2 m,呈斑块状分布,面积原有 67 hm^2,由于建设和养殖的原因,1996 年红树林剩 20 hm^2 左右,而到了 2002 年,红树林面积已不足 2 hm^2.涨潮时所有红树林林冠都可被潮水淹没.红树林带外为泥沙质滩涂,宽度 150～250 m,中潮区部分养殖缢蛏,低潮区有牡蛎养殖.

图 1 为 1996 和 2002 年厦门东屿红树林区生境变迁示意图,期间主要有两次较大的生境变迁,一是 1997 年的围海造陆,右图的荒废裸地就是当时形成的;另一次是 2001 年海沧滨海大道建设,海沧滨海大道建设从滩涂中间开出约 110 m 宽的公路,破坏了不少红树林,并把码头右上侧红树林围起来与滩涂隔离.另外由于滩涂养殖的发展,不少红树林变成养殖地也是当地红树林面积减少的一个原因.

* 国家自然科学基金(40276028),厦门大学科技创新项目(XDKJCX20041021)资助

原载于厦门大学学报(自然科学版),2007,46(1):104－108

除红树林破坏和生境隔离外,2002 年厦门东屿红树林区人为活动增多,主要有滨海大道建设期间大量的人员和车辆,旭日海湾房地产开发增加人流量,在荒废裸地和滩涂上挖鱼塘和进行渔业养殖的村民等.

1.2 调查方法

在鸟类迁徙季节(3～5 月,9～11 月)每月调查 2～3 次,其它月份每月调查 1 次. 调查采用路线调查和高位定点相结合的方法,对红树林带、堤岸、农田、鱼塘、荒废裸地等生境带采用路线调查法进行统计,对在泥滩和海面觅食、栖息的鸟类采用高位定点观察和直接进入泥滩观察的方法统计绝对数量[1,2].

1.3 数据处理

两次调查的数据进行比较时,若在一个月份内进行多次调查的用该月调查的平均值进行比较,相应的年累计数量用月平均值进行统计.

鸟类的优势度等级根据频度指数(*RB*)判定:

$$RB = \frac{d}{D} \times \frac{N}{D}$$

式中,*RB* 为频率指数; *d* 为遇见该种鸟类的天数; *N* 为遇见该种鸟类的总数量; *D* 为调查总天数.

*RB*10 以上为优势种(＋＋＋＋),5～10 常见种(＋＋＋),1～5 罕见种(＋＋),1 以下偶见种(＋).

2 结果与分析

2.1 厦门东屿红树林区鸟类物种数变迁

宋晓军于 1996 年 1 月～1997 年 1 月在厦门东屿红树林区调查到鸟类 64 种,我们在 2002 年 1 月～2003 年 1 月调查到鸟类 72 种,两次调查中共有种类 46 种,相异种类 44 种,其中宋晓军调查到而在本次调查中没有的 18 种,本次调查到而宋晓军没调查到的有 26 种(表 1).

两次调查种类差别主要由当地鸟类生境变迁引起. 厦门东屿红树林区中的荒废裸地是在 1997 年围海造陆以后才形成的,新形成的生境吸引了不少鸟类,与宋晓军的调查结果相比,该生境中新增加了 13 种鸟类,占新增加鸟类种数的 50 %. 在荒废裸地新出现的 13 种鸟类中有 8 种鸣禽,其中小云雀和理氏鹨主要裸地上活动,其他鸣禽主要出现在裸地周边的草丛中. 荒废裸地新出现的种类中还有 4 种鸻鹬类,其中流苏鹬和灰头麦鸡都只调查到 1 次 1 只,这 2 种鸟类比较罕见,在厦门的其它地方尚未有相关记录;黑翅长脚鹬调查到 1 次 2 只,主要在裸地上休息;蒙古沙鸻的记录次数比较多,但其数量少(1～2 只)而且混在大群铁嘴沙鸻中,由于两种沙鸻差别细微,不易辨识,这可能是在 1996 年 1 月～1997 年 1 月未记录到的原因.

在 1996 年出现而 2002 年未调查到的 18 种鸟类中有 10 种鸻鹬类,占了 55.56 %,鸻鹬类大部分为冬候鸟或过境鸟,它们主要在滩涂上进行觅食,涨潮滩涂淹没时飞到滩涂附近的浅水鱼塘或裸地上休息,由于滩涂面积的大量减少,鸻鹬类觅食地受到破坏,从而导致了该地区鸻鹬类种类大量减少.

2.2 厦门东屿红树林区鸟类数量变迁

(1) 优势种类数量变迁

厦门东屿红树林区鸟类在两次调查中共有鸟类优势种 12 种(表 2),其中本次调查比 1996 年 1 月～1997 年 1 月调查少了池鹭、大白鹭、绿翅鸭、红脚鹬、青脚鹬和红嘴鸥 6 种,多了铁嘴沙鸻和麻雀 2 种.

厦门东屿红树林区鸟类优势种的变化也主要是由当地生境变迁引起,红树林面积减少导致经常在红树林中活动的鹭科鸟类资源数量减少,表现在池鹭和大白鹭由优势种变为常见种和罕见种. 同时滩涂面积减少,鹭类觅食空间减小也是鹭类资源减少的重要原因.

绿翅鸭在 1996 年 1 月～1997 年 1 月间累计调查到 1 361 只,单次调查最大数量为 431 只,而本次调查只在 3 月份见到一次,数量为 30 只. 厦门东屿红树林区滩涂(涨潮时为水面)大面积减少和生境破碎化,是导致绿翅鸭数量锐减的主要原因. 根据相关报道雁鸭类在零散活动时与居民点距离为 500 m 以上,道路 100 m 以上,聚群活动与居民点距离为 1 500 m 以上,道路 400 m 以上[3],厦门东屿红树林区滨海大道建设和周边房地产开发决定了该地绿翅鸭只能进行零散活动,从而导致数量锐减.

鸻鹬类中青脚鹬和红脚鹬主要在滩涂生境中觅食,滩涂面积减少使得它们觅食空间缩小,从而导致青脚鹬和红脚鹬数量减少. 铁嘴沙鸻为旅鸟,但在厦门东屿红树林区全年都有调查到,在 6、7、8 月仍能见到不少个体(分别为 58、90、42 只)留下度夏,这种铁嘴沙鸻居留情况异常变化是导致其在厦门东屿红树林区成为优势种类的主要原因.

红嘴鸥在 1996 年 1 月～1997 年 1 月为优势种,到 2002 年 1 月～2003 年 1 月为罕见种,其数量减少原因可能跟滩涂面积减少,人为干扰比较严重有关.

麻雀在 1996 年 1 月～1997 年 1 月为偶见种,到 2002 年 1 月～2003 年 1 月为优势种. 麻雀通常结群活动于村庄附近,适应人居环境,厦门东屿红树林区的开发建设为麻雀创造了有利的生存空间,从而导致其数量剧增. 反过来,麻雀数量的增加也说明了厦门东屿红树林区正在逐步向人工环境转变.

表1 两次调查(1996年1月～1997年1月和2002年1月～2003年1月)中的鸟类物种变迁

Tab. 1 Changing of bird species from 1996-01～1997-01 to 2002-01～2003-01 in Dongyu mangrove areas

	种 类	优势等级[a]	优势等级[b]	累计数量[b]	出现生境[b]
1	珠颈斑鸠 *Streptopelia chinensis*	-	+	2	塘岸
2	戴胜 *Upupa epops*	+	-	-	-
3	蓝翡翠 *Halcyon pileata*	+	-	-	-
4	小白腰雨燕 *Apus affinis*	-	+	5	空中
5	白胸苦恶鸟 *Amaurornis phoenicurus*	+	-	-	-
6	黑水鸡 *Gallinula chloropus*	-	+ + +	23	鱼塘
7	扇尾沙锥 *Gallinago gallinago*	+	-	-	-
8	大杓鹬 *Numenius madagascariensis*	+	-	-	-
9	小杓鹬 *N. minutus*	+	-	-	-
10	鹤鹬 *Tringa erythropus*	+	-	-	-
11	斑尾塍鹬 *Limosa lapponica*	+	-	-	-
12	大滨鹬 *Calidris tenuirostris*	+	-	-	-
13	红腹滨鹬 *C. canutus*	+	-	-	-
14	青脚滨鹬 *C. temminckii*	+ + +	-	-	-
15	尖尾滨鹬 *C. acuminata*	+	-	-	-
16	弯嘴滨鹬 *C. ferruginea*	+	-	-	-
17	流苏鹬 *Philomachus pugnax*	-	+	1	荒废裸地
18	黑翅长脚鹬 *Himantopus himantopus*	-	+	2	荒废裸地
19	蒙古沙鸻 *Charadrius hiaticula*	-	+	12	荒废裸地、滩涂
20	灰头麦鸡 *Vanellus cinereus*	-	+	1	荒废裸地
21	中白鹭 *Mesophoyx intermedia*	-	+	1	鱼塘
22	牛背鹭 *Bubulcus ibis*	-	+	3	荒废裸地
23	绿鹭 *Butorides striatus*	-	+	2	红树林
24	夜鹭 *Nycticorax nycticorax*	+ +	-	-	-
25	黑脸琵鹭 *Platalea minor*	+	-	-	-
26	海鸥 *Larus canus*	-	+	5	滩涂、海面
27	小鸊鷉 *Tachybapus ruficollis*	-	+	4	鱼塘
28	红隼 *Falco tinnunculus*	-	+	5	空中
29	蓝矶鸫 *Monticola solitarius*	-	+	3	堤岸
30	乌鸫 *Turdus hortulorum*	-	+	1	堤岸
31	北红尾鸲 *Phoenicurus auroreus*	-	+	1	荒废裸地
32	灰背椋鸟 *Sturnus sinensis*	+	-	-	-
33	黑领椋鸟 *Sturnus nigricollis*	+	-	-	-
34	金腰燕 *Hirundo rustica*	-	+	2	空中
35	白喉红臀鹎 *Pycnonotus aurigaster*	-	+	3	塘岸
36	日本树莺 *Cettia diphone*	+	-	-	-
37	黑眉苇莺 *Acrocephalus bistrigiceps*	-	+	3	荒废裸地
38	东方大苇莺 *Acrocephalus orientalis*	-	+	1	荒废裸地
39	褐柳莺 *Phylloscopus fuscatus*	-	+	2	堤岸
40	小云雀 *Alauda gulgula*	-	+	37	荒废裸地
41	理氏鹨 *Anthus richardi*	-	+	6	荒废裸地
42	斑文鸟 *Lonchura punctulata*	-	+ +	63	荒废裸地
43	金翅[雀] *Carduelis sinica*	-	+	1	荒废裸地
44	黄胸鹀 *Emberiza aureola*	-	+	1	荒废裸地

调查时间:a. 1996-01～1997-01; b. 2002-01～2003-01

表 2 两次调查的厦门东屿红树林区鸟类优势种

Tab. 2 Dominant birds in Dongyu mangrove areas in 1996-01～1997-01 or 2001-01～2003-01

类群	种 类	1996-01～1997-01	2001-01～2003-01
鹭类	白鹭 *Egretta garzetta*	+ + + +	+ + + +
	苍鹭 *Ardea cinerea*	+ + + +	+ + + +
	池鹭 *Ardeola bacchus*	+ + + +	+ + +
	大白鹭 *Casmerodius albus*	+ + + +	+ +
鸻鹬类	环颈鸻 *Charadrius alexandrinus*	+ + + +	+ + + +
	黑腹滨鹬 *Calidris alpina*	+ + + +	+ + + +
	铁嘴沙鸻 *Charadrius leschenaultii*	+	+ + + +
	红脚鹬 *Tringa totanus*	+ + + +	+ + +
	青脚鹬 *Tringa nebularia*	+ + + +	+ + +
鸥类	红嘴鸥 *Larus ridibundus*	+ + + +	+ +
鸭类	绿翅鸭 *Anas crecca*	+ + + +	+
雀类	麻雀 *Passer montanus*	+	+ + + +

表 3 1996 和 2002 年厦门东屿红树林主要水鸟类群年累计数量比较

Tab. 3 Comparing of accumulative quantity of main waterbird groups between 1996 and 2002

	鸻鹬类	鹭类	鸭类	鸥类	合计
1996(ind.)	3088	2444	873	3023	9428
2002(ind.)	5478	352	10	73	5918
1996/2002(%)	177.40	14.40	1.15	2.41	62.84

表 4 2002 年 10～12 月厦门东屿荒废裸地和滩涂鸻鹬类数量

Tab. 4 Quantity of waders between dry-naked ground and marched beach in Dongyu mangrove areas from Oct. to Dec. in 2002

	10/6	10/14	10/31	11/11	11/22	12/13	合计
荒废裸地(ind.)	271	927	777	1363	1104	1191	5663
滩涂(ind.)	91	168	252	196	301	455	1463
滩涂/荒废裸地(%)	33.58	18.12	32.43	14.38	27.26	38.20	25.97

(2)主要水鸟类群的数量变迁

东屿红树林区的主要水鸟类群有鸻鹬类、鹭类、鸭类和鸥类,生境的变迁也导致这些主要水鸟类群数量的变迁(图 2 和表 3).

2002 年鸻鹬类全年累计数量要比 1996 年高,主要是由于围海造陆形成的荒废裸地是鸻鹬类理想的休息场所,大量红树林区外鸻鹬类也到该地休息,从而导致该区鸻鹬类数量增加. 根据 2002 年 10～12 月在红树林滩涂鸻鹬类的样方调查,红树林区滩涂见到的鸻鹬类数量只占荒废裸地数量的 14.38 %～38.2 %(表 4),说明在荒废裸地休息的鸻鹬类主要(74 %左右)还是在红树林区以外滩涂觅食,正是这些从其它地方迁来的鸻鹬类导致了鸻鹬类数量比 1996 年多. 2002 年 2～5 月厦门东屿鸻鹬类数量比 1996 年同期来得少,主要是此时滨海大道进行施工,同时有许多当地村民在荒废裸地中挖鱼塘,车辆与人为活动非常频繁,而聚群休息的鸻鹬类对人为干扰反应比较敏感,根据相关报道,在辽河三角洲湿地,鸻鹬类在聚群活动时的破碎化影响距离为居民点 500 m,道路 200 m[3]. 在此时期厦门东屿红树林区荒废裸地都在上述范围之内,因而只能见到个别的鸻鹬类鸟类.

2002 年鹭类和鸭类明显低于 1996 年(分别占 14.40 %,1.15 %),主要是由于红树林破坏和滩涂面积减少(本文 2.2(1)).

2002 年的鸥类数量急剧下降,只有 1996 年的 2.41 %,主要是由于红嘴鸥数量的大量减少引起,1996 年 1 月、12 月都调查到了上千只大群,而在 2002 年仅在 2 月调查到 6 只. 2002 年主要集中在 5～8 月,种类为夏候鸟白额燕鸥,累计数量 46 只,占 63.01 %.

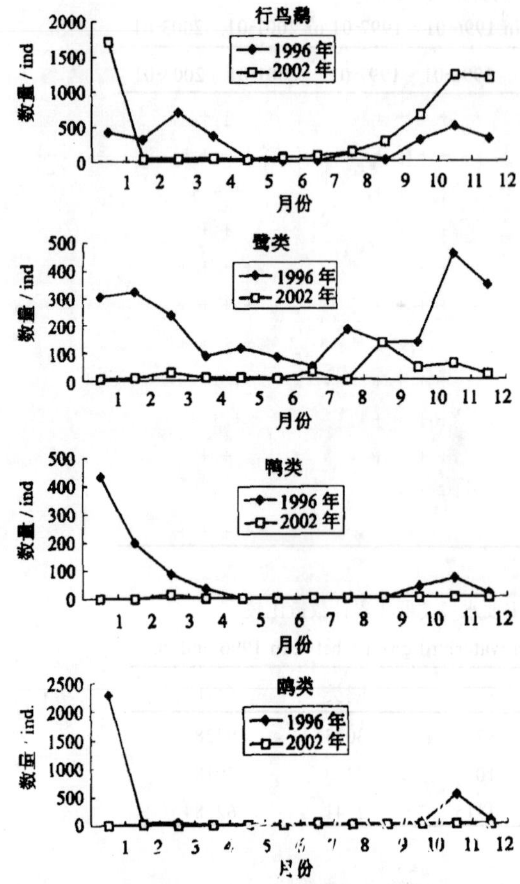

图2 1996和2002年主要水鸟类群数量月变动

Fig. 2 Monthly change of quantity of main waterbird group at Dongyu mangrove areas in 1996 and 2002

3 讨 论

东屿红树林区是厦门最重要的冬候鸟栖息地之一,滨海大道的建设填埋了大量的滩涂,破坏了大量的红树林,人为活动干扰增加,同时又增加了不同的生境类型,因而改变了当地鸟类的物种和数量结构组成.生境类型多样化一方面导致不同鸟种进入该地区活动,另一方面红树林的破坏和滩涂面积的减少导致了一些物种不再出现,特别是一些鸻鹬类的消失.

东屿红树林区的环境变迁对水鸟数量的影响最为明显,2002年的鸥类、鸭类、鹭类数量急剧下降,主要原因就是滩涂面积的减少、红树林大量被破坏及人为的干扰增加.而鸻鹬类在整个区域内的数量有所增加,这主要是由于新形成的荒废裸地生境为这些鸟类提供了理想的休息场所,许多在红树林区外觅食的鸻鹬类在涨潮时也到荒废裸地上休息,但在红树林区范围内觅食的鸻鹬类数量比1996年也下降了不少.由于觅食功能和栖息功能对鸟类的作用不一样,难以评价这样的生境改变是否对鸻鹬类有利.

参考文献:

[1] 宋晓军,林鹏.福建红树林湿地鸟类区系研究[J].生态学杂志,2002,21(6):5-10.

[2] 宋晓军.福建红树林区鸟类及其群落的空间结构和时间动态[D].厦门:厦门大学,1997.

[3] 肖笃宁,胡远满,李秀珍.环渤海三角洲湿地的景观生态学研究[M].北京:科学出版社,2001.

The Effect of Environment Changing on Birds at Dongyu Mangrove Area, Xiamen

LIN Qing-xian, CHEN Xiao-lin, LIN Peng

(School of Life Sciences, Xiamen University, Xiamen 361005, China)

Abstract: By comparing the data of birds between in 2002-01～2003-01 and in 1996-01～1997-01 at Dongyu mangrove areas, Xiamen, the effect of environment changing on bird's species, quantity was analyzed. The exploitation resulted in diversification of habitat type which increasing some bird species, on the other hand, mangrove and mudflat reducing resulted in disappear of some species. 64 bird species were recorded in 1996-01～1997-01 and 72 species were recorded in 2002-01～2003-01. There were 44 different species in the two indagations, among which 26 new species appearing and 18 species disappearing in 2002-01～2003-01. The habitat changing also resulted in the dominant species changing. *Ardeola bacchus*, *Casmerodius albus*, *Tringa tetanus*, *Tringa nebularia*, *Larus ridibundus* and *Anas crecca* were dominant species in 1996-01～1997-01, but they were not dominant species in 2002-01～2003-01, *Charadrius leschenaultia* and *Passer montanus* were new increased dominate species in 2002-01～2003-01. The quantity of some main water-bird flocks were changed sharply from 1996 to 2002. By comparing with 1996, the quantity of herons, ducks and gulls in 2002 was from 2 444, 3 023 and 873 reduced to 352, 73 and 10, but the quantity of chardrii was increased form 3 088 to 5 478. Some reasons that lead to the changing of birds were discussed in paper.

Key words: Dongyu; Xiamen; environment changing; bird

Are vegetated areas of mangroves attractive to juvenile and small fish? The case of Dongzhaigang Bay, Hainan Island, China*

Mao Wang[a] Zhenyuan Huang[a] Fushan Shi[a] Wenqing Wang[a,b]

(a. *Key Laboratory of Ministry of Education for Coastal and Wetland Ecosystems, School of Life Sciences, Xiamen University, Xiamen* 361005, *China*; b. *State Key Laboratory of Marine Environmental Science, Xiamen University, Xiamen* 361005, *China*)

A R T I C L E I N F O

Article history:
Received 2 March 2009
Accepted 11 August 2009
Available online 6 September 2009

Keywords:
mangrove
fish
species
habitat
seasonal distribution

A B S T R A C T

Well-developed aerial roots of mangroves make it difficult to study how fish utilize the mangrove forest as a habitat. In the present study, we compared the differences in fish assemblages in three major types of habitats of mangrove estuary (vegetated area, treeless mudflat, and creek) of a mangrove bay in Hainan Island, China, at different seasons during two consecutive years. Three types of gears, centipede net, gill net and cast net, were used in the different habitats of mangrove estuary and sampling efficiencies among gears were evaluated. Centipede nets were used in all the three types of habitats and cast nets and gill nets in treeless mudflats and creeks. Fish assemblages were dependent on gears used. Centipede net could efficiently catch fish occurring both inside and outside of vegetated areas efficiently. A total of 115 fish species in 51 families were collected. In terms of numbers of species per family, Gobiidae was the most diverse (17 species), followed by Mugilidae (5 species). Almost all of the fish were juvenile or small fish and few predators were recorded, implying low predation pressure in the bay. ANOVA analysis showed that significant seasonal and spatial variation existed in species richness, abundance, and biomass, which were less in the vegetated areas than those of treeless mudflats and creeks. The attraction of vegetated areas to fish was less than that of creeks and mudflats. Many species were specific to a particular habitat type, 4 species occurring exclusively in the creeks, 45 species occurring exclusively in the treeless mudflats, and 5 species occurring exclusively in the vegetated areas. The results indicated that mangrove estuaries were potentially attractive habitats for juvenile and small fish, but this attraction was accomplished by a connection of vegetated areas, treeless mudflats and creeks, not only by vegetated areas.

1. Introduction

Mangroves support high fish species richness and high numbers of individuals, some of which have a great commercial importance (Robertson and Duke, 1987, 1990; Blaber et al., 1989; Robertson and Blaber, 1992; Laegdsgaard and Johnson, 1995; Louis et al., 1995; Blaber, 1997; Faunce and Serafy, 2006), and mangroves are protected worldwide as a nursery ground of fishes (Spalding et al., 1997; Valiela et al., 2001; FAO, 2007). Two hypotheses have been proposed to explain why mangroves are so attractive to fishes (Laegdsgaard and Johnson, 2001): (1) the predator refuge hypothesis, which stresses that the structural complexity of mangrove pneumatophores and prop roots provides excellent shelter from predators for juvenile and small fishes by migrating into vegetated areas of mangroves particularly when the trees are inundated by water (Robertson and Duke, 1987; Abrahams and Kattenfeld, 1997; Kathiresan and Bingham, 2001; Laegdsgaard and Johnson, 2001); and (2) the feeding hypothesis, which asserts that there is a greater abundance of food within mangroves due to high productivity and the associated abundance of benthic fauna (Chong et al., 1990; Kathiresan and Bingham, 2001; Laegdsgaard and Johnson, 2001).

Vegetated areas (areas covered by mangrove trees), treeless mudflats and creeks are three main types of habitats of mangrove estuaries (Ikejima et al., 2003). The great variability in composition of fish assemblages in different habitats of reefs has been stressed by some researchers (Williams and Hatcher, 1983; Russ, 1984). Likely, mangrove estuaries vary greatly in levels of structural complexity among habitats (Rönnbäck et al., 1999, 2002). Different habitats within mangrove estuaries provide various levels of food and shelter for fish (Rönnbäck et al., 1999), which may influence the fauna assemblages across habitats (Satumanatpan and Keough, 2001; Ikejima et al., 2003). These highlight a need for spatially explicit sampling (Smith and Hindell, 2005). However, the differences in fish assemblage among habitats are ignored by most studies which treat

* From Estuarine, Coastal and Shelf Science, 2009, 85: 208—216

them as a single habitat because of higher ability of fish to swim between habitats (Hindell and Jenkins, 2004). In most studies, fish sampling were conducted at sites adjacent to vegetated areas such as channels (Chong et al., 1990), creeks (Robertson and Duke, 1987; Lin and Shao, 1999; He et al., 2001; He and Fan, 2002; Shinnaka et al., 2007), or along forest edges (Laroche et al., 1997; He, 1999; Hindell and Jenkins, 2004). Until today, only a few studies have investigated the heterogeneity of fish assemblages among habitats of mangrove estuaries (Blaber et al., 1989; Chong et al., 1990; Vance et al., 1996; Rönnbäck et al., 1999; Ikejima et al., 2003; Smith and Hindell, 2005).

Besides habitats, sampling gears used also have significant effects on the individual number and species number of fish catches (Tongnunui et al., 2002; Faunce and Serafy, 2006). Gear type, mesh size, and sampling frequency all have great influence on catches (Tongnunui et al., 2002; Faunce and Serafy, 2006). To date, few studies have assessed how fish samples change between different methods, or discussed the implications for sampling bias in shaping our understanding of small-scale faunal-habitat associations (Smith and Hindell, 2005). Differences in sampling methodologies prevent informative comparisons across studies (Connolly, 1994; Rozas and Minello, 1997), thereby masking broad-scale generality in findings (Smith and Hindell, 2005).

Gill net, trawl net and seine net are commonly used gears in mangrove-fish research (Tongnunui et al., 2002; Faunce and Serafy, 2006; Shinnaka et al., 2007). However, well-developed aerial roots of mangrove plants make it almost impractical to sample fish within vegetated areas of mangroves using these gears (Vance et al., 1996; Faunce and Serafy, 2006). Thus, most studies only collected fishes close to the mangrove fringe or in creeks (Vance et al., 1996; Beck et al., 2001; Huxham et al., 2004). Such sampling strategy, however, might not reflect real distribution of fish within vegetated areas of mangroves (Nagelkerken et al., 2000; Beck et al., 2001; Huxham et al., 2004). Although fish collection from vegetated areas has been conducted in some mangroves (Thayer et al., 1987; Morton, 1990; Vance et al., 1996; Ley et al., 1999; Rönnbäck et al., 1999; Huxham et al., 2004), only Thayer et al. (1987), Morton (1990), Nagelkerken et al. (2000), Huxham et al. (2004), Smith and Hindell, (2005) and Dorenbosch et al. (2007) made comparisons with adjacent habitats. Different gears used in different habitats make it difficult to compare fish from the three typical habitats of mangroves (Beck et al., 2001; Ikejima et al., 2003; Huxham et al., 2004; Smith and Hindell, 2005).

Fish assemblage structure of mangroves may also change seasonally (Ley et al., 1999; Ikejima et al., 2003; Huxham et al., 2004). Thus, the use of uniform gears to study fish distribution in the three typical habitats of mangroves is imperative in long-term research (Faunce and Serafy, 2006).

The aim of the present study was to compare the fish communities in a mangrove estuary, using the same sampling gear to investigate whether different habitats (vegetated areas, treeless mudflats and creeks) really do support different fish communities. We also compared the efficiency of the three most widely used gears (centipede net, cast net and gill net). Fish communities in the different habitats of mangroves were compared on the basis of species richness, abundance and biomass of fishes. We aimed to identify which habitat is more attractive to fish.

2. Materials and methods

2.1. Site description

The study was conducted in Dongzhaigang Bay (110° 32′–110°37′E, 19°51′–20°01′N), a mangrove estuary situated in the northeast part of Hainan Island, China. The bay experiences a tropical monsoon, with mean rainfall 1676 mm and average temperature 24.8 °C. It is a semi-enclosed, muddy bottom estuary with an area of 5400 hm^2, fed by 4 small rivers (Fig. 1). The bay is subjected to semidiurnal tides with mean tidal amplitude of 1.6–1.8 m. The vegetated area is approximately 1734 hm^2, dominated by *Bruguiera sexangula*, *Ceriops tagal*, *Avicennia marina*, *Rhizophora stylosa*, and *Aegiceras corniculatum*. Deforestation has ceased since 1986 when the bay was declared a national nature reserve (Dongzhaigang National Nature Reserve). It was one of the first 7 wetlands in China listed in the Ramsar List of Wetlands of International Importance (Wang and Wang, 2007).

2.2. Sampling methods

Three types of habitats, namely vegetated area of mangroves (M), treeless mudflat (B) and creek (C), were chosen and 9 sampling sites were established (Fig. 1). Each habitat had 3 sites (replicates). We established these sites according to the distribution of mangrove plant community and the suggestion of local fishermen (least artificial disturbance and easy processing). The mangrove forests at M1,

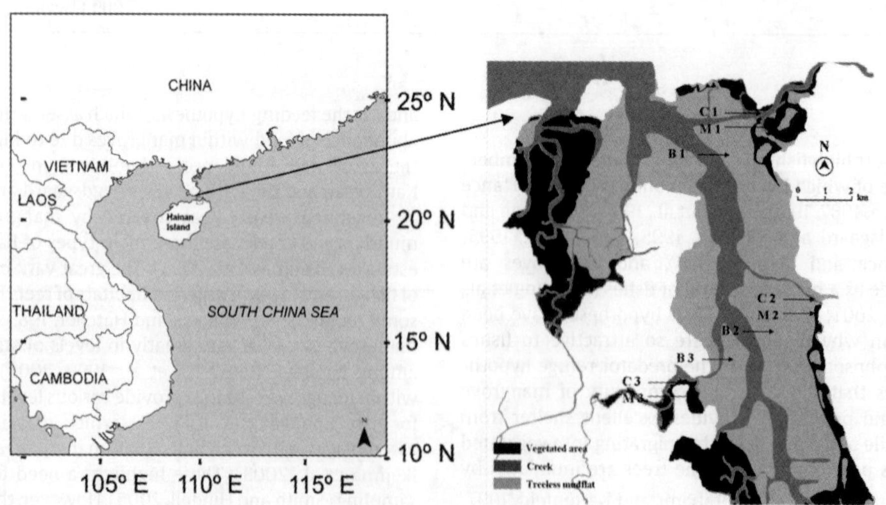

Fig. 1. The distribution of sampling sites in Dongzhaigang Bay, Hainan Island, China (white portion is higher land not inundated by tide). B: treeless mudflat, C: creek, M: vegetated area.

M2 and M3 were dominated by *Rhizophora stylosa*, *Avicennia marina* and *Bruguiera sexangula*, respectively. M1, M2 and M3 were 30 m landward away from the forest edges, B1, B2 and B3 were about 1000 m seaward away from the forest edges, and the C1, C2 and C3 were about 30–40 m seaward away from the forest edges. The water depth at the leading edge of vegetated areas at high tide was less than 1.2 m. In the rainy season, the vegetated areas are inundated at spring as well as neap high tide, and exposed completely at low tide. In the dry season, the vegetated areas are only inundated at spring high tide. The treeless mudflats are exposed completely at low tide also. The creeks are never completely dry, but are separated from the main water body at low tide.

Centipede net, cast net and gill net were used in this study since these gears are widely used by local fishers in the bay (Wang et al., 2007). Centipede net is a recently introduced but most widely used net in this bay. It is a prism made of a net unit (width 35 cm, height 25 cm, 10 m long, 8.5 mm mesh) supported by several rectangular iron frames (35 cm × 25 cm) every other 40 cm. The net acts as a kind of trap net with small lengthways, cupped and staggered holes on both sides. Each unit can be connected end to end. Fig. 2 shows a photograph of a centipede net unit arranged at the seaward fringe of *Rhizophora stylosa* + *Aegiceras corniculatum* forest. Cast net (coniform, height 3.12 m, girth 14.3 m, mesh 11.5 mm) and gill net (rectangle, height 1.2 m, length 200 m, mesh 18 mm) are traditional tools in this bay. Centipede nets were used in all the 9 sites, while cast nets and gill nets were only used in non-vegetated sites (B1, B2, B3, C1, C2, C3). To minimize the disturbance of near nets, the distance between two kinds of nets at the same site was more than 200 m.

Sampling was conducted seasonally, e.g. March (spring, late dry season), June (summer, early rainy season), September (autumn, late rainy season) and December (winter, early dry season) – from March 2004 to March 2006. At peak high tide, a centipede net group (30 units of centipede nets connected end to end) was arranged parallel to the forest edge at each site and left to catch fish. Therefore, there were three replicates for each habitat. About 4–6 h later, when the nets in vegetated areas were exposed to air, all the captured fishes were taken back to the lab, then sorted, identified, counted and weighed. Unknown specimens were preserved in 10% formalin and taken back to the lab for further identification. Fishes were categorized as juveniles or adults; all reproductively immature fish were classified as juveniles. The piscivory of each fish was identified according to South China Sea Fisheries Institute (1979). The vegetated areas (M1, M2 and M3) were only sampled in the rainy seasons (June and September) because these areas were rarely flooded in dry seasons. Water temperature and salinity at high tide were recorded simultaneously (Fig. 3).

Fig. 2. Centipede net arranged at the seaward fringe of mangroves in Dongzhaigang Bay, Hainan Island, China.

Cast nets and gill nets were used at high tide during the day. Gill nets were arranged vertically to the water-flow at each site. As for cast net, ten replicate nettings were performed at each site every day. The water depths at treeless mudflats (B1, B2 and B3) and creeks were about 1.1–1.5 m and 1.8–2.2 m, respectively. The sample period using these 2 gears was from March 2004 to March 2005.

2.3. Data analysis

The effects of gear type (centipede net, gill net and cast net) on species richness were tested using a one-way analysis of variance (ANOVA).

Only the data from centipede net samples were used in the following analyses. Species richness, abundance, and biomass were expressed by the mean numbers of species, individuals, and weights of each of the three replicates. The effects of habitat (vegetated area, treeless mudflat and creek) and season on fish assemblage properties (i.e. number of species, abundance and biomass) were tested using a two-way analysis of variance (ANOVA).

3. Results

3.1. Difference in catch of fish species among gears

A total of 115 fish species in 51 families were caught from March 2004 to March 2006 (Appendix 1). There were considerable differences in fish species composition among gears. The total species caught by centipede net, gill net and cast net were 114, 38 and 65, respectively. Except for *Chanos chanos*, which was caught only by gill net, all the fish recorded in Dongzhaigang Bay in this study were caught by centipede net. Of the 115 fish, about one-third (37 species) was caught exclusively by centipede net and only 20% species (25 species) were caught by all three gears.

Table 1 is a list of the 6 dominant species caught by different nets in Dongzhaigang Bay. As for centipede net, the dominant species were *Leiognathus brevirostros*, *Fugu alboplumbeus*, *Ambassis gymnocephalus*, *Sillago sihama*, *Acentrogobius caninus* and *Osteomngil ophuyseni*. Except for *O. ophuyseni*, the 6 dominant species caught by centipede net were not listed on the dominant species caught by gill net or cast net. The dominant species caught by gill net were similar to that caught by cast net.

3.2. Predator in mangroves

Few predators were caught at any time during the sample periods. Only 5 piscivorous fishes, *Onigocia macrolepis*, *Saurida*

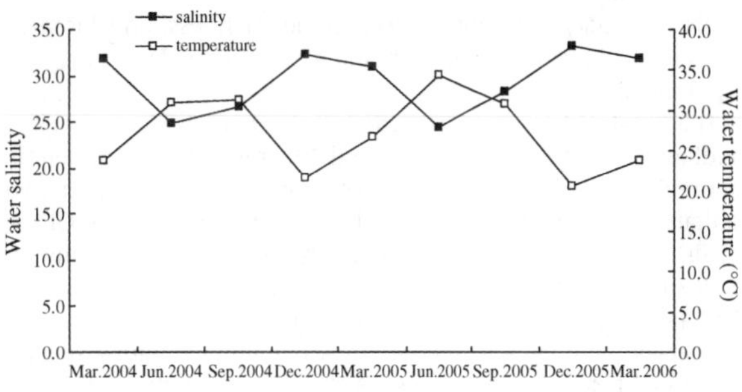

Fig. 3. Seasonal changes in water salinity and temperature in Dongzhaigang Bay, Hainan Island, China.

Table 1
Six dominant fish species caught by different gears from March 2004 to Decmber 2004 in Dongzhaigang Bay, Hainan Island, China, and the average weight and body length of the 6 dominant fish species using centipede net.

	Centipede net			Gill net	Cast net
	Fresh weight (g)	Body length (cm)	Maximal body length (cm)		
Leiognathus brevirostros	4.32	3.2	14	*Gerreomorpha japonica*	*Liza carinatus*
Fugu niphobles	20.74	13.5	15	*Gerres filamentosus*	*Mugil cephalus*
Ambassis gymnocephalus	1.36	3.1	8	*Gerres lucidus*	*Osteomugil ophuyseni*
Acentrogobius caninus	7.42	7.6	13	*Liza carinatus*	*Gerreomorpha japonica*
Sillago sihama	9.11	9.8	20	*Mugil cephalus*	*Gerres filamentosus*
Osteomngil ophuyseni	17.94	12.9	20	*Osteomugil ophuyseni*	*Hyporhamphus intermedius*

elongata, *Saurida undosquamis*, *Trachinocephalus myops* and *Epinephelus malabaricus* were recorded in Dongzhaigang Bay. All the 5 species did not belong to dominant species and were rarely recorded (Appendix 1). For example, only 12 individuals of *E. malabaricus* were caught during our 2-year investigation. Gut analysis showed that the prey species of the 5 piscivorous fishes, *Allanetta bleekeri* and *Stolephorus zollingeri* were also not listed on to the dominant species list and we found no fish that feed on the 6 dominant species of the bay (personal data, not published). Of the five piscivorous fishes, only a few individuals of *T. myops* were caught inside the vegetated areas, no piscivorous fishes entered the vegetated areas.

3.3. *Body size of fish in Dongzhaigang Bay*

Of the 114 fishes caught by centipede nets, 55 species (48.2% of total species) were represented only by juveniles, compared with 59 species comprising both juveniles and adults, and 1 species of adults only (Appendix 1). The results of average fresh weight and body length of the 6 dominant species also showed that fish in mangroves of Dongzhaigang Bay were juvenile and small fishes (Table 1). Additionally, except in September 2004, the average fresh weights in all the sampling seasons were not higher than 10 g and did not showed significant differences among habitats (4A). In September 2004, the average fresh weights of fishes in vegetated areas were higher than those of creeks and treeless mudflats ($P < 0.01$).

3.4. *Spatial and seasonal distribution of fish*

3.4.1. *Spatial changes*

ANOVA analysis of abundance, species richness, and biomass revealed significant seasonal and spatial variation in Dongzhaigang Bay (Table 2).

The total fish species recorded in vegetated areas, treeless mudflats and creeks were 45, 66 and 102, respectively (Appendix 1). Creeks had much higher species richness than the other two habitats, and vegetated areas had the least species richness. Of the 115 fish species recorded in the bay, only 35 species were common to all sites and most of them were specific to a particular habitat (Appendix 1). Five species were caught only in the vegetated areas, 45 species were caught only at treeless mudflats, and 4 species were caught exclusively at creeks. Although a few individuals were caught outside vegetated areas, most individuals of *Bostrichthys sinensis*, *Acentrogobius viridipunctatus*, *Ctenogobius brevirostris* and *Pisoodonophis boro* were caught inside vegetated areas. These 4 species were considered specific to vegetated areas. There were 20 species found both in treeless mudflats and creeks, while only 6 species occurred both at creeks and vegetated areas, and 2 species occurred both at treeless mudflats and creeks.

Although the number of species per sampling showed significant seasonal changes in creeks and treeless mudflats ($P < 0.01$), no consistent trends were found. The number of species per sampling in vegetated areas did not show any seasonal change, stabilizing in 11 species per sampling. There were significant differences in the number of species per sampling among habitats. The numbers of species per sampling in treeless mudflats in June 2004 and June 2005 were higher than those found in creeks. However, there were no significant differences in other seasons between the 2 habitats. The numbers of species per sampling in vegetated areas were lower than those of creeks and treeless mudflats ($P < 0.01$) (Fig. 4B). The seasonal average numbers of species per sampling in treeless mudflats, creeks and vegetated areas were 18.1, 17.1 and 10.5, respectively. In September 2004, the numbers of species per sampling in the three habitats were 20.0, 21.3 and 11.3, respectively.

Average individual number per sampling and fresh weight per sampling varied significantly among the seasons and habitats as well (Fig. 4C and D). In September 2004, the average individuals per sampling in treeless mudflats and vegetated areas were significantly higher than that of creeks. In September 2005, the average individuals per sampling in treeless mudflats were 172% and 80% higher than those of creeks and vegetated areas, respectively. Similarly, the fresh weight per sampling in September 2004 and September 2005 was higher than those of creeks and vegetated areas. In other seasons, these values had no significant differences among habitats ($P > 0.5$).

4. Discussion

4.1. *Sampling efficiency among gears*

In this study, the dominant species as well as the species numbers, were gear-specific (Appendix 1, Table 1). The total species caught by centipede net, gill net and cast net were 114, 38 and 65, respectively. Only 20% species (24 species) were caught by all three gears. Of the 115 fishes, about one-third (35 species) was caught exclusively by the centipede net. These suggested that centipede net provided a more complete picture of the ichthyofaunas associated with mangroves of Dongzhaigang Bay. As a result, the use of centipede net would increase the chance of catching rare species and therefore attain a better representation of the true community structure. In a summary of 111 mangrove-fish surveys published between 1955 and 2005, Faunce and Serafy (2006) claimed that the selection of gears was a more significant factor influencing fish assemblages. The manner by which fish samples are acquired from mangroves is of particular importance due, in part, to issues of species- and size-selectivity (Faunce and Serafy, 2006). They emphasized the importance of evaluating for a better understanding of fish assemblage in mangroves.

Dense prop roots and pneumatophores in vegetated areas of mangroves pose difficulties in the sampling of fish (Thayer et al., 1987; Laegdsgaard and Johnson, 1995; Faunce and Serafy, 2006). Most studies failed to sample fish in vegetated areas because either towed gears or general passive gears were difficult to place in dense mangrove prop roots (Faunce and Serafy, 2006). In some studies, gears were set at the periphery of vegetated areas (Laegdsgaard and Johnson, 1995; Laroche et al., 1997; He, 1999; Hindell and Jenkins, 2004). Laegdsgaard and Johnson (1995) collected fish amongst pneumatophores along the fringe of mangrove forests, and at the mouths of small tidal creeks draining vegetated areas. This lack of

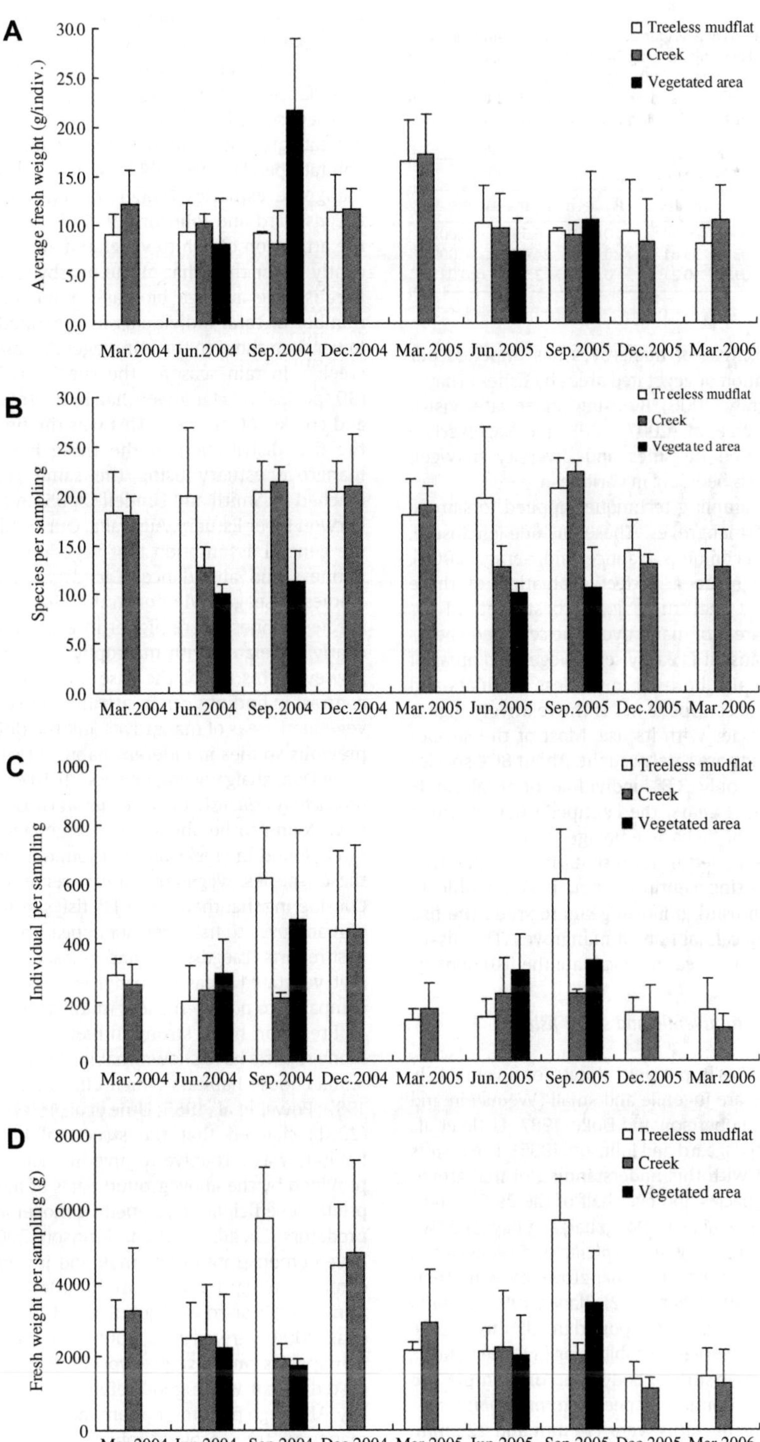

Fig. 4. Seasonal changes in the average fresh weight (A), species number per sampling (B), individual number per sampling (C) and fresh weight per sampling (D) using centipede net at different microhabitats in Dongzhaigang Bay, Hainan Island, China.

Table 2
Summary of two-factor (site × season) ANOVA for species richness, abundance, and biomass of fish seasonally collected in Dongzhaigang Bay, Hainan Island, China using centipede net from March 2004 to March 2006. Species richness was square root-transformed and abundance and biomass data were transformed to log (x). A posteriori tests were performed using the Tukey–Kramer method with a significance level of $P < 0.05$.

Variable	Microhabitat		Season		Interaction	
	F(df = 2)	P	F(df = 2)	P	F(df = 3)	P
Species richness	17.97	<0.0001	6.04	0.0081	6.59	0.0024
Abundance	3.98	0.0335	33.61	<0.0001	24.46	<.0001
Biomass	5.25	0.0137	10.2	0.0007	3.54	0.0312

sampling within vegetated areas of mangroves has impeded our understanding of the utilization of vegetated areas by fishes (Thayer et al., 1987; Faunce and Serafy, 2006). By using underwater visual census technique, Dorenbosch et al. (2007, MEPS) and Nagelkerken et al. (2000, ECSS) compared densities and diversity between mangroves, mudflats, seagrass beds, etc in Caribbean.

There have been some sampling techniques applied to sample fish in vegetated areas of mangroves. These include enclosure, entanglement and visual techniques (Faunce and Serafy, 2006). Faunce and Serafy (2006) made a perfect evaluation of these techniques. They concluded that there is no best method for sampling fish in vegetated areas of mangroves. The centipede net is advantageous not only because it is easily set in vegetated areas of mangroves, but that it can also be used in treeless mudflats and creeks. Another advantage of centipede net is the relatively higher chance of catching rare species with its use. Most of the species identified in the present study are rarely caught. About 80% species (90 species) accounted for only 3.8% individual of total catch. In comparison to the other 2 gears, the centipede net provided a more complete picture of fish assemblage associated with mangroves, and therefore a better representation of the true community structure. By using centipede net, it is possible to conduct long-term research using uniform gears to study the fish distribution in the three typical habitats of mangroves. The disadvantage of centipede net is that it cannot evaluate the fish density.

4.2. Attraction of mangroves to juvenile and small fishes

Mangroves are recognized as important habitats for juvenile fish, and most fish in mangroves are juvenile and small (Weinstein and Brooks, 1983; Wright, 1986; Robertson and Duke, 1987; Little et al., 1988; Chong et al., 1990; Laegdsgaard and Johnson, 1995). The results of this study are consistent with this understanding of mangroves (Tables 1 and 2). Fifty-five species and over half of the 29,836 individuals were found only as juveniles in Dongzhaigang Bay. The two most dominant species, *Ambassidae gymnocephalus* and *Leiognathids brevirostris*, were known to utilize the mangrove bays in their juvenile stages (Robertson and Blaber, 1992; Blaber, 1997; Ikejima et al., 2003). This phenomenon has been reported in other mangrove bays. Thus, Dongzhaigang Bay serves as habitat for juvenile fishes.

In addition, these fishes have smaller body size. For example, the maximal body lengths of the dominant species *Acentrogobius caninus* and *Acentrogobius viridipunctatus* were 114 mm and 123 mm, respectively. The maximal body length of *Bostrichthys sinensis* was 192 mm, but only a few individuals were of this length.

4.3. What attract fishes to mangroves?

Predator refuge hypothesis and feeding hypothesis have been proposed to explain why mangroves are attractive to juvenile fish (Robertson and Duke, 1987; Abrahams and Kattenfeld, 1997; Kathiresan and Bingham, 2001; Laegdsgaard and Johnson, 2001; Shinnaka et al., 2007). However, if we compare the contribution of different habitats of mangroves to high fish species richness and abundance, some questions arise.

Laegdsgaard and Johnson (1995) concluded that mangrove habitats are not equal in species richness and abundance. Distinct habitat-specific assemblages have been identified in some mangrove estuaries (Robertson and Duke, 1987; Morton, 1990; Laegdsgaard and Johnson, 1995). In this study, we concluded that the attraction of fish to vegetated areas of mangroves was significantly lower than that of the nearby treeless mudflats and creeks because the average biomass, abundance, species richness, and number of total individuals in vegetated areas were significantly lower than those in the non-vegetated areas (treeless mudflats and creeks). In rain seasons, the species richness in vegetated areas (30 species) was far lower than in the treeless mudflats (48 species) and creeks (60 species). This was the first report which compared the fish distribution in the three kinds of habitats in tropical mangrove estuary using the same gear. Similar results were reached by Smith and Hindell (2005) in a temperate estuary in the Barwon River Estuary, Australia. Our results are not consistent with the general assumption that vegetated areas have higher species richness and abundance (Laegdsgaard and Johnson, 1995). The studies on seagrass demonstrated that fish abundance, biomass and species richness were higher in vegetated areas, and were significantly correlated with macrophyte biomass (Lubbers et al., 1990). However, this is not the case in the mangroves we studied. The contrast in findings can be attributed to the lack of sampling within vegetated areas of mangroves and the different gears were used by previous studies in different habitats (Faunce and Serafy, 2006).

In Dongzhaigang Bay, only four fish (*Acentrogobius viridipunctatus*, *Bostrichthys sinensis*, *Ctenogobius brevirostris*, and *Pisoodonophis boro*) were found to be abundant in vegetated areas. These fishes were rarely found in creeks and not found at all in treeless mudflats. For these species, vegetated areas serve as their perfect habitats. Considering that there were 115 fish species in the bay, the attraction of mangroves to fish does not appear to be dramatic as described in past reports (Laegdsgaard and Johnson, 1995). These results inferred that vegetated areas of mangroves are not as attractive to fish as compared to nearby treeless mudflats and creeks.

Predation has a strong influence on the habitat choice of fish. Many reports have shown that in the presence of predators, aquatic fauna choose habitats which offer better shelter (Stein, 1977; Sih, 1982; Power et al., 1985; Utne et al., 1993). Laegdsgaard and Johnson (2001) claimed that the structural heterogeneity of mangrove habitats was attractive to juvenile fish. The structural complexity provided by the aboveground parts of mangrove trees may reduce predation efficiency by impeding movement or restricting vision of predators (Laegdsgaard and Johnson, 2001). Experimental studies have demonstrated that small and juvenile fishes are attracted to complex mangrove structure in order to reduce the risk of predation (Laegdsgaard and Johnson, 2001; Cocheret de la Morinière et al., 2004; Verweij et al., 2006). The rare abundance of piscivorous fish species and few piscivorous fish in Dongzhaigang Bay suggested that it was a good refuge for juvenile and small fish in the bay. Although pneumatophores and prop roots of mangrove trees of mangrove trees provide refuge from predation, fish species richness, abundance and biomass in vegetated areas were all lower than those of treeless mudflats and creeks (Fig. 4). These results suggest that the attraction of fish to mangroves cannot be attributed to low predation pressure in mangroves.

Mangroves support high fish species richness and high numbers of individuals (Faunce and Serafy, 2006). However, in this study, it can be concluded that the attraction of fish to vegetated areas of mangroves was significantly lower than that of the nearby treeless mudflats and creeks. There is no dramatic fish habitat advantage found in vegetated areas as compared with the treeless mudflats and creeks. The attraction of mangroves to fish is not fulfilled only by vegetated areas of mangroves. When we refer to mangroves, it should pertain to the whole mangrove ecosystem (or mangrove estuary) including vegetated areas, treeless mudflats and creeks, rather than merely vegetated areas of mangroves. In the context of the above considerations, the results of our study can then account for the attraction of mangroves to juvenile and small fish. The vegetated areas of mangroves are attractive to juvenile and small fish, only in so far as these are part of the entire ecosystem of mangroves, inclusive of the treeless mudflats and creeks.

Acknowledgements

We would like to thank missionary of Dongzhaigang National Nature Reserve for their help with fieldwork. This work was jointly funded by grants from the Natural Science Foundation of China (No. 40876046, No. 40376025) and Program for Innovative Research Team in Science and Technology in Fujian Province University.

Appendix 1. Relative abundance of fish species caught by centipede nets from different microhabitats of mangroves and fish species caught by different gears in Dongzhaigang Bay, Hainan Island, China.

Family	Species	Microhabitat type[a]			Percentage to total individuals	Living stage	Gear type		
		M	C	B			Centipede net	Gill net	Cast net
Aluteridae	*Monacanthus chinensis*			+	Δ	J	√		√
	Navodon septentrionalis			+	Δ	J	√		
Ambassidae	*Ambassis gymnocephalus*	++	++	+++	11.90	J	√	√	√
Anabantidae	*Anabas testudineus*	+	+	+	Δ	j	√		√
Aploactinidae	*Acanthosphex leurynnis*		+	++	Δ	J	√		√
	Hypodytes indicus			+	Δ	J	√		√
	Vespicula trachinoides		+	++	Δ	J	√		√
Apogonidae	*Apogonichthys striatus*			+	Δ	J	√		
Ariidae	*Arius thalassinus*		+	++	1.57	JA	√	√	√
Atherinidae	*Allanetta bleekeri*			++	0.49	J	√		√
Bembridae	*Bembrao japonicus*			+	Δ	JA	√		√
Belonidae	*Tylosurus strongylurus*	+	+	++	Δ	J	√	√	√
Bothidae	*Pseudorhombus malayanus*			+	Δ	J	√		
	Pseudorhombus cinnamoneus			+	Δ	J	√		
Callionymidae	*Callionymus flagris*		+	+	Δ	J	√		√
	Callionymus marisinensis		+	+	Δ	J	√		√
	Calliurichthys dorysus			+	Δ	J	√		√
	Calliurichthys fiamentosus			+	Δ	J	√		√
Carangidae	*Alectis indica*			+	Δ	J	√	√	
	Caranx sexfasciatus			+	Δ	J	√		
	Trachinotus ovatus			+	Δ	J	√		
Chanidae	*Chanos chanos*				Δ	J		√	
Clupeidae	*Clupanodon punctatus*		+	++	2.06	JA	√	√	√
	Clupanodon thrissa			+	Δ	JA	√	√	
	Nematalosa nasus	+	+	++	0.32	JA	√	√	√
	Sardinella jussieu			+	Δ	J	√		
Cynoglossidae	*Cynoglossus bilineatus*		+	+	Δ	J	√		√
	Paraplagusia blineata		+	+	Δ	J	√		√
Dasyatidae	*Dasyatis akajei*			+	Δ	JA	√		
Drepanidae	*Drepane punctata*			+	Δ	JA	√	√	
Echelidae	*Muraenichthys macropterus*		+		Δ	JA	√		
	Muraenichthys malabonensis		+		Δ	JA	√		
Eleotridae	*Bostrichthys sinensis*	+++	+		0.82	JA	√		
	Butis butis	++	+	+	Δ	JA	√		√
	Eleotris melanosoma	+			Δ	JA	√		
	Prionobutis koilomatodon		++	+	Δ	JA	√		√
Elopidae	*Elops saurus*			+	Δ	J	√		
Engraulidae	*Setipinna taty*	+		+	Δ	J	√	√	
	Stolephorus zollingeri	+	+	+	Δ	J	√	√	
	Trissa vitirostris	+	+	+	Δ	J	√	√	√
Formionidae	*Formio niger*			+	Δ	JA	√	√	
Gerridae	*Gerreomorpha japonica*	+	+++	++	3.81	J	√	√	√
	Gerres filamentosus	+	+++	++	1.11	J	√	√	√
	Gerres lucidus	+	++	+	0.66	J	√	√	√
Gobiidae	*Acentrogobius caninus*	++	++	+++	8.30	JA	√	√	√
	Acentrogobius chlorostigmatoides	+	++	++	0.37	JA	√		√
	Acentrogobius viridipunctatus	+++	+		5.82	JA	√		
	Amblygobius albimaculatus		+	+	Δ	JA	√		√
	Chaenogobius urotaenia			+	Δ	JA	√		√
	Ctenogobius brevirostris	+++	+		0.36	JA	√		√
	Crytocentrus pavoninoides			+	Δ	JA	√		√
	Glossogobius olivaceus	+	++	++	0.32	JA	√		√
	Glossogobius giuris	+	++	+	0.34	JA	√		√
	Mugilogobius abei	+	+		Δ	JA	√		√
	Oxyurichthys microlepis		++	++	1.39	JA	√		√

Appendix 1 (*continued*)

Family	Species	Microhabitat type[a]			Percentage to total individuals	Living stage	Gear type		
		M	C	B			Centipede net	Gill net	Cast net
	Oxyurichthys ophthalmonema	+	+	+	Δ	JA	√		√
	Oxyurichthys papuensis	+	++	+	Δ	JA	√		√
	Oxyurichthys tentacularis		++	+	Δ	JA	√		√
	Stigmatogobius javanicus	+	+	+	Δ	JA	√		√
	Triaenopogon barbatus			+	Δ	JA	√		√
	Tridentiger trigonocephalus		+	+	Δ	JA	√		√
Hemirhamphidae	*Hyporhamphus intermedius*	+	+++	++	0.62	J	√	√	√
Latidae	*Lates calcarifer*			+	Δ	J	√	√	
	Psammoperca waigiensis			+	Δ	J	√	√	
Leiognathidae	*Leiognathus brevirostros*	+	++	+++	16.70	J	√	√	√
Lethrinidae	*Lethrinus haematopterus*	+	+	+	Δ	J	√		
Lutianidae	*Lutianus johni*			+	Δ	J	√		
	Lutianus russelli		+	+	Δ	J	√		
	Lutianus spilurus			+	Δ	J	√		
Mugilidae	*Liza carinatus*	+++	+	++	3.66	JA	√	√	√
	Mugil cephalus	+++	++	++	3.34	JA	√	√	√
	Osteomngil ophuyseni	+++	+	++	6.60	JA	√	√	√
	Osteomngil strongylocephalus	++	+	+++	1.11	JA	√	√	√
	Sphyraena jello			+	Δ	JA	√	√	
Mullidae	*Upeneus bensasi*			+	Δ	J	√		
	Upeneus tragula			+	Δ	J	√		
Muraenidae	*Gymnomuraena concolor*		+	+	Δ	JA	√		
Muraenesocidae	*Muraenesox talabon*	+	++	+	Δ	J	√		
Ophichthyidae	*Myrichthys maculosus*			+	Δ	JA	√		
	Pisoodonophis boro	++	+		Δ	JA	√		
	Pisoodonophis cancrivorus		+		Δ	JA	√		
Periophthalmidae	*Boleophthalmus pectinirostris*	++	+	++	Δ	JA	√		
	Periophthalmus cantnensis	++	+	+	Δ	JA	√		
Platycephalidae	*Onigocia macrolepis*	+	+	++	0.56	JA	√	√	√
Plotosidae	*Platosus anguillaris*			+	Δ	A	√		√
Pomadasyidae	*Pomadasys hasta*			+	Δ	J	√		
	Pomadasys maculatus	+	+	+	Δ	J	√		
Scatophagidae	*Scatophagus argus*			+	Δ	J	√		√
Sciaenidae	*Umbrina russelli*	+	+		Δ	J	√	√	
Serranidae	*Epinephelus malabaricus*	+	+	+	Δ	JA	√		√
Siganidae	*Siganus fuscescens*			+	Δ	JA	√	√	√
	Siganus guttatus			+	Δ	JA	√	√	
	Siganus oramin		+	++	Δ	JA	√	√	√
Sillaginidae	*Sillago maculata*		+	+	Δ	JA	√	√	
	Sillago sihama	+++	++		7.88	JA	√	√	√
Soleidae	*Brachirus pan*		+	+	Δ	J	√		√
	Solea ovata			+	Δ	J	√		√
	Synaptura orientalis	+	+	++	0.34	J	√	√	√
	Zebrias zebra			+	Δ	J	√		√
Sparidae	*Sparus berda*	+	+	+	Δ	J	√		√
	Sparus latus	+	+	+	Δ	J	√		√
Syngnathidae	*Syngnathus acus*	+		+	Δ	JA	√		
	Syngnathus cyanospilus		+		Δ	JA	√		
Synodidae	*Saurida elongata*			+	Δ	JA	√		
	Saurida undosquamis			+	Δ	JA	√		
	Trachinocephalus myops			+	Δ	JA	√		
Taenioididae	*Taenioides anguillaris*			+	Δ	JA	√		
	Trypauchen vagina			+	Δ	JA	√		
Tetraodontidae	*Fugu alboplumbeus*			+	Δ	JA	√	√	√
	Fugu niphobles	+	+	+++	15.81	JA	√		√
	Fugu oblongus			+	Δ	JA	√		√
Theraponidae	*Pelates quadrilineatus*		+	+	0.54	J	√	√	√
	Therapon oxyrhynchus		+	+	Δ	J	√	√	√
	Therapon jarbua	+	+	+	Δ	J	√	√	√
Triacanthidae	*Triacanthus brevirostris*			+	Δ	J	√		√
	115	45	66	103			114	38	65

Δ: Individuals accounted to less 0.3% of the total fishes caught by centipede net at different habitats of mangroves.
Gill net and cast net were only used in creeks and treeless mudflats. In vegetated areas of mangroves, centipede nets were used only in wet seasons.
M: vegetated areas of mangroves, C: creeks, B: treeless mudflats.
Sampling time: from March 2004 to December 2004.
√: Caught by a net.
+: Incidential species, ++: common species, +++: dominant species.
J: juvenile, A: adult, JA: juvenile and adult.
[a] Caught by centipede net in different microhabitats.

References

Abrahams, M.V., Kattenfeld, M.G., 1997. The role of turbidity as a constraint on predator-prey interactions in aquatic environments. Behavioral Ecology and Sociobiology 40, 169–174.

Beck, M.W., Heck Jr., K.L., Able, K.W., Childers, D.L., Eggleston, D.B., Gillanders, B.M., Halpern, B., Hays, C.G., Hoshino, K., Minello, T.J., Orth, R.J., Sheridan, P.F., Weinstein, M.P., 2001. The identification, conservation, and management of estuarine and marine nurseries for fish and invertebrates. BioScience 51, 633–641.

Blaber, S.J.M., 1997. Fish and Fisheries of Tropical Estuaries. Chapman and Hall, London, UK, 367 pp.

Blaber, S.J.M., Brewer, D.T., Salini, J.P., 1989. Species composition and biomasses of fishes in different habitats of a tropical northern Australian estuary: their occurrence in the adjoining sea and estuarine dependence. Estuarine, Coastal and Shelf Science 29, 509–531.

Chong, V.C., Sasekumar, A., Leh, M.U.C., D'Cruz, R., 1990. The fish and prawn communities of a Malaysian coastal mangrove system, with comparisons to adjacent mud flats and inshore waters. Estuarine, Coastal and Shelf Science 31, 703–722.

Cocheret de la Morinière, E., Nagelkerken, I., Meij, H., Velde, G., 2004. What attracts juvenile coral reef fish to mangroves: habitat complexity or shade? Marine Biology 144, 139–145.

Connolly, R.M., 1994. Comparison of fish catches from a buoyant pop net and a beach seine in a shallow seagrass habitat. Marine Ecology Progress Series 109, 305–309.

Dorenbosch, M., Verberk, W.C.E.P., Nagelkerken, I., van der Veldel, G., 2007. Influence of habitat configuration on connectivity between fish assemblages of Caribbean seagrass beds, mangroves and coral reefs. Marine Ecology Progress Series 334, 103–116.

FAO, 2007. The Word's Mangroves 1980–2005. Forest Resources Assessment Working Paper No. 153. ftp://ftp.fao.org/docrep/fao/010/a1427e/a1427e00.pdf.

Faunce, C.H., Serafy, J.E., 2006. Mangroves as fish habitat: 50 years of field studies. Marine Ecology Progress Series 318, 1–18.

He, B.Y., 1999. Comparative study on the ecology of mangrove fishes between two bays of Guangxi. Marine Science Bulletin 18, 28–35 (in Chinese, with English Abstr.).

He, B.Y., Fan, H.Q., 2002. A study on seasonal dynamics of species diversity of fishes in tidal waters of creeks within the mangroves of Yinluo Bay, Guangxi. Biodiversity Science 10, 175–180 (in Chinese, with English Abstr.).

He, B.Y., Fan, H.Q., Mo, Z.C., 2001. Study on species diversity of fishes in mangrove area of Yinluo Bay, Guangxi Province. Journal of Tropical Oceanography 20, 74–79 (in Chinese, with English Abstr.).

Hindell, J.S., Jenkins, G.P., 2004. Spatial and temporal variability in the assemblage structure of fishes associated with mangroves (*Avicennia marina*) and intertidal mudflats in temperate Australian embayments. Marine Biology 144, 385–395.

Huxham, M., Kimani, E., Augley, J., 2004. Mangrove fish: a comparison of community structure between forested and cleared habitats. Estuarine, Coastal and Shelf Science 60, 637–647.

Ikejima, K., Tongnunui, P., Medej, T., Taniuchi, T., 2003. Juvenile and small fishes in a mangrove estuary in Trang province, Thailand: seasonal and habitat differences. Estuarine, Coastal and Shelf Science 56, 447–457.

Kathiresan, K., Bingham, B.L., 2001. Biology of mangroves and mangrove ecosystems. Advances in Marine Biology 40, 81–251.

Laegdsgaard, P., Johnson, C.R., 1995. Mangrove habitats as nurseries: unique assemblages of juvenile fish in subtropical mangroves in eastern Australia. Marine Ecology Progress Series 126, 67–81.

Laegdsgaard, P., Johnson, C.R., 2001. Why do juvenile fish utilize mangrove habitats? Journal of Experimental Marine Biology and Ecology 257, 229–253.

Laroche, J., Baran, E., Rasoanandrasana, N., 1997. Temporal patterns in a fish assemblage of a semiarid mangrove zone in Madagascar. Journal of Fish Biology 51, 3–20.

Ley, J.A., McIvor, C.C., Montague, V.L., 1999. Fishes in mangrove prop-root habitats of northeastern Florida Bay: distinct assemblages across an estuarine gradient. Estuarine, Coastal and Shelf Science 23, 701–723.

Lin, H.J., Shao, K.T., 1999. Seasonal and diel changes in a subtropical mangrove fish assemblage. Bulletin of Marine Science 65, 775–794 (in Chinese, with English Abstr.).

Little, M.C., Reay, P.J., Grove, S.J., 1988. Distribution gradients of ichthyoplankton in an east African mangrove creek. Estuarine, Coastal and Shelf Science 26, 669–677.

Louis, M., Bouchon, C., Bouchon-Navaro, Y., 1995. Spatial and temporal variations of mangrove fish assemblages in Martinique (French West Indies). Hydrobiologia 295, 275–284.

Lubbers, L., Boynton, W.R., Kemp, W.M., 1990. Variations in structure of estuarine fish communities in relation to abundance of submersed vascular plants. Marine Ecology Progress Series 65, 1–14.

Morton, R.M., 1990. Community structure, density and standing crop of fishes in a subtropical Australian mangrove area. Marine Biology 105, 385–394.

Nagelkerken, I., van der Velde, G., Gorissen, M.W., Meijer, G.J., van't Hof, T., den Hartog, C., 2000. Importance of mangroves, seagrass beds and the shallow coral reef as a nursery for important coral reef fishes, using a visual census technique. Estuarine, Coastal and Shelf Science 51, 31–44.

Power, M.E., Matthews, W.J., Stewart, A.J., 1985. Grazing minnows, piscivorous bass and stream algae: dynamics of a strong interaction. Ecology 65, 1448–1456.

Robertson, A.I., Blaber, S.J.M., 1992. Plankton, epibenthos and fish communities. In: Robertson, A.I., Alongi, D.M. (Eds.), Tropical Mangrove Ecosystems. American Geophysical Union Press, Washington DC, USA, pp. 173–224.

Robertson, A.I., Duke, N.C., 1987. Mangroves as nursery sites: comparisons of the abundance and species composition of fish and crustaceans in mangroves and other nearshore habitats in tropical Australia. Marine Biology 96, 193–205.

Robertson, A.I., Duke, N.C., 1990. Mangrove fish-communities in tropical Queensland, Australia: spatial and temporal patterns in densities, biomass and community structure. Marine Biology 104, 369–379.

Rönnbäck, P., Macia, A., Almqvist, G., Schultz, L., Troell, M., 2002. Do penaeid shrimps have a preference for mangrove habitats? Distribution pattern analysis on Inhaca Island, Mozambique. Estuarine, Coastal and Shelf Science 55, 427–436.

Rönnbäck, P., Troell, M., Kautsky, N., Primavera, J.H., 1999. Distribution pattern of shrimps and fish among *Avicennia* and *Rhizophora* microhabitats in the Pagbilao Mangroves, Philippines. Estuarine, Coastal and Shelf Science 48, 223–234.

Rozas, L.P., Minello, T.J., 1997. Estimating densities of small fishes and decapod crustaceans in shallow estuarine habitats: a review of sampling design with focus on gear selection. Estuaries 20, 199–213.

Russ, G., 1984. Distribution and abundance of herbivorous grazing fishes in the central Great Bamer Reef. I. Levels of variability across the entire continental shelf. Marine Ecology Progress Series 20, 23–34.

Satumanatpan, S., Keough, M.J., 2001. Role of larval supply and behaviour in determining settlement of barnacles in a temperate mangrove forest. Journal of Experimental Marine Biology and Ecology 260, 133–153.

Shinnaka, T., Sano, M., Ikejima, K., Tongnunui, P., Horinouchi, M., Kurokura, H., 2007. Effects of mangrove deforestation on fish assemblage at Pak Phanang Bay, southern Thailand. Fisheries Science 73, 862–870.

Sih, A., 1982. Foraging strategies and the avoidance of predation by an aquatic insect, *Notonecta hoffmanni*. Ecology 63, 786–796.

Smith, T.M., Hindell, J.S., 2005. Assessing effects of diel period, gear selectivity and predation on patterns of microhabitat use by fish in a mangrove dominated system in SE Australia. Marine Ecology Progress Series 294, 257–270.

South China Sea Fisheries Institute, 1979. The Fishes of the Islands in the South China Sea. Science Press, Beijing, China (in Chinese).

Spalding, M.D., Blasco, F., Field, C.D., 1997. World Mangrove Atlas. The International Society for Mangrove Ecosystems, Okinawa, Japan, p. 178.

Stein, R.A., 1977. Selective predation, optimal foraging and the predator-prey interaction between fish and crayfish. Ecology 58, 1237–1253.

Thayer, G.W., Colby, D.R., Hettler Jr., W.F., 1987. Utilization of the red mangrove prop root habitat by fishes in south Florida. Marine Ecology Progress Series 35, 25–38.

Tongnunui, P., Ikejima, K., Yamane, T., Horinouchi, M., Medej, T., Sano, M., Kurokura, H., Taniuchi, T., 2002. Fish fauna of the Sikao Creek mangrove estuary, Trang, Thailand. Fisheries Science 68, 10–17.

Utne, A.C.W., Aksnes, D.L., Giske, J., 1993. Food, predation risk and shelter: an experimental study on the distribution of adult two-spotted goby *Gobiusculus flavescens* (Fabricus). Journal of Experimental Marine Biology and Ecology 166, 203–216.

Valiela, I., Bowen, J.L., York, J.K., 2001. Mangrove forests: one of the world's threatened major tropical environments. BioScience 51, 807–815.

Vance, D.J., Haywood, M.D.E., Heales, D.S., Kenyon, R.A., Loneragan, N.R., Pendrey, R.C., 1996. How far do prawns and fish move into mangroves? Distribution of juvenile banana prawns *Penaeus merguiensis* and fish in a tropical mangrove forest in northern Australia. Marine Ecology Progress Series 131, 115–124.

Verweij, M.C., Nagelkerken, I., de Graaff, D., Peeters, M., Bakker, E.J., van der Velde, G., 2006. Structure, food and shade attract juvenile coral reef fish to mangrove and seagrass habitats: a field experiment. Marine Ecology Progress Series 306, 257–268.

Wang, M., Zhang, J.H., Shi, F.S., 2007. Investigation on fishing gear and fish catch in Dongzhai Harbor Mangrove Area, Hainan. Fisheries Science and Technology Information 27, 6–9 (in Chinese, with English Abstr.).

Wang, W.Q., Wang, M., 2007. The Mangroves of China. Science Press, Beijing, China, p. 186(in Chinese).

Weinstein, M.P., Brooks, V.H.A., 1983. Comparative ecology of nekton residing in a tidal creek and adjacent seagrass meadow: community composition and structure. Marine Ecology Progress Series 12, 15–27.

Williams, D.M., Hatcher, A.I., 1983. Structure of fish communities on outer slopes of inshore, mid-shelf and outer shelf reefs of the Great Barrier Reef. Marine Ecology Progress Series 10, 239–250.

Wright, J.M., 1986. The ecology of fish occurring in shallow water creeks of a Nigerian mangrove swamp. Journal of Fish Biology 29, 431–441.

截污后深圳河落马洲段大型底栖动物群落的恢复过程*

蔡立哲[1] 林 鹏[2] 历红梅[1]

(1. 厦门大学环境科学研究中心;2. 厦门大学湿地与生态工程研究中心 福建 厦门 361005)

摘 要 2000年 4月至 2002年 4月,对截污前后的深圳河落马洲段大型底栖动物进行两周年的季度监测.结果表明,截污前深圳河落马洲段没有发现大型底栖动物;截污后大型底栖动物种数、密度、生物量和种类多样性指数 (H′)有随时间推移呈增加的趋势;截污后底泥中硫化物、总氮、总磷随时间推移呈下降趋势.由此可见,截污后深圳河落马洲段大型底栖动物从无到有,种数、密度和生物量从少到多,群落结构处于向多样性恢复的过程中.图 6表 2参 11

关键词 大型底栖动物; 恢复过程; 截污; 深圳河

CLC X832 : Q958.894

Restoration Process of Macrofaunal Community at Luomazhou Section of the Shenzhen River after Pollution Interception

CAI Lizhe , LIN Peng & LI Hongmei

Abstract To find out the process of ecological restoration, macrofauna was seasonally investigated at Luomazhou section that was kept apart from polluted water in the Shenzhen River from April 2000 to April 2002. The results showed that only three species of oligochaetes and insects were collected in the early days after pollution interception, and 12 species of macrofauna (include polychaetes, oligochaetes, gastropods, bivalves and insects) were found within two years. After pollution interception, macrofaunal density, biomass and species diversity index were also found increasing, but sulphide, total nitrogen and total phosphorus decreasing in sediment. Fig 6, Tab 2, Ref 11

Keywords macrofauna; restoration process; pollution interception; Shenzhen River

CLC X832 : Q958.894

深圳河是深圳与香港的界河,北岸为深圳经济特区的罗湖区和福田区,南岸为香港特别行政区的新界地区.深圳河属雨源型河流,平时天然来水量不大,主要为流域内生活污水,为深圳市主要纳污河流,也是深圳市最重要的排洪入海河流.目前全市污水处理数量和深度都十分有限,深圳河中下游河水水质受到了严重污染,各项水质指标大大劣于国家地表水水质(GB3838-2002) V类水质标准,枯水期污染尤为严重[1].正是因为深圳河流域有大量的没有被截流的污水流入自净能力和稀释能力均较弱的深圳河,从而造成了深圳河严重的污染状况[2].1995年,深圳市治河办开始对深圳河进行拓宽、取直、挖深,主要目的是增加深圳河的排洪和航运能力.至 2000年初,开始将深圳河落马洲河段封闭,即在落马洲河段两端与深圳河连接处建造了水闸,隔断了落马洲河段与深圳河水的交换.这种隔断,相当于截断了污染物进入落马洲河段,使得落马洲河段水质有所改善.为了了解截污后落马洲河段大型底栖动物的恢复动态,受深圳市治河办的委托,作者从 2000年 4月开始,按季度对落马洲河段大型底栖动物进行了 2 a的监测.

1 材料与方法

深圳河落马洲段布设 3个大型底栖动物取样点. L1取样点靠近黄岗口岸, L2取样点靠近落马洲村, L3取样点靠近黄岗边检站和福田河与深圳河连接处 (图 1). L1和 L2取样点的沉积物为陈旧的黑泥,L3取样点沉积物为较硬的黄色粘土,带有细砂和小石块.每个取样点用深度为 20 cm、直径为 10 cm的塑料取样管随机采集 10管,管内的沉积物放入塑料桶内,加水

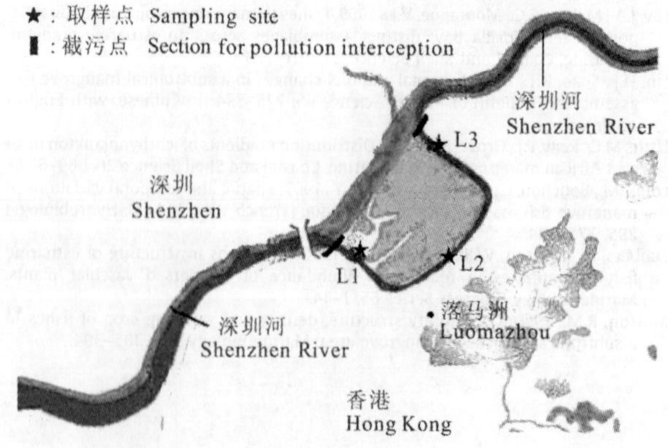

图 1 深圳河落马洲段大型底栖动物取样站 (★)示意图
Fig 1 Sampling sites (★) of macrofauna at Luomazhou section of the Shenzhen River

* 原载于应用与环境生物学报,2007,14(4):497－500

References

1 Li BH (李斌华), Yu XS (喻学山). Environmental problem and countermeasure for water in Shenzhen River Estuary. *Water Conservancy & Electricity of China Country* (中国农村水利水电), 2002(12): 51～52

2 Zhang JJ (张健君), He HB (何厚波), Hu JD (胡嘉东), Yang J (杨军). Control strategy of water pollution in the Shenzhen River. *Res Environ Sci* (环境科学研究), 2005, **18** (5): 40～44

3 Huang HY (黄汉禹). Characteristic of polluted soil on riverway in Shenzhen River and disposal step. *Acta Sci Nat Univ Sunyatseni* (中山大学学报自然科学版), 2001, **40** (Suppl 2): 5～9

4 Liu Y (刘玉), Vermaat JE, de Ruytered, de Kruijf Ham. Using macrofauna and ODP system to evaluate organic pollution of the Pearl River. *Chin J Appl Environ Biol* (应用与环境生物学报), 2003, **9** (2): 154～157

5 Hu BJ (胡本进), Yang LF (杨莲芳), Wang BX (王备新), Shan LN (单林娜). Functional feeding groups of macroinvertebrates in 1～6 order tributaries of the Changjiang River. *Chin J Appl Environ Biol* (应用与环境生物学报), 2005, **11** (4): 463～466

6 Sheng SC (盛世春), Gui HR (桂和荣), Zhang MQ (张明群), Tai Y (邰燕), Ni SG (倪仕钢). Ecological study on snails (Lymnaeidae) in Huainan section of the Huaihe River. *Chin J Appl Environ Biol* (应用与环境生物学报), 2005, **11** (5): 563～565

7 Den Besten PJ, Den Brink PJ. Bioassay responses and effects on benthos after pilot remediations in the delta of the rivers Rhine and Meuse. *Environ Poll*, 2005, **136** (2): 197～208

8 Moseman SM, Levin LA, Currin C, Forder C. Colonization, succession, and nutrition of macrobenthic assemblages in a restored wetland at Tijuana Estuary, California. *Estuarine, Coastal & Shelf Sci*, 2004, **60** (4): 755～770

9 Cai LZ (蔡立哲), Lin P (林鹏), Liu JJ (刘俊杰). Quantitative dynamics of three species of large individual polychaete and environmental analysis on mudflat in Shenzhen Estuary. *Acta Oceanol Sin*, 2000, **22** (3): 97～103

10 Cai LZ (蔡立哲), Li FX (李复雪), Zheng B (郑斌). Temporal and spatial distribution of Mollusks mudflat in Shenzhen Estuary. In: Chinese Society of Malacology (中国贝类学会) ed. Transactions of the Chinese Society of Malacology Ⅷ. Beijing (北京): Ocean Press (海洋出版社), 1999. 91～98

11 Liu YY (刘月英), Zhang WZ (张文珍), Wang YX (王耀先). Freshwater mollusks of the specific area of the Shenzhen, Guangdong Province, China. In: Chinese Society of Malacology (中国贝类学会) ed. Transactions of the Chinese Society of Malacology Ⅱ. Beijing (北京): Ocean Press (海洋出版社), 1986. 42～44

红树林区微生物资源*

龙 寒 向 伟 庄铁诚 林 鹏

(厦门大学生命科学学院,厦门 361005)

摘 要 随着工业的迅猛发展,工业废料源源不断地向海洋输出,污染日趋严重。人们在大力开发海洋微生物自净能力的同时,也对海岸线的绿色卫士——红树林给予了密切关注,积极展开红树林区微生物资源的开发利用。本文从红树林区微生物库的资源多样性、微生物在物质循环和能量流动中的作用、生理活性物质、代谢产物和污染治理等几个方面进行综述。

关键词 红树林,微生物,生理活性物质,代谢产物,污染处理

中图分类号 Q93 **文献标识码** A **文章编号** 1000-4890(2005)06-0696-07

Microorganism resource of mangrove ecosystems. LONG Han, XIANG Wei, ZHUANG Tiecheng, LIN Peng (*School of Life Science, Xiamen University, Xiamen* 361005, *China*). *Chinese Journal of Ecology*, 2005, **24** (6): 696~702.

With the rapid development of industry, more and more industrial waste is constantly exported to the ocean which causes serious pollution problems. Researchers are paying more attention to the mangrove, the important guard of the coastline, and the exploitation of mangrove microorganism resource has been expanded in many aspects, as well as developing the self-clean ability of ocean microorganism. This review summarized the diversity of mangrove microorganism resource, the role of microorganism in material recycle and energy flow, physiological active material, metabolic product and pollution management.

Key words mangrove, microorganism, physiological active matter, metabolic production, pollution management.

1 引 言

红树林在热带亚热带的海岸潮间带构建了一片片美丽的海上森林,虽然它占据的陆地面积不到全球陆地面积的千分之一,却以相当于亚马逊热带雨林的高生产力与珊瑚礁、上升流、海滨沼泽湿地组成了四大海洋高生产力生态系统。其特殊的生长环境创造了极为丰富又极具特色的微生物资源。近年来的红树林微生物研究工作中,已分离出了许多特殊功能菌,其中包括对各类污染物有较强降解能力的微生物,生理活性物质和代谢产物研究也相继展开。人们对于红树林微生物的认识,从"被忽视"的空白至"重要一环"并具有"多功能"的环境效应;从宏观深入到微观;从平面到立体;从群落结构到分子水平。人类已经意识到红树林对地球环境保护和生物资源开发具有的不可忽视作用。红树林微生物资源的开发是红树林资源开发不可缺少的部分,应当继续加强和拓宽其研究内容,为红树林生态系统更好的开发和利用提供更多的微生物资料。

2 红树林中微生物库的资源多样性

红树林生活在热带亚热带海岸潮间带,其生境具有强还原性、强酸性、高含盐量、营养丰富[12]等特征。因此,这里的微生物资源既丰富又不失特色,主要类群为细菌、真菌、放线菌、微型藻类等。其中已分离鉴定出的红树林真菌超过百种,成为海洋真菌第二大类群[58,92]。在热带红树林中,微生物的组成大致为:细菌和真菌占微生物资源总量的91%,藻类和原生动物分别占7%和2%[35,92]。目前,已发现并且分离出许多新的菌种,例如我国新发现的两个海生疫霉种(泡囊海疫霉、刺囊海疫霉)[31]、II型甲烷氧化菌等。正在研究中的红树林微生物有高效固氮菌、溶磷细菌、硫化还原菌、光合厌氧菌、产甲烷菌及红树林真菌。

在红树林生态系统中,固氮率的高低和植物体呼吸、根、树皮、林下土壤、蓝细菌、沉积物及覆于其

* 国家自然科学基金资助项目(30270272)
原载于生态学杂志,2005,24(6):696-702

上的腐解枝叶等因素有关。红树林内积累着大量生物降解产生的有机物,这些有机物为固氮提供大量的能量,所以,在红树林内发现高效固氮菌是很正常的。在红树林中,乙炔还原比率同有机物可利用性有着明显的相关性,林下非固氮微生物通过分解凋落物为固氮提供了足够能量,因此,凋落物分解过程中氮含量的增加不会受到外来碳源的影响。而在无红树林区,外来碳源的加入会显著提高固氮率[93,94]。从红树林区中分离的固氮微生物包括自生固氮菌、联合固氮菌和共生固氮菌,其中大部分蓝细菌均具有高效固氮活性。Sengupta 等[78]从不同种类红树林沉积物、根际及根表分离出的固氮菌分别归属于固氮螺菌属(*Azospirillum*)、固氮菌属(*Azotobacter*)、根瘤菌属(*Rhizobium*)、梭菌属(*Clostridium*)和克雷伯氏菌属(*Klebsiella*);Holguin 等[55]从墨西哥的大红树(*Rhizophora mangle*)、亮叶白骨壤(*Avicennia germinans*)和假红树(*Laguncularia racemosa*)林中也分离出 *Vibrio campbelli*, *Listonella anguillarum*, *Vibrio aestuarianus* 和 *Phyllobacterium* sp. 等固氮菌。另外,同位素示踪发现,这些固氮菌确能在红树植物体内定居并向植物根部提供无机氮,促进红树植物的生长发育[26,40,49,85]。在澳大利亚南部的红树林生态系中,凋落物及表面沉积物的固氮量能提供全年氮需求的 40 %[86];在佛罗里达的红树林,生物固氮能满足该生态系 60 %的氮需求量[93]。可见,固氮作用是红树林生态系统中细菌的重要功能之一。

以往红树林研究中,对微生物溶磷作用的研究不多,对溶磷细菌(Phosphate-solubilizing bacteria)的研究也就相对较少。在对墨西哥红树林的研究中,研究者们从黑红树(Black mangrove)根部分离出了 9 种溶磷细菌(*Bacillus amyloliquefaciens*, *Bacillus. atrophaeus*, *Paenibacillus macerans*, *Xanthobacter agilis*, *Vibrio proteolyticus*, *Enterobacter aerogenes*, *E. taylorae*, *E. asburiae*, 和 *Kluyvera cryocrescens*),从假红树中分离出 3 种溶磷细菌(*Bacillus. licheniformis*, *Chryseomonas luteola*, 和 *Pseudomonas stutzeri*)[87],其中,*Xanthobacter*, *Kluyvera* 和 *Chryseomonas* 是首次在红树林中发现,也是首次发现这 3 种细菌具有固氮能力。Vazquez 等[87]在添加了磷酸钙的培养基上对这些分离出来的细菌进行培养,发现在生长的菌落周围会出现透明圈,从而首次证明了这些种类细菌的溶磷能力。溶磷细菌为红树林植物提供了可溶性磷,在红树植物的生长发育中起着不可忽视的作用。

由于红树林沉积物的氧含量几乎为零,在这里发生的分解作用主要通过硫化还原进行。硫化还原菌(Sulfate-reducing bacteria)作为沉积物中有机物的主要分解者普遍存在于红树根际和林下沉积物中。例如,印度果阿红树林沉积物中硫化还原菌的种群密度为 10^3 cfu·g^{-1}[76];在佛罗里达大红树和亮叶白骨壤的根际,硫化还原菌的种群密度达到了 10^6 cfu·g^{-1}(鲜重)[93]。红树林生态系统含硫丰富及厌氧的土壤环境也为另一细菌群体的生存繁殖提供了很好的条件,这就是光合厌氧菌(Photosynthetic anoxygenic bacteria),包括紫色硫细菌(purple sulfur bacteria)、紫色非硫细菌(purple non-sulfur bacteria)和绿色非硫细菌(green non-sulfur bacteria)。可能由于这类细菌生长缓慢,难于在实验室培养,故红树林内这类细菌的研究报道并不多见,目前已分离鉴定的科属包括:着色菌科(Chromatiaceae)和红螺菌科(Rhodospirillaceae)[88,89],着色菌属(*Chromatium*)、贝氏硫细菌属(*Beggiatoa*)、板硫菌属(*Thiopedia*)、*Chloronema*、*Leucothiobacteria*[45,48]、*Rhodobacter*、*Rhodopseudomonas*[80]等。另外,硫化还原菌和一些光合厌氧菌也具有固氮功能[53,93]。在红树林生态系统中还生活着另一种重要的细菌群落——产甲烷菌,研究人员已分离出 *Methanococcoides methylutens*[65]和 4 种未鉴定耐热产甲烷菌[62]。产甲烷菌的数量会受到硫化还原菌的限制[70]。就研究最多的红树林真菌而言,目前已分离鉴定的红树林真菌种类超过了 100 种,红树植物区系多样性、群落年龄、群落周围陆栖树种以及微生境的差异造成了各个红树林区真菌种数的差异[56]。

红树林细菌除具有上述固氮、溶磷等功能特性外,还具有在一定盐胁迫下生长的特性,这是由红树林生态系统的环境特点决定的。红树林区为盐生环境[12],其中的微生物具有一定的耐盐和嗜盐性。红树林耐盐及嗜盐微生物资源的研究也正在起步中。嗜盐微生物是光能转化蛋白膜——紫膜提取的主要原料,是细菌视紫红质(BR)的生产者,二者均为应用前景极佳的生物分子光电材料和光存储材料[54,64]。嗜盐微生物的代谢产物多种多样,包括胞外多糖、维生素、抗生素胰岛素等等[17]。特别值得一提的是,嗜盐微生物所产生的酶在高盐条件下仍具有很高活性,因此它们在污染物降解——尤其是高盐环境中的污染物降解起到了非常重要的作用[61]。在医学上,嗜盐菌产生的具热塑性、生物降解和生物相容性的聚羟基丁酸可用于可降解生物材料的开发[7];在农业上,通过转基因技术将嗜盐微生物的一定基因片段转入农作物中,可提高作物的抗盐性[3]。此外,耐盐、嗜盐微生物在食品工业、石油开采等方面均有广泛的应用[7]。

3 红树林微生物在物质循环和能量流动中的作用

红树林生态系统的生产力极高[11~16],这归功于它所拥有的高效营养循环系统[2,4,12,60]。因为在红树林生态系统中,只有少量的营养保留,新的营养不断从腐烂的红树树叶中产生,微生物的活动在其中起到了非常主要的作用[36,38,55],是驱动红树林生态系统中营养转化的主要因素之一[9,59,69,70,90]。

生活在热带红树林沉积物中的细菌促使了大部分碳流的形成。它们推动了大部分的能量流动和营养流动,担当着碳沉降的角色[39]。红树林沉积物中的细菌群落能够消耗溶解在间隙水中的有机碳,通过这种方式防止了该种形式的碳流失到临近的生态系统中[35,36,37,43]。溶解在间隙水中的有机碳浓度高于沉积物上方水层中的有机碳浓度,而在这两个水层之间碳的净通量为零。具有很高活力和生产力的细菌群落在沉积物中大量繁殖。虽然间隙水中溶解的游离氨基酸与上层潮汐水中的氨基酸浓度具有很大的梯度,但在这两个水层之间没有发现氨基酸的流动[82]。所以说,在热带红树林中,细菌群落消耗绝大多数溶解在间隙水中的碳。在河口,大量的碳通过光合作用被固定,固定的碳堆积在沉积物中,被细菌在厌氧条件下矿化,细菌从这个过程获得能量用于自身生长,而代谢所产生的能量则通过碎屑食物链依次传递给无脊椎动物和鱼类。通过对河口淤泥和盐沼泥炭的测定表明,厌氧微生物的新陈代谢,特别是硫的循环,引起了大部分生态系统中大部分能量的流动[46],因此可以说在红树林中细菌作为循环者要比作为营养者更为出色[52]。

红树林生态系统中各类氮化物的比率和流动取决于系统的特性[52]。在墨西哥红树林中,研究者们在测定氮的流失时发现,由于反硝化作用而导致的氮流失是可以忽略的[74],这显示氮在被以氮气释放进大气前就已经被消耗了。细菌和植物对系统中可利用氮的竞争可能很强烈,沉积物中的由含氮有机化合物分解而来的硝酸盐可能在细菌的作用下转化成为铵离子,被细菌和植物吸收,这个过程保存了生态系统中的氮[72,73]。除此之外,异化的硝酸盐转化成铵(dissimilatory nitrate reduction to ammonium)[84]或铵可能为无氧氧化(the possible anaerobic oxidation of ammonium)[57]也可能为系统保存了氮。

Bano 等[39]研究发现,在热带和亚热带红树林生态系统中生活着多种多样的具有高生产力微生物群体,它们不断地把死掉的红树植物中的营养转化为能够被植物利用的氮、磷和其他形式的营养,与此同时,植物体也为生活在这个生态系统中的微生物提供食物来源。营养物质作为微生物和植物之间的桥梁使二者之间建立起非常亲密的关系,营养物质的循环和存储就是建立二者关系的机制[52]。

红树的叶和枝干在凋落后就为林下微生物占据并立即开始被分解。在分解过程中,腐烂的叶和枝干中的总氮、磷浓度以及蛋白质含量随时间而增加。van der Valk 等[86]研究白骨壤叶片分解,105 d 时,由于碳的流失导致氮浓度由 0.7 %增加到 1.2 %,而整个落叶层由于林下微生物的固氮作用使得氮由 41 %增加到 64 %;凋落的 *Rhizophora* spp. 枝干在分解的最初两个月内,氮含量增加了 500 %[75];巴拿马海湾中红树叶片在经 27 d 分解后,干重减少了 50 %,而氮浓度由 0.3 %增加到 2.9 %(干重),磷的浓度由 0.04 %增加到 0.13 %(干重)[47]。Odum 等[66]发现,刚凋落的红树叶片中蛋白质含量仅为 6 %,6 个月后,叶片蛋白质增加到 20 %,这可能是脂肪、碳水化合物转化的结果。

4 代谢产物的研究

随着人类对陆地资源的大力开发,继续从中寻找新的微生物药物及化合物资源已愈加困难。面对各种新型病毒的不断产生以及病毒抗性对传统药物的逐渐适应,人类开始把眼光投向海洋资源,而四大海洋生态系统之一的红树林也已经受到人们的关注。1989 年 Poch 等[67]从夏威夷红树林内生真菌中分离出 helascolidesA、B 和赭曲霉素;1991 年又从该地区内生真菌中分离到 AuranticinsA、B(其中 A 具有显著抗菌活性,可抑制枯草芽孢杆菌和金黄色葡萄球菌生长)[68];1998 年 Schlingmann 等[77]分离的 *Hypoxylon oceanicum* 可产生一系列脂肪缩酚肽和大环内脂化合物(其产生的脂肪缩酚肽 15G256γ 可抑制真菌细胞壁合成)。近年来,我国的研究者已经开展了不少有关红树林内生真菌代谢物的研究,分离了许多新型化合物和不少具抗菌活性、抗癌活性物质,其中包括首次在微生物中分离出的环(酪-脯)二肽[32],首次从海洋真菌中分离出的尿囊素、5-对羟基苯乙基-2,4-咪唑烷二酮和环(苯丙-丙苯)二肽、新的聚酮类、连烯类化合物及α型甘油单棕榈酸酯(α-glycerol monopal mitate)、piliformic 酸(2-hexylidene-3-methylsuccinic acid)、对羟基苯甲酸(p-hydroxy benzoic acid)、3,4-二羟基苯甲酸甲酯(pto-tocatechuic acid methyl ester)、2-甲氧基-4-甲基-乙酰苯(4-hydroxy-2-methoxy-acetophe-none)、麦角甾醇(ergosterol)等多种化合物,以及异香豆素、灰黄霉素和多种醌类抗生素等[1,6,18,20,27]。

研究人员在对红树林内生真菌抗菌、抗肿瘤活性物质的研究中发现,红树植物内生真菌中存在着一定数量的抗肿瘤活性物质产生菌。在我国福建省厦门地区红树林植物中分离出的125株内生真菌中,有8.8%的菌株对HL-60或KB有抑制作用,其中以拟青霉属和曲霉属为主。该研究分离的两株高效活性菌株HQ1和FQ1对KB、HL-60、HeLa、BGC等肿瘤细胞具有显著的细胞毒素作用。研究HQ1和FQ10代谢产物,从HQ1的发酵液中分离到具有抗肿瘤活性的化合物布雷菲德菌素A,从FQ10的发酵液体中分离到一个纯化合物环(酪-脯)二肽,该化合物是首次在微生物中发现[32]。

红树林以其生境的特殊性养育了大量具有特色代谢产物的微生物,肽类化合物(环肽类化合物)已越来越多地被从菌体和培养液中分离出来,例如,从南海红树林内生真菌1356发酵液中分离出来的多种鞘胺醇和环二肽(其中鞘胺醇A是一种新型的神经鞘胺醇葡萄糖苷,环Pro-Tyr可以诱导生物发光以及色素产生)[21,24],从真菌2524培养液的乙酸乙酯提取物中分离的5个环二肽(其中两个环二肽A、B氢谱具异常化学位移,环二肽E则具镇静作用)[5]。肽类化合物具有很强的生理活性,它们不仅是抗生素、毒素、免疫抑制剂、离子转移调节器、蛋白粘合抑制剂、酶抑制剂,它们还可以用来促进记忆、调节激素、调控能量代谢、抑制肿瘤细胞活性等。除了肽类、环肽类化合物外,对胞外多糖等一些大分子代谢产物的研究也在相继展开,我国研究者已在南海红树林内生真菌(endophytefungus)1356号的菌体中分离提取得到一种新的多糖W_{21},并通过甲醇解初步研究了该多糖的组成[28]。微生物胞外多糖在食品医药化工等众多领域广泛应用。早在80年代,黄单胞菌多糖(黄原胶)(xanthan gum)就已作为食品添加剂获得世界卫生组织和粮农组织在世界范围使用的批准,而美国Kelco公司更是在60年代初就已开始了大量商业性生产。此后Kelco公司又将结冷胶(gellan gum)投入生产,这种由沼假单胞菌ATCC31 431生产的线性阴离子杂多糖作为凝胶剂、增稠剂、悬浮剂和成膜剂在食品工业中的应用不逊于黄原胶。除此之外,它还被作为培养基凝固剂、胶囊、胶片、纤维等制品的制作材料广泛应用于工农业中。继黄原胶和结冷胶生产之后,凝结多糖(curdlan)又成为Kelco新推出的目标,该产品仍作为食品稳定剂、增稠剂致力于食品工业中的应用[34]。在临床医学中,胞外多糖能作为机体免疫增效剂,增强机体抗肿瘤、抗细菌、抗病毒、抗寄生虫的性能[19,29],有希望用以阻止癌症复发、微转移和HIV携带者症状的表现[23]。微生物胞外多糖已有很多种被成功的应用到人们的生产生活中。除此之外,有研究将胞外多糖用于污水处理收到良好的效果[33]。

5 红树林微生物酶资源

红树林生境非常特殊,因此微生物产酶情况也具有特殊性。在红树林内,凋落物非常丰富,在凋落物分解过程中微生物产生的酶以纤维素分解酶、木质素分解酶为主。实验表明,红树林区绝大多数真菌都能产生用于降解木质素、纤维素和其他植物成分的酶[44,50,51,81],某些地区的放线菌也能产生纤维素酶[90];另外,红树林内一些放线菌还产生如α-淀粉酶抑制剂、胰蛋白酶抑制剂、胃蛋白酶抑制剂以及几丁质酶等生物活性物质[26]。除主要的纤维素酶和木质素酶外,红树林微生物还生产如果胶酶、木聚糖酶、蛋白酶、葡萄糖酶、脂肪酶、淀粉酶和琼脂糖酶等用途广泛的酶类。例如,在台湾淡水河口的红树林真菌、放线菌大部分均能产生纤维素酶、果胶酶、蛋白酶、脂肪酶、琼脂糖酶[90]。还有人从台湾海峡红树林土壤中分离到一株产纤维素酶的短小芽孢杆菌S-27,该菌只产生葡聚糖内切酶[22]。在印度果阿的红树林中,异养细菌具有纤维素水解、果胶水解、淀粉水解和蛋白质水解活性[63],而红树植物的降解真菌具有果胶酶、蛋白酶、淀粉酶活性并具有降解木质纤维素化合物的能力[51]。红树林内微生物产酶多样化,这些酶类物质在医药卫生、工农业生产、生活资料加工上也都有着广泛的用途。纤维素酶的用途就极为广泛,可在纺织工业中作为生物整理剂(如靛蓝牛仔服的酶洗和棉、Tencel、粘胶、黄麻、亚麻及其混纺织物的生物整理)、饲料生产中作为生物添加剂、果蔬加工中提高果汁或药物的萃取率,另外在石油开采、发酵工业、造纸、中成药加工、洗涤剂工业、垃圾治理等方面,纤维素酶和木质素酶都有着非常广泛的用途。因此,大力开发红树林微生物酶资源对生产生活具有重要意义。

6 红树林微生物对污染的治理作用

红树林位于河口入海处,是阻止陆地污染向海洋生态系统扩散的一道坚固大门,是海洋污染净化工程的重要参与者,而红树林微生物则是肩负这一重要使命不可替代的"特种兵"。红树林污染生态学的研究早在20世纪70年代就开始了,而直到90年代微生物在红树林污染生态中扮演的角色才逐渐引起人们的关注。在微生物对污染物的降解研究中,人们已发现红树林微生物在处理沿海排放的城市废水上起了很大作用,它们可能将废水中的重金属离子吸附固定,并利用废水中的营养物质,从而达到净化废水的目的[83]。在对农药降解方面,经测定,红树林土壤微生物对甲胺磷具有较强的降解能力,某些细菌的降解率甚至高达70%以上[8,30]。在滨海油污的净化处理上,利用红树林微生物对海岸工业油污、船舶油污、原油泄漏等进行处理也收到了巨大的功效。研究人员发现红树林微生物对柴油的降解一个月内可达60%以上[10];红树林土壤微生物对多环芳烃等有机物污染也有显著的清除作用[25],研究者们还分离出了对油田钻井废液具降解活性的石油降解菌——产碱菌和微球菌[41,42]。另外,利用红树林土壤内好氧嗜热微生物对海洋淤泥进行初期发酵以防止海洋淤泥的富营养化也是目前正在进行的红树林微生物开发利用项目之一[91]。

红树林微生物对陆地、海洋污染的分解净化是多渠道多方面的,我们在不断分离高效的分解菌株的同时,应该利用分子技术和基因工程技术对现有菌株进行加工改造,以生产出分解效率更高、分解范围更广的工程菌。

参考文献

[1] 王 军,林永成,吴雄宇,等. 2001. 从红树林内生真菌No. 2533分离出新的异香豆素[J]. 中山大学学报(自然科学版),**40**(1):127~128.

[2] 云南大学生物系. 1980. 植物生态学[M]. 北京:人民教育出版社,310~311.

[3] 卢 青. 2000. 植物耐盐性的分子生物学研究进展[J]. 生物学杂志,**17**(4):9~11.

[4] 卢昌义,林 鹏. 1989. 两种红树植物叶分解速率的研究[J]. 厦门大学学报(自然科学版),**27**(6):679~683.

[5] 李厚金,林永成,刘晓红,等. 2002. 红树林内源真菌2524号的肽类成分(Ⅰ)[J]. 中山大学学报(自然科学版),**41**(1):110~112.

[6] 朱 峰,林永成,周世宁,等. 2003. 南海红树林内源真菌2534号代谢产物的研究[J]. 中山大学学报(自然科学版),**42**(1):52~54.

[7] 刘爱民. 2002. 嗜盐菌的研究进展[J]. 安徽师范大学学报(自然科学版),**25**(2):181~193.

[8] 庄铁诚,张瑜斌,林 鹏. 2000. 红树林土壤微生物对甲胺磷的降解[J]. 应用环境生物学报,**6**(3):276~280.

[9] 庄铁城,林 鹏. 1992. 九龙江口秋茄红树林凋落叶自然分解与落叶腐解微生物的关系[J]. 植物生态学与地植物学学报,**16**(1):17~25.

[10] 庄铁诚,林 鹏. 1995. 红树林下土壤微生物对柴油的降解[J]. 厦门大学学报(自然科学版),**34**(3):442~446.

[11] 林 鹏. 1990. 红树林研究论文集(第1集)[C]. 厦门:厦门大学出版社,23~30.

[12] 林 鹏. 1997. 中国红树林生态系[M]. 北京:科学出版社,23~30.

[13] 林 鹏,尹 毅,卢昌义. 1992. 广西红海榄群落的生物量和生产力[J]. 厦门大学学报(自然科学版),**31**(2):199~202.

[14] 林 鹏,卢昌义. 1985. 九龙江口红树林研究 Ⅰ. 秋茄群落的生物量和生产力[J]. 厦门大学学报(自然科学版),**24**(4):508~514.

[15] 林 鹏,卢昌义,王恭礼,等. 1990. 海莲红树林的生物量和生产力[J]. 厦门大学学报(自然科学版),**29**(2):209~213.

[16] 林 鹏,陈荣华. 1991. 红树林有机碎屑在河口生态系统中的作用[J]. 生态学杂志,**10**(2):45~48.

[17] 张永光,李文均,姜成林,等. 2002. 嗜盐放线菌的研究进展[J]. 微生物学杂志,**22**(4):45~48.

[18] 陈光英,刘晓红,温 露,等. 2003. 南海红树林内生真菌1893代谢产物研究[J]. 中山大学学报(自然科学版),**42**(1):49~54.

[19] 杜宇野. 1995. 香菇研究进展[J]. 中国食用菌,**14**(4):9~11.

[20] 吴雄宇,李曼玲,胡谷平,等. 2002. 南海红树林内生真菌2508代谢物研究[J]. 中山大学学报(自然科学版),**42**(3):34~36.

[21] 吴雄宇,林永成,冯 爽,等. 2001. 海南红树林内生真菌1356代谢产物的研究[J]. 热带海洋学报,**20**(4):80~86.

[22] 杨智源,陈荣忠,杨 丰,等. 2001. 短小芽孢杆菌葡聚糖内切酶基因的克隆及序列测定[J]. 微生物学报,**40**(1):76~81.

[23] 周卫东,刘如林,邢邦华,等. 1997. 深层发酵香菇水溶性胞外多糖的生物学活性[J]. 菌物系统,**16**(3):220~207.

[24] 周世宁,林永成,吴雄宇,等. 2002. 海洋真菌与细菌发酵物中的环二肽[J]. 微生物学通报,**29**(3):59~62.

[25] 郑天凌,庄铁城,蔡立哲,等. 2001. 微生物在海洋污染环境中的生物修复作用[J]. 厦门大学学报(自然科学版),**40**(2):524~534.

[26] 郑志成,周美英,姚炳新. 1989. 红树林根系放线菌的组成[J]. 厦门大学学报(自然科学版),**28**(3):306~310.

[27] 姜广策,林永成,周世宁,等. 2003. 中国南海红树林内生真菌No. 1403次级代谢物研究[J]. 中山大学学报(自然科学版),**39**(6):119.

[28] 胡谷平,佘志刚,吴耀文,等. 2002. 南海海洋红树林内生真菌胞外多糖的研究[J]. 中山大学学报(自然科学版),**41**(1):121~122.

[29] 胡承钰,王三英. 2001. 细菌胞外多糖复合应用的免疫增强作用[J]. 厦门大学学报(自然科学版),**40**(5):1129~1132.

[30] 郑天凌,庄铁城,蔡立哲,等. 2001. 微生物在海洋污染环境中的生物修复作用[J]. 厦门大学学报(自然科学版),**40**(2):524~534.

[31] 曾会才,郑服丛,贺春萍. 2001. 海南红树林生境中海疫霉种的分离与鉴定[J]. 菌物系统,**20**(3):310~315.

[32] 郑忠辉,缪 莉,黄耀坚,等. 2002. 红树植物内生真菌的抗肿瘤活性[J]. 厦门大学学报(自然科学版),**42**(4):513~516.

[33] 潘道东,陈 杰,韩正康,等. 2002. 胞外多糖 Pullulan 处理养猪场污水效果[J]. 畜牧与兽医,**34**(1):23~24.

[34] 魏培莲. 2002. 微生物胞外多糖研究进展[J]. 浙江科技学院学报,**14**(2):8~12.

[35] Alongi DM. 1988. Bacterial productivity and microbial biomass in tropical mangrove sediments[J]. *Microb. Ecol.*,**15**:59~79.

[36] Alongi DM. 1994. The role of bacteria in nutrient recycling in tropical mangrove and other coastal benthic ecosystems[J]. *Hydrobiologia*,**285**:19~32.

[37] Alongi DM,Boto KG,Tirendi F. 1989. Effect of exported mangrove litter on bacterial productivity and dissolved organic carbon fluxes in adjacent tropical nearshore sediments[J]. *Mar. Ecol. Prog. Ser.*,**56**:133~144.

[38] Alongi DM,Christoffersen P,Tirendi F. 1993. The influence of forest type on microbial-nutrient relationships in tropical mangrove sediments[J]. *J. Exp. Mar. Biol. Ecol.* **171**:201~223.

[39] Bano N,Nisa MU,Khan N,*et al*. 1997. Significance of bacteria in the flux of organic matter in the tidal creeks of the mangrove ecosystem of the Indus river delta, Pakistan[J]. *Mar. Ecol. Prog. Ser.*,**157**:1~12.

[40] Bashan Y,Puente ME,Myrold DD,*et al*. 1998. In vitro transfer of fixed nitrogen from diazotrophic filamentous cyanobacteria to black mangrove seedlings[J]. *FEMS Microbiol. Ecol.*,**26**(3):165~170.

[41] Benkacoker MO,Olumgin A. 1995. Waste drilling-fluid-utilising microorganisms in a tropical mangrove swamp oilfield location[J]. *Bioresour. Technol.*,**53**(3):211~215.

[42] Benkacoker MO,Olumgin A. 1996. Effects of waste drilling fluid on bacterial isolates from a mangrove swamp oilfield in the Niger Delta of Nigeria[J]. *Bioresour. Technol.*,**55**(3):175~179.

[43] Boto KG,Alongi DM,Nott ALJ. 1989. Dissolved organic carbon-bacteria interactions at sediment-water interface in a tropical mangrove system[J]. *Mar. Ecol. Prog. Ser.*,**51**:243~251.

[44] Bremer GB. 1995. Lower marine fungi (*Labyrinthulomycetes*) and the decay of mangrove leaf litter[J]. *Hydrobiologia*,**295**:89~95.

[45] Chandrika V,Nair PVR,Khambhadkar LR. 1990. Distribution of phototrophic thionic bacteria in the anaerobic and microaerophilic strata of mangrove ecosystem of Cochin[J]. *J. Mar. Biol. Assoc. India*,**32**:77~84.

[46] Day JW Jr eds. 1989. Microbial ecology and organic detritus in estuaries[A]. In: Day JW Jr,eds. Estuarine Ecology[C]. New York:John Wiley & Sons,Inc.,257~308.

[47] D'Croz L,Del Rosario J,Holness R. 1989. Degradation of red mangrove (*Rhizophora mangle*) leaves in the bay of Panama[J]. *Rev. Biol. Trop.* **37**:101~104.

[48] Dhevendaran K. 1984. Photosynthetic bacteria in the marine environment at Porto-Novo[J]. *Fish Technol. Soc. Cochin.*,**21**:126~130.

[49] Ellison AM,Farnsworth EJ,Twilley RR. 1996. Facultative mutualism between red mangrove and root-fouling sponges in Belizean mangal[J]. *Ecology*,**77**(8):2431~2444.

[50] Fell JW,Master IM,Wiegert RG. 1984. Litter decomposition and nutrient enrichment[A]. In: The Mangrove Ecosystem: Research Methods(Monograph on oceanographic methodology,8)[C]. Paris: UNESCO,239~251.

[51] Findlay RH,Fell JW,Coleman NK,*et al*. 1986. Biochemical indicators of the role of fungi and thraustochytrids in mangrove detrital systems[A]. In: Moss ST,ed. The biology of marine fungi [C]. Cambridge: Cambridge University Press,91~104.

[52] Gina H,Patricia V,Yoav B. 2001. The role of sediment microorganisms in the productivity, conservation, and rehabilitation of mangrove ecosystems: An overview[J]. *Biol. Fert. Soils*,**33**(4):265~278

[53] Gotto JW,Taylor BF. 1976. N_2 fixation associated with decaying leaves of the red mangrove (*Rhizophora mangle*) [J]. *Appl. Environ. Microbiol.*,**31**:781~783.

[54] Haronian D,Lewis A. 1991. Element of a unique bacteriorhodopsin neural network architecture[J]. *Appl. Optics*,**30**(5):597~608.

[55] Holguin G,Bashan Y,Mendoza-Salgado RA,*et al*. 1999. Microbiology of mangroves,forests in the frontier between land and sea [J]. *Ciencia Desarrollo*,**25**(144):26~35.

[56] Hyde KD,Lee SY. 1995. Ecology of mangrove fungi and their role in nutrient cycling: What gaps occur in our knowledge? [J]. *Hydrobiologia*,**295**(1~3):107~188.

[57] Jetten MSM,Strous M,van de Pas-Schoonen KT,*et al*. 1998. The anaerobic oxidation of ammonium[J]. *FEMS Microbiol. Rev.*,**22**:421~437.

[58] Jones EBG,Hyde KD. 1988. Methods for the study of marine fungi from the mangroves[A]. In: Agate AD,eds. Mangrove Microbiology: Role of Microorganism in Nutrient Cycling of Mangrove Soils and Waters[C]. Paris: UNDP/UNESCO,9~27.

[59] Kohlmeyer J,Bebout B,Volkmann KB. 1995. Decomposition of mangrove wood by marine fungi and teredinids in Belize[J]. *Mar. Ecol.*,**16**:27~39.

[60] Lu CY,Lin P. 1990. Studies on liter fall and decomposition of *Bruguier sexangula*(Lour) poir community on Hainan Island, China[J]. *Bull. Mar. Sci.*,**47**(1):139~148.

[61] Maltseva O,Oriel P. 1997. Monitoring of an alkaline 2,4,6-trichlorophenol-degrading enrichment culture by DNA fingerprinting methods and isolation of the responsible organism, *Haloakaliphilic Nocardioides* sp. strain M6[J]. *Appl. Environ. Microbiol.*,**63**:4145~4149.

[62] Marty DG. 1985. Description of four souches methanogenes thermotolerantes isolees of interial or marine sediments[J]. *C. R Acad. Sci.* Ⅲ,**300**:545~548.

[63] Matondkar SGP,Mahtani S,Mavinkurve S. 1981. Studies on mangrove swamps of Goa. I. Heterotrophic bacterial flora from mangrove swamps[J]. *Mahasagar Bull. Nat. Inst. Oceanogr*,**14**:325~327.

[64] Miyasaka T,Koyama K,Itoh I. 1992. Science quantum conversion and image detection by a bacteriorhodopsin-based artificial photoreceptor[J]. *Science*,**255**(1):342~344.

[65] Mohanraju R,Rajagopal BS,Daniels L,*et al*. 1997. Isolation and characterization of a methanogenic bacterium from mangrove sediments[J]. *J. Mar. Biotechnol.*,**5**:147~152.

[66] Odum WE,Heald EJ. 1975. Mangrove forests and aquatic productivity[A]. In: Hasler AD,ed. Coupling of Land and Water Systems: Ecological Study No. 10[C]. New York: Springer-Verlag,129~136.

[67] Poch GK,Gloer J. 1989. Helicascolides A and B: New lactones from the marine fungus *Helicascus kanaloanus*[J]. *J. Natl. Prod.*,**52**:257~260.

[68] Poch GK,Gloer J. 1991. Auranticins A and B: Two new depsidones from a mangrove isolate of the fungus *Preussia aurantiaca* [J]. *J. Natl. Prod.*,**54**:213~217.

[69] Raghukumar S,Sharma S,Raghukuma C,*et al*. 1994. Thraustochytrid and fungal component of marine detritus. 4. Laboratory studies on decomposition of leaves of the mangrove *Rhizophora apiculata* Blume[J]. *J. Exp. Mar. Biol. Ecol.*,**183**(1):113~131.

[70] Raghukumar S,Sathepatak V,Sharma S,*et al*. 1995. Thraus-

tochytrid and fungal component of marine detritus. 3. Field studies on decomposition of lesves of the mangrove *Rhizopphora apiculata*[J]. *Aquat. Microb. Ecol.*, **9**(2):117~125.

[71] Ramamurthy T, Raju RM, Natarajan R. 1990. Distribution and ecology of methanogenic bacteria in mangrove sediments of Pitchavaram, east coast of India[J]. *Indian J. Mar. Sci.*, **19**: 269~273.

[72] Rivera-Monroy VH, Day WJ, Twilley RR, *et al.* 1995a. Flux of nitrogen and sediment in a fringe mangrove forest in Terminos lagoon, Mexico[J]. *Estuar. Coast. Shelf Sci.*, **40**:139~160.

[73] Rivera-Monroy VH, Twilley RR, Boustany RG, *et al.* 1995b. Direct denitrification in mangrove sediments in Términos Lagoon, Mexico[J]. *Mar. Ecol. Prog. Ser.*, **126**:97~109.

[74] Rivera-Monroy VH, Twilley RR. 1996. The relative role of denitrification and immobilization in the fate of inorganic nitrogen in mangrove sediments (Términos Lagoon, Mexico) [J]. *Limnol. Oceanogr.*, **41**:284~296.

[75] Robertson AI, Daniel PA. 1989. Decomposition and the annual flux of detritus from fallen timber in tropical mangrove forests [J]. *Limnol. Oceanogr.*, **34**:640~646.

[76] Saxena D, Loka-Bharathi PA, Chandramohan D. 1988. Sulfate reducing bacteria from mangrove swamps of Goa, central west coast of India[J]. *Indian J. Mar. Sci.*, **17**:153~157.

[77] Schingmann G, Milne L, Williams DR, *et al.* 1998. Cell wall active antifungal compounds produced by the marine fungus *Hypoxylon oceanicum* LL-15G256. II. Isolation and structure determination[J]. *J. Antibiot.*, **51**(3):303~316.

[78] Sengupta A, Chaudhuri S. 1990. Halotolerant Rhizobium strains from mangrove swamps of the Ganges River Delta[J]. *Indian J. Microbiol.*, **30**:483~484.

[79] Sengupta A, Chaudhuri S. 1991. Ecology of heterotrophic dinitrogen fixation in the rhizosphere of mangrove plant community at the Ganges River Estuary in India[J]. *Oecologia*, **87**:560~564.

[80] Shoreit AAM, El-Kady IA, Sayed WF. 1994. Isolation and identification of purple nonsulfur bacteria of mangal and non-mangal vegetation of Red Sea Coast, Egypt[J]. *Limnologica*, **24**:177~183.

[81] Singh N, Steinke TD. 1992. Colonization of decomposing leaves of *Bruguiera gymnorrhiza* (Rhizopraceae) by fungi, and in vitro cellulolytic activity of the isolates[J]. *South Afric. J. Bot.*, **58**(6):525~529.

[82] Stanley SO, Boto KG, Alongi DM, *et al.* 1987. Composition and bacterial utilization of free amino acids in tropical mangrove sediments[J]. *Mar. Chem*, **22**:13~30.

[83] Tam NFY. 1998. Effects of wastewater discharge on microbial population and enzyme activites in mangrove soils [J]. *Environ. Poll.*, **102**(2~3):233~242.

[84] Tiedje JM. 1988. Ecology of denitrification and dissimilatory nitrate reduction to ammonium[A]. In: Zehnder AJB, ed. Biology of Anaerobic Microorganisms[C]. New York: Wiley, 179~244.

[85] Toledo G, Bashan Y, Soeldner A. 1995. In vitro colonization and increase in nitrogen fixation of seedling roots of black mangrove inoculated by a filamentous cyanobacteria[J]. *Can. Microbiol.*, **41**(11):1012~1020.

[86] van der Valk AG, Attiwill PM. 1984. Acetylene reduction in an *Avicennia marina* community in southern Australia[J]. *Aust. J. Bot.*, **32**:157~164.

[87] Vazquez P, Holguin G, Puente ME, *et al.* 2000. Phosphate-solubilizing microorganisms associated with the rhizosphere of mangroves in a semiarid coastal lagoon[J]. *Biol. Fertil. Soils*, **30**: 460~468.

[88] Vethanayagam RR. 1991. Purple photosynthetic bacteria from a tropical mangrove environment[J]. *Mar. Biol.*, **110**:161~163.

[89] Vethanayagam RR, Krishnamurthy K. 1995. Studies on anoxygenic photosynthetic bacterium *Rhodopseudomonas* sp. from the tropical mangrove environment[J]. *Indian J. Mar. Sci.*, **24**:19~23.

[90] Wu RY. 1993. Studies on the microbial ecology of Tansui Estuary[J]. *Bot. Bull. Acad. Sin.*, **34**(1):13~30.

[91] Yoshihiro A, Naoya M, Kazuyoshi Y, *et al.* 2001. Initial fermentation of sea sludge using aerobic and thermophilic microorganisms in a mangrove soil[J]. *Biores. Technol.*, **80**:83~85.

[92] Zhuang T, Cheng LP. 1998. Soil microbial function of *Kandelia candel* mangrove: degradation of diesel oil[A]. In: Morton B, ed. The Marine Biology of the South China Sea[C]. Hongkong: Hongkong University Press, 389~395.

[93] Zuberer DA, Silver WS. 1978. Biological dinitrogen fixation (Acetylene reduction) associated with Florida mangroves[J]. *Appl. Environ. Microbiol.*, **35**:567~575.

[94] Zuberer DA, Silver WS. 1979. N_2-fixation (acetylene reduction) and the microbial colonization of mangrove roots[J]. *New Phytol.*, **82**:467~471.

九龙江口红树林土壤微生物的时空分布*

张瑜斌 林 鹏 庄铁诚

(厦门大学生命科学学院,福建 厦门 361005)

摘要: 研究了福建九龙江口秋茄(*Kandelia candel*)林与白骨壤(*Avicennia marina*)林土壤及其相应的对照光滩 0~20 cm、20~40 cm 和 40~60 cm 3 个层次土壤好氧异养细菌、放线菌和丝状真菌 3 类微生物数量时空分布及其与土壤主要理化因子的关系.结果表明:秋茄林与白骨壤林土壤及其相应的对照光滩土壤的细菌、放线菌和丝状真菌和微生物总数在 3 个土壤层次中,随着土壤深度的增加而下降,差异显著;两个红树林及其对照光滩土壤的微生物数量均以夏季最多,冬季最少,春秋季位于夏冬两季之间,或高或低,但差值不大;红树林土壤各类群微生物数量多显著高于相应对照光滩;秋茄林土壤各类群微生物数量显著高于白骨壤林土壤.土壤微生物数量与主要理化因子间的多元回归分析显示:与两个红树林土壤微生物数量关系最密切的因子是全氮和全磷而不是有机质,这可能与微生物数量及其活性对氮和磷有着相对重要的调控作用有关.

关键词: 红树林;土壤;微生物数量;时空分布;九龙江口

中图分类号:Q 939;S 154.3 **文献标识码**:A **文章编号**:0438-0479(2007)04-0587-06

红树林是自然分布于热带、亚热带海岸的潮间带的木本植物群落,通常生长在港湾河口地区的淤泥质滩涂上,是海滩上特有的森林类型[1].许多学者研究了红树林土壤因子与红树植物的关系[2-3],土壤因子与动物的关系[4-5].微生物是生态系统的主要分解者,只有通过微生物的分解作用,才能最终推动生态系统的物质循环和能量流动.迄今为止,有关红树林沉积物或土壤微生物的时空变化研究多集中于热带地区[6-8],在亚热带地区的研究较少[9].亚热带地区,全年的温度和降雨变化明显,受其影响,植物生长和凋落物生产也有明显变化[1].受环境因子影响,土壤微生物也可能存在时空变化.福建九龙江口是我国红树林的主要典型分布区之一[1],尽管先前该区域在微生物学方面已有一些研究[10-12],但土壤微生物数量的时空动态研究尚未见报道,因此本文旨在研究该区域两个红树林及其对照光滩 0~60 cm 土壤层细菌、放线菌和丝状真菌的垂直分布和季节变化特征,为红树林的研究与管理提供依据.

1 材料与方法

1.1 样地概况

本研究样地的秋茄(*Kandelia candel*)林和白骨壤(*Avicennia marina*)林位于南亚热带的福建省九龙江口.秋茄林位于九龙江口南岸的龙海市浮宫镇草埔头村,24°29′N,117°23′E (图 1),秋茄林沿九龙江口南岸呈带状分布,宽度约 40 m,主要位于中、高潮带,林相整齐,郁闭度在 0.9 以上,树高 5.5~6 m.白骨壤林位于九龙江口厦门西海域海沧镇的东屿村,24°31′N,18°03′E (图 1),该群落为纯林,林冠整齐,群落结构简单,郁闭度 0.95 以上,高度 1.2 m,植株达 15 株/m^2,林中有大量繁生幼苗和指状呼吸根,整个白骨壤林滩面平整.

1.2 采样方法

土壤采样设在秋茄林内(标为站位 SA)和白骨壤林内(标为站位 SB)的中潮带,在距离秋茄林和白骨壤林 100 m 以远的同潮带的光滩面上各设置一个对照样地(秋茄林对照光滩样地标为站位 SCa,白骨壤林对照光滩样地标为站位 SCb),以作比较.土壤取样用内径 5 cm 的硬质聚氯乙稀管,在选定的样地内随机多点钻取 60 cm 深的土样,每一管钻取的土样分为上层(0~20 cm),中层(20~40 cm)和下层(40~60 cm)3 层,同一层次随机多点钻取的土样拣去根系并混匀后装入无菌袋,立刻带回实验室分析.所有取样用具均作无菌处理.土壤采样分春、夏、秋、冬 4 个季节进行,采样时秋茄林及其对照光滩春、夏、秋、冬表土温度依次为 20.5、31.5、25.0、13.0 ℃,白骨壤林及其对照光滩春、夏、秋、冬表土温度依次为20.0、30.0、25.5、13.5 ℃.

* 国家自然科学基金(30270272)资助

原载于厦门大学学报(自然科学版),2007,46(4):587-592

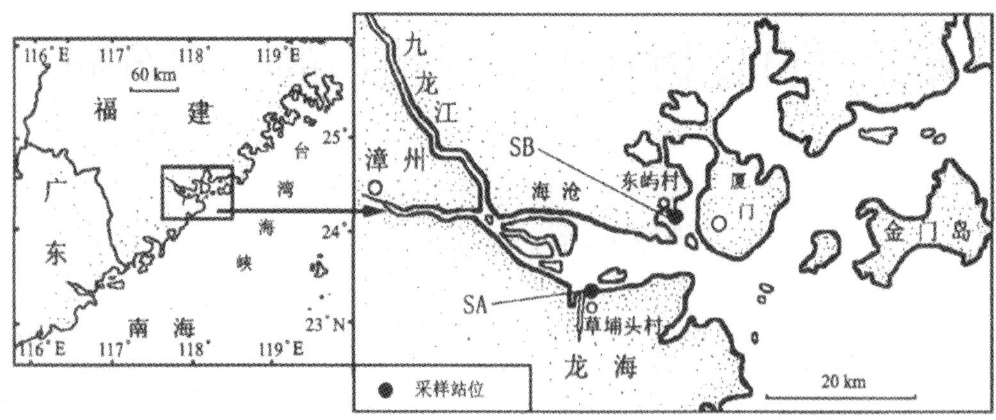

图 1 采样站位分布图

Fig. 1 A map illustrating geographical positions of the sampling stations

样地土壤的主要理化特征见表 1.

1.3 微生物数量分析方法

微生物数量测定采用稀释平板法[13]. 培养基:细菌为 2216E 好气异养菌培养基[13];放线菌为改良的高氏 1 号培养基[14],加入终浓度为 50 μg/g 的重铬酸钾作抑制剂;丝状真菌为马丁氏培养基[14],每 1 L 培养基添加 2 mL 医用氯霉素(0.25 g/2 mL)作细菌抑制剂. 根据作者的前期研究结果[15],上述培养基均使用淡水配制并添加 NaCl,保持 3 种培养基的 NaCl 终浓度依次为 15 ‰、5 ‰、8 ‰. 样品的稀释水细菌为陈海水,放线菌与丝状真菌为淡水. 陈海水取自厦门大学海洋系滨海水井(盐度 29.58). 样品梯度稀释后选取合适的梯度接种平板,每个梯度重复 3 板,接种后的平板于 28 ±1 ℃培养,定时观察、计数.

1.4 数据统计分析

本文中的差异显著性检验与多元回归等数据统计分析过程均使用 SPSS 10.0 统计软件完成.

2 结 果

2.1 红树林与其相应对照光滩土壤微生物数量的垂直分布与季节变化

秋茄林及其对照光滩土壤微生物数量的垂直分布中,无论是细菌、放线菌、真菌,春、夏、秋、冬每个季节及 4 个季节的平均值均是上层 > 中层 > 下层,表现为随土壤深度的增加而下降(表 2),且差异显著($p < 0.05$);秋茄林土壤微生物数量的季节变化中,无论是

表 1 土壤的主要理化特征*

Tab. 1 Main physi-chemical features of soil at sampling stations

站位	层次 (cm)	土壤质地	体积质量 (g/cm³)	有机质 (%)	全氮 (%)	全磷 (%)	全钾 (%)	全钠 (%)	全钙 (%)	全镁 (%)	盐度 (‰)	pH
SA	0~20	轻粘土	1.08	3.206	0.143	0.066	0.725	0.594	0.154	0.403	15.14	6.55
	20~40	中粘土	1.03	3.279	0.128	0.060	0.763	0.569	0.155	0.395	14.24	6.74
	40~60	中粘土	1.05	3.016	0.107	0.054	0.835	0.629	0.172	0.413	15.83	6.74
SCa	0~20	轻粘土	0.98	3.054	0.134	0.064	0.634	0.471	0.205	0.411	11.56	7.08
	20~40	中粘土	1.02	2.968	0.119	0.060	0.771	0.444	0.203	0.400	10.07	7.22
	40~60	中粘土	1.03	2.861	0.100	0.058	0.781	0.435	0.203	0.417	9.71	7.44
SB	0~20	轻粘土	0.93	3.600	0.148	0.057	0.851	0.828	0.337	0.596	24.30	7.00
	20~40	轻粘土	0.99	3.446	0.125	0.051	0.976	0.794	0.346	0.617	22.44	7.05
	40~60	中粘土	1.03	2.895	0.083	0.046	1.038	0.766	0.363	0.604	21.81	7.12
SCb	0~20	中粘土	1.01	2.139	0.117	0.055	1.121	0.791	0.393	0.624	21.25	7.18
	20~40	中粘土	1.03	1.896	0.076	0.054	1.022	0.731	0.393	0.609	19.16	7.21
	40~60	中粘土	1.04	1.784	0.054	0.053	0.952	0.700	0.337	0.606	18.33	7.29

*表中数据为 4 个季节的平均值.

细菌、放线菌还是真菌,在上、中、下3层均表现为夏季最多,冬季最少(下层冬季的真菌除外),春秋两季的数量位于夏冬两季之间.秋茄林对照光滩各类群土壤微生物数量的季节变化模式与秋茄林一致(表2).

白骨壤林土壤微生物数量的垂直分布模式也与秋茄林一致,随土壤层的加深而垂直下降(表3),垂直差异极显著($p<0.01$).白骨壤林对照光滩土壤微生物数量垂直分布也是如此,但真菌数在中下层的绝大多数季节没有检测到,这两层之间的差异极小.白骨壤林土壤微生物数量的季节变化也表现出细菌和放线菌在上中下3层均为夏季最多,冬季最少,春秋季位于夏冬季之间;真菌数量在上中层也是以夏季最多,冬季最少,但中层在春夏秋3季差异较小,下层每个季节都没有检测到真菌.白骨壤林对照光滩土壤细菌和放线菌的季节变化模式与白骨壤林一致,但真菌仅在上层表现出夏季最多,冬季最少,春秋季位于夏冬季之间,中层只在夏季分离到真菌,其他3个季节均没有检测到真菌,下层春夏秋冬4个季节也没有检测到真菌(表3).

2.2 红树林与相应对照光滩之间以及两个红树林之间土壤微生物数量的水平差异

秋茄林的细菌和放线菌多于其对照光滩,差异极显著($p<0.01$),而秋茄林土壤真菌数也多于其对照光滩,两者之间差异几近显著水平($p=0.0524$).白骨壤林土壤微生物数量多于对照光滩,差异显著($p<0.05$).

两个红树林土壤微生物数量之间也存在较大的水平差异,秋茄林的土壤细菌、放线菌、和真菌均多于白骨壤林,差异极显著($p<0.01$).

2.3 红树林及其对照光滩的土壤微生物数量与主要理化因子之间的关系

多元回归分析(表4)表明在所测定的土壤理化因子中,与秋茄林土壤微生物数量关系最为密切的因子是全氮和全磷;与秋茄林对照光滩土壤微生物数量关系最为密切的因子是全氮、全磷、C/N值和N/P值;与白骨壤林土壤微生物数量关系最为密切的因子是全磷和全氮;与白骨壤林对照光滩土壤微生物数量关系最为密切的因子是全镁、全氮、N/P值和全钠.总体看来,在所测定的土壤化学因子中,与两个红树林土壤微生物数量关系最为密切的是全氮和全磷.

3 讨 论

有关红树林土壤微生物数量的垂直分布研究较为少见.在本项研究中,4个站位土壤细菌、放线菌和丝状真菌三大类好氧异养微生物数量均有随土壤深度的

表2 秋茄林与对照光滩的土壤微生物数量*

Tab.2 Microbial densities of soil in *Kandelia candel* forest and controlled mudflat

站位与土层	微生物类群	春	夏	秋	冬	平均值 ±标准差
SA(0~20 cm)	细菌	26.25	33.12	22.81	13.61	23.95 ±8.12
	放线菌	44.35	47.87	46.88	40.82	44.98 ±3.14
	丝状真菌	47.06	52.78	38.92	32.13	42.72 ±9.07
SA(20~40 cm)	细菌	18.90	28.54	12.31	6.86	16.65 ±9.33
	放线菌	30.42	36.71	32.49	29.39	32.25 ±3.24
	丝状真菌	7.20	9.57	8.55	6.16	7.87 ±1.50
SA(40~60 cm)	细菌	5.98	9.80	4.76	4.67	6.30 ±2.41
	放线菌	9.50	10.23	8.84	8.82	9.35 ±0.67
	丝状真菌	0.00	4.30	4.25	0.87	2.36 ±2.25
SCa(0~20 cm)	细菌	6.63	9.26	7.41	2.67	6.49 ±2.78
	放线菌	42.90	44.20	41.73	28.27	39.28 ±7.41
	丝状真菌	36.08	41.50	33.15	24.83	33.89 ±6.96
SCa(20~40 cm)	细菌	3.01	5.25	3.50	1.77	3.38 ±1.44
	放线菌	27.08	35.84	29.44	21.24	28.40 ±6.04
	丝状真菌	6.20	8.15	7.36	5.31	6.76 ±1.25
SCa(40~60 cm)	细菌	1.76	2.05	1.89	1.59	1.82 ±0.20
	放线菌	8.10	8.89	8.26	7.79	8.26 ±0.46
	丝状真菌	3.52	3.76	3.44	2.66	3.35 ±0.48

*细菌 $\times10^4$ cfu/g(dm);放线菌 $\times10^3$ cfu/g(dm);真菌 $\times10^1$ cfu/g(dm).

表 3 白骨壤林与对照光滩的土壤微生物数量*

Tab. 3 Microbial densities of soil in *Avicennia marina* forest and controlled mudflat

站位与土层	微生物类群	春	夏	秋	冬	平均值 ±标准差
SB(0～20 cm)	细菌	20.91	30.84	17.14	13.07	20.49 ±7.61
	放线菌	30.43	38.41	28.17	22.04	29.76 ±6.77
	丝状真菌	7.25	1.92	6.9	5.85	6.98 ±0.86
SB(20～40 cm)	细菌	4.44	6.76	3.60	2.76	4.39 ±1.72
	放线菌	9.25	9.97	8.10	7.91	8.81 ±0.98
	丝状真菌	0.93	0.95	0.90	0.00	0.70 ±0.46
SB(40～60 cm)	细菌	2.72	2.99	1.91	1.77	2.35 ±0.60
	放线菌	2.17	2.29	1.91	1.59	1.99 ±0.31
	丝状真菌	0.00	0.00	0.00	0.00	0.00 ±0.00
SCb(0～20 cm)	细菌	7.94	12.16	8.56	3.89	8.14 ±3.39
	放线菌	21.55	27.86	18.41	15.36	20.80 ±5.35
	丝状真菌	5.67	6.55	5.58	2.78	5.15 ±1.64
SCb(20～40 cm)	细菌	2.70	3.81	2.05	1.75	2.58 ±0.91
	放线菌	3.06	5.54	3.76	1.40	3.44 ±1.71
	丝状真菌	0.00	0.87	0.00	0.00	0.22 ±0.44
SCb(40～60 cm)	细菌	1.59	1.90	1.27	1.13	1.47 ±0.34
	放线菌	0.35	0.87	0.51	0.32	0.51 ±0.25
	丝状真菌	0.00	0.00	0.00	0.00	0.00 ±0.00

*细菌 $\times 10^4$ cfu/g(dm);放线菌 $\times 10^3$ cfu/g(dm);真菌 ×10 cfu/g(dm).

表 4 土壤微生物数量与土壤主要理化因子之间的多元回归方程

Tab. 4 Mutiple regession equations between the microbial densities and main physi-chemical factors of soil

站位	微生物类群	回 归 方 程[a]	R	F	P	影响因子重要性排序[b]
SA	细 菌	$Y = -56.803 + 164.067 X_1 + 863.95 X_2$	0.935	31.448	0.000	氮$^+$ > 磷$^+$
	放线菌	$Y = -106.718 + 2262.77 X_2$	0.868	30.523	0.000	磷$^+$
	真菌	$Y = -137.17 + 2583.905 X_2$	0.800	17.749	0.002	磷$^+$
SCa	细菌	$Y = -4.195 + 68.931 X_1$	0.797	17.403	0.002	氮$^+$
	放线菌	$Y = -145.696 + 269.859 X_1 + 2309.2 X_2$	0.851	11.812	0.003	氮$^+$ > 磷$^+$
	真菌	$Y = -141.869 + 3115.011 X_2 - 2.037 X_3$	0.837	10.570	0.004	磷$^+$ > C/N$^-$
SB	细菌	$Y = -70.399 + 1560.879 X_2$	0.853	26.764	0.000	磷$^+$
	放线菌	$Y = -89.606 + 1745.248 X_2 + 120.366 X_1$	0.942	35.711	0.000	磷$^+$ > 氮$^+$
	真菌	$Y = -25.538 + 551.802 X_2$	0.855	27.190	0.000	磷$^+$
SCb	细菌	$Y = -137.785 + 227.126 X_6 + 34.008 X_1$	0.963	57.602	0.000	镁$^+$ > 氮$^+$
	放线菌	$Y = -78.645 + 9.729 X_4 + 97.447 X_5$	0.929	8.242	0.000	N/P$^+$ > 钠$^+$
	真菌	$Y = -87.228 + 114.483 X_6 + 29.614 X_1$	0.920	24.914	0.000	镁$^+$ > 氮$^+$

a. 回归方程中, X_1 为全氮含量, X_2 为全磷含量, X_3 为碳氮比, X_4 为氮磷比, X_5 为全钠含量, X_6 为全镁含量; b. 影响因子重要性排序中的"+"和"-"分别表示正影响和负影响.

增加而减少的趋势. 印度 Sunderban 红树林 0～63.5 cm 4 个层次的土壤真菌数量最多在表层,数量和发生频率随土壤深度增加而减少[6],本文的研究结果与其一致. 但在澳大利亚 Hinchinbrook 红树林 0～10 cm 的垂直剖面中,细菌密度变化反复无常,垂直分布模糊不清[8],与本文的结果不同,这可能是他们研究的土壤层太浅的缘故,因为 0～10 cm 还是属于活跃的表土层. 在温带沉积物中,细菌数量也随深度增加而呈下降趋势[8]. 在湿地中,水分和氧气含量是两个互为消长的因素,影响着微生物的数量,湿地土壤中的氧气含量随着深度增加下降,好氧异养微生物数量相应随之下降,氧气也就成为一个十分重要的因素[16].红树林区土壤好氧异养微生物数量的垂直分布极大程度上也受土壤 O_2 含量和氧化还原状况的影响,杨萍如等的研究显示海南琼山东寨港红树林土壤氧化还原势随土壤深度的增加而下降[17],虽然本项研究没有测定土壤 O_2 含量和氧化还原势,但先前的研究已证实本项研究的样地土壤还原性物质随土壤层的加深而增加[18],说明 O_2 含量随土壤层的加深而下降,还原性增强.因此,我们初步推测 O_2 含量和氧化还原势是影响土壤好氧微生物数量垂直分布的主导因子.

影响微生物数量季节变化的主导因素在热带与亚热带地区具有明显差异.在热带地区,季风降雨是红树林微生物季节变化的主导因子[19-21],因为季风降雨冲刷会影响沉积物有机碳与氮含量以及盐度变化.在亚热带地区,全年的季节和温度变化明显,温度对微生物活动的影响明显.台湾淡水河口湾红树林微生物总数随着季节变化,夏季数量最高,秋季次之,冬季最低[9],本项研究结果也表明两个红树林群落和相应光滩微生物数量夏季最多,冬季最少,春秋位于其间而或多或少,这与采样时的土温(见 1.材料与方法)夏季最高、冬季最低、春秋位于之间相符.秋季的土温高于春季,但微生物数量春秋季却是或高或底,这可能与春季温度上升促发红树根系和底栖动物的活动影响微生物数量有关.由此可见,在具有明显季节变化的亚热带区域,温度是影响红树林土壤微生物数量季节变化的主导因素.

本项研究的结果表明两个红树林群落土壤微生物数量均比相应的对照光滩要多,先前的研究也有一致的报道[22],这是因为红树林可以为其下的土壤提供荫蔽条件,减少风浪潮汐对土壤的冲击;同时,红树植物持续不断地将凋落物及碎屑加入红树林生态系统[23],使得红树林土壤有机质含量高于光滩土壤,土壤中红树植物根系的排泄物或分泌物也会促进土壤微生物的活动[3],根系将植物吸收的 O_2 释放入土壤中也会促进好氧异养微生物的活动[24].因此红树的存在,为林下的土壤微生物提供了良好的栖息环境.

秋茄林与白骨壤林之间土壤微生物数量的差异极大程度上是潮位差异决定的.本项研究中的秋茄林处于九龙江河口内部,而白骨壤林处于厦门西海域港湾内,所以白骨壤林每天浸潮时间长,土壤的含水量、还原性和盐分(表 1)均高于秋茄林[18],受其影响,白骨壤林土壤好氧异养微生物数量低于秋茄林.不少的研究表明,土壤细菌数量[19,25],真菌数量[6]或微生物数量[9]与有机质呈正相关,在有机质丰富的地方,微生物数量多.但在本项研究中,SB 站位的白骨壤林土壤有机质多于秋茄林,但微生物数量却刚好相反,说明在此有机质不是主要影响因子,盐度、淹水条件(O_2 含量)和氧化还原势应是主要的影响因子.Lee 等也认为红树林土壤营养水平、pH、盐度、淹水条件(O_2 含量与氧化还原势)可能是影响红树林沼泽土壤小型真菌特征和分布模式最重要的环境因素[26].由于白骨壤林土壤好氧异养微生物数量低于秋茄林,使得其对有机物的好氧分解要弱,相对可积累更多的有机质,因此白骨壤林土壤有机质也就高于秋茄林.

红树林湿地土壤化学因子毫无疑问会影响生存其中的微生物的丰度、活性与繁殖各个方面,反过来,微生物的分解活性及转化等作用也会对土壤化学性质起到调控作用.统计分析表明,在所测定的土壤化学因子中,与两个红树林土壤微生物数量关系最为密切的是全氮和全磷而不是有机质,说明有机质不会阻碍微生物的分解活性,而在此期间微生物丰度及其对有机质的分解、转化活性相对土壤氮和磷有着重要的调控作用.很少有研究报道镁可能与土壤细菌数量关联起来,本项研究的多元回归分析却显示与白骨壤林对照光滩土壤微生物数量关系最为密切的因子是全镁,其中的原因有待于进一步探讨.但迄今为止,有关红树林土壤微生物数量与主要理化因子关系的研究报道极少,有待于今后更深入和系统的研究.

参考文献:

[1] Lin P. Mangrove ecosystems in China[M]. Beijing and New York:Science Press,1999:1 - 7,142 - 167.

[2] Mckee K L. Soil physicochemical patterns and mangrove species distribution- reciprocal effects[J].Journal of Ecology,1993,81:477 - 487.

[3] Sherman R E,Fahey T J,Howarth R W. Soil-plant interactions in a neotropical mangrove forestiron,phosphorus and sulfur dynamics[J]. Oecologia,1998,115:553 - 563.

[4] Kristensen E M H,Bussarawit N. Benthic metabolism and sulfate reduction in south-east Asian mangrove swamp [J]. Marine Ecology Progress Series,1991,73:93 - 103.

[5] Smith T J,Boto K G,Frusher S D,et al. Keystone species and mangrove forest dynamics:the influence of burrowing by crabs on soil nutrient status and forest productivity [J]. Estuarine,Coastal and Shelf Science,1991,33:419 - 432.

[6] Garg K L. Vertical distribution of fungi in Sunderban mangrove mud[J]. Indian Journal of Marine Sciences, 1983,12:48 - 51.

[7] Mohanraju R,Natarajan R. Methanogenic bacteria in mangrove sediments[J]. Hydrobiologia,1992,247:187 - 193.

[8] Alongi D M ,Sasekumar A. Benthic communities[M]//Robertson A I,Alongi D M ,eds. Tropical mangrove ecosystems. Washington D C: American Geophysical Union , 1992:137 - 172.

[9] Wu R Y. Studies on the microbial ecology of Tansui Estuary[J]. Botanical Bulletin of Academin Sinica ,1993 ,34 (1) :13 - 30.

[10] 庄铁诚 ,林鹏. 红树林凋落物叶自然分解过程中土壤微生物的数量动态[J]. 厦门大学学报:自然科学版 ,1993 , 32(3) :365 - 370.

[11] 庄铁诚 ,林鹏. 红树林下土壤微生物对柴油的降解[J]. 厦门大学学报:自然科学版 ,1995 ,34(3) :442 - 446.

[12] 庄铁诚 ,张瑜斌 ,林鹏. 红树林土壤微生物对甲胺磷的降解[J]. 应用与环境生物学报 ,2000 ,6(3) :276 - 280

[13] 陈绍铭 ,郑福寿. 水生微生物实验法(上册) [M]. 北京:海洋出版社 ,1985 :34 - 35.

[14] 中国科学院南京土壤研究所微生物室. 土壤微生物研究法[M]. 北京:科学出版社 ,1985.

[15] 张瑜斌 ,林 鹏 ,邓爱英 ,等. 稀释平板技术应用于红树林区土壤微生物数量研究的方法学改进[C]//张洪勋 ,庄绪亮 ,主编. 微生物生态学研究进展(第五届微生物生态学术研讨会论文集) . 北京:气象出版社 ,2003 :85 - 93.

[16] Ulehlová B. The role of decomposers in wetlands[M]// Westlake D F , Květ J , Szczepański T L A ,eds. The production ecology of wetlands: the IBP synthesis. Cambridge:Cambridge University Press ,1998: 192 - 210.

[17] 杨萍如 ,何金海 ,刘腾辉. 红树林及其土壤[J]. 自然资源学报 ,1987 ,2(1) :32 - 37.

[18] 张银龙. 九龙江口红树林酶活性等性质及其细根的生态学研究[D]. 厦门:厦门大学 ,1996 : 27 - 40.

[19] Alongi D M. Bacterial productivity and microbial biomass in tropical mangrove sediments[J]. Microbial Ecology , 1988 ,15 :59 - 79.

[20] Matondkar S G P , Mahtani S , Mavinkurve S. Seasonal variations in the microflora from mangrove swamps of Goa[J]. Indian Journal of Marine Sciences ,1980 ,9 :119 - 120.

[21] Alongi D M. Microbial meiofunal interrelationships in some tropical mangrove sediments[J]. Journal of Marine Research ,1988 ,46 :349 - 365.

[22] 张瑜斌 ,庄铁诚 ,杨志伟 ,等. 海南东寨港红树林土壤微生物初探[J]. 生态学杂志 ,2001 ,20(1) :63 - 64.

[23] Jagtap T G. Seasonal distribution of organic matter in mangrove environment of Goa[J]. Indian Journal of Marine Sciences ,1987 ,16 :103 - 106.

[24] Boto K G,Wellington J T. Soil Characteristics and nutrient status in a Northern Australian mangrove forest[J]. Estuaries ,1984 ,7(1) :61 - 69.

[25] Alongi D M ,Christoffersen P , Tirendi F. The influence of forest type on microbial-nutrient relationships in tropical mangrove sediments[J]. Journal of Experimental Marine Biology and Ecology ,1993 ,171 :203 - 223.

[26] Lee B K H,Baker G E. An ecological study of the soil microfungi in a Hawaiian mangrove swamp[J]. Pacific Science ,1972 ,26 :1 - 10.

Temporal and Spacial Distribution of Microbial Densities of Soil in Mangrove Forest in Jiulongjiang Estuary

ZHANG Yu-bin , LIN Peng , ZHUANG Tie-cheng

(School of Life Sciences ,Xiamen University ,Xiamen 361005 ,China)

Abstract : Temporal and seasonal distribution of bacterial ,actinomycetes and fungal densities and the relation between microbial densities and physi-chemical factors in three soil layers (0～20 cm ,20～40 cm and 40～60 cm) were investigated in *Kandelia candel* forest , *Avicennia marina* forerst and their corresponding mudflats in Jiulongjiang Estuary of Fujian. The results indicated that the bacteria ,actinomycetes and fungi densities of soil declined with increase of soil depth in three soil layers in two mangrove forests and their corresponding mudflats ,and variation in microbial densities was significant among three layers. The most abundant microorganisms were presented in summer and the least in winter ;difference in densities between spring and autumn was little. Microbial densities of soil in two mangrove forests were significantly greater than that in their corresponding mudflats. Microbial densities were significantly greater in *K. candel* forest than in *A. marina* forest. The multiple regression analysis indicated that the relation between total nitrogen ,total phosphorus and microbial densities were the most outstanding ,which probably meant that microbial abundance and activity played an important role in controlling nitrogen and phosphorus.

Key words : mangrove ;soil ;microbial density ;temporal and spacial distribution ;Jiulongjiang Estuary

九龙江口红树林鹧鸪菜藻体自生固氮细菌*

张瑜斌[1,2] 林 鹏[2] 邓爱英[1] 庄铁诚[2]

(1. 广东海洋大学海洋资源与环境监测中心，广东 湛江 524088；

2. 厦门大学生命科学学院，福建 厦门 361005)

摘 要 初步研究了福建九龙江口秋茄(*Kandelia candel*)红树林红藻鹧鸪菜(*Caloglossa leprieurii*)藻体异养自生固氮菌数量的季节变化和微生物区系。结果表明:鹧鸪菜上异养自生固氮细菌数量以春季最多(1.033×10^4 cfu·g^{-1}),冬季最少(0.567×10^4 cfu·g^{-1}),固氮菌的季节变化模式表现为春季>秋季>夏季>冬季;鹧鸪菜藻体氮含量也以春季最高(22.08 g·kg^{-1}),冬季最低(16.63 g·kg^{-1}),二者差异显著($P<0.05$),且藻体含氮量的季节变化模式与固氮细菌数量一致;这与鹧鸪菜的生长和物质积累密切相关,鹧鸪菜与其藻体上的自生固氮菌可能存在着互惠互利的关系,这种关系同时也受到环境温度和水分等因子的综合影响;对9株固氮菌的初步鉴定结果显示,它们分属于固氮菌属(*Azotobacter*)与拜叶林克氏菌属(*Beijerinckia*)。

关键词 固氮菌;氮含量;季节变化;鹧鸪菜;红树林

中图分类号 X172; Q934.3 **文献标识码** A **文章编号** 1000-4890(2007)09-1384-05

Abiogenous azotobacter on the body of *Caloglossa leprieurii* growing in *Kandelia candel* mangrove forest in Jiulongjiang estuary of Fujian Province. ZHANG Yu-bin[1,2], LIN Peng[2], DENG Ai-ying[1], ZHUANG Tie-cheng[2] ([1]*Monitoring Center for Marine Resources and Environments, Guangdong Ocean University, Zhanjiang* 524088, *Guangdong, China*; [2]*School of Life Sciences, Xiamen University, Xiamen* 361005, *Fujian, China*). *Chinese Journal of Ecology*, 2007, **26**(9): 1384-1388.

Abstract: The study on the seasonal changes of heterotrophic abiogenous azotobacter on the body of *Caloglossa leprieurii* growing in the *Kandelia candel* mangrove forest in Jiulongjiang estuary indicated that the quantity of this kind of azotobacter was the most in spring (1.033×10^4 cfu·g^{-1}) and the least in winter (0.567×10^4 cfu·g^{-1}), with a seasonal pattern of spring > autumn > summer > winter. The nitrogen concentration of *C. leprieurii* also had the same seasonal pattern, being the highest in spring (22.08 g·kg^{-1}) and the lowest in winter (16.63 g·kg^{-1}), which was closely related to the growth and matter accumulation of the seaweed. There was probably a kind of special cooperative relation between the azotobacter and the seaweed, which was influenced simultaneously by the environmental factors such as temperature, water, and others. The preliminary identification of isolated nine strains azotobacter showed that they were belonged to *Azotobacter* or *Beijerinckia*.

Key words: azotobacter; nitrogen concentration; seasonal change; *Caloglossa leprieurii*; mangrove.

1 引 言

生物固氮是由固氮微生物完成的氮的生物地球化学循环中的重要一环。许多滨海海洋环境存在着生物固氮现象(Herbert, 1999)。Zuberer等(1975)首次报道了红树林区的固氮研究,随后陆续有不同红树林群落的氮固定报道(Potts, 1984; van der Valk & Attiwill, 1984; Toledo *et al*, 1995)。已发现在红树叶和根的凋落物、活根系以及周围的沉积物中有各种固氮生物存在(van der Valk & Attiwill, 1984)。氮的固定在红树林氮转化过程中研究得最详细(Alongi *et al*, 1992),近年来,红树林区有关固氮微生物的研究主要集中在固氮微生物与红树植物间的关系(Toledo *et al*, 1995; Bashan *et al*, 1998;

* 国家自然科学基金资助项目(30270272)
原载于生态学杂志,2007,20(9):1384—1388

Lugomela & Bergman, 2002; Ravikumar *et al*, 2004);固氮微生物与其他微生物的作用(Holguin & Bashan, 1996);以及固氮酶活性的变化(Sengupta, 1991; Woitchik *et al*, 1997),但这些研究多集中于能固氮的蓝细菌,对异养性的固氮细菌的研究仅见Ravikumar等(2004)的报道。在红树林区,大型藻类前人已有研究(Tanaka, 1987;林鹏等, 1997),但对大型藻类藻体上的微生物研究仅见于庄铁诚等(2000)的报道,对大型藻类上的自生固氮菌研究未见报道。在以往的研究中,庄铁诚等(2000)发现不同季节和红藻生长情况对藻体上异养微生物数量有明显的影响,红藻生长繁殖愈好,微生物数量愈多。本文研究了九龙江口秋茄林的红藻鹧鸪菜藻体上的异养自生固氮菌数量的季节变化,对藻体上的异养自生固氮菌作了初步的鉴定,并探讨了自生固氮菌与鹧鸪菜之间的关系,以期为进一步研究红树林生态系统内的藻菌关系以及藻-菌-红树植物之间的相互关系奠定基础。

2 研究地区与研究方法

2.1 样地概况

藻样采于福建省九龙江口南岸的龙海市浮宫镇草埔头村的秋茄(*Kandelia candel*)林,位于24°29′N, 117°23′E。样地的气候属南亚热带海岸气候,年平均气温为21.1 ℃,年降水量为1 475.2 mm,最冷月1月平均气温为13.0 ℃,气温平均年较差15.8 ℃,相对湿度79.5%,年日照时数为2 040.5 h。秋茄林沿九龙江口南岸呈带状分布,宽度约40 m,主要位于中、高潮带。该群落为1962年人工营造的秋茄纯林,林缘有少量桐花树(*Aegiceras corniculatum*)和白骨壤(*Avicennia marina*)伴生,林相整齐,郁闭度在0.9以上,树高5.5~6.0 m,在大洪潮时,外缘树干基本被淹没,仅露出林冠,植株密度大,且枝下高较高,林内具不发达的板状根,外缘有小型支柱根或板状根。

2.2 藻样的采集

藻样以红藻中的鹧鸪菜(*Caloglossa leprieurii*)为材料。鹧鸪菜属于红藻门(Rhodophyta),真红藻纲(Frorideophyceae),仙藻目(Ceraminales),红叶藻科(Declessericaceae),其藻体小型,呈紫红色或暗紫色,扁平窄长椭圆形,叉状分枝,在分枝处常缢缩,单轴型。鹧鸪菜是一种泛亚热带性藻类,在中国从浙江至广东沿海都有分布,尤喜生长在河口附近的高、中潮带(李伟新等, 1982)。在福建红树林区也是一类普遍存在的优势红藻,在本项研究的秋茄林内极为常见(林鹏等, 1997)。采集藻样时,在靠外滩的红树林边缘选择生长茂盛的红藻区为采样地段,用镊子或小刀采集近泥面处红树基部的附生鹧鸪菜(注意弃除树皮等杂物,尽可能不沾泥),装入无菌袋内混合均匀。所有采样用具都经无菌处理。采集的藻样立即带回实验室作微生物学分析,另取一部分藻样,80 ℃烘干研细,按实验要求过筛,贮存供分析藻样氮含量。另外,再取少量样品于105 ℃烘干至恒重,求含水量与干质量。藻样采集时间依次为秋季2000年11月6日,冬季2001年1月3日,春季2001年4月11日,夏季2001年7月9日。

2.3 藻体自生固氮菌数量测定

自生固氮菌数量测定采用稀释平板法。称取混合均匀的样品5 g,盛于无菌研钵中,加入少量无菌水(从已定容的三角瓶中倒出),充分研磨后,倒回三角瓶中(瓶中装有小玻璃珠),置旋涡混合器上振荡混匀,静置分层后,取上层悬浊液按要求进行梯度稀释,接种培养与计数。自生固氮细菌的分离培养基为阿须贝氏(Ashby)培养基,使用葡萄糖作碳源,用蒸馏水配制,添加NaCl使其浓度为15 $g \cdot kg^{-1}$,使用人工海水作无菌稀释水。自生固氮菌数量以每克干质量的平板菌落数计算($cfu \cdot g^{-1}$)。

2.4 藻样氮含量的测定

在检测自生固氮菌数量的同时,也测定了藻样的氮含量。藻样氮含量的测定采用硫酸过氧化氢消化,钠氏试剂比色法(鲁如坤, 2000)。

2.5 藻体异养固氮细菌的分类鉴定

在适当稀释平板上挑取生长良好的异养固氮细菌,转接于以葡萄糖为碳源,含NaCl 15‰的Burk培养基的试管斜面上,经纯化后参照相关文献初步鉴定到属(中国科学院微生物研究所细菌分类组, 1978;布坎南和吉本斯, 1984)。

2.6 数据处理

藻体含氮量春冬两季差异显著性的*T*检验采用SPSS 10.0统计软件完成。

3 结果与分析

3.1 藻体异养自生固氮细菌及藻体氮含量的季节变化

从红藻鹧鸪菜藻体上异养自生固氮菌的数量季节变化可知(图1),自生固氮菌季节变化模式为春

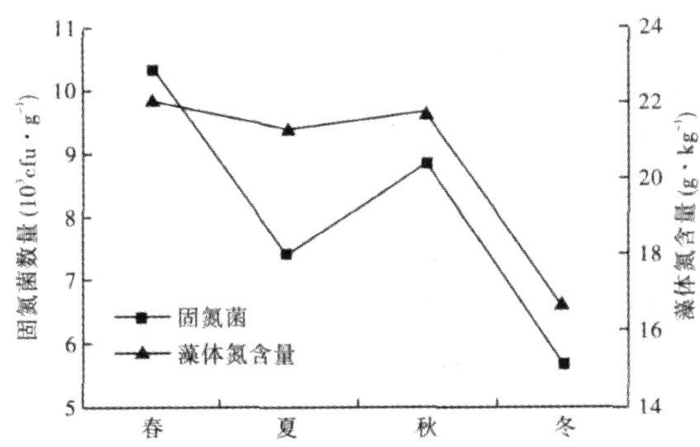

图 1 固氮菌和藻体氮含量的季节变化

Fig. 1 Seasonal change of nitrogen-fixing bacteria and nitrogen concentration in a lgal samples

季 >秋季 >夏季 >冬季,表现春季最多,为 1.033×10^4 cfu · g^{-1},冬季最少,为 5 670 cfu · g^{-1}。而鹧鸪菜藻体氮含量的季节变化模式也是春季 >秋季 >夏季 >冬季 (图 1),以春季的含氮量最高 (22.08 g · kg^{-1}),冬季最低 (16.63 g · kg^{-1}),两者差异显著 ($P = 0.0014 < 0.05$)。鹧鸪菜藻体异养自生固氮菌与藻体氮含量表现出一致的季节变化模式。

3.2 藻体异养固氮细菌的初步鉴定

分离纯化了 9株异养固氮细菌 (依次编号为 nf01—nf09),均对其进行了革兰氏染色、菌落特征、色素、菌体细胞形状与大小观察以及动力穿刺实验,并对其进行了一系列生理生化实验。在所鉴定的 9株菌株中,革兰氏染色均为阴性,呈杆状,大多数菌株在光学显微镜下同时可见到类球状细胞,所有菌株均有荚膜,除 nf06号菌株有单端单生鞭毛外,其他菌株均无鞭毛,异染粒 (Volutin)染色仅 nf01和 nf05呈阳性,其他为阴性,各菌株菌落特征不一;所有菌株过氧化氢酶反应呈阳性,不能分解淀粉,鉴定 9株菌的主要生理生化特征 (表 1)。

根据上述实验结果,参照"伯杰氏细菌鉴定手册"(布坎南和吉本斯, 1984),这 9株菌均属固氮菌科 (Azotobacteraceae)。nf01、nf02、nf03、f04和 nf05这 5株菌在形态特征上虽然有所差异,但它们已鉴定的生理生化特征一致,这些生理生化特征与固氮菌属内的维涅兰德固氮菌 (*Azotobacter vinelandii*)相符,但色素和鞭毛特征与该种有些不符,因此将这 5株菌初步定为固氮菌属未知种 (*Azotobacter* spp.)。nf06、nf07、nf08和 nf09或多或少兼有拜叶林克氏菌属 (*Beijerinckia*)各个种的大部分特征,初步定为拜叶林克氏菌属的未知种 (*Beijerinckia* spp.)。9株菌的种名确定有赖于进一步的表型和遗传特征。

表 1 固氮细菌的主要生理生化特征

Tab. 1 Main physiological and biochemical character istics of nitrogen-fixing bacteria

实验项目	菌株号								
	nf01	nf02	nf03	nf04	nf05	nf06	nf07	nf08	nf09
氧化酶	-	-	-	-	-	+	+	+	+
硝酸盐还原	+	+	+	+	+	-	-	-	-
甘露醇发酵	+	+	+	+	+	+	-	-	-
鼠李糖发酵	+	+	+	+	+	-	-	-	-
乳糖利用	+	+	+	+	+	+	-	+	+
蔗糖利用	+	+	+	+	+	+	-	-	W +
苯甲酸盐利用	+	+	+	+	+	W +	+	-	-
柠檬酸盐利用	+	+	+	+	+	+	-	-	-
钒盐代替钼盐	+	+	+	+	+	-	-	-	-

"+"为阳性反应, "-"为阴性反应, "W +"为弱阳性反应。

4 讨 论

Mann和 Steinke (1993)的研究表明,非洲 Beachwood红树林自然保护区能固氮的蓝绿藻夏季数量最多,冬季最少,季节变异明显,与本文所述的藻体异养自生固氮细菌春季数量最多不同。通常认为温度是季节变化的原因 (尤其是在寒冷的气候带),很少认为其他因素会起作用,但在南非 Swartkops河口湾表面沉积物秋季有较高的固氮活性,春夏季的固氮活性低是由于这些季节可利用的有机碳低的缘故 (Mann & Steinke, 1993)。

本研究中,温度也不是固氮菌数量季节变化的主导因子,因为鹧鸪菜上异养固氮细菌数量最多的是春季,而不是夏季。异养固氮细菌的季节变化模式受鹧鸪菜的生长与物质积累、温度、光照和水分的综合影响。鹧鸪菜与栖息其上的异养固氮细菌存在着密切的关系,鹧鸪菜一年四季的生长情况与藻体氮含量以及细菌的季节变化充分反映了这种关系(图 2)。鹧鸪菜以春、夏之交生长得最好,春季积累的物质最多,冬季生长得最差(林鹏等,1997),而藻体氮含量也是春季最多,冬季最少,固氮菌数量也是如此,藻体氮含量与固氮菌数量变化的季节模式趋势相近(图 2)。光合自养的鹧鸪菜与异养的固氮细菌可能存在着互惠互利的关系,鹧鸪菜提供给细菌碳源和能量物质,反过来,异养的固氮细菌提供给鹧鸪菜部分固定的氮源。这种现象在其他大型藻类也有(Head & Carpenter, 1975)。细菌-植物联合固氮共生体系(bacteria-plant associated N_2-fixing symbioses)在盐沼草的互花米草(*Spartina alterniflora*)(Patriquin, 1978),海草的海龟草(*Thalassia testudinum*)(Capone & Taylor, 1982),甚至浮游植物中也存在(Martinez *et al*, 1983)。

因此,鹧鸪菜上异养固氮细菌春季最多,是因为春季鹧鸪菜生长旺盛,可以提供较多的碳源和能量物质,加之温度也较适宜之故,夏季数量低于春季和秋季,可能是由于夏季温度高,日照强度大,蒸发量大,鹧鸪菜生长差,附生于红树茎基及根上的鹧鸪菜,容易被蒸发掉水分,这样也就影响到栖息其上的异养固氮细菌的生长发育,此时水分可能是一个重要的生态因子。Mann和 Steinke(1993)的研究中,最高的蓝绿藻数量与最多的月降雨量的时期一致,在所有采样期间,白骨壤(*Avicennia marina*)呼吸根在水浸没条件下比在暴露条件下明显有更高的乙炔还原速率,说明水分确实是一个重要的因素。冬季异养固氮细菌数量最少在较大程度上与此时的低温有关,因为低温既不利于鹧鸪菜的生长,也不利于细菌的活动。

综上所述,藻菌之间存在着密切的关系,这种关系又受到环境因子的影响,同时红树植物及其生境又会影响到藻和菌。因此,红树植物-藻-菌之间形成的复杂关系,是该系统内又一个很有吸引力的、需进一步研究的领域。

Herbert(1999)总结了海岸海洋生态系统中具有固氮能力的异养细菌。在红树林生态系统,异养的固氮细菌主要有梭菌属(*Clostridium*)、固氮菌属(*Azotobacter*)、芽孢杆菌属(*Bacillus*)和脱硫弧菌属(*Desulfovibrio*)(Herbert, 1999),Uchino等(1984)报道了木榄(*Bruguiera gymnorrhiza*)具瘤中的氮固定细菌,鉴定为阴沟肠杆菌(*Enterobacter cloacae*)、产气肠杆菌(*Enterobacter aerogenes*)和植生克雷伯氏菌(*Klebsiella planticola*)。在上述的属中,固氮菌属为好氧微生物(aerobes),梭菌属和脱硫弧菌属为厌氧微生物(anaerobes),芽孢杆菌属、肠杆菌属(*Enterobacter*)和克雷伯氏菌(*Klebsiella*)为兼性厌氧微生物(facultative anaerobes)。

本项研究所鉴定的菌株分别属于固氮菌属和拜叶林克氏菌属,二者均为好氧微生物,其中固氮菌属在红树林已有报道(Potts, 1984; Ravikumar *et al*, 2004),在海岸海洋沉积物中广泛存在(Herbert, 1999)。对于拜叶林克氏菌属,虽在红树林区尚未见报道,但在陆地生态系统广泛存在(Alexander, 1977),红树林生态系统位于陆海生态交错区,土壤生境同时具有陆地和海洋的性质,鹧鸪菜生长于树干基部,受潮水冲刷土壤的影响,鹧鸪菜上分离到拜叶林克氏菌为正常现象。

参考文献

李伟新,朱仲嘉,刘凤贤. 1982. 海藻学概论. 上海:上海科技出版社:82-83.

林 鹏,陈贞奋,刘维刚. 1997. 福建红树林区大型藻类的生态学研究. 植物学报,**39**(2):176-180.

鲁如坤. 2000. 土壤农业化学分析方法. 北京:中国农业出版社:107-147.

中国科学院微生物研究所细菌分类组. 1978. 一般细菌常用鉴定方法. 北京:科学出版社:98-194.

庄铁诚,张瑜斌,林 鹏. 2000. 红树林区红藻体上微生物初探. 厦门大学学报(自然科学版),**39**(2):227-234.

布坎南 RE,吉本斯 NE(中国科学院微生物研究所译). 1984. 伯杰细菌鉴定手册(第 8版). 北京:科学出版社:329-340.

Alexander M. 1977. Introduction to Soil Microbiology (2nd ed). New York: John Wiley & Sons Inc: 150-158.

Alongi DM, Boto KG, Robertson AI. 1992. Nitrogen and phosphorus cycles // Robertson AI, Alongi DM, eds. Tropical Mangrove Ecosystem. Washington DC: American Geophysical Union: 251-292.

Bashan Y, Puente ME, Myrold DD, *et al*. 1998. In vitro transfer of fixed nitrogen from diazotrophic filamentous cyanobacteria to black mangrove seedlings. *FEMS Microbiology Ecology*, **26**(3): 165-170.

Capone DG, Taylor EJ. 1982. N_2 fixation in the rhizosphere of *Thalassia testudinum*. *Canada Journal of Microbiology*, **26**: 998-1005.

Head WD, Carpenter EJ. 1975. Nitrogen fixation associated with macroalga *Codium fragile*. *Limnology Oceanogy*, **20**: 815-823.

Herbert RA. 1999. Nitrogen cycling in coastal marine ecosystems. *FEMS Microbiology Reviews*, **23**: 563-590.

Holguin G, Bashan Y. 1996. Nitrogen-fixation by *Azospirillum brasilense Cd is promoted when co-cultured with a mangrove* rhizosphere bacterium *(Staphylococcus* sp.). *Soil Biology & Biochemistry*, **28**(12): 1651-1660.

Lugomela C, Bergman B. 2002. Biological N_2-fixation on mangrove pneumatophores: Preliminary observations and perspectives. *Ambio*, **31**: 612-613.

Mann FD, Steinke TD. 1993. Biological nitrogen fixation (acetylene reduction) associated with blue-green algal (cynobacterial) communities in the Beachood Mangrove Nature Reserve. II. Seasonal variation in acetylene reduction activity. *South Africa Journal of Botany*, **59**(1): 1-8.

Martinez L, Silver MW, King JM, *et al.* 1983. Nitrogen fixation by floating diatom mats: A source of new nitrogen to oligotrophic ocean waters. *Science*, **221**: 152-154.

Patriquin DG. 1978. Nitrogen fixation (acetylene reduction) associated with cord grass, *Spartina alterniflora* Loisel. *Ecological Bulletin*, **26**: 20-27.

Potts M. 1984. Nitrogen fixation in mangrove forests // Por FD, Dor I, eds. Hydrobiology of the Mangal. The Hague: Dr W Junk: 155-162.

Ravikumar S, Kathiresan K, Ignatiammal STM, *et al.* 2004. Nitrogen-fixing azotobacters from mangrove habitat and their utility as marine biofertilizers. *Journal of Experimental Marine Biology and Ecology*, **312**(1): 5-17.

Sengupta AC. 1991. Ecology of heterotrophic dinitrogen fixation in rhizosphere of mangrove plant community at the Ganges river estuary in India. *Oecologia*, **87**(4): 560-564.

Tanaka J. 1987. Species composition and vertical distribution of macroalgae in brackish water of Japanese mangrove forests. *Ibid*, **13**(4): 141-150.

Toledo G, Bashan Y, Soeldner A. 1995. In vitro colonization and increase in nitrogen fixation of seedling roots of black mangrove inoculated by a filamentous cyanobacteria. *Canadian Journal of Microbiology*, **41**(11): 1012-1020.

Uchino F, Hambali GG, Yatazawa M. 1984. Nitrogen-fixing bacteria from warty lenticellate bark of a mangrove tree, *Bruguiera gymnorrhiza* (L.) Lamk. *Applied and Environmental Microbiology*, **47**(1): 44-48.

van der Valk AG, Attiwill PM. 1984. Acetylene reduction in an *Avicennia marina* community in Southern Australia. *Australia Journal of Botany*, **32**: 157-164.

Woitchik AF, Ohowa B, Kazungu JM, *et al.* 1997. Nitrogen enrichment during decomposition of mangrove leaf litter in an east African coastal lagoon (Kenya): Relative importance of biological nitrogen fixation. *Biogeochemistry*, **39**(1): 15-35.

Zuberer DA, Silver WS. 1975. Mangrove-associated nitrogen fixation // Walsh SS, Teas H, eds. Proceedings of International Symposium on Biology and Management of Mangroves. Gainesville: University of Florida: 643-653.

九龙江口红树林土壤微生物的类群及抗菌活性*

林　鹏[1]　张瑜斌[1,2]　邓爱英[1,2]　庄铁诚[1]

(1. 厦门大学生命科学学院,福建 厦门 361005;
2. 汕头大学水生生物技术与环境资源保护研究所,广东 汕头 515063)

摘要: 研究了九龙江口的秋茄(*Kandelia candel*)林和白骨壤(*Avicennia marina*)林两个红树林群落及其相应对照光滩土壤微生物的类群及抗菌活性. 对微生物类群的研究结果表明:九龙江口红树林区土壤细菌中,芽孢杆菌(*Bacillus*)是最占优势的属;放线菌以小单胞菌属(*Micromonospora*)最具优势;其次是链霉菌属(*Streptomyces*),从秋茄林到白骨壤林,由于潮位降低,小单胞菌比例增加,而链霉菌比例下降;丝状真菌以半知菌占绝对优势,木霉(*Trichoderma*)、曲霉(*Aspergillus*)和青霉(*Penicillum*)是最常见的属;随着土壤深度的增加,微生物的类群减少,但芽孢杆菌和小单胞菌的相对比例增加;红树林土壤微生物类群比对照光滩丰富,缘于林内土壤营养与微生物的栖息条件比光滩优越. 对抗菌活性研究表明:土壤真菌的抗菌活性低,抗菌谱窄;放线菌的抗菌活性高,抗菌谱宽,具有抗菌活性的放线菌多为小单胞菌,小单胞菌是一类值得重视的放线菌.

关键词:微生物区系;抗菌活性;土壤;红树林
中图分类号: Q939; S154. 3　　**文献标识码**: A　　**文章编号**: 0253-4193(2005)03-0133-09

1　引言

红树林是自然分布于热带、亚热带海岸的潮间带的木本植物群落,通常生长在港湾河口地区的淤泥质滩涂上,是海滩上特有的森林类型[1]. 生态系统中,微生物是主要的分解者,只有通过微生物的分解作用才能最终推动生态系统的物质循环和能量流动. 分解是各微生物类群综合作用的过程,因此对微生物的类群进行全面研究是很有必要的. 尽管在红树林地区的沉积物中存在着丰富而活跃的微生物[2,3],但是以往对红树林微生物类群的研究偏重于真菌类群[3],而对细菌和放线菌类群的研究较少[4,5]. 九龙江口是我国红树林的主要典型分布区之一,对该区域的红树林已有多方面的研究报道[1],尽管在微生物学方面已有一些研究[6~9],但对土壤微生物的类群和区系研究鲜见报道,因此,本文初步研究了该区域土壤好氧异养细菌、放线菌和丝状真菌3类微生物的类群.

微生物尤其是放线菌能产生许多具有各种价值的生物活性物质,并且仍然是筛选新的生物活性物质的重要来源. 天然的抗生素(包括医药用途的)大约2/3是来自放线菌[10]. 过去几十年来对陆栖土壤放线菌的筛选已越来越显露出再次发现和筛选到已经知道的生物活性物质的问题[11],解决这一问题的方法之一是从陆地土壤以外的环境寻找和扩展放线菌来源,因此海洋环境也就成为首选之一. Goodfellow等[12]认为海洋沉积物具有能分离到产生新的生物活性物质的放线菌的潜在价值,故此越来越引起人们的重视. 潮间带的红树林土壤处于海陆交错区,可能存在着能产生新颖活性物质的微生物. 为此,结合微生物类群的研究,我们对红树林土壤丝状真菌和放线菌的抗菌活性也进行了初步的研究.

---

* 国家自然科学基金资助项目(30270272)
原载于海洋学报,2005,27(3):133－141

2 材料和方法

2.1 样地概况

本项研究样地的秋茄(*Kandelia candel*)林和白骨壤(*Avicennia marina*)林位于福建九龙江口,该地的气候属南亚热带海岸气候.秋茄林位于九龙江口南岸的龙海市浮宫镇草埔头村,24°29′N,117°23′E(图1),秋茄林沿九龙江口南岸呈带状分布,宽度约为40 m,主要位于中、高潮带.该群落为1962年人工营造的秋茄纯林,林缘有少量桐花树(*Aegiceras corniculatum*)和白骨壤伴生,林相整齐,郁闭度在0.9以上,树高5.5~6 m.白骨壤林位于九龙江口厦门西海域海沧镇的东屿村,24°31′N,18°03′E(图1),该群落为纯林,林冠整齐,外貌呈褐绿色,群落结构简单,郁闭度在0.95以上,高度为1.2 m,植株密度大,达15株/m^2,基径约为2.7 cm,最大有4 cm,采样时很难穿行其内.林中有大量幼苗繁生,地表具大量指状呼吸根,高约10 cm,密度为35根/m^2,整个白骨壤林滩面平整,但样地仍选择在白骨壤林发育良好的地段.

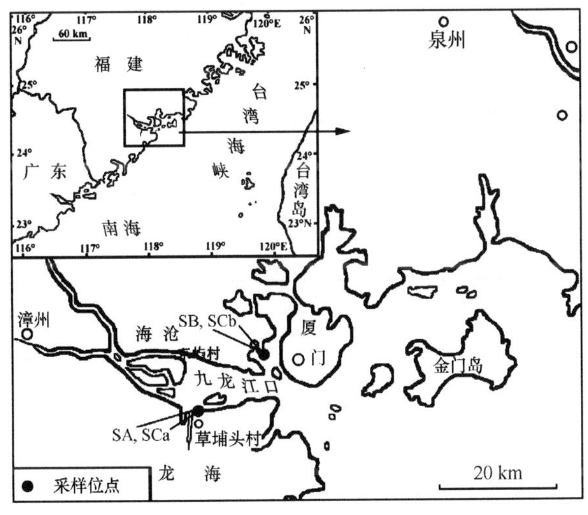

图1 采样位点分布

2.2 采样方法

土壤采样设在秋茄林内(标为样地SA)和白骨壤林内(标为样地SB)的中潮带,在距离秋茄林和白骨壤林100 m以远的同潮带的光滩面上各设置一对照样地(秋茄林对照光滩样地标为SCa,白骨壤林对照光滩样地标为SCb),以作比较.4个样地0~20 cm土壤主要理化参数见表1.土壤取样用内径5 cm的硬质聚氯乙稀管,在选定的样地内随机多点钻取20 cm深的土样,对随机多点钻取的土样拣去根系并混合均匀后装入无菌袋,立刻带回实验室处理.所有取样用具均作无菌处理.另在SA随机钻取了分为上层(0~20 cm)、中层(20~40 cm)和下层(40~60 cm)三层的土壤样品,用同样方法处理,以研究微生物的类群在土壤层中的垂直分布.

2.3 微生物的分离和鉴定

培养基:细菌为2216E好气异养菌培养基[13];放线菌为改良的高氏1号培养基[14],加入终浓度为50 μg/g的重铬酸钾作抑制剂;丝状真菌:马丁氏培养基[14],每1 dm^{3}培养基添加2 cm^{3}医用氯霉素(0.25 g/2 cm^3)作细菌抑制剂.根据前期的实验结果,对上述培养基用无菌淡水配制,添加NaCl,使其最终盐度分别为15,5和8.

分离方法:稀释平板法[14].根据前期的实验结果,把细菌用陈海水制作成稀释无菌水,把放线菌和丝状真菌用淡水制作成稀释无菌水.实验中使用的陈海水取自厦门大学海洋楼滨海水井(盐度为29.58),于密闭黑暗条件下放置15 d以上.对于放线菌,为减小细菌对其分离的影响,在所选择稀释度的前一稀释度添加1 cm^3 0.5%苯酚,振荡混匀,静置10 min后,再依次稀释至所需稀释度[14].

微生物的分类鉴定:选取适当稀释度的分离平板挑取菌株为优势菌株,转入相适应的试管斜面,经纯化后按有关文献鉴定到属[15~19].

表1 4个位点的土壤主要理化参数

位点	质地*	容重/g·cm^{-3}	有机质(%)	全氮(%)	全磷(%)	全钾(%)	全钠(%)	全钙(%)	全镁(%)	盐度	pH
SA	轻黏土	1.08	3.206±0.103	0.143±0.049	0.066±0.002	0.725±0.073	0.594±0.021	0.154±0.014	0.403±0.012	15.14±1.17	6.55±0.31
SCa	轻黏土	0.98	3.054±0.101	0.134±0.038	0.064±0.002	0.634±0.059	0.471±0.028	0.205±0.030	0.411±0.009	11.56±1.83	7.08±0.33
SB	轻黏土	0.93	3.600±0.109	0.148±0.029	0.057±0.004	0.851±0.091	0.828±0.070	0.337±0.027	0.596±0.012	24.30±1.49	7.00±0.13
SCb	中黏土	1.01	2.139±0.122	0.117±0.027	0.055±0.001	1.121±0.132	0.791±0.013	0.393±0.010	0.624±0.010	21.25±1.13	7.18±0.06

* 依据卡庆斯基土壤质地分类标准.

2.4 微生物的抗菌活性测定

本研究采用滤纸片液体扩散法[20]对土壤放线菌和丝状真菌进行了抗菌活性测定.供试菌摇瓶液体发酵培养基:放线菌为黄豆饼粉培养基[20];丝状真菌为 PDA 培养基[20],加入浓度为 0.8 %的 NaCl.摇瓶于 28～30 ℃摇床振荡培养 5 d,发酵液经离心分离后(丝状真菌发酵液先用超声波细胞破碎仪破碎细胞后再离心处理),取上清液供作测试.抗菌活性测定中所使用的指示菌为大肠杆菌(*Escherichia coli*, Ec)、金黄色葡萄球菌(*Staphylococcus aureus*,Sa)、枯草芽孢杆菌(*Bacillus subtilis*,Bs)、黑曲霉(*Aspergillus niger*, An)和白色假丝酵母(*Candida albicans*,Ca).本实验中所使用的滤纸片直径约为 6 mm,以抑菌圈直径大小衡量抗菌活性大小.

3 结果

3.1 土壤异养微生物类群的分布

表 2～4 分别是土壤细菌、放线菌和丝状真菌的主要属和相对密度.在秋茄林及其对照光滩和白骨壤林 3 个位点,共分离到 5 个属的细菌(表 2),为芽孢杆菌(*Bacillus*)、节细菌(*Anthrobacter*)、假单胞菌(*Pseudomonas*)、黄单胞菌(*Xanthomonas*)和链球菌(*Streptococcus*),其中秋茄林(SA 位点)0～20 cm 土壤层 5 个属均有出现,20～40 cm 层有芽孢杆菌、节细菌和假单胞菌 3 个属,40～60 cm 层只有芽孢杆菌和节细菌 2 个属,秋茄林区对照光滩(SCa 位点)0～20 cm 土壤层只有芽孢杆菌、假单胞菌、黄单胞菌和链球菌 4 个属,白骨壤林(SB 位点)0～20 cm 只有芽孢杆菌、假单胞菌和链球菌 3 个属.在所研究的 3 个位点中,芽孢杆菌在细菌属的组成中占有绝对优势,其相对密度均在 50 %以上(表 2).

在 4 个位点中,从放线菌共分离到 6 个属(表 3),秋茄林(SA 位点)0～20 cm 土壤层有链霉菌(*Streptomyces*)、小单胞菌(*Micromonospora*)、链轮丝菌(*Streptoverticillium*)、红球菌(*Rhodococcus*)、小多孢菌(*Micropolyspora*)和游动放线菌(*Actinoplanes*)6 个属,20～40 cm 层只有链霉菌和小单胞菌 2 个属,40～60 cm 层仅有小单胞菌 1 属,秋茄林对照光滩(SCa 位点)0～20 cm 土层有链霉菌、小单胞菌、红球菌、小多孢菌和游动放线菌 5 个属,白骨壤林(SB 位点)和其对照光滩(SCb 位点)0～20 cm 土层均只有链霉菌和小单胞菌 2 个属.放线菌均以小单胞菌最具优势,相对密度在 50 %以上,在白骨壤林和其对照光滩高达 90 %以上,其次为链霉菌,其他类型所占份额相对较少(表 3).

表 2 土壤异养细菌类群主要属组成和相对密度(%)*

菌 属	SA			SCa	SB
	0～20 cm	20～40 cm	40～60 cm	0～20 cm	0～20 cm
芽孢杆菌(*Bacillus*)	54.63	60	75	72.22	77.42
节细菌(*Anthrobacter*)	17.59	20	25	-	-
假单胞菌(*Pseudomonas*)	4.63	20	-	5.56	9.68
黄单胞菌(*Xanthomonas*)	16.67	-	-	16.66	-
链球菌(*Streptococcus*)	6.48	-	-	5.56	12.9

*相对密度(%)等于(该属总菌株数除以该区域细菌所有属的总菌株数)乘以 100,下同.

表 3 土壤放线菌主要属的组成和相对密度(%)

菌 属	SA			SCa	SB	SCb
	0～20 cm	20～40 cm	40～60 cm	0～20 cm	0～20 cm	0～20 cm
链霉菌(*Streptomyces*)	24.00	11.11	-	25.00	9.09	10.00
小单胞菌(*Micromonospora*)	56.00	88.89	100.00	62.50	90.91	90.00
链轮丝菌(*Streptoverticillium*)	4.00	-	-	-	-	-
红球菌(*Rhodococcus*)	8.00	-	-	4.17	-	-
小多孢菌(*Micropolyspora*)	4.00	-	-	4.17	-	-
游动放线菌(*Antinoplanes*)	4.00	-	-	4.17	-	-

表4 土壤丝状真菌主要属的组成和相对密度(%)

菌 属	SA			SCa	SB	SCb
	0～20 cm	20～40 cm	40～60 cm	0～20 cm	0～20 cm	0～20 cm
木霉(*Trichoderma*)	27.28	50.00	-	52.38	-	-
曲霉(*Aspergillus*)	22.22	-	-	42.86	16.67	-
青霉(*Penicillium*)	22.22	-	-	4.76	33.33	-
拟青霉(*Paecilomyces*)	16.67	-	-	-	-	-
腐殖霉(*Humicola*)	11.11	-	-	-	-	-
茎点霉(*Phoma*)	-	25.00	-	-	16.67	-
被孢霉(*Mortierella*)	-	25.00	-	-	-	-
无梗孢(*Trichocladium*)	-	-	-	-	33.33	-
链毛孢(*Streptothrix*)	-	-	-	-	-	50.00
未产孢(*Mycelia sterilia*)	-	-	-	-	-	50.00

研究了4个位点土壤层丝状真菌属的组成及其相对密度,共分离到9个属,另有一未产孢类型(表4),其中秋茄林0～20 cm土层有木霉(*Trichoderma*)、曲霉(*Aspergillus*)、青霉(*Penicillum*)、拟青霉(*Paecilomyces*)和腐殖霉(*Humicola*)5个属,20～40 cm层有茎点霉(*Phoma*)和被包霉(*Mortierella*)2个属,在40～60 cm层未检测到丝状真菌;秋茄林对照光滩0～20 cm土层有木霉、曲霉和青霉3个属,白骨壤林0～20 cm层有曲霉、青霉、茎点霉和无梗孢(*Trichocladium*)4个属,白骨壤林对照光滩0～20 cm土层有链毛孢(*Streptothrix*)1个属和一未产孢类型.在秋茄林内以木霉占优势,其次是曲霉和青霉;在秋茄林对照光滩也以木霉占优势,其次是曲霉;在白骨壤林占优势的是青霉和无梗孢;在白骨壤林对照光滩只分离到一个属和一未产孢类型(表4).

研究了SA位点的秋茄林土壤各类群微生物属的数目的垂直分布特征(见表2～4).细菌属的垂直分布中,在0～20 cm土层有5个属、在20～40 cm层有3个属、在40～60 cm层有2个属,随着土壤层的加深,细菌属的数目减少,但3个层次中均以芽孢杆菌最多,芽孢杆菌的比例随着土壤层的加深而增加.放线菌的垂直分布中,在0～20 cm土层有6个属,在20～40 cm层有2个属,在40～60 cm层只有1个属,同样,表现出随着土壤层的加深,放线菌属的数目也减少,但3个层次均以小单胞菌最多,并随土壤层的加深,小单胞菌所占比例增加.丝状真菌在0～20 cm土层有5个属,在20～40 cm层有3个属,在40～60 cm层没有分离到丝状真菌.在0～20和20～40 cm两个层次中均以木霉占优势,丝状真菌属的数目随土壤层的加深而下降.因此,各类群微生物属数均表现为随土壤深度的增加而减少.

比较秋茄林与其对照光滩土壤0～20 cm层各类群微生物属的数目,发现在秋茄林0～20 cm土壤层细菌、放线菌和丝状真菌属的数目分别为5属、6属和5属,而在其对照光滩却分别只有4属、5属和3属,秋茄林多于其对照光滩.尽管白骨壤林与其对照光滩的土壤放线菌都有2个属,但丝状真菌的属数却是白骨壤林多于其对照光滩,总体上仍然表现出红树林土壤微生物类群多于对照光滩的趋势。

3.2 土壤丝状真菌和放线菌的抗菌活性

土壤丝状真菌和放线菌的抗菌性能统计结果见表5,从表5可知,真菌的抗菌活性较小,拮抗菌株数的比例为11.54%,对5种指示菌拮抗比例也很低.相比之下,放线菌的抗菌活性要高许多,具有拮抗性能的菌株占供试菌株的57.81%,但放线菌的抗菌活性主要表现为对细菌的拮抗,对大肠杆菌(*Escherichia coli*)、金黄色葡萄球菌(*Staphylococcus aures*)和枯草芽孢杆菌(*Bacillus subtilis*)3种指示细菌的拮抗比例分别为31.25%,31.25%和20.31%,而对真菌的拮抗比例较小,对黑曲霉(*Aspergillus niger*)和白色假丝酵母(*Candida albicans*)的拮抗比例分别只有3.13%和4.69%.尽管如此,放线菌仍表现了广泛的抗菌性能.

表5 具拮抗性能的真菌和放线菌菌株数的百分比率*

微生物类群	供试菌株数	拮抗菌株数(%)	指示菌				
			Ec	Sa	Bs	An	Ca
真菌	26	3(11.54)	1(3.85)	2(7.69)	2(7.69)	0(0)	1(3.85)
放线菌	64	37(57.81)	20(31.25)	20(31.25)	13(20.31)	2(3.13)	3(4.69)

*括号内为具拮抗性能的菌株数占供试菌株总数的百分比率，指示菌种类见材料和方法.

表6是具拮抗性能的真菌和放线菌的抗菌谱.在具有拮抗性能的37株放线菌中,小单胞菌有27株,链霉菌有5株,游动放线菌有2株,小多胞菌有2株,链轮丝菌有1株,它们所占的百分比率分别为72.97%,13.51%,5.41%,5.41%和2.70%,这说明小单胞菌属是值得重视的一类放线菌.

4 讨论

在世界红树林区,对细菌类群的研究较少,但已报道的结果表明,各地细菌属的组成在沉积物(或土壤)中有一定的相似性,即均以芽孢杆菌占优势.Shome等[21]研究表明,在印度南Andaman红树林凋落物及其下表面沉积物的细菌区系中,芽孢杆菌属占50%的比例;胡承彪等[5]的研究也发现芽孢杆菌属是红树林土壤中的优势属.在本项研究中,芽孢杆菌也是土壤中的优势属(见表2),但在人畜粪便污染较重的红树林区,土壤中除芽孢杆菌外,还会伴随出现较多的病原细菌例如大肠杆菌(*E. coli*)、沙门氏杆菌(*Salmonella*)和志贺氏杆菌(*Shigella*)[4].

本项研究中的土壤放线菌类群与郑志成等[22]所研究的根际放线菌类群有一定的相似性.郑志成等[22]分离到链霉菌、链孢囊菌(*Streptosporangium*)、小单胞菌、小多胞菌、诺卡氏菌和游动放线菌(*Antinoplanes*)6个属,以链霉菌最多(占75.7%),小单胞菌其次.本项研究中,放线菌类群以小单胞菌最多,其次是链霉菌,这种差异可能缘于不同的小生境.在Wu[4]的研究中分离到链霉菌、链轮丝菌、小单胞菌、诺卡氏菌和红球菌5个属,其中多数为小单胞菌和链霉菌,与本文的结果相似.Pisano等[23]的研究结果表明,小单胞菌和链霉菌属是海洋沉积物放线菌的主要属,Jensen等[24]Chesapeake的研究中,小单胞菌属的数量最多,其次是链霉菌群切萨皮克,Takizawa等[25]对海湾沉积物放线菌的研究也是以小单胞菌占多数.本项研究也表明小单胞菌是红树林土壤中是最具优势的放线菌.在淡水湖泊沉积物中也是这样的结果,Jiang等[26]对云南12个湖泊沉积物的放线菌研究表明,小单胞菌属的菌株最占优势,链霉菌群是次于小单胞菌的第2大类群,Jiang等[26]认为这是湖泊水生放线菌类群与陆栖放线菌类群相比的突出特征,因为在陆地土壤中,优势放线菌是链霉菌群[27].本项研究的红树林土壤放线菌处于海陆交错的潮间带,有着周期性的潮汐浸润,该区域的放线菌也是以小单胞菌属的种类最多,链霉菌属其次.根据上述分析可以认为,无论是在淡水湿地还是在海岸湿地以及海洋沉积物中,小单胞菌是最占优势的放线菌.小单胞菌是一类水生放线菌,对营养的要求可能比链霉菌更为复杂,但对氧的要求较低,抗压能力强[28].Jensen等[24]和Takizawa等[25]以及本文的结果进一步证实了这一观点.Jensen等[24]的研究中,随着水深的增加,链霉菌群快速下降,游动放线菌群(其中绝大多数为小单胞菌)比例增加,但放线菌总数下降.Takizawa等在切萨皮克海湾的研究结果也表明,游动放线菌群(其中绝大多数为小单胞菌)在总放线菌类群中的比例朝着海湾的方向(海向)增加,在海湾口和远离海滨的站位,所有菌株均被鉴定为游动放线菌群,相反,其他放线菌的数量在海湾内部(陆向)的站位最高,朝着海湾口的方向下降[25].本项研究中,位于潮位较高的河口处的秋茄林(SA位点)及其对照光滩(SCa位点)20 cm土壤层中,小单胞菌所占的百分比分别为56.00%和62.50%,链霉菌的百分比分别为24.00%和25.00%,而位于潮位较低的海湾内的白骨壤林(SB位点)及其对照光滩(SCb位点)0~20 cm土壤层中,小单胞菌的比例分别上升为90.91%和90.00%,链霉菌的百分比分别下降为9.09%和10.00%,这也证实了小单胞菌是一类水生放线菌,对氧的要求较低,抗压能力强.

表6 具拮抗性能的真菌和放线菌的抗菌谱(以抑菌圈直径(mm)大小衡量)*

菌株号	菌 名	指 示 菌				
		Ec	Sa	Bs	An	Ca
AS203	*Humicola* sp.					6.5
AS205	*Penicillium* sp.	11.0	14.0	10.0		
BS201	*Trichocladium* sp.		7.0	7.0		
AS301	*Streptomyces* sp.	W		W		
AS304	*Micropolyspora* sp.	6.5	11.0	6.5		6.5
AS305	*Micromonospora* sp.			W		
AS307	*Micromonospora* sp.	6.5				
AS308	*Antinoplanes* sp.		6.5			
AS310	*Micromonospora* sp.	6.8	12.5	W		
AS311	*Micromonospora* sp.	7.0	21.0	16.0		
AS312	*Micromonospora* sp.		14.0			
AS313	*Micromonospora* sp.		19.0	14.0		
AS317	*Micromonospora* sp.	W				
AS319	*Streptomyces* sp.	W				
AS320	*Streptoverticillium* sp.		6.8			
AS321	*Streptomyces* sp.		12.0			
AZ305	*Micromonospora* sp.	6.5				
AZ306	*Micromonospora* sp.	7.0				
AX301	*Micromonospora* sp.			W		
AX302	*Micromonospora* sp.			W		
AcS301	*Micromonospora* sp.		17.0			
AcS303	*Antinoplanes* sp.	7.0		7.0		
AcS304	*Micromonospora* sp.			13.5		
AcS305	*Micromonospora* sp.	W	9.0	8.0		7.5
AcS307	*Micromonospora* sp.				9.5	
AcS313	*Streptomyces* sp.	8.5	6.5	6.5		
AcS314	*Streptomyces* sp.	W	W			
BS301	*Micromonospora* sp.		W			
BS302	*Micromonospora* sp.		W			
BS303	*Micromonospora* sp.	W			9.5	
BS305	*Micromonospora* sp.		7.3			
BS307	*Micromonospora* sp.	W	6.8			
BS308	*Micromonospora* sp.	W	6.5			
BS309	*Micromonospora* sp.		7.0			
BcS301	*Micromonospora* sp.		18.0	7.5		
BcS303	*Micromonospora* sp.	W				
BcS304	*Micromonospora* sp.	W				
BcS305	*Micromonospora* sp.	6.5				
BcS306	*Micromonospora* sp.		6.5			
BcS308	*Micromonospora* sp.	6.5				

*指示菌种类见材料和方法, W 表示微弱反应, 其抑菌圈直径不大于6.1 mm.

在红树林土壤丝状真菌中以半知菌占大多数,子囊菌和接合菌较少,鞭毛菌则更稀少,没有担子菌,许多的研究结果都证实了这一结果[29~32].本项研究中半知菌也占有绝对优势,除被包霉为接合菌和一未产孢类型外,其他均为半知菌,没有分离到子囊菌、鞭毛菌和担子菌(见表4),与上述一致.因此,可以说半知菌是红树土壤中最占优势的真菌类型,但在不同地域的红树林土壤丝状真菌中,常见属会有所差异,在夏威夷的欧湖岛红树林,木霉是最常见的属,其次是青霉、镰孢菌(*Fusarium*)和曲霉[29],在印度西孟加拉邦(West Bengal)的红树林,曲霉最多,青霉、毛霉次之[30,31],而在东非(Portuguese East Africa)伊尼亚卡岛的红树林,曲霉和青霉最多[32].在本项研究中(见表4),尽管SCb位点例外,但在SA位点,木霉是最常见的属,其次是曲霉和青霉,SCa位点最常见的属也是木霉,其次是曲霉,SB位点最常见的属为青霉、无梗孢,与前述结果一致.木霉、曲霉和青霉是红树林土壤丝状真菌中最常见的属.

本项研究中,红树林土壤微生物类群多于相应的对照光滩,这缘于红树林内的土壤生境比光滩优越,从表1可知,红树林土壤营养条件优于光滩,而且红树林可以为其下的土壤提供荫蔽条件,减少风浪潮汐对土壤的冲击,同时红树植物持续不断地将凋落物及碎屑加入红树林生态系统[33];土壤中红树植物根系的排泄物或分泌物也会促发土壤微生物的活动[34],根系将植物吸收的氧气释放入土壤中也会促进好氧异养微生物的活动[35].可以说,植被的存在改善了栖息环境,丰富了土壤微生物的类群.该区域先前的微生物学研究也表明红树林土壤中降解石油和农药的微生物比光滩土壤丰富[7,8],这说明红树林土壤中蕴藏的微生物资源值得重视.

有关红树林微生物的抗菌活性研究报道多见于放线菌[4,22],对真菌的抗菌活性研究较少,这可能缘于能产生较强生物活性物质的真菌极少.在郑志成等[22]的研究中,红树根际放线菌的拮抗比例为36.3%,拮抗菌株主要为链霉菌,在拮抗放线菌中,能抑制革兰氏阳性细菌者居多,抑制革兰氏阴性细菌者次之,能抑制真菌者最少.本项研究中,放线菌的拮抗比例(57.81%)要高于郑志成等[22]的结果,对革兰氏阳性和阴性细菌均有较好的拮抗性能,对真菌的抑制很小,即抗菌谱与他们的报道相似.Wu[4]的研究结果表明,许多放线菌能产生抗生素抑制黑曲霉、白色假丝酵母、枯草芽孢杆菌和藤黄八叠球菌(*Sarcina lutea*)的生长,而对金黄色葡萄球菌和大肠杆菌的抑制很少.在Pisano等[36,37]的海洋沉积物放线菌抗菌活性的研究中,尽管具拮抗性能的放线菌的比例有的低至15.8%[36],有的高至73%[37],但均表现为主要抑制革兰氏阳性细菌,其次是真菌,对革兰氏阴性细菌的拮抗最小.综上所述可以看出,在红树林及海洋环境不同研究者的报道之间存在两个差异:(1)具拮抗性能放线菌在供试菌株中所占比例的差异;(2)供试的放线菌抑制指示菌类型的差异.这些差异可能由下列原因引起:(1)培养基的性质及发酵条件将直接影响生物活性物质的分泌和抗菌活性,这一点已经被证实[37,38];不同研究者所使用的发酵培养基及发酵条件不同将直接引起他们之间结果的差异;(2)产生生物活性物质是菌株的特性而不是种的特性[39];(3)抗菌菌株的数量会随着指示微生物数的增加而增加[40].

参考文献:

[1] LIN P. Mangrove Ecosystems in China [M]. Beijing and New York: Science Press, 1999. 1-7, 142-167.

[2] ALONGD M, SASEKUMAR A. Benthic communities [A]. ROBERTSON A I, ALONGI D M. Tropical Mangrove Ecosystems[M]. Washington D C: American Geophysical Union, 1992. 137-172.

[3] JONES E B G, ALIAS S A. Biodiversity of mangrove fungi [A]. Biodiversity of Tropical Microfungi [M]. Hong Kong: Hong Kong University Press, 1996. 86.

[4] WU Rong-yang. Studies on the microbial ecology of Tansui Estuary [J]. Botanical Bulletin of Academia Sinica, 1993, 34(1): 13-30.

[5] 胡承彪,梁秀棠.合浦滨海海滩森林土壤微生物区系及生化活性[J].热带林业科技,1987,(1):1-7.

[6] 庄铁诚,林 鹏.红树林凋落物叶自然分解过程中土壤微生物的数量动态[J].厦门大学学报(自然科学版),1993,32(3):365-370.

[7] 庄铁诚,林 鹏.红树林下土壤微生物对柴油的降解[J].厦门大学学报(自然科学版),1995,34(3):442-446.

[8] 庄铁诚,张瑜斌,林 鹏.红树林土壤微生物对甲胺磷的降解[J].应用与环境生物学报,2000,6(3):276-280.

[9] 庄铁诚,张瑜斌,林 鹏.红树林区红藻体上微生物初探[J].厦门大学学报(自然科学版),2000,39(2):227-234.

[10] OKAMI Y, HOTTA K. Search and discovery of new antibiotics [A]. GOODFELLOW M, WILLIAMS S T, MORDARSKI M. Actinomycetes in Biotechnology [M]. San Diego, Calif.: Academic Press Inc, 1988. 33-67.

[11] NOLAN R D, CROSS T. Isolation and screening of actinomycetes [A]. GOODFELLOW M, WILLIAMS S T, MORDARSKI M. Actinomycetes in Biotechnology [M]. San Diego, Calif.: Academic Press Inc, 1988. 1-32.

[12] GOODFELLOW M, HANYNESJ A. Actinomycetes in marine sediments [A]. ORTIZ-ORTIZ L, BOJALIL L F, YAKOLEFF V. Biological, Biochemical, and Biomedical Aspects of Actinomycetes [M]. Orlando, Florida: Academic Press Inc, 1984. 453-472.

[13] 陈绍铭,郑福寿.水生微生物实验法上册[M].北京:海洋出版社,1985.

[14] 中国科学院南京土壤研究所微生物室. 土壤微生物研究法[M]. 北京:科学出版社, 1985.

[15] 布坎南 R E, 吉本斯 N E. 伯杰细菌鉴定手册(第八版)[M]. 中国科学院微生物研究所译. 北京:科学出版社, 1984.

[16] 阎逊初. 放线菌的分类与鉴定[M]. 北京:科学出版社, 1992.

[17] 姜成林,徐丽华,许宗雄. 放线菌分类学[M]. 昆明:云南大学出版社, 1995.

[18] 魏景超. 真菌鉴定手册[M]. 上海:上海科学技术出版社, 1979.

[19] 巴尼特 H L,亨特 B B. 半知菌属图解[M]. 沈崇尧译. 北京:科学出版社, 1977.

[20] 周德庆. 微生物学实验手册[M]. 上海:上海科技出版社, 1986.

[21] SHOME R, SHOME B R, MANDAL A B, et al. Bacterial flora in mangroves of Andaman: Part I. Isolation, identification and antibiogram studies [J]. Indian Journal of Marine Sciences, 1995, 24: 97–98.

[22] 郑志成,周美英,姚炳新. 红树林根际放线菌的组成及生物活性[J]. 厦门大学学报(自然科学版), 1989, 28(3): 306–310.

[23] PISANO M A, SOMMER M J, BRANCACCCIO L. Isolation of bioactive actinomycetes from marine sediments using rifampicin [J]. Applied Microbiology and Biotechnology, 1989, 31: 609–612.

[24] JENSEN P R, DWIGHT R, FENICAL W. Distribution of actinomycetes in nearshore tropical marine sediments [J]. Applied and Environmental Microbiology, 1991, 57(4): 1 102–1 108.

[25] TAKIZAWA M, COLWELL R R, HILL R T. Isolation and diversity of actinomycetes in the Chesapeake Bay [J]. Applied and Environmental Microbiology, 1993, 59(4): 997 - 1 002.

[26] JIANG Cheng-lin, XU Li-hua. Diversity of aquatic actinomycetes in lakes of the middle plateau, Yunnan, China [J]. Applied and Environmental Microbiology, 1996, 62: 249–253.

[27] GOODFELLOW M, WILLIAMS S T. Ecology of actinomycetes [J]. Annual Review Microbiology, 1983, 37: 189–216.

[28] 姜成林,徐丽华. 微生物资源学[M]. 北京:科学出版社, 1997, 99: 144–149.

[29] LEE B K H, BAKER G E. An ecological study of the soil microfungi in a Hawaiian mangrove swamp [J]. Pacific Science, 1972, 26: 1–10.

[30] RAI J N, TEWARI J P, MUKERJI K G. Mycoflora of mangrove mud [J]. Mycopthologia et Mycologia Applicata, 1969, 3: 17–31.

[31] MISRA J K. Fungi from mangrove mud of Andama - Nicobar Islands [J]. Indian Journal of Marine Sciences, 1986, 15: 185–186.

[32] SWART H J. An investigation of the mycoflora in the soil of some mangrove swamps [J]. Acta Botanica Neerlandica, 1958, 7: 741–768.

[33] JAGTAP T G. Seasonal distribution of organic matter in mangrove environment of Goa [J]. Indian Journal of Marine Sciences, 1987, 16: 103–106.

[34] SHERMAN R E, FAHEY T J, HOWARTH R W. Soil - plant interactions in a neotropical mangrove forest: iron, phosphorus and sulfur dynamics [J]. Oecologia, 1998, 115: 553–563.

[35] BOTO K G, WELLINGTON J T. Soil characteristics and nutrient status in a northern Australian mangrove forest [J]. Estuaries, 1984, 7(1): 61–69.

[36] PISANO M A, SOMMER M J, LOPEZ M M. Application of pretreatments for isolation of bioactive actinomycetes from marine sediments [J]. Applied Microbiology and Biotechnology, 1986, 25: 285–288.

[37] PISANO M A, SOMMER M J, TARAS L. Bioactivity of chitinolytic actinomycetes of marine origin [J]. Applied Microbiology and Biotechnology, 1992, 36: 553–555.

[38] OKAZAKI T, OKAMI Y. Studies on marine microorganisms: II. Actinomycetes in Sagami Bay and their antibiotic substances [J]. Journal of Antibiotic, 1972, 25: 461–466.

[39] OKAMI Y. Marine microorganisms as a source of bioactive agents [J]. Microbial Ecology, 1986, 12: 65–78.

[40] TAKIZAWA M, COLWELL R R, HILL R T. Isolation and diversity of actinomycetes in the Chesapeake Bay [J]. Applied and Environmental Microbiology, 1993, 59(4): 997–1 002.

Microflora and antimicrobial activities of soil microorganisms in mangrove forests in the Jiulong Estuary, China

LIN Peng[1], ZHANG Yu-bin[1,2], DENG Ai-ying[1,2], ZHUANG Tie-cheng[1]

(1. *School of Life Sciences, Xiamen University, Xiamen* 361005, *China*; 2. *Institute of Aquatic Biotechnology and Environmental Resources Protection, Shantou University, Shantou* 515063, *China*)

Abstract: Microflora and antimicrobial activities of soil microorganisms were investigated in *Kandelia candel* forest, *Avicennia marina* forest and their corresponding mud-flats in the Jiulong Estuary, Fujian of China. The studies on microflora indicate that Bacillus is dominant genus in bacterial genera; *Micromonospora* is dominant genus followed by *Streptomyces* in actinomycete genera; increase of percentage of *Micromonospora* and decrease of that of *Streptomyces* is presented with descending tidal level from *K. candel* forest to *A. marina* forest; the percentage of imperfect fungi is absolutely dominant in isolated fungi, of which *Trichoderma, Aspergillus* and *Penicillum* is dominant and conmon genera; microflora declins with increasing soil depth, however, the percentage of *Bacillus* and *Micromonospora* increases relatively with it; the reason why microflora of soil is more in mangrove forest than in mud-flats attributed to more abundant soil nutrition and better habitat in mangrove forests. Studies on antimicrobial activity reveals that antimicrobial activity of soil fungi is weaker than that of soil actinomycetes, so antibiogram of soil fungi is narrow while that of soil actinomycetes is wide; the genus *Micromonospora* should be worth paying attention to owing to its strong antimicrobial activities.

Key words: microflora; antimicrobial activities; soil; mangrove

Four newly recorded species of Bacillariophyta from the mangroves in China*

CHEN Chang-Ping GAO Ya-Hui LIN Peng

(*School of Life Sciences, Xiamen University, Xiamen* 361005, *China*)

Abstract Four species of diatoms from the mangroves in Fujian Province and Shenzhen City of China are described. They are *Cymbella cucumis* A. Schmidt, *Navicula elegantoides* Hustedt, *N. platyventris* Meister, and *N. tenera* Hustedt. They represent new records for China. Detailed description of the taxonomic characters of the four species and of their ecological behavior is given. *Cymbella cucumis* was defined as a freshwater and brackish water species for it occurred, though occasionally, where water salinity was more than 15.

Key words Diatom, mangroves, new record, China.

Mangroves are woody plant communities which occur in the intertidal zones of tropical and subtropical coastlines of the world (Lin, 1984). Mangrove algae constitute a significant food source for various organisms in the mangrove ecosystem (Nicholas et al. , 1988). Taxonomic studies of diatoms from mangrove environments have been carried out in many countries (Foged, 1979; Navarro, 1982; Nagumo & Hara, 1990; Sequeiros-Beltrones & Castrejon, 1999). In China, however, such studies have been less reported (Du & Jin, 1983), especially for benthic diatoms (Fan et al. , 1993; Chen et al., 2005).

Benthic diatoms serve as a common food for certain mangrove fishes (Beumer, 1978) and diatoms were abundant in phytoplankton in mangroves (Liu & Chen, 1997). In this paper, four newly recorded species of diatoms from the mangroves in China were reported. They are *Cymbella cucumis* A. Schmidt, *Navicula elegantoides* Hustedt, *N. platyventris* Meister and *N. tenera* Hustedt.

1 Material and methods

Samples were collected from water and mudflat in mangroves in Fujian Province and Shenzhen City, China, respectively. All samples were treated with 10% HCl to remove the calcareous matter, and treated with 30% H_2O_2 to destroy the organic material. Each sample was diluted by adding distilled water until no acid was left in the sample. Treated samples were identified and photographed under an Olympus BH-2 microscope (1000×) and a JEM-100CX II transmission electron microscope (TEM).

2 Description of species

1. Cymbella cucumis A. Schmidt in A. Schmidt et al., Atlas Diatomaceenkunde 9, pl. 9, figs. 21, 22. 1885; Cleve, Kongliga Svenska Vetenskaps-Akademiens Handlingar 26: 165. 1894; Hustedt in A. Schmidt et al., Atlas Diatomaceenkunde 375. 1931.

瓜形桥弯藻 Fig. 1

* From: Acta Phytotaxonomica Sinica, 2006, 44(1): 95—99

Valve broad, with convex dorsal and almost straight or slightly convex ventral margin. Ends rostrate-truncate, 70 μm long, 22 μm wide (79–90 μm long and 24 μm wide in Cleve (1894)). Axial area narrow, slightly dilated around the central nodule. Raphe slightly arcuate. Striae 9 (dorsal) to 10 (ventral) in 10 μm; areolae 12–14 in 10 μm.

This species is similar to *Cymbella lata* Grunow and *C. ehrenbergii* Kützing, but *C. lata* has dense areolae, approximately 30 in 10 μm, valve 40–60 μm long, 16–18 μm wide, raphe slightly eccentric, slightly curved, with dorsally reflexed apical fissures, and occurs in fresh and slightly saline water. *Cymbella ehrenbergii* has narrow, slightly protracted ends, almost straight raphe, axial area moderately wide, lanceolate, roundishly widened around the central nodule, valve 50–220 μm long, 19–50 μm wide, and occurs in freshwater, brackish water and sea water (Cleve, 1894).

Habitat: Freshwater and brackish water. Cleve (1894) pointed out that *Cymbella cucumis* is a freshwater species, but we found it occurred where water salinity ranged from 5.0 to 26.1 sampled from January 2001 to January 2003.

Distribution: Our samples were collected from mudflat in mangroves in Yunxiao County, Fujian Province, China. This species has been previously found in Bengal and Cameroon (Cleve, 1894).

2. Navicula elegantoides Hustedt in A. Thienemann, Die Binnengewasser 16 (2): 76, fig. 142. 1942; Prowse in Garden's Bull., Singapore 19: 42. 1962.

拟优美舟形藻 Fig. 2

Valve elliptical-lanceolate with rostrate, produced apices, 60 μm long, 23 μm wide (60–85 μm long and 22–26 μm wide in John (1983)). Axial area broad, narrowing towards the apices, central area broadly lanceolate, slightly asymmetrical. Raphe branches broad. Striae costate, radiate towards the middle, becoming convergent towards the apices, 7 in 10 μm.

This species is similar to *Navicula yarrensis* Grunow and *N. elegans* W. Smith, but valve of *N. yarrensis* has 4–4.5 striae in 10 μm, 80–162 μm long, 26–35 μm wide, with the striae radiate in the middle, parallel to slightly convergent at the apices, and that of *N. elegans* has 9–11 striae in 10 μm, approximately parallel margins and large orbicular central area, 75–80 μm long, 18–21 μm wide, with the striae strongly curved, radiate towards the middle and parallel to convergent towards the apices (John, 1983).

Habitat: Freshwater and brackish water, benthic and planktonic. Water salinity ranges from 2.2 to 35.6 in John (1983) and from 5.0 to 26.1 in our samples, respectively.

Distribution: Our samples were collected from mudflat in mangroves in Yunxiao County, Fujian Province, China. This species has been previously found in Sri Lanka (Foged, 1976) and estuarine of Swan River in Australia (John, 1983).

3. Navicula platyventris Meister in Bibl. Diatom. 44: 95, fig. 33. 1935.

侧偏舟形藻 Fig. 3

Valve long elliptical, 2-rostrate, 17 μm long and 4.5 μm wide (11–22 μm long and 5–7 μm wide in Navarro (1982)). Striae radiate with short bar areolae, 27 in 10 μm. Axial area narrowing towards the apices, central area orbicular, with a short stria on each side.

This species is different from *Navicula rhaphoneis* (Ehr.) Grunow by its short striae and areolae (Cheng et al., 1993). In *N. rhaphoneis*, the valve is 11.5–35 μm long and 5–11 μm wide, the apical endings turn to the same side, the central area is expanded, with 2 short striae on each side (only 2 areolae), striae radiate towards the middle, becoming parallel or slightly convergent near the apices, 11–12.5 in 10 μm.

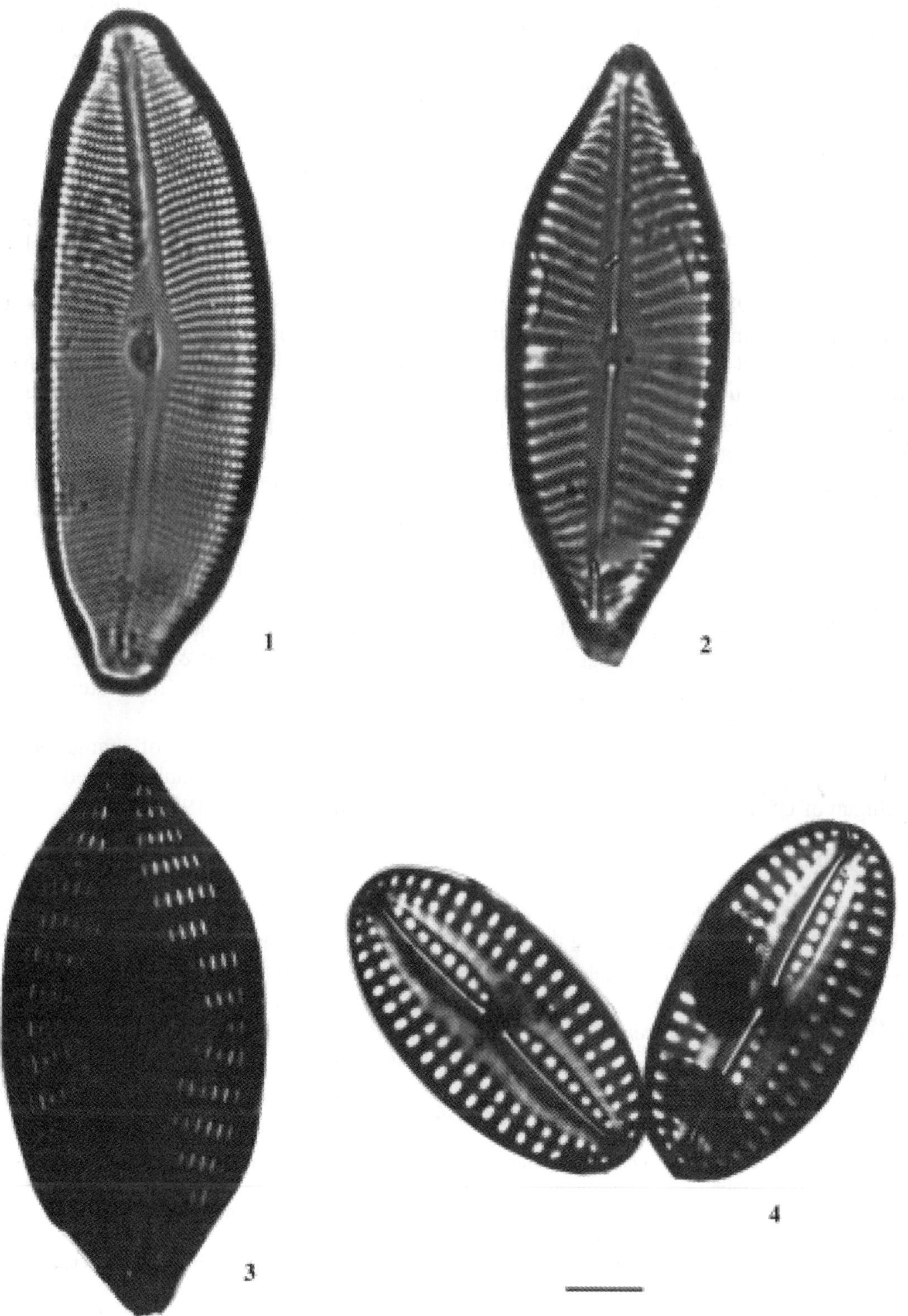

Figs. 1–4. Photomicrographs of four newly recorded species of Bacillariophyta. **1.** *Cymbella cucumis* (LM). **2.** *Navicula elegantoides* (LM). **3.** *Navicula platyventris* (TEM). **4.** *Navicula tenera* (TEM). Scale bar=3 μm for Figs. 3, 4 and 10 μm for Figs. 1, 2.

Habitat: Sea water and brackish water, planktonic. Water salinity ranges from 25 to 40 in Navarro (1982) and from 17.6–21.8 in mangroves in Shenzhen City, China, respectively.

Distribution: Our samples were collected from water in mangroves in Shenzhen City, China. This species has been previously found in Florida, USA (Navarro, 1982).

4. Navicula tenera Hustedt in A. Schmidt et al., Atlas Diatomaceenkunde 405, pl. 405. 1936; et in Archiv Hydrobiologie, suppl.-Bd. 15 (Tropische Binnengewasser, Bd. 7): 259. 1937; Prowse in Garden's Bulletin, Singapore 19: 48. 1962.

柔弱舟形藻 Fig. 4

Valve elliptical with broadly rounded poles, 12 μm long, 5.5 μm wide (9–14.5 μm long and 4–6.5 μm wide in Archibald (1983), 9–27 μm long and 4–9 μm wide in Krammer & Lange-Bertalot (1986)). The raphe is slightly arcuate, and the axial area is a wide lanceolate region made asymmetrical on account of the row of isolated pores on one side of the raphe. The axis rib is out of the axial area, with fine areolae. The transapical striae are relatively short extending a third of the valve width inwards from the margin. A curved longitudinal costa on either side of the raphe divides the striae into a smaller pore on the inner side of the costa, with a larger areole on the outer side.

The species has a row of isolated pores on one side of the raphe in the axial area. In its closely related species, such as *Navicula insociabilis* Krasske and *N. monoculata* Hustedt, the pores lie just outside the axial rib or slightly indented into it. The valve of *N. monoculata* is 6.5–11 μm long, 2.5–4.7 μm wide, with 30 axis ribs in 10 μm, and the apical endings turn to the same side.

Habitat: Freshwater and brackish water, planktonic and benthic. Archibald (1983) reported *Navicula tenera* to occur in estuarine of river, though Wujek & Rupp (1980) and Foged (1976) pointed out that it is a freshwater species. In our samples the water salinity ranged from 17.6–21.8 in mangroves of Shenzhen City, China.

Distribution: Our samples were collected from mudflat and water of mangroves in Shenzhen City, China. This species has been previously recorded in Sri Lanka (Foged, 1976), Michigan of USA (Wujek & Rupp, 1980) and South Africa (Archibald, 1983).

3 Discussion

Former studies have ascribed *Cymbella cucumis* to only freshwater species (Cleve, 1894; Prowse, 1962). We observed that it occurred, though occasionally, where water salinity was more than 15, suggesting that it was also a brackish water species. *Navicula elegantoides*, *N. platyventris* and *N. tenera* occurred in both freshwater and brackish water. Though *N. elegantoides* and *N. platyventris* were rare, *N. tenera* was common, ranging from 2.0×10^3–2.6×10^5 cells/L from Jan. 2001 to Jan. 2003 in the samples.

References

Archibald R E M. 1983. The diatoms of the Sundays and Great Fish Rivers in the Eastern Cape Province of South Africa. Bibliotheca Diatomologica 1: 1–362.

Beumer J P. 1978. Feeding ecology of four fishes from a mangrove creek in north Queensland, Australia. Journal of Fishery Biology 12: 475–490.

Chen C-P, Gao Y-H, Lin P. 2005. Biomass, species composition and diversity of benthic diatoms in mangroves of the Houyu Bay, China. Acta Oceanologica Sinica 24 (2): 145–154.

Cheng Z-D (程兆第), Gao Y-H (高亚辉), Liu S-C (刘师成). 1993. Nanodiatom from Fujian coast (福建沿岸微型硅藻). Beijing: China Ocean Press. 1–91.

Cleve P T. 1894. Synopsis of the naviculoid diatom. Kongliga Svenska Vetenskaps-Akademiens Handlingar 26: 1–194.

Du Q (杜琦), Jin D-X (金德祥). 1983. Studies on the epiphytic diatoms in the intertidal zones of the Jiulong River Estuary, Fujian, China. Journal of Oceanography in Taiwan Strait (台湾海峡) 2 (2): 76–97.

Fan H-Q (范航清), Cheng Z-D (程兆第), Liu S-C (刘师成), Gao Y-H (高亚辉). 1993. Species of benthic diatoms in Guangxi mangrove habitats. Journal of the Guangxi Academy of Sciences (广西科学院学报) 9 (2): 37–42.

Foged N. 1976. Freshwater diatoms in Sri Lanka (Ceylon). Bibliotheca Phycologica 23: 1–112.

Foged N. 1979. Diatoms in New Zealand, the North Island. Bibliotheca Phycologica 47: 1–224.

John J. 1983. The diatom flora of the Swan River Estuary, western Australia. Bibliotheca Phycologica 64: 1–358.

Krammer K, Lange-Bertalot H. 1986. Bacillariophyceae, Teil 1: Naviculaceae. In: Ettl H, Gerloff J, Heynig H, Mollenhauer D eds. Süsswasserflora von Mitteleuropa (Begründet von A. Pascher). New York, Stuttgart: Gustav Fischer Verlag. 2 (1): 1–876.

Lin P (林鹏). 1984. Mangrove Vegetation (红树林). Beijing: China Ocean Press. 1–102.

Liu Y (刘玉), Chen G-Z (陈桂珠). 1997. Study on community structure and ecology of algae in mangrove area in Futian, Shenzhen. Acta Scientiarum Naturalium Universitatis Sunyatseni (中山大学学报(自然科学版)) 36 (1): 101–106.

Nagumo T, Hara Y. 1990. Species composition and vertical distribution of diatoms occurring in a Japanese mangrove forest. The Japanese Journal of Phycology 38: 333–343.

Navarro J N. 1982. Marine diatoms associated with mangrove prop roots in the Indian River, Florida, USA. Bibliotheca Phycologica 61: 1–151.

Nicholas W L, Stewart A C, Marples T G. 1988. Field and laboratory studies of *Desmodora cazca* Gerlach, 1956 (Desmodoridae: Nematoda) from mangrove mud-flats. Nematologica 34: 331–349.

Prowse G A. 1962. Diatoms of Malayan Freshwaters. Garden's Bulletin, Singapore 19: 1–104.

Sequeiros-Beltrones D A, Castrejon E N. 1999. Structure of benthic diatom assemblages from a mangrove environment in a Mexican subtropical lagoon. Biotropica 31: 48–70.

Wujek D E, Rupp R F. 1980. Diatoms of the Tittabawassee River, Michigan. Bibliotheca Phycologica 50: 1–100.

红树林下中国新记录的四种硅藻

陈长平　高亚辉　林　鹏

(厦门大学生命科学学院　厦门 361005)

摘要　报道了来自福建和深圳红树林下中国首次记录的4种硅藻，即瓜形桥弯藻*Cymbella cucumis* A. Schmidt、拟优美舟形藻*Navicula elegantoides* Hustedt、侧偏舟形藻*N. platyventris* Meister 和柔弱舟形藻*N. tenera* Hustedt，同时描述了每个种类的细胞形态特征和生态分布特点。作者认为淡水硅藻瓜形桥弯藻*C. cucumis*在半咸水和海水的环境中也有分布(盐度>15)，应属于淡水和半咸水种。

关键词　硅藻；红树林；新记录；中国

盐度对稀释平板法研究红树林区土壤微生物数量的影响*

张瑜斌[1,2] 林 鹏[2] 魏小勇[3] 庄铁诚[2]

(1. 广东海洋大学海洋资源与环境监测中心，湛江 524088；
2. 厦门大学生命科学学院，厦门 361005；
3. 广州中医药大学基础医学院，广州 515045)

摘要：在使用稀释平板法分离潮间带红树林及其对照光滩土壤微生物以及计数时，多数情况下使用陈海水制作培养基和稀释水，很少考虑培养基和稀释水的盐度对最终计数结果的影响。使用稀释平板法研究了盐度对福建九龙江口红树林区与深圳福田红树林保护区土壤微生物平板计数的影响，结果表明培养基与稀释水盐度对微生物数量有明显的影响。统计分析显示细菌的海水稀释效果优于淡水，而放线菌与真菌则刚好相反(P＜ 0.05，一个例外)。海水不适合配制红树林区土壤微生物平板计数的培养基，从 0～35，高盐度的平板培养基会降低微生物的数量，尤其是放线菌的数量，尽管培养基的盐度对真菌影响无规律，但细菌数量在低盐度时比在高盐度和不加氯化钠时要多。根据盐度效应，提出了稀释平板技术应用于潮间带的红树林及其相应光滩时的优化方法，认为细菌应该用海水作无菌稀释水，而放线菌和真菌则应用淡水作稀释水；包括光滩在内的红树林区土壤微生物分离与计数的培养基宜控制较低盐度范围。

关键词：盐度；微生物数量；土壤；红树林；潮间带

文章编号：1000－0933 (2008)03-1287-09　中图分类号：Q939；S154.3　文献标识码：A

Effect of salinity on microbial densities of soil in the dilution platetechnique applied in mangrove areas

ZHANG Yu-Bin[1,2], LIN Peng[2], WEIXiao-Yong[3], ZHUANG Tie-Cheng[2]

1 *Monitoring Center forMarine Resource and Environment, Guangdong Ocean University, Zhanjiang* 524088, *China*

2 *School of Life Science, Xiamen University, Xiamen* 361005, *China*

3 *School of Basic Medicial Science, Guangzhou University of TCM, Guangzhou* 510405, *China*

Acta Ecologica Sinica, 2008, 28 (3):1287～1295.

Abstract: When the soilmicrobial densities are determined in mangroves and correspondingmudflat at the same tidal level by the dilution plate method, the agarmedia and dilution water are generallymade up of aged seawater in most

* 国家自然科学基金资助项目(30270272)
原载于生态学报，2008，28(3)：1287－1295

cases, and effects of salinity in agar media and dilution water on the enumeration of microbes is seldom taken into consideration. The effects of salinity on soil microbial counting from the samples in mangrove areas in J iulongjiang Estuary of Fujian, and FutianMangrove Nature Reserve of Shenzhen, China, were tested by dilution plate technique. The results showed that the soil microbial densities in mangroves and mudflat were significantly influenced by the salinity of dilution water and agar media. For the bacteria, the seawater served as sterilized dilution water was significantly ($P<0.05$)more benefic to the enumeration on the plates than the freshwater, but in reverse for the actinomycetes and fungi. The increasing salinity of media within 35 significantly decreased microbial colonies on the plates, especially for the actinomycetes, in spite of the fact that the effect of salinity ofmedia on fungal numberswas not indefinite. The bacterial colonieswere more abundant on the agar plates with low salinity than with high salinity or without any NaCl. It was proposed that some methodological improvements were needed when the dilution plate technique was applied to microbial counting in the samples of mangrove forest and mudflat at the same tidal level in inter-tidal zone. The sterilized dilution water should be prepared with seawater for the bacteria, but with freshwater or low saline water for the actinomycetes and fungi. The salinity of agarmedia should be low for the microbial isolation and enumeration of soil samples from the mangrove areas including mudflats.

KeyWords: salinity; microbial densities; soil; mangroves; inter-tidal zone

1 Introduction

Mangroves are woody plant communities in the inter-tidal zone[1]. Mangroves and their corresponding mudflat occupy a littoral habitat, characterized almost invariably by salt or brackish water and coastal silt. The salinity is one of the most important ecological factors in this coastal wetland, which is significantly different from other aquatic environments. Many researchers have addressed the effect of salinity in mangrove ecosystem[2-4]. Beyond all doubt, the microbes in this habitatmust be able to tolerate the conditions characteristic in this kind of special ecosystem. The salinity influences generation time, morphological and physiological characters of bacteria and fungi in aquatic saline habitat[5]. Similarly, microbial distribution is easily influenced by the habitat salinity in mangrove ecosystem[6]. A study from Leano etal[7] revealed that three isolates from mangroves identified as *Halophy tophthora vesicular* grew at a salinity range of 0—60 normally, but vigorously at 15—25 and the optimum salinity levels for sporulation were different among them. Nakagiri and Ito[8] also reported that the ascospore appendages formed by *Aniptodera salsuginosa* isolated from mangroves were functional only when they were submerged in brackish water. Moreover, there was a gradual decrease in the populations of soil nitrogen-transforming microorganisms in Bhitarkanika mangrove forest due to deforestation and adverse effect of increased salinity in deforested land and salty land respectively[9]. It is obvious that the salinity is an extensive ecological factor controlling microbes in mangrove ecosystem.

The dilution plate method was, whether for isolation or enumeration of bacteria and fungi, often utilized in previousmicrobiological investigations in mangrove ecosystem[10-12]. However, in most cases, the salinity of dilution water and plate media was hardly considered in this method applied in the mangrove areas, although Garg[10] discovered that the colonies and occurrence frequency of fungi decreased with the increasing salinity of media at a gradient of 0, 30, and 60. The few salinity gradients and last two higher ones probably led to his experimental results because the high salinity of media above the habitat one by far would stress the fungal growth and development. Besides a few of bacteria, no actinomyctes and fungi were detected when dilution plate technique was utilized to investigate microbial distribution in Dongzhaigang Mangroves Reserve, Hainan of China[13], in which the dilution water and plate media were prepared with the seawater. The similar results were obtained when the same method was used to investigate microbial community in Xiamen West Sea Area, China[14]; Rare actinomyctes and fungi in actual habitat and the investigative method might be responsible for it. However, a great number of actinomycete colonies appeared on the agar plates as soon as the media salinity declined[14], implying that the salinity of agarmedia could influence the final number of colony. The inter-tidal zone includingmangrove area is an ecotone from the land to the sea, where soil habitat is both terrestrial and marine, but is not completely different from any one of them in some degrees. Should this area be taken as the terrestrial or marine one, and what range of salinity should be controlled when the dilution plate technique is applied to measure the microbial density of soil? Based on the findings above2mentioned, in this study, the effects of salinity of dilution water and agar media on colony counting of heterotrophic bacteria, actinomycetes and fungi on the plates were examined when this technique was used to investigate soilmicrobial densities in two mangrove areas of China.

2 Materials and methods

2.1 Sampling stations

Two mangrove areas investigated in this study locate at J iulongjiang Estuary, Fujian Province, China, and FutianMangrove Nature Reserve, Guangdong Province, China, respectively, where soil sampleswere collected from *Kandelia candel* forests, *Avicennia marina* forests and their corresponding mudflats. In J iulongjiang Estuary, sampling stations J1 (*K. candel* forest) and J2 (mudflat) locate at Caoputou village (24°24′N, 117°55′E), Fugong Town, Longhai City, and J3 (*A. marina* forest) and J4 (mudflat) at Dongyu Village (24°31′N, 118°03′E), Haicang Town, Xiamen City. The mudflat stations are at least 100 m away from their corresponding *K. candel* forest and A. m arina forest at the same tidal level and without any vegetation. In FutianMangrove Nature Reserve, sampling stations F1 (*K. candel* forest) and F2 (*A. marina* forest) locate at a bird observation zone near Chegongmiao (114°05′E, 22°32′N), which has been described by Chen et al. [15] and L in et al. [16], and no sample was taken from their corresponding mudflats because of administerial restriction. The climate in J iulongjiang Estuary belongs to southern subtropical maritime climate. The *K. candel* forest in J iulongjiang Estuary mainly occupiesmiddle and high inter-tidal zone, and is a zonal distribution (about 40 m in width) along the southern coast of this estuary, with a small quantity of *Aegiceras corniculatum* trees and *A. marina* trees at the forest fringe. The forest form is tidy with the tree height from 5.5 m to 6.0 m and crown density over 0.9. The *A. marina* forest locating at Dongyu village is also a pure forest, with the tidy canopy, green physiognomy, and simple community structure. The average tree height is about 1.2 m in *A. marina* forest, but the density of tree is so great (reaching 15 indiv · m^{-2}) that it is difficult to go through when sampling there. In addition, there are plenty of seedlings in this forest (above 31 ind · m^{-2}), and many finger · like pneumatophores (about 35 ind · m^{-2}) rise out of the surface of mudflat about 10 cm high. The entire tidal flat is smooth.

2.2 Collection of soil samples

Soil sampleswere aseptically collected at random from top soil of 20 cm depth at the middle inter-tidal zone with a sterile PVC pipe at low tide. Multiple samples collected at each plot were mixed together uniformly and kept in sterile polyethylene bags after the roots were removed. The mixed samples were then taken to the laboratory for analysis immediately.

2.3 Media

(1)Bacterialmedium (2216E agar by Zobell), peptone 5 g, yeast extract 1 g, $FePO_4$ 0.01 g, agar 15 g, water 1000 ml, and pH 7.2—7.4; (2)Actinomycetesmedium: soluble starch 20 g, casein 1 g, KNO_3 1 g, K_2HPO_4 0.5 g, $MgSO_4$ $7H_2O$ 0.5 g, agar 15 g, trace elemental solution 1 ml, water 1000 ml, pH 7.4, and $K_2Cr_2O_4$ solution as inhibitor supplemented into medium after autoclave at final concentration of 50 ($\mu g \cdot g^{-1}$); (3)Fungal medium: glucose 10 g, peptone 5 g, K_2HPO_4 1 g, $MgSO_4$ $7H_2O$ 0.5 g, agar 15 g, Bengal red solution (0.5%)0.66 ml, water 1000 ml, medical chloramphenicol (as inhibitor)2 ml, and pH 6.6.

2.4 Measurement ofmicrobial densities

In this study, microbial densities of soil sampleswere determined using dilution plate method. Soil sample of 10 g wasmixed with sterilized water of 90 ml. The mixture was vigorously shaken on a swirlmixer and then settled for 10 min. The overlying suspension containingmicrobeswas further diluted with the sterilized water at a rate of 1∶10. Then, the suspension was plated out on the agarmedia, respectively. Three appropriate serial dilutionswere selected for the bacteria, actinomycetes and fungi respectively. Each dilution replicated three times.

The samples from J1 and J3 were diluted with aged seawater (SW, salinity 29.6) and freshwater (FW), respectively, to compare the effect of salinity on the final colony counting on the plates, but the samples from J2, J4, F1 and F2 were diluted only with freshwater. To evaluate the effect of salinity in the media on microbial enumeration, the diluted suspensions were plated out on the agar plates above2mentioned with different NaCl concentration from 0 to 35.

The plates were incubated at (28±1)℃. The formation of colonies was observed regularly every day and the enumeration was performed after incubation for 223 d for the bacteria, 10 d for the actinomycetes and 3—5 d for the fungi. CFU (colony2forming unit)per gram of wet soil was calculated for each sample, but the final results were expressed as CFU per gram dry weight ($CFU \cdot g^{-1} \cdot dw$) based on the water coefficient of samples.

2.5 Analysis of main physical-chemical characteristics of soil

The soil samples were air dried naturally, grounded to fine powder and sifted. The sifted samples were used to analyze the physical-chemical characteristics, including soil texture by the densimetermethod, organic matter (OM) by wet oxidation with potassium dichromate, total nitrogen (TN) by Kjeldahl nitrogen, total phosphorus (TP) by standard colorimetric methods after acid digestion, total potassium (TK) by inductively coupled plasma atom emission spectrum (ICP-AES), salinity by electric conductivity, pH value by potentiometric analysis with a pH meter according to the methods described in related reference byNanjing Institute of Soil Science, Chinese Academic of Sciences[17].

2.6 Statistical analysis

Differences in microbial densities between the seawater dilution and freshwater dilution and between stations were assessed using nonparametricW ilcoxon test. The statistical analysis was performed in the SPSS for windows.

3 Results

3.1 Main physical-chemical characteristics of soil

In J iulongjiang Estuary, the concentrations of OM and TN were lower at J1 than at J3, but TP was higher at J1 and its correspondingmudflat J2 than at J3 and its correspondingmudflat J4, respectively (Tab. 1). The soil salinity and pH value were correspondingly higher at J3 and J4 than at J1 and J2. In FutianMangrove Nature Reserve, the concentrations of OM, TN, and TP, and salinitywere lower at F1 than at F2, while the pH value was higher at F1 than at F2.

Table 1 Ma in physical-chemical character istics of soil at sampling stations in two mangrove areas

Mangrove area	Station	Soil texture	OM(%)	TN(%)	TP (%)	TK(%)	Salinity	pH
J iulongjiang Estuary	J1	Light clay	3.206	0.143	0.066	0.725	15.14	6.55
	J2	Light clay	3.054	0.1344	0.064	0.634	11.56	7.08
	J3	Light clay	3.600	0.148	0.057	0.851	24.30	7.00
	J4	Middle clay	2.139	0.117	0.055	1.121	21.25	7.18
FutianMangrove Nature Reserve	F1	Light clay	4.09	0.139	0.086	—	15.25	6.50
	F2	Light clay	6.72	0.363	0.099	1.25*	19.30	5.88

* Cited from Lin *et al.* [16]

3.2 Effect of salinity of dilution water

The salinity of dilution water significantly influenced the final numbers of colony for the samples from J1 and J3 (Table 2). The statistical analysis indicated that the bacterial colonies were significantly ($p<0.05$)more abundant when the seawater was used as dilution water, but the actinomycetes and fungi were significantly ($p<0.05$)more when the samples were diluted with the freshwater, except the fungi at J3 (Table 3). Therefore, the seawater was more suitable for dilution of bacterial suspension than the freshwater, with the reverse results for the actinomycetes and fungi.

Table 2 Microbial densities of soil under salinity gradient of media in two mangrove areas*

Station	Microbial group	Dilution water	Concentration of NaCl in the media							
			0	5	10	15	20	25	30	35
J1	Bacteria	SW	68.17	77.61	79.57	69.06		23.32	19.76	16.73
J1	Bacteria	FW	13.88	11.93	9.79	11.21		9.79	8.90	6.76
J1	Actinomycetes	SW	22.07	18.33	14.95	9.97		5.52	1.96	0.71
J1	Actinomycetes	FW	39.69	23.5	16.38	12.99		11.93	3.92	1.25
J1	Fungi	SW	6.23	8.9	8.9	8.01		7.12	3.56	3.56
J1	Fungi	FW	38.27	38.27	32.04	35.6		31.15	32.93	26.7
J2	Bacteria	FW	26.38	32.40	26.38	22.12		21.92	11.83	13.58
J2	Actinomycetes	FW	28.13	25.41	19.01	10.86		7.95	3.10	0.97
J2	Fungi	FW	26.19	21.34	20.37	28.13		13.58	13.58	13.58
J3	Bacteria	SW	19.02	21.74	19.44	19.65		19.44	17.14	16.3
J3	Bacteria	FW	19.96	14.21	8.78	7.73		4.6	3.76	2.72
J3	Actinomycetes	SW	26.75	24.45	20.06	7.11		2.72	1.46	0.21
J3	Actinomycetes	FW	27.38	26.13	21.95	15.05		9.82	3.76	1.25
J3	Fungi	SW	5.23	8.36	5.23	8.36		6.51	4.18	5.23
J3	Fungi	FW	9.41	8.36	8.36	8.36		4.18	10.45	10.45
J4	Bacteria	FW	5.48	7.37	4.35	4.73		3.02	1.51	0.76
J4	Actinomycetes	FW	16.63	22.30	20.98	13.42		7.56	4.73	0.95
J4	Fungi	FW	0.95	3.78	0.95	4.73		3.78	1.89	3.78
F1	Bacteria	FW	11.64	14.32	14.32	6.27		9.85	6.27	3.58
F1	Actinomycetes	FW	20.59	14.86	11.10	6.44		3.22	0.72	0.18
F1	Fungi	FW	8.06	7.16	7.16	6.27		6.51	4.34	6.51
F2	Bacteria	FW	34.72	56.42	39.06	44.49		24.96	22.79	8.64
F2	Actinomycetes	FW	30.16	25.39	20.18	14.11		8.03	2.39	1.09
F2	Fungi	FW	45.57	51	27.22	39.06		56.42	52.08	56.42

* Bacteria 10^4 CFU · g^{-1} dw; Actinomycetes 10^3 CFU · g^{-1} dw; Fungi 10 CFU · g^{-1} dw; SW: seawater; FW: fresh water

Table 3 Nonparametr icW ilcoxon test on var ia tion of m icrob ia l den sities in d ilution wa ter type

Microbial groups	Station	Comparison of dilution water	Z value of Wilcoxon test	p value
Bacteria	J1	FW-SW	−2.521	0.012*
	J3	FW-SW	−2.380	0.017*
Actinomycetes	J1	FW-SW	−2.380	0.017*
	J3	FW-SW	−2.366	0.018*
Fungi	J1	FW-SW	−2.527	0.012*
	J3	FW-SW	−1.753	0.080

* Significant at the level of $P<0.05$

3.3 Effect of salinity ofmedia

The colony numbers on the plates were significantly influenced by the salinity of media. The most bacterial colonies from the sample of J1 were found on the plates containing 10 salinity and the least at 35, and the colonies were significantlymore abundant at low salinity (0−15) than at high salinity (20−35) when the seawaterwas used as dilution water. However, such variation was not found when the freshwaterwas used as dilution water (Table 2). At J2, the bacterial densities decreased graduallywith increasing salinity ofmedia from 0 to 35 when the freshwaterwas used as dilution water, with the most densities at 5. At J3, the bacterial colonieswere the most abundant at 5, and then decreased with increasing salinity of media when the samples were diluted with the fresh water (Table 2), but the difference between low and high salinitieswasmuch smaller at J3 than at J1. The bacterial densities at J3 sharply decreased with increasing salinity of media from 0 to 35 when in the freshwater dilution system. However, the bacterial densities decreased a little with the increasing salinity of media at J4 although the variation tendency was similar to that at J3 (Table 2)to some extent, which resulted in the fact thatmore bacterial colonieswere counted at J3 than at J4.

For the actinomycetes in J iulongjiang Estuary, the highest densities were almost found at 0, only with one exception at J4 in which the highest one emerged at 5 (Table 2). The actinomycetes densities at all stations declined sharply with the increasing salinity, and down to the leastwhen the salinity rose up to the highest (35). Moreover, some blank plateswere present at the highest salinity of 35. When the freshwater was used as dilution water, the difference in densities of actinomycetes between J1 and J2, and between J3 and J4 was not as large as that of the bacteria.

The densities of fungi in J iulongjiang Estuary did not show well2regulated variation with the increasing salinity of media, except at J1 and J2 the fungi colonies were more abundant at low salinities than at high salinities when the samples were diluted with the freshwater (Table 2). However, much more fungi were counted in the samples from two mangrove stations than those from two corresponding mudflat stations when the freshwater was used as dilution water.

In FutianMangrove Nature Reserve, the significant decreasing colony densities with the increasing salinity of media were also found in the bacteria and actinomycetes. However, the variation tendency of fungi was indefinite since at F2, the densities of fungi seemed to be higher at high salinity than those at low salinity, nor did the fungi at F1 (Table 2).

3.4 Difference in microbial densities between mangrove stations

The soilmicrobeswere significantly ($p<0.05$, with the exception of fungi)more abundant at J1 than at J2 in J iulongjiang Estuary when the samples were diluted with suitable water. In Futian Mangrove Nature

Reserve, for three groups ofmicrobes, the higher microbial densities were all measured at F2 than at F1 ($p < 0.05$), and the biggest difference was found in the fungi (Table 2).

4 Discussion

Although the plate countingmethod remarkably underestimates true microbial densities, it is continually used for the isolation and enumeration of microbes, especially for the actinomycetes and fungi, in mangrove areas[6,10-13,18]. The bacterial density is generally determined by ep iflourescence microscopy due to its accurateness, but the bacteria are also isolated and their densities are measured sometimes by dilution plate method[12,13,18], therefore few researcher cared about the effect of salinity of dilution water and media on bacterial abundance in this method. However, this study indicated that the bacterial colonieswere more at the low salinity range of agarmedia whether in J iulongjiang Estuary or in Futian Mangrove Nature Reserve, and the effects of seawater dilution were significantly better than the fresh water. Therefore, it is concluded that the bacteria are terrestrial in the mangrove areas including mudflats, but they probably adapt to the salt stress from these areas easily during a long evolution, and their growth and development are more vigorous under the certain salinity than without any salt stimulation.

The regulation that the abundance of fungal colony varied with the salinity ofmedia was apparently indefinite in this study, which was to some extent relative to the fungal nature. Gray *et al.* considered thatmanymangrove fungi were able to tolerate the great variations in the salinity of media with ease[19], which was in agreement with this investigation, while Garg discovered that the numbers and occurrence frequency of fungi decreased with the increasing salinity ofmedia at a gradient of 0, 30, and 60[10]. We consider that the few salinity gradients and last two higher onesmay be key factors leading to his finding. The statistical analysis showed that the fungal colonieswith the fresh water dilution were significantlymore than with the seawater dilution in this study; we think it is relative to the terrestrial origin of fungi and their nature with the resistant spore. It was confirmed that the fungi found in mangrove mud are all "terrestrial"species with individual exceptions[20].

The actinomycetes seemed to be very sensitive to the salinity by reason of that their abundance on the plates declined rapidly with the increasing salinity of agar media, moreover, much more colonies were counted as the samples were diluted with the fresh water. So, the low salinity in dilution plate method is significantlymore effective and helpful to the isolation and enumeration of actinomycetes. Takizawa et al.[21] reported that the actinomycetes densities reached $1.8 \times 10^2 - 1.4 \times 10^5$ CFU · ml^{-1} sedimentwhen the media-starch casein agarwas supplemented with NaCl at final concentration of 0.5% , and all sediment samples were diluted with sterilized 0.5% saline water. The same effect occurred at both mudflat stations and mangrove ones in this investigation, asmeant that thismethod could identically be applied to the mangrove areas including mudflats. We think that the actinomycetes are also "terrestrial"like the bacteria and fungi in the mangrove areas includingmudflats, as a result, itwasmore effective for the soil samples in the mangrove areas to be treated in a non-marine way or a terrestrial way. There were few reports involved in the effect of salinity of dilution water and agar media on the microbial enumeration on the plates in the previous studies in spite of the finding by Garg[10]. Neither the salinity of dilution water nor that of agarmedia wasmentioned or described clearly in a few of reports[10-12,22] despite the reliable results were obtained in them. In practice, no or scarce colonies of soil actinomycetes[13,22,23] and fungi[11,13,22] were counted in some mangrove stations, which had to some extent relevant to this effect of salinity of dilution water and plate media in the majority of these reports. This study confirmed that the higher salinity would cut microbial numbers down with ease. Therefore, it should be considered carefully for the salinity of dilution water and agar media when the dilution plate method was employed in mangrove areas includingmudflats.

Kohlmeyer found that the fungal densities decreased with the increasing habitat salinity from the inland zone to the seaward zone in mangrove forest[24]. The same phenomenon involved in the fungi, bacteria and actinomycetes occurred in mangrove area of J iulongjiang Estuary, where although the concentration of OM and TN was higher at J3 (*Avicenn ia m arina* forest) than at J1 (*Kandelia candel* forest) (Table 1), soil microbes were more abundant at J1 than at J3, similarly, the microbial densitieswere more at J2 than J4, because J3 and J4 were more seaward than J1 and J2. As a result, the time submerged by tide was longer, the soil salinity was higher (Table 1), and the redox potential①was lower at J3 and J4 than at J1 and J2 respectively. Lee and Baker[6] considered that organic matter content, nutrient level, oxygen content (equivalent of redox potential), pH and salinity were probably the most important environmental factors controlling the nature and distribution pattern ofmicrofungi in mangrove swamp soil. But the time submerged by tide, soil salinity and redox potential seemed to be more important than other factors such as organic matter content, nutrient level among these stations in J iulongjiang Estuary. However, this phenomenon did not occurred at two mangrove stations in the FutianMangrove Nature Reserve, in respect thatmicrobial densities were least at F1 (*K. candel* forest) than at F2 (*A. marina* forest), which was without question associated with special habitat. The concentration ofOM and nutrient level in soilwas lower at F1 than at F2, so did soil salinity, in spite of the fact that two stations have same tidal level[15,16]. Therefore, the organic matter and nutrient level should be considered as the leading factor at these two stations. Microbial distribution and horizontal difference is, in virtually, a complicated ecological question involved in habitat salinity.

On the basis of this study, some optimum improvements are proposed as follows when the dilution plate technique is employed to investigate the microbial density of soil in mangrove areas including correspondingmudflats: ① How to prepare sterile dilution water. Sterilized dilution water should be prepared with the seawater for the bacteria, and with the freshwater or low saline water (for instance 5) for the actinomycetes and fungi respectively at least in this study.

②The seawater is not suitable for preparing the agarmedia. The suitable salinity of bacterialmedia is at the low range of salinity (approximately 10 in this study) and approp riate salinity ofmedia can be adjusted easily in the light of actual salinity in the habitat. The salinity of agarmedia for the actinomycetes should not be over 5, and that for the fungi should also be in low salinity although the effect of salinity of fungalmedia was indefinite.

③A better strategy is to measure the habitat salinity and provide preliminary experiments for some samples following the method supplied by authors. In this way, the salinity of media and the types of dilution water may finally be controlled and selected with ease.

If the study aims at obtaining halophilic microbes or realmarine ones, both media and dilution water should be

① Zhang Y L. Ecological studies on enzymatic activity and fine roots of soil in mangrove forest in Jiulongjiang Estuary. Doctor Thesis in Xiamen University, Xiamen, China, 1996. 27—40

prepared with the seawater, and even the salinity ofmedia should further be raised when necessary.

Refrences:

[1] Lin P. Mangrove Ecosystem in China. Beijing, and New York:Science Press, 1999. 1.

[2] Garg K L. Effect of salinity on cellulolytic activity of some pneumatophore inhabiting fungi of Sunderban mangrove swamps. Indian Journal of Marine Sciences, 1982, 11:339—340.

[3] Sun Q and Lin P. Wood structure of *Aegiceras corniculatum* and its ecological adaptations to salinities. Hydrobiologia, 1997, 352:61—66.

[4] Naidoo G, Tuffers A V, Willert D J. Changes in gas exchange and chlorophyll fluorescence characteristics of two mangroves and a mangrove associate in response to salinity in the natural environment. Trees2Structure and Function, 2002, 16:140—146.

[5] Rheinheimey G. Aquatic microbiology (third edition). John Wiley & Sons Incorporation, Chichester • New York • Brisbane • Toronto, 1985,108—111.

[6] Lee B K H, Baker G E. An ecological study of the soilmicrofungi in a Hawaiian mangrove swamp. Pacific Science, 1972, 26:1—10.

[7] Leano EM, Virijmoed L L P, Jones E B G. Physiological studies on *Halophytophthora vesicula* (straminpilous fungi) isolated from fallen mangrove leaves from Mai Po Hong Kong. BotanicaMarina, 1998, 41 (4):411—419.

[8] Nakagiri A, Ito T. *Aniptodera salsuginosa*, a new mangrove-inhabiting Ascomycete, with observation on the effect of salinity on ascospore appendage morphology. Mycological Research, 1994, 98 (8):931—936.

[9] Routray T K, Satapathy G G, Mishra A K. Seasonal fluctuation of nitrogen transforming microorganisms in *Bhitarkanika mangrove* forest. Journal of Environmental Biology, 1996, 17:325—330.

[10] Garg K L. Vertical distribution of fungi in Sunderban mangrove mud. Indian Journal ofMarine Sciences, 1983, 12:48—51.

[11] Misra J K. Fungi from mangrove mud of Andama-Nicobar Islands. Indian Journal ofMarine Sciences, 1986, 15:185—186.

[12] Shome R, Shome B R, Mandal A B, *et al*. Bacterial flora in mangroves of Andaman-Part I:Isolation, identification and antibiogram studies. Indian Journal of Marine Sciences, 1995, 24:97—98.

[13] Zhang Y B, Zhuang T C, Yang ZW, *et al*. Microbial study of mangrove soil at Dongzhaigang Harbor in Hainan. Chinese Journal of Ecology, 2001, 20 (1):63—64.

[14] Zhang Y B, Wang W Q, Zhuang T C, *et al*. Microbial amounts variation in mudflat soil at southwest in west Xiamen Harbor. Journal of Oceanography in Taiwan Strait, 2000, 19 (1):54—59.

[15] Chen G Z, Miao S Y, Zhang J H. Ecologic Study on the Mangrove Forest in Futian Nature Reserve, Shenzhen, China. Acta Scientiarum Naturalium Universitatis Sunyatseni, 1996, 35 (Suppl.):294—300.

[16] Lin P, ZhengW J, Li Z J, *et al*. Accumulation and distribution of K, Na, Ca and Mg in *Avicennia marina* mangrove community in Shenzhen. In:Lang H Q, Lin P, Lu J J, eds. Conservation & Research of Wetlands in China. Shanghai: East China NormalUniversity Press, 1998. 273—278.

[17] Nanjing Institute of Soil Science, Chinese Academic of Sciences. Analysis for Soil physics and Chemistry. Shanghai: Shanghai Science and Technology Press, 1978.

[18] Matondkar S G P, Mahtani S, Mavinkurve S. Seasonal Variations in the microflora from mangrove swamps of Goa. Indian Journal of Marine Sciences, 1980, 9:119—120.

[19] GrayW, Pinto P U C, Pathak S G. Growth of fungi in seawatermedium. Applied Microbiology, 1963, 11:501—505.

[20] Swart H J. Further investigations of the mycoflora in the soil of some mangrove swamps. Acta Botanica Neerlandica, 1963, 12:98—111.

[21] TakizawaM, Colwell R R, Hill R T. Isolation and diversity of actinomycetes in the Chesapeake Bay. Applied and

Environmental Microbiology, 1993, 59 (4):997—1002.

[22] Hu C B, Liang X T. A study on themicrobiological flora and the biochemical activities of forest soil in the coastal beach of Hepu. Tropical Forestry(Science and Technology), 1987 (1):127.

[23] Jensen P R, Dwight R, Fenical W. Distribution of actinomycetes in near-shore tropical marine sediments. Applied and Environmental Microbiology, 1991, 57 (4):1102—1108.

[24] Kohlmeyer J. Ecological notes on fungi in mangrove forests. Transaction of BritishMycology Society, 1969, 53:237—250.

说明:培养基 NaCl 浓度单位为 g/kg The unit of concentration of NaCl in the media is g/kg.

Biomass, species composition and diversity of benthic diatoms in mangroves of the Houyu Bay, China*

CHEN Changping[1] GAO Yahui[1] LIN Peng[1]

(*School of Life Sciences, Xiamen University, Xiamen* 361005, *China*)

Abstract

The biomass, species composition and diversity of benthic diatom assemblages in mud-flat soils in *Kandelia candel* (L.) Druce communities with and without vegetation were studied seasonally at the Houyu Bay in Fuding City, Fujian Province, China. a total of 103 taxa were identified (including varieties). Eighty-four taxa were found in the mud-flat with vegetation and 74 taxa in the mud-flat without vegetation, while the biomass was large in January and April and decreased from July to October. The most abundant species in the mud-flat with vegetation are *Nitzschia cocconeiformis*, *Gyrosigma scalproides* and *N. fasciculata*, compared with *G. scalproides* and *N. obtusa* var. *scalpelliformis* in the mud-flat without vegetation. High *H'* values at 2 sites during all seasons suggest that diatom assemblages in the sediments of the Houyu Bay represent an original environment. Multi-dimensional scaling of diatom assemblages from mud-flats with and without vegetation shows that a slight seasonal change and only a single association occur in the mangroves.

Key words: diatom, mangrove, Houyu Bay, biomass, benthic

1 Introduction

Algae in mangroves constitute a significant food resource for various organisms in the mangrove ecosystem (Leh and Sasekumar, 1986; Leija-Tristan and Sanchez-Vargas, 1988; Nicholas et al., 1988; Vicente, 1989). Benthic diatoms serve as a common food for certain mangrove fishes (Milward, 1974; Beumer, 1978). Abundant pennate diatoms were found within the phytoplankton of mangrove environments (Lugomela and Semesi, 1995; Damroy, 1995), which suggests that benthic diatoms are an important source of enriched phytoplankton.

Taxonomic studies of diatoms in mangrove environments have been carried out mainly in the New World (Reyes-Vasquez, 1975; Sullivan, 1981; Navarro, 1982; Maples, 1983;Sequeiros-Beltrones and Castrejon, 1999), and there are few studies in the Australasian group (Foged, 1979) and Indo-Malesia group (Wah and Wee, 1988; Nagumo and Hara, 1990).

In China, only qualitative study on benthic diatom in the mangroves was done by Fan et al.(1993). No quantitative research work exists so far. This study aims to quantitatively report the benthic diatom assemblages from mangrove environments at the Houyu Bay in Fuding City, Fujian Province, China, which will enrich the database of mangrove ecosystem, and further our understanding of ecological functions of benthic diatoms in the energy flow and food webs of mangrove ecosystems.

The Houyu Bay is located in the northeast of Fujian Province, China (27°20′N and 120°18′ E) (Fig. 1). It is the northernmost area in nature for mangroves in China. The yearly average temperature is 18.5 ℃, and the lowest average temperature (January) is 8.4 ℃. The location is estuarine. Mangroves at the Houyu Bay

* From Acta Ocean Logica Sinica, 2005, 24(2): 1—13

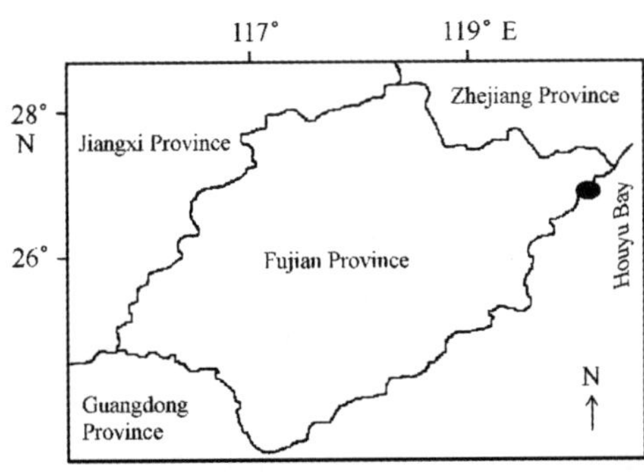

Fig. 1. Study site at the Houyu Bay, Fujian Province.

are *Kandelia candel* communities. The average height is less than 2 m. The communities are in a healthy state, damaged little by pests.

2 Materials and methods

Three replicated mud-flat soil samples were collected seasonally [winter (January), spring (April), summer (July), autumn (October)] in mud-flat in *K. candel* communities from January 2001 to January 2003, and in mud-flat without vegetation (50 m out of the *K. candel* communities) from July 2001 to January 2003 during low tide. These were taken at a depth of 1, 10 and 10 cm in length and breadth respectively, and from the exposed substratum approximately at the same point. The samples were collected in the Petri dish to be transported to the laboratory. Salinity and pH in the mud-flat soil were measured using a refractometer and a pH-meter.

In the laboratory, samples were diluted to 400 cm^3 with synthetic seawater in order to reduce the amount of sediment particles in the sample, and preserved in 30 cm^3 of 40% formaldehyde. The samples were prepared based on the method described by Hakansson (1984). All samples were treated with 10% HCl to remove calcareous matter, and treated with 30% H_2O_2 to destroy organic material. Then, each acid was diluted by adding distilled water and centrifuged until no acid was left in the sample.

An aliquot (0.1 cm^3) of each sample solution was placed on a slide with a grid, and the number of individuals was counted. Analysis of the samples was carried out with an Olympus BH-2 microscope, and 100 objective (enlargement 1000) for diatom identification. More than 300 diatom valves were counted and the density (cells/cm^3) was estimated for each sample.

The taxa of benthic diatoms were determined based on the studies of Clark and Rushforth (1977), Czarnecki and Blinn (1978), Foged (1979, 1980, 1981), Archibald (1983), Jin et al. (1982), Jin et al. (1986), Podzorski and Hakansson (1987) and Sun and Liu (2002).

3 Analysis of the assemblages

The organic carbon content of diatom was estimated according to the Strathmann (1967) equation, $\log_{10}C=0.892\times\log_{10}PV-0.610$, where C is the carbon content, and PV is the plasma volume, which was calculated based on the cell shape and size (Sun et al., 2000 Sun and Liu, 2003).

The species diversity of benthic diatoms was calculated using the Shannon–Weaver Index (H') (Shannon and Weaver, 1963). The degree of distribution of individuals among the species (equitability) was estimated using Pielou's (1969).

As an expression of dominance in the sample, the biological value index (*BVI*) (Sanders, 1960) was estimated for those species that represented 85% of the individuals in each sample. The maximum hierarchical value per sample was 10 and since there were 9 and 7 samples in mud-flats with and without vegetation, the maximum *BVI* possible was 90 and 70, respectively.

An ordination of the samples based on diatom data was obtained by carrying out multi-dimensional scaling (MDS), and the statistical program SPSS version 10.0 was used (Jackson, 1993). Between-groups linkage and Pearson correlation were applicable for the numerical analysis.

4 Results

4.1 Salinity and pH

Salinity in the mud-flat soil in *K. candel* communities varied from 16 to 24 (Table 1), compared with 11 and 21 in the mud-flat without vegetation. Salinity in the mud-flat with vegetation was higher than that in the mud-flat without vegetation, except during April 2002.

Table 1. Salinity and pH in mud-flats with and without vegetation

		Jan. 2001	Apr. 2001	Jul. 2001	Oct. 2001	Jan. 2002	Apr. 2002	Jul. 2002	Oct. 2002	Jan.2003
Mud-flat without vegetation	pH	—	—	6.92	7.05	7.17	7.10	6.94	7.27	7.35
	salinity	—	—	16	13	15	21	17	15	11
Mud-flat with vegetation	pH	6.40	6.32	6.29	6.15	6.21	6.54	6.40	6.76	6.58
	salinity	18	19	20	18	19	16	24	17	17

pH in the mudflat with vegetation was lower than 7, ranging from 6.15 in October 2001 to 6.76 in October 2002, and that in the mud-flat without vegetation varied from 6.92 in July 2002 to 7.35 in January 2003. The mud-flat with vegetation was more acid than that without vegetation.

4.2 Seasonal change of species composition, density and biomass

A total of 103 taxa (see Table 2) including varieties in 36 genera were identified. Eighty-four taxa occurred in the mud-flat with vegetation, and the 3 dominant genera, in terms of the number of taxa encountered, were *Nitzschia* (16), *Gyrosigma* (10) and *Navicula* (9). Seventy-four taxa were identified in the mud-flat without vegetation, and the 3 dominant genera were *Nitzschia* (15), *Navicula* (9) and *Gyrosigma* (8).

The highest densities in mud-flats with and without vegetation were observed in spring and winter, respectively, while the lowest corresponded to that in autumn (Fig. 2). *Nitzschia fasciculata*, *Gyrosigma scalproides* and *Navicula* sp. 3 were the dominant species in spring in the mud-flat with vegetation, but *Navicula* sp. 3 was few in other seasons. *Nitzschia obtusa* var. *scalpelliformis* had the highest density in winter in the mud-flat without vegetation, and it was common in other seasons.

A different pattern of seasonal change was observed for the biomass of diatom in mud-flats with and without vegetation (Fig. 2). The highest biomass in the mud-flat with vegetation was observed in winter (January 2002), not in spring (April 2002). Though the density in April 2002 was the highest, the dominant species such as *Nitzschia fasciculata* and *Gyrosigma scalproides* are few. The increase of larger cells such as *Pleurosigma major* and *Gyrosigma balticum* var. *sinicum* in January 2002 led to the higher biomass. This also accounted for the highest biomass in the mud-flat without vegetation and was observed in spring, not in winter.

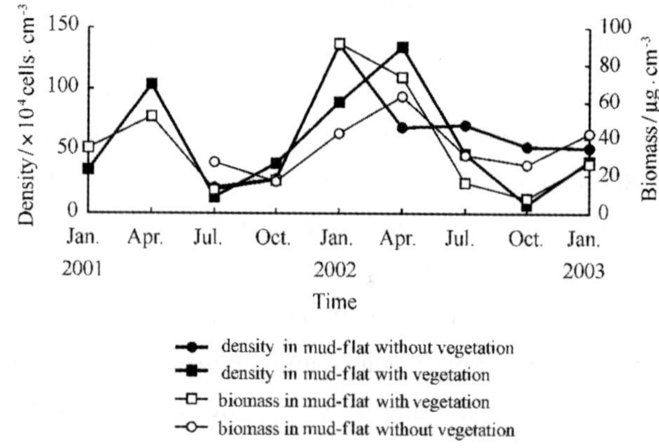

Fig. 2. Seasonal change of density and biomass of diatoms in mud-flats with and without vegetation.

The two-way analysis of variance indicated that there were no significant differences between the densities (or the biomasses) due to salinity and pH.

Table 2. List and abundance of benthic diatoms in mud-flats with and without vegetation at the Houyu Bay

Species	Mud-flat with vegetation	Mud-flat without vegetation	*PV*	Carbon content
Achnanthes lanceolata (Bréb.) Grunow	U	—	919	108
A. brevipes var. *intermedia* (Kutz.) Cleve	R	—	2532	267
A. crenulata Grunow	U	—	3514	357
A. javanica var. *subcontricta* Meister	C	C	4775	470
A. longipes Agardh	U	U	13246	1167
Actinocyclus ellipticus Grunow	U	U	11315	1014
Actinoptychus splendens (Shadb.) Ralfs	R	—	38204	3001
A. senarius (Ehrenberg) Ehrenberg	U	U	57311	4309
Amphora angusta Gregory	—	C	1604	177
A. angusta var. *chinensis* Skvortzow	C	C	1918	208
A. angusta var. *diducta* (A.S.) Cleve	U	—	960	112
A. coffeaeformis (Ag.) Kutzing	C	A	589	73
Amphora sp.1	U	R	655	80
A. sp.2	—	U	589	73
Aulacoseira moniliformis (O. F. Mull.) Simonsen	C	U	226	31
A. islandica (Mull.) Simonsen	—	U	353	46
Bacillaria paxillifera (Muller) Hendey	—	U	10773	970
Caloneis elongata (Greg.) Boyer	—	R	25120	2065
C. oregonica (Ehr.) Patrick	C	U	18149	1545
C. sp.1	U	—	95568	6799
Campylodiscus ecclesianus Greville	U	—	51145	3893
Cocconeis placentula var. *euglypta* (Ehr.) Cleve	C	U	9068	832
C. scutellum var. *parva* Grunow	U	—	2653	278
C. notata Petit	C	U	699	85
Coscinodiscus asteromphalus Ehrenberg	—	U	1607680	84320
C. excentricus Ehrenberg	U	U	6133	587
C. oculatus (Fauv.) Petit	—	U	14507	1265
C. subconcavus var. *tenuior* Rattray	—	U	48081	3684
Cyclotella meneghiniana (Ehr.) Kutzing	U	U	1077	124
C. striata (Kutz.) Grunow	C	C	43407	3363
C. stylorum Brightwell	R	R	43407	3363
Cymbella tumida (Bréb.) Van Heurck	U	—	5341	519
Diploneis bombus Ehrenberg	—	U	38341	3010
D. grundleri (A.S.) Cleve	R	—	13319	1172
D. smithii (Bréb.) Cleve	C	U	8167	758
D. smithii var. *dilatata* (M. Per.) Terry	U	C	9734	886
Eunotia major (W. Sm.) Rabenhorst	—	U	318.5	42
Gomphonema subtilis Ehrenberg	U	—	946	111
G. sp.1	U	—	479	60
G. sp.2	U	—	530	66
G. sp.3	U	—	362	47
Grammatophora marina (Lyngbye) Kutzing	U	—	1452	162
Gyrosigma acuminatum (Kutz.) Rabenhorst	U	—	4719	465
G. balticum var. *sinensis* (Ehr.) Cleve	C	U	8349	773
G. balticum var. *sinicum* Chin et Liu	C	U	32400	2591
G. fasciola var. *tenuirostris* (Grun.) Cleve	U	U	9000	826

Tobe continued

Table 2 Continued

Species	Mud-flat with vegetation	Mud-flat without vegetation	*PV*	Carbon content
G. obliquum (Grun.) G. West	C	C	5626	543
G. scalproides (Rab.) Cleve	A	A	1728	190
G. spencerii (W.Sm.) Griffith et Henfrey	U	—	7200	677
G. strigilis (W.Sm.) Griffith et Henfrey	—	U	40000	3126
Gyrosigma strigilis var. *excentriraphe* Chin et Liu	U	U	50000	3815
G. wormleyi (Sull.) Boyer	U	C	7350	690
G. sp.1	U	—	10000	908
Hantzschia amphioxys (Ehr.) Grunow	U	—	1476	165
Luticola mutica (Kü tzing) Mann	U	U	2276	242
Navicula cancellata Donkin	C	C	15272	1324
N. rhaphoneis (Ehr.) Grunow	C	U	2669	279
N. rhynchocephala Kutzing	—	U	2983	309
N. scopulorum Brébisson	C	C	10163	921
N. sp.1	U	C	40694	3175
N. sp.2	C	U	1526	170
N. sp.3	A	U	20347	1711
N. sp.4	U	U	2374	252
N. sp.5	A	A	621	76
Nitzschia acuminata (W Smith) Grunow	U	U	594	73
N. amphibia Grunow	C	U	99	15
N. cocconeiformis Grunow	A	C	4050	405
N. constricta (Greg.) Grunow	C	U	1815	198
N. fasciculata Grunow	A	A	1212	138
N. lanceolata var. *incrustans* Grunow	—	U	696	84
N. lorenziana var. *densestriata* (Per.) A Schmidt	U	U	918	108
N. obtusa var. *scalpelliformis* Grunow	C	A	1404	158
N. obtusa W Smith	C	C	15120	1313
N. panduriformis var. *minor* Grunow	U	U	5595	541
N. sigma (Kutz.) W Smith	C	C	1404	158
N. subtilis Grunow	C	U	3168	326
N. tryblionella Hantzsch	C	—	26875	2193
N. tryblionella var. *victorica* Grunow	—	—	2450	259
N. sp.1	U	U	8167	758
N. sp.2	U	A	6936	655
N. sp.3	U	—	342	45
N. sp.4	U	U	648	79
Paralia sulcata (Ehrenberg) Cleve	U	C	1692	186
Petrodictyon gemma Ehrenberg	C	C	6782	642
Petroneis granulata (Bailey) Mann	—	U	39611	3099
P. marina (Ralfs in Pritchard) Mann	U	—	71389	5241
Pinnularia sp.1	U	—	1368	154
P. sp.2	U	—	8591	793
Pleurosigma major Liu et Chin	C	U	128675	8865
Psammodictyon corpulentum (Hendey) Mann	C	—	1149	132
Rhopalodia gibberula (Ehr.) O Muller	—	U	1158	133
Stauroneis sp.	—	U	3989	400

Tobe continued

Table 2 Continued

Species	Mud-flat with vegetation	Mud-flat without vegetation	*PV*	Carbon content
Surirella fastuosa Ehrenberg	U	—	56972	4286
S. kurzii Grunow	U	U	109525	7678
S. voigti Skvortzow	U	C	29186	2360
Thalassionema nithzchioides Grunow	U	C	604	74
Thalassiosira anguste-lineata (Schmidt) Fryxell & Hasle	—	U	522	65
Trachyneis antillarum Cleve	—	U	7231	680
Tryblionella granulata (Grunow) Mann	C	C	3468	353
T. punctata (Smith) Mann	U	U	5508	533
T. navicularis (Bréb. et Kutz.) Mann	U	U	2812	293
Tryblioptychus cocconeiformis (Cleve) Hendey	U	—	13319	1172
Vanheurckia lewisiana (Greville) Brébisson	U	—	8888	817

Notes: Plasma volume (*PV*, μm^3) is calculated based on cell shape and size. Carbon content (pg/cell) is estimated according to the Strathmann equation (1967). A represents abundant (>5%), C common (0.5%~5.0%), U uncommon (0.05%~0.50%) and R rare (<0.05%).

4.3 Species richness, diversity and equitability

Species richness in the mud-flat with vegetation was the highest in spring (April 2001 and April 2002), compared with that in winter (January 2002 and January 2003) in the mud-flat without vegetation, while species richness was the lowest in summer and autumn in the mud-flat with vegetation, compared with that in autumn in the mud-flat without vegetation (Table 3). The increase of species in *Nitzschia* and *Navicula* genera resulted in the high species number in mud-flats with and without vegetation.

H′ values of assemblages in mud-flats with and without vegetation varied little during the sampling period (Table 3). The two-way analysis of variance indicated that there were no significant differences between the *H′* values due to the sampling site or date. High *H′* values at two sites during all seasons suggested that diatom assemblages in the sediments of the Houyu Bay represented an original environment.

Higher *J* values in the mud-flat with vegetation showed the higher equitability in the mud-flat with vegetation than that in the mud-flat without vegetation.

4.4 Biological value index (*BVI*)

The estimated values for the *BVI* corresponded with the numerical importance of the dominant species, in space and time (see Table 4). In the mud-flat with vegetation, *Nitzschia cocconeiformis* (*BVI*=63) was by far the dominant species throughout the sampling period. Its minimum abundance occurred in the April 2001 and April 2002 samples, respectively, which is due to the increase of other species such as *Gyrosigma scalproides* (*BVI*=62) and *Nitzschia fascic*

Table 3. Species number, diversity index (*H′*) and equitability (*J*) of assemblages in mud-flats with and without vegetation

Site		Jan. 2001	Apr. 2001	Jul. 2001	Oct. 2001	Jan. 2002	Apr. 2002	Jul. 2002	Oct. 2002	Jan. 2003
mud-flat without vegetation	*H′*	—	—	4.119	3.732	3.666	3.651	3.776	3.515	4.335
	J	—	—	0.887	0.878	0.721	0.777	0.813	0.788	0.845
	species umber	—	—	25	19	34	26	25	22	35
mud-flat with vegetation	*H′*	3.618	4.055	3.288	3.826	3.973	3.970	3.659	3.881	4.112
	J	0.852	0.791	0.842	0.870	0.827	0.756	0.847	0.914	0.865
	species number	19	35	15	21	28	38	20	19	27

Table 4. Biological value index (*BVI*)

Species in the mud-flat with vegetation	*BVI*	Species in the mud-flat without vegetation	*BVI*
Nitzschia cocconeiformis	63	*Gyrosigma scalproides*	49
Gyrosigma scalproides	62	*Nitzschia obtusa* var. *scalpelliformis*	41
Nitzschia fasciculata	36	*Navicula* sp. 5	39
Tryblionella granulata	36	*Nitzschia fasciculata*	32
Navicula sp. 5	28	*Nitzschia subtilis*	23
Amphora angusta var. *chinensis*	25	*Amphora coffeaeformis*	22
Amphora coffeaeformis	24	*Achnanthes javanica* var. *subconstricta*	17
Navicula sp. 3	22	*Amphora angusta* var. *chinensis*	17
Nitzschia sigma	22	*Nitzschia cocconeiformis*	17
Cyclotella striata	20	*Gyrosigma obliquum*	16
Nitzschia obtusa	19	*Navicula scopulorum*	16
Gyrosigma balticum var. *sinensis*	14	*Nitzschia sigma*	16
Navicula sp. 2	11	*Navicula cancellata*	10
Nitzschia constricta	10	*Navicula* sp. 1	9
Navicula scopulorum	9	*Nitzschia obtusa*	9
Nitzschia subtilis	9	*Gyrosigma strigilis*	7
Cocconeis placentula var. *euglypta*	8	*Amphora angusta*	6
Navicula sp. 1	7	*Paralia sulcata*	5
Thalassionema nithzchioides	7	*Tryblionella navicularis*	5
Amphora sp. 1	6	*Cyclotella striata*	4
Cocconeis notata	6	*Petrodictyon gemma*	4
Thalassiosira anguste-lineatus	6	*Eunotia major*	3
Diploneis smithii	5	*Gyrosigma strigilis* var. *excentriraphe*	3
Navicula cancellata	5	*Tryblionella granulata*	3
Nitzschia amphibia	5	*Surirella voigti*	3
Nitzschia obtusa var. *scalpelliformis*	5	*Thalassionema nithzchioides*	3
Petrodictyon gemma	5	*Actinoptychus senariuss*	2

Notes: *BVI* values were estimated considering 85% of the abundance in each sample.

ulata (*BVI*=36).

In the mud-flat without vegetation, *Gyrosigma scalproides* (*BVI*=49) was the most dominant species in many samples though it did not occur in the July 2002 sample. *Nitzschia obtusa* var. *scalpelliformis* (*BVI*=41) was observed in all the samples and dominant in the January 2002 sample.

4.5 Diatom assemblages in mud-flats with and without vegetation

Diatom samples were positioned in a similar place to the species in the MDS ordination diagram (see Fig. 3). Samples with similar diatom assemblages are located close to each other in the MDS ordination diagram, and they can therefore be grouped according to their mutual distances. Multi-dimensional scaling of diatom assemblages from mud-flats with and without vegetation showed that a slight seasonal change and only a single association occurred in the mangroves. Assemblages in ka2 and mj2 had different scores from the rest of the data, and discrete assemblages were shown in some seasons.

4.6 Discussion and conclusions

Benthic diatom flora in the mangroves were more diverse than those in the estuaries without mangrove forests on the main island of Japan (Nagumo and Hara, 1990). Sequeiros-Bentrones and Castrejon (1999) found that there were nonsignificant variations among benthic diatom assemblages for the mangroves in a

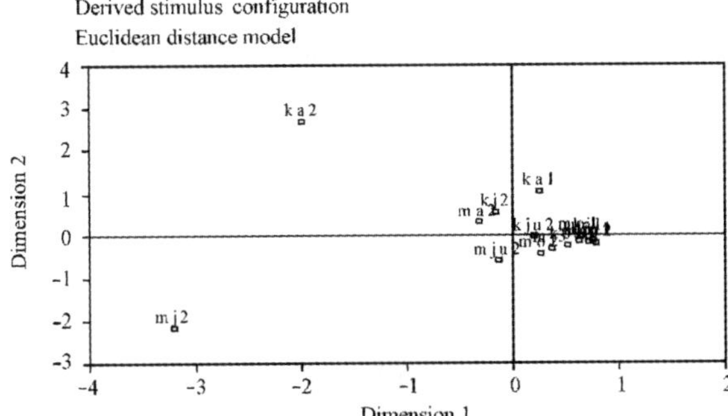

Fig. 3. MDS of assemblages due to site and date (stress equals 0.051). kj1 represents mud-flat with vegetation in January 2001. ka1 mud-flat with vegetation in April 2001, kju1 mud-flat with vegetation in July 2001, ko1 mud-flat with vegetation in October 2001, kj2 mud-flat with vegetation in January 2002, ka2 mud-flat with vegetation in April 2002, kju2 mud-flat with vegetation in July 2002, ko2 mud-flat with vegetation in October 2002, kj3 mud-flat with vegetation in January 2003. mju1 mud-flat without vegetation in July 2001, mo1 mud-flat without vegetation in October 2001, mj2 mud-flat without vegetation in January 2002, ma2 mud-flat without vegetation in April 2002, mju2 mud-flat without vegetation in July 2002, mo2 mud-flat without vegetation in October 2002, and mj3 mud-flat without vegetation in January 2003.

Mexican subtropical lagoon. Multi-dimensional scaling of diatom assemblages from mud-flats with and without vegetation showed that only a single association occurred in the mangroves in our results. Thus differences between the two sites, which are located within 50 m, may be too subtle for detection against the distribution of common benthic diatoms.

Density and species richness of assemblages in the mud-flat with vegetation were lower than those in the mud-flat without vegetation, except during October 2001 and April 2002, though the total species number identified in the mud-flat with vegetation was more than that in the mud-flat without vegetation. The abundant organisms diet of benthic diatoms count for the low density of assemblages in the mangrove forest (Leija-Tristan and Sanchez-Vargas 1988; Vicente, 1989). Moreover, low light intensity in the mangroves due to the low density of assemblages.

Biomass based on cell volume rather than cell census was needed, considering food chains and nutrient dynamic analyses. Contribution of the same species in different stain to the biomass varied with cell shape and size, usually influenced by nutrient and region. Standing stock expressed as a cell census may overestimate the contribution of small cells and under estimate that of large cells (Sun et al., 2000). Larger cells such as *Pleurosigma major* and *Gyrosigma balticum* var. *sinicum* resulted in the highest biomass in January 2002 in the mud-flat with vegetation and in April 2002 in the mud-flat without vegetation, respec tively, though the highest density occurred in April 2002 in the mud-flat with vegetation and in January 2002 in the mud-flat without vegetation, respectively.

Some researches on benthic diatoms in lagoons and rivers (Facca et al., 2002; Winter and Duthie, 2000) showed a slight seasonal change, which also oc curred in our study by multi-dimensional scaling of diatom assemblages from the mangroves. Deviation of two samples (Fig. 3) (the mud-flat with vegetation in April 2002 and the mud-flat without vegetation in Jan uary 2002) indicated instability of assemblages in some seasons.

The dominant genera, *Nitzschia* and *Navicula*, found in this study were the same as those in other mangroves reported by Fan et al. (1993), Nagumo and Hara (1990), Sequeiros-Beltrones and Castrejon (1999). Differing from that in other mangrove environments, *Gyrosigma* was abundant in the mangroves at the Houyu Bay, though the species richness was low in the mud-flat without vegetation. Most species at the Houyu Bay were of tropical to arctic nature, and the temperate species such as *Gyrosigma acuminatum*, *Gyrosigma fasciola* var. *tenuirostris*, *Gyrosigma spencerii* and *Gyrosigma strigilis* were common (Chin, 1981). The dominant species such as *Nitzschia fasciculata* and *Gyrosigma scalproides* were also in a temperate form. It corresponded to the situation that the Houyu Bay was located in the temperate region.

Since only a single association occurred in the

mangroves, the most dominant species in the mangroves at the Houyu Bay were *Gyrosigma scalproides* (*BVI*=111), *Nitzschia cocconeiformis* (*BVI*=80) and *Nitzschia fasciculata* (*BVI*=68) (the maximum *BVI* is 160). *Amphora coffeaeformis* (*BVI*=46) and *Nitzschia obtusa* var. *scalpelliformis* (*BVI*=41) were common in some seasons such as spring and autumn. The species diversity and equitability of assemblages from mangrove environments were high and varied little, which showed an original environment of mangroves at the Houyu Bay.

Acknowledgements

We thank Professor Cheng Zhaodi for his constructive comments on the identification of benthic diatoms. This study was supported by the National Natural Science Found ation of Fujian Province under contract No: D0210003.

References

Archibald R E M. 1983 The Diatoms of the Sundays and Grent Fish Rivers in the Eastern Cape Province of South Africa. Bibliotheca Diatomologica, 1~362

Beumer J P. 1978. Feeding ecology of four fishes from a mangrove creek in north Queensland, Australia. Journal of Fish Biology, 12: 475~490

Chin T G, Jin D X. 1981. The Geographical Distribution of the Marine Diatoms in China. Nova Hedwigia. 763~792

Clark R L, Rushforth S R. 1977. Diatom Studies of the Headwaters of Henrys Fork of the Snake River, Island Park, Idaho, USA. Bibliotheca Phycologica, 33: 1~204

Czarnecki D B, Blinn D W. 1978. Diatoms of the Colorado River. Bibliotheca Phycologica, 38: 1~181

Damroy S. 1995. Studies on mangrove ecology of Chouldari area, South Andaman. Journal of Andaman Science Association, (1—2): 29~33

Fan H Q, Cheng Z D, Liu S C, et al. 1993. Species of benthic diatoms in Guangxi mangrove habitats. Journal of the Guangxi Academy of Sciences (in Chinese), 9(2): 37~42

Facca C.A, Sfriso & G. Socal, 2002. Temporal and Spatial Distribution of Diatoms in the Surface Sediments of the Venice Lagoon. Botanica Marina. 45:170~183

Foged N. 1979. Diatoms in New Zealand, the North Island. Bibliotheca Phycologica, 47: 1~224

Foged N. 1980. Diatoms in Oland, Sweden. Bibliotheca Phycologica, 49: 1~192

Foged N. 1981. Diatoms in Alaska. Bibliotheca Phycologica, 53: 1~316

Hakannsson H. 1984. The recent diatom succession of Lake Havgardssjon, south Sweden. In: Mann D G, ed Proceedings of the Seventh International Diatom Symposium. Koeltz, Koenigstein, 411~429

Jackson D A. 1993. Multivariate analysis of benthic invertebrate communities: the implication of choosing particular data standardizations, measures of association and ordination methods. Hydrobiologia, 268: 9~26

Jin D X, Cheng Z D, Lin J M, et al. 1982. Marine benthic diatoms in China, v1(in Chinese). Beijing: China Ocean Press, 256

Jin D X, Cheng Z D, Lin J M, et al. 1986. Marine benthic diatoms in China, v2(in Chinese). Beijing: China Ocean Press, 313

Leh C. M. U, Sasekumar A. 1986. The food of sesarmid crabs in Malaysian mangrove forests. Malayan Nature Journal, 39(2): 135~145

Leija–Tristan A. Sanchez–Vargas D P. 1988. Biology and ecology of the common mud shrimp *Upogebia dawsoni* (Crustacea: Thalassinoidea) of Manglar Requeson, Bahia Concepcion and Estero Rio Mulege, Baja California Sur, Mexico. Revista de Biologia Tropical, 36(1): 107~114

Lugomela C V, Semesi A K. 1995. Spatial and temporal variations of phytoplankton in Zanzibar nearshore waters. Symposium on the Biology of Microalgae, Macroalgae and Seagrasses in he Western Indian Ocean, Reduit (Mauritius). 3~10

Maples R S. 1983. Community struture of ditoms epiphytic on pneumatophores of the black mangrove *Avicennia germinans* in a Louisiana USA salt marsh. Gulf Research Reports, 7(3): 255~260

Milward N E. 1974. Studies on the taxonomy, ecology, and physiology of Queenland mudskippers:[dissertation]. Univer Queensland, Australia, 276

Nagumo T, Hara Y. 1990. Species composition and vertical distribution of diatoms occurring in a Japanese mangrove forest. The Japanese Society of Phycology, 38(4): 333~343

Navarro J N. 1982. Marine diatoms associated with mangrove prop roots in the Indian River, Florida, USA. Bibliotheca Phycologica, 61:1~151

Nicholas W L, Stewart A C, Marples T G. 1988. Field and laboratory studies of *Desmodora cazca* Gerlach, 1956 (Desmodoridae: Nematoda) from mangrove mud-flats, Nematologica, 34 (3):331~349

Pielous E. C. 1969. An Introduction to Mathematical Ecology. N Y Wiley Interscience, 268

Podzorski A C, Hakansson, H. 1987. Freshwater and marine diatoms from Palawan. Bibliotheca Diatomologica, 13: 1~244

Reyes–Vasquez G. 1975. Littoral Diatoms of the Family Naviculaceae of Laguna La Restinga, Margarita Island. Biology Institution Oceanography, University Oriente, 14(2): 199~225

Sanders H L. 1960. Benthic studies in Buzzards Bay: III. The structure of soft-bottom community.Limnology and Oceanography, 5:138~153

Sequeiros–Beltrones D A, Castrejon E N. 1999. Structure of benthic diatom assemblages from a mangrove environment in a Mexican subtropical lagoon. Biotropica, 31(1): 48~70

Strathmann R R. 1967. Estimating the organic carbon content of phytoplankton from cell volume or plasma volume. Limnology and Oceanography, 12:411~418

Shannon C E, Weaver W. 1963. The Mathematical Theory of Communication. Ubana: University of Illinois Press

Sullivan M J. 1981. Community structure of diatoms epiphytic on mangroves and Thalassia in Bimini Harbour, Bahamas. In: Ross R, ed. Proc Sixth Symp. Living and Fossil Diatoms, Budapest, 1980. Koenigstein: Koeltz Science Publication, 385~398

Sun J, Liu D Y. 2002. The Preliminary notion on nomenclature of common phytoplankton in China seas waters. Oceanologia et Limnologia Sinica, 33(3):271~286.

Sun J, Liu D Y. 2003. Geometric models for calculating cell biovolume and surface area for phytoplankton. Journal of Plankton Research, 25(11):1331~1346

Sun J, Liu D Y, Qian S B. 2000. Estimating biomass of phytoplankton in the Jiaozhou Bay: I. Phytoplankton biomass estimated from cell volume and plasma volume. Acta Oceanologica Sinica. 19(2): 97~110.

Vicente H J. 1989. Monthly population density fluctuation and vertical distribution of meiofauna community in tropical muddy substrate. Asian Fisheries Forum, Tokyo (Japan), 4: 17~22

Wah T T, Wee Y C. 1988. Diatoms from mangrove environments of Singapore and Southern peninsular Malaysia, Botanica Marina, 31: 317~327

Winter J G, Duthie H C. 2000. Stream epilithic, epipelic and epiphytic diatoms: habitat fidelity and use in biomonitoring. Aquatic Ecology, 34: 345~353

福建漳江口红树林保护区浮游植物群落的季节变化研究*

陈长平　高亚辉　林　鹏
(厦门大学生命科学学院,福建 厦门 361005)

摘要: 于2001年1月~2003年1月对福建省漳江口红树林区水体浮游植物群落的季节变化进行了研究。结果表明，漳江口红树林区浮游植物以硅藻门种类为主，优势种为长菱形藻(*Nitzschia longissima*)和菱形藻（*Nitzschia* sp.1）等，同时出现多种裸藻、绿藻和甲藻。本次调查共鉴定到浮游植物31属87种（包括变种），其中硅藻门23属75种(包括变种)，蓝藻门3属3种，绿藻门1属4种，金藻门1属1种，甲藻门2属2种，裸藻门1属2种。浮游植物密度变化范围为2.78×10^4~1.14×10^6个/L，平均为3.51×10^5个/L，季节变化为双峰型。出现大量的底栖硅藻和淡水性藻类是该水域浮游植物的一个特点。浮游植物的组成和结构表明该水域水质较好。

关键词: 红树林；浮游植物；硅藻；漳江口
中图分类号: Q948.1　**文献标识码:** A　**文章编号:** 1000-3096（2007）07-0025-07

红树林区拥有丰富的饵料和适宜的栖息环境，是许多海洋动物发育生长的重要场所，浮游植物是红树林生态系统初级生产者之一，在食物链中占有重要地位，一些浮游植物如骨条藻等具有指示环境的作用[1]。研究红树林中的浮游植物群落结构可以更好地说明红树林生态系统的独特生态功能及在河口海岸中的重要性。目前对中国红树林区浮游植物的研究比较缺乏，特别是在浮游植物群落结构的季节变化方面[2~4]。作者通过对福建漳江口红树林区浮游植物群落结构和组成季节变化的研究，了解浮游植物在红树林内的生态作用，对红树林生态系统基础资料的补充和生态环境的保护具有重要的意义。

1　材料和方法

福建漳江口红树林自然保护区在2002年成为国家级自然保护区。保护区距离云霄县城10 m，地处117°24′~117°30′ E，23°53′~23°56′ N，海拔-6~8 m。保护区在漳江口石矶塔以西广阔的滩涂湿地，面积2360 hm^2, 属亚热带海洋性气候，气候温暖湿润，光、热、水资源丰富，年均温21.2 ℃，年均降水1 714.5 mm。该保护区是中国红树林自然分布最北的大面积重要湿地类型保护区。主要红树植物有秋茄（*Kandelia candel*）、白骨壤(*Avicennia marina*)、桐花树(*Aegiceras corniculatum*)，另外还有少量木榄(*Bruguiera gymnorrhiza*)、老鼠簕(*Acanthus ilicifolius*)及藤本植物鱼藤(*Derris trifoliata*)等。红树林分布在中高潮带，面积约260 hm^2，宽度25~150 m，高度1.5~6 m，主要红树植物中秋茄最高，在核心区可达6m，白骨壤高1.5~3 m，桐花树高2~3m，红树林外貌整齐，林相郁闭度高，很难在林中穿行。林外为泥质滩涂，宽度50~400 m，主要为养殖地，滩涂分布的其他植物有卡开芦(*Phragmites karka*)，短叶茳芏(*Cyperus malaccensis*)和铺地黍(*Panicum repens*)等，为斑块分布[5]。

作者于2001年1月~2003年1月对漳江口红树林区水体浮游植物群落的季节变化进行研究，分别于1月（冬）、4月（春）、7月（夏）、10月（秋）现场采样。高潮时在林内外和林缘采集表层海水共3L，用

* 福建省自然科学基金资助项目(2006J0145;D0210003)
原载于海洋科学,2007,31(2):25－31

Lugol's solution 固定，静置，沉淀，逐步浓缩到 50 mL。取 0.1 mL 于显微镜下用浮游植物计数框进行种类鉴定和计数。

2 结果和讨论

2.1 浮游植物的种类组成

福建漳江口红树林区水体浮游植物的种类组成和密度见表 1。共计浮游植物 6 门 31 属 87 种(包括变种)，其中硅藻门 23 属 75 种，蓝藻门 3 属 3 种，绿藻门 1 属 4 种，金藻门 1 属 1 种，甲藻门 2 属 2 种，裸藻门 1 属 2 种。

表 1 福建漳江口红树林区水体浮游植物的种类组成和密度

Tab.1 Species composition and density of phytoplankton in water under mangrove forest at the estuary of the Zhangjiang River, Fujian Province, China

种名	密度（$\times10^3$ 个/L）								
	2001 年				2002 年				2003 年
	1 月	4 月	7 月	10 月	1 月	4 月	7 月	10 月	1 月
硅藻门									
翼茧形藻（*Amphiprora alata*）			1.6		0.4	2.7			8.0
狭窄双眉藻（*Amphora angusta*）		3							
狭窄双眉藻中国变种（*Amphora angusta* var. *chinensis*）								2.5	
狭窄双眉藻分离变种（*Amphora angusta* var. *diducta*）							2.5		
咖啡形双眉藻（*Amphora coffeaeformis*）			1.6						0.3
双眉藻（*Amphora* sp.）	0.2					2.0	2.5		
奇异棍形藻（*Bacillaria paradoxa*）	1.7				0.8	0.2	2.5		
双突角毛藻（*Chaetoceros didymus*）			1.1						
角毛藻（*Chaetoceros* sp.）									16.5
盾卵形藻（*Cocconeis scutellum*）				0.8			2.5		
卵形藻（*Cocconeis* sp.）									0.3
星脐圆筛藻（*Coscinodiscus asteromphalus*）						0.2			
线形圆筛藻（*Coscinodiscus lineatu* ）						0.2			
梅里小环藻（*Cyclotella meneghiniana*）						61	45.0		
条纹小环藻（*Cyclotella striata*）	0.5			10.0			2.5		1.0
柱状小环藻（*Cyclotella stylorum*）						0.7		2.5	
新月细柱藻（*Cylindrotheca closterium*）		0.2	1.1			0.7			
地中海指管藻（*Dactyliosolen mediterraneus*）									1.0
史密斯双壁藻（*Diploneis smithii*）			1.1				2.5		0.3
异极藻（*Gomphonema* sp.1）					0.4				
异极藻（*Gomphonema* sp.2）			0.3			0.2			
簇生布纹藻（*Gyrosigma fasciola* var. *arcuata*）	0.2				0.4				

表 1（续）

种名	密度（$\times10^3$ 个/L）								
	2001 年				2002 年				2003 年
	1 月	4 月	7 月	10 月	1 月	4 月	7 月	10 月	1 月
刀形布纹藻（*Gyrosigma scalproides*）	0.2			0.8					
澳立布纹藻（*Gyrosigma wormleyi*）			0.5					2.5	
布纹藻（*Gyrosigma* sp.）						0.7			
丹麦细柱藻（*Leptocylindrus danicus*）									38.5
具槽直链藻（*Melosira sulcata*）	0.3		0.3						
尤氏直链藻（*Melosira juergensi*）	0.3				1.6				
货币直链藻（*Melosira moniliformis*）				6.0					
冰岛直链藻（*Melosira island*）				4.0					
变异直链藻（ *Melosira varians*）							20		
侏儒舟形藻（*Navicula pygmaea*）			1.1				2.5		
缝舟舟形藻（*Navicula rhaphoneis*）			39.5			1.3	2.5		
岩石舟形藻（*Navicula scopulorum*）						0.2			
舟形藻（*Navicula* sp.1）	0.3				0.8	7.3	2.5		11.5
舟形藻（*Navicula* sp.2）		0.7	3.7	6.0		3.3	35.0	5.0	
新月菱形藻（*Nitzschia closterium*）			0.5			0.7		5.0	2.0
卵形菱形藻（*Nitzschia cocconeiformis*）		0.2		4.0					
簇生菱形藻（*Nitzschia fasciculata*）		0.7	6.9	0.8	0.8	11.0	25.0	10.0	
碎片菱形藻（*Nitzschia frustulum*）		0.7	8.0	0.8		0.2	2.5	2.5	0.3
颗粒菱形藻（*Nitzschia granulata*）				10.0		0.2	2.5		
杂菱形藻（*Nitzschia hybrida*）						0.7			
披针菱形藻（*Nitzschia lanceolata*）						13.0			
长菱形藻（*Nitzschia longissima*）	0.3	2.0	2.7			409.0		342.5	4.0
洛氏菱形藻（*Nitzschia lorenziana*）		0.7	1.1		4.0	1.3			1.0
较大菱形藻线形变种（*Nitzschia majuscula* var. *lineata*）			1.1						
钝头菱形藻（*Nitzschia obtusa*）							10.0		
钝头菱形藻刀形变种（*Nitzschia obtusa* var. *scalpelliformis*）			0.3	0.8	1.6	0.7	2.5		0.3
琴式菱形藻小型变种（*Nitzschia panduriformis* var. *minor*）			1.1				2.5		0.3
具点菱形藻（*Nitzschia punctata*）						0.2			
弯菱形藻（*Nitzschia sigma*）	1.2				1.2				
拟螺形菱形藻（*Nitzschia sigmoides*）	0.3								
细弱菱形藻（*Nitzschia subtilis*）	0.2								
盘形菱形藻维多变种（*Nitzschia tryblionella* var. *victorica*）			1.6						

表 1（续）

种名	密度（×10³ 个/L）								
	2001 年				2002 年				2003 年
	1 月	4 月	7 月	10 月	1 月	4 月	7 月	10 月	1 月
菱形藻（*Nitzschia* sp.1）		111.4		1 030.0				17.5	
菱形藻（*Nitzschia* sp.2）			0.5		2.0	12.0			3.3
伊氏斜纹藻（*Pleurosigma aestuarii*）	0.5	0.7							
相似斜纹藻（*Pleurosigma affine*）					0.4				0.3
宽角斜纹藻（*Pleurosigma angulatum*）		2.3	0.3		3.2				
柔弱斜纹藻（*Pleurosigma delicatulum*）					14.0				
美丽斜纹藻（*Pleurosigma formosum*）	4.2	0.7	5.3	58.0		2.7	2.5		
海洋斜纹藻（*Pleurosigma pelagicum*）	1.2				0.4				0.3
斜纹藻（*Pleurosigma* sp.）									0.3
中肋骨条藻（*Skeletonema costatum*）	15.0								
针杆藻（*Synedra* sp.）									0.3
菱形海线藻（*Thalassionema nitzschioide*）	0.8	0.2	2.1		0.4	2.7	20		
海链藻（*Thalassiosira* sp.）						17.0			32.0
安蒂粗纹藻（*Trachyneis antillarum*）		0.2							
硅藻未知种 sp.1							30.0	15.0	
硅藻未知种 sp.2			0.5						
硅藻未知种 sp.3							2.5		
蓝藻门									
念珠藻（*Nostoc* sp.）			15.5						
点形平裂藻（*Merismopedia punctata*）			8.5						
二角盘星藻纤细变种（*Pediastrum duplex* var. *gracillimum*）								25.0	
绿藻门									
被甲栅藻（*Scenedesmus armatus*）								10.0	
二形栅藻（*Scenedesmus dimorphus*）								10.0	
斜生栅藻（*Scenedesmus obliquus*）				8.0				10.0	
四尾栅藻(*Scenedesmus quadricauda*)								14.0	
金藻门									
六异刺硅鞭藻（*Distephanus speculum*）									0.3
甲藻门									
亚历山大藻（*Alexandrium* sp.）								5.0	

表1（续）

种名	密度（×10³个/L）								
	2001年				2002年				2003年
	1月	4月	7月	10月	1月	4月	7月	10月	1月
原多甲藻（*Protoperidinium* sp.）									0.3
裸藻门									
静裸藻（*Euglena denses*）								55.0	
密盘裸藻（*Euglena wangi*）	0.2			0.8	0.8	0.2	2.5	262.5	27.5
种类数(种)	18	14	27	16	17	29	24	17	25
总细胞密度(×10³个/L)	28.0	123.7	108.0	1141.0	33.0	552.0	228.0	782.5	163.9

2.2 浮游植物密度的季节变化

漳江口红树林保护区水体浮游植物密度的季节变化呈双峰型（图1），2001~2003年浮游植物细胞密度的最高值均出现在秋季（10月），细胞密度近 10^6 个/L，但优势种不一样，2001年秋季优势种是菱形藻（*Nitzschia* sp.1），占总细胞密度的90.3%；2002年秋季优势种是长菱形藻（*Nitzschia longissima*）和近轴裸藻(*Euglena wangi*)，分别占总细胞密度的43.7%和33.5%。次高峰均出现在春季，以2002年春季的细胞密度较高。2001~2003年最低的细胞密度均出现在冬季（1月）。

2.3 浮游植物优势种和种类数的季节变化

漳江口红树林区浮游植物种类数的季节变化没有一定的规律性，2002年春季种类数最高，为29种；2001年春季种类数最低，为14种（表1）。

不同季节浮游植物的优势种有很大的不同（表2），这与底栖硅藻大量参与到浮游植物中有密切的关系，如底栖硅藻菱形藻(*Nitzschia* sp.1)在2001年春季和秋季形成绝对的优势种，这与潮汐和风浪关系密切。底栖硅藻在2001年夏季、2002年冬季和2002年夏季也成为优势种。浮游性种类如中肋骨条藻在浮游植物中也占有重要地位。

2.4 浮游植物在红树林生态系统中的生态作用及特征

（1）国内外某些红树林区水体浮游植物调查的报道表明浮游植物以硅藻门种类为主[2~4, 6, 7]。漳江口红树林区浮游植物亦是以硅藻门种类为主，但在一些季节如2002年秋季和2003年冬季密盘裸藻成为优势种之一，同时出现多种绿藻和甲藻，表现出与以往水体浮游植物不同的群落结构，水体可能受到一定程度的污染。

（2）相对于其它红树林区（表3），云霄漳江口红树林区浮游植物的种类数和细胞密度较高，这样可以为各种生物提供较多的饵料和食物，对生态系统食

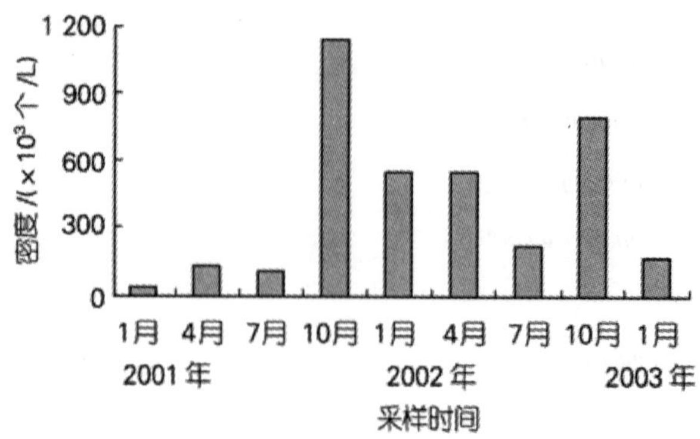

图1 福建漳江口红树林区水体浮游植物密度的季节变化

Fig.1 Seasonal changes of phytoplankton density at waters of mangroves at the estuary of the Zhangjiang River in Fujian Province, China

表 2 福建漳江口红树林区水体浮游植物优势种密度的季节变化

Tab.2 Seasonal changes of dominant species and their percentages in phytoplankton at the waters of the mangroves at the estuary of the Zhangjiang River in Fujian Province, China

采样时间		优势种	比例（%）*
2001 年	1 月	中肋骨条藻（*Skeletonema costatum*）	54.8
	4 月	菱形藻（*Nitzschia* sp.1）	85.6
	7 月	缝舟形舟形藻（*Navicula rhaphoneis*）	36.7
	10 月	菱形藻（*Nitzschia* sp.1）	90.3
2002 年	1 月	柔弱斜纹藻（*Pleurosigma delicatulum* ）	41.5
	4 月	长菱形藻（ *Nitzschia longissima*）	74.1
	7 月	梅里小环藻（*Cyclotella meneghiniana*）	19.8
		舟形藻（*Navicula* sp.2）	15.4
	10 月	长菱形藻	43.8
		密盘裸藻（*Euglena wangi*）	33.6
2003 年	1 月	丹麦细柱藻（*Leptocylindrus danicus* ）	23.6
		密盘裸藻	16.8

注：*占总细胞密度的百分率（%），其它非优势种未列出。

表 3 中国不同红树林区水体浮游植物的比较

Tab.3 Comparison of phytoplankton communities between different mangrove areas in China

红树林区	采样时间（年-月）	种类组成	密度变化（$\times10^3$ 个/L）	平均密度（$\times10^3$ 个/L）	主要优势种
云霄	2001-01~2003-01	6 门 31 属 87 种（包括变种）	27.8~1 140.8	350.8	中肋骨条藻、菱形藻、密盘裸藻、长菱形藻等
福鼎[10]	2001-01~2003-01	3 门 32 属 77 种（包括变种）	1.1~1 156.0	32.1	长菱形藻、中肋骨条藻等
广西英罗港[2]	1992-06，1992-11	2 门 97 种（包括变种）	0.1~3.4	1.7	窄隙角毛藻等角毛藻变种、短孢角毛藻、扁面角毛藻等
深圳福田[4]	1994-12	2 门 11 属 34 种	2 374	-	矮小胸隔藻

物链的贡献也较大[8]。过高的密度也可能是富营养化造成的，这与当地滩涂大量的蛏苗养殖关系密切。

（3）漳江口红树林区浮游植物的组成特点是出现大量的羽纹纲硅藻。由于在红树林阻挡下，林前冲刷的潮汐和风浪的影响，底栖硅藻极易悬浮于水体中，从而大量地参与到浮游植物中，起到丰富浮游植物的作用，云霄漳江口红树林区浮游植物中底栖性的种类占总种类数的 50%以上，同时一些附着性的硅藻如卵形藻等也出现在浮游植物中。

（4）每个季节的浮游植物均出现具有典型淡水性质的种类，如密盘裸藻、被甲栅藻和四尾栅藻等共 13 种，其中密盘裸藻在 2002 年秋季成为优势种类，表明水质属咸淡水性质，这与红树林区处于漳江口的位置关系密切，而大洋性的种类如角毛藻等仅有少量的种类和数量。

（5）云霄红树林区浮游植物中也出现一些赤潮藻，如中肋骨条藻、原多甲藻和亚历山大藻等种类，但是出现的次数和细胞密度都较低，这表明该红树林区出现赤潮的可能性较小。

（6）对云霄红树林区浮游植物的调查发现较多的微型硅藻（<20 μm），如梅里小环藻、舟形藻(*Navicula* sp.1)和菱形藻(*Nitzschia* sp.2)等，其中梅里

小环藻在 2002 年夏季成为优势种类，而舟形藻(*Navicula* sp.1)和菱形藻(*Nitzschia* sp.2)也在多个季节出现。因此有必要对红树林区的微型硅藻开展研究，以补充和完善硅藻集群[9]。

3 结论

漳江口红树林区水体浮游植物以硅藻门种类为主，种类丰富，赤潮藻种类少，密度低，表明水体质量较好。但是某些季节如 2002 年秋季和 2003 年冬季出现大量的裸藻，而秋季浮游植物密度也达到了 10^6 个/L，说明水体质量有下降的可能，对此应加以注意。出现大量的底栖、附着性硅藻和淡水性藻类是该水域浮游植物的一个特点，而微型硅藻在该水域浮游植物中也占有重要的地位。

参考文献：

[1] 林鹏．中国红树林生态系[M]．北京：科学出版社，1997．

[2] 陈坚，范航清，陈成英．广西英罗湾红树林区水体浮游植物种类组成和数量分布的初步研究[J]．广西科学院学报，1993, **9**（2）：31-33．

[3] 陈长平，高亚辉，林鹏．红树林区硅藻研究进展[J]．海洋科学，2002，**26**（3）：17-19．

[4] 刘玉，陈桂珠．深圳福田红树林区藻类群落结构和生态学研究[J]．中山大学学报（自然科学版），1997，**36**（1）：102-106．

[5] 林鹏．福建漳江口红树林湿地自然保护区综合科学考察报告[M]．厦门：厦门大学出版社，2001．

[6] Damroy S． Studies on mangrove ecology of Chouldari area, South Andaman[J]． **J Andaman Sci Assoc**， 1995． **11**(1-2): 29-33．

[7] Roy S D． Mangrove ecology of Alexandra Island and Manjera area of South Andaman[J]． **J Andaman Sci Assoc**， 1995, **11**(1-2): 58-61．

[8] Leija-Tristan A, Sanchez-Vargas D P． Biology and ecology of the common mud shrimp *Upogebia dawsoni* (Crustacea: Thalassinoidea) of Manglar Requeson, Bahia Concepcion and Estero Rio Mulege, Baja California Sur, Mexico[J]． **Revista de Biologia Tropical**，1988. **36**(1): 107-114．

[9] 程兆第，高亚辉，刘师成．福建沿岸微型硅藻[M]．北京：海洋出版社，1993．

[10] 陈长平，闽粤沿海几个红树林区硅藻的生态分布和 6 种重金属对底栖硅藻胞外产物的影响[D]. 厦门：厦门大学，2004,51-72.

Seasonal change of phytoplankton community in waters of mangrove in the estuarine of the Zhangjiang River, Fujian Province, China

CHEN Chang-ping，GAO Ya-hui，LIN Peng

(School of Life Sciences, Xiamen University, Xiamen 361005, China)

Received：Nov., 10,2005

Key words: mangrove; phytoplankton; diatom; the estuarine of the Zhangjiang River

Abstract: Structure of phytoplankton communities in waters of mangrove natural reserve in the estuarine of the Zhangjiang River, Fujian Province, China was studied seasonally from Jan. 2001 to Jan. 2003. Water samples were collected every three months and observed under a light microscope. The results showed that dominant species were diatoms in each season, such as *Nitzschia longissima* and *Nitzschia* sp.1. Some taxa belonging to Euglenophyta,Pyrrophyta and Chlorophyta also occurred in the water. Totally 87 taxa (including variety) in 31 genera were identified, in which 75 taxa in 23 genera belonged to Bacillariophyta, and 3 taxa in 3 genera for Cyanophyta, 4 taxa in 1 genus for Chlorophyta, 1 taxon in 1 genus for Chrysophyta，2 taxa in 2 genera for Pyrrophyta, 2 taxa in 1 genus for Euglenophyta, respectively. Cell density of phytoplankton ranged from 2.78×10^4 to 1.14×10^6 cells/L, and average density was 3.51×10^5 cells/L. The communities were characterized by abundant estuarine species and benthic diatom related with erosion by tide and wind wave, which distributed in the inter-tide in the mangroves. Nanodiatoms are an important component in phytoplankton. Water quality in mangrove area in the estuarine of the Zhangjiang River was in good condition according to composition and structure of phytoplankton communities.

福建省福鼎市后屿湾红树林区水体浮游植物群落动态研究*

陈长平 高亚辉 林 鹏

(厦门大学生命科学学院,福建 厦门 361005)

摘要: 于 2001 年 1 月～2003 年 1 月对福建省福鼎市后屿湾红树林区水体浮游植物种类和数量的季节变化等进行了研究. 结果表明:浮游植物的优势种除了 2001 年 1 月是裸藻类的密盘裸藻(*Euglena wangi*)外,其余季节优势种均为硅藻类,长菱形藻(*Nitzschia longissima*)和中肋骨条藻(*Skeletonema costatum*)是常出现的优势种,还有一些甲藻和绿藻偶尔出现. 本次调查共鉴定浮游植物 32 属 77 种(包括变种),其中硅藻门 29 属 74 种(包括变种),甲藻门 1 属 1 种,绿藻门 1 属 1 种,裸藻门 1 属 1 种. 浮游植物密度变化范围为 1.14×10^4～1.16×10^6 个/L,平均为 3.21×10^5 个/L,季节变化为单峰型. 福鼎红树林区水体中浮游植物的群落组成以河口性种类为主,红树林潮间带底栖硅藻在浮游植物中占有较大的比例,与潮汐和风浪冲刷有关.

关键词: 红树林;浮游植物;硅藻;福鼎

中图分类号: Q 948.1 **文献标识码**: A **文章编号**: 0438-0479(2005)01-0118-05

红树林是热带、亚热带海洋潮间带的木本植物群落,是海岸和河口湿地生态系统重要的第一生产者之一[1]. 红树林区水体中的浮游植物是红树林生态系重要的食物链组成部分,是海洋动物,尤其是海洋动物幼虫和幼体的直接饵料[2]. 通过对福鼎红树林区水体的浮游植物群落组成及其生态学的研究,了解红树林区的浮游植物组成特征及其生态作用,为研究整个红树林生态系统的作用、功能提出科学的、基础的资料.

1 材料和方法

福鼎市地处福建东北沿海,位于北纬 26°55′～27°26′,东经 119°55′～120°43′之间,属亚热带海洋性季风气候. 南距省会福州 250 km,北邻浙江温州 114 km,东至台湾基隆港 2.63×10^2 km,是闽东南通浙江乃至长江三角洲的"北大门". 福鼎市是中国红树林天然分布的北界. 年均气温 18.5 ℃,月最低气温 8.4 ℃(1 月). 福鼎后屿湾红树林为秋茄群落,地处河口,面积约为 8.8 hm^2,平均树高不及 2.0 m,长势良好. 林外为泥质滩涂,主要为鱼塘和滩涂养殖.

于 2001 年 1 月～2003 年 1 月对福鼎后屿湾红树林区水体浮游植物的种类和密度的季节变化进行调查研究,采集时间为 1 月(冬)、4 月(春)、7 月(夏)、10 月(秋). 高潮时在林内、林外和林缘的位点,各采集海水 1 L 后混合,用 Lugol's solution 固定,静置,沉淀,逐步浓缩到 50 mL. 于显微镜下取 0.1 mL 用浮游植物计数框进行种类鉴定和计数[4,5].

2 结果与讨论

2.1 浮游植物的种类组成

福鼎红树林区水体浮游植物的种类组成和密度见表 1. 共鉴定到浮游植物 32 属 77 种(包括变种),其中硅藻门 29 属 74 种,甲藻门 1 属 1 种,绿藻门 1 属 1 种,裸藻门 1 属 1 种.

2.2 浮游植物密度的季节变化

福鼎红树林区水体浮游植物密度变化范围为 1.14×10^4～1.16×10^6 个/L,平均 3.21×10^5 个/L,季节变化呈单峰型(图 1),2001～2002 年密度的最高值出现在春季(1.13×10^6 个/L),而 2002～2003 年密度的最高值出现在冬季(1.16×10^6 个/L). 2001～2003 年中出现最大密度的季节是不同的,但优势种都是长菱形藻(*Nitzschia longissima*),而且都是绝对优势种(2001 年 4 月和 2002 年 1 月分别占总密度的 82.2 % 和 76.4 %). 夏季浮游植物的密度均最低.

* 福建省自然科学基金(D0210003)资助

原载于厦门大学学报(自然科学版),2005,44(1):118－122

表 1　福鼎红树林区水体浮游植物的种类组成和密度($\times 10^3$ 个/L)

Tab. 1　Species composition and cell density of phytoplanktons from mangrove area in Fuding City, Fujian Province, China

种　名	2001 年				2002 年				2003 年
	1 月	4 月	7 月	10 月	1 月	4 月	7 月	10 月	1 月
硅藻门									
爪哇曲壳藻亚缢变种 *Achnanthes javanica* var. *subconstricta*		3.0		0.2					
翼茧形藻 *Amphiprora alata*	1.8	25.0			71.3	0.1		5.1	30.0
截端双眉藻 *Amphora terroris*				1.0					
双眉藻 *Amphora* sp.	0.1		0.3			0.3		0.1	0.5
尤氏直链藻 *Aulacoseira juergensi*							0.2		
货币直链藻 *Aulacoseira moniliformis*							0.1		
具槽直链藻 *Aulacoseira sulcata*				0.2			0.6		0.3
奇异棍形藻 *Bacillaria paradoxa*	3.0		0.3		1.3	0.4		0.3	
洛伦角毛藻 *Chaetoceros lorenziana*							0.5		
假弯角毛藻 *Chaetoceros pseudocurvisetus*							3.3		
角毛藻 *Chaetoceros* sp.								0.1	
卵形藻 *Cocconeis* sp.									0.3
中心圆筛藻 *Coscinodiscus centralis*					1.3				
线形圆筛藻 *Coscinodiscus lineatus*				1.0	1.3		0.2		
小眼圆筛藻 *Coscinodiscus oculatus*				0.2					
圆筛藻 *Coscinodiscus* sp.			0.6						
小环藻 *Cyclotella* sp.							0.2		
条纹小环藻 *Cyclotella striata*	0.1	2.0	0.3		1.3		0.3		
新月细柱藻 *Cylindrotheca closterium*		0.5	0.3					0.1	
膨胀桥弯藻 *Cymbella tumida*		0.5							
蜂腰双壁藻 *Diploneis bombus*						0.1	0.2		
史密斯双壁藻 *Diploneis smithii*				1.3	1.3				
史密斯双壁藻扩大变种 *Diploneis smithii* var. *dilatata*	0.1								
波罗的海布纹藻中华变种 *Gyrosigma balticum* var. *sinicum*							0.1		
优美布纹藻 *Gyrosigma exoticus*	0.1								
簇生布纹藻弧形变种 *Gyrosigma fasciola* var. *arcuata*					3.8				0.3
簇生布纹藻薄喙变种 *Gyrosigma fasciolaca* var. *tenuirostris*						0.1		0.5	
斜布纹藻 *Gyrosigma obliquum*		0.5							
刀形布纹藻 *Gyrosigma scalproides*		0.5	0.3	0.7		1.5		0.4	
布纹藻 *Gyrosigma* sp.			0.6			0.1			
斯氏布纹藻 *Gyrosigma spencerii*			0.9						
双尖菱板藻 *Hantzschia amphioxys*	0.1								
丹麦细柱藻 *Leptocylindrus danicus*									24.5
短楔形藻 *Licmophora abbreviata*			0.3			0.1			
岩石舟形藻 *Navicula scopulorum*		0.5							
舟形藻 *Navicula* spp.	19.2	12.0	0.6	5.3	1.3	4.9	0.4	2.6	6.0
新月菱形藻 *Nitzschia closterium*	0.3	4.0	0.3	0.3				1.3	
卵形菱形藻 *Nitzschia cocconeiformis*						0.1	0.1	2.0	
缢缩菱形藻 *Nitzschia constricta*		3.0							
簇生菱形藻 *Nitzschia fasciculata*		2.0		0.2		0.5			

续表 1

种名	2001年				2002年				2003年
	1月	4月	7月	10月	1月	4月	7月	10月	1月
碎片菱形藻 *Nitzschia frustulum*							0.1		
颗粒菱形藻 *Nitzschia granulata*		0.5	0.6	0.7					
长菱形藻 *Nitzschia longissima*	41.2	930.0	0.3	5.3	882.5	0.5	0.3	1.0	5.5
洛氏菱形藻 *Nitzschia lorenziana*	3.2						0.1		0.3
洛伦菱形藻密条变种 *Nitzschia lorenziana* var. *densestriata*							0.1		
较大菱形藻线形变种 *Nitzschia majuscula* var. *lineata*	0.1		0.3		62.5			0.1	
舟形菱形藻 *Nitzschia navicularis*				0.2					
钝头菱形藻 *Nitzschia obtusa*	0.1								
钝头菱形藻镰刀变种 *Nitzschia obtusa* var. *scalpelliformis*	3.5	15.0	0.3	2.0	2.5			0.9	3.0
具点菱形藻 *Nitzschia punctata*		13.0		0.7					
弯菱形藻 *Nitzschia sigma*	0.1	30.0	0.3		3.8				2.5
菱形藻 *Nitzschia* spp.	2.1	72.0	0.9	6.7	106.3	1.0	0.1	0.1	3.0
细弱菱形藻 *Nitzschia subtilis*	0.1	5.0	1.5	0.2		0.9	0.1	0.6	11.5
盘形菱形藻维多变种 *Nitzschia tryblionella* var. *victoriae*					1.3				
伊氏斜纹藻 *Pleurosigma aestuarii*					1.3			0.1	
相似斜纹藻 *Pleurosigma affine*							0.1		
宽角斜纹藻 *Pleurosigma angulatum*	0.1					0.1			0.3
宽角斜纹藻镰刀变种 *Pleurosigma angulatum* var. *falcatum*		5.0							
长斜纹藻中华变种 *Pleurosigma elongatum* var. *sinica*	0.1			27.0					
美丽斜纹藻 *Pleurosigma formosum*	0.1		0.3	6.0	3.8			0.1	
海洋斜纹藻 *Pleurosigma pelagicum*		4.0							1.0
斜纹藻 *Pleurosigma* sp.							0.1		
翼根管藻 *Rhizosolenia alata*	0.1								
中肋骨条藻 *Skeletonema costatum*	3.2			29.3	6.3	13.5	2.8	0.6	208.0
芽形双菱藻 *Surirella gemma*									0.3
沃氏双菱藻 *Surirrela voigti*			0.3			0.1			
双菱藻 *Surirella* sp.		2.0							
菱形海线藻 *Thalassionema nithzchioides*	0.7		1.2			0.6	0.4	0.4	
圆海链藻 *Thalassiosira rotula*							0.3	0.1	
佛氏海毛藻 *Thalassiothrix frauenfeldii*						0.1		1.4	
长海毛藻 *Thalassiothrix longissima*					1.3				
卵形褶盘藻 *Tryblioptychus cocconeiformis*				0.2					
硅藻未知种 sp. 1	1.7								
硅藻未知种 sp. 2				1.0					
裸藻门									
密盘裸藻 *Euglena wangi*	62.5	2.0	0.6	0.7	1.3	0.4	0.1	0.4	6.5
甲藻门									
四刺原多甲藻 *Protoperidinium quinquecorne*							0.7		
绿藻门									
四尾栅藻 *Scenedesmus quadricauda*							0.3		
种数合计(种)	25	23	22	23	19	20	26	22	19
密度合计($\times 10^3$ 个/L)	143.7	1131.7	11.4	90.4	1155.0	25.4	11.5	18.3	303.8

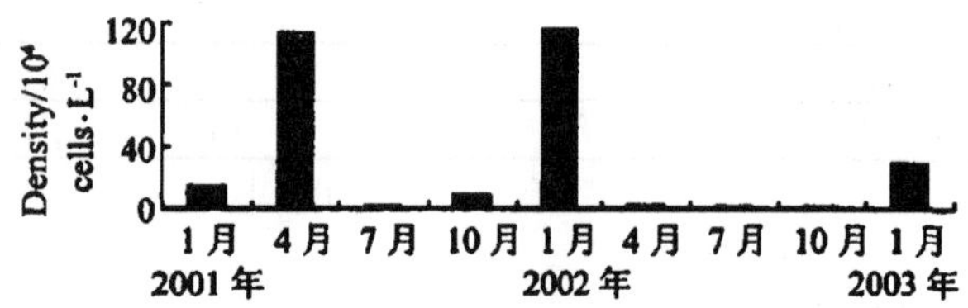

图1 福鼎红树林水体中浮游植物密度的季节变化

Fig. 1 Seasonal change of density of phytoplankton from mangroves in Fuding City in Fujian Province, China

2.3 浮游植物优势种和种类数的季节变化

福鼎红树林区水体浮游植物优势种在不同季节是不同的(表2). 除了2001年冬季优势种是裸藻类的密盘裸藻外,其他季节优势种均是硅藻. 浮游植物的优势种以长菱形藻(*Nitzschia longissima*)和中肋骨条藻(*Skeletonema costatum*)为多次出现的种类. 底栖硅藻也在浮游植物中占有重要的地位, 如2001年夏季, 纤细菱形藻(*Nitzschia subtilis*)成为优势种类.

福鼎红树林区水体浮游植物种类数的季节变化没有一定的规律性(表1),变化范围在19～26种之间. 这与底栖硅藻大量参与到浮游植物中有一定的关系.

2.4 红树林区水体浮游植物的生态作用及特征

1) 福鼎红树林区水体浮游植物以硅藻门种类为主,密度(除了2001年冬季为密盘裸藻外)和种类分别占浮游植物群落的50%和90%以上. 国内外某些红树林区水体浮游植物调查的报道也表明浮游植物以硅藻门种类为主[3~6]. 这是由于红树林区有丰富的硅(Si)和氮(N),有利于硅藻的繁殖.

2) 福鼎红树林区水体浮游植物的组成特点是水体中出现大量的原本为底栖硅藻的羽纹纲硅藻. 由于在红树林阻挡下,林前冲刷的潮汐和风浪的影响,底栖硅藻极易悬浮于水体中,起到丰富浮游植物的作用,福鼎后屿湾红树林区浮游植物集群中底栖性的种类占总种类数的50%以上,同时一些在红树植物、大型藻类上附着的硅藻如爪哇曲壳藻亚缢变种、卵形藻等也出现在浮游植物集群中.

3) 每个季节的浮游植物均出现具有典型淡水性质的种类如密盘裸藻、膨胀桥弯藻和四尾栅藻,其中密盘裸藻在2001年冬季成为优势种类,表明水质属咸淡水性质,这与后屿湾处于河流上游位置相符合,而大洋性的种类如角毛藻等仅有少量的种类和数量.

4) 赤潮生物在每个季节的浮游植物中均有出现,如中肋骨条藻(*Skeletonema costatum*)和四刺原多甲藻(*Protoperidinium quinquecorne*)等,但密度都不高,这说明福鼎后屿湾红树林区赤潮出现的可能性很小.

5) 前些时期红树林浮游植物研究大多集中小型浮游植物(20～200 μm)(包括小型硅藻),然而目前人们发现海洋微型浮游生物(<20μm)在生态系统中占有相当重要的位置,特别是在初级生产力中占有很重要的地位[7~9],微型食物网的特点是营养盐的快速更

表2 2001年1月～2003年1月福鼎红树林区水体浮游植物的优势种

Tab. 2 Dominant species and their percentage in phytoplankton assemblages from mangrove area in Fuding City, Fujian Province, China from Jan. 2001 to Jan. 2003

		优势种	百分比/%*
2001年	1月	裸藻 *Euglena wangi*	43.5
		长菱形藻 *Nitzschia longissima*	28.7
	4月	长菱形藻 *Nitzschia longissima*	82.2
	7月	纤细菱形藻 *Nitzschia subtilis*	13.2
		菱形海线藻 *Thalassionema nitzschioides*	10.5
	10月	中肋骨条藻 *Skeletonema costatum*	32.5
		宽角斜纹藻中华变种 *Pleurosigma angulatum var. sinica*	29.9
2002年	1月	长菱形藻 *Nitzschia longissima*	76.4
	4月	中肋骨条藻 *Skeletonema costatum*	52.9
	7月	假弯角毛藻 *Chaetoceros pseudocurvisetus*	28.8
		中肋骨条藻 *Skeletonema costatum*	24.5
	10月	翼茧形藻 *Amphiprora alata*	27.9
2003年	1月	中肋骨条藻 *Skeletonema costatum*	68.5

*占总密度的百分率(%),其它非优势种之和未列出.

新,能量的高速转换,与碎屑紧密联系,以及具有复杂的摄食类型,包括吞噬营养[10]. 因此对红树林区微型藻类包括微型硅藻的研究有助于补充和完善浮游植物群落的组成,对于深入了解微型浮游植物包括微型硅藻在整个浮游植物群落中的地位有重要意义.

3 结 论

福鼎红树林区水体中浮游植物的群落组成以硅藻为主,赤潮生物少量出现,表明该区域水体质量较好. 水体中较丰富的浮游植物种类和较高的密度可以维持较高的初级生产力,对红树林生态系统次级生产的贡献也相对较大. 红树林潮间带底栖硅藻在浮游植物中占有较大的比例,红树植物上的附着硅藻也有少量出现,与潮汐和风浪冲刷有关,表明了该水域浮游植物群落组成和结构上的特殊性.

参考文献:

[1] 林鹏. 红树林[M]. 北京:海洋出版社,1984. 1 - 102.

[2] Leija-Tristan A, Sanchez-Vargas D P. Biology and ecology of the common mud shrimp *Upogebia dawsoni* (Crustacea: Thalassinoidea) of Manglar Requeson, Bahia Concepcion and Estero Rio Mulege, Baja California Sur, Mexico [J]. Revista de Biologia Tropical, 1988, 36(1): 107 - 114.

[3] 陈坚,范航清,陈成英. 广西英罗湾红树林区水体浮游植物种类组成和数量分布的初步研究[J]. 广西科学院学报,1993,9(2):31 - 33.

[4] Damroy S. Studies on mangrove ecology of Chouldari area, South Andaman[J]. J. Andaman Science Association, 1995, (1 - 2): 29 - 33.

[5] Roy S D. Mangrove ecology of Alexandra Island and Manjera area of South Andaman[J]. J. Andaman Science Association, 1995, 11(1 - 2): 58 - 61.

[6] 刘玉,陈桂珠. 深圳福田红树林区藻类群落结构和生态学研究[J]. 中山大学学报(自然科学版),1997,36(1):102 - 106.

[7] Takahashi M, Bienfang P K. Size structure of phytoplankton biomass and photosynthesis in subtropical Hawaiian waters[J]. Marine Biology, 1983, 76(2): 203 - 211.

[8] Weber L H, El-sayed S Z. Contributions of the net, nano-, and picoplankton to the phytoplankton standing crop and primary productivity in the Southern Ocean[J]. Journal of Plankton Research, 1987, 9: 973 - 994.

[9] 高亚辉,金德祥,程兆第. 厦门港微型浮游植物叶绿素的分布与作用[J]. 海洋与湖沼,1994,25(1):87 - 93.

[10] 程兆第,高亚辉,刘师成. 福建沿岸微型硅藻[M]. 北京:海洋出版社,1993. 1 - 91.

Dynamics of Phytoplankton Community in Mangrove Waters in Fuding City, Fujian Province, China

CHEN Chang-ping, GAO Ya-hui, LIN Peng

(School of Life Sciences, Xiamen University, Xiamen 361005, China)

Abstract: Species composition and cell density of phytoplankton in Mangrove waters in Fuding City, Fujian Province, China was studied seasonally from Jan. 2001 to Jan. 2003. Water samples were collected every three months and observed under light microscope. The results showed that dominant species was diatoms in each season, such as *Nitzschia longissima* and *Skeletonema costatum*, except during Jan. 2001 which was *Euglena wangi*. Some taxa belonging to Pyrrophyta and Chlorophyta also occurred occasionally in the waters. Totally 77 taxa in 32 genera were identified, in which 74 taxa in 29 genera belonged to Bacillariophyta, and 1 taxa in 1 genera for Euglenophyta, Pyrrophyta and Chlorophyta, respectively. Cell density of phytoplankton ranged from 1.14×10^4 to 1.16×10^6 cells/L, and average density was 3.21×10^5 cells/L. The communities were characterized by abundant estuarine species and benthic diatoms concerned with erosion by tide and wind wave, which distributed in the inter-tide in the mangroves.

Key words: mangrove; phytoplankton; diatom; Fuding City

深圳福田红树林保护区浮游植物群落的季节变化及其生态学研究*

陈长平　高亚辉　林　鹏

(厦门大学生命科学学院,福建 厦门 361005)

摘要：对深圳福田红树林保护区浮游植物群落结构和季节变化进行研究.共鉴定到浮游植物5门25属75种,其中硅藻门21属68种,蓝藻门1属1种,甲藻门1属1种,绿藻门1属3种,裸藻门1属3种.密度的季节变化范围为1.0 ×10⁶～5.0 ×10⁶个/L,平均密度为2.7 ×10⁶个/L.赤潮藻和耐污染特征的种类如威氏海链藻和微小小环藻等是浮游植物的主要成分.底栖性、附着性和淡水性的种类在浮游植物中经常出现.与以前的研究比较,福田红树林区水体浮游植物群落朝着种类个体变小、种类数减少、密度增加、耐污染种类增加的方向变化,这反映了该红树林区水体富营养化程度高,水质持续恶化.

关键词：福田红树林保护区;浮游植物;硅藻;底栖

中图分类号：Q 948.8　　**文献标识码**:A　　**文章编号**:0438-0479(2005)Sup-0011-05

红树林区的浮游植物是红树林生态系统中除红树植物、底栖藻类外的又一重要生产者,是海洋动物尤其是海洋动物幼虫和幼体的直接饵料[1].一些浮游植物如骨条藻等可作为“三废指标”的指示种,也是海水富营养化、发生赤潮的主要生物之一[2~4].浮游植物的大量繁殖与红树林区某些浮游动物的大量爆发有着密切关系[5].研究红树林中的浮游植物群落结构可以更好地说明红树林生态系统的独特生态功能及在河口海岸中的重要性,而且可以指示红树林区水体质量.目前对我国红树林区浮游植物的研究比较缺乏,特别是在浮游植物群落结构的季节变化方面[6,7].本文通过对深圳福田红树林区浮游植物群落结构和生态学的研究,了解浮游植物在红树林内的分布和生态作用,对红树林生态系统基础资料的补充和在监测保护红树林环境方面具有重要的意义.

1　材料与方法

1.1　取样地点

深圳湾东北岸的福田红树林鸟类自然保护区位于北纬22°32′,东经114°03′～114°05′之间,该保护区东起深圳河口皇岗,西至车公庙,北到广深高速公路,南达海滩外5.1 km处,保护面积304 hm^2,其中红树林面积66.7 hm^2,以深圳河为界,与香港米埔红树林自然保护区隔水相望.年均温22.5 ℃,最热月均温(7月)28.7 ℃,最冷月均温(1月)15.0 ℃,年均降水量1 926.80 mm,年均相对湿度79 %.红树林生长在水陆交界的潮间带,林外是广阔平坦的滩涂湿地,淤泥的平均厚度达3 m,海岸线曲折,小河或排水沟通过红树林并在滩涂上形成潮沟.深圳湾的潮汐属不规则的半日潮,由低低潮上涨至低高潮的涨潮差约1 m,由高低潮到高高潮的涨潮差约2 m.

1.2　研究方法

于2001年4月到2003年1月对深圳福田红树林区水体浮游植物群落的季节变化进行研究,分别于1月(冬)、4月(春)、7月(夏)、10月(秋)现场采样.高潮时在林内外和林缘采集表层海水共3 L,用Lugol′s solution固定,静置,沉淀,逐步浓缩到50 mL.取0.1 mL于显微镜下用浮游植物计数框进行种类鉴定和计数[8].

2　结果与讨论

2.1　浮游植物的种类组成

深圳福田红树林保护区鉴定到浮游植物5门25属75种,其中硅藻门21属68种,占总种类数的90.7 %,蓝藻门1属1种,甲藻门1属1种,绿藻门1属3种,裸藻门1属3种(表1).

* 福建省自然科学基金(D0210003),厦门大学科技创新项目(XD KJCX20041018)资助
原载于厦门大学学报(自然科学版),2005,44(Sup.):11－15

表 1 深圳福田红树林水体浮游植物群落的种类组成和密度

Tab. 1 Composition and density of phytoplankton community in water under mangrove forest in Futian of ShenZhen City, China

门类	种类	2001 年			2002 年			2003 年	
		4	7	10	1	4	7	10	1
硅藻门	短柄曲壳藻				0.8				
	咖啡形双眉藻	2.7	2.7	0.6					
	具槽直链藻	53.3							
	尤氏直链藻					0.6		0.2	
	念珠直链藻								0.5
	奇异棍形藻		1.3	0.1	0.8			0.1	0.1
	美壁藻未知种 1					0.2			
	卵形藻未知种 1					0.6			
	弓束圆筛藻			0.1					
	琼氏圆筛藻					0.1			
	线形圆筛藻						4.0	0.1	0.1
	辐射圆筛藻		1.3	0.1	0.1				
	圆筛藻未知种 1					0.1			
	极微小环藻		1133.3	315.0					
	微小小环藻		1200.0	135.0			2880.0		
	梅里小环藻	2.7			2.4				
	条纹小环藻		1.3					2.0	14.0
	条纹小环藻双斑点变种		222.2						
	柱状小环藻	2.7		0.1	0.1				
	小环藻未知种 1					0.3			0.5
	新月细柱藻			0.6					
	细柱藻未知种 1			0.1				0.5	
	地中海指管藻								0.5
	史密斯双壁藻					0.1		0.1	
	尖布纹藻								3.0
	波罗的海布纹藻中华变种					0.1			
	簇生布纹藻薄喙变种						4.0		
	刀形布纹藻		5.3			0.2	8.0	0.1	
	布纹藻未知种 1					0.3			
	布纹藻未知种 2						4.0		
	斯氏布纹藻					0.4			
	丹麦细柱藻			0.1					
	内实舟形藻					0.1			
	截端舟形藻				2.4				
	缝舟形舟形藻			0.1	2.4			0.1	0.1
	喙头舟形藻			0.3					
	柔弱舟形藻		269.4	165.0	1.6		4.0		
	舟形藻未知种 1	2.7		0.4		0.8	4.0	0.5	3.0
	舟形藻未知种 2	29.3	2.7		0.8				
	尖锥菱形藻		2.7						
	新月菱形藻					0.1			

表 1(续)

门类	种类	2001年			2002年			2003年	
		4	7	10	1	4	7	10	1
	簇生菱形藻	10.7	2.7		8.0	0.2	4.0		0.1
硅藻门	碎片菱形藻	45.3	10.7	2.2	101.6		780.0	0.1	43.0
	颗粒菱形藻					0.2			
	长菱形藻				0.8				2.0
	洛氏菱形藻	2.7			4.0			0.5	3.0
	较大菱形藻线形变种					0.3		0.5	
	钝头菱形藻					0.3			
	钝头菱形藻刀形变种	2.7			1.6			0.5	
	琴式菱形藻	5.3		0.3	1.6	2.1			0.1
	琴式菱形藻小型变种		2.7						
	弯菱形藻	26.7		0.3	0.1	1.2	4.0		
	纤细菱形藻	2.7				3.8			
	菱形藻未知种 1		5.3	0.1		0.1	4.0		5.0
	菱形藻未知种 2			0.5					
	伊氏斜纹藻				0.8	1.0		0.5	
	美丽斜纹藻	2.7	2.7			0.3			0.5
	海洋斜纹藻	5.3				2.1			
	斜纹藻未知种 1					0.3	8.0		
	斯氏根管藻								22.0
	中肋骨条藻	24.0		4.0	9.6			53.5	3211.5
	芽形双菱藻	2.7				0.1		0.1	
	华丽针杆藻					0.1			
	针杆藻未知种 1					0.1			0.5
	菱形海线藻	2.7	10.7	0.2		0.2		0.5	0.1
	太平洋海链藻								0.5
	威氏海链藻	613.3	133.0	402.2	1395.0	1720.0	8.0	3690.0	1483.0
	硅藻未知种 1					0.1			
蓝藻门	颤藻未知种 1	320.0			120.0	240.0			
甲藻门	裸甲藻未知种 1	18.7							
绿藻门	被甲栅藻		10.7						
	二形栅藻						8.0		
裸藻门	静裸藻	8.0							
	密盘裸藻	2.7							
	裸藻未知种 1				5.6	0.1		0.5	182.0
	种类数(种)	24	19	22	21	34	14	19	23
	总密度	1189.6	3020.7	1027.4	1660.1	1976.6	3724	3750.4	4975.1

2.2 浮游植物密度的季节变化

深圳福田红树林水体浮游植物密度的变化范围为 $1.0\times10^6\sim5.0\times10^6$ 个/L,平均密度为 2.7×10^6 个/L(图 1).2001～2002 年 4 各季节中夏季密度最高,为 3.0×10^6 个/L,2002～2003 年个季节中冬季密度最高,为 5.0×10^6 个/L.每个季度浮游植物密度均已达到富营养化的水平(10^6 个/L),2002～2003 年平均密度比 2001～2002 年平均密度高出近二倍,说明水体富营养化加剧,水质趋向恶化.

2.3 浮游植物优势种和种类数的季节变化

深圳福田红树林保护区水体浮游植物的优势种见表 2.微型硅藻和耐污染种类在浮游植物群落中占有

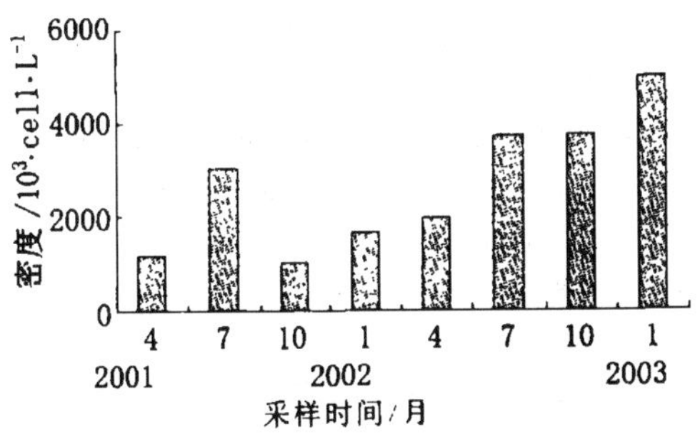

图 1 福田红树林区水体浮游植物密度的季节变化

Fig. 1 Seasonal changes of phytoplankton community density at water of mangroves in Futian of Shenzhen City, China

优势地位. 威氏海链藻和小环藻是浮游植物的主要种类,中肋骨条藻在 2003 年冬季成为优势种类. 有污染特征的蓝藻门种类颤藻在 2001 年夏季也是主要种类之一.

浮游植物种类数较低,变化范围为14~24种,平均种类数为 22 种,春季种类数较高,达 24 种,夏季种类数较低,仅为 14 种.

表 2 深圳福田红树林水体浮游植物的优势种及其占总密度的百分比

Fig. 2 Dominant species and their percentage in phytoplankton in water of mangroves in Futian of Shenzhen City, China

	月份	优势种	百分比/ % *
2001 年	4	威氏海链藻	51.6
		颤藻未知种 1	26.9
	7	微小小环藻	39.7
		极微小环藻	37.5
	10	威氏海链藻	39.2
		极微小环藻	30.7
2002 年	1	威氏海链藻	80.0
	4	威氏海链藻	87.0
	7	微小小环藻	77.3
	10	威氏海链藻	98.4
2003 年	1	中肋骨条藻	64.6

* 占总密度的百分率(%),其它非优势种之和未列出.

2.4 浮游植物群落结构分析

深圳福田红树林水体浮游植物群落多样性指数变化范围为 0.134~2.244,平均多样性指数为 1.268,均匀度变化范围为 0.031~0.496,平均均匀度为 0.289,优势度变化范围为 0.290~0.968,平均优势度为 0.569. 每个季节浮游植物的多样性指数均小于 3,而且多样性指数和均匀度逐渐减小,优势度逐渐增加(表 3),说明水体环境恶化.

2.5 福田红树林保护区浮游植物的生态作用和特征

1) 浮游植物以硅藻门种类为主,分别占总种类数和密度的 90.7 %和 95.7 %. 同时出现多种绿藻、裸藻等. 有污染特征性的蓝藻门种类(颤藻)、裸藻门种类在水体中经常出现,并且颤藻密度较高,最高可达 3.2×10^5 个/L.

2) 底栖性和附着性的硅藻在浮游植物中大量存在,约占总种类数的 40 %和 10 %. 由于在红树林阻挡下,林前冲刷的潮汐和风浪的影响,底栖硅藻极易悬浮于水体中,起到丰富浮游植物的作用.

3) 淡水性的种类在浮游植物群落中经常出现,如颤藻、裸藻等种类,说明水质属咸淡水性质.

4) 赤潮藻和耐污染特征的种类是浮游植物的主要成分. 赤潮藻如威氏海链藻、中肋骨条藻等在多个季节成为绝对优势种,耐污染的藻类如微小小环藻、极微小环藻、颤藻、裸藻等在浮游植物中大量存在. 浮游植物的密度在每个季度均达到富营养化的水平,并且有继续增加的趋势. 有毒的赤潮藻如裸甲藻等也偶然出现.

5) 与以前的研究比较[7],福田红树林区水体浮游植物群落朝着种类个体变小、种类数减少、密度增加、耐污染种类增加的方向变化,这反映了该红树林区水体富营养化程度高,水质持续恶化.

2.6 小 结

深圳福田红树林保护区浮游植物以硅藻门种类为

表 3 深圳福田红树林水体浮游植物集群结构分析

Tab. 3 Structure of phytoplankton assemblages in waters of mangroves in Futian of Shenzhen City, China

	2001 年			2002 年				2003 年
	4	7	10	1	4	7	10	1
优势度	0.344	0.314	0.290	0.715	0.772	0.642	0.968	0.507
多样性指数	2.244	2.033	1.957	0.959	0.635	0.920	0.134	1.264
均匀度	0.496	0.479	0.439	0.218	0.125	0.242	0.031	0.279

主,并有多种绿藻、裸藻等.大量的底栖和附着硅藻以及淡水性藻类是该红树林区浮游植物组成的特点.赤潮藻和耐污染的藻类在浮游植物群落中占有优势地位.浮游植物的组成和结构反映了该红树林区水体富营养化程度高,水质趋向恶化.

参考文献:

[1] 林鹏.红树林[M].北京:海洋出版社,1984.1-102.

[2] 李少菁.厦门港几种海洋浮游桡足类的食性与饵料成分的初步研究[J].厦门大学学报(自然科学版),1964,11(3):93-109.

[3] 郑重,张松踪,李松.中国海洋浮游桡足类(上卷)[M].上海:上海科学技术出版社,1965.144-156.

[4] Vicente H J. Monthly population density fluctuation and vertical distribution of meiofauna community in tropical muddy substrate[J]. Asian Fisheries Forum, Tokyo (Japan),1989,(4):17-22.

[5] Nicholas W L, Stewart A C, Marples T G. Field and laboratory studies of Desmodora cazca Gerlach, 1956 (Desmodoridae: Nematoda) from mangrove mud-flats[J]. Nematologica 1988,34(3):331-349.

[6] 陈坚,范航清,陈成英.广西英罗湾红树林区水体浮游植物种类组成和数量分布的初步研究[J].广西科学院学报,1993,9(2):31-33.

[7] 刘玉,陈桂珠.深圳福田红树林区藻类群落结构和生态学研究[J].中山大学学报(自然科学版),1997,36(1):102-106.

[8] 金德祥,陈金环,黄凯歌.中国海洋浮游硅藻类[M].上海:上海科学技术出版社,1965.1-230.

Study on the Seasonal Changes of Phytoplankton Community and Its Ecology in Futian Mangrove Reserve of Shenzhen, China

CHEN Chang-ping, GAO Ya-hui, LIN Peng

(School of Life Sciences, Xiamen University, Xiamen 361005, China)

Abstract: Seasonal changes and ecology of phytoplankton in mangrove reserves in Futian, Shenzhen were studied. Totally 75 species belonging to 25 genera in 5 phyta were identified, among which 68 taxa in 21 genera belonged to Bacillariophyta, 1 taxa in 1 genera for Cyanophyta and Pyrrophyta, and 3 taxa in 1 genera for Euglenophyta and Chlorophyta, respectively. Density varied from 1.0×10^6 to 5.0×10^6 ind./L, and average density was 2.7×10^6 ind./L. Redtide causive species and pollution-tolerance species such as Thalassiosira weissflogii and Cyclotella caspia were dominant in phytoplankton. Benthic, epiphytic and freshwater species occurred frequently. Comparison with former study, phytoplankton community in mangrove reserves of Futian were characterized by smaller species, less species number, increasing density and more pollution-tolerance species, showing eutrophic and deteriorated water.

Key words: mangrove reserves in Futian; phytoplankton; diatom; benthic

盐度和 pH 对底栖硅藻胞外多聚物的影响*

陈长平 高亚辉 林 鹏

(厦门大学生命科学学院,福建 厦门 361005)

摘要: 研究了盐度和 pH 值对底栖硅藻新月筒柱藻(*Cylindrotheca closterium* (Ehr.) Reimann et Lewin)增殖、蛋白质含量和胞外多聚物(Extracellular Polymeric Substances, EPS)的影响. 结果表明新月筒柱藻最适生长的盐度和 pH 值分别是 15 和 8, 属半咸水性生活. 高盐度(>15)和低 pH 值(<pH8)的胁迫促进了胞外多聚物(EPS)的积累, 说明 EPS 的存在可能有利于缓解外界的不利条件. 胶体 EPS 和附着 EPS 对盐度和 pH 值的响应不同, 反应了两种 EPS 功能上的差异. 盐度和 pH 值对新月筒柱藻胞内碳水化合物的影响不显著.

关键词: 底栖硅藻; 胞外多聚物; 盐度; pH

中图分类号: Q949.27 **文献标识码**: A **文章编号**: 0253-4193-(2006)05-0123-08

1 引言

底栖微藻是海洋潮间带滩涂生境的主要初级生产者, 以底栖硅藻为主要组成部分[1,2], 是海洋动物幼虫和幼体的直接饵料, 也是一些经济软体动物的主要饵料[3,4]. 底栖硅藻通常根据潮汐变化、滩涂特点等在滩涂的表层进行不同程度的运动, 同时分泌一类被称为胞外多聚物(Extracellular Polymeric Substances, EPS)的黏性物质[5]. EPS 不仅是细菌和底栖动物等的碳源, 在河口生态系统的碳循环中起重要作用[6,7]; 而且能通过不同的方式与沉积物颗粒结合[8], 提高潮间带沉积物的稳定性, 这一过程对受潮汐和河流影响而显得脆弱的潮间带尤为重要[9,10]. EPS 还具有稳定群落组成[8]、抗失水、防止硅质壁溶解以及减少渗透压等功能[11].

河口海岸生境复杂多变, 受海潮和河流的影响, 环境因素如营养盐、光照、盐度、pH 值、温度等变化较大. 底栖硅藻主要集中在土层表面, 会分泌大量的 EPS, 以达到结合有机或无机营养盐的作用[6], 适应贫营养的环境[12], 特别是针对其大量繁殖、高密度生长造成的营养盐缺乏[13~15]. 底栖硅藻的 EPS 还与温度、光照、光合作用及其运动规律有关[16,17].

目前还没见到盐度和 pH 值对底栖硅藻胞外多聚物影响的相关报道. 本文研究了盐度和 pH 值对从潮间带滩涂中分离出来的一种底栖硅藻——新月筒柱藻(*Cylindrotheca closterium*)胞外多聚物产量的影响, 为了解底栖硅藻如何适应复杂的潮间带滩涂环境提供了科学依据, 并有利于进一步研究底栖硅藻的 EPS 在海洋潮间带生态系统及碳循环中的重要作用, 在评估潮间带滩涂养殖容量, 水产增养殖业的合理布局方面有一定的参考价值.

2 材料和方法

底栖硅藻藻种为羽纹纲菱形藻科筒柱藻属新月筒柱藻(*C. closterium*), 分离自福建九龙江口潮间带滩涂. f/2 培养液培养, 温度 20 ℃, pH8.0, 盐度 15, 光强 1 500~2 500 lx, 光照采用白色日光灯, 光照周期 12D:12L.

研究了 3 个盐度梯度即 15, 25 和 35; 3 个不同 pH 值即 pH6.0, 7.0 和 8.0 对新月筒柱藻增殖、蛋

* 厦门大学科技创新项目(XMDX KJ CX20051018); 福建省自然科学基金(D020031)

原载于海洋学报, 2006, 28(5): 123－129

白质和胞外多糖的影响.每个梯度 3 个重复.

2.1 生长过程细胞密度的测定

在无菌条件下取一定量的藻液后,吸取0.1 mL,在光学显微镜下计数,计算得到藻细胞密度.

2.2 胞外多糖、胞内碳水化合物和蛋白质含量的测定

胞外多糖含量的测定采用苯酚硫酸法[18].

参考 Wolfstein 和 Stal[16] (2002) 的方法,把胞外多糖分为 2 种组份:胶体 EPS(Colloidal EPS, CEPS) 和附着 EPS(Attached EPS, AEPS). CEPS 是指溶解在培养基中的多糖,AEPS 是指黏附在细胞上的多糖,两者均为水溶性的多糖.提取方法见文献[16].

分别以 CEPS, CEPS/ Pr., CEPS/ Cell, AEPS, AEPS/ Pr.和 AEPS/ Cell 作为不同的参数比较了不同条件下新月筒柱藻分泌胞外多糖的差异,其中 CEPS和 AEPS 是指每毫升培养基中含有的 CEPS 和 AEPS 含量(mg/ dm^3), CEPS/ Pr.和 AEPS/ Pr.是指每微克蛋白分泌的 CEPS 和 AEPS 含量(μg/ μg), CEPS/ Cell 和 AEPS/ Cell 是指每个细胞分泌的 CEPS 和 AEPS 含量(10^{-6} mg/ 个).

胞内碳水化合物和蛋白质含量分别采用硫酸苯酚法和考马斯亮蓝法[19]测定.

2.3 数据统计分析

本文中的数据统计过程使用 SPSS 10.0 For Windows 统计软件完成[20]

3 结果

3.1 细胞的增殖和蛋白质含量

结果表明,盐度对细胞增殖的影响不显著($p>0.05$)(图 1).最高的细胞密度均出现在第 10 天,比第 1 天增加了近 3 倍.细胞蛋白质含量也在第 10 天最大,比第 1 天的细胞蛋白质含量,盐度 15,25 和 35 实验组分别增加了 2.3,1.8 和 1.7 倍.随着盐度的增加,细胞蛋白质含量显著下降($p<0.05$).

与 pH7 和 pH8 实验组相比,pH6 显著促进了细胞的增殖($p<0.05$),而 pH7 和 pH8 两个实验组差异不显著($p>0.05$).不同 pH 值实验组细胞的蛋白质含量均在第 10 天最高,比第 1 天的细胞蛋白质含量,pH6,pH7 和 pH8 实验组分别增加了 1.8, 1.5和 2.6 倍. pH8 实验组细胞蛋白质含量显著大于 pH6 和 pH7 实验组的蛋白质含量($p<0.05$),而 pH6 实验组细胞蛋白质含量显著大于 pH7 实验组的细胞蛋白质含量($p<0.05$).

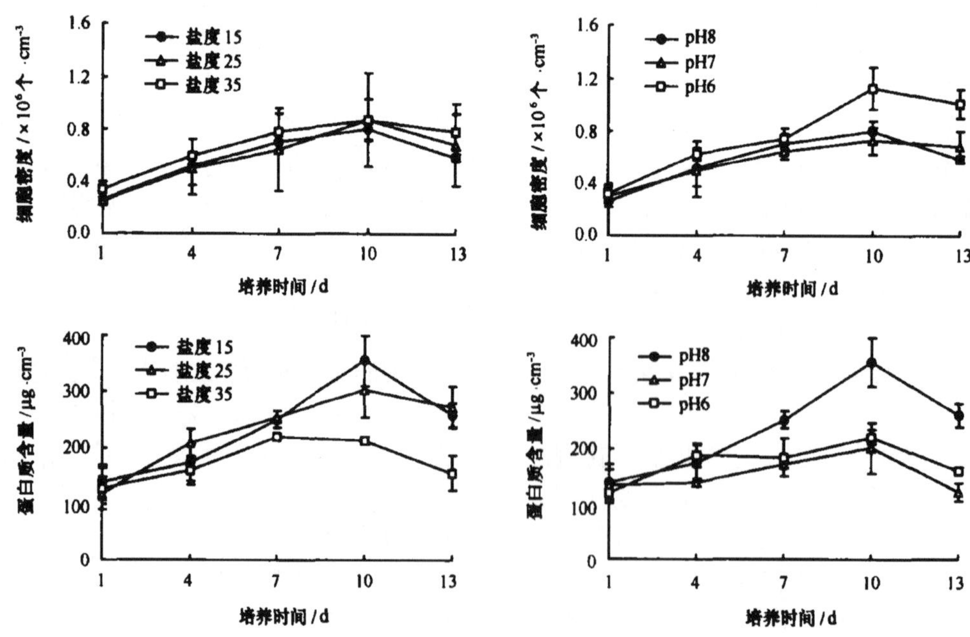

图 1 盐度和 pH 值对新月筒柱藻(*C. closterium*)细胞增殖和蛋白质含量的影响

3.2 胞外多糖

3.2.1 胶体胞外多糖(CEPS)

不同盐度和pH值对新月筒柱藻胶体胞外多糖的影响见图2.结果表明,在实验早期,不同处理的CEPS含量都相对较低,在实验后期或末期达到最高值,最高可增加近5倍(盐度15和pH8实验组).盐度25实验组的CEPS含量显著高于盐度15实验组的CEPS含量($p<0.05$),而盐度35实验组与盐度15和25实验组之间无显著差异($p>0.05$).pH7实验组CEPS含量显著大于pH6和pH8实验组的CEPS含量($p<0.05$),而pH6和pH8实验组差异不显著($p>0.05$).

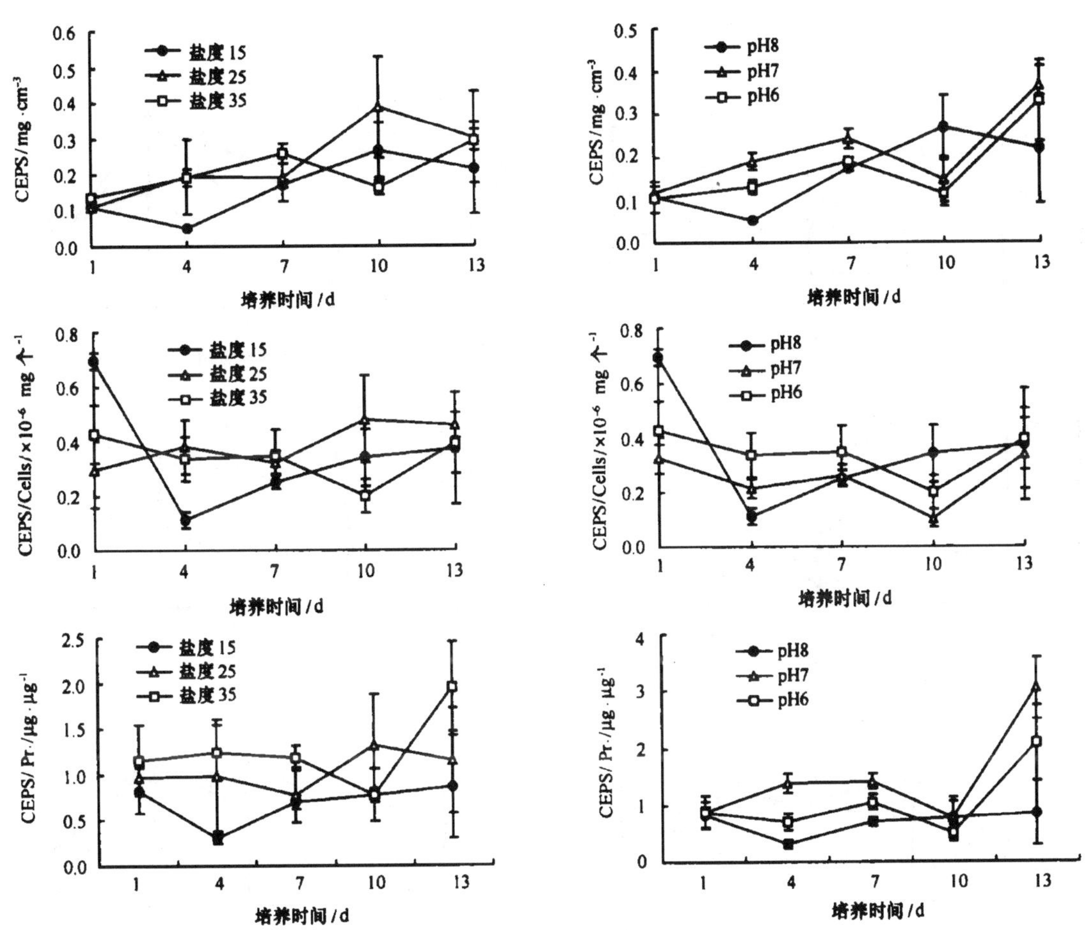

图2 盐度和pH值对新月筒柱藻(*C. closterium*)CEPS,CEPS/Cell和CEPS/Pr.的影响

CEPS/Cell和CEPS/Pr.值在实验后期有减小的可能,这是因为细胞密度或蛋白质含量在实验后期较高,其他的研究也有类似的现象[2,13].盐度对CEPS/Cell含量的影响不显著($p>0.05$).pH6实验组CEPS/Cell含量显著大于pH 7和pH8实验组($p<0.05$),而pH 7和pH8实验组之间没有显著差异($p>0.05$).

盐度35实验组CEPS/Pr.含量显著大于盐度15和25实验组的CEPS/Pr.含量($p<0.05$),而盐度15和25实验组之间没有显著差异($p>0.05$).pH7实验组CEPS/Pr.含量显著大于pH6和pH8实验组的CEPS/Pr.含量($p<0.05$),而pH6实验组CEPS/Pr.含量显著大于pH8实验组的CEPS/Pr.含量($p<0.05$).

3.2.2 附着胞外多糖(AEPS)

不同盐度和pH值对新月筒柱藻AEPS含量的影响见图3.结果表明,AEPS在实验早期含量很小,在第10天时AEPS含量达到最大值,与CEPS相比,AEPS含量在前期更低,但在生长后期含量高于CEPS含量.盐度25实验组AEPS含量显著大于盐度15和35实验组的AEPS含量($p<0.05$),而盐度15和35实验组之间没有显著差异($p>0.05$).pH7实验组的AEPS含量显著大于pH8实验组的AEPS含量($p<0.05$),而pH6实验组和pH7,pH8实验组之间差异不显著($p<0.05$).

盐度对AEPS/Cell含量的影响不显著($p<0.05$).pH7实验组AEPS/Cell含量显著大于pH6实验组和pH8实验组AEPS/Cell含量($p<0.05$).而pH6实验组和pH8实验组之间无显著差异($p>0.05$).

盐度15实验组AEPS/Pr.含量显著小于盐度25和35实验组的AEPS/Pr.含量($p<0.05$),盐度25和35实验组之间没有显著差异($p>0.05$).pH值对AEPS/Pr.含量的影响与CEPS/Pr.含量的影响是一样的,pH7实验组AEPS/Pr.含量显著大于pH6和pH8实验组的AEPS/Pr.含量($p<0.05$),而pH6实验组AEPS/Pr.含量显著大于pH8实验组的AEPS/Pr.含量($p<0.05$).

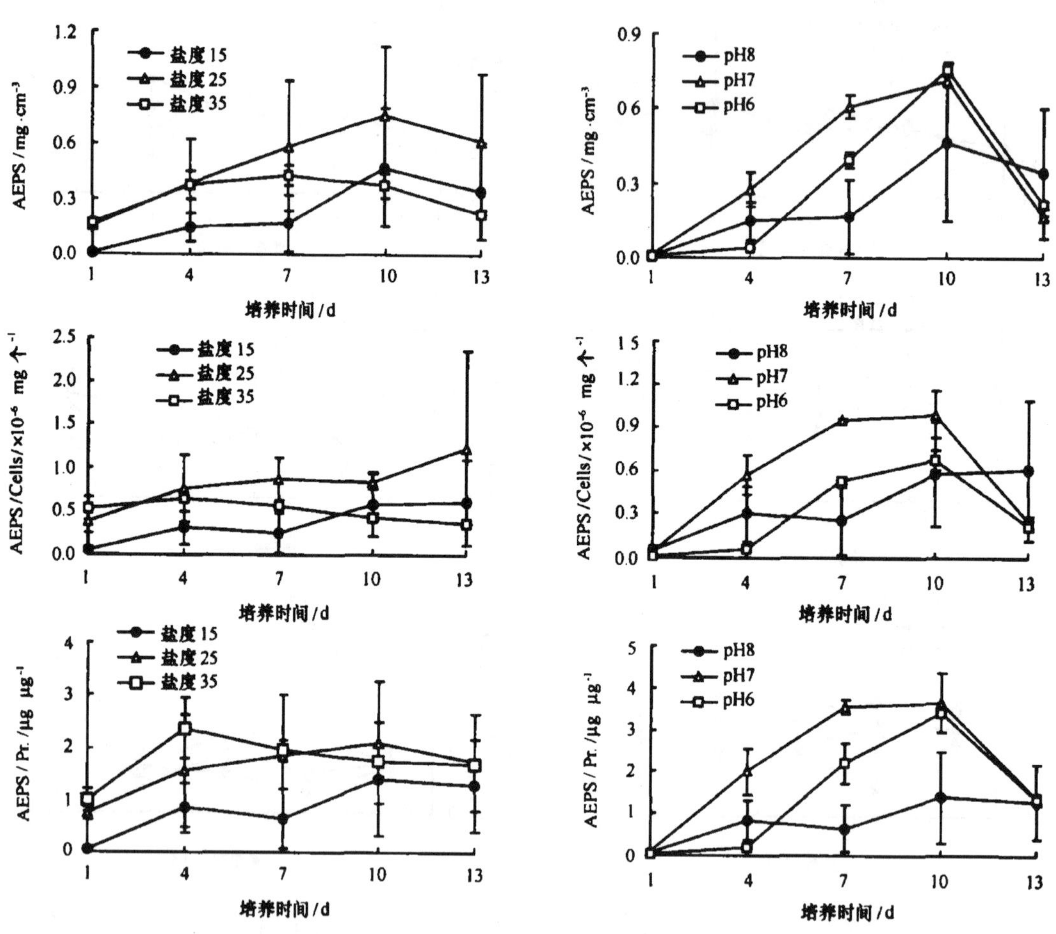

图3 盐度和pH值对新月筒柱藻(*C. closterium*)分泌AEPS,AEPS/Cell和AEPS/Pr.的影响

3.3 胞内碳水化合物(Intracellular carbohydrate, ICH)

细胞ICH含量在生长后期较高,积累速度明显低于EPS(见图4).盐度对ICH含量和ICH/Cell含量没有显著的影响($p>0.05$).盐度35实验组的ICH/Pr.含量显著于盐度15和25实验组ICH/Pr.含量($p<0.05$),而盐度15和25实验组之间无显著差异($p>0.05$).pH6实验组ICH含量显著高于pH7实验组的ICH含量($p<0.05$),而pH8实验组与pH6,pH7实验组之间无显著差异($p>0.05$).pH6,pH7和pH8实验组对ICH/Cell含量没有显著的影响($p>0.05$).而pH8实验组的ICH/Pr.含量显著低于pH6和pH7实验组的ICH/Pr.含量($p<0.05$).

3.4 总胞外多糖(Total EPS, TEPS)

TEPS在生长前期含量较低,到第10天含量达到最大(见图5).盐度25实验组的TEPS显著高于盐度 15 实验组和 35 实验组的 TEPS 含量($p < 0.05$),而盐度 15 实验组和 35 实验组之间没有显著差异($p > 0.05$). pH7 实验组的 TEPS 显著大于 pH6 实验组和 pH8 实验组的 TEPS($p < 0.05$),而 pH6 实验组和 pH8 实验组之间无显著差异($p > 0.05$).

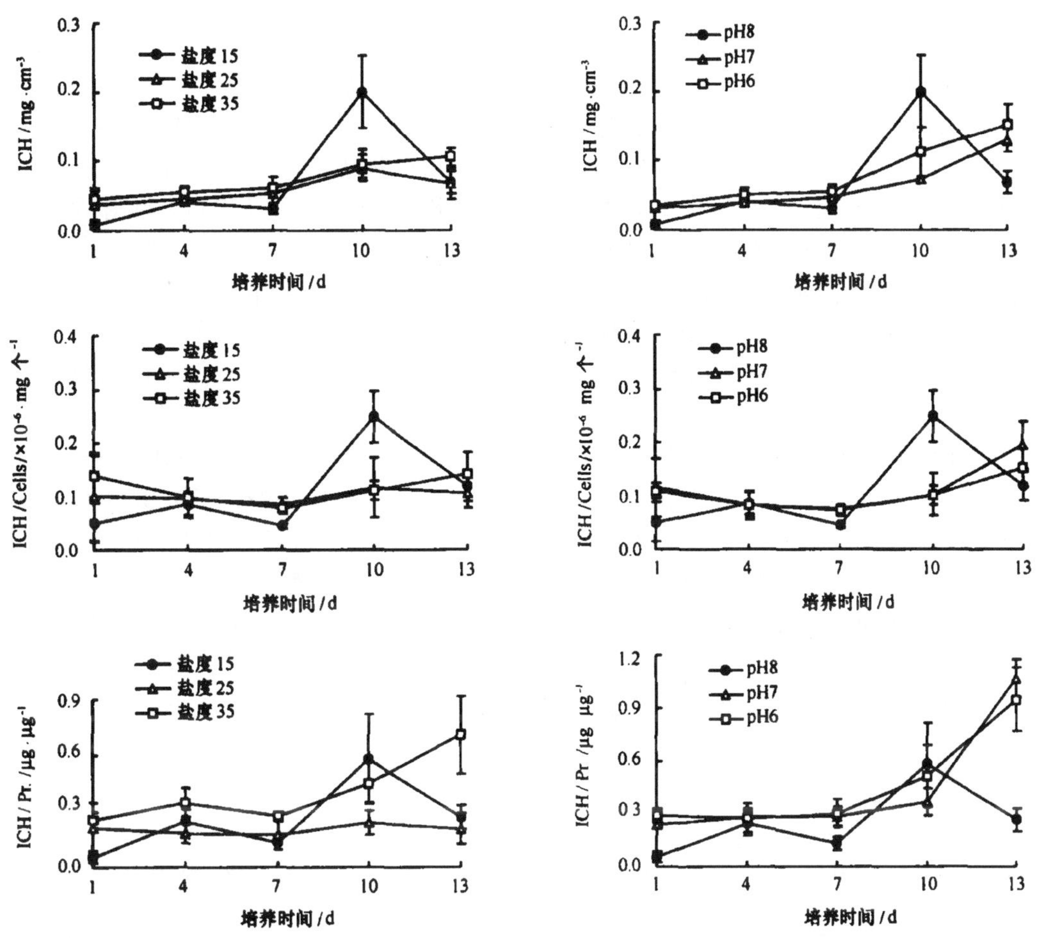

图 4 盐度和 pH 值对新月筒柱藻(*C. closterium*)分泌 ICH, ICH/Cell 和 ICH/Pr. 的影响

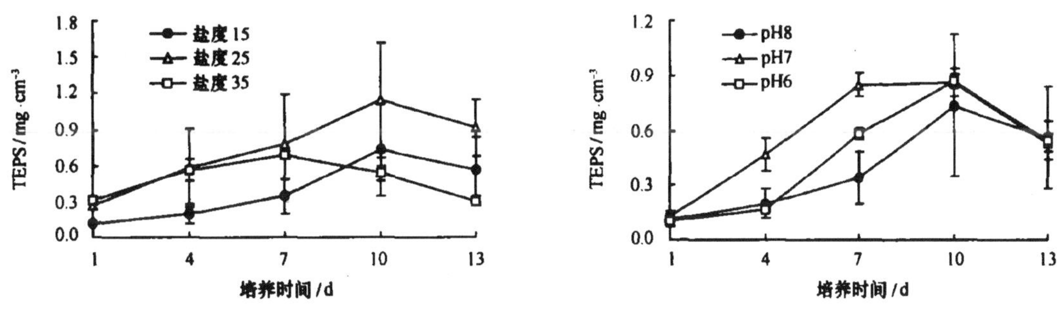

图 5 不同盐度和 pH 值对新月筒柱藻(*C. closterium*)总胞外多糖的影响

4 讨论与结论

低盐度和高pH显著促进新月筒柱藻细胞的蛋白质含量,细胞增殖的速度也较快,这表明该藻株是适宜咸淡水性和高pH生活的种类,而高盐度和低pH则被认为是一种胁迫.高盐度和低pH的胁迫均促进了新月筒柱藻CEPS,CEPS/Cell,CEPS/Pr.,AEPS,AEPS/Cell,AEPS/Pr.和TEPS的积累.这与一些研究的结论一致,即认为不利的生长条件是硅藻大量分泌EPS的主要原因之一[12~15].也有不同的观点,认为最适的生长条件下底栖硅藻EPS的产量最高,这是因为胁迫超过了阈值,生长后期限制了细胞的光合作用,使得底栖硅藻光合作用合成的碳源不能满足其细胞活动所需分泌EPS的碳源[21],导致EPS的减少.但在实验早期,胁迫下EPS的产量还是高于最适条件下EPS的产量[16].

盐度和pH值对细胞ICH的影响没有显著的差异,特别是实验前期阶段,细胞ICH的积累速度与细胞CEPS和AEPS的积累速度相差也很大,这说明细胞调控EPS和ICH的机制是不同的.

在不同季节,5~35 ℃范围内均可观察到硅藻的迁移,AEPS被认为与底栖硅藻的运动关系密切,并且可作为储藏物质供细胞在黑暗下使用[16].而CEPS有助于底栖硅藻获得更多的营养[22].因此AEPS与CEPS对底栖硅藻的作用是不一样的.这种差异也反应在同样条件下两种EPS产量的不同,在所有的实验组中,AEPS含量一直高于CEPS的含量,培养过程中AEPS含量增加了近2倍,而CEPS含量仅增加了近1倍.

在潮间带沉积物上,底栖硅藻是主要的微型光合生物,低潮时,底栖硅藻通常运动到表层进行光合作用,涨潮时,底栖硅藻迁移到土壤内,防止被流水冲走,以及被浮游动物等摄食.白天底栖微藻的光合作用导致了周围微环境pH值的增加,而夜间的呼吸作用则降低了周围微环境的pH值,并导致沉积物的稳定性发生不同变化,与pH值呈负相关关系[23].结合EPS和沉积物稳定性的正相关关系[24]及本研究的结论,也可说明低pH值导致底栖微藻分泌大量的EPS,从而提高了沉积物的稳定性,反之,高pH值条件下,沉积物的稳定性低.

分泌EPS是底栖硅藻适应不良环境的机制之一,并具有多种生态功能.我们的研究认为高盐度和低pH值的胁迫是半咸水性的新月筒柱藻产生大量的EPS的主要原因.

参考文献:

[1] De WINDER B,STAATS N,STAL L J,et al. Carbohydrate secretion by phototrophic communities in tidal sediments[J]. Journal of Sea Research,1999,42:131 - 146.

[2] SMITH D J,UNDERWOOD G J C. Exopolymer production by intertidal epipelic diatoms[J]. Limnol Oceanogr,1998,43(7):1 578 - 1 591.

[3] 尤仲杰,王一农,徐海军. 泥螺 *Bullacta exarata* (Philippi)生态的初步观察[J]. 浙江水产学院学报,1994,(13)4:245 - 250.

[4] 陈品健. 尖刀蛏食性和食料的研究[J]. 台湾海峡,1998,17 (sup.):39 - 43.

[5] PERKINS E J. The diurnal rhythm of the littoral diatoms of the river Ouse estuary, Fife[J]. J Ecol,1960,48:725 - 728.

[6] DECHO A W. Microbial exopolymer secretions in ocean environments: their role(s) in food webs and marine processes[J]. Oceanogr Mar Biol Annu Rev,1990,28:73 - 153.

[7] DECHO A W,MORIARTY D J W. Bacterial exopolymer utilization by a Harpacticoid Copepod —a methodology and results[J]. Limnol Oceanogr,1990,35:1 039 - 1 049.

[8] GREENLAND D J,LINDSTROM B G,QUIRK J P. Role of polysaccharides in stabilization of natural soil aggregates[J]. Nature,1961,191:1 283 - 1 284.

[9] HOAGLAND K D,ROSOWSKI J R,GRETZ M R,et al. Diatom extracellular polymeric substances: function, fine structure, chemistry and physiology[J]. J Phycol,1993,29:537 - 566.

[10] SUTHERLAND T F,GRANT J,AMOS C L. The effect of carbohydrate production by the diatom Nitzschia curvelineata on the erodibility of sediment[J]. Limnol Oceanogr,1998,43:65 - 72.

[11] PETERSON C G. Influences of flow regime on development and desiccation response of lotic diatom communities [J]. Ecology,1987,(68):946 - 954.

[12] STAATS N,LUCAS J S,LUUC R M. Exopolysaccharide production by the epipelic diatom Cylindrotheca closterium: effects of nutrient conditions[J]. J Exp Mar Bio Eco,2000,249:13 - 27.

[13] ADMIRAAL W. Influence of various concentrations of orthophosphate on the division rate of an estuarine benthic diatom, Navicula arenaria, in culture [J]. Mar Biol, 1977, 42: 1 - 8.

[14] SULLIVAN M J, DAIBER F C. Light, nitrogen and phosphorus limitation of edaphic algae in a Delaware salt marsh[J]. J Exp Mar Biol Ecol, 1975, 18: 79 - 88.

[15] DAVID J S, UNDERWOOD G J C. Exopolymer production by intertidal epipelic diatoms[J]. Limnol Oceanogr, 1998, 43(7): 1 578 - 1 591.

[16] WOLFSTEIN K, STAL L J. Production of extracellular polymeric substances (EPS) by benthic diatoms: effect of irradiance and temperature[J]. Mar Eco Progr Ser, 2002, 236: 13 - 22.

[17] HILLEBRAND H, SOMMER U. Response of epilithic microphytobenthos of the Western Baltic Sea to in situ experiments with nutrient enrichment[J]. Mar Ecol Progr Ser, 1997, 160: 35 - 46.

[18] 范 晓,严小军,韩丽君. 海藻化学分析方法[M]. 北京:学苑出版社,1996.

[19] 蔡武城,袁厚积. 生物物质常用化学分析法[M]. 北京:科学出版社,1982. 1 - 183.

[20] 三味工作室. 世界优秀统计软件 SPSS V10.0 For Windows 实用基础教程[Z]. 北京:北京希望电子出版社,2001.

[21] RUDDY G, TURLEY C M, JONES T E R. Ecological interation and sediment transport on an intertidal mudflat. II. An experimental dynamic model of the sediment-water interface [A]. BLACK K S, PETERSON D M, CRAMP A. Sedimentary Processes in the Intertidal Zone[C]. London: Geological Society, 1998. 149 - 166.

[22] STAATS N, STAL L J, deWINDER B, et al. Oxygenic photosynthesis as driving process in exopolysaccharide production of benthic diatoms [J]. Mar Ecol Prog Seri, 2000, (193): 261 - 269.

[23] MONTAGUE C L. Influence of biota on erodability of sediments[A]. MEHTA A J. Estuarine Cohesive Sediment Dynamics: Lecture Notes on Coastal and Estuarine Studies[C]. New York: Springer-Verlag, 1986. 14: 251 - 269.

[24] HOLLAND A F, ZINGMARK R G, DEAN J M. Quantitative evidence concerning the stabilization of sediments by marine benthic diatoms[J]. Marine Biology, 1974, 27: 191 - 196.

Production of extracellular polymeric substances (EPS) by benthic diatom: effect of salinity and pH

CHEN Chang-ping[1], GAO Ya-hui[1], LIN Peng[1]

(1. *School of Life Sciences, Xiamen University, Xiamen* 361005, *China*)

Abstract: Influences of salinity and pH value on reproduction, protein content and extracellular polymeric substances (EPS) of benthic diatom *Cylindrotheca closterium* (Ehr.) Reimann et Lewin were studied. Salinity and pH value fit to C. closterium were 15 and 8, respectively, showing brackish water species. Abundant EPS were secreted by *C. closterium* characterized by brackish water species under high salinity (> 15) and low pH (< 8) stress, showing functions in abnormal conditions. The fraction of EPS that was closely bound to the cells (attached EPS) and the soluble fraction (colloidal EPS) were produced in different amounts at the different salinity and pH value. This suggested that the production of the 2 operationally defined fractions of EPS might serve different functions. No significant differences were found on intracellular carbohydrate production of *C. closterium* under different salinity and pH value.

Key words: benthic diatoms; extracellular polymeric substances (EPS); salinity; pH

中国红树林湿地物种多样性及其形成*

何斌源[1,2] 范航清[1] 王 瑁[2] 赖廷和[1] 王文卿[2]

(1. 广西红树林研究中心,北海 536007;2. 厦门大学生命科学学院,厦门 361005)

摘要:目前中国红树林湿地共记录了 2854种生物,包括真菌 136种、放线菌 13种、细菌 7种、小型藻类 441种、大型藻类 55种、维管束植物 37种、浮游动物 109种、底栖动物 873种、游泳动物 258种、昆虫 434种、蜘蛛 31种、两栖类 13种、爬行类 39种、鸟类 421种和兽类 28种。这些动物中有 8种国家一级保护动物,75种二级保护动物。中国红树林湿地是中国濒危生物保存和发展的重要基地,并在跨国鸟类保护中起着重要作用。中国红树林湿地单位面积的物种丰度是海洋平均水平的 1766倍。从初级生产物质基础、食物关系多样性、宏观尺度和微观尺度的空间异质性、生境利用的时序性等方面分析了中国红树林湿地物种多样性极其丰富的原因。

关键词:红树林湿地;生物多样性;初级生产力;食物关系多样性;生境多样性;空间异质性;时序性

文章编号:1000-0933(2007)11-4859-12 **中图分类号**:Q146,Q948 **文献标识码**:A

Species diversity in mangrove wetlands of China and its causation analyses

HE Bin-Yuan[1,2], FAN Hang-Qing[1,*], WANG Mao[2], LAI Ting-He[1], WANG Wen-Qing[2]

1 *Guangxi Mangrove Research Center, Beihai 536007, China*

2 *School of Life Sciences, Xiamen University, Xiamen 361005, China*

Acta Ecologica Sinica, **2007, 27(11): 4859~4870.**

Abstract: To date, total of 2854 species of organisms were recorded in Chinese mangrove wetlands, including 136 species of fungi, 13 species of actinobacteria, 7 species of bacteria, 441 species of microalgae, 55 species of macroalgae, 37 species of vascular plants, 109 species of zooplankton, 873 species of macrobenthos, 258 species of nektons, 434 species of insects, 31 species of spiders, 13 species of amphibians, 39 species of reptiles, 421 species of birds and 28 species of mammals. Among them, 8 species belonged to the category 1 of Chinese national protected animals and 75 species belonged to the category 2. Chinese mangrove wetlands are very important bases for the conservation and development of the endangered species to China, and playing a critical role in the international activities for protecting the migrating birds. The species abundance in Chinese mangrove wetlands was 1766 times as much as that for the averaged species abundance in Chinese sea fields. The prolific species diversity in Chinese mangrove wetlands can be attributed to their high primary productivity, high diversity in their consumers' food preferences, high spatial heterogeneity at macroscopical and microscopic scale levels, and their dynamic temporal sequence in habitat utilization.

---

* 国家自然科学基金资助项目(40676050);联合国环境规划署(UNEP)全球环境基金(GEF)资助项目;广西科学基金资助项目(0640014)

原载于生态学报,2007,27(11):4859－4869

Key Words: Mangrove wetland; species diversity; primary productivity; food preference diversity; habitat diversity; spatial heterogeneity; temporal sequence

红树林湿地作为一种特殊的生态交错带,立足于狭长的海岸潮间带滩涂,潮汐循环往复地改变着基质状况,植物群落异质地向水平和垂直延伸,构建起景观上与其它生态系统迥异的三维复合体。这种生物地貌特征极其鲜明,构造复杂多样,开放性、包容性极强,相似性和特异性并存,汇集承载了生境需求、饵料选择、形体大小、营养级别、功能角色上千差万别的各种海洋和陆地生物类群,同时并行着海岸护卫前沿、有机物质生产车间、碎屑食物链源端、饵料场、繁殖地、越冬场所、栖息地、幼苗库、中途加油站、避难所等许多性质各异而又共存的结构和功能载体。

中国红树林现有面积仅约 220km^2,无论与陆地森林还是海洋相比都微不足道。中国红树林自然分布于海南、广东、广西、福建、台湾、香港和澳门等省区,从最南端的海南省三亚市 (18°13′N)到最北端的福建省福鼎市 (27°20′N),并人工引种至浙江省乐清县 (28°25′N),跨越纬度 10°12′。红树林广泛生长在我国东南沿海的海湾和河口,但多数林带狭窄,群落低矮,组成简单,发育保存良好者少。尽管如此,我国红树林湿地仍具有较高的经济和生态价值,据估算广西红树林湿地在木材、果实、近海渔业、减少风暴潮损失、维护海堤、保护耕地、防止侵蚀、保持肥力及绿肥、产生氧气、净化空气水体、维持林区动物等方面的经济效益为 59458.98元 $\cdot hm^{-2}\cdot a^{-1}$[1]。红树林以有限的群体,有力地支撑我国东南沿海的近海海洋生态安全和可持续发展。近年来我国各地开展了大量的红树林生物多样性调查研究,然而尚缺乏国家层次的更新数据。本文总结了中国红树林湿地在物种多样性研究方面的最新结果,为正确、全面评价中国红树林湿地生态价值提供科学依据。

2 材料与方法

尽可能完整地收集中国有关红树林湿地物种多样性的资料和文献,按类群归属和分类系统整理排序,形成名录。分析总结各生物类群的一般性规律。从食物关系和生境两大方面探讨中国红树林湿地物种多样性丰富的原因。引用除正式出版文献外,一些重要的未正式出版的参考资料有:广东湛江建立红树林鸟类国家级自然保护区的综合报告 (1995),广西北仑河口国家级海洋自然保护区本底研究报告 (2000),海南省东寨港国家级红树林保护区考察报告 (2001),全国红树林资源调查报告 (2002)。

几乎全部已有中国红树林湿地生物多样性研究的区域范围仅涉及红树林植株、林地,及林间面积狭小的潮沟、裸滩、临时或长期的水体等。《全国湿地资源调查与监测技术规程》对于红树林沼泽的定义为:以红树植物群落为主的潮间沼泽。本文认为红树林湿地与红树林沼泽同义,其范围可取 2002年全国红树林资源调查中的郁闭度 >0.2天然林、未成林造林地和天然更新林地三者之和,面积约 239km^2。

3 结果与分析

红树林湿地是具有复杂完整结构和功能的生态系统,包含了三大生物功能类群:生产者、消费者和分解者。

3.1 中国红树林湿地中的生产者

3.1.1 小型藻类

本文所称的小型藻类包括浮游植物和底栖硅藻。由于红树林水较浅,在风浪和潮汐作用下,底栖硅藻被大量带入水体中,进而影响浮游植物的群落组成[2,3]。本文将两者合并,避免生物多样性编目重复。本文统计了中国红树林湿地小型藻类 441种,包括硅藻门 408种,裸藻门 18种,甲藻门 9种,蓝藻门和绿藻门各 3种。硅藻是红树林湿地的小型藻类的优势类群[2~11],种数较多的属有:菱形藻属 *Nitzschia*和舟形藻属 *Navicula*,均 46种;角毛藻属 *Chaetoceros*和圆筛藻属 *Coscinodiscus*同为 25种;双眉藻属 *Amphora*有 21种。

浮游植物和底栖硅藻都是红树林湿地初级生产的补充力量,在光合放氧、有机物转换、营养元素循环、改变土壤的 pH值和氧化还原电位、吸收重金属等方面起着一定的作用。小型藻类中有赤潮、污染监测指示种,如硅藻门的威氏海链藻 *Thalassiosira weissflogii*、中肋骨条藻 *Skeletonema costatum*,蓝藻门的颤藻及裸藻门的种类。广东深圳福田红树林湿地水体中裸藻门多达 18种,某些赤潮种在个别月份数量激增,成为优势种[7]。

3.1.2 大型藻类

红树林湿地大型藻类生长在潮沟、滩面或低矮的红树根系枝干上。林益明等[12]报道我国红树林区大型藻类有 4门 55种,其中蓝藻门 17种、红藻门 13种、褐藻门 2种、绿藻门 23种。红树林湿地大型藻类的主要优势属有:红藻门的鹧鸪菜属 *Caloglossa*、卷枝藻属 *Bostrychia*和节附链藻属 *Catenella*,绿藻门的绿球藻属 *Chlorococcum*、根枝藻属 *Rhizoclonium*、无隔藻属 *Vaucheria*、浒苔属 *Enteromorpha*。数量上红树林海藻以红藻为主要优势种,绿藻次之。红藻较喜荫蔽潮湿的环境,而绿藻适生在光照条件较好生境[13]。

3.1.3 红树植物

红树植物的定义虽然趋于统一,但对于中国红树植物种数有多种不同表述,代表性观点有:郑德璋[14]报道真红树 27种 (含引种成功的无瓣海桑 *Sonneratia apetala*)和半红树 8种,范航清[1]认为有真红树 26种和半红树 11种,林鹏[15]归纳了真红树 28种 (含无瓣海桑)和半红树 11种,Wang等[16]主张真红树 19种和半红树 7种。归属出入较大的有卤蕨属、老鼠簕属和银叶树 *Heritiera littoralis*。卤蕨属或被认为是草本,或强调茎的木质化。老鼠簕属有时被列入半红树[16]。银叶树被认为分布在几乎只有特大潮才能波及的地带,生境趋同于许多伴生种类。

本文列出中国现存的原生真红树 12科 14属 24种 (含 1变种),包括了:卤蕨 *Acrostichum aureum*、尖瓣卤蕨 *Acrostichum speciosum*、木榄 *Bruguiera gymnorrhiza*、海莲 *B. sexangula* 、尖瓣海莲 *B. s* var. *rhynochopetala*、角果木 *Ceriops tagal*、秋茄 *Kandelia obovata*、红树 *Rhizophora apiculata*、红海榄 *R. stylosa*、小花老鼠簕 *Acanthus ebracteatus*、老鼠簕 *Acanthus ilicifolius*、红榄李 *Lumnitzera littorea*、榄李 *L. racemosa*、海漆 *Excoecaria agallocha*、木果楝 *Xylocarpus granatum*、桐花树 *Aegiceras corniculatum*、水椰 *Nypa fruticans*、瓶花木 *Scyphiphora hydrophyllacea*、杯萼海桑 *Sonneratia alba*、海桑 *S. caseolaris*、海南海桑 *S. hainanensis*、卵叶海桑 *S. ovata*、拟海桑 *S. gulngai*、白骨壤 *Avicennia marina*。中国原生真红树种数占世界总种数 (70种)的 34.3%。

我国所有原生真红树种类都可在地处热带的海南省找到,广东广西均有 11种,香港 9种,台湾 8种,福建 7种,澳门 5种。浙江引种秋茄成功。

中国红树林湿地中常见的半红树植物有 12种:玉蕊 *Barringtonia racemosa* 、海芒果 *Cerbera manghas*、海滨猫尾木 *Dolichandrone spathacea*、阔苞菊 *Pluchea indica*、莲叶桐 *Hernandia nymphiifolia*、水黄皮 *Pongamia pinnata*、水芫花 *Pemphis acidula*、黄槿 *Hibicus tiliaceus*、杨叶肖槿 *Thespesia populnea*、银叶树、苦朗树 *Clerodendrum inerme*和钝叶臭黄荆 *Premna obtusifolia*。

综上,中国红树林湿地中生长红树植物 37种 (包括引种归化的无瓣海桑)。

3.2 中国红树林湿地中的消费者

红树林是滨海湿地,容纳了大量海洋动物,它们可划分为浮游动物、底栖动物和游泳动物等 3大生态类群,每个类群包含了丰富多样的分类层次和营养级别。同时,红树林湿地也具有陆地森林性质,生活着昆虫、蜘蛛、两栖类、爬行类、鸟类和兽类。

3.2.1 浮游动物

红树林湿地浮游动物是植食性食物链的重要中间链结,是生态系统物质流动和能量转化的关键一环。据资料[4,5,8,11]统计表明中国红树林区水体的浮游动物记录了 4门 109种,包括腔肠动物门 49种 (绝大部分为水母),节肢动物门 48种,毛颚动物门 9种,尾索动物门 3种。

3.2.2 底栖动物

底栖动物是中国红树林湿地最为丰富多样的生物类群,研究最充分[4,6~8,11,17~35]。本文统计了 13门 873种,分别为腔肠动物门 8种,扁形动物门 3种,线形动物门 29种,纽形动物 4种,环节动物门 142种,星虫动物门 11种,螠虫动物门 3种,软体动物门 348种,甲壳动物门 250种,腕足动物门 1种,棘皮动物门 28种,尾索动物门 3种,脊索动物门 43种。

红树林湿地底栖动物群落多以珠带拟蟹守螺为优势种,这种贝类经济价值不高,常被用作养殖虾蟹的新鲜蛋白补充。但它和众多以小型藻类为食的拟蟹守螺、滩栖螺一样,是食物链的重要中间环节,支撑起更高级别的营养类群。底栖动物中经济种类很多,如可口革囊星虫 *Phascolosoma esculenta*、裸体方格星虫 *Sipunculus mudus*、团聚牡蛎 *Ostrea glomerata*、缢蛏 *Sinonovacula constricta*、红树蚬 *Gelolna coaxans*、文蛤 *Meretrix meretrix*、青蛤 *Cyclina sinensis*、脊尾白虾 *Exopalaemon carinicauda*、锯缘青蟹 *Scylla serrata* 和各种底栖鰕虎鱼、弹涂鱼等。到红树林湿地赶小海捕底栖动物是红树林沿岸村民的重要副业之一。

3.2.3 游泳动物(鱼类)

中国红树林湿地的游泳鱼类记录有 258种,其中软骨鱼纲 4种,硬骨鱼纲 254种[36~39]。红树林湿地的游泳鱼类以小型鱼类为主,生长期以幼苗为主;区系组成以暖水性种占绝对优势,生态类型上底层鱼类十分丰富,尤其是鰕虎鱼科种类。游泳鱼类数量的季节变化明显,优势种突出且季节间差异很大。

3.2.4 昆虫和蜘蛛

根据文献[7,11,40~42]统计了中国红树林湿地昆虫 434种。红树林湿地的昆虫在种类上与沿岸灌草丛、农田作物上的差异不大。一般而言,昆虫的飞行距离不远,种类和数量呈从靠岸林带向靠海林带减少的趋势。捕食性和寄生性天敌、授粉昆虫对红树植物的保护和发展起着重要作用,但害虫的数量爆发对红树植物造成破坏。卷蛾科 *Lasiognatha* sp. 猖獗造成广西钦州港大面积桐花树受害[40]。螟蛾科害虫几乎每年夏天都侵害深圳福田的白骨壤植株[7]。2004年广西沿海受广州小斑螟 *Oligochroa cantoonella* 等虫害的白骨壤林面积累计达到 700hm^2,受害严重的植株 90%以上的叶片干枯,约 45%的枝条枯死[42]。

颜增光等[43]初步研究表明广西英罗港红树林区蜘蛛群落由 12科 31种组成,以圆蛛科和肖蛸科的种类占优势。红树林蜘蛛群落数量呈由靠陆林带到向海林带递减,以昆虫为食的蜘蛛与昆虫在种类和数量分布上规律一致。

3.2.5 两栖类、爬行类和兽类

中国红树林湿地两栖类、爬行类和兽类调查[6,8,18,19,44]相对很少且研究不系统,还远不能反映出我国红树林湿地这些类群多样性的真实状况。

两栖类统计出 5科 6属 13种,均为无尾目种类,虎纹蛙 *Rana rugulosa* 为国家二级保护动物。

爬行类计有 11科 39种,包括龟鳖目 3科 8属 8种,蜥蜴目 3科 4属 5种,蛇目 5科 22属 26种。其中国家一级保护动物 1种蟒 *Python molurus*,二级保护动物有太平洋丽龟 *Lepidochelys olivacea* 等 5种。野生爬行动物几乎都可食用和药用,大多面临危险境地。

兽类记录了 15科 24属 28种,中华白海豚 *Sousa chinensis* 为国家一级保护动物,小灵猫 *Viverricula indica* 等 6种为二级保护动物。红树林沿岸人口密度较大、经济活动频繁,林区相对简单狭窄,兽类的正常活动容易受到干扰,种类自然较其他森林类型贫乏。

3.2.6 鸟类

我国东南沿海红树林湿地位于重要的鸟类迁徙通道上:亚洲东部沿海鸟类迁徙路线和中西伯利亚—中国中部的内陆鸟类迁徙路线在这一带交汇后,再继续往南延伸至东南亚和澳大利亚。在迁徙季节,大量候鸟途经红树林湿地,它们往往在这里歇息取食、休整一段时间后再继续迁飞。

中国重要的红树林湿地分布区均开展过鸟类调查[7,11,18,45~55],已记录有 19目 58科 421种,占我国 1331种鸟类[56]的 31.6%。国家一级保护鸟类有 6种:黑鹳 *Ciconia nigra*、白鹳 *Ciconia ciconia*、东方白鹳 *Ciconia boyciana*、中华秋沙鸭 *Mergus squamatus*、白肩鵰 *Aquila heliaca* 和遗鸥 *Larus relictus*,二级保护鸟类有黑脸琵鹭 *Platalea minor* 等 63种。

红树林湿地是水鸟和陆鸟共存的生境,421种鸟类中水鸟 177种、陆鸟 224种,分别占 42%和 58%。红树林湿地给需求不一的鸟类提供了适宜的觅食区、栖息地和繁殖地,尽管基于鸟类的生活习性和安全性选择,这几种生活分区有时鸟类仅选择其中一种,它们强大的空间移动能力保证它们在不同生活分区内畅通无阻。红

树林湿地既有长期或临时的水域,又有经常出露的滩涂,适宜水鸟生活。红树林也具有陆地森林性质,同时由于人类活动干扰,陆鸟的原有生境被压缩而被迫向红树林转进。红树林湿地鸟类中,大多数是往来迁徙的候鸟。种类和数量呈现出明显的季节变动,春、秋两季为候鸟迁徙季节,鸟类种类和数量急骤大幅增多,呈现出两个显著高峰。红树林湿地鸟类的种间竞争导致群落组成发生动态变化,尤其是在发育良好并僻静的林区,一些种类数量逐年上升而栖息地扩张,而有些种类被排挤边缘化。

3.3 中国红树林湿地中的还原者

微生物是红树林湿地的最主要还原者,在凋落物和有机碎屑的分解转换中处于先锋地位,对红树林湿地的物质循环和能量流动起了重要的推动作用。同时,微生物会引发煤污病、炭疽病等[57,58],导致红树叶片脱落、枝梢枯萎,甚至植株死亡。目前红树林微生物研究趋向在抗菌、抗肿瘤方面有活性的菌株的筛选[59~61]。

红树林湿地微生物包括细菌、放线菌和真菌等类群,数量上以细菌类群占绝对优势,其中芽孢杆菌属为突出优势属,放线菌和真菌极少[6,62,63]。王伯荪等[7]总结了广东、香港和澳门红树林湿地的真菌区系,共计有 76种;优势种有 *Trichocladium linderii*、*Marinosphaera mangrovei*、*Lignincola laevis*、*Hypoxylon oceanicum* 等。本文根据文献记录,汇总到至少真菌 136种、放线菌 13种和细菌 7种。由于微生物研究的特殊性,鉴定到种的占总种数的比例很低,远不能反映红树林湿地微生物的全貌。对于海洋细菌,王岳坤等[64]指出不经微生物分离培养步骤,直接从土壤中抽提总 DNA,分析其中 16S rDNA 的序列多态性,以此反映微生物的种群构成,是近 10a来逐步发展起来的方法;此研究方法所揭示的土壤微生物种群结构较传统方法更加复杂多样。

3.4 中国红树林湿地所有生物类群的物种多样性

目前中国红树林湿地动物已记录了 14门 31纲 2165种 (表 1),加上藻类、高等植物和微生物 789种,所有生物类群的总种数达 2854种。在总面积约为 300万 km^2的中国海洋国土,记载了海洋生物物种为 20278种[8]。中国红树林湿地以 239 km^2的狭长地带,繁衍生息着至少 2854种生物,红树林湿地单位面积的物种丰度是海洋平均水平的 1766倍;有 8种国家一级保护动物,75种二级保护生物,把中国红树林湿地作为栖息地、饵料场、中途加油站和避难所。此外,红树林湿地鸟类中还有属于中日、中澳双边协定共同保护的候鸟超过 150种。1992年至 2001年,海南东寨港、香港米埔、台湾淡水河口、广西山口和广东湛江等 5块红树林湿地先后被纳入国际重要湿地之列,彰显了中国红树林湿地在全世界濒危生物保存和发展的重要地位。

4 讨论

中国红树林湿地丰富的物种多样性首先得益于以红树林为主角的初级物质生产,形成完整而多样的食物关系,诸多景观单元提供了数量众多的细化生境,基质的节律性动态变化又增益了生境变数,而它的高度开放性则给物种交流提供了非常便捷的通道。

4.1 强大的初级生产物质基础

红树植物是红树林湿地最主要的初级生产者,底栖硅藻、浮游植物和大型藻类通过光合作用为系统补充部分有机物。红树林具有"三高"特性:高生产率、高归还率和高分解率,枝干叶片等凋落物构成滨海湿地食物网的深厚基础。表征中国红树林区域性特征的海莲、红海榄和秋茄的初级生产力分别为 29.49、15.37 $t \cdot hm^{-2} \cdot a^{-1}$和 23.46 $t \cdot hm^{-2} \cdot a^{-1}$,归还率为 12.55、6.31 $t \cdot hm^{-2} \cdot a^{-1}$和 9.21 $t \cdot hm^{-2} \cdot a^{-1}$,半分解期为 20~45 d,20~71 d和 18~56 d[18]。近年来大规模种植的无瓣海桑生产力也较高,在深圳 6龄树林净生产力达 16.92 $t \cdot hm^{-2} \cdot a^{-1}$[65]。范航清等[11]估算广西山口保护区红树林的地上部总生产力 4.58 $t \cdot hm^{-2} \cdot a^{-1}$,理论上每年足以支撑的植食性海洋动物 1.22 t鲜重。初级生产力向各营养级消费者转化的过程逐级锐减,红树林湿地的高生产力增加食物链得以延长的可能性,从而使位于不同营养层次的生物类群多样化。

4.2 食物网结构的完整和多样性

通过食物关系,红树林湿地各类群生物形成复杂的食物网结构,这些具有不同营养特点的生物对红树林湿地具有十分重要的贡献,它们导致了能流和物流的多样化过程,以及生物之间复杂的相互关系。

红树植物归还的凋落物经分解后形成可溶性养分和颗粒状有机碎屑,开启了两种性质不同的食物关系走向:①可溶性有机和无机物是底栖硅藻和浮游植物等低等植物初级生产的基础原料,初级生产物由此被浮游动物利用;而底栖硅藻、浮游植物、浮游动物是大部分大型底栖动物和鱼类的饵料。②颗粒状有机碎屑被许多虾蟹和杂食性鱼类直接食用而进入更高的营养级。刘劲科等[66]研究发现生活在红树林区的鱼类的肠胃饱满度较高,说明红树林区具有丰富的饵料资源。

表 1 中国红树林湿地动物所属门、纲种数

Table 1 Numbers of species, class and phylum in the fauna of Chinese mangrove wetlands

门 Phylum	纲 Class	种数 Species number
腔肠动物门 Coelenterata	珊瑚虫纲 Anthozoa	7
	水螅虫纲 Hydrozoa	50
扁形动物门 Platyhelminthes	涡虫纲 Turbellaria	3
线形动物门 Nematoda	线虫纲 Nematoda	29
纽形动物门 Nemertea	无针纲 Anopla	4
环节动物门 Annelida	多毛纲 Polychaeta	138
	寡毛纲 Oligochaeta	4
星虫动物门 Sipuncula	革囊星虫纲 Phascolosomatidea	8
	方格星虫纲 Sipunculidea	3
螠虫动物门 Echiura	螠纲 Echiurida	3
软体动物门 Mollusca	双壳纲 Bivalvia	180
	腹足纲 Gastropoda	161
	头足纲 Cephalopoda	6
节肢动物门 Arthropoda	肢口纲 Merostomata	3
	甲壳纲 Crustacea	295
	蛛形纲 Arachnoidea	31
	昆虫纲 Insecta	434
毛颚动物门 Chaetognatha	矢虫纲 Sagittoidea	9
腕足动物门 Brachiopoda	无关节纲 Inarticulata	1
棘皮动物门 Echinodermata	海星纲 Asteroidea	4
	海胆纲 Echinoidea	3
	蛇尾纲 Ophiuroidea	16
	海参纲 Holothuroidea	5
尾索动物门 Urochordata	有尾纲 Appendiculata	1
	海鞘纲 Ascidiacea	2
脊索动物门 Chordata	软骨鱼纲 Chondrichthyes	4
	硬骨鱼纲 Osteichthyes	260
	两栖纲 Amphibia	13
	爬行纲 Reptilia	39
	鸟纲 Aves	421
	兽纲 Mammalia	28
所有类群动物的种数 Total species of fauna		2165

同时,昆虫直接取食鲜活的红树植物有机物和器官,如汁液、叶片和胚轴等,为肉食性鸟类和蜘蛛制造食饵。两栖爬行类在退潮出露的林地捕食贝类、虾蟹类和底栖鱼类。鸟类尤其是水鸟捕食滩涂上底栖动物、鱼类甚至两栖爬行类。兽类动物在食物链中虽处于较高层次,作用却远小于鸟类。

4.3 红树林湿地生境的多样性

红树林湿地以生长着红树林而区别于其它类型的湿地,红树林湿地生境本质上首先是红树林能够适应的生境,其次为红树林、其它生物类群与物理环境共同构建的生物环境。红树林湿地生境有着高度的异质性和复杂性,诸多不同尺度的景观单元提供了物种多样性空间分布的支体。

4.3.1 红树林湿地宏观生境多样性

宏观上,红树林湿地生物能适应丰富多样的气候类型、水文类型、岸滩类型和基质类型。气候类型从大尺度上影响红树林湿地生物多样性,水文类型、岸滩类型和基质类型则在中小尺度上起作用。

气候类型的影响:红树植物是热带起源的喜热型木本植物群落,在我国分布范围从热带延续到亚热带,在纬度上跨越了 10°12′,年均气温 18.5～25.5℃,最冷月均温 11～21℃ [67]。温度是制约红树植物分布的主导因素,红树植物多样性随纬度增高而递减。从南到北,底栖动物和游泳动物的区系性质由完全的暖水性种组成向出现一些暖温性种的转化,不过暖水性种仍占绝对优势,这与红树林湿地分布在湿热的华南沿海有关。温度决定了鸟类的迁徙和繁殖,由于候鸟的迁徙空间尺度较大,各红树林湿地间候鸟的种类差异相对较小,但留鸟和繁殖鸟的种类受温度因素影响而差异较大。地带性差异导致整体上中国红树林湿地生物在物种水平上的多样性十分丰富。

水文类型的影响:水文因素中海水盐度被认为是影响红树林分布的主要宏观因素之一;通过比较发现我国主要红树林分布区的海水盐度有一定的差异;海水盐度决定着生长的红树种类[18]。海水盐度是限制某些物种传播的天然障碍,可能导致形成独有的地方性种。同时由于淡水流的梯度变化,细化了适应生境类型和适应能力不同的物种的适宜生境,在河口湾红树林湿地由河向海依次出现了淡水种、河口种、海洋性种。在一些海水盐度很低的红树林海区,出现了如细鳊 *Rasborinus lineatus*、鲢 *Hypophthalmichthys molitrix*、鲮 *Cirrhinus molitorella*等典型的淡水种,高盐度海区鱼类则为完全海洋性质的种[39]。

岸滩(波能)类型的影响:郑德璋等[68]根据地貌、风浪等生境特点将红树林划分为前缘浪击型、内湾型和河流型,不同类型树林里红树植物种类差异较大。范航清等[69]按地貌和波能特点分类将红树林海岸分为开阔性海岸、隐蔽性海湾和河口海岸;红树林污损动物忠实地反映岸滩类型不同的红树林湿地之间的差异,物种丰度、种群密度和生物量表现为:开阔海岸型 >河口型 >海湾型。

基质性质的影响:红树林生长的基质复杂多样,淤泥质、泥质至沙质之间的各种过渡类型、砾石滩、甚至仅有很薄的沉积物覆盖且稳定性很差的基岩上均可着生。粗基质湿地沉积物贫瘠,多着生养分要求低的白骨壤、老鼠簕、榄李和桐花树等,细基质湿地养分丰富但同时透气性差、硫化物含量高、氧化还原电位低,适合红树、红海榄、海桑、木榄等。有时在很小的局部林带就出现多种生境基质的显著交替变化。广西大冠沙红树林的林带宽度不超过 300m,但底质类型从向海林带的纯砂质变化到向陆林带的深厚淤泥质;由于人工筑堤引起水文条件的巨大改变,在局部林区发生沙丘入侵,破碎了原本比较均一的红树林滩涂,引起利用不同生境的大型底栖动物的物种丰度、群落组成和数量分布的变化[27]。

4.3.2 红树林湿地微观生境多样性

红树林湿地宏观的空间异质性已很多样化,微生境的差异更增益空间异质性,丰富较小尺度的景观单元种类和数量。

红树林的种类组成和群落结构比陆地森林简单得多,但它有丰富的形态结构和地面微景观地貌,这些细化的生境对生物多样性的丰富是至关重要的。红树植物类型丰富形态各异的器官直接参与构建了众多细化的微生境,叶层、枝杈、气生根、支柱根、表面根、呼吸根与空气、滩涂、水体互相交叉渗透。这些不同尺度的景观单元给不同形体大小的动物类群提供了适合的微生境,并且动物类群也直接参与生境的构建活动,如蟹类挖穴、贝类构建滤食水管体系等,从而使红树林湿地的结构和功能更为复杂化。生活在滩涂基质内部和表面的底栖动物充分利用各种细化的生境,按生活类型可分为匍匐生活型、底内生活型、附着生活型、固着生活型、凿穴生活型、穴居生活型、游泳生活型、寄居生活型和管栖生活型等[18]。红树林湿地滩涂层次和结构的复杂性,降低底栖动物被捕食的几率,红树根围的动物种类和数量远大于林地滩涂和林外裸滩。叶层多样性高的红树林,空间异质性增加,表现为树种相对丰富,叶层结构较复杂,垂直层次多,不但提供了良好的隐蔽条件,而且还意味着更加多样的小生境和食物资源以及更大的取食面积,允许更多的鸟类共存[70]。红树林湿地里丰富的饵料,吸引黄毛鼠 *Rattus losea exiguus*等在不被潮水淹到的树冠上用红树枝叶在筑巢[71]。

4.3.3 异质性的时序变动多样性

空间异质性和复杂性的时序变动多态性在丰富红树林湿地生物多样性上也起着重要作用,它提高了动物对红树林湿地生境的重复利用率。

潮汐节律性的影响 红树林湿地基质的物理化学性质随时间变数很大。潮汐基本上是有规律的,循环往复地进出红树林湿地,湿地基质表面轮流被淹没和出露,湿地性质出现海洋和陆地性质的交替。在河口地区,淡水流在潮汐退去后独自影响红树林湿地,基质间隙水体盐度变化剧烈。需要不同生活基质被不同生物类群按时序利用,也可以同时利用。忠实定居者按时作息:滤食性底栖动物摄食随潮汐而来的浮游生物饵料,退潮后则龟息;底栖硅藻食性动物则在退潮滩涂出露时行走觅食。生活分区多的物种按潮汐规律而在红树林湿地分时段出没:红树拟蟹守螺 *Cerithidea rhizophorarum* 等在涨潮时爬上树干,退潮时又回到滩涂上。游泳动物乘潮水进入红树林觅食,又乘之退出;翠鸟等俯冲捕食鱼类多发生在潮水充满湿地潮沟、滩涂的时候,而水鸟则逐潮而乘机捕食尚未来得及深潜的多毛类、甲壳类、贝类和鱼类。

生物物候的影响 植物的物候变化明显地影响不同昆虫的觅食行为,取食嫩叶和汁液的昆虫往来频繁或定居;取食花蜜的种类则在红树植物的花期出现,然而红树植物的花期不整齐而延长,客观上增加这些昆虫在红树林湿地出现的时间。文蛤、锯缘青蟹等在生活史的不同阶段在红树林湿地潮间带和潮下带浅海之间移动。某些鱼类幼年期生长在红树林,长成迁出。冬、夏候鸟的往来具有明显的季节性,鸟类在红树林里繁殖期也较严格地限定在一年中的某些月份。

4.4 红树林湿地的开放性

红树林湿地可通过水体、空气、土壤界面与浅海、入海河流及陆地生态系统建立起交流通道,形成开放性的系统。广西红树林岸线 1243.18 km,红树林面积为 8374.9 hm^2 (平均林带宽度 67.4 m),面积不大于 5.0 hm^2的斑块数累计占 70.4%,现有红树林分布比较零星,大部分连片面积较小[72]。中国红树林湿地的高度开放性与林带狭窄有关,边缘地带基质的理化性质趋同,客观上增加了更多新的海洋动物进入红树林湿地的机会。

红树林的海洋属性基本上由潮汐来表征。潮汐是各种形式物质流动的廉价载体,涨退之间完成了双向的物质交换、物种交流。主动性移动能力强的游泳动物趁机进出;各种浮游植物、浮游动物、浮游阶段幼体被动地随潮汐移动。空气的流通性给昆虫、鸟类等活动能力强的动物以自由进出。红树林湿地土壤与其它生态系统连接,是一种更为固化的通道。红树植物在潮间带滩涂上带状分布与潮汐水文、土壤理化性质及动物利用等影响因素有关[73]。红树林湿地底栖动物的分布决定于其抗逆境能力和饵料来源,不同种群生长的关键生态因子及生态位有所差异。

References:

[1] Fan H Q. Coastal guarder—mangrove. Nanning: Guangxi Science and Technology Press, 2002. 126.

[2] Liu Y, Chen G Z. Study on community structure and ecology of algae in mangrove areas in Futian, Shenzhen. Acta scientiarum naturalium universitatis sunyatsen, 1997, 36(1): 102-106.

[3] Chen C P, Gao Y H, Lin P. Progress in the studies of the diatoms in mangrove environment. Marine Science, 2002, 26(3): 17-19.

[4] Jiang J X, Li R G, Lu L, *et al*. A study on the biodiversity of mangrove ecosystem of Dongzhai Harbor in Hainan Province. In: Proceedings of the ECOTONE VI, Beihai, 1997. 161-183.

[5] Jiang J X, Li R G, Lu L, *et al*. A study on the biodiversity of mangrove ecosystem of Qinglan Harbor in Hainan, China. Acta Oceanologica Sinica, 22(sup.): 261-271.

[6] Lin P, Xie S Z, Lin Y M, eds. Comprehensive survey report of Zhangjiang Estuary Mangrove Wetland Nature Reserve in Fujian Province. Xiamen: Xiamen University Press, 2001. 114.

[7] Wang B S, Liao B W, Wang Y J, *et al*. Mangrove forest ecosystem and its sustainable development in Shenzhen Bay. Beijing: Science Press, 2002. 362.

[8] Huang Z G, Li J J, Lin Y Y, *et al*. Biodiversity on marine estuarine wetland. Beijing: Ocean Press, 2004. 426.

[9] Chen C P, Gao Y H, Lin P. Dynamics of phytoplankton community in mangrove waters in Fuding City, Fujian Province, China. Journal of Xiamen University (Natural Science), 2005, 44(1): 118-122.

[10] Chen C P, Gao Y H, Lin P. Study on the seasonal changes of phytoplankton community and Its ecology in Futian Mangrove Reserve of Shenzhen, China. Journal of Xiamen University (Natural Science), 2005, 44(Sup.): 11-15.

[11] Fan H Q, Chen G H, He B Y, *et al*. Coastal wetland and management of Shankou mangroves. Beijing: Ocean Press, 2005. 126.

[12] Lin Y M, Lin P. Species, diversities, functions and protections of plants in mangrove ecosystem in China. Transactions of Oceanology and Limnology, 2001, (3): 8-16.

[13] Liu W G, Lin Y M, Chen Z F, *et al*. Distribution and seasonal change of algae in Fujian mangrove areas. Acta Oceanologica Sinica, 2001, 23(3): 78-86.

[14] Zheng D Z, Zheng S F, Liao B W, *et al*. The utilization, protection and afforestation on mangrove wetland. Forest Research, 1995, 8(3): 322-328.

[15] Lin P. A review on the mangrove research in China. Journal of Xiamen University (Natural Science), 2001, 40(2): 592-603.

[16] Wang B S, Liang S C, Zhang W Y, *et al*. Mangrove flora of the world. Acta Botanica Sinica, 2003, 45 (6): 644-653.

[17] He M H. Studies on ecology of polychaeta in mangrove in Jiulong river estuary. Marine Science Bulletin, 1991, 10(3): 56-62.

[18] Lin P. Mangrove ecosystem in China. Beijing: Science Press, 1997. 342.

[19] Zhang H D, Chen G Z, Liu Z P, *et al*. Studies on Futian mangrove wetland ecosystem, Shenzhen. Guangzhou: Guangdong Science and Technology Press, 1997.

[20] Cai L Z, Tam N F Y, Wong Y K. Characteristics of quantitative distribution and species composition of macro zoobenthos in mangrove stands in Eastern Hong Kong. Journal of Xiamen University (Natural Science), 1998, 37(1): 115-121.

[21] He B Y, Fan H Q, Zhang Z R. A preliminary study on the ecology of mangrove macrobenthos in Pearl Bay, Guangxi. In: He Q R ed. Exploitation and research on the resources in Southern China Sea. Guangzhou: Guangdong Economics Press, 1998. 1036-1048.

[22] Lai T H, He B Y. Studies on the macrobenthos species diversity for Guangxi mangrove areas. Guangxi Sciences, 1998, 5(3): 166-172.

[23] Li R G, Jiang J X. Ecological study of mollusca in Puyuzhou mangrove near Daya Bay nuclear power station. Studia Marina Sinica, 1998, (39): 115-122.

[24] Zhang Y Z, Chen C Z, Wang Y Y, *et al*. The ecology of benthos in Fujian mangrove swamps. Acta Ecologica Sinica, 1999, 19(6): 896-901.

[25] Zou F S, Song X J, Chen K, *et al*. The research on benthic macrofauna of swamp in Qinglangang mangrove, Hainan. Ecologic Science, 1999, 18(12): 42-45.

[26] Zou F S, Song X J, Chen W, *et al*. The diversity of benthic macrofauna on mud flat in Dongzhaigang Mangrove Reserve, Hainan. Chinese Biodiversity, 1999, 7(3): 175-180.

[27] Fan H Q, He B Y, Wei S Q. Influences of sand dune movement within the coastal mangrove stands on the Macrobenthos *in situ*. Acta Ecologica Sinica, 2000, 20(5): 722-727.

[28] Lui T H, Lee, S Y, Sadovy Y. Macrobenthos of a tidal impoundment at the Mai Po Marshes Nature Reserve, Hong Kong. Hydrobiologia, 2002, 468: 193-212.

[29] Han W D, Cai Y Y, Liu J K, *et al*. Molluscs of mangrove areas in Leizhou Peninsula, China. Journal of Zhanjiang Ocean University, 2003, 23(1): 1-7.

[30] Xiao H H, Li F M. The composition and ecological distribution of intertidal crabs of Naozhou Island in Guangdong. Guizhou Science, 2003, 21(4): 59-62.

[31] Gao A G, Chen Q Z, Zeng J N, *et al*. Macrofauna community in the mangrove area of Ximen Island, Zhejiang. Journal of Marine Sciences, 2005, 23(2): 33-40.

[32] Hong Y B, Lu X M, Chen L, *et al*. Benthos on mangrove wetland and smooth cordgrass (*Spartina alteriflora*) wetland in Jiulongjiang Estuary. Journal of Oceanography in Taiwan Strait, 2005, 24(2): 189-194.

[33] Liang C Y, Zhang H H, Ji X Y, *et al*. Study on biodiversity of mangrove benthos in Leizhou Peninsula. Marine Science, 2005, 29(2): 18-25, 31.

[34] Tang Y J, Lin W, Chen J F. Species diversity of benthic mollusc in different habitats of intertidal zone in Shangchuan Island. Biomagnetism, 2005, 5(1): 4-7.

[35] Tang Y J, Yu S X. Community structure of macrofauna in Zhanjiang Mangrove Nature Reserves. Progress in Modern Biomedicine, 2006, 6(7): 7-11.

[36] Fan H Q, Wei S Q, He B Y, *et al*. The seasonal dynamics of nekton assemblages in mangrove-fringed tidal waters of Yingluo Bay, Guangxi. Guangxi Science, 1998, 5(1): 45-50.

[37] He B Y. Comparative study on the ecology of mangrove fishes between two bays of Guangxi Marine Science Bulletin, 1999, 18(1): 28-35.

[38] He B Y, Fan H Q. A study on seasonal dynamics of species diversity of fishes in tidal waters of creeks within mangroves of Yingluo Bay, Guangxi Biodiversity Science, 2002, 10(2): 175-180.

[39] He X L, Ye N, Xuan L Q. Investigation of fishes in mangrove areas of Leizhou Peninsula Journal of Zhanjiang Ocean University, 2003, 23(3): 3-10.

[40] Jiang G F, Zhou Z Q. A preliminary study on the insect community and its diversity in mangrove of Qinzhou Bay. Journal of Guangxi Academy of Science, 1996, 12(3,4): 50-53.

[41] Jiang G F, Yan Z G, Cen M. Insect community and its diversity in Mangrove forest at Yingluo Bay of Guangxi Chinese Journal of Applied Ecology, 2000, 11(1): 95-98.

[42] Fan H Q, Qiu G L. Insect pests of *Avicennia marina* mangroves along the coast of Beibu Gulf in China and the research strategies Guihaia, 2004, 24(6): 558-562.

[43] Yan Z G, Jiang G F, Zhang Y Q. A preliminary survey of spider communities in Yingluo Bay mangrove area, Guangxi Journal of Guangxi Academy of Science, 1998, 14(4): 5-7.

[44] Wang Y J, Zan Q J. Studies and protection on amphibians and reptiles of Shenzhen Bay wetlands Ecologic Science, 1998, 17(1): 90-94.

[45] Chang H, Bi X F, Chen G Z, *et al* Composition and Avifauna of Birds in Dongzhaigang National Nature Reserve, Hainan Island Ecologic Science, 1999, 18(2): 53-61.

[46] Zhou F, Fang H L, Zhang H X. The birds of mangroves in the north coastal area of Beibu-bay. In: China Ornithological Society, eds China Animal Sciences Beijing: Chinese Forestry Press, 1999. 257-265.

[47] Zou F S, Song X J, Chen K, *et al* A preliminary study on avian ecology in mangrove wetland of Qinglangang, Hainan Chinese Biodiversity, 2000, 7 (3) : 175-180.

[48] Zou F S, Song X J, Chen K, *et al* Avian diversity in the mangrove wetland of Dongzhaigang Chinese Journal of Ecology, 2001, 20 (3) : 20-23.

[49] Fang B Z The dynamic and protection of winter residents in mangrove forests in Zhangjiangkou, Fujian Journal of Fujian Forestry Science and Technology, 2002, 29(3) : 65-68.

[50] Song X J, Lin P. Bird communities in four mangrove wetlands in Fujian Chinese Journal of Ecology, 2002 , 21 (6) : 5-10.

[51] Wu S B, Ke Y Y, Wu G S, *et al* A preliminary study on the fauna composition and ecological distribution of waterfowl of Leizhou Peninsula wetland Chinese Journal of Zoology, 2002, 37 (2): 58-62.

[52] Zhou F, Fang H L, Zhang H X, *et al* The waterbirds of mangroves area in the coastal area of Guangxi Journal of Guangxi Agriculture and Biology Science , 2002, 21(3) : 145-150.

[53] Lin Q X, Chen X L, Lin P. Investigation of avifauna and It s annual fluctuation in mangroves in Xiamen , China Journal of Xiamen University (Natural Science) , 2002, 41(5): 634-640.

[54] Lin Q X, Chen X L, Lin P. Bird Resources and its distribution in mangroves at Dongyu Islet , Xiamen , China Journal of Xiamen University (Natural Science) , 2005, 44(Sup): 37-42.

[55] Zhou F, Han X J, Lu Z, *et al* The birds of wetland of Nanliu river estuary. Guangxi Sciences, 2005, 12(3) : 221-226.

[56] Zheng G M. A checklist on the classification and distribution of the birds of China Beijing: Science Press, 2005. 426.

[57] Huang Z Y, Zhou Z Q. Anthracnose of mangrove in Guangxi Guangxi Sciences, 1997, 4 (4) : 319-324.

[58] Huang Z Y, Zhou Z Q. A preliminary observation on pathogenic fungi of sooty molds parasitized *Aegiceras corniculatum* and disease occurrence characteristics Guangxi Sciences, 1998, 5 (4) : 314-317.

[59] Zeng C H, Zheng F C, He C P. Isolation and identification of *Halophytophthora* species from mangrove habitats in Hainan Island Mycosystema, 2001, 20(3): 310-315.

[60] Lin X, Huang Y J, Hu Z Y, *et al* Study on antimicrobial activities of marine *Lignicolous* fungi Journal of Oceanography in Taiwan Strait, 2004, 23(3): 308-313.

[61] Lin A Y, Xing X K, Guo S X, *et al* Study on isolation of endophytic fungi from four medicinal semi-mangrove plants and Its antimicrobial activity. Chinese Pharmacy Journal, 2006, 41 (12): 892-894.

[62] Wang G W, Li H Y. Preliminary study on the fungal endophytes in the root of mangrove plants at Qinzhou Bay, China Guangxi Forestry Science, 2003, 32(3): 121-124.

[63] Lin P, Zhang Y B, Deng A Y, *et al* Microflora and antimicrobial activities of soil microorganisms in mangrove forests in the Jiulong Estuary, China Acta Oceanologica Sinica, 2005, 27(3): 133-141.

[64] Wang Y K, Hong K Mangrove soil community analysis using DGGE of 16S rDNA V3 fragment polymerase chain reaction products Acta Microbiologica Sinica, 2005, 45(2): 201-204.

[65] Zan Q J, Wang Y J, Liao B W, *et al* Biomass and Net Productivity of *Sonneratia apetala*, *S. caseolaris* mangrove man-made forest Journal of Wuhan Botanical Research, 2001, 19 (5) : 391-396.

[66] Liu J K, Xuan L Q. Study on the composition of fishery captures in mangrove areas of Leizhou Peninsula China Fisheries, 2005, (2): 74-75, 77.

[67] Lin P. Ecological notes on mangroves in southeast coast of China including Taiwan Province and Hainan Island Acta Ecologica Sinica, 1981, 1 (3): 283-290.

[68] Zheng D Z, Liao B W, Zheng S F, *et al* Mangrove plants' adaptive ability to habitat and their horizontal distribution in Qinglan harbour, Hainan Island Forest Research, 1995, 8(1): 67-72.

[69] Fan H Q, Chen J, Li J L. Species composition and distribution of fouling macro fauna which are attached to mangroves of Guangxi Journal of Guangxi Academy of Science (China), 1993, 9(2), 58-62.

[70] Zhou F, Fang H L, Zhang H X. Diversity of birds in mangroves of Shankou Guangxi Sciences, 2000, 7 (2) : 154-157.

[71] Zhao S X. On the preliminary investigation of Turkestan rat, Rattus rattoides exiguous Howell, in sea beach of red-woods Zoological Research, 1982, 3(1): 18.

[72] Li C G Distribution and forest structure of mangrove in Guangxi Journal of Nanjing Forestry University (Natural Sciences Edition), 2003, 27 (5): 15-19.

[73] Ellison A M, Farnsworth E J. Seedling survivorship, growth, and response to disturbance in Belizean mangal American Journal of Botany, 1993, (80) : 1137-1145.

参考文献:

[1] 范航清. 海岸环保卫士——红树林. 南宁:广西科技出版社, 2000. 126.

[2] 刘玉,陈桂珠. 深圳福田红树林区藻类群落结构和生态学研究. 中山大学学报(自然科学版), 1997, 36(1): 102~106.

[3] 陈长平,高亚辉,林鹏. 红树林区硅藻研究进展. 海洋科学, 2002, 26(3): 17~19.

[5] 江锦祥,李荣冠,鲁琳,等. 海南省清澜港红树林生态系生物多样性. 海洋学报, 22(sup.): 261~271.

[6] 林鹏,谢绍舟,林益明,编. 福建漳江口红树林湿地自然保护区综合科学考察报告. 厦门:厦门大学出版社, 2001. 114.

[7] 王伯荪,廖宝文,王勇军,等. 深圳湾红树林生态系统及其持续发展. 北京:科学出版社, 2002. 362.

[8] 黄宗国,李经建,林永源,等编. 海洋河口湿地生物多样性. 北京:海洋出版社, 2004. 426.

[9] 陈长平,高亚辉,林鹏. 福建省福鼎市后屿湾红树林区水体浮游植物群落动态研究. 厦门大学学报(自然科学版), 2005, 44(1): 118~122.

[10] 陈长平,高亚辉,林鹏. 深圳福田红树林保护区浮游植物群落的季节变化及其生态学研究. 厦门大学学报(自然科学版), 2005, 44(sup.): 11~15.

[11] 范航清,陈光华,何斌源,等. 山口红树林滨海湿地与管理. 北京:海洋出版社, 2005. 126.

[12] 林益明,林鹏. 中国红树林生态系统的植物种类、多样性、功能及其保护. 海洋湖沼通报, 2001, (3): 8~16.

[13] 刘维刚,林益明,陈贞奋,等. 福建红树林区海藻的分布及季节变化. 海洋学报, 2001, 23(3): 78~86.

[14] 郑德璋,郑松发,廖宝文,等. 红树林湿地的利用及其保护和造林. 林业科学研究, 1995, 8(3): 322~328.

[15] 林鹏. 中国红树林研究进展. 厦门大学学报(自然科学版), 2001, 40(2): 592~603.

[17] 何明海. 九龙江口红树林海岸潮间带多毛类生态研究. 海洋通报, 1991, 10(3): 56~62.

[18] 林鹏. 中国红树林生态系. 北京:科学出版社, 1997. 342.

[19] 张宏达,陈桂珠,刘治平,等编. 深圳福田红树林湿地生态系统研究. 广州:广东科技出版社, 1997.

[20] 蔡立哲,谭凤仪,黄玉山. 香港东部红树林区大型底栖动物种类组成与数量分布特点. 厦门大学学报(自然科学版), 1998, 37(1): 115~121.

[21] 何斌源,范航清,张振日. 珍珠港红树林大型底栖动物生态的初步研究. 见:何其锐主编,南海资源开发研究. 广州:广东经济出版社, 1998. 1036~1048.

[22] 赖廷和,何斌源. 广西红树林区大型底栖动物种类多样性研究. 广西科学, 1998, 5(3): 166~172.

[23] 李荣冠,江锦祥. 大亚湾核电站邻近埔渔洲红树林区软体动物生态研究. 海洋科学集刊, 1998, (39): 115~122.

[24] 张雅芝,陈灿忠,王渊源,等. 福建红树林区底栖生物生态研究. 生态学报, 1999, 19(6): 896~901.

[25] 邹发生,宋晓军,陈康,等. 海南清澜港红树林滩涂大型底栖动物初步研究. 生态科学, 1999, 18(12): 42~45.

[26] 邹发生,宋晓军,陈伟,等. 海南东寨港红树林滩涂大型底栖动物多样性的初步研究. 生物多样性, 1999, 7(3): 175~180.

[27] 范航清,何斌源,韦受庆. 海岸红树林地沙丘移动对林内大型底栖动物的影响. 生态学报, 2000, 20(5): 722~727.

[29] 韩维栋,蔡英亚,刘劲科,等. 雷州半岛红树林海区的软体动物. 湛江海洋大学学报, 2003, 23(1): 1~7.

[30] 肖汉洪,李方满. 广东硇洲岛潮间带蟹类的组成及生态分布. 贵州科学,2003,21(4):59～62.
[31] 高爱根,陈全震,曾江宁,等. 西门岛红树林区大型底栖动物的群落结构. 海洋学研究,2005,23(2):33～40.
[32] 洪荣标,吕小梅,陈岚,等. 九龙江口红树林湿地与米草湿地的底栖生物. 台湾海峡,2005,24(2):189～194.
[33] 梁超愉,张汉华,颉晓勇,等. 雷州半岛红树林滩涂底栖生物多样性的初步研究. 海洋科学,2005,29(2):18～25,31.
[34] 唐以杰,林炜,陈结芬. 上川岛潮间带不同生境底栖软体动物物种多样性初步研究. 生物磁学,2005,5(1):4～7.
[35] 唐以杰,余世孝. 湛江红树林保护区大型底栖动物的群落结构. 现代生物医学进展,2006,6(7):7～11.
[36] 范航清,韦受庆,何斌源,等. 英罗港红树林缘潮水中游泳动物的季节动态. 广西科学,1998, 5(1):45～50.
[37] 何斌源. 广西两港湾红树林鱼类生态对比研究. 海洋通报,1999,18(1):28～35.
[38] 何斌源,范航清. 广西英罗港红树林潮沟鱼类多样性季节动态研究. 生物多样性,2002,10(2):175～180.
[39] 何秀玲,叶宁,宣立强. 雷州半岛红树林海区的鱼类种类调查. 湛江海洋大学学报,2003,23(3):3～10.
[40] 蒋国芳,周志权. 钦州港红树林昆虫群落及其多样性初步研究. 广西科学院学报,1996,12(3、4):50～53
[41] 蒋国芳,颜增光,岑明. 英罗港红树林昆虫群落及其多样性的研究. 应用生态学报,2000,11(1):95～98.
[42] 范航清,邱广龙. 中国北部湾白骨壤红树林的虫害与研究对策. 广西植物,2004,24(6):558～562.
[43] 颜增光,蒋国芳,张永强. 广西英罗港红树林蜘蛛群落初步调查. 广西科学院学报,1998,14(4):5～7.
[44] 王勇军,昝启杰. 深圳湾湿地两栖爬行动物及其保护. 生态科学,1998,17(1):90～94.
[45] 常弘,毕肖峰,陈桂珠,等. 海南岛东寨港国家级自然保护区鸟类组成和区系的研究. 生态科学,1999,18(2):53～61.
[46] 周放,房慧伶,张红星. 北部湾北部沿海红树林的鸟类. 见:中国动物学会. 中国动物科学研究. 北京:中国林业出版社,1999. 257～265.
[47] 邹发生,宋晓军,陈康,等. 海南清澜港红树林湿地鸟类初步研究. 生物多样性,2000,7(3):175～180.
[48] 邹发生,宋晓军,陈康,等. 海南东寨港红树林湿地鸟类多样性研究. 生态学杂志,2001,20(3):20～23.
[49] 方柏州. 福建漳江口红树林冬候鸟动态及保护. 福建林业科技,2002,29(3):65～68.
[50] 宋晓军,林鹏. 福建红树林湿地鸟类区系研究. 生态学杂志,2002,21(6):5～10.
[51] 吴诗宝,柯亚永,吴桂生,等. 雷州半岛湿地水鸟区系组成及生态分布的初步研究. 动物学杂志,2002,37(2):58～62.
[52] 周放,房慧伶,张红星,等. 广西沿海红树林区的水鸟. 广西农业生物科学,2002,21(3):145～150.
[53] 林清贤,陈小麟,林鹏. 厦门凤林红树林区鸟类组成和年变动研究. 厦门大学学报(自然科学版),2002,41(5):634～640.
[54] 林清贤,陈小麟,林鹏. 厦门东屿红树林湿地鸟类资源及其分布. 厦门大学学报(自然科学版),2005,44(sup.):37～42.
[55] 周放,韩小静,陆舟,等. 南流江河口湿地的鸟类研究. 广西科学,2005,12(3):221～226.
[56] 郑光美. 中国鸟类分类与分布名录. 北京:科学出版社, 2005. 426.
[57] 黄泽余,周志权. 广西红树林炭疽病研究. 广西科学,1997,4(4):319～324.
[58] 黄泽余,周志权. 桐花煤污病的病原菌和病害发生特点初步观察. 广西科学,1998,5(4):314～317.
[59] 曾会才,郑服丛,贺春萍. 海南红树林生境中海疫霉种的分离与鉴定. 菌物系统,2001,20(3):310～315.
[60] 林昕,黄耀坚,胡志钰,等. 海洋木栖真菌抗菌活性的初步研究. 台湾海峡,2004,23(3):308～313.
[61] 林爱玉,邢晓科,郭顺星,等. 4种药用半红树植物内生真菌的分离及其抗菌活性研究. 中国药学杂志,2006,41(12): 892～894.
[62] 王桂文,李海鹰. 钦州湾红树植物根部内生真菌初步研究. 广西林业科学,2003,32(3):121～124.
[63] 林鹏,张瑜斌,邓爱英,等. 九龙江口红树林土壤微生物的类群及抗菌活性. 海洋学报,2005,27(3):133～141.
[64] 王岳坤,洪葵. 红树林土壤细菌群落 16S rDNA V3片段 PCR产物的 DGGE分析. 微生物学报,2005,45(2):201～204.
[65] 昝启杰,王勇军,廖宝文,等.无瓣海桑、海桑人工林的生物量及生产力研究. 武汉植物学研究,2001,19(5):391～396.
[66] 刘劲科,宣立强. 2005. 雷州半岛红树林海区渔获组成研究. 中国水产,(2):74～75,77.
[67] 林鹏. 中国东南部海岸红树林的类群及其分布. 生态学报,1981,1(3):283～290.
[68] 郑德璋,廖宝文,郑松发,等. 海南岛清澜港红树树种适应生境能力. 林业科学研究,1995,8(1):67～72.
[69] 范航清,陈坚,黎建玲. 广西红树林大型固着污损动物的种类组成及分布. 广西科学院学报,1993,9(2):58～62.
[70] 周放,房慧伶,张红星. 山口红树林鸟类多样性初步研究. 广西科学,2000,7(2):154～157.
[71] 赵善贤. 海滩红树林中黄毛鼠生态学的初步研究. 动物学研究,1982,3(1):18.
[72] 李春干. 广西红树林资源的分布特点和林分结构特征. 南京林业大学学报(自然科学版),2003,27(5):15～19.

三 红树林植物化学

PART Ⅲ MANGROVE PHYTOCHEMISTRY

中国海洋红树林药物的研究现状、民间利用及展望*

林　鹏　林益明　杨志伟　王湛昌

（厦门大学生命科学学院，福建 厦门 361005）

中图分类号：Q945　　**文献标识码：**A　　**文章编号：**1000-3096(2005)09-0076-04

1　红树林药物研究的历史、现状

海洋是一个巨大的药源宝库。海洋生物活性物质主要包括生物信息物质、各类活性成分、海生毒素、生物功能材料等，研究海洋生物活性物质是海洋药物研究的主导方向[1]。20 世纪 60 年代以来，从海洋动物、植物及微生物中已分离获得新型化合物 1 万多种，其中 1/2 以上具有抗肿瘤、抗菌、抗病毒、抗凝血等药理活性；这些新型化合物为药物设计提供了可贵的分子模型，为海洋药物的开发提供了重要的先导化合物库[2]。但与来源于陆生植物的 15 万种天然产物相比，海洋天然药物至今才 1 万多种，从资源研究上看，目前用于研究的海洋生物仅几千种，海洋生物具有巨大潜力等待开发与研究[3]。

海洋药物的应用在我国有着悠久的历史，是中医药科学宝库的重要组成部分。海洋药物对功能性疾病、自身免疫性疾病、老年性疾病及多种疑难病症具有优于化学药物和传统中药的疗效。因此，海洋药业对现代中药产业的发展将起到决定性的作用。

海洋药物由于具有药理的稳定性、强效性和特异性等优点，决定了海洋药物是具有明显优势的新药品种。目前用于临床的海洋药物主要有：藻酸双酯钠、甘糖酯、精母注射液、龙珠口服液等，基本完成研制的有：珍珠贝胶囊和聚甘古酯抗艾滋病注射液等[2]。

我国的药物研究，极少有来源于海洋生物活性先导化合物的原创性研究和专利。到目前为止，我国仅研究了约 1 000 个海洋化合物，鉴定了约 500 个海洋生物化合物的结构，具有生理活性的化合物更少。而国外发现了约 9 000 种海洋生物化合物，其中 50% 具有活性[4,5]。

红树林是分布在热带亚热带海岸潮间带的木本植物群落，中国有红树植物 12 科 15 属 26 种（含 1 变种），以及半红树植物 9 科 10 属 11 种[6]。在我国 37 种红树植物和半红树植物中，已发现具有药用价值的 18 种，说明一半的红树植物种类是已知具有民间药物利用传统的。其中，老鼠簕（*Acanthus ilicifolius*）1 种真红树植物，黄槿（*Hibiscus tilisceus*）和海芒果（*Cerbera manghas*）2 种半红树植物被《全国中草药汇编》所收录。

红树植物药的有效成分研究很少。具有药用价值的有毒植物海漆（*Excoecaria agallocha*）[7,8] 和海芒果（*Cerbera manghas*）[9]，主要是利用其各种有毒成分，而其中多种植物的药用，很大程度上与红树植物单宁含量较高有关。单宁具有收敛、止血、解毒和防腐等药学性质[10]。

海莲（*Bruguiera sexangula*）树皮的提取物能有效地抑制 2 种类型的肿瘤，肉瘤 180（sarcoma 180）和刘易斯肺癌（Lewis lung carcinoma）[11]。

老鼠簕是国际上研究较为深入的药用红树植物之一。在我国及亚太地区其它国家的红树林海岸居民都有传统的药物利用经验。泰国学者 Kokpol[11] 研究报道，在老鼠簕的根提取液中发现的成分如苯并噁唑啉（benzoxazoline-2-ene）因其对中枢神经系统具有抑制作用而具有较高的医药利用价值，可作止痛药、退热剂、抗惊厥药和安眠药及具有肌肉松弛活性；苯并唑啉能抗真菌病害，这种糖的核糖衍生物具有抗癌和抗病毒活性；鼠类试验中证明老鼠簕的根具有抗白血病（leukemia）的活性。Kanchanapoom 等[12,13] 从老鼠簕的地上部分分离得到多种化学成分。Babu 等[14] 从老鼠簕叶的提取物进行鼠类实验中发现具有抗肿瘤效应。迄今为止从老鼠簕（*Acanthus ilicifolius*）中

* 福建省自然科学基金重点资助项目（2003Y036）
原载于海洋科学，2005，29(9)：76－79

分离到了约 30 个化合物，主要有：2-喹啉羧酸，2-喹啉羧酸、酮类化合物甲基芹菜素-7-O-β-D-葡萄糖醛酸苷(Methylapigenin-7-O-β-D-glucuronate)、槲皮素-3-O-β-D-葡萄糖酸苷(quercetin-3-O-β-D-glucopyranoside)，生物碱 7-氯-(2R)-2-O-β-D-吡喃型葡萄糖基-2 氢 1,4-氧氮杂奈酮-3-(4 氢)-(7-chloro-(2R)2-O-β-D-glucopyranosyl-2H-1,4benzoxazin-3(4H)-one)，三萜系列化合物三萜苷和三萜皂苷(齐墩果烷、羽扇豆醇、乌苏烷和乌苏酸等)。国内外对慢性肝炎还没有特效药物和特效疗法。比较有前景的是干扰素和基因疗法。干扰素价格昂贵，基因疗法很不成熟。而在中国民间，红树植物老鼠簕根捣碎水煮加上蜂蜜后口服是治疗乙型肝炎的特效药，这是与它的独特的化学成分有关的。

海漆(*Excoecaria agallocha*)为有毒红树植物，迄今为止从中分离到的化合物有二萜、三萜、查尔酮和吡啶生物碱，主要以二萜化合物为主，其中 Laphnane(瑞香烷)型 22 种化合物，它们具有相同的母核与致癌物质佛波酯相同，这类化合物已被证实是对人的皮肤造成刺激损伤的原因；此外，从海漆中分离了 19 种 Labdane 型化合物和 5 种 Beyerane 型化合物。

Erickson 等[15]报道了红树植物海漆的提取物的抗艾滋病原理。最近，印度学者 Babu 等[14,16]报道了红树植物老鼠簕的乙醇提取物(浓度为 250 500 mg/kg)能有效地抑制肿瘤的生长和致癌物诱导在老鼠皮层瘤的生成等；日本学者 Konoshima 等[8]采用 12-O-四癸酰基-佛波-13-乙酸酯诱导的 EBV-EA(非洲淋巴细胞瘤病毒)活化的体外肿瘤模型，对从海漆中分离的 8 个 Labdane 型二萜化合物初步活性筛选，其中化合物 7 在肿瘤催进剂 TPA(12-O-四癸酰基-佛波-13-乙酸酯)和激动剂 DMBA(7,12-二甲基苯并蒽)协同作用的双阶段小鼠肿瘤模型中，该化合物显示出显著的抗肿瘤活性。

值得注意的是，某些与海洋动植物共生或附生的微生物可产生不同于陆生生物所产生的生物活性物质，其中有些成分以前还被认为是动植物宿主所产生的，如从海藻中分离的 *Flavobacteium uliginosum* 产生可抗肿瘤的 Marinactan；20%～50%海鞘和海参体内的微生物可产生具细胞毒性和杀菌的化合物；相当部分以前被认为是动植物产生的毒素已被证明是其共生的细菌所产生[17,18]。目前，研究与海洋动植物共生或附生的微生物所产生的生物活性物质已成为海洋药物资源研究和开发的新兴领域，在这方面已经进行了有益的探索[19,20]。

从国际上海洋药物研究的进展看，近几年来，天然药物化学家从红树植物中分离得到大量有良好抗肿瘤活性的结构新颖的化合物[21,22]。但目前进入临床及临床前研究的抗肿瘤海洋药物主要有膜海鞘素(Didemnin B)、海兔素肽(Doladtatin 10)、苔藓虫素(Bryostatin 1)等[1]。它们多为脂类、多肽或萜类，且多从海洋动物中提取。从海洋高等植物——红树林中得到抗肿瘤药物用于临床的未见报道。可见大力开发红树植物药物十分紧迫，争取早日获得我国有自主知识产权的红树海洋植物药物。

2 中国海洋红树林植物药物的民间利用

从中国的自然地理条件和红树林的生态习性看来，中国历史上一直有红树林的自然分布。从我国海岸开发史来看，东南沿海红树林海岸居民利用红树林生态系统也至少有几百年历史。因此，红树林海岸居民积累了极为丰富的红树植物的药物利用知识。

(1) 正红树(*Rhizophora apiculata* Bl.)(红树科)

民间利用正红树作为治疗结石、尿路结石的有效药物。通常摘取正红树的气生根顶端嫩尖，长约 1 cm，水煮口服，能化解结石，使泌尿系统得以畅通。在海南琼山县，正红树的树皮民间用于治疗烧伤、烫伤。采集方法是刮去最外层老树皮，剥取第二层树皮，捣碎磨烂之后外敷于患处，不仅能防止患处化脓感染，而且促进伤口较快愈合，新皮再生。当地部队 162 医院，经临床试验，证明其对烧伤治疗有特效，正在进一步研究开发中。

(2) 木榄(*Bruguiera gymnorrhiza* (L.) Lamk.)(红树科)

木榄果(胚轴)捣碎，水煮口服，民间作为腹泻的收敛剂[23]。在海南琼山市民间还利用木榄胚轴来治糖尿病。

(3) 海莲(*Bruguiera sexangula* (Lour.) Poir.)(红树科)

海莲的树叶，水煮熬汁口服，可以用来治疗痢疾。

(4) 角果木(*Ceriops tagal* (Perr.) C. B. Rob.)(红树科)

角果木的树皮捣碎外敷能止血，治恶疮；种子榨油外敷能止痒，治疥癣和冻疮。其叶熬汁可以作为奎宁的代用品，能治疗痢疾[23]。

(5) 秋茄(*Kandelia candel* (L.) Druce)(红树科)

在福建省福鼎县鲎屿村，民间将秋茄的根挖掘捣碎，水煮口服，能够治疗风湿性关节炎，疗效较好。当地居民说能根治风湿性关节炎慢性病。

(6) 白骨壤(*Avicennia marina* (Forsk.) Vierh.)

（马鞭草科）

白骨壤的叶，捣烂外敷，可治脓肿，其树皮胶可外用作为避孕药品[23]。

（7）海漆（*Excoecaria agallocha* L.）（大戟科）

海漆为有毒的红树植物，分泌乳状汁液，有刺激性，可使皮肤肿胀，倘误入眼睛，为害更大[24]，少量可引起暂时失明，民间有用作箭毒或毒鱼的[25]。

（8）老鼠簕（*Acanthus ilicifolius* L.）（爵床科）

在海南澄迈县，民间广泛将其用作药物。其根捣碎水煮，加上蜂蜜后口服，是治疗乙型肝炎的特效药。一些乙型肝炎患者，用许多其它药物无法治愈，就专程到红树林海岸，挖掘老鼠簕的根煎药服用，几个月后可治愈；另一种用途是将根捣碎后外敷于患处，有消炎作用，治无名肿痛。

老鼠簕目前已被《全国中草药汇编》收录[26]。以全株或根人药，全年可采，洗净晒干备用。性味功能：微咸、凉，清热解毒，能消肿散结、止咳平喘。外敷可治瘰疬等。老鼠簕的药用仍限于民间，国家法定药典《中华人民共和国药典》中药部分中尚未记载。

海南民间用老鼠簕消肿、解毒、止痛、治淋巴结肿大、急性肝脾疼痛、黄疸、胃痛和哮喘。广西民间运用其根治神经痛、腰肌劳损、祛痰等。另有报道其叶和根混合捣碎成糊状，可治蛇伤。根叶捣碎可作为毛发防腐剂。茎叶可消肿，种子捣碎可治脓肿。林鹏[23]在福建龙海调查了解到民间还用老鼠簕治疗男子不育症。该种在福建已濒临灭绝，各地适宜种植和利用相结合，以保护种质资源。

（9）小花老鼠簕（*Acanthus ebrecteatus* Vahl.）（爵床科）

小花老鼠簕的果实捣碎外敷，可治疗疔[27]。

（10）海桑（*Sonneratia caseolaria*（L.）Engler）（海桑科）

海桑的果实捣烂成糊状，可以治扭伤[23]。其果实、叶和花作为内科用药。

（11）杯萼海桑（*Sonneratia alba* Sm.）（海桑科）

杯萼海桑果实榨汁发酵，可用于止溢血[23]。

（12）榄李（*Lumnitzera racemosa* Presl.）（使君子科）

榄李的树叶熬汁，可治鹅口疮、雪口病[27]。

（13）银叶树（*Heritiera littoralis*（Drgand.）Ait.）（梧桐科）

银叶树的树皮，水煮熬汁内服，可治疗血尿病[23]。银叶树也用作治疗腹泻和赤痢，其种仁被认为是一种滋补品[27]。

（14）海芒果（*Cerbera manghas* L.）（夹竹桃科）

海南民间用海芒果的叶、树皮、乳汁、有催吐、下泻之效，但有毒，用量须慎重。广西民间用海芒果做泻下剂[27]。

海芒果为《全国中草药汇编》收录的红树植物，药用部分为其树液，具有催吐、泻下之效用，属少用中草药类。

海芒果所含的海芒果甙具有强心作用，国内已从海芒果中提取海芒果甙。属一种显效快，正性肌力作用强，持续时间短的强心甙，用于治疗心力衰竭的急性病症。

海芒果为中国红树林中的有毒植物[25]。其树叶和果均有毒，核仁毒性最强。因含氢氰酸和海芒果甙（cerberin）等，其茎呈生物碱和酸性物质反应。种子可毒鱼[26]。在海南琼山县调查了解到，民间有吃食海芒果果实自杀者。误食其果实中毒时，民间用灌鲜羊血、饮椰子水解毒。

（15）玉蕊（*Barringtonia racemosa* Roxb.）（玉蕊科）

海南民间将玉蕊作为药用植物，其根可退热，果可以止咳。玉蕊有一定毒性，民间有误食中毒发生[25]。

（16）黄槿（*Hibiscus tiliaceus* L.）（锦葵科）

黄槿为《全国中草药汇编》中收录的红树林药用植物。药用部分为叶、树皮和花，其性甘、淡、微寒、具清热解毒、散瘀和消肿之效用。民间用来治木薯中毒。方法是采摘鲜花或嫩叶 50～100 g，捣烂取汁冲白糖水口服。严重者可每日服 2～3 剂。另外，能治疮疖肿痛，方法是采摘嫩叶或鲜树皮捣烂外敷。黄槿属于少用中草药类，但尚未见于国家法定的《中华人民共和国药典》中药部分。

广西民间用黄槿治痈疮肿痛、解除木薯中毒。黄槿树叶掺水磨汁，可作为祛痰剂和利尿剂[23]。

（17）杨叶肖槿（*Thespesia populnea*（L.）Soland. ex Correa）（锦葵科）

杨叶肖槿的果实可捣烂制药膏，能去虱，其树叶水煮熬汁，可治头痛和疥癣[23]。

（18）木果楝（*Xylocarpus granatum* Koenig）（楝科）

树皮可治赤痢，种仁用作滋补品[23]。在海南琼山县和三亚市调查了解到，来自东南亚的游客常采摘木果楝的果实，据说是一种很好的滋补品。

3 展望

红树林民间药物利用的调查，目前已经十分困难。随着海岸居民医疗条件的改善，大多数以往用红树林植物治疗的疾病，现在都已采用西药和其他中成药。民间积累的丰富宝贵的药物利用经验正在迅速被遗忘。红树林民间药物利用调查虽然十分困难，但也十分紧迫和必要。

海洋生物似乎能提供无穷尽的新化合物。尤其是近年来海洋生物中新化合物发现的速度和数量超出了人们的想象,并且不断有全新骨架的海洋化合物被报道。从海洋生物中寻找新的活性天然产物依然是当前海洋天然产物研究的主要内容。从研究的海洋生物物种看,主要是海绵,其次是珊瑚,其他物种还有贝囊类动物、苔藓虫、棘皮动物、海藻、海星等。而分布于热带亚热带海岸潮间带的红树林天然产物的研究是海洋天然产物研究领域的新热点,将扮演越来越重要的角色。

同时,开发其药物价值,促进海岸居民把其作为药用资源来加以保护和种植,才可能使红树林避免进一步遭受破坏。把经济效益与生态效益相结合,是保护我国红树林的关键。

参考文献:

[1] 关美君,林文翰,丁源. 海洋药物——二十一世纪中国药学研究的新热点[J]. 中国海洋药物, 2001, 1:1—5.

[2] 管华诗,耿美玉,王长云. 21世纪——中国的海洋药物[J]. 中国海洋药物, 2000, **19**(4):44.

[3] 林文翰. 海洋生物——中国天然药物研究的新领域[A]. 田乃旭,屠鹏飞主编. 药物学研究与展望[C]. 北京:科学出版社, 1999. 18—27.

[4] Abda M J , Bermejo P. Bioactive natural products from marine sources[J]. **Studies in Natural Products Chemistry**, 2001,25:683—756.

[5] Faulkner D J. Marine natural products[J]. **Nat Prod Rep**, 2000,17:72.

[6] 林鹏. 中国红树林生态系[M]. 北京:科学出版社, 1997.

[7] Anjaneyulu A S R. , Lakshmana R V. Five diterpenoids (agallochins A-E) from the mangrove plant *Excoecaria agallocha* Linn[J]. **Phytochemistry**, 2000,**55**(8):891—901.

[8] Konishi T, Takasaki M, Tokuda H, *et al*. Anti-tumor-Promoting Activity of Diterpenes from *Excoecaria agallocha* [J] . **Bio Pharm Bull**, 1998, **21**(9):993—996.

[9] Chang L C, Gills J J, Bhat K P L, *et al*. Activity—guided isolation of constituents of *Cerbera manghas* with antiproliferative and antiestrogenic activities[J]. **Bioorganic and Medicinial Chemistry Letters**, 2000, **10**(21):243.

[10] 林鹏,傅勤. 中国红树林环境生态及经济利用[M]. 北京:高等教育出版社, 1995.

[11] Kokpol U. Chemistry of natural products from mangrove plants, UNDP/UNESCO, Training course on life history of selected species of flora and fauna in mangrove ecosystem[J]. **Thailand**, 1985, 159—169.

[12] Kanchanapoom T, Kamel M S, Kasai R, *et al*. Lignan glucosides from *Acanthus ilicifolius* [J]. **Phytochemistry**,2001, **56**(4): 369—372.

[13] Kanchanapoom T, Kamel M S, Kasai R, *et al*. Benzoxazinoid glucosides from *Acanthus ilicifolius* [J]**Phytochemistry**, 2001, **58**(4):637—640.

[14] Babu B H, Shylesh B S, Padikkala J, *et al*. Tumour reducing and anticarcinogenic activity of *Acanthus ilicifolius* in mice[J]. **Journal of Ethnopharmacology**, 2002,**79**(1):27—33.

[15] Erickson K L , Beutler J A , Cardellina II J H, *et al*. A novel phorbol ester from *Excoecaria agallocha* [J]. **J Natl Prod**, 1995, 769—972.

[16] Babu B H, Shylesh B S, Padikkala J, *et al*. Antioxidant and hepatoprotective effect of *Acanthus ilicifolius*[J]. **Fitoterapia**, 2001, **72**(3):272—277.

[17] Gil-Turnes M. S, Hay M E, Fenical W, *et al*. Symbiotic marine bacteria chemically defined crustacean embryos from a pathogenic fungus [J]. **Science**, 1989,246:116.

[18] Rouhiainen L, Sivonen K. , Buikema WJ, *et al*. Characterization of toxin — producing cyanobacteria by using an oligonucleotide probe containing a tandemly repeated heptamer [J]. **J Bacteriol**, 1995, 117:6 021.

[19] Huang Y J, Wang J F, LI G L, *et al*. Antitumor and antifungal activities inendophytic fungi isolated from pharmaceutical plants *Taxus mairei*, *Cephalataxus fortunei* and *Torreya grandis*. **FEMS Immunology and Medical Microbiology**, 2001, **31**(2):163—167.

[20] Zheng Z H, Wei Z, Huang Y J, *et al*. Detection of antitumor and antimicrobial activities in marine organism associated actinomycetes isolated from the Taiwan Strait, China. [J]. **FEMS Microbiology Letters**, 2000, **188**(1):87—91.

[21] Anjaneyulu A S R, Rao V L. Seco diterpenoids from *Excoecaria agallocha* L[J] . **Phytochemistry**,2003, **62** ,585—589.

[22] Konishi T, Yamazoe K, Konoshima T, *et al*. Seco-labdane type diterpenes from Excoecaria agallocha. [J]. **Phytochemistry**,2003, **64** ,835—840.

[23] 林鹏. 我国药用的红树林植物[J]. 海洋药物, 1984, 4:45—51.

[24] 华南植物研究所. 海南植物志[M]. 北京: 科学出版社, 1964—1974.

[25] 陈冀胜,郑硕. 中国有毒植物[M]. 北京: 科学出版社, 1987. 257—258.

[26] 《全国中草药汇编》编写组. 全国中草药汇编(下册)[M]. 北京: 人民卫生出版社, 1978. 231—232,756,804.

[27] 林鹏. 红树林[M]. 北京: 海洋出版社, 1984. 96.

红树林单宁的研究进展*

林益明　向　平　林　鹏

（厦门大学生命科学学院，福建 厦门 361005）

中图分类号：Q948.885.3　**文献标识码**：A　**文章编号**：1000-3096(2005)03-0059-05

1　植物单宁的概念及性质

植物单宁（vegetable tannin），又称植物多酚（plant polyphenol）是一类广泛存在于植物体内的多元酚化合物，在维管植物中的含量仅次于纤维素、半纤维素和木质素，主要存在于植物的皮、根、叶、果中，含量可达20%[1, 2]。作为皮革的一种传统鞣剂，单宁一般指的是相对分子质量为500~3 000的多酚[3, 4]。Haslam[5]提出了植物多酚这一术语，它包括了单宁及相关化合物（如单宁的前体化合物和单宁的聚合物）。根据化学结构的不同，植物多酚分为水解单宁（棓酸酯类多酚）和缩合单宁（黄烷醇类多酚或原花色素）。前者主要是棓酸及其衍生物与多元醇以酯键或甙键形成，可细分为棓单宁和鞣花单宁两类；后者主要是羟基黄烷醇类单体的缩合物，单体间以C—C键相连[6]；缩合单宁和水解单宁之间的这种结构的差异引起了这2种化合物在植物体内的功能的不同[7]。

在被子植物中，单宁具有潜在的提供源信息的潜力，这与木质素和角质是互补的[8]。例如单子叶植物不能被木质素的结构所分开，仅仅能被角质略微分开，一种缩合单宁的单体（*ent*-epicatechin），只特异地存在单子叶植物中的，并且含原天竺葵的聚合物在单子叶植物中的分布比在双子叶植物中的分布要更普遍[9]。相反，水解单宁仅仅在双子叶植物中发现[10, 11]，这易与单子叶植物相区别；单宁相关的化合物也能用来区分被子植物和裸子植物。如黄烷醇主要是在被子植物中发现的。除此以外，缩合单宁的二、三聚合物在被子植物中的出现包含了更多种类依赖的分类学信息[5]。

在维管束植物含量最丰富的组分中，单宁是继纤维素、半纤维素和木质素之后，排第4。在快速循环中的叶（包括针叶）中，单宁的含量高达20%[1, 12]。因而，除作为潜在的生物标记物外，单宁还大量地存在于有机物的特性，包括颜色、收敛性和反应性。单宁与蛋白质结合的能力称之为收敛性或涩性。单宁与生物碱和多糖也可发生与单宁衆蛋白质结合相似的复合反应。收敛性是单宁多种生理活性的基础。如多种红树植物的民间药用，很大程度上与红树植物单宁含量较高有关[13]。Miyamoto分析了单宁的结构及其抗肿瘤之间的关系，表明单宁的抗肿瘤活性与其收敛性相关。

植物单宁的多元酚结构赋予其一系列独特的化学性质。单宁与蛋白质的结合是其最重要的特征。在过去，对单宁研究的兴趣来源于单宁的反应过程中与蛋白氮结合的能力。从地球化学循环角度，单宁对氮的潜在结合与对氮的固定是十分有意义和有趣的。例如，在红树林沼泽，在沉积物和浸没的叶子中氮的固定机制（外生氮的结合到有机物中）还是一个未被了解的过程[2]，单宁的研究也许会对这个关键的过程提供一个线索。

2　植物单宁测定的方法

在植物多酚的应用及相关领域，多酚的含量或纯度测定是非常重要的也是最常遇到的问题[14, 15]。例如，植物中鞣剂的单宁含量决定了其使用质量；

---

* 国家自然科学基金资助项目(40376026)

原载于海洋科学，2005，29(3)：59－63

而高粱、豆类中多酚的含量影响着其营养价值；木材、农作物中多酚含量与其抗虫、抗病菌有关，在食品中有时要求一定量的多酚以保持恰当的风味。所有这些关系到植物多酚含量的测定，因此如何快速、简便、准确地测定出各类样品中多酚的含量更是急需解决的问题[16]。

由于植物多酚与黄酮类、蒽醌类、简单酚和木质素等在植物体内共存并且性质相近，其本身更是一类结构和性质都极为相似的混合物，并且多酚性质较为活泼，很容易发生缩合或者降解，因此对其进行精确定量比较困难，特别是分离提纯的方法对其进行“绝对”的定量测定是不可能的。目前虽然有数十种多酚的定量方法，但这些方法都是相对的、有局限性的，几乎没有一种可以通用的测定方法。最常用的几种方法为：Folin-Denis (FD)法、Prussian Blue (PB)法、香草醛法、正丁醇-盐酸法、Bovine Serum Albumin (BSA)沉淀法[4, 15, 17]。

测定植物样品中的总酚含量，可采用FD法（或改进的FC法）和PB法[18]；如果测定缩合单宁含量，可采用香草醛法和正丁醇-盐酸法[14, 19]；香草醛法和正丁醇法联合起来，可用于测定缩合单宁的聚合度，在不能用复杂的波谱技术测定的情况下，可用这种方法粗略确定缩合单宁的分子质量，它特别适用于比较植物不同生长期缩合单宁的含量；当测定由多酚涩味性所带来的生物活性及相关特性时，可采用BSA沉淀法；当需要着重了解样品中具有捕捉自由基、络合金属离子等生物活性的多酚含量时，可相应采取还原法或络合法等测定方法对多酚进行定量[4, 6, 15~17, 20]。

在实际测定中，不同方法的测定结果通常具有较大的差异。经常采取几种不同的方法对同一样品进行测定，从而对样品所含的总酚、特定结构的酚、单宁与非单宁酚含量得到一个综合的表征和认识。对于特定的体系，可以将总酚量近似看成单宁量，尤其在比较同一类样品时较为合适[4]。

植物多酚含量的测定中最为关键的一步是多酚的提取。提取时样品的状况和提取条件都可能导致多酚量的很大变化。植物原料的贮存、干燥、粉碎，提取溶剂、温度都可能改变多酚的化学结构和提取效率，从而改变了多酚的化学、物理化学和生物活性，使测定值与真实情况有很大出入[21, 22]。当测定植物原料中多酚和单宁的含量时，贮存通常使提取率降低；因此尽可能采取新鲜材料，首先应采取短时间（2~5 min.）的水蒸气加热，使样品中多酚氧化酶PPO等酶丧失活性，避免对原料成分的改变，否则应对样品进行干燥后才能短时间的贮存，最好是冷冻干燥，避免高温。样品提取前需要粉碎成粉末，最适合的尺寸在100目左右。水虽然是植物多酚的良好溶剂，但并非最适合多酚的提取，有机溶剂和水的复合体系（有机溶剂占50%~70%）最适合多酚的提取，有机溶剂的提取顺序为：丙醇<乙醇<甲醇<丙酮<四氢呋喃。其中应用最多的是丙酮-水体系[3, 4, 15]。

3 红树植物单宁含量、组成及其分布

目前，我国红树植物的单宁含量尚缺乏全面系统的测定。现有的资料主要是从鞣料开发利用的角度对我国主要的红树植物树皮的单宁含量所作的测定[23]。

红树植物的单宁含量随植物种类的不同而不同，真红树的树皮的单宁含量，通常在1%~30%之间，有开发利用的价值。但白骨壤除外，单宁含量仅0.3%；红树植物单宁的含量与植物种类、树龄、产地、生态环境的关系，以及在植物体内不同部位的分布规律，都还无系统的分析研究。

Hsu等[24]通过对来自秋茄树皮中的缩合单宁的酸催化苄硫醇硫解，应用色谱和^{1}H-NMR、^{13}C-NMR对秋茄树皮中的缩合单宁（包括单体）进行了结构的研究，不仅发现有原天竺葵定（propelargonidin）二聚物、原花青定（procyanidin）三聚物这些与其他物种共有的原花色素（缩合单宁），还发现了2种新的原花色素二聚物：秋茄素(kandelins)A-1、A-2，和4个三聚物：秋茄素B-1、B-2、B-3、B-4，他们聚合链上最上端是一个苯丙基取代的黄烷-3-醇（金鸡纳因）。以苄硫醇-醋酸对秋茄素B-1进行完全硫解时，生成表儿茶素、儿茶素-4β-苄硫醇及金鸡纳因I$_a$-4β-苄硫醇。金鸡纳因(cinchonains)含于红金鸡纳树皮内[2]。对秋茄素B-1进行局部硫解时，在反应产物中能找到原花青定B-2及金鸡纳因II$_a$-4β-苄硫醇。在秋茄树皮中含有的黄烷三醇有：阿福豆素(afzelechin)、儿茶素（catechin)、表儿茶素(epicatechin）和醅儿茶素（gallocatechin)。Hernes等[2]对采自巴哈马群岛的大红树叶单宁进行酸催化降解，然后用GS-NMR对其进行分析，发现组成大红树叶单宁的黄烷醇单元有：儿茶素、表儿茶素、醅儿茶素、阿福豆素、表阿福豆素、表醅儿茶素和醅酸。在该文献中没有提到在大红树叶的单宁的组成单元

中有金鸡纳因。

4 单宁与红树植物的抗盐适应

红树林是分布在热带亚热带海岸潮间带的木本植物群落。单宁对红树植物的重要性表现在其所处的特殊的物理环境、化学环境和生物环境等各方面都有生态适应意义[25，26]。在红树林海滩，低潮带的盐度可达20~35，即使在高潮带，一般也有5~15。红树林采取拒盐性、泌盐性和聚盐性3种机制适应高盐度环境，其中以依靠单宁的聚盐性抗盐机制是最重要的机制[25]。红树林区共生着许多鸟类、昆虫；潮间带分布着许多甲壳动物、软体动物；海水中存在许多微生物。在这样的环境中，较高的单宁含量为红树植物提供了自我保护的能力。在红树植物与外界接触的界面—树皮和果皮往往单宁含量最高，形成一个有效的保护层①。由于单宁有涩味，避免或减轻了动物的直接啃食，红树植物白骨壤易受病虫害，可能与其单宁含量较低有关。同时，单宁有抑制微生物活动、杀灭病原菌的效能，增强了红树植物的抗病能力和抗海水腐蚀的能力。

在红树植物繁殖体的发育过程中，单宁同样起重要的作用。未成熟的果实，如海桑、老鼠簕的果实，单宁含量高，避免了动物的啃食。特别是显胎生植物，其胚轴的外皮单宁含量很高，减轻了鸟类的啄食和昆虫的危害。更重要的是，在种子传播过程中，胚轴能漂洋过海行程数百海里，存活数月，正是富含单宁使其不受海水腐蚀和海洋动物的啃食，保证种质得以广泛传播。

5 红树林单宁的地球化学循环

红树林具有高归还率和高分解率的特性，凋落物的分解过程是红树林生态系统的物流和能流的关键环节，红树林凋落物富含单宁，单宁的降解无疑对其分解过程，进而对生态系统的物流和能流产生影响[25]。红树植物凋落叶进入水生环境的N固定是被经常注意，但还未被人们所了解的现象。单宁与N的相互关系（特别是与基本氨基酸的关系）是研究N固定的一个关键领域，通过单宁的分子结构、氨基酸的数据和^{13}C-NMR的数据，提供凋落物腐殖化过程N固定的分子水平的研究证据[2]。

当维管束植物材料在水环境的分解过程中，研究者一直认为最初的淋溶阶段伴随着N含量的减少，而研究表明，淋溶过程伴随着N含量绝对增高[1，27，28]，N增高来源的困惑在以下水环境中进行探究，包括海岸盐沼[29~33]、红树林[1，31~34]、溪[35]、河[28，36]。绝大多数调查者都推崇微生物作用的N固定，而直接的分子或同位素证据却没有。许多研究表明，分解物中与微生物生物量直接相关的N仅占总N的小部分。N在分解残留物中存在的主要形式有植物蛋白、微生物蛋白和植物成分与微生物成分结合而成的复杂含N缩合物[37]。Rice [31]，Rice 等[33], Poutanen 等[38], Melillo 等 [28], Harrison [37] 都认为，凋落物分解中N含量的提高是一种生物物理化学过程，微生物产生胞外酶，这些酶引起诸如多酚化合物和木质纤维等大分子的降解，降解后的部分产物（如“反应性酚类物质”）与含N化合物（主要来自微生物）缩合，这种降解—缩合过程贯穿于整个分解期，最终形成富含N、稳定而不易被分解的大分子化合物，从而导致分解残留物中N含量的相对量或绝对量的提高。Bradley 等 [39]研究认为，缩合单宁通过结合和隐蔽有机N而减少了矿化N的非生物循环。Odum 等曾经研究发现，陆地、淡水和河口维管植物分解叶总N中高达30%的N为非蛋白N，并认为这些N以难分解、稳定的几丁质或腐殖酸类化合物形式存在。Bradley 等[39]研究表明缩合单宁对腐殖质中N循环有负效应。对红树林湿地凋落物分解过程中N含量提高机理的探讨，将成为进一步开展红树林凋落物分解研究的努力方向。

红树林湿地代表了一种重要的连接陆地和海洋的生态系统。掉落的红树叶是一种重要的对于河口食物网提供碳、氮和其他营养的重要的来源。应用液态的^{13}C-NMR（碳13-核磁共振）和Folin—Denis分析，Benner 等[1]对大红树（*Rhizophora mangle*）研究表明，大量的大红树叶的碳是存在于单宁中。

Hernes 等[2]认为，关于单宁的生物地球化学循环的进展就好像30年前对木质素的研究一样。它们的相似性很大的：都是酚类、都是聚合物、都是维管束植物所特有；并且都已经对天然产物进行了很好的研究。但是它们的差异也是十分明显的：木质素的聚合物以任意方式高度分支，而单宁大分子则通常对应于它们的功能，靠着一些已经推测的锁－钥匙的机制高度地结构化[7]；木质素在木材组织中占主导，而单宁则通常在叶片（包括针叶）里面含量丰富；而且，木质素是相当稳定的，而单宁趋向于许多反应，这后一点是非常重要的：单宁在某种程度上可以提供一个潜在的去研究有机物质反应过程，而这是木质素，即结构性多糖做不到的[30]。

6 红树林单宁的化感作用

单宁还是一种重要化感作用物质，化感作用对红树植物群落的种类组成以及群落演变都具有重要的生态意义。单宁是一种能与蛋白质和其他化合物形成交联能力的酚化合物。许多森林植物都产生叶单宁，这种单宁凌乱地渗出到森林地面上，进而影响凋落物的降解率、腐殖质的形成、N的循环和最终的植物营养。Schimel 等[40]从*Populus balsamifera* L.中分离出次生代谢物质（小分子质量的酚和单宁），并且发现小分子质量的酚促进土壤的呼吸而单宁却抑制土壤的呼吸。一些陆地森林研究表明，*Kalmia* 属植物可能通过产生单宁而影响腐殖质中营养的循环而引起云杉生长的非生物性阻碍[41]; *Kalmia* 属植物控制的地域是引起松类植物幼苗生长抑制的原因[39，42]。

在环境胁迫下，植物通过释放化感作用物质的方式抑制周围其它植物的生长，从而增强其对养分、水分的竞争力；另一方面，某些化感物质（如酚类、酸类物质）有助于吸收N、P以及金属离子等营养成分[43，44]，提高抗逆性等生理作用（如多胺的增加，能显著提高植物的抗寒力），从而增加植物在逆境条件下的相对竞争力，对其他植物产生间接的抑制作用，后者更可能是环境胁迫条件下，植物化感物质增多，化感作用增强的主要原因。

7 结语

实际上，由于单宁的测定具有分析上的挑战性，大量的关于单宁的文献集中在天然产物的单宁分子结构研究，而关于单宁在红树植物的代谢过程及其在沉积物中的地球化学循环研究是红树林研究的前沿领域，国际上正开始相关研究。开展红树林此领域的研究不仅具有重要的理论意义，而且对红树林区的渔业和水产养殖业也具有特别重要的实践价值。

参考文献:

[1] Benner R, Weliky K, Hedges J I. Early diagenesis of mangrove leaves in a tropical estuary: Molecular-level analyses of netural sugars and lignin-derived phenols [J]. **Geochimica et Cosmochimica Acta**, 1990, 54: 1 991-2 001.

[2] Hernes P J, Benner R, Cowie G L, *et al.* Tannin diagensis in mangrove leaves from a tropical estuary: A novel molecular approach[J]. **Geochimica et Cosmochimica Acta**, 2001, 65:3 109-3 122.

[3] 孙达旺. 植物的单宁化学[M]. 北京: 中国林业出版社, 1992.

[4] 石碧, 狄莹. 植物多酚[M]. 北京: 科学出版社, 2000.

[5] Haslam E. Plant Polyphenols-Vegetable Tannins Revisited [M]. Cambridge: Cambridge University Press, 1989.

[6] Hemingway R W, Karchesy J J. Chemistry and Significance of Condensed Tannins[M]. New York and London: Plenum Press, 1989.

[7] Zucker W V. Tannins: Does structure determine function? An ecological perspective[J]. **Am Nat**, 1983, 121:335-365.

[8] Goni M A, Hedges J I. Potential applications of cutin-derived CuO reaction products for discriminating vascular plant sources in natural environments[J]. **Geochim Cosmochim Acta**, 1990, 54:3 073-3 081.

[9] Ellis C J, Foo L Y, Porter L J. Enantiomerism: A characteristic of the proanthocyanidin chemistry of the monocotyle-donae[J]. **Phytochemistry**, 1983, 22:483-487.

[10] Okuda T, Yoshida T, Hatano T. Hydrolyzable tannins and related polyphenols[J]. **Prog Ch Org Nat Prod**, 1995, 66:1-117.

[11] Salminen J P, Ossipov V, Haukioja E, *et al.* Seasonal variation in the content of hydrolysable tannins in leaves of *Betula pubescens*[J]. **Phytochemistry**, 2001, 57:15-22.

[12] Hedges J I, Weliky K. Diagenesis of conifer needles in a coastal marine environment[J]. **Geochim Cosmochim Acta**, 1989, 53:2 659-2 673.

[13] 林鹏, 傅勤. 中国红树林环境生态及经济利用[M]. 北京: 高等教育出版社, 1995.

[14] Waterman P G, Mole S. Analysis of phenolic plant metabolites[M]. Oxford: Blackwell Scientific Publications, 1994.

[15] Mueller-Harvey I. Analysis of hydrolysable tannins [J]. **Animal Science and Technology**, 2001, 91: 3-20.

[16] Hernes P J, Hedges J I. Tannin geochemistry of natural systems: Method development and application. Ph.D. thesis[M]. Seattle, WA: University of Washington,1999.

[17] Schofield P, Mbugua D M, Pell A N. Analysis of condensed tannins: a review[J]. **Animal Science and Technology**, 2001, 91:21-40.

[18] Hyder P W, Fredrickson E L, Estell R E, *et al.* Distribution and concentration of toatal phenolics, condensed tannins, and nordihydroguaiaretic acid (NDGA) in creosotebush (*Larrea tridentata*) [J]. **Biochemical Systematics and Ecology**, 2002,30: 905-912.

[19] Giner-Chavez B, Van Soest P J, Robertson J B, *et al.* A method for isolating condensed tannins from crude

plant extracts with trivalent ytterbium[J]. **J Sci. Food Agric**, 1997, 74:359-368.

[20] Makkar H P S, Gamble G, Becker K. Limitation of the butanol-hydrochloric acid-iron assay for bound condensed tannins[J]. **Food Chemistry**, 1999, 66: 129-133.

[21] Palmer B, Jones R J, Wina E, *et al.* The effect of sample drying conditions on estimates of condensed tannin and fibre content, dry matter digestibility, nitrogen digestibility and PEG binding of *Calliandra calothyrsus*[J]. **Animal Feed Science and Technology**, 2000, 87: 29-40.

[23] Chavan U D, Shahidi F, Naczk M. Extraction of condensed tannins from beach pea (*Lathyrus maritimus* L.) as affected by different solvents[J]. **Food Chemistry**, 2001,75:509-512.

[24] 王宗训. 中国资源植物利用手册[M]. 北京:中国科学技术出版社, 1989.

[25] Hsu F L, Nonaka G I, Nishioka I. Tannins and related compounds. XXXI. Isolation and characterization of proanthocyanidins in *Kandelia candel* (L) [J] .**Druce Chem Pharm Bull**, 1985, 33:3 142-3 152.

[26] Lin P, Fu Q. Environmental Ecology and Economic Utilization of Mangroves in China[R]. China Higher Education Press Beijing and Springer-Verlag Berlin Heidelberg, 2000.

[27] Kovacs J M. Assessing mangrove use at the local scale [J]. **Landscape and Urban Planning**, 1999, 43: 201-208.

[28] Fell J W, Master I M. The association and potential role of fungi in mangrove detrital systems[J]. **Bot Mar**, 1980, 23:257-263.

[29] Melillo J M, Naiman R J, Aber J D, *et al.* Factors controlling mass loss and nitrogen dynamics of plant litter decaying in northern Streams[J]. **Bull Mar Sci**, 1984, 35:341-356.

[30] White D S, Howes B L. Nitrogen incorporation into decomposing litter of *Spartina alterniflora*[J]. **Limnol Oceanogr**, 1994, 39:133-140.

[31] Benner R, Fogel M L, Sprague E K. Diagenesis of belowground biomass of *Spartina alterniflora* in saltmarsh sediments[J]. **Limnol Oceanogr**, 1991, 36: 1 358-1 374.

[32] Rice D L. The detritus nitrogen problem: new observations and perspectives from organic geochemistry [J]. **Mar Ecol Prog Ser**, 1982, 9:153-162.

[33] Rice D L, Tenore K R. Dynamics of carbon and nitrogen during the decomposition of detritus derived from estuarine macrophytes[J]. **Estuar Coast Shelf**, 1981, 13:681-690.

[34] Rice D L, Hanson R B. A kinetic model for detritus nitrogen: Role of the associated bacteria in nitrogen accumulation[J]. **Bull Mar Sci**, 1984, 35:326-340.

[35] Zieman J C, Macko S A, Mills A L. Role of sea grasses and mangroves in estuarine food webs: Temporal and spatial changes in stable isotope composition and amino acid content during decomposition[J]. **Bull Mar Sci**, 1984, 35:380-392.

[36] Qualls R G. The role leaf litter nitrogen immobilization in the nitrogen budget of a swamp stream[J]. **J Environ Qual**, 1984, 13:640-644.

[37] Bowden W B. Nitrification, nitrate reduction, and nitrogen immobilization in a tidal freshwater marsh sediment [J]. **Ecology**, 1986, 67:88-99.

[38] Harrison P G. Detrital processing in seagrass systems: a review of factors affecting decay rates, remineralization and detritivory[J]. **Aquat Bot**,1989, 23:263-288.

[39] Poutanen E L, Morris R J. A study of the formation of high molecular weight compounds during the decomposition of a field diatom population[J]. **Estuar Coast Shelf Sci**, 1983, 17:189-196.

[40] Bradley R L, Titus B D, Preston C P. Changes to mineral N cycling and microbial communities in black spruce humus after additions of $(NH_4)_2SO_4$ and condensed tannins extracted from *Kalmia angustifolia* and balsam fir [J]. **Soil Biology & Biochemistry**, 2000, 32:1 227-1 240.

[41] Schimel J P, Van Cleve K, Cates R G, *et al.* Effects of balsam poplar (*Populus balsamifera*) . tannins and low molecular weight phenolics on microbial activity in taiga floodplain soil: implications for changes in N cycling during succession[J]. **Canadian Journal of Botany**, 1996,74:84-90.

[42] Kuiters A T. Role of phenolic substances from decomposing forest litter in plant-soil interactions[J]. **Acta Botanica Neerlandica**, 1990, 39:329-348.

[43] Bradley R L, Titus B D, Fyles J W. Nitrogen acquisition and competitive ability of *Kalmia angustifolia* L , paper birch (*Betula papyrifera* Marsh) and black spruce (*Picea mariana* (Mill.)B.S.P.) Seedlings grown on different humus forms[J]. **Plant and Soil**, 1997, 195:209-220.

[44] Chapin F S III. New cog in the nitrogen cycle[J]. **Nature**, 1995, 377:199-200.

[45] Northup R R, Zengshou Y, Daligren R A, *et al.* Polyphenol control of nitrogen release from pine litter[J]. **Nature**, 1995, 377:227-229.

Effects of Adduct Ions on Matrix-assisted Laser Desorption/Ionization Time of Flight Mass Spectrometry of Condensed Tannins: A Prerequisite Knowledge*

Xiang Ping[1] Lin Yiming[1,2] Lin Peng[1,2] Xiang Cheng[1]

(1. *Department of Biology, School of Life Sciences, Xiamen University, Xiamen* 361005, *China*;
2. *Research Center for Wetlands and Ecological Engineering, Xiamen University, Xiamen* 361005, *China*)

Abstract: Three different condensed tannins were analyzed by MALDI-TOF MS, with Cs^+ and Na^+ being the cationization reagents. When Na^+ was used as the cationization reagent, good quality MALDI-TOF spectra of condensed tannins were obtained, accompanied by the possibility of overevaluation of the gallocatechin/epigallocatechin units because of the influence of the remaining K^+. It is critical to investigate the appearance of gallocatechin/epigallocatechin units in condensed tannins from special plant tissues for taxonomic information and for the better understanding of plant metabolism and other ecological properties. Cs^+ is a suitable substitute for this situation. The more complicated MALDI-TOF spectrum of the complex condensed tannins from *Bruguiera gymnorrhiza* leaves was detected by the selection of Cs^+ as the cationization reagent in the presence of Na^+ and K^+. Using MALDI-TOF without deionization or with deionization and selection of Cs^+ as the cationization reagent rather than with deionization and selection of Na^+, condensed tannin polymers of higher polymerization degree (PD) were observed. Meanwhile, the polymer with the highest intensity ion peak changed with the ion adducts used.

Key Words: Condensed tannins; Matrix-assisted laser desorption/ionization time-of-flight mass; Flavan-3-Ol unit; Polymerization degree

1 Introduction

Condensed tannins are a group of typical polyphenolic compounds ranging in molecular weight from 500 to 3000 Daltons[1]. In addition to preventing damage from herbivory, they play an important role in a number of ecological processes, including litter decomposition, nutrient cycling, nitrogen sequestration, microbial activity, humic acid formation, metal complexation, and pedogenesis[2]. To understand the effects of condensed tannins on the physiological and ecological processes in plants, it is a prerequisite to analyze the chemical structure of condensed tannins. Because of its soft ionization energy and high ion transmission yield, Matrix-assisted laser desorption/ionization time-of-flight mass spectrometry (MALDI-TOF MS) has gained wide acceptance for the detection of biomacromolecule and synthetic polymers and it has been shown to be an available method for the characterization of polydispersed vegetable tannins[3,4].

When MALDI-TOF MS is used to characterize vegetable tannins, in addition to the matrix, solvent, and the preparation of samples, the mass spectra of condensed tannins are notably affected by naturally abundant Na^+ or K^+ and the added single charge ions[5]. Selection of an inappropriate ion as the cationization reagent for MALDI makes it difficult to accurately characterize the condensed vegetable tannins using MALDI-TOF MS from special plant tissues to obtain taxonomic information and other ecological properties. When the results obtained from the MALDI-TOF MS spectrum cannot be confirmed using additional methods, the effects of

* From Chinese Journal of Analytical Chemistry, 2006, 34(7): 1019－1222

adduct ions on MALDI-TOF MS of condensed tannins must be considered. In this study, three different condensed tannins were chosen and the adduct ions effects on the MALDI-TOF MS spectra of condensed tannins were investigated. The condensed tannins/matrix mixtures were either deionized and spiked with a solution containing Na^+ or Cs^+ or applied directly to a steel target.

2 Experimental

2.1 Instrument and reagents

Bruker Reflex III MALDI-TOF Mass Spectrometer was purchased from Bruker Daltonics (Bremen, Germany). Rotatory evaporator and Freeze drier were purchased from Beijing Detianyou Technology Development Co. Ltd, China. All solvents including acetone, methanol, and chloroform were of analytical reagent grade. Amberlite IRP-64 cation-exchange resin was purchased from Sigma-Aldrich and Sephadex LH-20 was from Amersham. Water was purified on a Millipore Milli-Q apparatus. Condensed tannins were extracted from the hypocotyls of *Kandelia candel,* and flowers and leaves of *Bruguiera gymnorrhiza* and were purified on Sephadex LH-20.

2.2 Methods

2.2.1 Sample preparation

Mature hypocotyls of *K. candel* and leaves and flowers of *B. gymnorrhiza* were collected from a mangrove forest in the Jiulong River Estuary (24°24′N, 117°55′E), Fujian, China. All samples were taken to the laboratory immediately after collected and cleaned with distilled water. Fresh materials were weighed and extracted thrice with 7:3 (v/v) acetone/water solution at 5℃. After each extraction, the samples were centrifuged and the supernatant was combined. The acetone was completely removed by rotary evaporation at 30℃. The tannin-containing aqueous solution was filtered to remove the non-tannin debris, followed by extraction thrice with 300 ml hexane and chloroform to remove lipids, pigments, and monomer phenols. The extracted residue was freeze-dried, resolubilized with a small amount of methanol/water (50:50), and purified using chromatography on Sephadex LH-20[6,7]. The purified tannins were freeze-dried and stored at –20℃ before for analysis by MALDI-TOF MS.

2.2.2 MALDI-TOF MS

The spectra were recorded on a Bruker Reflex Ⅲ. The irradiation source was a pulsed nitrogen laser with a wavelength of 337 nm, and the duration of the laser pulse was 3 ns. In the positive reflectron mode, an accelerating voltage of 20.0 kV and a reflectron voltage of 23.0 kV were used. The spectra of condensed tannins were obtained from a sum of 100–150 shots and were calibrated using Angiotensin II (1046.5 MW), Bombesin (1619.8), ACTHclip18–39 (2465.2), and Somatostatin28 (3147.47 MW) as external standards. 2,5-Dihydroxy benzoic acid (DHB, 10 mg ml^{-1} aqueous solution) was used as the matrix. The sample solutions (7.5 mg/ml aqueous) were mixed with the matrix solution at a volumetric ratio of 1:3. The mixture (1 μl) was applied to the steel target. Amberlite IRP-64 cation-exchange resin (Sigma-Aldrich), equilibrated in deionized water, was used to deionize the analyte/matrix solution thrice. NaCl (1 mg ml^{-1}) or cesium trifluoroacetate (1 mg ml^{-1}) was mixed with the analyte/matrix solution at the 1:3 volumetric ratio to promote the formation of a single type of ion adduct ($[M+Na]^+$ or $[M+Cs]^+$)[4,8].

3 Results and discussion

3.1 Assigning structures to MALDI-TOF MS

Condensed tannins are polydispersed oligomers and polymers of flavan-3-ol, which are linked by C–C bonds most commonly between positions 4 and 8, but may also involve positions 4 and 6 of the monomer. In addition to the usual C4–C8 or C4–C6 bond, another infrequent type of linkage is that between the C2 of the upper unit and the oxygen-bearing C7 of the lower unit (A-type linkage). The catechin/epicatechin (C/EC) and gallocatechin/epigallocatechin (GC/EGC) are the most common flavan-3-ol units occurring in condensed tannins. GC/EGC, besides the two hydroxyl groups at positions 3′ and 4′ of the B ring as the C/EC, contains an additional hydroxyl group at position 5′ of the B ring.

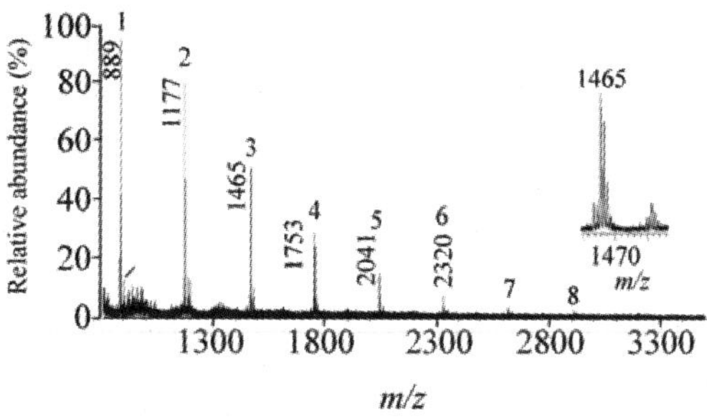

Fig. 1 MALDI-TOF mass spectrum $[M+Na]^+$ of condensed tannins extracted from hypocotyls of *K. candel* and the enlarged spectrum of the polyflavan-3-ol pentamer

In the case of a condensed tannin mixture from hypocotyls of *K. candel* (Fig. 1), the following equation was applied to predict the mass distribution and the flavan-3-ol unit composition, $M + Ca^+ = Ca + 2.0 + 288.0\ n + 304.0\ m$ [4], where n is the number of C/EC, m is the number of GC/EGC units, Ca corresponds to the molecular weight of the added cations (m/z 23 to Na^+), m/z 2.0 is the number of H in the end groups (corresponding to 2), m/z 288.0 corresponds to the molecular weight of one C/EC unit, m/z = 304.0 to GC/EGC unit, and m + n is the degree of polymerization (DPs) contributed by the repeated C/EC and GC/EGC units occurring in oligomers and polymers of condensed tannins.

It can be seen from Fig. 1 that clear spectra of the condensed tannins from *K. candel* hypocotyls were obtained, showing the oligomers and polymers of building units from trimer(m/z 889.3, 905.3) to decamer (m/z 2905.6, 3011.6) and the oligomer or polymer series with masses of the repeat units of 288.0 Da corresponding to a mass difference of one C/EC between each oligomer or polymer. For each multiplet, a substructure with mass increments of 16 Da appears. These masses have been identified as heteropolymers of repeating flavan-3-ol units containing an additional hydroxyl group (Δm 16 Da) at the 5′ position of the B-ring as GC/EGC. It can be seen from the spectra that most masses are homopolymers consisting of three to ten C/EC units in trimer to decamer. The heteropolymers containing only one GC/EGC unit were also detected with low intensity.

3.2 Na⁺ as the cationization reagent associated with overevaluation of the GC/EGC units

Another explanation for the detection of 16 Da higher in the mass spectrum of *K. candel* hypocotyls was that the signal from each condensed tannin oligomer and polymer was split and detected as both $[M + Na]^+$ and $[M + K]^+$. The atomic mass difference between Na^+ and K^+ (Δm = 15.9739) is almost equal to the molecular weight difference of C/EC and GC/EGC differing by one hydroxyl group substitution (Δm = 15.9949)[4].

Fig. 2 A, B, and C show the enlarged MALDI-TOF mass spectra (pentamer) of the condensed tannins from the hypocotyls of *K. candel* in the case of no deionization or addition of cations, deionization and addition of Na^+, and deionization and addition of Cs^+, respectively.

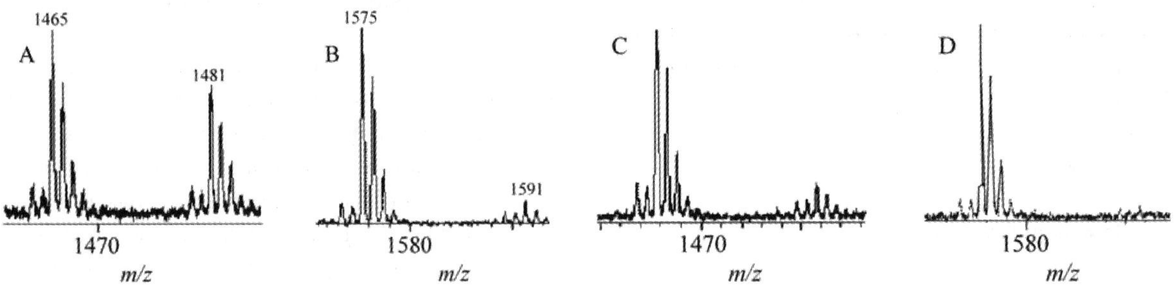

Fig. 2 The enlarged MALDI-TOF mass spectra of the polyflavan-3-ol pentamer in condensed tannins purified from *K. candel* hypocotyls (A, B and C) or *B. gymnorrhiza* flowers (D)

A. no-deionization or addition of cation to the matrix/analyte before deposition on the target; B. deionization and addition of Na^+; C, D. deionization and addition of Cs^+

According to the spectrum in the case of deionization and addition of Cs^+ (Fig. 2C), the condensed tannins from the hypocotyls of *K. candel* contain a minimal amount of heteropolymers, whereas in the case of no deionization or addition of cation, the mass spectra (Fig. 2A) are notably affected by naturally abundant K^+ ($[M + K]^+$ formation), which leads to the overevaluation of the heteropolymers (as m/z 1481 in pentamer) in each multiplet. This problem can be partly solved by deionization and selection of Na^+ as the cationization reagent for MALDI (as shown in Fig. 2B).

In the case of condensed tannins from *B. gymnorrhiza* flowers, no heteropolymers containing the GC/EGC unit were detected by deionization and addition of Cs^+ (Fig. 2D). The analyte/matrix was cautiously deionized and Na^+ was added, but the mass spectrum of the condensed tannins from *B. gymnorrhiza* flowers was similar to that of the hypocotyls of *K. candel*. According to the mass spectrum by deionization and addition of Na^+, an inaccurate result regarding the condensed tannins from *B. gymnorrhiza* flowers containing the GC/EGC unit will be obtained. Hence, it is critical to investigate the appearance of GC/EGC units in condensed tannins from special plant tissues for taxonomic information and for the better understanding of plant metabolism and other ecological properties.

To solve the problem of cationization with Na^+ and K^+, Krueger *et al.* selected cesium trifluoroacetate as the cationization reagent for MALDI. However, some researchers did not consider the potential problem of the mass spectrum of condensed tannins being notably affected by naturally abundant K^+ [3,9,10].

3.3 Cs⁺ as cationization reagent associated with the quality of MALDI-TOF MS spectrum

Condensed vegetable tannins are diverse in structure and vary with plant species and tissues. The complexity of individual condensed tannins results from the increase in the number of heteropolymers and substitutions with aliphatic or carbohydrate moieties and makes it difficult to characterize these condensed tannins by MALDI-TOF MS.

In the case of condensed tannins from *B. gymnorrhiza* leaves, a more complicated mass spectrum was obtained by deionization and addition of Cs^+ (Fig. 3C). In this spectrum, the series of $[M+Na]^+$ and $[M+K]^+$ molecular ions appear because the naturally abundant Na^+ and K^+ in various vessels could not be completely removed. It was difficult to obtain the accurate structure information including the number of A-type linkage and the number of glycosides of the condensed tannins from the complex mass spectra. However, when the analyte/matrix mixture was deionized and Na^+ was added, a high quality MALDI-TOF spectrum of the condensed tannins from *B. gymnorrhiza* leaves was obtained (Fig. 3B). The mass spectrum without deionization or addition of cations provided less accurate information (Fig. 3C). Hence, it is recommended that two MALDI-TOF spectra obtained by deionization and addition of Cs^+ and by deionization and addition of Na^+ be compared to accurately characterize the complex condensed tannins such as those from *B. gymnorrhiza* leaves.

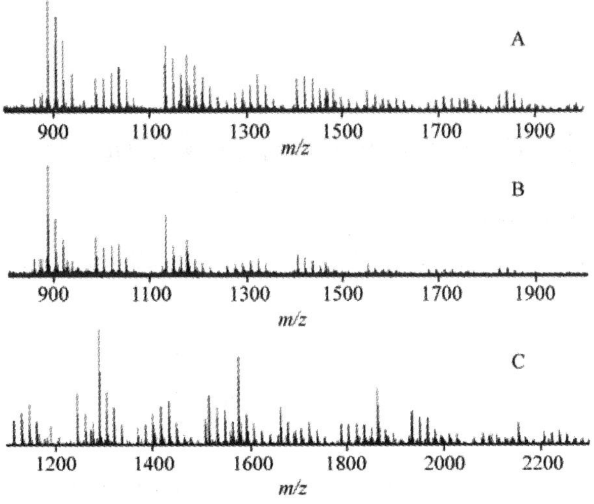

Fig. 3 Partly enlarged MALDI-TOF mass spectra of complex condensed tannins purified from *B. gymnorrhiza* leaves

A. no-deionization or addition of cation to the matrix/analyte before deposition on the target; B. deionization and addition of Na^+;
C. deionization and addition of Cs^+

3.4 Effects of adduct ions on the measurement of mass distribution and DPs

The mass distribution and DPs are important structural properties of condensed tannins. MALDI-TOF MS is an attractive method to measure the mass distribution and DPs of condensed tannins[3]. With deionization and addition of Na^+, the MALDI-TOF mass spectra of condensed tannins from *B. gymnorrhiza* leaves indicated the presence of oligomers to the maximum of hexamers, and polymers more than 1900 Da were not detected (Fig. 3B). The octamers still appeared on the mass spectrum of these condensed tannins in the case of no deionization or addition of cations, and deionization and addition of Cs^+ (Fig. 3A, C). The oligomers with highest intensity were tetramers in MALDI-TOF by deionization and addition of Cs^+ but were trimers by selection of Na^+ as the adduct ion.

Similar results were observed from the MALDI-TOF mass spectra of condensed tannins from *K. candel* hypocotyls and *B. gymnorrhiza* flowers. These phenomena suggest that it is still a challenge to measure the mass distribution and DPs of condensed tannins by MALDI-TOF MS. However, under conditions of identical sample preparation and selection of the same adduct ion, MALDI-TOF MS should be a useful strategy for measuring the mass distribution and DPs of condensed tannins.

References

[1] Shi B, Di Y. *Plant Polyphenols*. Beijing: Science Press, **2000**: 5

[2] Kraus T E C, Dahlgren R A, Zasoski R J. *Plant. Soil,* **2003**, 256: 41–66

[3] Ohnishi-kameyama M, Yanagida A, Kanda T, Nagata T. *Rapid Commun. Mass Spectrom.,* **1997**,11: 31–36

[4] Krueger C G, Vestling M M, Reed J D. *J. Agric. Food Chem.,* **2003**, 51: 538–543

[5] He M Y. *Modern Organic and Biologic Mass Spectrum*, Beijing: Beijing University Press, **2002**: 175–197

[6] Foo L Y, Porter L J. *J. C. S. Perkin.,* **1978**, I : 1186–1190

[7] Maie N, Behrens A, Knicker H. *Soil Biol. Biochem.*, **2003**, 35: 577–589

[8] Krueger D G, Dopke N C, Treichel P M, Folts J, Reed J D. *J. Agric. Food Chem*., **2000**, 48: 1663–1667

[9] Maie N, Behrens A, Knicker H, Kögel-Knabner I. *Soil Biol. Biochem.*, **2003**, 35: 577–589

[10] Rösch D, Mügge C, Fogliano V, Kroh L W. *J. Agric. Food Chem.*, **2004**, 52: 6712–6718

药用红树植物老鼠簕(*Acanthus ilicifolius*)乙醇提取物的化学成分研究*

张亮亮[1] 王湛昌[1] 陈俊德[1] 林 鹏[1,2] 杨志伟[1] 林益明[1,2]

(1. 厦门大学生命科学学院生物学系,福建 厦门 361005;
2. 厦门大学湿地与生态工程研究中心,福建厦门 361005)

摘 要:**目的** 探讨老鼠簕的化学成分。**方法** 将老鼠簕全株粉碎后,用95%乙醇浸提3次,合并浓缩,浸膏的石油醚部位经反复硅胶柱层析分离,结合波谱学数据鉴定化合物结构。**结果** 从石油醚部位分离得到8个化合物,分别为β-谷甾醇(Ⅰ)、豆甾醇(Ⅱ)、正十六烷酸(Ⅲ)、正二十八烷酸(Ⅳ)、4,22-二烯-3-酮豆甾烷(stigmasta-4,22-dien-3-one)(Ⅴ)、4-烯-3-酮豆甾烷(stigmast-4-en-3-one)(Ⅵ)、5,22-二烯-7-酮-3β-羟基豆甾烷(3β-hydroxystigmasta-5,22-dien-7-one)(Ⅶ)、2-唑啉酮(2-benzoxazolinone)(Ⅷ)。**结论** 其中Ⅴ,Ⅵ,Ⅶ为首次从该植物中分离得到。

关键词:红树植物;老鼠簕;化学成分;2-苯并噁唑啉酮

中图分类号:R284.2,Q949 **文献标识码**:A **文章编号**:1002-3461(2007)06-0005-05

Studies on chemical constituents in ethanolic extract from *Acanthus ilicifolius* as a pharmaceutic mangrove

ZHANG Liang-liang[1], WANG Zhan-chang[1], CHEN Jun-de[1], LIN Peng[1,2], YANG Zhi-wei[1], LIN Yi-ming[1,2]

(1. Department of Biology, School of Life Sciences, Xiamen University, Xiamen 361005, *China;* 2. *Research Centre for Wetlands and Ecological Engineering, Xiamen University, Xiamen* 361005, *China)*

Abstract:**Objective** To study the chemical constituents of *Acanthus ilicifolius*. **Methods** The air-dried and powdered plant material was extracted with 95% ethanol. Silica gel and Sephadex LH-20 column chromotography were used in the isolation from the ethanol extracts of the whole plant, the compounds were determined on the basis of various modern spectroscopic analyses and physical constants. **Results** Eight compounds were isolated from the petroleum ether soluble portion, identified as β-sitosterol(Ⅰ), stigmasterol (Ⅱ), n-hexadecane acid (Ⅲ), n-octacosanic acid(Ⅳ), stigmasta-4,22-dien-3-one(Ⅴ), stigmasta-4-en-3-one (Ⅵ), 3β-hydroxystigmasta-5,22-dien-7-one (Ⅶ) and 2-benzoxazolinone (Ⅷ). **Conclusion** Compound Ⅴ, Ⅵ, Ⅶ were obtained from the genus for the first time.

Key words:mangrove plant;*Acanthus ilicifolius*;chemical constituents;2-benzoxazolinone

老鼠簕(*Acanthus ilicifolius*)民间又称 老鼠怕,软骨牡丹,为爵床科老鼠簕属一种带

* 国家自然科学基金(30671646);福建省科技计划资助项目(2003Y036)
原载于中国海洋药物杂志,2007,26(6):5—9

刺的药用红树植物，在我国主要分布于广东、广西、福建、海南和台湾沿海地区[1]。该植物为直立稍分枝灌木，茎圆柱形，高 0.5～1.5m，淡绿色。老鼠簕是国际上研究较为深入的药用红树植物之一。在我国及亚太地区其它国家的红树林海岸居民都有传统的药物利用经验。泰国学者 Kokpol[2]研究报道，在老鼠簕的根提取液中发现的成分如苯并噁唑啉(benzoxazoline-2-ene)因其对中枢神经系统具有抑制作用，可用作止痛药、退热剂、抗惊厥药和安眠药及具有肌肉松弛活性；苯并噁唑啉能抗真菌病害，这种糖的核糖衍生物具有抗癌和抗病毒活性；鼠类试验中证明老鼠簕的根具有抗白血病的作用。Kanchanapoom 等[3,4]从老鼠簕的地上部分分离得到多种化学成分。Babu 等[5]从老鼠簕叶的提取物进行鼠类实验中发现具有抗肿瘤效应。迄今为止，从老鼠簕(*Acanthus ilicifolius*)中分离到了约 30 个化合物，主要有：2-喹啉羧酸、酮类化合物甲基芹菜素-7-O-β-D-葡萄糖醛酸苷(4 氢)-(7-chloro-(2R) 2-O-β-D-glucopyranosyl-2H-1，4benzoxazin-3(4H)-one)，三萜系列化合物三萜苷和三萜皂苷(齐墩果烷、羽扇豆醇、乌苏烷和乌苏酸等)。

为了深入探讨老鼠簕的化学成分，并从老鼠簕的醇提浸膏中分离获得活性物质，我们对老鼠簕醇提浸膏进行了化学成分分离和鉴定。

1 实验部分

1.1 样品

老鼠簕样品于 2005 年 7 月采自海南东寨港，种属由厦门大学生命科学学院侯学良博士鉴定，标本存放于厦门大学生命科学学院生物系标本馆。

1.2 仪器和试剂

熔点用 X-5 型显微熔点测定仪测定(温度未矫正)；质谱(ESI-MS)用 Bruker Bio TOF-Q 型质谱仪测定；核磁共振谱用 Unity 500MHZ 超导核磁共振仪(美国 Varian 公司)测定，TMS 作内标；柱色谱用 Sephadex LH 3/20 为 Pharmacia 公司产品；薄层色谱及柱色谱用硅胶均为青岛海洋化工厂产品；其他试剂均为分析纯。

1.3 提取分离

老鼠簕剪碎，取 4kg 用 95%乙醇浸泡 3 次，浸提液浓缩成浸膏(150g)分散于适量水中，依次用石油醚、乙酸乙酯、正丁醇萃取。石油醚部分(15g)经石油醚-乙酸乙酯梯度洗脱，馏分经反复硅胶柱层析，得化合物 Ⅰ(103mg)、Ⅱ(26.1mg)、Ⅲ(12.8mg)、Ⅳ(15.7mg)、Ⅴ(21.5mg)、Ⅵ(17.4mg)、Ⅶ(10.5mg)、Ⅷ(7.8mg)。

2 结构鉴定

化合物 Ⅰ 无色针状结晶，mp：136.0～138.2 ℃，用 TLC 检测，20%硫酸显色显紫红色，且多个溶剂系统均与β-谷甾醇对照品 R_f 值相同，Liebermann-Burchard 反应呈阳性。其光谱数据与文献[6]数据基本一致，鉴定为β-谷甾醇。

化合物 Ⅱ 无色针状结晶，mp：139～140 ℃，ESI-MS m/z：413 $(M+H)^+$，Liebermann-Burchard 反应呈阳性。^{1}H-NMR (500MHz，$CDCl_3$，TMS) δ：5.34 (1H，m，6-H)，5.14 (1H，m，22-H)，5.02 (1H，m，23-H)，3.52 (1H，m，3-H)，1.03 (3H，s，CH_3)，1.01 (3H，d，CH_3)，0.85 (3H，t，CH_3) 0.84 (3H，d，CH_3)，0.79 (3H，d，CH_3)，0.70 (3H，s，CH_3)。以上光谱数据与文献[7]数据基本一致，鉴定为豆甾醇。

化合物 Ⅲ 白色粒状固体，mp：62～63 ℃，ESI-MS m/z：257 $(M+H)^+$，分子式为 $C_{16}H_{32}O_2$。^{1}H-NMR (500MHz，$CDCl_3$，TMS) δ：0.88 (3H，t，J = 6.96Hz，16-H)，2.35 (2H，t，J = 7.74Hz，2-H)，2.02 (2H，m，3-H)，1.30 (4～15-H，m)；^{13}C-NMR

(125MHz, $CDCl_3$, TMS) δ: 14.1 (C-CH_3), 22.7(C-15), 24.7(C-14), 29.0(C-13), 29.3 (C-12), 29.4(C-11), 29.5(C-10), 29.6(C-9), 29.7(C-4～8), 31.9(C-3), 34.0(C-2), 179.7(COOH)。以上光谱数据与文献[8]数据基本一致,鉴定为正十六烷酸。

化合物 Ⅳ 无色粒状结晶, mp: 86～88 ℃, ESI-MS *m/z*: 425(M + H)$^+$, 分子式为 $C_{28}H_{56}O_2$。^{1}H-NMR (500MHz, $CDCl_3$, TMS) δ: 0.88(3H, t, 28-H), 1.26(4～27-H, m), 1.64(2H, m, 3-H), 2.35(2H, t, 2-H)。以上光谱数据与文献[9]数据基本一致,鉴定为长链脂肪酸正二十八烷酸。

化合物 Ⅴ 无色针状结晶, mp: 115～116 ℃, ESI-MS *m/z*: 411(M + H)$^+$, 分子式为 $C_{29}H_{46}O$。Liebermann-Burchard 反应呈阳性。^{1}H-NMR(500MHz, $CDCl_3$, TMS) δ: 5.72(1H, s, H-4), 5.14(1H, dd, J = 15.1, 8.8Hz, 22-H), 5.02(1H, dd, J = 15.1, 8.8 Hz, 23-H), 2.36(1H, m, 2a-H) 2.26(1H, m, 2b-H,), 2.01(1H, m, 1a-H), 1.53(1H, m, 1b-H)。1.18(3H, s, 19-CH_3), 1.01(3H, d, J = 6.6Hz, 21-CH_3), 0.84 (3H, d, J = 6.6 Hz, 27-CH_3), 0.80(3H, t, J = 7.4Hz, 29-CH_3), 0.79(3H, d, J = 7.3Hz, 26-CH_3), 0.72(3H, s, 18-CH_3)。^{13}C-NMR (125MHz, $CDCl_3$, TMS) 信号的归属见表 1, 以上光谱数据与文献[10]数据基本一致, 鉴定为 4,22-二烯-3-酮豆甾烷。

化合物 Ⅵ 白色粉末, mp: 174～176 ℃, ESI-MS *m/z*: 413(M + H)$^+$, 分子式为 $C_{29}H_{48}O$。Liebermann-Burchard 反应呈阳性。^{1}H-NMR (500MHz, $CDCl_3$, TMS) δ: 0.71 (3H, s, 18-H), 1.15 (3H, s, 19-H), 0.90 (3H, d, J = 6.5Hz, 21-H), 0.82(3H, d, J = 6.7Hz, 26-H), 0.83(3H, d, J = 6.7Hz, 27-H), 0.84(3H, d, J = 6.7Hz, 29-H), 5.72 (1H, s, 4-H)。^{13}C-NMR (125MHz, $CDCl_3$, TMS) 信号的归属见表 1. 以上光谱数据与参考文献基本一致[11]。所以确定该化合物为 4-烯-3-酮豆甾烷。

化合物 Ⅶ 无色针状结晶, mp: 148～151 ℃, ESI-MS *m/z*: 427(M + H)$^+$, 分子式为 $C_{29}H_{46}O_2$。Liebermann-Burchard 反应呈阳性。^{1}H-NMR (500MHz, $CDCl_3$, TMS) δ: 5.69(1H, s, 6-H), 5.17(1H, dd, J = 15.4, 8.8Hz, 22-H), 5.02(1H, dd, J = 15.4, 8.8 Hz, 23-H), 3.68(1H, m, 3-H), 2.51 (1H, m, 4a-H), 2.38(1H, m, 4b-H), 2.25(1H, dd, J = 12.4, 10.9Hz, 8-H), 1.34(1H, m, 9-H), 1.21 (3H, s, 19-CH_3), 1.02(3H, d, J = 6.6Hz, 21-CH_3), 0.85(3H, d, J = 6.6Hz, 27-CH_3), 0.81 (3H, t, J = 7.3Hz, 29-CH_3), 0.79(3H, d, J = 7.3Hz, 26-CH_3), 0.70(3H, s, 18-CH_3)。^{13}C-NMR (125MHz, $CDCl_3$, TMS) 信号的归属见表 1。以上光谱数据与参考文献基本一致[12]。所以确定该化合物为 5,22-二烯-7-酮-3β-羟基豆甾烷。

表 1 化合物 Ⅴ, Ⅵ和 Ⅶ的^{13}C-NMR(125MHz)数据

Tab 1 ^{13}C-NMR(125MHz) data for compounds Ⅴ, Ⅵ and Ⅶ

碳位	Ⅴ	Ⅵ	Ⅶ
1	35.6	35.7	36.3
2	33.0	33.9	31.1
3	199.7	199.8	70.5
4	123.7	123.7	41.8
5	171.7	171.8	165.1
6	32.9	33.0	126.1
7	32.0	32.1	202.4
8	35.6	35.6	45.4
9	53.8	53.8	49.9
10	38.6	38.6	38.3
11	21.0	21.0	21.2
12	35.6	39.6	38.5
13	42.2	42.4	42.9
14	55.8	55.9	50.0
15	24.2	24.2	26.4

碳位	V	VI	VII
16	28.9	28.2	29.1
17	55.9	56.0	54.6
18	12.2	12.0	12.2
19	17.4	17.4	17.3
20	40.5	36.1	40.3
21	21.0	18.7	21.4
22	138.1	34.0	138.1
23	129.4	26.0	129.4
24	51.2	45.8	51.2
25	32.0	29.1	31.9
26	19.0	19.8	18.9
27	21.0	19.0	21.1
28	25.4	23.1	25.4
29	12.2	12.0	12.2

化合物 VIII 白色粉末,mp:139～140 ℃,ESI-MS *m/z*:136(M + H)$^+$,分子式为 $C_7H_5O_2N$,^{1}H-NMR(500MHz,$CDCl_3$,TMS)δ:10.03(1H,br,NH),7.04-7.23(4～7-H,m);^{13}C-NMR(125MHz,$CDCl_3$,TMS)δppm:156.4(C-2),143.9(C-8),129.4(C-9),124.2(C-5),110.1(C-4),110.1(C-7),122.7(C-6)。以上光谱数据与文献[13,14]数据基本一致,鉴定为2-唑啉酮。

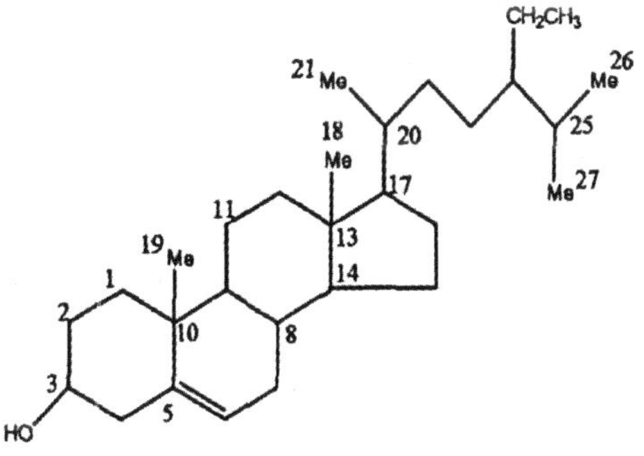

I. β-谷甾醇(β-sitosterol)

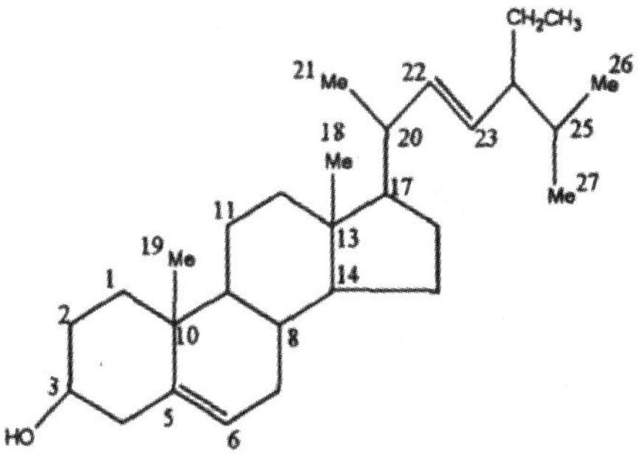

II. 豆甾醇(stigmasterol)

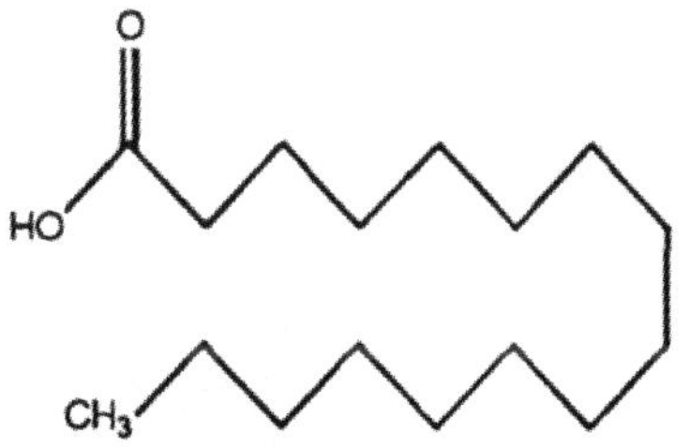

III. 正十六烷酸

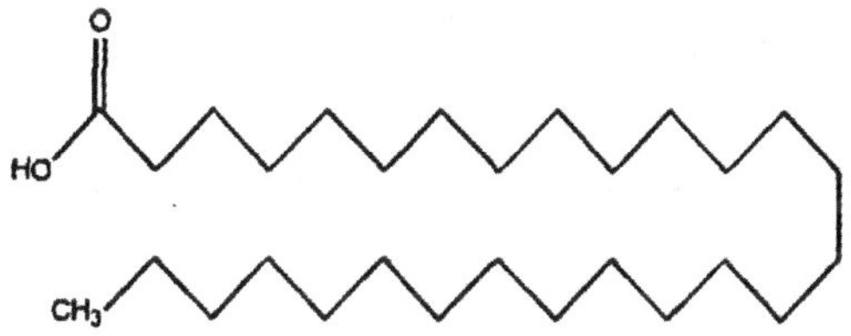

IV. 正二十八烷酸

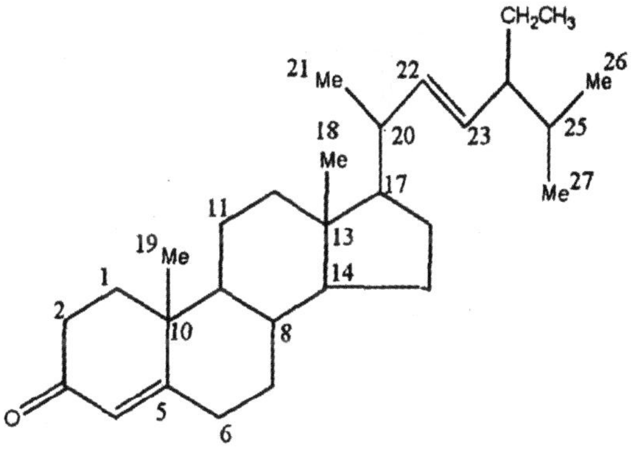

V. 4,22-二烯-3-酮豆甾烷(stigmasta-4,22-dien-3-one)

3 结论

从石油醚部位分离得到8个化合物,分别为β-谷甾醇(I)、豆甾醇(II)、正十六烷酸(III)、正二十八烷酸(IV)、4,22-二烯-3-酮豆甾烷(stigmasta-4,22-dien-3-one)(V)、4-烯-3-酮豆甾烷(stigmast-4-en-3-one)(VI)、5,22-二烯-7-酮-3β-羟基豆甾烷(3β-hydroxystigmasta-5,22-dien-7-one)(VII)、2-唑啉酮(2-benzoxazolinone)(VIII)。其中V、VI、VII为首次从该植物中分离得到。

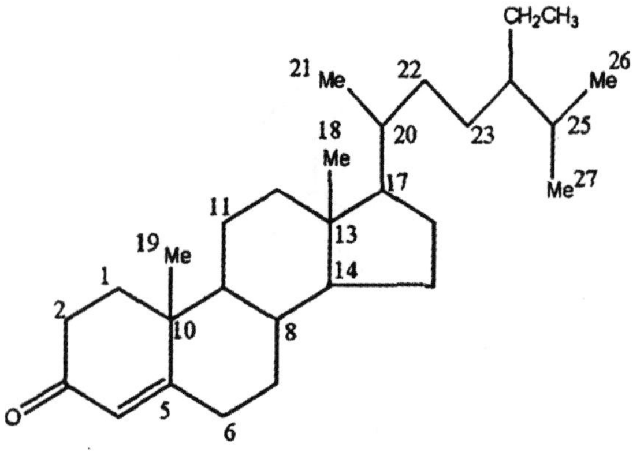

Ⅵ. 4-烯-3-酮豆甾烷(stigmasta-4-en-3-one)

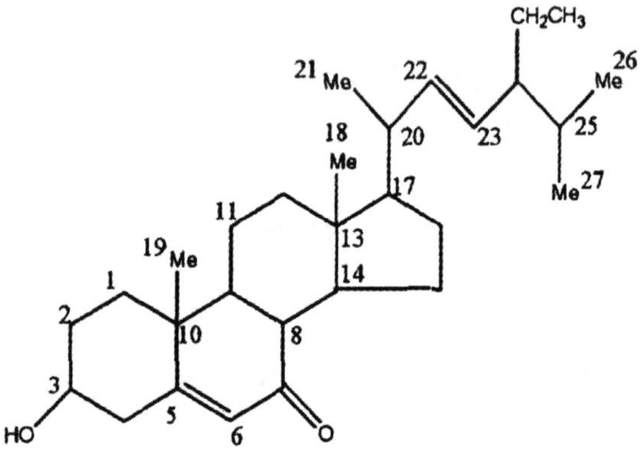

Ⅶ. 5,22-二烯-7-酮-3β-羟基豆甾烷
(3β-hydroxystigmasta-5,22-dien-7-one)

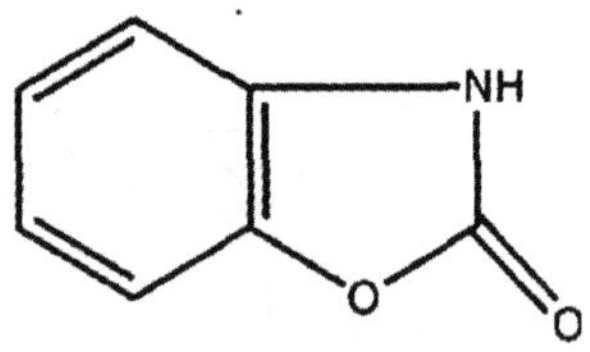

Ⅷ. 2-唑啉酮(2-benzoxazolinone)

图1 8个化合物的化学结构式

Fig 1 The chemical structures of the eight compounds

参考文献:

[1] 林鹏. 中国红树林生态系[M]. 北京:科学出版社,1997.

[2] Kokpol U. Chemistry of natural products from mangrove plants, UNDP/UNESCO, Training course on life history of selected species of flora and fauna in mangrove ecosystem[J]. *Thailand*,1985:159.

[3] Kanchanapoom T, Kamel MS, Kasai R, *et al*. Lignanglucosides from *Acanthus ilicifolius*[J]. *Phytochemistry*,2001,56(4):369.

[4] Kanchanapoom T, Kamel MS, Kasai R, *et al*. Benzoxazinoid glucosides from *Acanthus ilicifolius* [J]. *Phytochemistry*,2001,58(4):637.

[5] Babu BH, Shylesh BS, Padikkala J, *et al*. Tumour reducing and anticarcinogenic activity of *Acanthus ilicifolius* in mice[J]. *Journal of Ethnopharmacology*,2002,79(1):27.

[6] 袁珂,吕洁丽,贾安. 含羞草化学成分的研究[J]. 中国药学杂志,2006,41(17):1293.

[7] Kojima H, Sato N, hatano A, *et al*. Sterol glucosides from *Prunella vulgaris*[J]. *Phytochemistry*,1990,29(7):2351.

[8] 杨念云,钱士辉,段金廒,等. 野马追地上部分的化学成分研究(Ⅰ)[J]. 中国药科大学学报,2003,34(3):220.

[9] 向兰,范国强,郑俊华,等. 窄叶大黄非蒽醌类成分研究[J]. 中国中药杂志,2001,26(8):551.

[10] 刘安,田景奎,邹忠梅,等. 乌藤化学成分研究[J]. 中草药,2002,33(3):205.

[11] Elvira M M. Gaspar and highinald: steroidal constituent from Mature wheat Straw[J]. *Phytochemistry*,1993,34(2):523.

[12] Natro G, Piccialli V, Sica D. New steroidal hydroxyketones and closely related diols from the marine sponge *Cliona copiosa*[J]. *J Nat Prod*,1992,55:1588.

[13] Werner C, Hedberg C, Lorenzi-Riatsch, *et al*. Accumulation and metabolism of the spermine alkaloid, ahelandrine in roots of *Aphelandra tetragona*[J]. *Phytochemistry*,1993,33(5):1033.

[14] Udom K, Vallapa C. Chemical constituents of the roots of *Acanthus ilicifolius*[J]. *J Nat Prod*,1986,49(2):355.

福建浮宫红海榄(*Rhizophora stylosa*)次生代谢产物研究*

王湛昌[1] 林文翰[2] 张亮亮[1] 林 鹏[1,3]

(1. 厦门大学生命科学学院，福建 厦门 361005；2. 北京大学天然药物及仿生药物国家重点实验室，北京 100083；
3. 近海海洋环境科学国家重点实验室(厦门大学)，福建厦门 361005)

摘要：研究红海榄的化学成分.将红海榄的细枝粉碎后用 95 %乙醇浸提、回流热提，合并浓缩，浸膏的乙酸乙酯萃取物，经柱色谱分离，并应用 EI MS，^{1}H NMR，^{13}C NMR 等谱学方法确定其结构.从福建浮宫九龙江河口采集的红海榄中分离得到 9 个化合物，分别为β-谷甾醇(1)，胡萝卜甙(2)，Taraxerol(3)，Careaborin(4)，*Cis*-careaborin(5)，2，6-二甲氧基-对羟基苯甲醛(6)，异香草酸(7)，原儿茶酸(8)，2，4，6-三甲氧基苯酚(9).其中化合物(6)，(7)，(8)，(9)为首次从该种植物中分离获得.

关键词：红海榄；三萜；结构鉴定

中图分类号：Q 945　　**文献标识码**：A　　**文章编号**：0438-0479(2006)06-0873-04

红树植物是热带、亚热带海洋潮间带的木本植物群落.红海榄(*Rhizophora stylosa*)是红树科(Rhizophraceae)的红树植物，在我国海南、香港、广东、广西、台湾、福建沿海有分布[1].红海榄生长在海岸泥滩上，为小乔木或灌木，树皮富含单宁，可入药.红树植物有广泛的民间药用，《全国中草药汇编》(1978)，收录了 3 种红树林植物[2].国外学者对红树林植物进行过化学成分的研究，得到一系列结构新颖和具有生物学活性的化合物，对我国开发利用红树林资源，寻找有药用前景的天然产物有启发作用[3].我们选取了福建浮宫的红树科植物红海榄进行了深入的化学研究，从中得到 1 个甾体、1 个甾体皂苷、3 个三萜类化合物以及 4 个芳香化合物.这些芳香族的酚类化合物大多具有抗氧化、消炎等活性.化合物(8)是许多中药如五灵脂、丹参、接骨木的有效成分，具有抗氧化(保护生物膜过氧化损害)、抗血栓(抑止血小板聚集)，抗菌等活性[4~7].

1 试验部分

1.1 样 品

红海榄(*Rhizophora stylosa*)于 2002 年 7 月采自福建龙海浮宫镇草浦头，种属由厦门大学生命科学学院张宜辉博士鉴定，标本存放于北京大学医学部天然药物及仿生药物国家重点实验室.

1.2 仪器与试剂

质谱(EI MS)用 AEI-MS-50 型质谱仪测定；^{1}H NMR、^{13}C NMR 用 Bruker Abence-500FT 核磁共振仪(TMS 作外标)测定；层析硅胶：160～200 目，200～300 目，青岛海洋化工有限公司；其他试剂均为分析纯(北京化工二厂).

1.3 提取与分离

红海榄小枝 8.5 kg 用乙醇冷浸两次，每次 3 d，提取液减压浓缩回收溶剂得粗提物 490 g.浸膏用水分散后用乙酸乙酯、正丁醇萃取，得到乙酸乙酯部分 40 g.并经反复硅胶柱色谱分离，得到化合物(1) 140 mg，(2) 6.8 mg，(3) 14.7 mg，(4) 3.6 mg，(5) 1.5 mg，(6) 7.8mg，(7) 9.4 mg，(8) 8.6 mg，(9) 12 mg.

2 结构鉴定

化合物(1)：为白色针晶(甲醇)，mp155～156 ℃，Liberman-Burchard 反应呈阳性. EIMS：m/z 414 $[M]^+$，396，381，329，303，273，255，231，145，43，数据与文献[4]报道基本一致.^{1}H NMR($CDCl_3$) δ：5.38 (1H，d，J = 5.0)，3.54 (1H，m)，2.28 (2H，m)，1.89 (2H，m)；^{13}C NMR($CDCl_3$) δ：11.9，12.0，18.8，19.0，19.4，19.8，21.1，23.1，24.3，26.1，28.3，29.2，31.7，31.9，31.9，34.0，36.2，37.3，39.8，42.3，45.9，50.2，56.1，56.8，71.8，121.7，140.8.以上光谱数据与文献[8]报道基本一致，鉴定该化合物为β-谷甾醇.

化合物(2)：白色无定粉形末(甲醇)，mp298～301 ℃，Liberman-Burchard 反应呈阳性.^{1}H NMR

* 福建省科技计划项目(2003Y036)资助
原载于厦门大学学报(自然科学版)，2006，45(6)：873－876

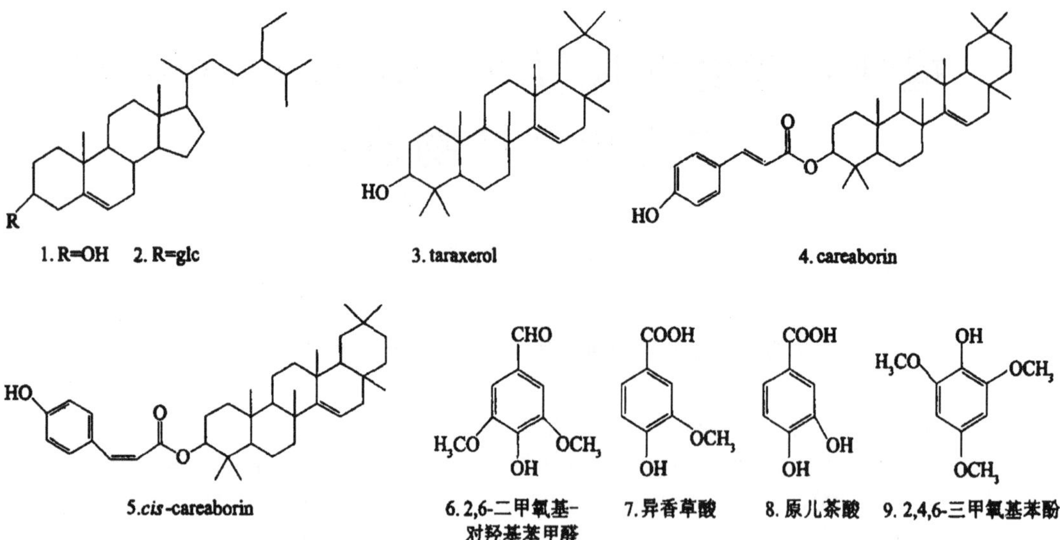

图 1 9 个化合物的化学结构式

Fig. 1 The chemical structures of the nine compounds

(C_5D_5N)δ:0.65(3H,s),0.92(3H,s),5.06(1H,d,J=7.5HZ,glc H-1),5.34(1H,m);^{13}C NMR(C_5D_5N) δ:11.9,12.1,19.0,19.2,19.4,19.9,21.2,23.3,24.5,26.3,28.5,29.4,30.2,32.0,32.1,34.1,36.3,37.0,37.5,39.3,39.9,42.4,46.0,50.3,56.2,56.8,62.8,71.6,75.3,78.0,78.5,78.6,102.5,121.9,140.8.以上光谱数据与文献[9]报道基本一致,鉴定该化合物为胡萝卜甙($C_{35}H_{60}O_6$).

化合物(3):为白色无定形粉末,Liberman-Burchard 反应呈阳性. EIMS:m/z 426 $[M]^+$,409,393,218,204,189,135.^{1}H NMR ($CDCl_3$)δ:5.53(1H,dd,J=3.1,8.1Hz),3.20(1H,dd,J=4.6,11.1Hz),1.08(3H,s,H-26),0.98(3H,s,H-23),0.95(3H,s,H-29),0.93(3H,s,H-25),0.91(6H,s,H-27,H-30),0.82(3H,s,H-28),0.80(3H,s,H-24).^{13}C NMR ($CDCl_3$)数据见表 1.根据 HMQC、HMBC 确定碳、氢化学位移的归属,并纠正文献[10],7、12、16、17、19、21、22、26、27 等碳的归属错误,鉴定该化合物为 taraxerol($C_{30}H_{50}O$).

化合物(4):为白色无定形粉末,Liberman-Burchard 反应呈阳性. EIMS:m/z 572 $[M]^+$,393,218,204,189,147,135.^{1}H NMR($CDCl_3$)δ:7.61(1H,d,J=15.9 Hz,H-3′),7.43(2H,d,J=8.6Hz,H-6′,8′),6.85(2H,d,J=8.6Hz,H-5′,9′),6.30(1H,d,J=15.9Hz,H-2′),5.54(1H,dd,J=3.0,8.1Hz,H-15),4.60(1H,dd,J=5.3,11.0Hz,H-3),1.10(3H,s),0.98(3H,s),0.95(6H,s),0.90(6H,s),0.89(3H,s),0.82(3H,s).^{13}C NMR($CDCl_3$)数据见表 1.此数据和文献[11]基本一致,鉴定该化合物为 careaborin ($C_{39}H_{56}O_3$).

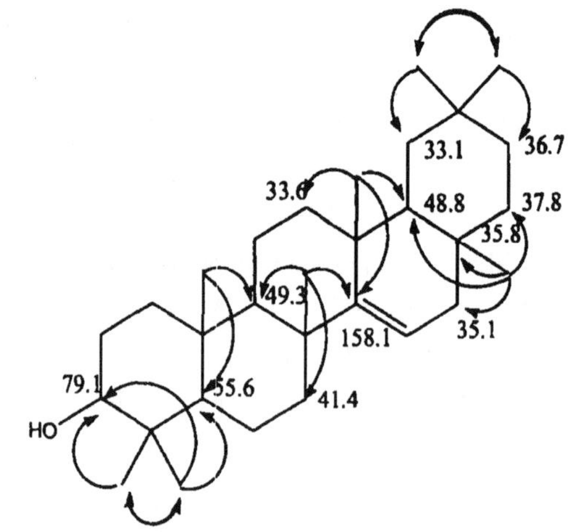

图 2 化合物(3)的 HMBC 相关

Fig. 2 The HMBC of compound(3)

化合物(5):为白色无定形粉末,Liberman-Burchard 反应呈阳性. EIMS:m/z 572 $[M]^+$,393,218,204,189,147,135.^{1}H NMR($CDCl_3$)δ:7.60(2H,d,J=8.7Hz,H-6′,8′),6.84(1H,d,J=12.6Hz,H-3′),6.77(2H,d,J=8.7Hz,H-4′,9′),5.84(1H,d,J=12.6Hz,H-2′),5.54(1H,dd,J=3.0,8.1Hz,H-15),4.52(1H,dd,J=4.9,11.0Hz,H-3),1.08(3H,s),0.95(6H,s),0.91(6H,s),0.87(3H,s),0.84(3H,s),0.81(3H,s).氢谱数据纠正了文献[8]中 H-2′,3′,5′,6′,8′,9′的归属错误.^{13}C NMR($CDCl_3$)数据见表 1,数据和文献[12]基本一致,鉴定该化合物为 *cis*-careaborin($C_{39}H_{56}O_3$).

表1 化合物3～5的^{13}C NMR化学位移值

Tab.1 ^{13}C NMR spectra of compound 3～5

碳位编号	化合物3	化合物4	化合物5	碳位编号	化合物3	化合物4	化合物5
1	37.8	37.5	37.4	21	36.7	36.7	36.7
2	27.2	23.6	23.4	22	37.8	37.4	37.8
3	79.1	81.1	81.2	23	28.0	28.1	28.0
4	38.8	37.9	37.9	24	15.5	16.8	16.6
5	55.6	55.7	55.6	25	15.5	15.5	15.5
6	18.8	18.7	18.7	26	25.9	25.9	25.9
7	41.4	41.2	41.2	27	29.9	29.9	29.9
8	39.0	39.0	38.9	28	29.8	29.8	29.8
9	49.3	49.2	49.1	29	33.7	33.7	33.7
10	38.0	37.9	37.5	30	21.3	21.3	21.3
11	17.5	17.5	17.5	1′		167.3	167.3
12	33.6	33.7	33.7	2′		116.3	117.8
13	37.6	37.6	37.8	3′		144.0	143.9
14	158.1	158.0	158	4′		127.3	129.9
15	116.9	116.9	117	5′		129.9	132.4
16	35.1	35.1	35.1	6′		115.9	115.9
17	35.8	35.8	35.8	7′		157.8	157.7
18	48.8	48.8	48.8	8′		115.9	115.9
19	33.1	33.1	33.1	9′		129.9	132.4
20	28.8	28.8	28.8				

化合物均在$CDCl_3$中测定.

化合物(6):为白色无定形粉末,EIMS:m/z 182 [M]$^+$,167,139,111,96,65.^{1}H NMR($CDCl_3$)δ:9.84 (1H,s,1-CHO),7.18(2H,s,H-2,H-6),6.11(1H,s,4-OH),4.00(6H,s,3-OCH_3,5-OCH_3).^{13}CNMR ($CDCl_3$)δ:191.2(1-CHO),147.8(C-3),147.8(C-5),141.2(C-4),128.8(C-1),107.1(C-2),107.1(C-6),56.9(3-OH),56.9(5-OH).以上数据和文献[13]基本一致,鉴定该化合物为2,6-二甲氧基-对羟基苯甲醛.

化合物(7):无色针状结晶(甲醇),mp248～250℃,EIMS:m/z 168[M]$^+$,153,151,97.^{1}HNMR (DMSO)δ:12.49(1H,s,1-COOH),9.87(1H,s,3-OH),7.45(1H,d,J=8.0Hz,6-H),7.44(1H,s,2-H),6.85(1H,d,J=8.0Hz,5-H),3.81(3H,s,4-OCH_3).^{13}CNMR(DMSO)δ:168.0(-COOH),151.8 (C-3),148.0(C-4),124.3(C-6),122.4(C-1),115.8 (C-5),113.6(C-2),56.4(4-OCH_3).此数据和文献[14]基本一致,鉴定该化合物为异香草酸.

化合物(8):无色针状结晶(甲醇),mp198～199℃,EIMS:m/z 154[M]$^+$,137,109,81,63,53.^{1}HNMR(DMSO)δ:7.36(1H,d,J=1.8Hz,2-H),7.31(1H,dd,J=1.7Hz,8.2Hz,6-H),6.80(1H,d,J=8.2Hz).此数据和文献[15]基本一致,鉴定该化合物为原儿茶酸.

化合物(9):无色针状结晶(甲醇),EIMS:m/z 184[M]$^+$,169,141,111,69,53.^{1}H NMR($CDCl_3$)δ:3.81(6H,s,2-OCH_3,6-OCH_3),3.80(3H,s,4-OCH_3),6.11(2H,s,3-H,5-H),5.40(1H,s,1-OH).^{13}CNMR($CDCl_3$)δ:153.8(C-2),153.8(C-6),152.4(C-4),131.8(C-1),92.9(C-3),92.9(C-5),61.1 (4-OCH_3),56.0(2-OCH_3),56.0(6-OCH_3).以上数据和文献[13]基本一致,鉴定该化合物为2,4,6-三甲氧基苯酚.

致谢:感谢北京师范大学分析测试中心邓志威教授在波谱测试方面给予的大力帮助.

参考文献:

[1] 林鹏.中国红树林生态系[M].北京:科学出版社,1979.

[2] 林鹏.中国红树林环境生态及经济利用[M].北京:高等教育出版社,1995.

[3] Bandaranayake W M. Traditional and medicinal uses of mangroves[J]. Mangroves and Salt Marshes,1998,2:133 - 148.

[4] 陈月开,吴弢,王美素.五灵脂活性成分单体原儿茶酸的分析测定[J].山西大学学报:自然科学版,2000,23(3):260-262.

[5] 彭永芳,马银海,郭亚东,等.微柱高效液相色谱法测定丹参中的几种有效成分[J].分析实验室,2005,24:19-21.

[6] 刘耕陶,张铁梅,王保恩,等.丹参的7种酚类成分对生物膜过氧化损伤的保护作用[J].中国药理学和毒理学杂志,1992,6(1):77.

[7] 杨序娟,黄文秀,王乃利,等.接骨木种酚酸类化合物及其对大鼠类成骨细胞UMR106增殖及分化的影响[J].中草药,2005,36(11):1604-1607.

[8] 张雁冰,李玲,刘宏民,等.马桑化学成分研究[J].郑州大学学报:理学版,2005,37:75-77.

[9] 关永霞,杨小生,佟丽华,等.大驳骨化学成分研究(II)[J].天然产物与开发,2004,16:516-517.

[10] Nobuko S, Yoshikatsu Y, Takao I. Triterpenoids from *Myrica rubra*[J]. Phytochemistry,1987,26 (1):217-219.

[11] Talapatra B,Basak A,Talapatra S K. Triterpenoids and related compounds: patr XX. careaborin, a new triterpene ester from the leaves of *Careya arborea*[J].J. Indian Chem. Soc. ,1981,58:814-815.

[12] Kokplo U,Chavasiri W,Chittawong V,et al. Taraxeryl *cis*-p-hydroxycinnamate, a novel traxeryl from *Rhizophora apiculata*[J].Journal of Natural Products,1990,53(4):953-955.

[13] 于德全,杨峻山.分析化学手册(第七分册)[M].北京:化学工业出版社,1999.

[14] 解军波,李萍.四季青酚酸类化学成分研究[J].中国药科大学学报,2002,33(1):76-77.

[15] 李良琼,李美蓉,朱爱江.锈毛寄生化学成分的研究[J].中国中药杂志,1996,21:34-36.

Study on the Chemical Constituents of *Rhizophora stylosa*

WANG Zhan-chang[1],LIN Wen-han[2],ZHANG Liang-liang[1],LIN Peng[1,3]

(1. School of Life Sciences,Xiamen University,Xiamen 361005,China;

2. National Research Laboratories of Natural and Biomimetic Drugs,Peking University,Beijing 100083,China;

3. State Key Laboratory of Marine Environmental Science,Xiamen University,Xiamen 361005,China)

Abstract: The chemical constituents of *Rhizophora stylosa*,which occurs in the intertidal zones of tropical and subtropical coastlines were studied. The twigs of *Rhizophora stylosa* were collected from Fugong Fujian in July 2002. The air-dried and powdered plant material(8.5 kg) was estracted with 95 % ethanol. Removal of the sovlent from the combined 95 % ethanol extracts under reduced pressure gave a residue(490 g). The extract was taken up in H_2O and treated with EtOAc. The EtOAc fraction(40 g) was subjected to column chromatography over a column of silica gel(200～300 mesh) using solvent from petroleum ether to acetone. All fractions were collected and the fractions showing similar spots were combined. The residues were subjected to rechromatography over silica gel or sephadex LH-20 to yield 9 pure compounds. Their structures were indentified by EIMS,^{1}H NMR and ^{13}C NMR analysis and the spectral data were compared with those reported in literatures. Nine compounds are isolated from *Rhizophora stylosa* of Fugong,Fujian province by using column chromatography. The nine compounds are β-stitosterol(1),dausterol(2),taraxerol(3),careaborin(4),*cis*-careaborin(5),2,6-dimethoxy-*p*-hydroxybenzal-dehyde(6),2,4,6-trimethoxy-phenol(7),protocatechuic acid(8). Compound(6),(7),(8),(9) are isolated from this genus for the first time.

Key words: *Rhizophora stylosa*;triterpenoids;preparative chromatography

HPLC测定木榄繁殖器官内源ABA和GA_3含量*

王 洁 李 敏 张宜辉 杨盛昌
(厦门大学生命科学学院,福建 厦门 361005)

摘要:淹水、高盐和土壤缺氧等不良环境因子导致了红树植物组织富含多酚、色素、黏多糖等次生代谢物,激素测定有一定困难.本文以胎生红树植物——木榄(*Bruguiera gymnorrhiza*(L.) Lamk)为研究对象,探讨红树植物内源激素——脱落酸(ABA)和赤霉素(GA_3)的提取纯化方法以及合适的HPLC色谱测定条件.结果表明:采用等度洗脱、35 %甲醇(含0.15 % 0.1 mol/L H_3PO_4)流动相、流速1.0 mL/min,分别在245和208 nm下检测,ABA和GA_3分离效果理想,回收率分别达到99.8 %和95.5 %.整个过程简单、准确、易操作.比较木榄的花蕾、种子、幼胚轴、成胚轴中ABA和GA_3含量变化,种子和幼胚轴ABA含量较低,而GA_3含量较高,这可能是影响红树植物胎生现象发生的重要原因之一.

关键词:红树植物;木榄;液相色谱法;内源激素
中图分类号:Q 945.4 **文献标识码**:A **文章编号**:0438-0479(2008)05-0752-05

红树植物是生长在热带、亚热带海岸潮间带的木本植物,淹水、高盐、土壤缺氧和潮水冲击等不良环境因子导致了红树植物在形态、生理和生态方面的特异性,尤其是其独特的胎生现象,越来越多地引起人们的关注.对于非红树植物的胎生机理,研究很多[1],但对于红树植物如何适应潮间带生境及其胎生现象的机理仍不清楚.

脱落酸(ABA)和赤霉素(GAs)在红树植物生长发育过程中尤其是胎生期间种子萌发时期有很重要的调节作用,ABA能诱导种子休眠,抑制早萌发生,例如:用ABA合成抑制剂(fluridone)处理玉米早期胚胎,可诱发种子提前萌发[2-3],而对缺乏ABA的突变体和成熟胚胎外施ABA,可诱发并延长它对干旱的耐受性[4].GAs则能促进胚萌发,Leon-Kloosterziel等研究表明,增强对GAs的敏感性可使拟南芥种子萌发提前[5],而Khan研究表明通过增加GAs抑制剂可使非休眠种子进入休眠[6].从目前的研究看,探讨胎生植物繁殖器官ABA、GAs含量变化对于解释种子休眠具有重要作用.

测定植物激素的常见方法有酶联免疫法(ELISA)和放射免疫法(RIA)等,他们对激素提取液的纯度要求不高,但准确度低,重复性差.自20世纪70年代中期开始使用高效液相色谱法(HPLC)测定植物内源激素以来[7],HPLC法在测定植物激素的研究领域得到了广泛应用,它具有灵敏度高、重复性好、专一性强及分析速度快等特点,但对所测样品的激素纯度要求高.由于红树植物生长于高盐、潮汐等特殊的生境,其组织富含多酚、色素、黏多糖等次生代谢物,严重干扰了激素的准确测定,使得红树植物激素的测定有一定困难.对红树植物内源激素的分析,除应用RIA法[8]和ELISA法[9]分别测定了美国红树(*Rhizophora mangle* L.)的ABA和GAs含量外,目前采用HPLC法测定红树植物内源激素的研究还未有报道.故本文作者以胎生红树植物木榄(*Bruguiera gymnorrhiza* (L.) Lamk)的花蕾、种子、幼胚轴(种子萌发早期)和成胚轴(种子萌发后期)为研究对象,探讨HPLC法测定内源激素ABA和GA_3含量的具体过程,为准确分析红树植物繁殖器官中的内源激素含量变化提供技术保障.

1 材料和方法

1.1 样品的采集与保存

木榄的花蕾、种子、幼胚轴(伸出果皮3 cm)、成胚轴(即将离开母体的最成熟的胚轴)采自福建省龙海市浮宫镇,置于-80 ℃保存.

1.2 仪器设备

Agilent 1100高效液相色谱仪(美国),Hypersil ODS柱(4.6 mm i.d.(250 mm,5 μm),SDS紫外检测器,KQ-120型超声波清洗器,UV 310-Fix/scan紫外分光光度计,RE 52-3冷冻干燥机,RE 52-3旋转蒸发器,SHZ-ⅢB循环水真空泵.

1.3 试 剂

ABA和GA_3标样、非水溶性聚乙烯聚吡咯烷酮(PVPP)均为美国Sigma公司产品,色谱甲醇为上海

* 福建省自然科学基金(B0410001)资助项目
原载于厦门大学学报(自然科学版),2008,47(5):752—756

化学试剂研究所产品,其他试剂均为国产优质分析纯.

1.4 内源激素的提取与纯化

样品的处理参照文献[10 - 13]并加以改进.基本流程:称取 5 000 mg 植物材料,冰浴研磨后加入 30 mL 的 80 %冷甲醇,放入超声波仪,4 ℃超声 1 h 后,8 000 r/min 冷冻离心 15 min,转移上清液,残渣再加入 30 mL 的 80 %冷甲醇,4 ℃,过夜搅拌,合并上清.30 ℃减压浓缩至原体积的 1/3,用 8 mL pH 8.0 的磷酸缓冲液冲洗旋转蒸发瓶,混匀.加入饱和石油醚(体积比 0.8:1)萃取至有机相无色为止,再用 Na_2HPO_3 水溶液调水相pH值为 8.0 并加入 1.5 g PVPP,4 ℃搅拌 30 min,抽滤得到上清液,用冰乙酸调 pH 值为 3.0,加入适量饱和乙酸乙酯(体积比 0.8:1)萃取 3 次,合并 3 次酯相减压浓缩至干,用 3 mL 混合溶液(V(乙腈):V(水) = 1:4)复溶,调节样品液 pH 值为 3.0,过 C_{18}小柱(C_{18}小柱使用前先用 4 mL 甲醇反相冲洗活化柱,之后用 2 mL pH 3.0 的水正向冲洗两次,样液过柱后,先用 pH 3.0 的冰醋酸水溶液 2 mL 冲洗,之后用含 2 %冰乙酸的 40 %的甲醇 4 mL 洗脱 3 次并收集).样液冷冻干燥备用.整个实验过程中注意避光操作.

1.5 色谱条件

Hypersil ODS 色谱柱[4.6 mm i.d. (250 mm,5 μm)],测定 ABA 和 GA_3 的流动相,流速为 35 %甲醇(含 0.15 % 0.1 mol/L H_3PO_4),1.0 mL/min,检测波长分别为 245 nm 和 208 nm,进样量均为 40 μL,柱温 30 ℃.

2 结果与分析

2.1 ABA 和 GA_3 标准工作曲线的测定

用外标峰面积法测定.配制一系列质量浓度的标准液,用峰面积(y)对质量浓度(μg/μL)作线性回归曲线,得到的 ABA 和 GA_3 标准工作曲线分别为,ABA:

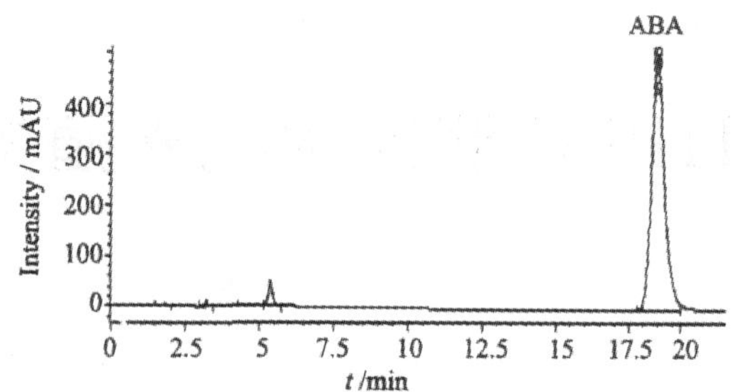

图 1 ABA 标准样品的图谱分析

Fig.1 Chromatogram of standard ABA

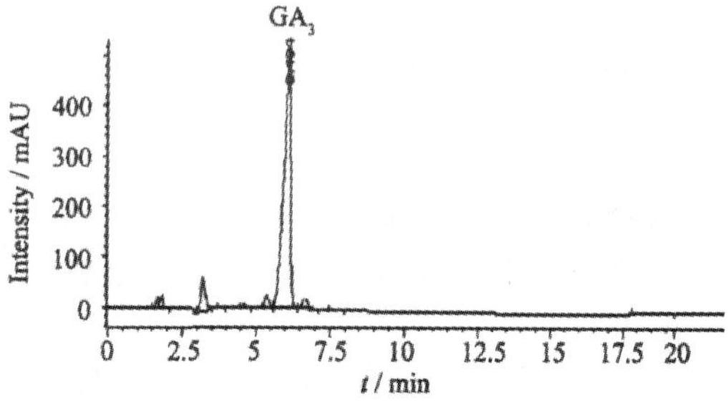

图 2 GA_3 标准样品的图谱分析

Fig.2 Chromatogram of standard GA_3

$y = 138\ 974x + 3.390\ 8$ ($r^2 = 0.999\ 5$, $p < 0.01$);GA_3:$y = 63\ 544x - 59.772$ ($r^2 = 0.996\ 6$, $p < 0.01$),其标准曲线如图 1、2,植物样品色谱图见图 3.

2.2 精密度和回收率实验

平行测定 ABA 和 GA_3 标准样品 6 次,得出 ABA 和 GA_3 保留时间的相对标准偏差分别为 0.13 %和 0.16 %,ABA 和 GA_3 峰面积的相对标准偏差为 2.17 %和 5.67 %.分别将 0.10 μg ABA 和 0.10 μg GA_3 标准溶液加入样品(木榄胚轴)中,进行 6 次平行测定,测得回收率分别为 ABA 99.8 %,GA_3 95.5 %.

2.3 木榄不同繁殖器官内源 ABA 和 GA_3 含量测定

对木榄不同繁殖器官的两种内源激素同时测定,

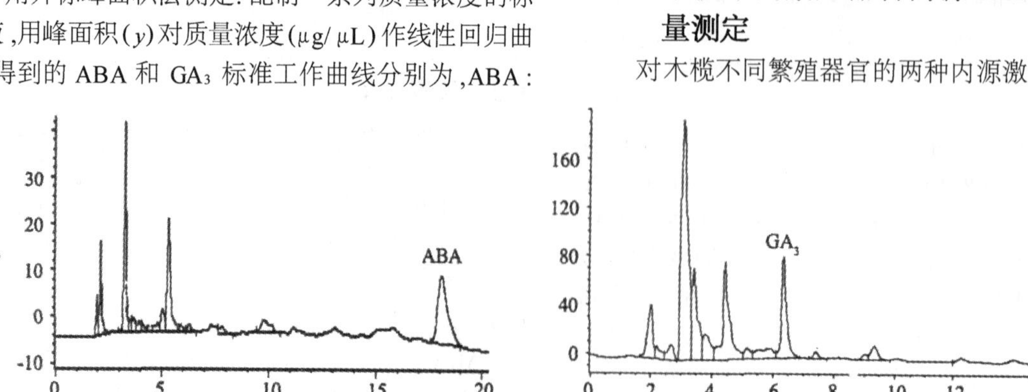

图 3 木榄成胚轴的 HPLC 图谱分析

Fig.3 HPLC chromatogram of mature hypocotyle of *Bruguiera gymnorrhiza*

表1　木榄不同繁殖器官内源 ABA 和 GA_3 含量

Tab.1　The hormone content in different reproductive organs of *Bruguiera gymnorrhiza*　(μg/g)

激素种类	木榄不同繁殖器官内源激素含量			
	花蕾	种子	幼胚轴	成胚轴
ABA	0.37 ±0.07^{a}	0.24 ±0.03^{bc}	0.21 ±0.01^{b}	0.32 ±0.04^{ac}
GA_3	2.80 ±0.24^{c}	18.55 ±0.99^{a}	5.33 ±0.53^{b}	2.60 ±0.19^{c}

注:不同字母表示显著性差异($p<0.05$,LSD).

结果见表1.由表可知:ABA在木榄花蕾中含量为0.37 ±0.07 μg/g,在种子中含量显著降低,降为0.24 ±0.03 μg/g,幼胚轴与种子 ABA 含量变化不大,在成胚轴中 ABA 含量又显著增至 0.32 ±0.04 μg/g. GA_3 在4个繁殖器官内与 ABA 变化趋势相反,尤其是在种子中,ABA 含量较低,而 GA_3 含量达到最高.

3　讨　论

3.1　提取纯化方法的优化

1)生长在特殊生境下的红树植物含有大量单宁等多酚杂质[14],所以在红树植物的激素提纯过程中最关键的一步是去除高含量的单宁等多酚杂质的影响.目前能有效吸附单宁等多酚杂质的是 PVPP 粉末或 PVPP 柱.本实验采用 PVPP 粉末振荡吸附的方法,具有简单、方便、易操作的特点.一般样品中 PVPP 用量为0～0.2 g/g fm[10,15-16],与长春花、卵叶韭、苹果叶片等材料相比,红树植物各器官中单宁含量要高很多,本实验分别用0.1、0.2、0.3、0.4g/g fm PVPP 处理木榄样品,从处理后上清液的颜色来看,0.1、0.2 g/g fm PVPP 处理后的上清液颜色为红褐色,吸附多酚杂质不完全,HPLC 测定时杂峰太多,影响激素峰的分离;0.3、0.4 g/g fm PVPP 处理后的上清液颜色相差不大,颜色为淡红色,接近无色.考虑到 PVPP 对激素有一定的吸附,所以选用0.3 g/g fm PVPP,HPLC 实测时也发现杂峰影响少,分离结果满意.为了兼顾除杂效果和回收率,我们采用少量多次去除杂质的方法,搅拌20 min后抽滤,重复3次,提高了 HPLC 的检测效果.

2)提取红树植物胚轴的内源激素时,可能是材料本身的原因,极易在分液除杂时产生乳化层,影响定量的准确性.为避免乳化层的出现,一是要用饱和的石油醚和乙酸乙酯溶液抽提,其次采取少量多次的原则,即按照0.8:1的体积比,让有机相的体积微少于溶液体积,同时在萃取过程中注意轻摇并充分静置,若已形成乳化层,也可采取低温离心的方法;对于氯仿形成的乳化层,可加少许乙醇促进乳化液分层;对于乙酸乙酯形成的乳化层,可加少许氯化钠.此外采用低温离心的方法代替常规的抽滤法获得甲醇初提液上清,因为红树植物(尤其是胚轴)含有较多黏多糖,使得抽滤无法正常进行.

3)杨世民[17]在研究中发现 C_{18} 小柱能够很好的去除干扰激素测定的叶绿体色素和酚类氧化物等小分子色素.在实验操作中为了保证高的回收率和激素纯化率,需注意两点,一是上样溶液的 pH 值应调至3.0;二是洗脱液应选择最佳洗脱浓度.

游离态激素在 pH 值为3.0时与 C_{18} 小柱具有极强的吸附能力[18],许多实验则忽略了这点[11-13],在实验中我们以木榄花蕾为材料,分别测定了未调节上样溶液 pH 值(pH = 5.1)和调节上样溶液 pH 值(pH = 3.0)后激素的含量,结果见表2.从表可知:未调 pH 值所测得的 ABA、GA_3 含量仅为调节 pH 值所测激素含量的56.76%和86.43%,因此调节上样溶液的 pH 值至3.0可以减少激素测定时的损失.

表2　C_{18} 柱上样前调节 pH 值对得率的影响

Tab.2　Influence of sample pH in C_{18} Sep-pak cartridge on measurement results

pH 值	ABA 含量/(μg·g^{-1})	GA_3 含量/(μg·g^{-1})
5.1	0.21 ±0.04^{b}	2.42 ±0.09^{b}
3.0	0.37 ±0.07^{a}	2.80 ±0.24^{a}

注:不同字母表示显著性差异($p<0.05$,LSD).

C_{18} 小柱萃取条件的选择实际是洗脱液浓度的选择,但目前对于洗脱液和洗脱浓度的选择差别较大,一般有50%的甲醇[19]、60%的甲醇[16]、65%的甲醇[10]、100%的乙腈[13]等,但需注意的是有机溶剂含量越高,色素等杂质越易被洗脱下来.

在本实验中,分别上样0.1 mg/mL 的混合标样1 mL,然后分别用0%～100%甲醇系列各4 mL 洗脱,收集洗脱液,进行 HPLC 测定,结果见表3.从表中得知:50%和40%的甲醇能分别将 ABA 和 GA_3 几乎完全洗脱.但在处理样品时发现50%以上的甲醇浓度洗脱时,一部分色素也会被洗脱,杂峰影响较大.所以综合考虑,选取能将大部分激素洗脱的最低甲醇浓度为40 %并洗脱 3 次,计算此洗脱条件下激素的萃取回收率,结果为 99.98 %.

表 3 C_{18}小柱的最佳甲醇洗脱浓度

Tab. 3 Optimization of methanol concentration in C_{18} Sep-pak cartridge

甲醇浓度/ %	ABA 的峰面积	GA_3 的峰面积
0	0.00 ± 0.00^f	0.00 ± 0.00^e
10	57.89 ± 1.09^e	20.42 ± 1.15^d
20	107.23 ± 1.10^d	84.26 ± 1.56^c
30	847.93 ± 2.23^c	398.65 ± 1.43^b
40	2885.78 ± 2.36^b	1515.68 ± 4.42^a
50	3465.21 ± 2.4^a	1520.49 ± 3.98^a
60	3468.28 ± 2.67^a	1521.46 ± 3.56^a
70	3464.89 ± 2.51^a	1526.68 ± 4.02^a
80	3466.57 ± 3.01^a	1524.98 ± 3.98^a
90	3464.92 ± 2.21^a	1526.24 ± 5.05^a
100	3466.68 ± 2.76^a	1523.52 ± 5.52^a

注:不同字母表示显著性差异($p<0.05$,LSD).

3.2 色谱条件的筛选

GA_3 在较远的紫外区域(如 200 nm 附近)接近其最大吸收波长. 而甲醇和一些化学物质在 210 nm 处有较大吸收,所以一些研究将 GA_3 的检测波长设定为 250 nm 以上[10,17]. 但在这些波长下,GA_3 的吸收大大降低,造成 GA_3 的检测灵敏度下降,甚至一些 GA_3 含量甚微的样品根本无法检测出. 在我们的实验中发现 0.1 mg/ mL 的 GA_3 标样 40 μL 在 208 nm 和 245 nm 下测得的含量分别为 4.0 μg 和 0.29 μg. 本研究均在 GA_3 和 ABA 的最大吸收波长下进行扫描检测,为了避免甲醇本底的影响,在溶解样品时用乙腈(V(乙腈) :V(水) = 1 :4),确定了两种检测波长下同时检测 ABA 和 GA_3 的色谱分析条件——即流动相为 35 %甲醇(0.15 % 0.1 mol/ L H_3PO_4),流速为 1.0 mL/ min,实验结果也显示良好的提取纯化条件配以合适的色谱条件使得在此检测波长下激素与杂质分离和测定结果理想(图 3).

3.3 内源激素含量的变化及其生物学意义

木榄不同繁殖器官——花蕾、种子、幼胚轴(种子萌发早期)、成胚轴(种子萌发后期)内源 GA_3 和 ABA 的含量不同,其中木榄 GA_3 的含量在种子成熟期达到最高,之后急剧下降,木榄 ABA 的含量一直处较低水平,尤其在成熟种子中最低. 低含量的 ABA 和高含量的 GA_3 共同作用使得胎生红树植物种子休眠解除,促进萌发的提前发生. 至于激素如何调控红树植物胎生的机制,还需进一步研究.

参考文献:

[1] Vertucci C W,Farrant J M. Seed development and germination[M]. New York:Marcel Dekker,1995.

[2] Fong F,Smith J D,Koehler D E. Early events in maize seed development: 1-Methyl-3-phenyl-5-(3-[trifluoromethyl]phenyl)-4-(1H)-pyridinome induction of vivipary [J]. Plant Physiol,1983,73:899 - 901.

[3] Oishi M Y,Bewley J D. Premature drying,fluridone-treatment,and embryo isolation during development of Maize Kernels(*Zea mays* L.) induce germination,but the protein synthetic responses are different. potential regulation of germination and protein synthesis by abscisic acid[J].J Exp Bot,1992,43:759 - 767.

[4] Meurs C,Basra A S,Karssen C M,et al. Role of abscisic acid in the induction of desiccation tolerance in developing seeds of *Arabidopsis Thaliana*[J]. Plant Physiol,1992,98:1484 - 1493.

[5] Leon-Kloosterziel K M,Gil M A,Ruijs GJ,et al. Isolation and characterization of abscisic acid-deficient Arabidopsis mutants at two new loci[J]. Plant J,1996,10:655 - 661.

[6] Khan A A. Induction of dormancy in non-dormant seeds [J].J Am Hort Soc,1994,119:408 - 413.

[7] Ciha A J,Brenner M L,Brun W A. Rapid separation and quantification of abscisic acid from plant tissues using high performance liquid chromatography[J]. Plant Physiol,1977,59:821 - 826.

[8] Farnsworth E J,Farrant J M. Reductions in abscisic acid are linked with viviparous reproduction in mangroves[J]. Am J Bot,1998,85:760 - 769.

[9] Smith S M,Yang Y Y,Kamiya Y,et al. Effect of environment and gibberellins on the early growth and development of the red mangrove,*Rhizophora mangle* L. [J]. Plant Growth Regul,1996,20:215 - 223.

[10] 赵晓菊,唐中华,郭晓瑞,等. 固相萃取富集——高效液相色谱法测定长春花种的 3 种内源激素[J]. 色谱,2006,24:534.

[11] 陈小鹏,王秀峰,孙小镭,等. 高效液相色谱测定黄瓜瓜条中赤霉素和脱落酸含量[J]. 山东农业科学,2005(1):65 - 67.

[12] 马海燕,王美丽,张震文. 葡萄新梢生长过程中内源激素含量的动态变化[J]. 西北农业学报,2007,16:177 - 179.

[13] 雷蕾,康庆华,张晓波,等. 反相高效液相色谱法分离和测定亚麻植株中植物激素[J]. 黑龙江农业科学,2001,(6):21 - 22.

[14] 林益明,向平,林鹏. 红树林单宁的研究进展[J]. 海洋科

学,2005,29:59 - 63.

[15] 陈远平,杨文钰.卵叶韭休眠芽中 GA_3、IAA、ABA 和 ZT 的高效液相色谱法测定[J].四川农业大学学报,2005,23:498 - 450.

[16] 吴耕西,毕桂红.高效液相色谱测定苹果叶片中的吲哚乙酸和脱落酸[J].山东农业大学学报,1994,25:51 - 55.

[17] 杨世民,何科学,赵彤.植物激素提纯中去除色素的有效方法研究[J].四川农业大学学报,1996,14:613 - 615.

[18] 曾建明,马志超.茶树激素提取方法的改进[J].中国茶叶,1999,21:28 - 29.

[19] 王若仲,萧浪涛,蔺万煌,等.亚种间杂交水稻内源激素的高效液相色谱测定法[J].色谱,2002,20:148 - 150.

HPLC Analysis of ABA and GA_3 in Reproductive Organs of *Bruguiera gymnorrhiza*

WANG Jie, LI Min, ZHANG Yi-hui, YANG Sheng-chang

(School of Life Sciences, Xiamen University, Xiamen 361005, China)

Abstract: Mangrove plant produces high levels of secondary metabolites (polyphenolics, pigment and mucopolysaccharide, etc.) in response to environmental factors such as hypersaline stress, waterlogging and low oxygen level in sediment, which hinder the extraction of hormones. In this paper, the extraction and purification methods and the operation condition of reversed-phase high performance liquid chromatography (RP-HPLC) were studied for a quantitative analysis of abscisic acid (ABA) and gibberellin acid (GA_3) in *Bruguiera gymnorrhiza* Lamk, a viviparous mangrove species. Under the conditions of isocratic concentration 35% methanol with 0.15% 0.1 mol/L H_3PO_4, flow rate 1.0 mL/min, UV wavelength 245 nm for ABA and 208 nm for GA_3, two mangrove plant hormones recoveries were high up to 95.5% and 98.8% respectively. This HPLC quantitative analysis method is convenient, fast and accurate. Compared with flower buds and mature hypocotyles, seeds and young hypocotyles had lower level of ABA and higher level of GA_3, which might contribute to the viviparity of mangrove plant.

Key words: mangroves; *Bruguiera gymnorrhiza*; HPLC; endogenous hormone

红树植物繁殖器官发育过程中铁钼锌元素的含量变化*

李旷达 王 洁 杨盛昌

(厦门大学生命科学学院,福建 厦门 361005)

摘要:探讨了显胎生红树植物木榄 (*Bruguiera gymnorrihiza*)、秋茄 (*Kandelia obovata*)、隐胎生红树植物白骨壤 (*Avicennia marina*)、桐花树 (*Aegiceras corniculatum*)和非胎生红树植物木果楝 (*Xylocarpus grantum*)的繁殖器官发育过程中铁 (Fe)、钼 (Mo)、锌 (Zn)元素的含量变化.结果表明:在花蕾期,除白骨壤外,其他种类 Fe元素含量相近,在 0.035 2~0.045 4 μg/mg之间.隐胎生红树植物白骨壤、桐花树中 Fe元素含量在种子期最高,分别为 0.090 2和 0.073 3 μg/mg,萌发后呈现降低的趋势,在胚轴成熟期达到最低.在种子期,Mo元素含量在胎生红树植物中差异不大,在 0.288 2~0.302 0 ng/mg之间,平均为 0.294 5 ng/mg,明显高于非胎生红树植物木果楝 0.105 6 ng/mg.种子萌发后,胎生红树植物中 Mo元素含量下降.在同一时期,胎生红树植物种子的 Mo元素含量明显大于木果楝.在木果楝的繁殖器官发育过程中,Zn元素含量逐渐下降,且比同期的胎生红树植物低.

关键词:红树植物;胎生;繁殖器官;元素含量

中图分类号:Q 945.5 **文献标识码**:A **文章编号**:0438-0479(2008)S2-0169-04

红树林是热带、亚热带海岸潮间带的木本植物群落,在海岸河口生态系统中占有重要地位.胎生现象是红树植物适应生境的一种特殊繁殖方式.一些红树植物,如木榄 (*Bruguiera gymnorrihiza*)、秋茄 (*Kandelia obovata*)等果实在离开母树前种子就萌发成棒状胚轴,有利于其固着在滩涂淤泥上或扩展分布范围,为显胎生.另一些红树植物,如桐花树 (*Aegiceras corniculatum*)、白骨壤 (*Avicennia marina*)等的种子在果实内萌发,形成具有幼苗雏型的胚体,但不形成长筒形胚轴突出果实之外,称为隐胎生.

围绕红树植物胎生现象,在形态结构、抗盐适应性以及能量动态等方面作了大量的研究工作[1-5].但有关 Fe、Mo、Zn等元素的影响还未见报道.本文通过测定红树植物从花蕾到胚轴萌发这一发育过程中 Fe、Mo、Zn等元素含量的变化,初步探究其对繁殖器官的发育的影响,加深对红树植物胎生特性的了解.

1 材料与方法

1.1 材料采集和预处理

胎生红树植物木榄、秋茄、白骨壤和桐花树的花蕾、种子 (败花)、胚轴幼期 (幼果)、胚轴成熟期 (成果)等于 2006年 4月至 7月采自福建省龙海浮宫红树林保护区.非胎生红树植物木果楝的花蕾、种子等采自海南东寨港红树林保护区,以种子萌发 3 d和 20 d后的材料对应于胎生红树植物胚轴幼期和胚轴成熟期.

为避免生境差异可能导致元素含量变化的影响,采用定点定株的方式进行样品采集.植物样品用无离子水洗净,80 ℃烘干 24 h,高速药物粉碎机 (细度 50~250目,山东青州精诚机械有限公司)研磨粉碎,称量待用.

1.2 主要仪器和试剂

电感耦合等离子体质谱仪 (ICP-MS) (PE DRC-e型)为美国 PE公司产品;程控箱式电炉 (马福炉,SXL-1016型)为上海精宏实验设备有限公司产品.

浓硝酸和无水乙醇为国产优级纯;超纯水配制标准液,其中 Fe元素所用标准液单位为:(1 ±0.005) μg/mL (5% HNO_3 + Trace HF),Fe元素浓度是 100倍单位浓度.Mo、Zn元素所用标准液单位为:(10 ±0.005) μg/mL (2% HNO_3 + Trace HF),浓度均为 1倍单位浓度.

1.3 测试方法

消解:称取 0.100 0 g植物干粉样品,转入 10 mL坩埚中.加入 200 μL的无水乙醇湿润样品,盖好盖子.放入马弗炉灰化,程序为:200 ℃,1 h;300 ℃,1 h;

* 国家基础科学人才培养基金项目(J0630649)资助。

原载于厦门大学学报(自然科学版),2008,47(2):169-172

400 ℃, 1 h; 525 ℃, 6 h. 取出灰化样品,每个坩埚加入 1 mL的浓硝酸,过夜消化,直到观察到坩埚内为澄清液体.

定容:向坩埚中加入 2/3体积的超纯水,转入 PET瓶中,再用超纯水清洗坩埚 4~5次,洗液均转入 PET瓶,用超纯水定容到约 50.000 g,记录重量,此溶液为原液.

稀释: ICP-M对待测溶液中所测离子浓度要求不超过 2×10^{-6},所以原液要进一步稀释才能符合浓度要求. 取原液 2 mL倒入 PET瓶中,加入超纯水使最终重量约为 50.000 g,即稀释了约 25倍.

测定:用 ICP-M测定 Fe、Mo、Zn等 3种元素的含量. 制作标准曲线时,将标准液分别稀释 500倍, 1 000倍, 2 000倍.

每个样品重复 3次以上. 计算样品中 3种元素的相对质量分数,单位为 μg/mg或 ng/mg

2 结果与分析

2.1 Fe元素的含量变化

红树植物繁殖器官发育过程中铁 (Fe)元素的含量变化见图 1.

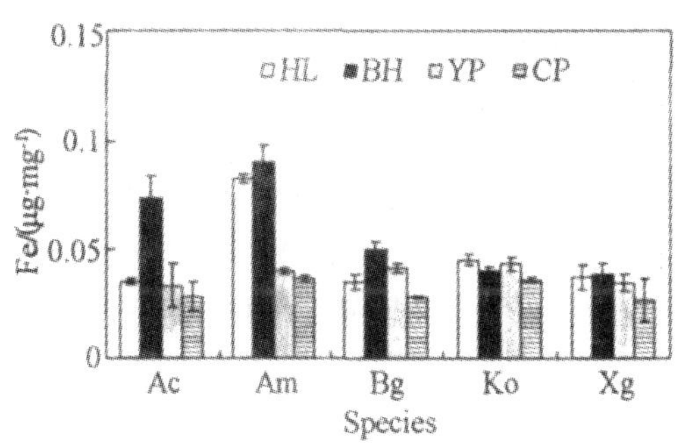

图 1 红树植物繁殖器官发育过程中铁 (Fe)元素的含量变化比较

Fig 1 Changes of Fe element content in the development of mangrove plant's propagative organs (HL, BH, YP and CP indicate Flower phase, Seed phase, Young hypocotyl phase and Adult hypocotyl phase, respectively; Ac = *Aegiceras corniculatum*, Am = *Avicennia marina*, Bg = *Bruguiera gymnorrihiza*, Ko = *Kandelia obovata*, Xg = *Xylocarpus grantum*)

在花蕾期,除白骨壤中 Fe元素含量明显较高外,其他种类含量接近,在 0.035 2~0.045 4 μg/mg之间. 在种子期,胎生红树植物中 Fe元素含量显著比非胎生红树植物木果楝高,其中隐胎生红树植物白骨壤、桐花树又高于显胎生红树植物木榄和秋茄. 在胚轴幼期和胚轴成熟期, Fe元素含量在不同种类的植物中差异较小.

比较不同发育时期的 Fe元素含量变化可以发现,种子萌发后呈现下降趋势,隐胎生红树植物白骨壤和桐花树在萌发初期尤为显著.

2.2 Mo元素的含量变化

红树植物繁殖器官发育过程中钼 (Mo)元素的含量变化见图 2.

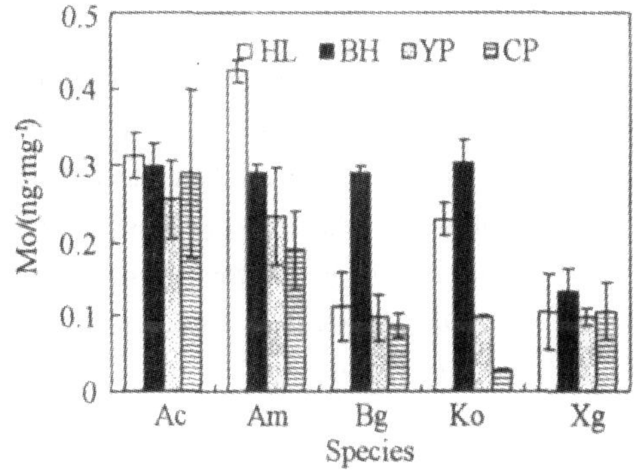

图 2 红树植物繁殖器官发育过程中钼 (Mo)元素的含量变化比较

Fig 2 Changes of Mo element content in the development of mangrove plant's propagative organs (HL, BH, YP and CP indicate Flower phase, Seed phase, Young hypocotyl phase and Adult hypocotyl phase, respectively; Ac = *Aegiceras corniculatum*, Am = *Avicennia marina*, Bg = *Bruguiera gymnorrihiza*, Ko = *Kandelia obovata*, Xg = *Xylocarpus grantum*)

在花蕾期,Mo元素含量在隐胎生红树植物白骨壤和桐花树中较高,非胎生红树植物木果楝较低.

种子期的 Mo元素含量在胎生红树植物中差异不大,在 0.288 2~0.302 0 ng/mg之间,平均为 0.294 5 ng/mg,明显高于非胎生红树植物木果楝 0.105 6 ng/mg

在胚轴幼期和胚轴成熟期,Mo元素含量在隐胎生红树植物白骨壤和桐花树中较高,显胎生红树植物和非红树植物含量较低,且差异较小.

比较不同发育时期的 Mo元素含量变化可以发现,隐胎生红树植物白骨壤和桐花树随发育进程呈现逐渐降低趋势,而显胎生红树植物中,种子期含量最高,萌发后呈现下降趋势. 非红树植物木果楝的 Mo元素含量随发育进程呈现略微增加趋势.

2.3 Zn元素的含量变化

红树植物繁殖器官发育过程中锌 (Zn)元素的含量变化见图 3.

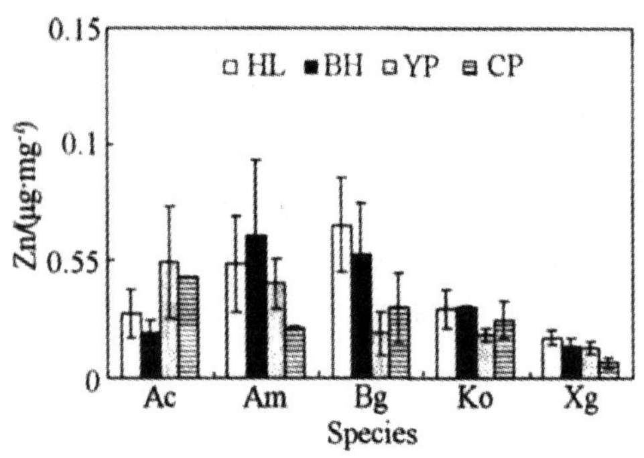

图 3 红树植物繁殖器官发育过程中锌(Zn)元素的含量变化比较

Fig 3 Changes of Zn element content in the development of mangrove plant's propagative organs (HL, BH, YP and CP indicate Flower phase, Seed phase, Young hypocotyl phase and Adult hypocotyl phase, respectively; Ac = *Aegiceras corniculatum*, Am = *Avicennia marina*, Bg = *Bruguiera gymnorrihiza*, Ko = *Kandelia obovata*, Xg = *Xylocarpus grantum*)

在花蕾期,Zn元素含量在非胎生红树植物木果楝中最低,为 0.017 3 μg/mg 胎生红树植物较高,在 0.027 4～0.065 7 μg/mg之间,平均为 0.0408 8 μg/mg

在种子期、胚轴幼期和胚轴成熟期,Zn元素含量均在非胎生红树植物木果楝中最低,胎生红树植物较高.其中,在胚轴幼期,隐胎生红树植物白骨壤和桐花树明显高于显胎生红树植物.

比较不同发育时期的 Zn元素含量变化,可以发现除木果楝随发育进程呈现下降趋势外,胎生红树植物无统一变化趋势.

3 讨 论

Fe是植物生长必需的微量营养元素,它是细胞色素蛋白、铁氧还蛋白、铁硫基蛋白等的重要辅助成分,在维持细胞内物质的正常代谢、叶绿体发育、酶活性等方面发挥重要作用[6]. Fe在植物中的流动性很小,不能向新生组织转移,因此它不能再度利用[6].本实验结果表明,在种子期,胎生红树植物中 Fe元素含量显著比非胎生红树植物木果楝高,而种子萌发后均呈现下降,说明胎生过程早期对 Fe元素有着较高的需求.

Mo元素存在于生物催化剂的组成之中,在植物体内的生理功能主要表现在氮素代谢方面;Mo还能促进光合作用的强度以及消除酸性土壤中活性铝在植物体内累积的毒害作用[7].刘鹏 (2006)报道,钼可能对植物生殖器官并无直接作用,但可通过对营养器官的影响间接作用于生殖器官[7].本实验发现,在花蕾期和种子期,Mo元素含量在胎生红树植物中明显高于非胎生红树植物木果楝,隐胎生红树植物白骨壤和桐花树随发育进程呈现逐渐降低趋势,而显胎生红树植物中,均以种子期含量最高,萌发后呈现下降.非胎生红树植物木果楝的 Mo元素含量随发育进程呈现略微增加趋势.推测 Mo元素对红树植物胎生早期的生长发育也起到促进作用.

Zn元素也是多种氧化酶活性的核心,参与电子的接受与传递,在植物体内的氧化还原反应中起重要作用,与叶绿素的形成以及碳水化合物、蛋白质的合成有密切关系[6].实验结果表明,同期相比,非胎生红树植物木果楝 Zn元素含量均胎生红树植物,反映了 Zn元素在红树植物胎生过程中可能发挥一定作用.

红树胎生是一个具有种特异性的、由激素调控的内在过程,并且该过程受到环境因素的一定影响[5,8]. Farnsworth & Farrant(1998)通过比较胎生和非胎生红树植物的内源 ABA含量,证实胎生红树植物成熟种子缺少休眠阶段与其内源 ABA含量低有关[9].在高等植物中,ABA的生物合成可能是通过以类胡萝卜素为前体的间接途径进行的,因此,类胡萝卜素的生物合成受到抑制,或从类胡萝卜素向 ABA的转化受抑制,都将导致种子中 ABA缺乏或低含量,从而促进种子胎生[10]. Fe、Mo、Zn元素可能通过影响 ABA生物合成的某些酶的活性而与胎生现象关联.

不同种红树植物虽然生长环境与生理特性相似,但种间亲缘较远,因此 3种元素的含量变化有所不同.至于 Fe、Mo、Zn元素在红树植物生殖器官中分布的动力学机理以及对繁殖器官的生长发育的作用方式还有待深入研究.

参考文献:

[1] 赵胡,郑文教.红树植物桐花树生长发育过程的元素动态与抗盐适应性 [J].海洋科学,2004,28: 1 - 5.

[2] 张宜辉,王文卿,池敏杰,等.显胎生红树植物木榄胎生胚轴发育 [J].海洋学报,2006,28: 121 - 127.

[3] 王文卿,王瑁.中国的红树林 [M].北京:科学出版社,2007.

[4] Farnsworth E J. Hormones and shifting ecology throughout plant development[J]. Ecology, 2004, 85: 5 - 15.

[5] Tomlinson P B, Cox P A. Systematic and functional anatomy of seedlings in mangrove Rhizophoraceae: vivipary explained? [J]. Botanical Journal of the Linnean Society, 2000, 134: 215 - 231.

[6] 武维华,主编.植物生理学 [M].北京:科学出版社,2003: 86 - 98.

[7] 刘鹏.大豆钼、硼营养研究进展 [J].中国农学通报,2001,

6: 41 - 44.

[8] Farnsworth E J. The ecology and physiology of viviparous and recalcitrant seeds[J]. Annual Review of Ecology and Systematics, 2000, 31: 107 - 138.

[9] Farnsworth E J, Farrant J M. Reductions in abscisic acid are linked with viviparous reproduction in mangroves[J]. American Journal of Botany, 1998, 85: 760 - 769.

[10] Kermode A R. Role of abscisic acid in seed dormancy[J]. Journal of Plant Growth Regulation, 2005, 24: 319 - 344.

Changes of Fe, Mo and Zn Elements in the Development of Mangrove Plant's Propaga tive Organs

LI Kuang-da, WANG Jie, YANG Sheng-chang

(*School of Life Sciences*, *Xiamen University*, *Xiamen* 361005, *China*)

Abstract: Changes of Fe, Mo and Zn elements were studied in the development of propagative organs from five mangrove species, including two viviparous species *Bruguiera* gymnorrihiza and *Kandelia obovata*, two cryp to vivipary species *Avicenniamarina* and *Aegiceras corniculatum* and one non-viviparous species *Xylocarpus grantum*. In flower phase, mangrove plants had similar Fe element content with 0.035 2 ~ 0.045 4 μg/mg except *Avicennia marina*. *Avicennia marina* and *Aegiceras corniculatum* had highest Fe contents in seed phase with 0.090 2 and 0.073 3 μg/mg, respectively, then decreased after germ ination, and reached lowest in adult hypocotylphase. In seed phase, four viviparousm angrove species had similar Moelement content with 0.288 2~0.302 0 ng/mg, further more than non-viviparous X. grantum with 0.105 6 ng/mg. After germ ination, viviparousm angrove species decreased Mocontent. At same phase, X. grantum had lowest Mocontent. In the development of propagative organs, X. grantum gradually decreased Zn element content, but was always lower than viviparous mangrove species at same phase.

key words: mangrove plant; viviparity; propagative organs; element

铝胁迫对海莲幼苗保护酶及脯氨酸含量的影响*

马　丽　杨盛昌

(厦门大学生命科学学院滨海湿地生态系统教育部重点实验室,福建 厦门 361005)

摘　要:为探讨 Al^{3+} 胁迫对海莲的影响,研究了 10~50 mmol/L Al^{3+} 处理下海莲幼苗叶片和根系的过氧化物酶(POD)、过氧化氢酶(CAT)、超氧化物歧化酶(SOD)、抗坏血酸过氧化物酶(APX)的活性以及可溶性蛋白质、丙二醛(MDA)和游离脯氨酸(Pro)含量的变化。结果表明,海莲幼苗能耐受 50 mmol/L 的 Al^{3+} 胁迫处理,具有较高的耐铝性。但在 50 mmol/L Al^{3+} 处理时,海莲幼苗叶片和根系的质膜系统膜脂过氧化加重,MDA 含量增加;细胞活性氧代谢失衡。在保护酶系统中,Al^{3+} 处理促进了叶片中 APX 和 POD 活性的提高,降低了 CAT 的活性,SOD 的活性呈下降趋势;海莲根部 POD 和 SOD 活性均显著提高,而 CAT 活性下降。25~50 mmol/L Al^{3+} 处理下,海莲叶片和根部可溶性蛋白质含量均显著下降;Pro 的含量在叶片和根均有显著增加。

关键词:海莲;铝胁迫;保护酶系统;脯氨酸

中图分类号:Q945　　**文献标识码**:A　　**文章编号**:1000-3142(2009)05-0648-05

Effect of aluminium on protection enzyme system and proline of *Brugiera sexangula* seedlings

MA Li, YANG Sheng-Chang

(*Key Laboratory of Ministry of Education for Coastal and Wetland Ecosystem, School of Life Sciences, Xiamen University, Xiamen* 361005, *China*)

Abstract: The activity of POD, CAT, SOD, APX, the content of soluble protein, MDA and proline in the leaves and roots of *Brugiera sexangula* were studied under 10—50 mmol/L $AlCl_3$ treatments. The results showed that, *B. sexangula* had strong tolerance to high concentrations of Al^{3+}, even 50 mmol/L of Al^{3+}, but under this concentration, the membrane system in the leaves and roots were seriously damaged due to peroxidation, while the content of MDA increased and the active oxygen metabolism of the cells were unbalanced. In protection enzyme system, Al^{3+} treatments concentrated on the raising of APX and POD activity, reducing of CAT and SOD activity in leaves; for the roots, the POD and SOD activity were significantly increased, while the CAT activity were decreased. Under treatments of 25—50 mmol/L Al^{3+}, in both leaves and roots, the soluble protein content decreased evidently but the Pro content increased.

Key words: *Brugiera sexangula*; Al stress; protection enzyme system; proline

铝是自然界中丰度最高的金属元素,占地壳质量7%左右,无机离子态的铝在酸性条件下大量渗出,污染了生态环境,严重危害了农作物和森林树种的生长(孔繁翔等,2000)。铝离子对植物的伤害反映在:影响植物营养成分的吸收与代谢,破坏细胞膜的结构和功能,影响保护酶的活性,使植物生长停

* 福建省自然科学基金(B0510003)资助

原载于广西植物,2009,29(5):648—652

止，根系、茎部和叶片功能坏死等，这在小麦（Dark等，2004）、水稻（Ma 等，2002）、马尾松（曹洪法等，1992）、茶树（陆建良等，1997）、龙眼（肖祥希等，2003）、荞麦（王芳等，2006）植物中已得到研究。

红树林是生长于热带和亚热带海岸和河口潮间带的木本植物群落，具有防风固堤，净化环境和维护生态平衡等作用。特殊的生境造就了红树植物耐盐和抗水淹等生理生态学特性。已有的研究表明：红树植物对汞和镉等重金属元素有较强的富集作用和耐受特性（林鹏等，1989；吴桂容等，2006；杨盛昌等，2003）。但到目前为止，有关红树植物的耐铝特性、铝离子对红树植物的生理伤害等方面的研究还鲜有报道。因此，本文以海莲（*Brugiera sexangula*）幼苗为材料，通过砂培和不同浓度 Al^{3+} 盐处理，探讨 Al^{3+} 胁迫对红树植物生长和保护酶系统等的影响，为研究红树植物的耐铝特性及其适应机制提供依据。

1 材料与方法

1.1 材料培养及 Al^{3+} 盐处理

海莲成熟胚轴采于海南省文昌清澜港红树林自然保护区，平均长度为 8.23±1.04 cm、平均重量为 12.2±0.68 g。采用随机分组，将海莲胚轴于厦门大学生命科学学院温室内进行砂培，砂砾粒径 2～4 mm，经自来水反复冲洗后装入塑料网盆中，每盆重约 2.5 kg，每盆种植海莲胚轴 5～6 棵。培养液采用 Hoagland's 液体培养基，并混合 3‰ NaCl 以保证海莲幼苗的生理需盐，每隔 5 d 更换 1 次培养基，每天补充因蒸发损失的水分。1 个月后进行铝胁迫处理，氯化铝（Al^{3+}）浓度分别为 10、25、50 mmol/L，以未添加铝盐的培养基作为对照。每一处理重复 4 次，为期 2 个月。

1.2 方法

1.2.1 采样　采取海莲第 3～4 对叶片和完整根系，经自来水及重蒸水反复清洗，晾干。

1.2.2 酶液的制备　取 1 g 新鲜叶片或根组织，剪碎，分次加入 10 mL 62.5 mmol/L 磷酸缓冲液（PBS pH 为 7.8，含 1%PVP）冰浴研磨，15 000 g，4 ℃离心 20 min，取上清。酶液中的蛋白质含量以 Brandford（1976）考马斯亮蓝 G-250 方法测定，以牛血清蛋白作标准曲线。

1.2.3 过氧化物酶（POD）活性的测定　取 100 μL 酶液（用 PBS 代替酶液做空白），加入 1.5 mL 反应混和液（62.5 mmol/L，pH7.8 PBS，20 mmol/L 愈创木酚），混匀，25 ℃温浴 5 min，加 10 μL 过氧化氢启动反应于 470 nm 波长处做时间扫描，扫描曲线率为酶反应速率，酶活性以 ΔA_{470}/g · Fw · min 表示（Jiang 等，2003）。

1.2.4 过氧化氢酶（CAT）活性的测定　CAT 活性的测定参考郝再彬等（2004）并略加改进，反应体系为 100 μL 酶液加 900 μL 62.5 mmol/L 磷酸缓冲液（含 1% H_2O_2）启动反应，于 240 nm 波长处做时间扫描，扫描曲线率为酶反应速率，酶活性以 ΔOD_{240}/g · Fw · min 表示。

1.2.5 超氧化物歧化酶（SOD）活性的测定　SOD 的活性采用氮蓝四唑（NBT）光照氧化还原显色反应的方法测定，以每单位时间内抑制光还原 50%的氮蓝四唑（NBT）为一个酶单位，酶活性以 U/gFw。

1.2.6 抗坏血酸过氧化物酶（APX）活性的测定　以 Nakano & Asada（1981）的方法进行，290 nm 波长处做时间扫描，根据单位时间内 OD_{290} 减少的值，计算 APX 的活性。

1.2.7 丙二醛（MDA）含量测定　取上述上清液 2.0 mL 加入 2.0 mL 10%三氯乙酸（TCA）（含 0.5%硫代巴比妥酸钠（TBA））混匀，煮沸 15 min 后快速冷却，4 000 r/min 离心 20 min，以 10% TCA（含 0.5%TBA）为参比，上清液在 532 和 600 nm 波长处测定 OD 值。

1.2.8 游离脯氨酸（Pro）含量的测定　采用郝再彬等（2004）茚三酮显色法，测定植物组织游离脯氨酸含量的变化。

1.2.9 统计方法　用 SPSS 11.0 统计软件处理所有结果，采用单因素方差分析法（One-way ANOVA）对实验数据进行统计分析。

2 结果与分析

2.1 对海莲幼苗生长的影响

100 mmol/L 和 75 mmol/L 铝处理的海莲植株，分别在处理 7 d 和 15 d 后停止生长，最终因根系坏死导致植株枯死。在 10～50 mmol/L 铝浓度范围内，海莲幼苗随铝浓度增加，生长明显减慢，与对照组抽条高度 18.4±1.4 cm 相比，分别为 16.2±2.7 cm、15.4±1.1 cm 和 12.3±1.7 cm。叶片数量和叶面积也相应减少。对根系观察发现，处理组的根系侧根和根毛较对照组明显减少,且显得松脆。

2.2 对海莲幼苗叶片和根系丙二醛(MDA)含量的影响

丙二醛(MDA)是脂质过氧化产物,其含量高低反映了膜质的破损程度。表 1 结果显示,低浓度铝盐处理下,海莲叶片和根的 MDA 含量变化不显著,但随铝浓度增加到 50 mmol/L,叶片和根的 MDA 含量均显著增强,较对照组分别增加了 1.5 倍和 1.9 倍,说明 50 mmol/L 铝浓度引发了海莲叶片和根细胞的膜质过氧化,破坏了脂膜的完整性。

表 1 Al^{3+}胁迫对海莲幼苗叶片和根部丙二醛(MDA)含量的影响

Table 1 Effect of Al^{3+} stress on MDA in leaves and roots of *Brugiera sexangula* seedlings

铝处理 Al^{3+} treatment (mmol/L)	叶片 MDA 含量 MDA in leaves (nmol/gFw)	根 MDA 含量 MDA in roots (nmol/gFw)
0	4.716±0.116Aa	1.657±0.124Aa
10	4.699±0.530Aa	1.574±0.211Aa
25	4.620±0.333Aa	2.010±0.375Aa
50	5.987±0.107Bb	3.215±0.433Bb

注:同一栏内字母不同者为差异显著,大小字母分别表示 $P=0.01$ 和 $P=0.05$ 的显著水平。下同。

Note: values in same column followed by different capital or small letter are significantly different at $P=0.01$ or $P=0.05$ level, respectively. The same below.

2.3 对海莲幼苗叶片和根部保护酶系统的影响

不同铝浓度处理下,海莲幼苗叶片和根部的某些保护酶活性呈现不同的变化趋势。

2.3.1 对 POD 和 CAT 活性的影响 从表 2 可知,与对照相比,铝盐处理下海莲叶片和根部的 POD 活性显著增强。在根部,随着铝盐浓度增加,POD 活性相应提高,50 mmol/L 处理组较对照组增加 2.8 倍。与对照组相比,海莲叶片和根系的 CAT 活性随铝浓度增加呈现下降趋势,但叶片 CAT 活性在各处理组之间差异并不显著。

表 2 Al^{3+}胁迫对海莲幼苗叶片和根中 POD 和 CAT 活性的影响

Table 2 Effect of Al^{3+} stress on POD and CAT in leaves and roots of *Brugiera sexangula* seedlings

铝处理 Al^{3+} treatment (mmol/L)	叶 POD 活性 POD in leaves (ΔA_{470}/g·Fw·min)	叶 CAT 活性 CAT in leaves (ΔA_{240}/g·Fw·min)	根 POD 活性 POD in roots (ΔA_{470}/g·Fw·min)	根 CAT 活性 CAT in roots (ΔA_{240}/g·Fw·min)
0	8.53±0.32Aa	2.73±0.45Aa	5.30±0.68Aa	16.02±1.67Aa
10	10.93±0.51Ab	1.57±0.47Bb	7.33±0.58Aa	11.55±0.49Bb
25	20.77±0.94Bc	1.23±0.31Bb	12.87±2.70Bb	10.34±0.76Bb
50	11.73±0.61Ab	1.15±0.41Bb	20.10±2.92Cc	7.20±0.85Bc

表 3 Al^{3+}胁迫对海莲幼苗叶片和根部 SOD 和 APX 活性的影响

Table 3 Effect of Al^{3+} stress on SOD and APX activity in leaves and roots of *Brugiera sexangula* seedlings

铝处理 Al^{3+} treatment (mmol/L)	叶 SOD 活性 SOD in leaves (U/g·Fw)	根 SOD 活性 SOD in roots (U/g·Fw)	叶 APX 活性 APX in leaves (ΔA_{290}/g·Fw·min)	根 APX 活性 APX in roots (ΔA_{290}/g·Fw·min)
0	118.99±6.3Aa	96.80±4.70Aa	4.87±0.74Aa	9.70±0.15Aa
10	106.51±21.8Aa	113.31±5.14Ab	9.06±1.21Bb	14.50±0.11Bb
25	97.81±11.05Aa	124.88±4.66Bc	10.10±1.01Bb	8.76±0.16Aa
50	85.88±6.13Bb	127.71±9.25Bc	11.42±1.18Bb	9.10±0.1Aa

2.3.2 对 SOD 和 APX 活性的影响 表 3 结果显示,海莲叶片 SOD 活性在 10~50 mmol/L 铝处理下表现降低趋势,但 10 mmol/L、25 mmol/L 两个处理对 SOD 活性影响不显著;铝处理水稻幼苗的研究(石贵玉,2004)中,也发现水稻叶片 SOD 活性随着铝浓度的上升呈下降趋势的类似情况。在 10~50 mmol/L 铝处理下,海莲根中 SOD 的活性均显著增加。海莲叶片的 APX 活性较对照组增加了 0.86~1.34 倍;而海莲根部的 APX 活性除 10 mmol/L 浓度处理下有所增加外,其他浓度处理的结果与对照组相比差异不显著。SOD 是植物抵御胁迫伤害,清除活性氧的第一道防线。长期铝胁迫下,海莲叶片的 SOD 活性下降,可能因为 SOD 酶只在一定时间范围内比较活跃,作用持续的时间短;细胞内产生的活性氧可能超出了 SOD 酶的清除范围,活性氧代谢的平衡遭到破坏,产生了一定程度的氧化伤害。而海莲叶片中 APX 活性的增加,可能弥补了 SOD 对膜质过氧化损伤修复作用的减弱趋势。根部

SOD 活性得到增高，表明海莲幼苗可以通过提高根部 SOD 活性来适应铝盐胁迫；而 APX 在海莲根部细胞器中作用并不活跃。

2.4 对海莲幼苗不同部位蛋白质、脯氨酸含量的影响

表 4 结果显示，海莲叶片和根部的可溶性蛋白质含量在 25～50 mmol/L 两个浓度 Al^{3+} 处理下较对照组均显著降低，在铝胁迫对龙眼幼苗蛋白质和核酸含量的影响的研究中(肖祥希等，2006)，龙眼叶片和根中可溶性蛋白质总量也随铝浓度的增加呈现降低的变化趋势；对此的解释有：蛋白酶活性升高导致蛋白质水解；逆境胁迫导致活性氧增加，膜脂过氧化加剧，蛋白质合成受到抑制。

脯氨酸是植物蛋白质的组分之一，以游离状态广泛地存在于植物体中。正常条件下，高等维管植物体中游离脯氨酸的含量并不多，但在逆境条件下(干旱、盐渍、冷冻等)植物体内游离脯氨酸可增加 10～100 倍(张显强等，2004)。

铝胁迫处理导致海莲叶片和根部游离脯氨酸的大量积累，且随着铝处理浓度的增加，脯氨酸含量呈现极显著的递增趋势。海莲叶片脯氨酸含量增加了 7.2～14.4 倍，根部脯氨酸含量增加了 10.7～30.1 倍。

表 4 Al^{3+} 胁迫对海莲幼苗叶片和根部可溶性蛋白和游离脯氨酸含量的影响

Table 4 Effect of Al^{3+} stress on soluble protein and Pro in leaves and roots of *Brugiera sexangula* seedlings

铝处理 Al^{3+} treatment (mmol/L)	叶蛋白质含量 Protein in leaves (mg/g · Fw)	根蛋白质含量 Protein in roots (mg/g · Fw)	叶 Pro 含量 Pro in leaves (μg/g · Fw)	根 Pro 含量 Pro in roots (μg/g · Fw)
0	9.27±0.28Aa	8.54±0.64Aa	21.3± 7.83Aa	13.6±1.7Aa
10	9.03±0.86Aa	8.42±0.22Aa	34.12±8.55Aa	36.97±2.4Aa
25	6.24±1.01Bb	5.52±0.14Bb	174.72±7.83Bb	159.26±14.5Bb
50	5.42±0.46Bb	4.96±0.21Bb	328.28±3.56Cc	422.8±19.49Cc

3 讨论

当 Al^{3+} 盐浓度在 50 mmol/L 时，海莲幼苗能够生长存活，说明海莲对 Al^{3+} 盐有很强的耐受性。但当浓度大于 50 mmol/L 时，海莲生长受到严重抑制，并发生枯死。

海莲叶片和根的 MDA 含量在 10～25 mmol/L Al^{3+} 浓度处理时变化不显著，当浓度达到 50 mmol/L 时，MDA 含量在叶片和根两个部位均显著增加，说明高浓度的铝盐使细胞膜脂过氧化作用增强，细胞膜的结构和功能受到破坏。

在海莲叶片的活性氧清除系统中，各种保护酶与抗坏血酸(AsA)和谷胱甘肽(GSH)等叶绿体中的重要抗氧化剂协调作用。APX 是叶绿体中专一地清除 H_2O_2 的关键酶，与谷胱甘肽还原酶(GR)等酶协同作用，通过抗坏血酸一谷胱甘肽循环移去叶绿体中的 H_2O_2，同时使 AsA 和 GSH 再生(肖祥希等，2003)。铝胁迫下，海莲叶片 APX 活性呈显著增加的趋势，可能是叶绿体系统对铝胁迫的适应。

CAT 和 POD 也是植物体内清除 H_2O_2 的主要保护酶，在铝胁迫下，海莲叶片的 CAT 活性显著下降，引起了 H_2O_2 的积累，但由于 POD 活性显著增强，加剧了 H_2O_2 的降解，从而降低细胞内的 H_2O_2 的水平。在 50 mmol/L 铝盐处理时，POD 的活性有所下降，可能是由于叶片积累了较多的铝而抑制了 POD 的活性。

此外，SOD 作为活性氧清除系统中的第一道防线，与其他保护酶协调作用，有效控制细胞内活性氧水平，阻止脂质过氧化物的积累。在高铝浓度下，海莲叶片的 SOD 活性降低，表明铝胁迫抑制了 SOD 活性，这也可能是导致膜脂过氧化程度加深的原因之一。

根系是植物面对土壤铝胁迫的首要器官，其生理代谢功能直接决定了植物对铝胁迫的耐受程度。但有关根系保护酶系统的特性研究相对较少。在本研究中，海莲根部的保护酶系统的活性变化与叶片有些不同：随铝处理浓度的增加，根 POD 和 SOD 活性显著增强，但 APX 的活性变化并不明显。根系 POD 和 SOD 活性的增强，表明根系保护酶系统的防御能力增强，对活性氧的清除作用更为活跃。究其原因，可能与铝胁迫因子诱导了根系 POD 和 SOD 酶基因的大量表达有关。另一种可能是，在铝胁迫条件下，根系能合成包括有机酸在内的特殊物质，这些物质的存在有助于稳定根系 POD 和 SOD 酶基因的转录产物或者促进了酶活力提高。因此，从保护酶系统的变化来看，红树植物根系和叶片可能存在不同的铝胁迫适应机制。

铝胁迫导致海莲幼苗叶片和根部蛋白质含量下降,这与高盐浓度培养海莲(郑海雷等,1998),随着盐分的增加,幼叶的蛋白质含量在总体上呈下降的趋势的结果相似。铝胁迫下蛋白质含量下降的主要原因目前认为有:植物为了适应和抵御胁迫,提高了蛋白酶的活性,从而加剧了蛋白质的水解;胁迫条件下植物细胞内RNA、DNA的含量减少,降低了蛋白质的合成速率;逆境胁迫导致活性氧增加,膜脂过氧化加剧(肖祥希等,2003),抑制蛋白质合成。

脯氨酸作为一种重要的渗透保护物质,在植物的抗性生理中发挥着重要的作用。最近有关脯氨酸合成和代谢的研究表明了脯氨酸具有多种功能,即脯氨酸可作为一种迅速利用的能源,氮源和碳源(张显强等,2004)。此外,脯氨酸代谢中间产物具有诱导基因表达作用(Iyer 等,1998)及降低渗透胁迫所造成的氧伤害作用。

从海莲叶片和根部的游离脯氨酸积累情况看,铝胁迫下,两个部位的游离脯氨酸含量均随铝浓度的增加有显著增加。但是,作为渗透胁迫的有机渗透物质积累,脯氨酸含量变化与铝胁迫是否直接相关还存在很多争议,仍需要进一步验证。

参考文献:

林鹏,陈荣华. 1989. 九龙江口红树林对汞的循环和净化作用[J]. 海洋学报,**11**(2):242－247

郝再彬,苍晶,徐仲. 2004.《植物生理实验》[M]. 哈尔滨工业大学出版社

Cao HF(曹洪法),Gao JX(高吉喜),Shu JM(舒俭民). 1992. Study on the response of *Pinus massoniana* seedling to aluminum(铝对马尾松幼苗影响的研究)[J]. *Acta Ecol Sin*(生态学报),**12**(3):239－246

Darko E,Ambrus H. 2004. Aluminum toxicity,Al tolerance and oxidative stress in an Al-sensitive wheat genotype and in Al-tolerant lines developed by *in vitro* microspore selection[J]. *Plant Sci*,**166**:583－591

Iyer S,Caplan A. 1998. Products of proline catabolism can induce osmoticalay regulated geres in rice[J]. *Plant Physiol*,**116**:203－211

Jiang,M,Zhang J. 2003. Cross-talk between calcium and reactive oxygen species originated from NADPH oxidase in abscisic acid-induced antioxidant defence in leaves of maize seedlings[J]. *Plant Cell Environ*,**26**:929－939

Kong FX(孔繁翔),Sang WL(桑伟莲),Jiang X(蒋新),*et al*. 2000. Aluminum toxicity and tolerance in plants(铝对植物毒害及植物抗铝作用机理)[J]. *Acta Ecol Sin*(生态学报),**20**(5):855－862

Lu JL(陆建良),Liang YR(梁月荣). 1997. Effect of aluminium stress on superoxide dismutase in tea and other plants(铝对茶树等植物超氧化物歧化酶的影响)[J]. *J Tea Sci*(茶叶科学),**17**:197－200

Ma JF,Shen RF. 2002. Response of rice to Al stress and identification of quantitative trait Loci for Al tolerance[J]. *Plant Cell Physiol*,**43**(6):652－659

Nakano Y,Asada K. 1981. Hydrogen peroxide is scavenged by ascorbate specific peroxidase in spinach chloroplasts[J]. *Plant Cell Physiol*,**22**:867－880

Shi GY(石贵玉). 2004. Effect of aluminium on growth and some physiological function of rice seedlings(铝对水稻幼苗生长和生理的影响)[J]. *Guihaia*(广西植物),**24**(1):77－80

Wang F(王芳),Liu P(刘鹏),Xu GD(徐根娣),*et al*. 2006. Effects of aluminium amount in soil on the root growth of buckwheat(铝对荞麦根系的影响)[J]. *Guihaia*(广西植物),**26**(3):321－324

Wu GR(吴桂容),Yan CL(严重玲). 2006. Effects of Cd on the growth and osmotic adjustment regulation contents of *Aegiceras conrniculatum* seedlings(镉对桐花树幼苗生长及渗透调节的影响)[J]. *Ecol Environ*(生态环境),**15**(5):1 003－1 008

Xiao XX(肖祥希),Yang ZW(杨宗武),Xiao H(肖晖),*et al*. 2003. Effect of aluminum stress on active oxygen metabolism and membrane system of Longan(*Dimocarpus longan*) leaves(铝胁迫对龙眼叶片活性氧代谢及膜系统的影响)[J]. *Sci Silv Sin*(林业科学),**39**:52－57

Xiao XX(肖祥希),Liu XH(刘星辉),Yang ZW(杨宗武),*et al*. 2006. Effect of aluminum stress on the content of protein and nucleic acid of longan (*Dimocarpus longan*) seedlings(铝胁迫对龙眼幼苗蛋白质和核酸含量的影响)[J]. *Sci Silv Sin*(林业科学),**42**(10):24－30

Yang SC(杨盛昌),Wu Q(吴琦). 2003. Effect of Cd on growth and physiological characteristics of *Aegiceras conrniculatum* seedlings(Cd对桐花树幼苗生长及某些生理特性的影响)[J]. *Marine Environ Sci*(海洋环境科学),**22**(1):38－42

Zhang XQ(张显强),Luo ZQ(罗在柒),Tang JG(唐金刚). 2004. Effect of high temperature and drought stress on free proline content and soluble sugar content of *Taxiphyllum taxirameum*(高温和干旱胁迫对鳞叶藓游离脯氨酸和可溶性糖含量的影响)[J]. *Guihaia*(广西植物),**24**(6):570－573

Zheng HL(郑海雷),Lin P(林鹏). 1998. Effect of salinity on membrane protection system for *B. sexangula* and *B. gymnorrhiza* seedling(培养盐度对海莲和木榄幼苗膜保护系统的影响)[J]. *J Xiamen Univ*(*Nat Sci*)(厦门大学学报·自然科学版),**37**(2):278－282

四 红树林植物形态学

PART IV MANGROVE PLANT MORPHOLOGY

中国红树植物生态解剖学研究综述*

李元跃[1,2] 林 鹏[1,2]

(1. 厦门大学生命科学学院,福建 厦门 361005;2. 厦门大学湿地与生态工程研究中心,福建 厦门 361005)

中图分类号:Q94 文献标识码:A 文章标号:1000-3096(2006)04-0069-05

植物与其生长的环境是一个整体,环境对植物的生长作用影响了形态构成,使植物形成了适应环境的形态结构,研究植物结构与生态环境关系的学科就称为"植物生态解剖学",它既是植物解剖学的分支学科,也是植物解剖学与生态学的交叉学科。随着环境问题的日益突出,这一学科有了较快的发展,近些年来,人们从植物个体、组织、细胞、亚细胞水平各个层次上来研究植物与生态环境或生态因子之间的关系,以及利用形态解剖学指标来指示、评价环境质量[1]。

红树林是指热带海岸潮间带的木本植物群落。对红树林的研究具有多方面的意义,其中最主要的是维护海岸生态平衡的特殊生态系以及林鹏[2]提出的"三高理论",即高生产率、高归还率、高分解。红树林是热带河口海湾生态系重要的第一性生产量的贡献者,它的生产量的形成和变化具有独特的生态学规律,也间接影响到河口海湾水产业和渔业的正常发展。红树植物是专指生长在红树林中的木本植物,是红树林群落中的重要物种,因此,红树植物的研究对于沿海海岸的防风固堤、渔业发展、湿地保护等具有重要的作用。

根据林鹏[3]的研究,中国红树植物在海南、广西、广东、福建、台湾五省有自然分布,浙江省引种一种秋茄成功。红树植物的种类除海南岛有较多外,各省(区)随纬度提高而种类逐渐减少,具体分布见表1[4]。

近年来,对中国红树植物的研究主要集中在能流物流、生理生态学、污染生态学、物种多样性、分子多样性等几方面,而在红树植物的形态解剖学以及生态解剖学的研究较少。作者试图对近些年来对中国红树植物的生态解剖学的研究情况做一些简要的介绍。

1 红树植物形态结构的研究

国内外对红树植物形态解剖学方面研究得较少,叶庆华等[5]对桐花树叶片的盐腺系统进行了研究,结果表明:盐腺系统5个部分,从叶肉向叶表排列是:收集细胞、基细胞、分泌细胞、收集室、盐腺盖。林鹏[2]对我国红树植物根、茎、叶等营养器官的结构及其对生境的适应性作了概括性讨论。研究表明:(1)红树植物根生长在含有高水分、高盐分、缺乏氧气而含有大量还原性物质的土壤中,经过长期的自然选择,形成了各种各样的适应性结构。其中主要包括具有复杂结构的周皮,含发达通气组织的皮层以及适应特殊生境的各种异常次生构造。(2)红树植物为多年生植物,具有木质化的茎。茎常具有发达的周皮,它包括木栓层、木栓形成层和栓内层三部分。红树植物茎皮层中的显著特征就是皮层薄壁细胞通常排列疏松,通气组织发达。许多红树植物的皮层还有一些异常结构,常见的有皮层维管束与皮层木栓腔;各种红树植物的树皮中都含有丰富的单宁。红树植物生长在高盐度的环境中,茎一般生长缓慢,次生木质部材质坚硬。表现在次生木质部导管排列为单管孔或复管孔,管孔直径略小但数量偏多。导管分子穿孔板甚斜,具有数量少或厚的横条。导管间的纹孔梯状排列;导管与木射线间的纹孔大多一侧开口,在木射线一边的纹孔大,呈轴或斜向的椭圆形延长,包着导管壁的2至数个小的圆形至卵圆形的纹孔。木射线宽且高,木射线细胞大多含有深色树胶物和单晶体。(3)红树植物叶总的来说表现出旱生叶的构造。一般包括表皮、下皮、叶肉组织与叶脉四部分。红树

* 国家自然科学基金资助项目(40276028);教育部博士点基金资助项目(20030384007)
原载于海洋科学,2006,30(4):69－73

科名	种名	分布							
		海南	香港	澳门	广东	广西	台湾	福建	浙江
卤蕨科	1. 卤蕨(*Acrostichum aureum*)	+	+	+	+	+	+	+	
(Acrostichaceae)	2. 尖叶卤蕨(*A. speciosum*)	+			+	+			
红树科	3. 柱果木榄(*Bruguiera cylindrica*)	+							
(Rhizophoraceae)	4. 木榄(*B. gymnorrhiza*)	+	+		+	+	+	+	
	5. 海莲(*B. sexangula*)	+							
	6. 尖瓣海莲(*B. s. var. rhynchopetala*)	+							
	7. 角果木(*Ceriops tagal*)	+	+		+	+	+		
	8. 秋茄(*Kandelia candel*)	+	+	+	+	+	+	+	+
	9. 红树(*Rhizophora apiculata*)	+							
	10. 红海榄(*R. stylosa*)	+	+		+	+	+	+	
爵床科	11. 小老鼠簕(*Acanthus ebractearas*)	+			+				
(Acanthaceae)	12. 老鼠簕(*A. ilicifolius*)	+	+	+	+	+	+	+	
使君子科	13. 厦门老鼠簕(*A. xiamenensis*)							+	
(Combretaceae)	14. 红榄李(*Lumnitzera littorea*)	+							
大戟科	15. 榄李(*L. racemosa*)	+	+		+	+	+		
(Euphorbiaceae)	16. 海漆(*Excoecaria agallocha*)	+	+		+	+	+	+	
楝科 (Meliaceae)	17. 木果楝(*Xylocarpus granatum*)	+							
紫金牛科 (Myrsinaceae)	18. 桐花树(*Aegiceras corniculatum*)	+	+	+	+	+	+	+	
棕榈科 (Palmaceae)	19. 水椰(*Nypa fruticans*)	+							
茜草科 (Rubiaceae)	20. 瓶花木(*Scyphiphora hydrophyllacea*)	+							
海桑科 (Sonneratiaceae)	21. 杯萼海桑(*Sonneratia alba*)	+							
	22. 海桑(*S. caseolaris*)	+							
	23. 海南海桑(*S. hainanensis*)	+							
	24. 大叶海桑(*S. ovata*)	+							
	25. 拟海桑(*S. paracaseolaris*)	+							
梧桐科	26. 无瓣海桑(*S. apetala*)	+			+			+	
(Sterculiaceae)	27. 银叶树(*Heritiera littoralis*)	+			+				
马鞭草科 (Verbenaceae)	28. 白骨壤(*Avicennia marina*)	+	+	+	+	+	+	+	
合计		27	10	5	14	11	10	11	1

注:作者 1997 年到中国台湾省考察鉴定结果,原台湾定的红茄苳,台湾称五梨蛟(*R. mucronata*)应改为红海榄(*R. stylosa*),海峡两岸同属一个种;无瓣海桑已引种成功

植物叶表皮上具有的显著特征之一是表皮细胞外平周壁上有厚的连续分布的角质层。角质层的存在有助于减少叶内水分的散失。表皮上还有气孔器的结构;一些红树植物适应海滩盐渍环境,在叶表皮上还分化了各种各样的排水、泌盐的分泌结构,其中的一种重要分泌结构就是盐腺,它可以排出体内过剩盐分。下皮是红树植物与陆生植物叶的另一显著差别的结构。下皮是表皮层细胞通过平周分裂向内侧产生的细胞形成的,下皮的细胞通常较大,且一般为薄壁细胞,细胞内含有大量的水分,因而下皮的主要功能就是贮水作用。红树植物的叶肉组织具有栅栏组织和海绵组织的分化。红树植物叶脉的脉梢通常是由数个直径大、长度短的管胞分子组成。此外,大多数红树植物的叶中具有一般植物罕见的两种特殊结构:木栓瘤及皮孔排水器。黄桂玲和黄庆昌[6~8]对中国10科13属16种红树植物的叶、根、茎的形态结构进行了描述,这些植物的主要特征有:(1) 叶:角质层通常较厚,气孔一般仅分布于下表皮;大多数植物的叶具有木栓层产生的木栓瘤及皮孔排水器;贮水组织特别发达;普遍具有各种各样的分泌结构;脉梢的管胞通常由数个直径大而短的管状分子组成。(2) 根:在木栓形成前有1至多层外皮;木栓细胞壁薄,有的栓内层分化出石细胞,有的木栓细胞呈马蹄形增厚;气生根皮孔大而多;支柱根机械组织发达;根的导管直径大,数量多,常含侵填体;多数根不具凯氏带加厚的内皮层。(3) 茎:树皮富含单宁;周皮组成多样;皮层排列疏松;具各种各样的异常次生结构;机械组织发达;多数红树树干基部具大型密集的皮孔。

2 生态因子对中国红树植物形态结构的影响

2.1 光因子

光因子对植物的结构和形态建成有重要影响,这在其他植物研究上有许多报道,比如 Osborn 和 Taylor[9]通过颤毛栎(*Quercus velutina*)的形态结构、角质膜的显微及亚显微结构观察,总结出向阳叶和遮阴叶的特征,指出向阳叶比起遮阴叶来,叶片较小,叶缘浅裂较深,气孔多,叶肉厚,角质膜较厚。光因子对红树植物形态结构影响的研究较少,集中表现在光因子对红树植物光合作用和蒸腾作用的影响。林鹏[10]和雷泽湘[11]等对红海榄和秋茄的蒸腾作用与生态因子的关系进行了研究,结果表明:蒸腾作用的变化在不同天气条件下都与光照强弱有关,光照越强,其蒸腾强度也越大。杨盛昌等[12]等研究了红树植物对光环境的光合适应,研究结果表明,阳叶较阴叶有更大的光合能力或净光合生产。

2.2 温度因子

温度关系着红树植物的生长发育。林鹏[3]研究了红树植物种类分布与温度的关系,指出随纬度升高而温度下降是影响红树植物分布和生长的主要因子。在不受人类干扰的自然分布区,随纬度升高,温度下降,红树植物种类减少,生产力下降。并根据中国红树林区内最低月均气温的变化,划分出中国红树植物耐寒性等级序列。详见表2[13]。

表2 中国红树植物耐寒性等级序列

等级	最低月均温度(℃)	地域	种类
Ⅰ	8~10	闽东北沿海的福鼎至莆田之间	秋茄
Ⅱ	10~12	闽中沿海的莆田至厦门之间	桐花树、白骨壤、老鼠簕、黄槿(半红树)
Ⅲ	12~14	厦门以南至汕头沿海,台湾北部	
Ⅳ	14~16	广西沿海和广东汕头以南(不包括雷州半岛南端)	木揽、海漆、厦门老鼠簕
Ⅴ	16~18		红海揽、角果木、榄李、杨叶肖槿和海芒果(半红树植物)
Ⅵ	18~20	广东雷州半岛南端,海南岛北部(包括东寨港)	海莲、尖瓣海莲、小花老鼠簕、银叶树、玉蕊
Ⅶ	20~22	海南岛东岸(包括清澜港)和西岸,台湾岛西南海岸	海桑、大叶海桑、海南海桑、瓶花木、红树
		海南岛东南岸端(包括三亚、陵水)及热带珊瑚岛(包括西沙群岛,台湾岛以南海域小岛)	红揽李、水芫花(半红树植物,仅分布于岛礁、海岸或珊瑚岛)

根据该序列,中国耐寒性最强的红树植物是秋茄,自然分布纬度最高,达到福建福鼎,最低月均温度为8.4 ℃。秋茄等耐寒性强的红树植物,在低纬度炎热地区同样有分布,但生长并不高大,受优势种排挤,而多呈零星分布而不成纯林。因此,红树植物的分布广度取决于其耐寒性,越耐寒的种类,其自然分布范围越大,即其纬度跨度越大。

在抗寒性方面,杨盛昌等[14]对龙海浮宫红树植物树冠不同部位叶片的抗寒力进行了比较研究,结果表明:上层叶片接受的光辐射多,叶片革质化程度高,同时光合作用强,光合产物累计较多,因此抗寒力较强;内部叶片接受的光辐射少,叶片革质化程度低,同时光合作用弱,光合产物累计少,因此抗寒力较弱。李银鹏等[15]通过秋茄幼苗对低温的反应研究指出,秋茄苗在低温胁迫时总叶绿素含量不断下降。

林益明等[16]对秋茄次生木质部生态解剖学的比较研究表明:随着纬度的升高,温度下降,秋茄各分布区导管分子长度和导管直径呈现下降的趋势,穿孔板横隔的数目也随温度的下降而减少,射线分布频率随温度的下降而减少。

2.3 盐因子

红树植物是一类生活在周期性遭受海水浸淹的潮间带环境的植物类型,因此其生理特征和形态结构受盐度的影响较大。关于这一方面的研究也较多。

林鹏[3]研究了土壤盐度对桐花树叶片形态结构的影响,结果表明:(1) 叶面上单位面积的气孔数随土壤盐度提高而减少;(2) 角质层厚度随土壤盐度的提高而增厚;(3) 栅栏组织随土壤盐度的提高而增厚;(4) 海绵组织随土壤盐度的提高而细胞间隙加大;(5) 叶片厚度随土壤盐度的提高而加厚。叶庆华等[17,18]有关海滩盐度对秋茄和桐花树叶肉细胞超微结构影响的研究表明:生长在高盐度海滩的比生长在低盐度海滩的叶肉细胞的质膜皱缩厉害;细胞膜与细胞壁之间的间隙增大;叶绿体的基粒和基粒片层都显著膨胀,基粒与基粒片层之间的界限模糊,以至好象没有基粒;同时它们的线粒体的脊结构也更不清楚。

在盐度对红树植物生长的影响的研究中,郑文教等[19,20]在盐度对秋茄和海莲幼苗的生长影响研究表明:低盐度对红树植物高度生长以及对根、茎、叶和总生物量具有正刺激效应;高盐度起抑制作用,随着盐度的提高,幼苗生长量降低,叶片变小,叶片叶绿素含量随盐度提高而提高,相反的叶片可溶性糖含量则降低。

林益明等[16]对秋茄次生木质部生态解剖学的比较研究表明:盐度高造成秋茄低导管分布频率、低单管孔率、低双管率。邓传远[21]通过对红海榄和海桑木材的解剖证明,在研究的土壤盐度范围内,导管密度和管孔直径没有显著影响,但随盐度增大,导管的聚合度增大,单孔率下降。

2.4 污染物

由于工业发展,环境污染问题已日益严重,许多污染物对植物的形态结构产生了不同程度的影响,引起了植物学家的关注。在中国有关红树植物污染的研究目前主要有以下几个方面[2]:(1) 红树植物对重金属污染物的吸收与抗性;(2) 红树植物对有机农药的吸收、累计及抗性研究;(3) 红树植物对油污染的生物监测作用;(4) 城市污水对红树植物的生态生理特性影响等。污染物对红树植物形态结构的影响的研究则是空白。因此,这一方面内容可成为今后红树植物污染生态学研究的新课题。

总的来看,虽然中国红树植物的研究已在多方面取得良好的成果,但在红树植物生态解剖学的研究还是远远不够。红树植物生态解剖学作为红树植物解剖学与生态学的交叉学科,承担着各种红树植物的形态结构的研究以及各生态因子对红树植物形态结构影响的研究,为各种红树植物的分类、生境、系统进化地位、移栽以及湿地保护等方面提供重要的依据。因此红树植物生态解剖学可作为红树植物研究中新的重要领域之一。建议从以下几方面加以研究:(1) 在显微和亚显微水平上研究各种红树植物的形态结构;(2) 单主导生态因子对红树植物形态建成的影响;(3) 在显微和亚显微水平上定量化研究生态因子对红树植物形态结构变化的影响;(4) 利用红树植物形态结构指标指示环境质量和变化;(5) 应用陆海生红树科种类的形态解剖的变化,研究陆海植物进化的规律。

参考文献:

[1] 王勋陵. 植物生态解剖学研究进展[J]. 植物学通报, 1993, **10**(增刊):1-10.

[2] 林鹏. 中国红树林生态系[M]. 北京:科学出版社, 1997. 1-108.

[3] 林鹏. 红树林[M]. 北京:海洋出版社, 1984. 1-48.

[4] 林鹏. 中国红树林研究进展[J]. 厦门大学学报(自然科学版), 2001, **40**(2):592-603.

[5] 叶庆华, 章菽, 林鹏. 福建九龙江口桐花树叶片的盐腺系统[J]. 台湾海峡, 1998, **7**(3):264-267.

[6] 黄桂玲, 黄庆昌. 中国红树植物的营养器官结构与生态适应 I[J]. 生态科学, 1989, 1:100-105.

[7] 黄桂玲, 黄庆昌. 中国红树植物的营养器官结构与生态

适应 Ⅱ[J]. 中山大学学报(自然科学版), 1990, **29**(2):94-101.

[8] 黄桂玲,黄庆昌. 中国红树植物的营养器官结构与生态适应 Ⅲ. 茎[J]. 植物学通报,1991,**8**(3):41-44.

[9] Osborn J M, Taylor T N. Morphological and ultrastructural studies of plant cuticular membranes. I. Sun and shade leaves of *Quercus velutina* (Fagaceae)[J]. **Bot Gaz**,1990,**151**(4):465-476.

[10] 林鹏,陈荣华,雷泽湘. 红海榄红树林的蒸腾作用与生态因子的关系[J]. 华南植物学报,1992,1:101-106.

[11] 雷泽湘,林鹏. 秋茄蒸腾作用日变化及其与生态因子的相关分析[J]. 湖北农学院学报,1998,**18**(3):204-208.

[12] 杨盛昌,林鹏,中须贺常雄. 红树林的光合作用[J]. 植物学通报,1996,**13**(增刊):33-38.

[13] 林鹏,傅勤. 中国红树林环境生态及经济利用[M]. 北京:高等教育出版社,1995. 40-43.

[14] 杨盛昌,林鹏. 红树植物秋茄幼苗抗低温特性的初步研究[A]. 林鹏. 红树林研究论文集,第四集[C]. 厦门:厦门大学出版社,2000. 80-86.

[15] 李银鹏,林鹏,杨盛昌. 秋茄幼苗对低温的反应及钙的效应[J]. 台湾海峡,1998,**17**(3):324-329.

[16] 林益明,林建辉,林鹏. 红树植物秋茄次生木质部生态解剖学的比较[J]. 台湾海峡,1998,**17**(2):219-223.

[17] 叶庆华,林鹏. 海滩盐度对两种红树叶肉细胞超微结构影响的研究[A]. 范航清,梁仕楚. 中国红树林研究与管理[C]. 北京:科学出版社,1995. 65-70.

[18] 叶庆华,林鹏. 两种盐度下桐花树叶肉细胞结构变化[J]. 厦门大学学报(自然科学版), 1995. **34**(1):104-108.

[19] 郑文教,林鹏. 盐度对秋茄幼苗的生长和水分代谢的效应[J]. 厦门大学学报(自然科学版),1990,**29**(5):575-579.

[20] 郑文教,林鹏. 盐度对红树植物海莲幼苗的生长和某些生理生态特征的影响[J]. 应用生态学报, 1992,**3**(1):9-14.

[21] 邓传远. 几种红树植物的木材解剖学研究[D]. 厦门大学,博士论文,2001,73-85.

三种红树植物叶片的比较解剖学研究*

李元跃 林 鹏
(厦门大学生命科学学院，湿地与生态工程研究中心，福建 厦门 361005)

摘要：研究了采自福建九龙江口的3种红树植物，秋茄(*Kandelia candel*)、木榄(*Bruguiera gymnorrhiza*)和红海榄(*Rhizophora stylosa*)的叶片结构并探讨了其生态学意义。结果表明，这3种红树植物叶片都具有适应海生环境的结构，较厚的角质层，表皮之内有内皮层，内皮层属贮水组织；气孔都分布在下表皮，下陷，并有大的孔下室；中脉有发达的维管束，其导管粗大。从叶片的横切面来看，秋茄叶片具有对称的结构，为等面叶；木榄和红海榄的叶片结构不对称，为异面叶。3种植物叶内都含有较丰富的单宁，以秋茄最高，红海榄次之，木榄最少。这些结构差异可为物种鉴定提供依据。
关键词：红树植物；叶片结构；比较解剖；秋茄；木榄；红海榄
中图分类号：Q944.56 文献标识码：A 文章编号：1005-3395(2006)04-0301-06

Anatomical Characteristics of Leaves in Three Mangrove Species

LI Yuan-yue, LIN Peng
(School of Life Science, Research Centre for Wetland and Ecological Engineering, Xiamen University, Xiamen 361005, China)

Abstract: The leaves of three mangrove species, *Kandelia candel* (L.) Druce, *Bruguiera gymnorrhiza* (L.) Blume and *Rhizophora stylosa* Griff., collected from the access to Jiulong River, Fujian, were studied anatomically. Results showed that the three species were adapt to aquatic environment in structure: having thicker cuticle, epidermis with hypodermis as aquiferous tissue, stomata sunk in and distributed in lower epidermis, with large substomatal chambers, developed vascular bundles with enlarged vessels in mid rib. Cross section views indicated that leaves of K. candel was isolateral with symmetric structure, whereas those of B. gymnorrhiza and R. stylosa were dorsi-ventral and asymmetric. Leaves of all the 3 species were abundant in tannin that was richest in K. candel, followed by R. stylosa and B. gymnorrhiza. The differences in leaf structure could help specific identification in taxonomy.
Key words: Mangrove; Leaf structure; Comparative anatomy; Kandelia candel; Bruguiera gymnorrhiza; Rhizophora stylosa

红树林是生长在热带和亚热带海滩的木本植物群落，通常生长在港湾河口地区的淤泥质滩涂上，是滨海湿地特有的森林类型。红树林生态系统处于海洋与陆地的动态交界面，周期性遭受海水浸淹的潮间带环境，使其结构和功能上具有既不同于陆地生态系统也不同于海洋生态系统的特性，作为独特的海陆边缘生态系统在自然生态平衡中起着特殊的作用。

红树科植物是红树植物的重要组成类型，Schimpe[1]、Chapman[2,3]和林鹏[4,5]等对红树林的植物生态学、植物生理学、植物形态学等有较详细研究；Walsh[6]、林鹏[7]、邓传远等[8]对红树植物的木材和根的解剖特征也进行了研究；Stace [9]研究了红树科4属间叶片的不同解剖特征；Rao[10,11]研究了 Aegiceras

* 教育部博士点基金项目(20030384007)；国家自然科学基金项目(40276028)资助
原载于热带亚热带植物学报，2006，14(4)：301－306

和 Scaevola 的叶片和石细胞的特征，随后 Rao 和 Hugh[12]研究了生长在新加坡的 16 种红树植物的叶片结构及其生态适应性；除红树植物外，Waisel[13]和 Rao[11,14]对其他一些海滨植物叶片的含水量、表皮毛和石细胞的特征也做了详细的描述。

本文主要对红树科的 3 种红树植物－秋茄(Kandelia candel)、木榄(Bruguiera gymnorrhiza)和红海榄(Rhizophora stylosa) 的叶片结构进行研究，探讨其生态学意义，为红树科植物的分类、系统进化和移栽提供依据。

1 材料和方法

秋茄(Kandelia candel)、木榄(Bruguiera gymnorrhiza)和红海榄(Rhizophora stylosa)的叶片均采自福建九龙江口龙海市浮宫镇草埔头村(24°29' N，117°55' E)，选取正常植株上的完整成熟叶片(顶芽下第三对叶片)10 片。

用火棉胶溶液直接涂抹于 3 种植物的叶片，取其胶膜装片观察，并用 100 目网格测微尺计算单位面积的气孔数和测量气孔的大小。

剪取成熟叶片中脉两侧约 5 mm×5 mm 的小块，用 FAA 固定，系列酒精脱水，石蜡包埋，Leica-2235 切片机切片，厚度 8-10 μm，番红 - 固绿对染，中性树胶封片，制成永久切片，显微测微尺测量，OLYMPUS 显微镜观察拍片。切片经番红 - 固绿对染后，部分细胞中具有被染成红褐色的小体，红褐色的小体表明了单宁的存在，因单宁细胞中的单宁化合物可氧化成褐色和红褐色的鞣酐[15]。每个实验数据各为 15 个数值的平均值。

叶片面积的测量采用剪纸法计算；叶片的含水量以烘干法，计算其水分占叶片鲜重的百分比。

用 SPSS 软件进行方差分析。

2 结果

2.1 叶片的形态特征

秋茄的上下表皮细胞外的角质层较厚，各具有 2 层的内皮层细胞(图版 I：1，2)，第一层细胞较小，不含单宁，第二层细胞较大，含有大量的单宁(图版 I：3)，栅栏组织分化为上下栅栏组织两部分，中间是海绵组织，栅栏组织细胞排列较紧密，且上栅栏组织比下栅栏组织厚，海绵组织排列疏松，无规则，在栅栏组织和海绵组织细胞中均含有单宁(图版 I：3)；气孔只位于下表皮，气孔下陷，气孔下有大的孔下室，并与海绵组织的空隙连在一起组成气道；中脉维管束发达，木质部导管多列，径向放射排列，呈类圆形(图版 I：4)。

木榄的上下表皮细胞外具有很厚的角质层，且表皮细胞外壁加厚，上表皮下具有一层内皮层，不具有下内皮层，在上表皮细胞和内皮层中具有大量的紫红色的单宁(图版 I：5，6)，栅栏组织排列紧密，海棉组织排列疏松，也含有单宁(图版 I：7)，靠近下表皮的海绵组织排列较紧密和有规律，气孔仅分布在下表皮，气孔也下陷，并有大的孔下室(图版 I：8)；中脉有一个大的维管束，维管束不发达，呈新月形或半圆形(图版 I：9)。

红海榄的上下表皮细胞外的角质层也较厚，上下表皮细胞较小，细胞外壁不见加厚(图版 I：10，11)，上表皮下具 5-7 层的内皮层细胞，且第一层含有大量的单宁，逐层减少，且细胞由小到大，并嵌入栅栏组织，使栅栏组织成山峰状排列(图版 I：10-12)，海绵组织排列疏松，不规则，具有大的细胞通道，具一层下内皮层，且含单宁(图版 I：13)，气孔也只位于下表皮并下陷，气孔下也有大的孔下室(图版 I：10)；中脉维管束略发达，木质部导管径向放射排列，且木质部偏下表皮方向发达，近上表皮方向不发达，呈横向长圆形(图版 I：14)。

2.2 叶片的形态数量性状

3 种植物叶片的形态特征数据见表 1、2。方差分析表明，3 种植物叶的组织结构和指标间的数量都存在显著差异($P< 0.01$)。

3 种植物叶片横切面上，都具有上内皮层，其中红海榄最厚，有 4-7 层细胞，占叶片总厚度的 34.2%；秋茄次之，有 1-2 层细胞，占叶片总厚度的 7.4%；木榄最薄，只有一层细胞，占叶片总厚度的 5.3%。但秋茄和红海榄同时具有下内皮层，分别占叶片厚度的 5.4%和 4.1%，且秋茄在下内皮层的内侧还分化出下栅栏组织，占 13.5%。因此，从叶片的横切面上看，只有秋茄具有对称的结构，为等面叶，而木榄和红海榄为异面叶。

秋茄成熟叶片厚度最大，达 666.60 μm，红海榄次之，为 596.50 μm，木榄最小，为 582.73 μm。叶片厚度排列顺序与它们的维管束发达程度(秋茄 > 红

海榄 > 木榄)是一致的,说明维管束的发达程度与叶片厚度有关,这是由于维管束越发达,水分及营养输送则越丰富,细胞分裂和生长越旺盛,叶片则越厚。叶片面积大小依次为红海榄 37.5 cm^2)、木榄(35.0 cm^2)、秋茄 24.0 cm^2),可见叶片的面积和厚度不存在正相关关系。

木榄叶片的含水量最高,为 72.1%,红海榄次之,为 70.5%,秋茄最小,为 62.8%,3 种植物的含水量和叶片的内皮层厚度无关。3 种植物都含有较多的单宁,从显色效应看,3 种植物的单宁含量是秋茄 > 红海榄 > 木榄,这和林鹏[16]的研究结果一致。3 种植物的气孔都分布在下表皮,气孔密度较高,气孔密度木榄 > 秋茄 > 红海榄,其气孔面积大小与气孔密度成反比。

表 1 3 种红树植物叶片各组织的厚度和层数

Table 1 Thickness and number of tissue layers in leaves of mangrove plants

		秋茄 *Kandelia candle*	木榄 *Bruguiera gymnorrhiza*	红海榄 *Rhizophora stylosa*
上角质层 Upper cuticle	厚度 Thickness (μm)	6.87 ±0.46 (1.0)	13.17±0.93 (2.3)	6.10±0.68 (1.0)
上表皮 Upper epidermis	厚度 Thickness (μm)	13.53 ±1.50 (2.0)	21.53±2.10 (3.7)	12.73±1.32 (2.1)
上内皮层 Upper hypodermis	厚度 Thickness (μm)	49.60±6.91 (7.4)	31.13 ±4.78 (5.3)	203.30±29.41 (34.2)
	层数 No. of layers	1–2	1	4–7
上栅栏组织 Upper palisade tissue	厚度 Thickness (μm)	158.53±14.91 (23.8)	174.90±21.4 (30.0)	110.00±32.27 (18.4)
	层数 No. of layers	4–5	4–9	2–7
海绵组织 Spongy tissue	厚度 Thickness (μm)	294.00±14.13 (44.1)	320.40±44.29 (55.0)	219.13±28.58 (36.8)
	层数 No. of layers	13–16	15–19	10–15
下栅栏组织 Lower palisade tissue	厚度 Thickness (μm)	89.53±5.23 (13.5)	—	—
	层数 No. of layers	2–3		
下内皮层 Lower hypodermis	厚度 Thickness (μm)	35.87 ±1.38 (5.4)	—	24.60±2.11 (4.1)
	层数 No. of layers	2		1
下表皮 Lower epidermis	厚度 Thickness (μm)	14.07±1.15 (2.1)	15.70 ±2.32 (2.7)	14.40±3.01 (2.4)
下角质层 Lower cuticle	厚度 Thickness (μm)	4.60±0.56 (0.7)	5.90±1.16 (1.0)	5.73±1.18 (1.0)
叶片 Leaf	厚度 Thickness (μm)	666.60	582.73	596.5

括号中数字指该组织的厚度占叶片厚度的百分率。Values in parentheses indicate percentage of tissue thickness to total leaf thickness.

表 2 3 种红树植物叶片面积、含水量、气孔密度及气孔大小

Table 2 Leaf area, moisture, stomatal densitry and size in 3 plants

植物 Species	叶片面积 Leaf area (cm^2)	含水量 Water content (%)	气孔密度 No. of stomata per mm^2	气孔长度×宽度 Mean length× width (μm)
秋茄 *Kandelia candel*	24.0±5.8	62.8	195.7±32.6	36.1 ±3.9×17.1±1.9
木榄 *Bruguiera gymnorrhiza*	35.0±7.1	72.1	250.7±26.5	30.8 ±2.4×17.8±1.6
红海榄 *Rhizophora stylosa*	37.5±4.6	70.5	104.1 ±29.9	44.6 ±3.5×29.1±2.1

3 讨论

红树植物通常分布于热带海岸潮间带，土壤盐渍化，通气不良，富有机质的淤泥海滩，受着潮汐的影响。为适应特殊的生境，红树植物的营养器官显示出与陆生植物不同的形态、结构与功能。从3种红树科植物的叶片研究可以看出，红树科海生植物作为红树植物的重要组成类型，其形态结构和海生环境相适应。

3种红树植物叶片的解剖结构说明，红树植物叶片为适应环境，主要是出现了较厚的角质层和贮水组织等旱生及抗盐结构，这和Rao和Hugh[12]的研究是一致的。另一方面，红树植物的厚角质层及内皮层的形成与其水生环境关系密切。叶片结构中具有厚的角质层也是红树植物区别于其他中生植物的重要特征[17,18]。在红树植物的生态生理学研究中，叶片结构中的贮水组织对于植物的蒸腾作用和调节水分平衡的功能具有重要意义[6,13]，因此，内皮层的结构对于红树植物适应海滩环境具有重要的意义。

黄桂玲和黄庆昌[19]的研究表明，秋茄叶片具有木栓瘤结构，红海榄叶片有皮孔排水器。本研究没有发现类似的结构，说明叶片是否有木栓瘤和排水器的结构还和植物本身生长的环境有关。

从3种红树植物的叶片横切面上看，三者的叶片解剖结构特征具有显著的差异，只有秋茄同时具有上下内皮层和上下栅栏组织，具有对称的结构，为等面叶；红海榄只有上下内皮层和上栅栏组织，没有下栅栏组织，木榄则只有上内皮层和上栅栏组织，没有下内皮层和下栅栏组织，因此红海榄和木榄同为异面叶。以上结构特征可作为种间鉴别依据。

Powman认为，美洲红树(Rhizophora mangle)叶的内皮层的数目与土壤水分中盐的含量有关[20]，而其他学者认为可能是一个年龄现象[2]。本实验表明，内皮层数目和红树植物种类也有一定关系。

参考文献

[1] Schimper A F W. Die Indo-Malaysche Strandflora [J]. Bot Mit Trop, 1891, 3:1-204.

[2] Chapman V J. Mangrove Vegetation [M]. Heidelberg: Strauss & Cramer Gmbt, 1975. 10-301.

[3] Chapman V J. Wet coastal formations of Indo-Malesia and Papua New Guinea [A]. In: Chapman V J. Ecosystems of the World 1: Wet Coastal Ecosystems [M]. Amsterdam: Elsevier, 1977. 261-

[4] Lin P, Chen R H. Study on the mangrove ecosystem of the Jiulongjiang River Estuary in China. III. Accumulation and biological cycle of calcium and magnesium in Kandelia candel Community [J]. Acta Ocean Sin, 1986, 5(3):447-455.

[5] Lin P, Wang W Q. Changes in the leaf composition, leaf mass and leaf area during leaf senescence in three spescis of mangroves [J]. Ecol Engin, 2001, 16(3):415-424.

[6] Walsh G E. Mangroves: A review [A]. In: Reimold R J. Ecology of Halophytes [M]. New York: Acad Press, 1974. 51-74.

[7] Lin P(林鹏), Lin Y M(林益明), Lin J H(林建辉). The ecological secondary xylem anatomy of the mangrove Aegiceras corniculatum and Sonneratia caseolaris [J]. Sci Sil Sin(林业科学), 2000, 36(2): 125-128.(in Chinese)

[8] Deng C Y(邓传远), Lin P(林鹏), Li Z B(黎中宝). Study on comparative anatomy of secondary xylem in six Sonneratia species of mangrove [J]. J Xiamen Univ (Nat Sci) (厦门大学学报自然科学版), 2001, 40(5):1100-1106.(in Chinese)

[9] Stace C A. The use of epidermal characters in phylogenetic considerations [J]. New Phytol, 1966, 65:304-318.

[10] Rao A N. Morphology of foliar sclereids in Aegiceras corniculatum [J]. Israel J Bot, 1971, 20:124-132.

[11] Rao A N. Sclereid variations in Scaevola frutescens from two different habitats [J]. Flora, 1977, 166:111-116.

[12] Rao A N, Tan H. Leaf structure and its ecological significance in certain mangrove plants [A]. In: Soepadmo E, Rao AN, Macintosh D J. Mangrove Environment Research and Management [C]. Kuala Lumpur: Organised and sponsored by university of Malaya and Unesco, 1984.183-194.

[13] Waisel Y. Biology of Halophytes [M]. New York: Acad Press, 1972. 16-28.

[14] Rao A N. Sclereid variations in Scaevola frutescens from two different habitats [J]. Flora, 1977, 166:111-116.

[15] 伊稍K. 李正理译. 种子植物解剖学 [M]. 上海: 上海人民出版社, 1973. 158.

[16] 林鹏. 红树林 [M]. 北京: 海洋出版社, 1984. 26-34.

[17] Sharma G K, Dunn D B. Effect of environment on the cuticular features in Kalanchoe fedchenkoi [J]. Bull Torrey Bot Club, 1968, 95:464-473.

[18] Sharma G K, Dunn D B. Environmental modifications of cell differentiations of leaf surface traits in Datura stramonium [J]. Can J Bot, 1969, 47:1211-1216.

[19] Huang G L(黄桂玲), Huang Q C(黄庆昌). The structure of vegetative organs and ecological adaptation of mangrove plants in China I [J]. Ecol Sci(生态科学), 1989, (1):100-105.(in Chinese)

[20] Powman. 张宏达译. 热带雨林 [M]. 北京: 科学出版社, 1976. 334-349.

图版说明

图版 I

织；S: 气孔。

1. 秋茄的叶片横切面; ×100
2. 秋茄的叶片横切面(描);
3. 秋茄叶片的上表皮、上内皮层及栅栏组织; ×400
4. 秋茄叶片的中脉维管束; ×100
5. 木榄的叶片横切面; ×100
6. 木榄的叶片横切面(描);
7. 木榄叶片的上表皮、上内皮层及栅栏组织; ×400
8. 木榄叶片下表皮的气孔; ×1 000
9. 木榄叶片的中脉维管束; ×100
10. 红海榄的叶片横切面; ×100
11. 红海榄的叶片横切面(描);
12. 红海榄叶片的上表皮及上内皮层; ×400
13. 红海榄叶片的下表皮、下内皮层及海绵组织; ×400
14. 红海榄叶片的中脉维管束; ×100

Explanation of plate

Plate I

Cu: Cuticle; Ep: Epidermis; Hy: Hypodermis; Pt: Palisade tissue; St: Spongy tissue; S: Stoma.

1. Transverse section of the leaf of Kandelia candel; ×100
2. Transverse sections of the leaf of Kandelia candel (drawing);
3. Upper epidermis upper hypodermis and palisade tissue of the leaf of Kandelia candel; ×400;
4. A vascular bundle in midrib of Kandelia candel; ×100
5. Transverse section of the leaf of Bruguiera gymnorrhiza; ×100
6. Transverse section of the leaf of Bruguiera gymnorrhiza (drawing);
7. Upper epidermis, upper hypodermis and palisade tissue of the leaf of Bruguiera gymnorrhiza; ×400
8. Stomata in lower epidermis of the leaf of Bruguiera gymnorrhiza; ×1 000
9. A vascular bundle in midrib of Bruguiera gymnorrhiza; ×100
10. Transverse section of the leaf of Rhizophora stylosa; ×100
11. Transverse section of the leaf of Rhizophora stylosa (drawing);
12. Upper epidermis and upper hypodermis of the leaf of Rhizophora stylosa; ×400
13. Lower epidermis. Lower hypodermis and spongy tissue of the leaf of Rhizophora stylosa; ×400
14. A vascular bundle in midrib of Rhizophora stylosa. ×100

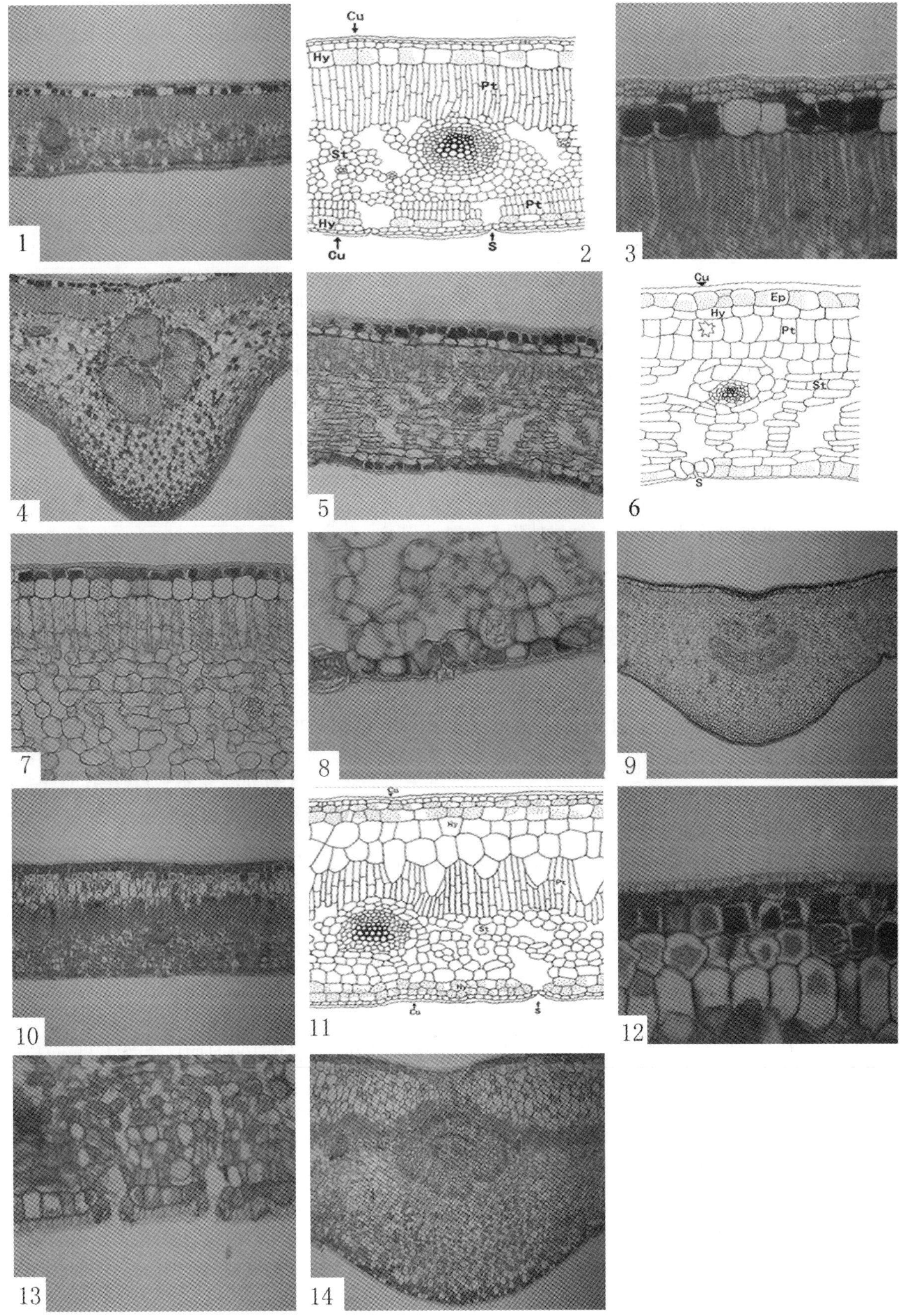

李元跃等:图版 I

LI Yuan-yue et al.: Plate I

三种红树植物叶片结构及其生态适应*

李元跃[1,2]　林　鹏[1,2]

(1. 厦门大学生命科学学院,福建 厦门 361005;2. 厦门大学湿地与生态工程研究中心,福建 厦门 361005)

摘要：研究了采自福建九龙江口龙海县浮宫镇草埔头村（24° 29′ N，117° 55′ E）的 3 种红树植物——桐花树(*Aegiceras corniculatum*)、白骨壤(*Avicennia marina*)和秋茄(*Kandelia candel*)的叶片结构及其生态学意义。这 3 种红树植物叶片结构中，都具有适应海滩环境的结构——较厚的角质层，表皮之内有内皮层，内皮层是贮水组织的一种。3 种红树植物的叶片结构表明，秋茄具等面叶，桐花树和白骨壤是异面叶。从 3 种红树植物叶片横切面的染色状况可判断，3 种红树植物的单宁含量，秋茄最高，桐花树第二，白骨壤最少。

关键词：叶片结构；红树植物；解剖；适应

中图分类号：Q944.5　**文献标识码**：A　**文章编号**：1000-3096（2006）07-0053-05

红树林是指热带海岸潮间带的木本植物群落。对红树林的研究具有多方面的意义，其中最主要的是维护海岸生态平衡的特殊生态系以及林鹏[1]提出的“三高理论”，即高生产率、高归还率、高分解率，红树林为热带河口海湾生态系重要的第一性生产量的贡献者，它的生产量的形成和变化具有独特的生态学规律，也间接影响到河口海湾水产业和渔业的正常发展。红树植物是专指生长在红树林中的木本植物，是红树林群落中的重要物种，因此，红树植物的研究对于沿海海岸的防风固堤、渔业发展、湿地保护等具有重要的作用。

对红树植物的研究国内外已取得很多重要的成果。Schimper[2]、Chapman[3,4]和林鹏[5~7]等对红树植物的生态学、生理学和形态学都有详细的讨论；Walsh[8]和林鹏等[9,10]对红树植物的木材解剖也做了很好的研究。而对红树植物叶片解剖的研究则较少，Stace[11]对红树科的 4 个种类的叶片的结构和生态做了研究，Rao 等[12]对新加坡的 16 种红树植物的叶片结构和生态意义做了比较。在国内对红树植物叶片的解剖学及其生态学的意义的研究也不多[13]。作者主要对中国 3 种广泛分布的红树植物——桐花树(*Aegiceras corniculatum*)、白骨壤(*Avicennia marina*）和秋茄（*Kandelia candel*）的叶片结构及其生态学意义做了研究，为 3 种红树植物的分类、系统进化和移栽提供一定的依据。

1 材料与方法

供试材料均采自福建九龙江口龙海市浮宫镇草埔头村（24° 29′ N，117° 55′ E)，选取正常植株上的完整成熟叶片（顶芽下第三对叶片）。

用火棉胶溶液直接涂抹于植株上的叶片，取其胶膜装片观察，并用 100 目网格测微尺计算单位面积的气孔数和测量气孔的大小。

剪取成熟叶片中脉两侧约 5mm×5mm 的小块，用 FAA 固定，系列酒精脱水，石蜡包埋，Leica-2235 切片机切片，厚度 8~10μm 番红-固绿对染，中性树胶封片，制成永久切片，显微测微尺测量，OLYMPUS 显微镜观察拍片。切片经番红-固绿对染后，部分细胞中具有被染成红褐色的小体，因单宁

* 国家自然科学基金资助项目(40276028);教育部博士点基金资助项目(20030384007)
原载于海洋科学,2006,30(7):53－57

细胞中的单宁化合物可氧化成褐色和红褐色的鞣酐[14]，所以将此红褐色的小体定为单宁。每个实验数据各为 10 个数值的平均值。

叶片面积的测量采用剪纸法计算；叶片的含水量是以烘干法求算其水分占叶片鲜质量的百分比来衡量。

2 结果

2.1 3 种红树植物叶片解剖特征

(1) 桐花树具有较厚的角质层，上下表皮内都具有内皮层，上内皮层 2～4 层细胞（图 1-1），下内皮层 1～2 层细胞（图 1-2），细胞中含有大量染成紫红色的单宁，栅栏组织呈柱状排列，具一定的细胞间隙（图 1-1），海绵组织排列无规则，具有较大的细胞空隙（图 1-2），气孔只位于下表皮（图 1-3），各层的细胞都含有单宁（图 1-4）。

(2) 白骨壤的角质层相对较薄，上表皮内具有 4～7 层的内皮层细胞（图 1-5），无下内皮层，栅栏组织呈柱状排列，紧密而有规律，海绵组织较疏松，无规则，细胞间的空隙较大（图 1-6），下表皮的外面具有表皮毛，气孔只位于下表皮，但由于表皮毛的遮盖，观察较困难，只在栅栏组织和海绵组织中含有一定量的单宁（图 1-7）。

(3) 秋茄的角质层也较厚，上下表皮内各具有 2 层的内皮层细胞（图 1-8），第一层内皮层细胞较小，不含单宁，第二层内皮层细胞较大，含有大量的单宁，栅栏组织分化为上下栅栏组织两部分，中间是海绵组织，栅栏组织细胞排列较紧密，且上栅栏组织比下栅栏组织厚，海绵组织排列疏松，无规则，在栅栏组织和海绵组织中均含有一定量的单宁（图 1-9），气孔也只位于下表皮。

2.2 3 种红树植物叶片解剖结果比较

3 种红树植物叶片的解剖结果和数据见表 1 和表 2。

表 1 3 种红树植物叶片各组织的厚度及其占叶片总厚度的百分率

Tab.1 Thickness and percentage of tissue layers in leaves of mangrove plants

组织名称	厚度（μm）		
	桐花树	白骨壤	秋 茄
上角质层	6.70±0.56 (1.7)	4.33±0.67 (1.3)	6.87±0.46 (1.0)
上表皮	18.47±1.76 (4.9)	12.00±0.33 (3.7)	13.53±1.50 (2.0)
上内皮层	76.67±16.71 (20.3)	109.67±32.89 (33.5)	49.60±6.91 (7.4)
上栅栏组织	61.40±14.48 (16.3)	109.67±13.00 (33.5)	158.53±14.91 (23.8)
海绵组织	161.00±20.13 (42.6)	79.33±9.78 (24.1)	294.00±14.13 (44.1)
下栅栏组织	-	-	89.53±5.23 (13.5)
下内皮层	23.93±4.20 (6.5)	-	35.87±1.38 (5.4)
下表皮	23.67±2.36 (6.3)	9.33±0.67 (2.8)	14.07±1.15 (2.1)
下角质层	5.83±0.89 (1.4)	3.50±0.83 (1.1)	4.60±0.56 (0.7)
叶片	377.67 (100)	327.83 (100)	666.60 (100)

注：括号中数字指该组织的厚度占叶片总厚度的百分率

(1) 3 种红树植物叶片横切面上，都具有上内皮层，其中白骨壤最厚，有 4～7 层细胞，占叶片总厚度的 33.5%，桐花树次之，占 20.3%，秋茄最薄，占 7.4%，但桐花树和秋茄同时具有下内表皮，分别占叶片厚度的 6.5%和 5.4%，且秋茄在下内皮层的内侧还分化出下栅栏组织，占 13.5%。因此，从 3 种红树植物的叶片横切面上看，只有秋茄具有对称的结构，为等面叶，而桐花树和白骨壤为异面叶。

(2) 3 种红树植物叶片厚度，秋茄最大，达 666.6 μm，桐花树第二，为 377.7 μm，白骨壤最小，为 327.8 μm；相应的叶片面积从大到小也依次为秋茄（24.0 cm^2）、桐花树（14.9 cm^2）、白骨壤（12.3 cm^2）。有趣的是，3 种红树植物的叶片厚度与面积的比例非常接近，在 26 左右。

(3) 白骨壤由于叶片下表皮密布皮毛，所以在光学显微镜下比较难观察到气孔，秋茄和桐花树则较容易观察到气孔，秋茄的气孔密度为 195.7 个/mm^2，气孔面积为 617.3 μm^2，桐花树的气孔密度则较大，为 305.6 个/mm^2，气孔面积和秋茄差不多，为 621.6 μm^2。3 种红树植物叶片的含水量比较接近，最大的是白骨壤，占鲜质量的 67.0%，其次是秋茄，为 62.8%，最低的是桐花树，为 58.3%。

(4) 3 种红树植物叶片中都含有一定量的单宁，从叶片横切面解剖图可看出：秋茄叶片中的单宁被染成红褐色，且分布在各层细胞中，所以含量应最多（图 1-8）；桐花树和白骨壤叶片中的单宁则被染成红色，且桐花树叶片中的单宁也分布在各层细胞中，含量第 2（图 1-4），而白骨壤叶片中的单宁只分布在栅栏组织和海绵组织的细胞中，含量最少（图 1-6）。

表 2 3 种红树植物叶片面积、含水量、气孔密度及气孔大小

Tab.2 Leaf area,water content,stomatal density and dimensions in the three kinds of mangrove plants

种名	叶片面积(cm^2)	含水质量分数(%)	气孔密度(个/mm^2)	气孔大小(μm)
桐花树	14.9±2.2	58.3±3.0	305.6±40.7	33.6±2.6×18.5±1.3
白骨壤	12.3±1.2	67.0±2.9	ND	ND
秋茄	24.0±5.8	62.8±3.2	195.7±32.6	36.1±3.0×17.1±1.1

注：ND 表示不能观察到

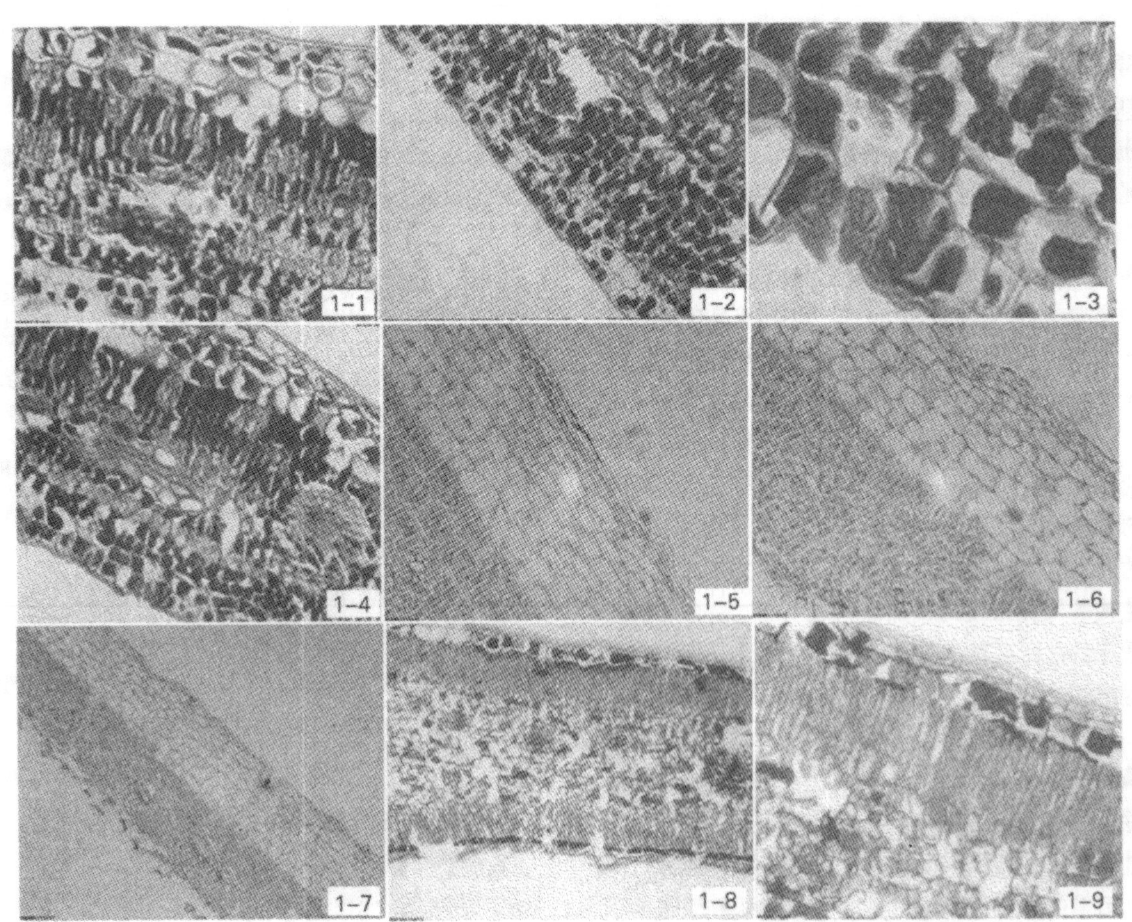

图 1 桐花树、白骨壤和秋茄叶片结构示意图

Fig.1 Sketch map of leaf structure of *A.corniculatum*, *A. marina* and *K. candel*

1-1.桐花树叶片的上表皮、上内皮层及栅栏组织；1-2.桐花树叶片的下表皮、下内皮层及海绵组织；1-3.桐花树叶片下表皮

的气孔；1-4.单宁在桐花树叶片横切面各组织的分布；1-5.白骨壤叶片的上表皮、上内皮层和栅栏组织；1-6.白骨壤叶片的栅栏组织和海绵组织；1-7.具表皮毛的白骨壤叶片横切面；1-8 秋茄的叶片横切面；1-9.秋茄叶片的上表皮、上内皮层及栅栏组织

1-1. Upper epidermis. upper hypodermis and palisade tissue of the leaf of *A.corniculatum*;1-2 .Lower epidermis. Lower hypodermis and spongy tissue of the leaf of *A.corniculatum*;1-3. Stomata in lower epidermis of the leaf of *A.corniculatum*;1-4. Distribution of tannin in each tissue of transverse sections of leaves of *A.corniculatum*;1-5. Upper epidermis. upper hypodermis and palisade tissue of the leaf of *A. marina*;1-6. Palisade tissue and spongy tissue of the leaf of *A. marina*;1-7. Trichomes in transverse sections of the leaf of *A. marina*;1-8. Transverse sections of the leaf of *K. candel*;1-9 Upper epidermis. upper hypodermis and palisade tissue of the leaf of *K. candel*

3 讨论

红树植物通常分布于热带海岸潮间带，土壤盐渍化，通气不良，富有机质的淤泥海滩，受着潮汐的影响。为适应特殊的生境，红树植物的营养器官显示出与陆生植物不同的形态、结构与功能。

(1)从3种红树植物叶片的解剖结构可证明，红树植物叶片适应环境，主要是出现了较厚的角质层、表皮毛和贮水组织等旱生结构及抗盐适应。从研究的3种红树植物叶片解剖结构看，3种类叶片结构中都出现了较厚的角质层，在白骨壤的叶片结构中下表皮还出现了表皮毛，这些结构都可以防止水分的过度蒸发；另外，从研究的3种红树植物叶片解剖结构看，最明显的特征是出现了内皮层的结构，内皮层具有贮水组织功能。在红树植物的生态生理学研究中，叶片结构中的贮水组织对于红树植物的蒸腾作用和调节水分平衡的功能具有重要意义[8,15]，因此，内皮层的结构对于红树植物适应海滩环境具有重要的意义。

(2)研究的3种红树植物中，白骨壤和桐花树是泌盐植物，秋茄是拒盐植物[16]，而3种红树植物叶片的含水量白骨壤最大，桐花树最小，说明红树植物的叶片含水量与红树植物盐分代谢的形式无关。桐花树的气孔密度比秋茄高将近 1/3，是否与桐花树适应低盐区蒸腾率较大有关，有待进一步研究才能得出结论。

(3)单宁是红树植物对环境的另一适应，它跟红树植物的抗腐蚀具密切关系[16]。单宁对红树植物的生存、对污染的抗性和净化作用，以及药物利用和各种工业用途方面具有极为重要的影响和意义[17]。在红树植物叶片结构中，不同种类的红树植物单宁分布部位是不同的，而且单宁含量的多少具有一定的显色效应，因此，单宁的含量可作为判别红树植物种类的依据之一。

参考文献:

[1] 林鹏.中国红树林生态系[M].北京：科学出版社，1997.1-8.

[2] Schimper A F W. Die Indo-Malaysche strandflora[J]. **Bot Mit Trop**, 1891,3: 1-204.

[3] Chapman V J. Mangrove Vegetation[M]. Heidelberg: Strauss & Cramer Gmbt, 1975.10-42.

[4] Chapman V J. Wet coastal formations of Indo-Malesia and Papua New Guinea[A].Chapman V J. Ecosystems of the World 1: Wet Coastal Ecosystems[C]. Amsterdam: Elsevier, 1977. 261-270.

[5] 林鹏.中国东南部海岸红树林的类群及其分布[J].生态学报,1981,**1**(3):283-290.

[6] 林鹏,陈德海,李钨金.两种红树叶的几种酶的生理特性和海滩盐度的相关性初探[J].植物生态学与地植物学丛刊.1984,**8**(3):222-227.

[7] Lin P, Wang W Q. Changes in the leaf composition, leaf mass and leaf area during leaf senescence in three species of mangroves[J]. **Ecological Engineering**, 2001,**16**(3): 415-424.

[8] Walsh G E. Mangroves: A review[A]. Reimold R J, Queen W H .Ecology of Halophytes[C]. New York: Acad Press,1974.51-74.

[9] 林鹏,林益明,林建辉.红树植物次生木质部的结构与进化[J].海洋学报,1998,**20**(4):108-114.

[10] 林鹏,林益明,林建辉.桐花树和海桑次生木质部的生态解剖[J].林业科学,2000,**36**(2):125-128.

[11] Stace C A. The use of epidermal characters in phylogenetic considerations[J]. **New Phytol**, 1966, 65: 304-318.

[12] Rao A N,Tan H. Leaf Structure and its ecological significance in certain mangrove plants[A]. Soepadmo E, Rao A N. Mangrove Environment Research and Management[C]. Kuala Lumpur: Organised and Sponsored by University of Malaya and Unesco,1984.183-194.

[13] 黄桂玲.黄庆昌，中国红树植物的营养器官结构与生态适应 I [J].生态科学，1989，1：100-105.

[14] 伊稍 K.种子植物解剖学[M].李正理译.上海:上海人民出版社,1973.158.

[15] Waisel Y. Biology of Halophytes[M]. New York:Acad Press, 1972.16-28.

[16] 林鹏.红树林[M].北京:海洋出版社,1984.21-34.

[17] Lin P, Fu Q. Environmental Ecology and Economic Utilization of Mangrove in China[M]. Beijing and Heidelberg:China Higher Education and Springer -Verlag Press, 2000.128-148.

Leaf structure and its ecological adaptability in three species of mangroves

LI Yuan-yue[1,2], LIN Peng[1,2]

(1.*School of Life Science, Xiamen University*, Xiamen 361005, China;2.*Research Centre for Wetland and Ecological Engineering,Xiamen University*,Xiamen 361005,China)

Received: Jan.,10,2005

Key wards: leaf structure; mangroveplant; anatomy; adapt

Abstract: The anatomy of the leaves and its ecological adaptation of three species of mangrove plants—— *Aegiceras corniculatum*, *Avicennia marina* and *Kandelia candel* were studied. They are collected from Fugong village, Longhai , Fujian（24° 29′ N，117° 55′ E）.Their relatively thick cuticle and hypodermis in leaf structure of three species of mangrove plants make them adapt to the environment of foreshore, and hypodermis is aqueous tissue. *K. candel* showed isobilateral leaf structure and the remaining two species dorsiventral. It can be referred that according to the degree of stained and tannin distribution in different tissues of leaf, the tannin in *K. candel* is most, that in *A. corniculatum* is second, that in *A. marina* is least among three species of mangroves.

Leaf and stem anatomical responses to periodical waterlogging in simulated tidal floods in mangrove *Avicennia marina* seedlings*

Yan Xiao[a,c] Zuliang Jie[a] Mao Wang[a] Guanghui lin[a,b] Wenqing Wang[a,b]

(a. *Key Laboratory of Ministry of Education for Coastal and Wetland Ecosystems*, *School of Life Sciences*, *Xiamen University*, Xiamen 361005, PR China; b. *State Key Laboratory of Marine Environmental Science*, *Xiamen University*, Xiamen 361005, Fujian, PR China; c. *School of Life Science*, *Nanjing University*, Nanjing 210093, PR China)

ARTICLE INFO

Article history:
Received 23 November 2008
Received in revised form 2 July 2009
Accepted 7 July 2009
Available online 14 July 2009

Keywords:
Avicennia marina
Waterlogging
Mangrove
Leaf
Stem
Anatomy

ABSTRACT

This study was aimed to evaluate anatomical responses to waterlogging of mangrove seedlings (*Avicennia marina* (Forsk.) Vierh.) grown in experimentally simulated semidiurnal tides. The following treatments were used: 0, 2, 4, 6, 8, 10 and 12 h submergence period with two daily tidal cycles. With increasing waterlogging duration, the leaf thickness, mesophyll thickness, palisade parenchyma thickness, palisade–spongy ratio and hypodermis thickness decreased, but the mesophyll to leaf thickness ratio, stem and pith diameter, and cortex thickness increased. The tangential vessel diameter, vessel wall thickness in stem and leaf and fiber wall thickness in stem showed a similar tendency in response to waterlogging, remaining constant between 0 and 4 h waterlogging duration, but decreasing with more prolonged waterlogging. When the waterlogging duration exceeded 4 h, no sclerenchyma cells in leaves or gelatinous fibers in stems were observed. The response of these leaf and stem features indicated that water transport and mechanical support could remain relatively stable in the 0–4 h waterlogging duration, but they would be negatively influenced by longer flooding. Tissues for gas exchange were stimulated by waterlogging, while the functions of water storage, photosynthesis, mesophyll conductance were weakened with increasing waterlogging.

1. Introduction

The grey mangrove *Avicennia marina* (Forsk.) Vierh. is probably the best known mangrove because of its distinctive root system and widespread distribution (Wang and Wang, 2007). It is commonly found close to the seaward side of mangroves and was reported as a pioneer, where its roots are submerged during high tides (Lin, 1999; He et al., 2007). It is therefore subjected to high wave energy and has well-developed pencil-like erect aerial roots (Tomlinson, 1986), which also give it a firm foothold against wind and waves. Numerous studies indicated that the duration of periodical waterlogging was a limiting factor for the survival of mangrove seedlings (Komiyama et al., 1996; Kitaya et al., 2002; Chen et al., 2004; He et al., 2007). Low seedling survival rate was common in mangrove reforestation (Wang et al., 2000; Thampanya et al., 2006; Wang and Wang, 2007). Compared with other mangrove species, *A. marina* seedlings exhibit relatively stronger waterlogging tolerance, including the ability to oxidize the rhizosphere and conserve oxygen to maintain aerobic metabolism for longer periods during submergence in glasshouse (Youssef and Saenger, 1996), a fast recovery of physiological function after flooding under natural conditions (Sayed, 1995), and a high survival rate over a wide tidal range (He et al., 2007).

The anatomical and morphological characteristics of living plants are commonly correlated with the particular combination of environmental conditions under which individual plants established and grew (Arens, 1997). Mature *A. marina* has specialized roots, which are important for gas exchange in anaerobic substrata (Kathiresan and Bingham, 2001). Curran et al. (1986) demonstrated that the conductance of pneumatophores in *A. marina* was sufficiently large to resupply the root internal gas space during a normal low tide when the pneumatophores were exposed to the atmosphere. Curran et al. (1996) measured the gas space in the roots of this species and found it to be 40–50% by volume. On exposure at low tide, pressure recovers immediately to atmospheric level, but oxygen slowly rises to a plateau below the concentration in the atmosphere. The changes in oxygen concentration are consistent with oxygen supply by diffusion (Allaway et al., 2001). Conductance to oxygen of pneumatophore in *A. marina* was found to depend on the number of lenticels (Hovenden and Allaway, 1994). However, small young seedlings generally lack aerial roots and lenticels. Although horizontal

* From Aquatic Botany, 2009, 91: 231—237

structures are likely to represent a significant pathway for admission of oxygen, especially in rapidly growing roots where the tip region lacks lenticels (Hovenden and Allaway, 1994), young seedlings are more sensitive to flooding injury than older seedlings and adult plants (Kozlowski, 1997). In seedlings, small size results in a limited access to air due to partial or total submergence, and absence of aerial roots and undeveloped internal ventilation systems results in a low efficiency of oxygen transport (McKee et al., 1988; McKee, 1993). Factors affecting early stages of mangrove establishment are important in determining mangrove species' distribution in this environmental setting (McKee, 1993; Delgado et al., 2001). Sediment burial, vegetation density, and water movement could affect *Avicennia* species and relatively latter species (*Sonneratia* and *Rhizophora*) differently (Thampanya et al., 2002a,b).

Based on extensive studies on anatomical and morphological adaptations to waterlogged and saline environment in mature mangroves (Yáñez-Espinosa et al., 2001, 2008; Verheyden et al., 2004, 2005; Schmitz et al., 2006) and previous findings in *Bruguiera gymnorrhiza* seedlings (Wang et al., 2007), we hypothesized that the anatomical variations of *A. marina* seedlings would respond positively to moderate flooding, but would be negatively influenced by excessive flooding. Thus, to test this hypothesis, a variety of anatomical measurements were made on individuals of *A. marina* seedlings growing in experimental equipment that simulated semidiurnal tides.

2. Materials and methods

2.1. Experimental design, plant material and culture conditions

The experimental design, plant material and culture conditions were as described by Chen et al. (2004, 2005) and Wang et al. (2007). Seven plastic tanks (65 cm × 50 cm × 50 cm) acting as artificial-tidal tanks simulating the semidiurnal tides were arranged as shown in Fig. 1. It took 2 h to fill a tank with diluted seawater through pipes. After Tank A was full, the water flowed over into Tank B, and so on. After Tank F was full, all water in Tanks A, B, C, D, E and F was unloaded by timer-controlled valves at the bottom of each tank. It took 5 min for the water to drain out of the tank. Therefore, Tank A was full of water for 12 h per tidal cycle, and the inundation duration in the other tanks was 10, 8, 6, 4, 2 and 0 h, respectively. There were two tide cycles each day. There were four pots in each tank, each being 25 cm tall and 25 cm in diameter, with a small hole at the bottom to allow rapid drainage while water in the tanks was drained away. Each pot was filled with washed river sand (diameter = 1 mm). Each set had seven tanks and a water container (Fig. 1), three sets of equipment acted as three replicates, and one plant per replicate was collected for anatomical analysis.

Healthy and mature hypocotyls of *A. marina* were collected from the Jiulong River Estuary at Fugong Town, Longhai County, Fujian Province of China (24°29′N, 117°55′E). The average seawater salinity was 17‰ (Lin, 1999). Five hypocotyls were planted in each pot and the seedlings were periodically submerged under artificial

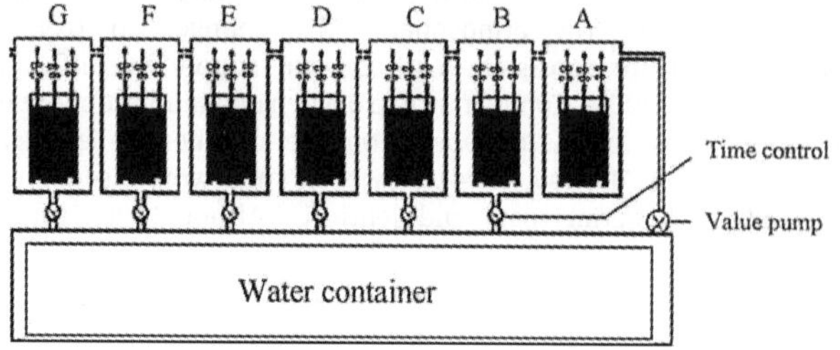

Fig. 1. Arrangement of simulated semidiurnal tides tanks.

seawater with a salinity of 15‰ (seawater from the west coast of Xiamen of 22–28‰ in salinity was diluted by tap water). Tap water was added daily to compensate for evaporation losses and the seawater was renewed weekly. All seedlings were grown in a greenhouse with an air temperature of 27–32 °C. Seedlings were flooded at 'high tide' to a maximum depth of 60 cm above the bottom of the tank, and at 'low tide' the water was slightly below the sand level.

In previous studies, waterlogging treatment was conducted only by soil flooding, and the shoots and leaves of plants remained exposed to the air during the treatments (Ellison and Farnsworth, 1997; Ye et al., 2003). However, in the early stages of development, seedlings are so small that they are often completely submerged by flood water in the field. In our simulation experiment, the flooding time for a semidiurnal tidal cycle was averaged into seven classes, according to the length of flooding during a tidal cycle at different locations in the inter-tidal zones. Twelve hours treatment meant that seedlings were inundated all the time in a tidal cycle (12 h), which denoted the flooding time at the lowest tidal level, while 0 h treatment equaled the highest tidal level where seedlings were not inundated. All plants were inundated twice daily except in the case of 0 h treatment (Chen et al., 2005). When the tanks were full of diluted seawater, the mangrove plants would be completely submerged under controlled mesocosm conditions.

2.2. Measurement of anatomical features of stems

After 70 days in culture, mature leaves (second pair from the shoot top) and stems (second node from the shoot top) of *A. marina* seedlings were fixed in formalin–70% alcohol–glacial acetic acid (5:90:5). Stem and leaf samples were dehydrated in an alcohol series (70–100%), cleared in xylene and embedded in paraffin (56–58 °C). Transverse sections of 10 μm thickness for leaf and 20 μm thickness for stem were obtained using a rotary microtome. These slides were deparaffinized, rehydrated and stained with safranin-fast green (1% aqueous safranin and 0.5% fast green in 95% alcohol). Sections of samples were photographed under a light microscope (Olympus BX41, Japan) and digitized (Olysia BioReport software) to determine anatomical parameters. Thickness of leaf, epidermis, hypodermis, palisade parenchyma, spongy parenchyma and mesophyll, as well as the number of vessels in midrib were measured on 30 random field of 456 μm × 341 μm. Vessel wall thickness and tangential vessel diameter in midrib of leaf, as well as fiber and vessel wall thickness and tangential vessel diameter in stem were measured in 30 fields of 182 μm × 136 μm at random, whereas vessel density and gelatinous fiber ratio was calculated based on 15 random measurements. Gelatinous fiber is a type of fiber present in the secondary xylem of dicotyledons. In such types the innermost layer of the secondary wall contains much α-cellulose and is poor in lignin (Fahn, 1983). The gelatinous fibers were obvious when the inner wall layer was stained with fast green. The gelatinous fiber ratio was calculated based on the equation, gelatinous fiber ratio = the number of gelatinous fibers/the number of total fibers. Stem diameter, pith diameter and cortex thickness were randomly determined in 15 fields of 4563 μm × 3410 μm.

2.3. Statistical analysis

All statistical analyses were performed with SPSS 11.0 software. One-way ANOVA followed by Bonferroni multiple comparison method was used to analyze the difference among the seven treatments and linear regression analysis was applied to evaluate association between anatomical features and waterlogging duration.

3. Results

The upper and lower epidermis thickness to leaf thickness ratio and mesophyll to leaf thickness ratio increased progressively with prolonged waterlogging duration (Fig. 2B and E). In contrast, prolonged waterlogging duration had significant negative effects on hypodermis thickness, hypodermis to leaf thickness ratio, leaf thickness, mesophyll thickness, palisade parenchyma thickness, palisade–spongy ratio, tangential vessel diameter and vessel wall thickness in leaf tissue (Figs. 2A, B, D, F and 3A–C, Table 1). Spongy parenchyma thickness first increased with prolonged waterlogging, reached a maximum value at $75 \pm 9\ \mu m$ under the 6 h treatment, and then tended to decline (Fig. 2C). Under a light microscope, sclerenchyma cells around the main vein only existed in short-term waterlogging duration and no waterlogging conditions (waterlogging duration less than 4 h) (Fig. 3D–F). However, the number of vessels in midrib was not significant throughout all treatments ($F = 1.752$, $P > 0.05$) (data not shown).

Positive correlations were observed in stem diameter, cortex thickness, pith diameter and cortex thickness to stem diameter ratio, and waterlogging duration (Fig. 4A and B, Table 1). Tangential vessel diameter, vessel and fiber wall thickness decreased significantly with prolonged waterlogging duration (Fig. 4C and D, Table 1), whereas no significant difference in vessel density was observed throughout all the treatments ($F = 1.616$, $P > 0.05$) (data not shown). The highest gelatinous fiber ratio occurred in the 2 h treatment (80%), followed by 0 h (71%) and 4 h (54%). When waterlogging duration was longer than 4 h, no gelatinous fiber was observed (Figs. 4D and 5A–C).

4. Discussion

Similar to our earlier findings in *B. gymnorrhiza* (Wang et al., 2007), the upper and lower epidermis to leaf thickness ratio increased progressively with prolonged waterlogging duration in the present study (Fig. 2E), indicating an adaptation in response to flooding. Previous study confirmed that the water storage tissue, in the form of hypodermis, is prominently present in the leaves of *Avicennia*, which played an important function in the regulation of water loss (Rao and Tan, 1984). Camilleri and Ribi (1983) suggested that thicker leaves had more water storage than thinner leaves. The decrease of hypodermis to leaf thickness ratio and leaf thickness with prolonged waterlogging duration may be due to the decreased requirement of water storage in waterlogged conditions. No sclerenchyma cells were observed when waterlogging duration

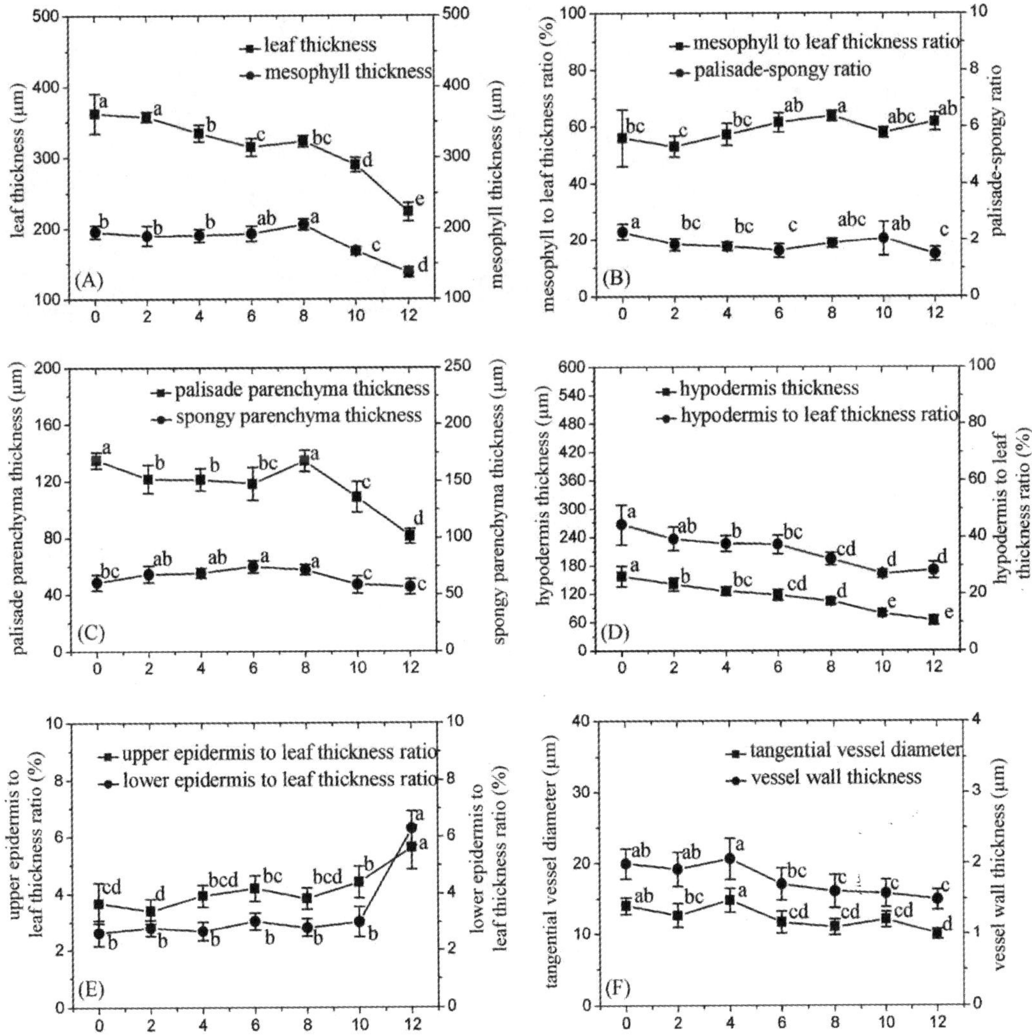

Fig. 2. Leaf anatomical characteristics of *Avicennia marina* seedlings under periodical waterlogging grown for 70 days. (A–E) Tissue thickness; (F) vascular system. Values are means ± S.E. Means followed for different letters were statistically different at $P < 0.05$.

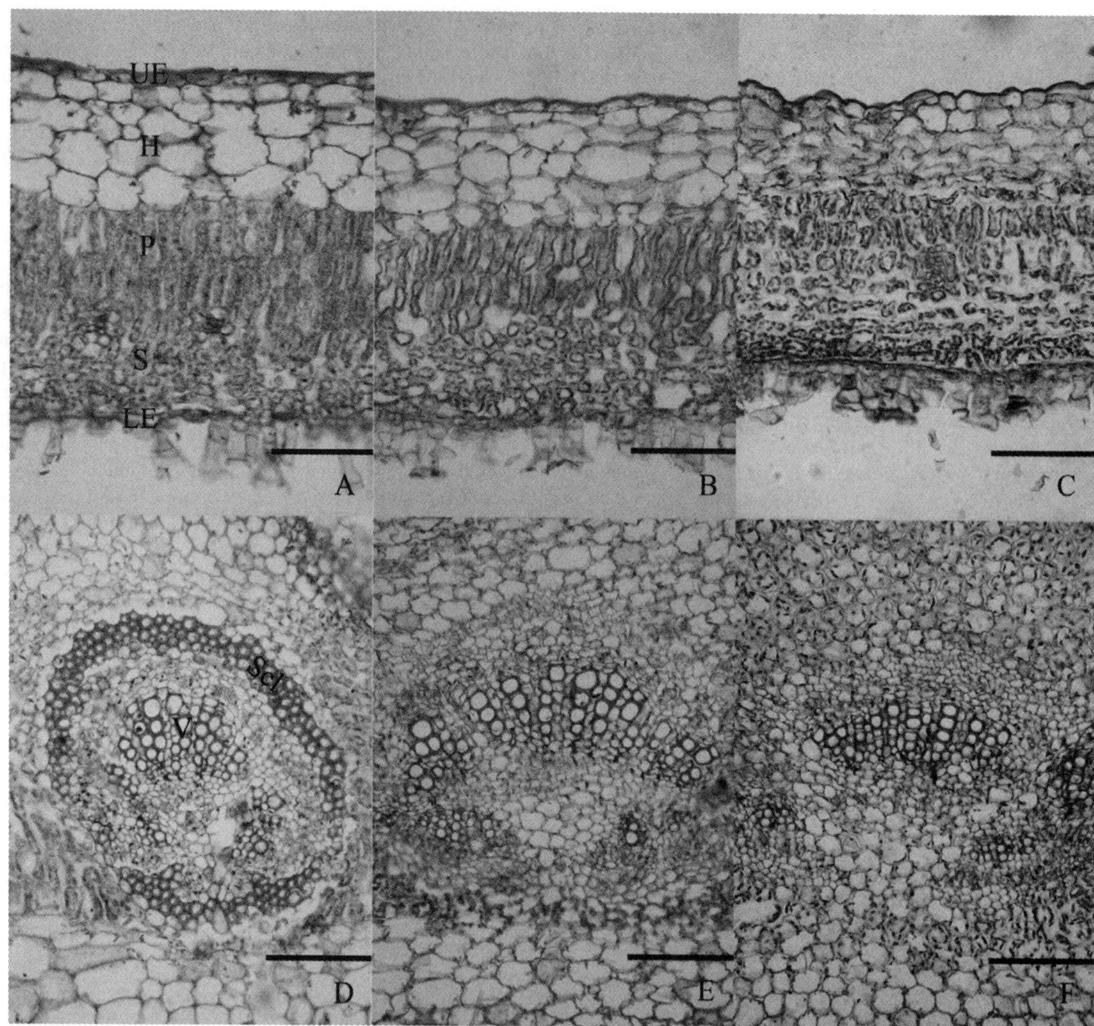

Fig. 3. Leaf blade cross-sections showing mesophyll and midrib details in *A. marina* seedlings under periodical waterlogging grown for 70 days. (A) Mesophyll, 0 h; (B) mesophyll, 6 h; (C) mesophyll, 12 h; (D) midrib, 0 h; (E) midrib, 6 h; (F) midrib, 12 h. Bars, 100 μm. UE, upper epidermis; H, hypodermis; P, palisade parenchyma; S, spongy parenchyma; LE, lower epidermis; V, vessel; Scl, sclerenchyma cells.

exceeded 4 h (Fig. 3E and F), which resulted in the reduction of support and protection to leaves.

Leaf thickness, mesophyll thickness, palisade parenchyma thickness and palisade–spongy ratio showed similar tendency as earlier study in *B. gymnorrhiza* (Wang et al., 2007), decreasing with prolonged waterlogging duration (Fig. 2A–C). In *Kandelia candel*, photosynthetic rate declined progressively with increasing duration of waterlogging (Chen et al., 2005). High palisade–spongy ratio was considered as an adaptation for light capture (Fahn, 1983); consequently, the low palisade–spongy ratio might partly result in reduction of utilization of light. Earlier controlled experiments have proved that salt-induced decrease of mesophyll thickness might have contributed to the reduction in photosynthesis and mesophyll conductance in some mangrove species (e.g. *B. parviflora*, *B. gymnorrhiza*, *Excoecaria agallocha*, *Heritiera fomes*, *Phoenix paludosa* and *Xylocarpus granatum*) (Parida et al., 2004; Nandy et al., 2007). Mesophyll thickness being inversely proportional to mesophyll density is otherwise positively correlated to conductance and supports the direct relationship between leaf porosity and mesophyll conductance (Loreto et al., 1992). Mesophyll to leaf thickness ratio was used as a criterion to compare the abundance of mesophyll and reflected mesophyll conductance (Nandy et al., 2007). Nandy et al. (2007) assumed that the higher the ratio, the less would be the mesophyll conductance and vice versa. Maximum mesophyll to leaf thickness ratio occurred in the 8 h treatment (Fig. 2B), implying the possibility of decrease in mesophyll conductance under long-term waterlogging duration. Thinner leaves (i.e., increased specific leaf area) of *Rumex palustris* and *R. crispus* achieved a significant reduction of diffusion resistance and an enhanced gas exchange between the leaf and the surrounding water (Vervuren et al., 1999; Mommer et al., 2005). Thus, we concluded that leaf anatomical features benefited gas exchange with increasing waterlogging duration at the cost of loss of photosynthesis, water storage, support and protection.

Soil waterlogging is considered to increase the parenchymatous tissue in some plants (Kozlowski, 1997). In stems, aerenchyma can occur in the cortex and in the pith cavity (Armstrong, 1979). It was suggested that the increased stem and pith diameter, as well as cortex thickness accompanying prolonged waterlogging duration (Fig. 4A and B) would be in response to anoxic conditions.

Generally, mangrove woods have narrow and densely distributed vessels that help create high tensions in the xylem since a slight decrease in vessel diameter produces a disproportionally

Table 1
Linear regression between anatomical characteristics of *Avicennia marina* seedlings and periodical waterlogging.

Variable	Slope ± S.E.	R^2	P
Leaf			
Upper epidermis to leaf thickness ratio	0.126 ± 0.018	0.197	0.000
Lower epidermis to leaf thickness ratio	0.170 ± 0.020	0.272	0.000
Hypodermis thickness	−7.569 ± 0.389	0.657	0.000
Hypodermis to leaf thickness ratio	−1.430 ± 0.124	0.402	0.000
Leaf thickness	−9.388 ± 0.550	0.596	0.000
Mesophyll thickness	−3.118 ± 0.412	0.225	0.000
Palisade parenchyma thickness	−2.743 ± 0.342	0.245	0.000
Spongy parenchyma thickness	−0.376 ± 0.225	0.014	0.096
Palisade–spongy ratio	−0.030 ± 0.011	0.039	0.005
Mesophyll to leaf thickness ratio	0.614 ± 0.157	0.071	0.000
Tangential vessel diameter	−0.292 ± 0.046	0.169	0.000
Vessel wall thickness	−0.048 ± 0.007	0.177	0.000
Stem			
Stem diameter	0.024 ± 0.003	0.279	0.000
Cortex thickness	10.255 ± 0.745	0.516	0.000
Pith diameter	0.010 ± 0.002	0.189	0.000
Cortex thickness to stem diameter ratio	0.276 ± 0.035	0.254	0.000
Tangential vessel diameter	−0.666 ± 0.095	0.216	0.000
Vessel wall thickness	−0.045 ± 0.008	0.142	0.000
Fibre wall thickness	−0.062 ± 0.007	0.307	0.000
Gelatinous fibre ratio	−4.398 ± 1.057	0.200	0.000

large increase in flow resistance (Scholander et al., 1965; Kathiresan and Bingham, 2001). Vessel diameter increments increase the efficiency of water conduction dramatically, whereas smaller and more numerous vessels indicate increased hydraulic safety (Carlquist, 2002). Thicker vessel walls would increase the mechanical strength and prevent vessel deformation when strong tensions occurred in vessels (Carlquist, 2002). Sperry et al. (1988) assumed that thick vessel walls in *R. mangle* could substantially reduce the probability of air seeding when these walls were in contact with air-filled extracellular spaces. In mature mangrove *Rhizophora mucronata*, higher vessel densities and narrower vessels are found in the dry season in comparison to those in the rainy season (Verheyden et al., 2004, 2005; Schmitz et al., 2006). Tolerance to salinity in mangrove species was related with less efficient water transport and more conservation water use (Sobrado, 2000, 2001, 2007). It has been found that hydraulic conductance was decreased by waterlogging in young deciduous trees (*Quercus robur*, *Q. petraea*, *Fagus sylvatica*) (Schmull and Thomas, 2000). In this experiment, Figs. 2F and 4C revealed that tangential vessel diameter in stem was dramatically higher than that in leaf. However, both vessel wall thickness and tangential vessel diameter in leaf and stem exhibited similar responses to prolonged waterlogging, tending to decline in the 6 h treatment (Figs. 2F and 4C). From these results, it was predicted that water-transporting capacity and mechanical strength of *A. marina* in leaf tissue and stem might be stable in 0–4 h waterlogging duration, but could be negatively affected when waterlogging duration was longer than 4 h. For *B. gymnorrhiza*, tangential vessel diameter and vessel wall thickness in leaf tissue declined dramatically under the 2 h treatments (Wang et al., 2007), suggesting that leaf vascular system of *B. gymnorrhiza* was more sensitive to waterlogging than that of *A. marina*.

Gelatinous fibers also play an important role in support (Fisher and Tomlinson, 2002). Our results showed that short-term waterlogging duration (2 h) stimulated the formation of gelatinous fibers (Fig. 4D). But gelatinous fibers decreased and were even absent with prolonged waterlogging, in accordance with a previous study in *Laguncularia racemosa*, which reported that highest percentage of gelatinous fibers was present at sites with a low flooding level in estuarine environment (Yáñez-Espinosa et al., 2004). The lack of gelatinous fiber and thinner fiber wall in the long-term waterlogging would impact protection against waterlogging stress.

Although constant vessel diameter, vessel and fiber wall thickness between 0 and 4 h waterlogging duration did not

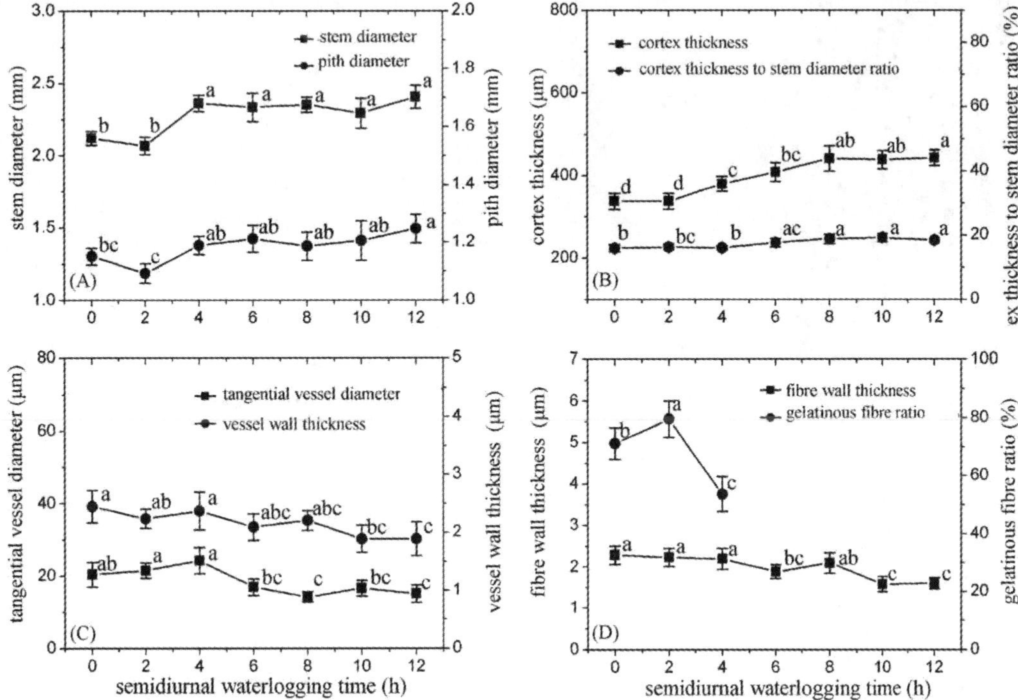

Fig. 4. Stem anatomical characteristics of *A. marina* seedlings under periodical waterlogging grown for 70 days. (A and B) Stem and pith diameter, cortex thickness; (C) vessel; (D) fibre. Values are means ± S.E. Means followed for different letters were statistically different at $P < 0.05$.

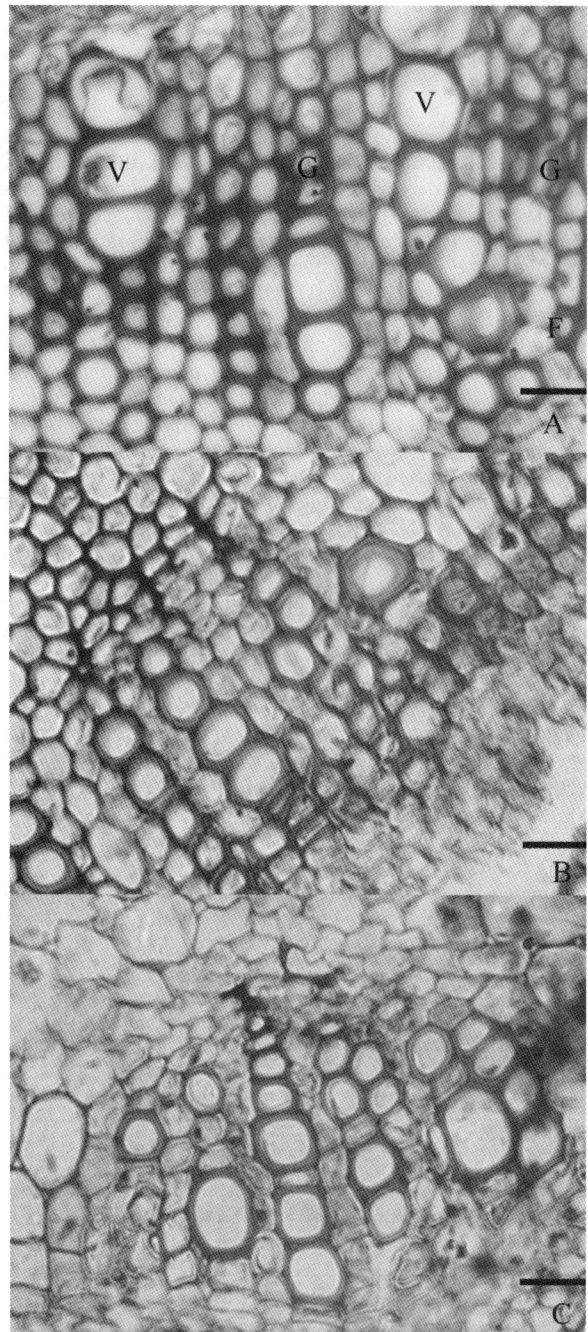

Fig. 5. Stem cross-sections showing the xylem details in *A. marina* seedlings under periodical waterlogging grown for 70 days. (A) 0 h; (B) 6 h; (C) 12 h. Bars, 20 μm. V, vessel; G, gelatinous fibre; F, fibre.

support our hypothesis that *A. marina* seedlings could respond positively to short-term flooding, when waterlogging durations exceeded 4 h, the decreased vessel diameter, vessel and fiber wall thickness, lack of sclerenchyma cells in leaf and gelatinous fibers in stem supported our hypothesis that this species could be negatively influenced by excessive flooding. However, thinner leaves and increased parenchymatous tissue in stem (e.g., pith and cortex) contributed to the function of gas exchange with increasing inundation duration, indicating positive responses throughout all treatments. But lower hypodermis to leaf thickness ratio, palisade-spongy ratio and higher mesophyll to leaf thickness ratio were observed with increasing waterlogging time, suggesting reductions of water storage, photosynthesis and mesophyll conductance. Taken together, these results showed that anatomical characteristics associated with different functions responded differently with increasing waterlogging duration. Nevertheless, additional measurements of hydraulic properties and mechanical strength should be made in future studies to obtain further information.

Acknowledgements

The project was jointly supported by Natural Science Fund of China (Nos. 40776046 and 30200031), the National Foundation for fostering talents of basic science from NSFC (No. J0630649) and Natural Science Foundation of Fujian Province (No. 2006J0146).

Reference

Allaway, W.G., Curran, M., Hollington, L.M., Ricketts, M.C., Skelton, N.J., 2001. Gas space and oxygen exchange in roots of *Avicennia marina* (Forssk.) Vierh. var. *australasica* (Walp.) Moldenke ex N.C. Duke, the Grey Mangrove. Wetlands Ecol. Manage. 9, 211–218.

Arens, N.C., 1997. Responses of leaf anatomy to light environment in the tree fern *Cyathea caracasana* (Cyatheaceae) and its application to some ancient seed ferns. Palaios 12, 84–94.

Armstrong, W., 1979. Aeration in higher plants. Adv. Bot. Res. 7, 225–332.

Camilleri, J.C., Ribi, G., 1983. Leaf thickness of mangroves (*Rhizophora mangle*) growing in different salinities. Biotropica 15, 139–141.

Carlquist, S., 2002. Comparative Wood Anatomy. Springer-Verlag, Berlin.

Chen, L.Z., Wang, W.Q., Lin, P., 2004. Influence of waterlogging time on the growth of *Kandelia candel* seedlings. Acta Oceanol. Sin. 23, 149–158.

Chen, L.Z., Wang, W.Q., Lin, P., 2005. Photosynthetic and physiological responses of *Kandelia candel* L. Druce seedlings to duration of tidal immersion in artificial seawater. Environ. Exp. Bot. 54, 256–266.

Curran, M., Cole, M., Allaway, W.G., 1986. Root aeration and respiration in young mangrove plants *Avicennia marina* (Forsk.) Vierh. J. Exp. Bot. 37, 1225–1233.

Curran, M., James, P., Allaway, W.G., 1996. The measurement of gas spaces in the roots of aquatic plants—Archimedes revisited. Aquat. Bot. 54, 255–261.

Delgado, P., Hensel, P.F., Jimenez, J.A., Day, J.W., 2001. The importance of propagule establishment and physical factors in mangrove distributional patterns in a Costa Rican estuary. Aquat. Bot. 71, 157–178.

Ellison, A.M., Farnsworth, E.J., 1997. Simulated sea level change alters anatomy, physiology, growth, and reproduction of red mangrove (*Rhizophora mangle* L.). Oecologia 112, 435–446.

Fahn, A., 1983. Plant Anatomy. Pergamon Press, New York.

Fisher, J.B., Tomlinson, P.B., 2002. Tension wood fibers are related to gravitropic movement of red mangrove (*Rhizophora mangle*) seedlings. J. Plant Res. 115, 39–45.

He, B.Y., Lai, T.H., Fan, H.Q., Wang, W.Q., Zheng, H.L., 2007. Comparison of flooding-tolerance in four mangrove species in a diurnal tidal zone in the Beibu Gulf. Estuar. Coast. Shelf Sci. 74, 254–262.

Hovenden, M.J., Allaway, W.G., 1994. Horizontal structures on pneumatophores of *Avicennia marina* (Forssk.) Vierh.—a new site of oxygen conductance. Ann. Bot. 73, 377–383.

Kathiresan, K., Bingham, B.L., 2001. Biology of mangroves and mangrove ecosystems. Adv. Mar. Biol. 40, 81–251.

Kitaya, Y., Jintana, V., Piriyayotha, S., Jaijing, D., Yabuki, K., Izutani, S., Nishimiya, A., Iwasaki, M., 2002. Early growth of seven mangrove species planted at different elevations in a Thai estuary. Trees 16, 150–154.

Komiyama, A., Santiean, T., Higo, M., Patanaponpaiboon, P., Kongsangchai, J., Ogino, K., 1996. Microtopography, soil hardness and survival of mangrove (*Rhizophora apiculata* BL) seedlings planted in an abandoned tin-mining area. For. Ecol. Manage. 81, 243–248.

Kozlowski, T.T., 1997. Responses of woody plants to flooding and salinity. Tree Physiol. Monogr. 1, 1–29.

Lin, P., 1999. Mangrove Ecosystem in China. Science Press, Beijing.

Loreto, F., Harley, P.C., Di Marco, G., Sharkey, T.D., 1992. Estimation of mesophyll conductance to CO_2 flux by three different methods. Plant Physiol. 98, 1437–1443.

McKee, K.L., 1993. Soil physicochemical patterns and mangrove species distribution—reciprocal effects? J. Ecol. 81, 477–487.

McKee, K.L., Mendalssohn, I.A., Hester, M.W., 1988. Reexamination of pore water sulfide concentrations and redox potentials near the aerial roots of *Rhizophora mangle* and *Avicennia germinans*. Am. J. Bot. 75, 1352–1359.

Mommer, L., de Kroon, H., Pierik, R., Bogemann, G.M., Visser, E.J.W., 2005. A functional comparison of acclimation to shade and submergence in two terrestrial plant species. New Phytol. 167, 197–206.

Nandy, P., Das, S., Ghose, M., Spooner-Hart, R., 2007. Effects of salinity on photosynthesis, leaf anatomy, ion accumulation and photosynthetic nitrogen use efficiency in five Indian mangroves. Wetlands Ecol. Manage. 15, 347–357.

Parida, A.K., Das, A.B., Mittra, B., 2004. Effects of salt on growth, ion accumulation, photosynthesis and leaf anatomy of the mangrove, *Bruguiera parviflora*. Trees 18, 167–174.

Rao, A.N., Tan, H., 1984. Leaf structure and its ecological significance in certain mangrove plants. In: Soepadmo, E., Rao, A.N., Mclntosh, D.J. (Eds.), Proc. As. Symp. Mangr. Environ.—Res. Manage. UNESCO, pp. 183–194.

Sayed, O.H., 1995. Effects of the expected sea level rise on *Avicennia marina* L: a case study in Qatar. Qatar Univ. Sci. J. 15, 91–94.

Schmitz, N., Verheyden, A., Beeckman, H., Kairo, J.G., Koedam, N., 2006. Influence of a salinity gradient on the vessel characters of the mangrove species *Rhizophora mucronata*. Ann. Bot. 98, 1321–1330.

Schmull, M., Thomas, F.M., 2000. Morphological and physiological reactions of young deciduous trees (*Quercus robur* L., *Q. petraea* [Matt.] Liebl., *Fagus sylvatica* L.) to waterlogging. Plant Soil 225, 227–242.

Scholander, P.F., hammel, H.T., Bradstreet, E.D., Hemmingsen, E.A., 1965. Sap pressure in vascular plants. Science 148, 339–346.

Sobrado, M.A., 2000. Relation of water transport to leaf gas exchange properties in three mangrove species. Trees 14, 258–262.

Sobrado, M.A., 2001. Hydraulic properties of a mangrove *Avicennia germinans* as affected by NaCl. Biol. Plant. 44, 435–438.

Sobrado, M.A., 2007. Relationship of water transport to anatomical features in the mangrove *Laguncularia racemosa* grown under contrasting salinities. New Phytol. 173, 584–591.

Sperry, J.S., Tyree, M.T., Donnelly, J.R., 1988. Vulnerability of xylem to embolism in a mangrove vs an inland species of Rhizophoraceae. Physiol. Plant. 74, 276–283.

Thampanya, U., Vermaat, J.E., Duarte, C.M., 2002a. Colonization success of common Thai mangrove species as a function of shelter from water movement. Mar. Ecol.-Prog. Ser. 237, 111–120.

Thampanya, U., Vermaat, J.E., Sinsakul, S., Panapitukkul, N., 2006. Coastal erosion and mangrove progradation of Southern Thailand. Estuar. Coast. Shelf Sci. 68, 75–85.

Thampanya, U., Vermaat, J.E., Terrados, J., 2002b. The effect of increasing sediment accretion on the seedlings of three common Thai mangrove species. Aquat. Bot. 74, 315–325.

Tomlinson, P.B., 1986. The Botany of Mangrove. Cambridge University Press, New York.

Verheyden, A., De Ridder, F., Schmitz, N., Beeckman, H., Koedam, N., 2005. High-resolution time series of vessel density in Kenyan mangrove trees reveal a link with climate. New Phytol. 167, 425–435.

Verheyden, A., Kairo, J.G., Beeckman, H., Koedam, N., 2004. Growth rings, growth ring formation and age determination in the mangrove *Rhizophora mucronata*. Ann. Bot. 94, 59–66.

Vervuren, P.J.A., Beurskens, S., Blom, C., 1999. Light acclimation, CO_2 response and long-term capacity of underwater photosynthesis in three terrestrial plant species. Plant Cell Environ. 22, 959–968.

Wang, W., Wang, M., 2007. The Mangroves of China. Science Press, Beijing.

Wang, W., Zhao, M., Deng, C., Lin, P., 2000. Species and its distribution of mangroves in Fujian coastal area. J. Oceanogr. Taiwan Strait 19, 534–540 (in Chinese).

Wang, W.Q., Xiao, Y., Chen, L.Z., Lin, P., 2007. Leaf anatomical responses to periodical waterlogging in simulated semidiurnal tides in mangrove *Bruguiera gymnorrhiza* seedlings. Aquat. Bot. 86, 223–228.

Yáñez-Espinosa, L., Terrazas, T., Angeles, G., 2008. The effect of prolonged flooding on the bark of mangrove trees. Trees 22, 77–86.

Yáñez-Espinosa, L., Terrazas, T., López-Mata, L., 2001. Effects of flooding on wood and bark anatomy of four species in a mangrove forest community. Trees 15, 91–97.

Yáñez-Espinosa, L., Terrazas, T., López-Mata, L., Valdez-Hernandez, J.I., 2004. Wood variation in *Laguncularia racemosa* and its effect on fibre quality. Wood Sci. Technol. 38, 217–226.

Ye, Y., Tam, N.F.Y., Wong, Y.S., Lu, C.Y., 2003. Growth and physiological responses of two mangrove species (*Bruguiera gymnorrhiza* and *Kandelia candel*) to waterlogging. Environ. Exp. Bot. 49, 209–221.

Youssef, T., Saenger, P., 1996. Anatomical adaptive strategies to flooding and rhizosphere oxidation in mangrove seedlings. Aust. J. Bot. 44, 297–313.

五 红树林分子生态学

PART V MANGROVE MOLECULAR ECOLOGY

木榄属三种红树植物的遗传变异和亲缘关系分析*

潘 文[1,2] 周涵韬[1,3] 陈 攀[1] 林 鹏[1,2]

(1. 厦门大学生命科学学院,福建 厦门 361005;2. 厦门大学湿地与生态工程研究中心,福建 厦门 3361005;

3. 福建省农业科学院 生物技术中心,福建 福州 350003)

摘要:用随机扩增多态性DNA(RAPD)和inter-简单重复序列(ISSR)分子标记技术对木榄属(*Bruguiera*)3种红树木榄(*Bruguiera gymnorrhiza*)、海莲(*B. sexangula*)、尖瓣海莲(*B. sexangula var. rhynchopetala*)进行遗传亲缘关系研究。12个RAPD引物和10个ISSR引物分别扩增出240和191条带,多态位点百分率分别为38.75%和52.88%,ISSR检测到的多态位点率高于RAPD。运用Nei指数法计算木榄-海莲、木榄-尖瓣海莲、海莲-尖瓣海莲之间的遗传距离,RAPD分析结果为0.47、0.36、0.29,ISSR分析结果为0.62、0.41、0.32。同时运用UPGMA统计法进行聚类分析,结果显示,海莲和尖瓣海莲聚为一组,木榄单独一组。结合宏观形态和等位酶资料,作者把尖瓣海莲确定为海莲的变种。

关键词:木榄属;ISSR;RAPD;分子标记

中图分类号:Q37 **文献标识码**:A **文章编号**:1000-3096(2005)05-0023-06

DNA序列多态性的检测和利用是分子生物学最主要的进展之一[1]。目前,已发展了多种以DNA多态性为基础的遗传标记,其中RAPD(Random Amplified Polymorphic DNA)[2]和ISSR(Inter-Simple Sequence Repeat)[3]标记,在生物种属鉴定、种质资源的遗传多样性、基因定位、分子连锁图构建[4-7]等方面得到了迅速而广泛地运用。相对于组织蛋白和等位酶分析而言,RAPD和ISSR具有多态性高、无需活材料,能实现全基因组无偏取样和无组织器官特异等优点;与微卫星分析相比,二者不要求预知基因组序列信息,大大减少了多态性分析的预备工作。

红树林是分布于热带海岸潮间带的木本植物群落,是热带海岸重要的植被类型,是维护海岸生态平衡的特殊生态系。由于不合理的开发利用,全世界的红树林都面临着资源枯竭的严重境地,为保护和可持续利用这一独特资源,各有关国家开展了许多研究。但研究多集中在生理学、生态学、生物化学[8]等领域,而分子生物学的研究较少。与其他生物保护一样,用分子标记的方法研究红树植物遗传多样性是红树植物种质资源保护及开发利用的基础。目前,红树植物某些种属的分类关系仍不确定。红树植物中红树科木榄属的植物在中国共有4种[9]:柱果木榄、木榄、海莲、尖瓣海莲。柱果木榄在海南有记录,现已灭绝。木榄在我国分布较广,海莲、尖瓣海莲均分布在海南岛。3种红树植物在形态学上的差异较小,主要集中在花器的结构上,这给引种、造林研究造成一定困难。通常,证实种间关系主要依靠形态学、次生化合物、染色体、地理分布和实验杂交等方法,但这些方法都有其局限性。相比之下,分子标记技术是一种有效的分析手段,对于任何一个类群来说,要么含有某个基因位点,要么不含该位点,不存在形态性状中难以判断的中间性状。作者利用RAPD和ISSR两种DNA分子标记技术对木榄属3种红树植物木榄、海莲、尖瓣海莲的进行遗传亲缘关系研究。

1 材料和方法

1.1 材料

木榄(*Bruguiera gymnorrhiza*)、海莲(*B. sexangula*)、尖瓣海莲(*B. sexangula var. rhynchopetala*)均采集于海

* 教育部科技重点项目(104105);福建省青年科技项目(2001J033)资助

原载于海洋科学,2005,29(5):23-28

南岛东寨港红树林保护区内。选择生长良好，无病虫害，胸径 4 cm以上的母树。每隔 5～10 m随机选取一株，每个种取 5 个个体，采集幼嫩叶片单独做好标记。叶片采集后置于 - 20 ℃冰箱或液氮储存备用。

RAPD 和 ISSR 引物，以及 Taq 酶等药品均购自上海生工生物工程技术服务有限工司。

1.2 方法

1.2.1 红树植物 DNA 的提取

参照周涵韬[8]等的方法进行总 DNA 的提取。

1.2.2 RAPD - PCR 反应

从 100 个 RAPD 随机引物 (上海生工) 中筛选出 12 个能获得清晰多态性条带，反应稳定引物。RAPD - PCR 扩增总体积为 25 μL。包括 Tris - HCl 10.0 mmol/L (pH8.0)，KCl 50.0 mmol/ L，0.1 % Tritonx - 100，2.5 mmol/ L $MgCl_2$，0.1 mmol/ L dNTPs，0.4 μmol/ L 引物，80 ng 的 DNA 模板，1UTaq 聚合酶。PCR 循环设置为：94 ℃变性 1 min，36 ℃退火 1 min，72 ℃延伸 2 min，共 40 个循环，然后 72 ℃延伸 7 min。反应产物在含有 EB 的 1.4 %琼脂糖凝胶中电泳检测，电压为 5 V/ cm，2 h，电泳结束后，在紫外检测仪上观察，并在凝胶成像系统保存图像。

1.2.3 ISSR - PCR 反应

按照与 RAPD 相同的筛选策略从 30 个 ISSR 引物中选出 10 个用于 ISSR 的 PCR 反应。ISSR - PCR 扩增总体积为 20 μL。包括 Tris - HCl 10.0 mmol/ L (pH8.0)，KCl 50.0 mmol/ L，0.1 %Tritonx - 100，2 mmol/ L $MgCl_2$，0.2 mmol/ L dNTPs，0.4 μmol/ L 引物，80 ng 的 DNA 模板，1UTaq 聚合酶，无菌水 6.6 μL。PCR 循环设置为：94 ℃变性 5 min 后，94 ℃变性 30 s，52 ℃退火 45 s，72 ℃延伸 2 min，共 45 个循环，然后 72 ℃延伸 7 min。反应产物在含有 EB 的 1.5 %琼脂糖凝胶中电泳检测，电压为 5 V/ cm，2 h 记录和拍照同 RAPD。

1.2.4 数据统计分析

在 RAPD 和 ISSR 扩增结果电泳图谱中，有带计为“1”，无带计为“0”，强带和弱带均计为“1”。

遗传一致度、遗传距离及聚类分析：根据 RAPD 扩增结果所统计的数据，遗传距离 (D) 和遗传一致度 (F) 的计算运用 Nei 指数法[10]，$F = 2Nxy/ (Nx + Ny)$，其中 Nxy 为两个个体共同享有的 RAPD 标记数，Nx，Ny 分别为 X 和 Y 个体分别拥有的 RAPD 标记数，再经 $D = 1 - F$ 计算相应的遗传距离。聚类分析采用 UPGMA (unweighted pair group mean average) 进行[11]。

2 结果与分析

2.1 RAPD 分析结果

在运用 RAPD 引物对 3 个供试材料的 DNA 扩增过程中。需对模板 DNA 的浓度做梯度实验。模板浓度在 50～150 ng 时都有扩增，最后选定为 80 ng / μL。在 PCR 扩增中，酶是最主要的影响因素，由于不同厂家不同批次酶的活性不同，对镁离子浓度要求不同，因此，每次买新酶必须做镁离子梯度实验，以确定最佳镁离子浓度。本实验最后选定镁离子浓度为 2.5 mmol/ L。

从 100 个 RAPD 引物中筛选出 12 个能获得清晰条带，反应稳定的引物(重复 2～3 次)进行研究。引物号序列以及 RAPD 扩增情况见表 1。RAPD 反应扩增片段大部分集中在 0.35～3.5 kb 范围内，如图 1。12 个 RAPD 引物在木榄属 3 种红树植物中共扩增出 290 条带，平均每个引物扩增出 20 条带，其中具有多态性的扩增带有 93 条，占总扩增带的 38.75 %，平均每个引物可扩增出 7.75 条多态性带。

根据 RAPD 扩增结果，运用 Nei 指数法计算木榄属 3 种红树植物的遗传距离和遗传一致度，结果见表 2。木榄 - 海莲，木榄 - 尖瓣海莲，海莲 - 尖瓣海莲的遗传距离分别为 0.47、0.34、0.29。平均遗传距离为 0.36。并运用 UPGMA 统计分析对木榄属 3 种植物的亲缘关系进行聚类分析，结果见图 2。木榄与海莲、尖瓣海莲的亲缘关系较远单独聚为一组，海莲与尖瓣海莲的关系较近。两者聚为一组。

表 1 木榄属 3 种红树植物 RAPD - PCR 扩增情况

Tab.1 Amplification of 3 mangrove species of *Bruguiera* by RAPD - PCR

引物号	引物序列	扩增条带	多态条带	多态百分率(%)
S124	GGTGATCAGG	13	4	30.77
S125	CCGAATTCCC	17	14	82.35
S129	CCAAGCTTCC	29	14	48.28
S130	GGAAGCTTGG	26	8	30.77
S141	CCCAAGGTCC	25	4	16
S143	CCAGATGCAC	26	11	42.31
S144	GTGACATGCC	22	7	31.82
S148	TCACCACGGT	14	8	57.14
S149	CTTCACCCGA	18	6	33.33
S152	TTATCGCCCC	16	4	25
S155	ACGCACAACC	21	9	42.86
S156	GGTGACTGTG	13	4	30.77
平均		240	93	
总计		20	7.75	38.75

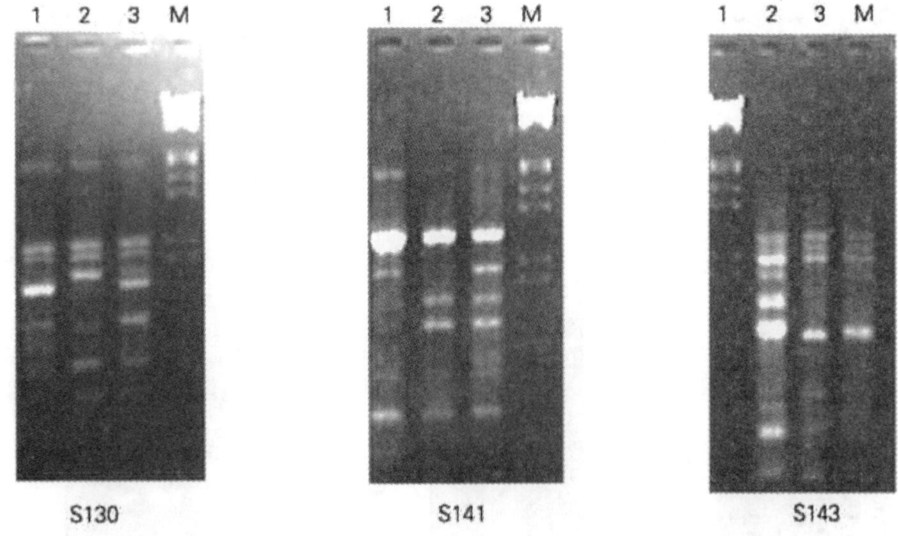

图 1 木榄属 3 种红树植物 RAPD 引物扩增图谱

Fig. 1 genomic DNA fingerprints of 3 mangrove species in *Brugiera* by RAPD

1. 木榄;2. 海莲;3 尖瓣海莲,M 为 λDNA EcoR Ⅰ/Hind Ⅲ分子量标记

1. *Bruguiera gymnorrhiza*;2. *B. sexangula*;3. *B. sexangula var. rhynchopetala*, M:λDNA EcoR Ⅰ/Hind Ⅲ

表 2 木榄属 3 种红树植物 RAPD 分析的遗传距离(对角线下)及遗传一致度(对角线上)

Tab. 2 Similarity matrit and genetic distance of 3 species of mangrove in *Bruguiera* by RAPD

木 榄	海 莲	尖瓣海莲
0	0.53	0.66
0.47	0	0.71
0.34	0.29	0

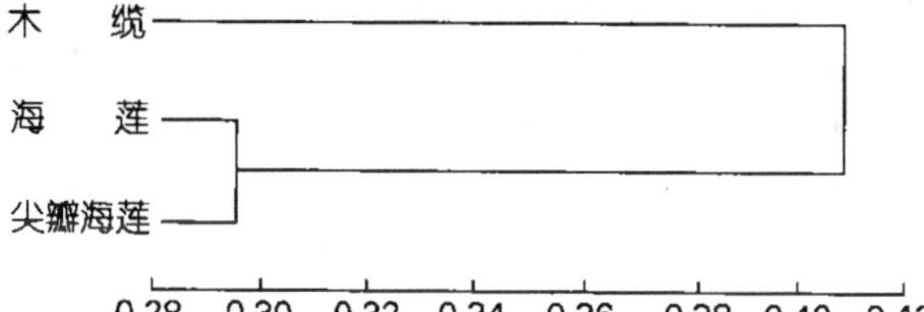

图 2 木榄属 3 种红树植物分子分类系统(RAPD 分析)

Fig. 2 DNA molecular dendrogram of 3 mangrove species in *Brugiera* by RAPD

2.2 ISSR 分析结果

在 RAPD 体系的基础上，发现 ISSR 体系模板 DNA 的浓度与 RAPD 差别不大,同时也对镁离子进行梯度实验,以获得最适合酶活性的镁离子浓度。实验结果以 Taq 酶用量 1 U,镁离子浓度为 2 mmol/L,

从 30 个 ISSR 引物中筛选出 10 个扩增效果好的引物进行遗传差异研究(表 3)。ISSR 反应扩增片段大部分集中在 0.2～2.0 kb 范围内，如图 3。10 个 ISSR 引物在木榄属 3 种红树植物中共扩增出 191 条带，平均每个引物扩增出 19.1 条，多态性条带有 101 条，占总扩增带的 52.88%。平均每个引物可扩增出 10.1 条多态性带，ISSR1 引物扩增多态性条带百分率最高达 85%。运用 Nei 指数法，根据 ISSR 扩增结果，计算木榄属 3 种红树植物的遗传距离和遗传一致度(表 4)，木

表 3 木榄属 3 种红树植物 ISSR-PCR 扩增情况

Tab. 3 Amplification of 3 mangrove species of *Bruguiera* by ISSR-PCR

引物号	引物序列	扩增条带	多态性条带	多态性条带百分率(%)
ISSR1	$GC(AC)_4$	20	17	85
ISSR2	$CCC(GT)_6$	18	9	50
ISSR3	$CGC(GA)_6$	19	7	36.84
ISSR4	$CCA(GTG)_4$	7	1	14.29
ISSR5	$GCG(AC)_6A$	29	14	48.27
ISSR6	$(AG)_8T$	28	16	57.14
ISSR7	$(GA)_8T$	18	9	50
ISSR8	$(GA)_8GC$	21	9	42.86
ISSR9	$(AC)_8CA$	23	14	60.87
ISSR10	$(CT)_8AC$	8	5	62.5
总计		191	101	
平均		19.1	10.1	52.88

榄-海莲、木榄-尖瓣海莲、海莲-尖瓣海莲的遗传距离分别为。0.62、0.42、0.31,平均值为0.45。运用UPGMA统计分析对木榄3种红树植物的亲缘关系进行聚类分析,结果见图4。木榄与海莲、尖瓣海莲的亲缘关系较远单独聚为一组,海莲与尖瓣海莲的关系较近,两者聚为一组。

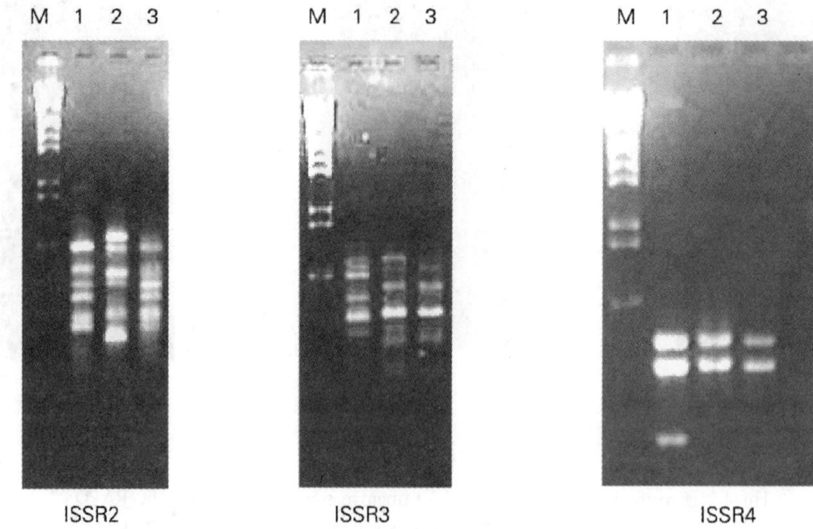

图3 木榄属3种红树植物ISSR引物扩增图谱

Fig. 3 genomic DNA fingerprints of 3 mangrove species in *Brugiera* by ISSR

1. 木榄;2. 海莲;3尖瓣海莲,M为λDNA EcoR Ⅰ/Hind Ⅲ分子量标记

1. *Bruguiera gymnorrhiza*; 2. B. *sexangula*; 3. B. *sexangula var. rhynchopetala*, M:λDNA EcoR Ⅰ/Hind Ⅲ

表4 木榄属3种红树植物ISSR分析的遗传距离(下三角)及遗传一致度(上三角)

Tab.4 Similarity matrit and genetic distance of 3 species of mangrove in *Bruguiera* by ISSR

	木 榄	海 莲	尖瓣海莲
木 榄	0	0.38	0.59
海 莲	0.62	0	0.68
尖瓣海莲	0.41	0.32	0

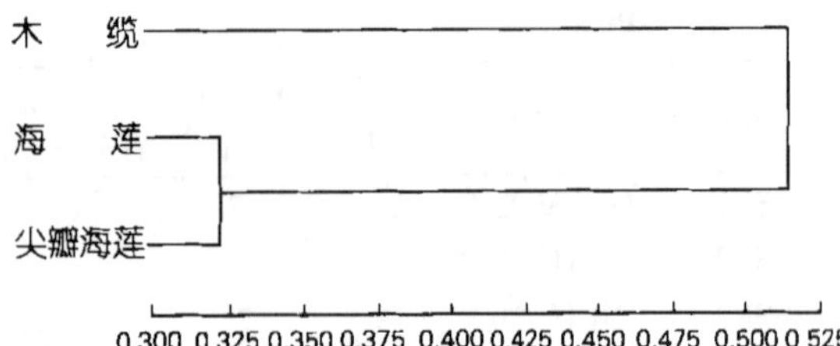

图4 木榄属3种红树植物分子分类系统(ISSR分析)

Fig.4 DNA molecular dendrogram of 3 mangrove species in *Brugiera* by ISSR

3 讨论

多态位点百分率是衡量物种遗传变异水平高低的一个重要指标,是度量遗传多样性的重要参数。Hamrick等[12]用等位酶对449种植物进行遗传变异研究,首次在种水平上对植物的遗传多样性做出估计,并且和在种群水平上的遗传多样性做了对比,揭示出种多态位点平均水平为50.5%,种群多态位点平均水平为34.2%。本研究用RAPD方法检测木榄属3种红树植物的平均多态百分率为38.75%,低于449个种的平均水平,高于种群平均水平。ISSR方法检测的平均多态百分率为52.88%。说明ISSR比RAPD检测到更多的遗传多态位点。显然,这与它们在对基因组进行扩增时引物结合位点的不同有直接的关系。在小麦和水稻的研究中也证明ISSR检测遗传多态性的能力比RAPD更高[5,13]。RAPD和ISSR分析所获得的遗传距离和UPGMA聚类树状图结果均比较一致。说明这两种分子标记技术可以有效地应用于象红树植物这类遗传多态性低物种的个体鉴定和遗传多样性评价研究。

遗传变异是进化的动力,而进化过程必然伴随或大或小的遗传变异,这是新物种形成的关键一步[14]。一般来说,随着分类或进化水平的提高,分类群间遗传一致度下降,遗传距离加大。Gottlieb[15]对28种植物的亲缘关系用等位酶进行了研究,发现种内种群间遗传距离很低,平均值为 0.05。种间遗传距离明显升高,平均值为 0.33。在本研究中,RAPD 和 ISSR 方法所获得木榄、海莲、尖瓣海莲平均的遗传距离分别为 0.36 和 0.45。表明 3 种红树植物为属内种间关系。从分子聚类图上,海莲、尖瓣海莲聚为一组,而木榄单独聚为一组,表明海莲与尖瓣海莲亲缘关系较近,二者与木榄的亲缘关系较远。电泳图谱上,尖瓣海莲与海莲的指纹图谱极为相似,而与木榄的指纹图谱差异较大。林鹏[9]对中国红树植物的形态学分类描述中,木榄与海莲、尖瓣海莲在形态有较大的差别,而海莲与尖瓣海莲仅存在极小差异(表 5),但这些差异都不是辨别种的主要特征,因此,把尖瓣海莲定位海莲的变种。葛菁萍[16]用等位酶方法对木榄属 3 种红树植物进行了种间亲缘关系分析,共检测了 5 个酶系统,7 个酶位点和 17 个等位基因,结果表明,木榄 - 海莲的遗传一致度为 0.796 6,海莲 - 尖瓣海莲的遗传一致度为 0.915 0,尖瓣海莲 - 木榄的遗传一致度为 0.889 5,表明海莲与尖瓣海莲之间的亲缘关系要近于木榄,尖瓣海莲是海莲的变种。综合上述研究,作者认为把尖瓣海莲处理为海莲的一个变种似乎更为妥当。

作者用 RAPD 和 ISSR 两种分子标记有效地将亲缘关系接近的尖瓣海莲和海莲区分开来。可看出分子分类对于红树资源的鉴定、分类及遗传多样性评价是一种有效手段。它应用于红树植物的分子分类研究是可行的。分子分类技术结合传统的分类方法必将进一步促进红树资源的合理开发、保护和利用。

表 5 木榄属三种植物的形态学差异

Tab.5 Morphlogic variance of 3 species in *Bruguiera*

项目	海 莲	尖瓣海莲	木 榄
1. 花萼裂片	9~12 枚,通常 10 枚	9~12 枚,通常 10 枚	10~14 枚,通常 12 枚
2. 花萼/萼管	结果时长于萼管	结果时长于萼管	结果时短于萼管
3. 花萼管	有纵棱	有纵棱	平滑
4. 花瓣	边缘被粗毛	边缘被粗毛	基部被密毛,上部秃净
5. 花瓣裂片	裂片顶端具刺毛 1~2 条,裂口 1 条	裂片顶端具刺毛 1~2 条裂口 1 条	顶端锐具刺毛 2~4 条,裂口 1 条明显超出花瓣顶端
	1. 刺毛:1 条	2 条	
	2. 花药:不具喙	具喙	

参考文献:

[1] 阎文昭,蔡平钟,宣 朴,等. RAPD 技术在生物学研究中的应用[J]. 世界科技研究与发展,1999,2:83 - 84.

[2] Williams J G L. DNA polymorphisms amplified by arbitrary primers are useful as genetic markers [J]. **Nuc Aci Res**, 1990 ,**18**(22) :6 531 - 6 535.

[3] Ziekiewicz E,Rafalski A,Labuda D. Genome fingerprinting by simple sequence repeat (SSR) - anchored polymerase chain reaction amplification [J]. **Genomica**, 1994, 20: 176 - 183.

[4] 王建波. ISSR 分子标记及其在植物遗传学研究中的应用[J]. 遗传,2002,**24**(5) :613 - 615.

[5] 杜金昆,姚颖垠,倪中福,等. 普通小麦、斯卑尔脱小麦、密穗小麦和轮回小麦选择后代材料 ISSR 分子标记遗传差异研究[J]. 遗传学报,2002,**29**(5) :44 - 48.

[6] 陈 洪,朱立煌,徐吉臣,等. 水稻 RAPD 分子连锁图谱的构建[J]. 植物学报,1995,37:677 - 684.

[7] Penner G A, Chong J ,Levesque L M, *et al*. Identification of a RAPD marker linked to the oat stem rust gene Pg3[J]. **Theor Appl Genet**,1993,85:702 - 705.

[8] 周涵韬,林 鹏. 中国红树科 7 种红树遗传多样性分析[J]. 水生生物学报,2001,**25**(4) :362 - 369.

[9] 林 鹏. 红树林[M]. 北京:海洋出版社,1984.

[10] Nei M, Li W. A mathematical model for studying genetic variation in the terms of restriction endonucleases[J]. **Proc Natl Acad Sci USA**, 1979,17:5 269 - 5 273.

[11] Vierling R A, Nguyen H T. Use of RAPD markers to determine the genetic diversity of diploid, wheat genotyes [J]. **Theor Appl Genet**,1992,84:835 - 838.

[12] Hamrick J L, Godt MJ W. Alloayme diversity in plant species [A]. Brown A H D, Clegg M T, Kahler A L, *et al*. eds. Plant population genetics, breeding, and genetic resources[C].

Sunderland Mass:Sinauer,1989. 43 - 46.

[13] 钱伟,葛颂,洪德元. 采用RAPD和ISSR标记探讨中国疣粒野生稻的遗传多样性[J]. 植物学报,2000,**42**(7):741 - 750.

[14] Grant V. Plant speciation [M]. New York: Columbia University. Press,1981. 149 - 172.

[15] Gottlieb L D. Electrophoretic evidence and plant systematics [J]. **Ann Missouri Bot Gard**,1977,64:161 - 180.

[16] 葛菁萍. 几种红树植物遗传变异和生态分化的研究[D]. 厦门大学博士论文,1999.

Genetic variation and relationship of 3 Bruguiera species by RAPD and ISSR

PAN Wen[1,2], ZHOU Han - tao[1,3], CHEN Pan[1], LIN Peng[1,2]

(1. School of Life Science, Xiamen University, Xiamen 361005, China; 2. Research Centre for Wetlands and Ecological Engineering, Xiamen University, Xiaman 361005, China; 3. Biotechnology Centre, FuJian Academy of Agricultural Science, Fuzhou 350003, China)

Received: Feb.,10,2004

Key word: *Bruguiera*; RAPD; ISSR; molecular marker

Abstract: Random amplified polymorphic DNA (RAPD) and inter - simple sequence repeat (ISSR) methods were applied to detect genetic variation and relationship of 3 mangrove species of *Bruguiera*(*Bruguiera gymnorrhiza*, *B. sexangul*, *B. sexangula var. rhynchopetala*). 12 RAPD and 10 ISSR primers generated 240 and 191 bands, of which 91 and 101 were polymorphic, respectively. Percentages of polymorphic bands of RAPD and ISSR were 38.75% and 52.88%. Comparison of the two marker systems shows that ISSR was better than RAPD in terms of reproducibility and ability of detecting genetic polymorphism. It was found that the average genetic distance among different 3 mangrove species of *Bruguiera* by RAPD and ISSR were 0.37 and 0.45, respectively. Based on UPGMA cluster analysis, it showed that *Bruguiera sexangula* and *B. sexangula var. rhynchopetala* were classified into same group, while *Bruguiera gymnorrhiza* in an other group. Comparing with the results of morphology and allozyme by other reports, *B. sexangula var. rhynchopetala* could be regarded as a variety of *Bruguiera sexangula*.

不同地区桐花树种群的分子遗传变异分析*

潘 文[1,2] 周涵韬[1,3] 陈 攀[1] 林 鹏[1,2]

(1. 厦门大学生命科学学院,2. 厦门大学湿地与生态工程研究中心,福建 厦门 361005;

3. 福建省农业科学院生物技术中心,福建 福州 350003)

摘要:利用RAPD和ISSR两种分子标记,对4个不同纬度桐花树种群的遗传多样性和遗传分化进行探讨,根据RAPD和ISSR数据计算遗传距离并进行聚类分析.结果表明,4个种群桐花树遗传变异性低,4个纬度桐花树种群分为南北两个类群,海南和广西为1个类群,广东和福建为1个类群.利用Shannon's指数计算4个种群的遗传多样性,广西种群多样性指数最高,福建种群最低.大部分的遗传变异存在于种群内(64 %),小部分的遗传变异存在种群间(36 %).本文同时探讨了种群遗传保护和引种策略.

关键词:桐花树;遗传多样性;RAPD;ISSR

中图分类号:Q 943 **文献标识码**:A

DNA序列多态性的检测和利用是分子生物学重要的进展之一[1].目前,已发展了多种以DNA多态性为基础的遗传标记,其中RAPD(Random Amplified Polymorphic DNA)[2]和ISSR(Inter-Simple Sequence Repeat)[3]标记,在生物种属鉴定、种质资源的遗传多样性、基因定位、分子连锁图构建[4~7]等方面得到了迅速而广泛地运用.

桐花树属紫金牛科桐花树属隐胎生泌盐红树植物,它广泛分布到亚洲沿岸和东太平洋群岛,大洋洲沿岸.我国主要分布在海南、香港、澳门、广东、广西、台湾、福建[8].红树植物在维护海岸生态平衡,防风减灾、护堤保岸等方面发挥重要的作用.由于不合理的开发利用,全世界的红树林都面临着资源枯竭的严重境地,为保护和可持续利用这一独特资源,各有关国家开展了许多研究.但研究多集中在生理学、生态学、生物化学[9]等领域,而分子生物学的研究较少.与其它生物保护一样,用分子标记的方法研究红树植物遗传多样性是红树植物种质资源保护及开发利用的基础.

本实验以桐花树种群为代表,用RAPD和ISSR两种标记方法对4个不同纬度分布的桐花树种群开展遗传多样性和遗传分化研究,旨在探讨桐花树种群遗传多样性和遗传分化的变化规律.

1 材料与方法

1.1 材 料

桐花树分别采集福建龙海红树林保护区、广西山口红树林保护区、海南东寨港红树林保护区、广东深圳福田红树林保护区,种群特征情况见表1,桐花树的采集均利用种群取样的方法进行.选择胸径4 cm以上,无病虫害的母树,每隔10~20 m取1株,每个种群取10个个体,采集幼叶,叶片采集后置于-20 ℃冰箱或液氮储存备用.

RAPD和ISSR引物,以及Taq酶等药品均购自上海生工生物工程技术服务有限公司.引物及引物序列见表2.

1.2 方 法

红树植物DNA的提取,采用改进的CTAB(Cetyl trimethyl ammoniumbromide,十六烷基三甲基溴化铵)法提取[9].

RAPD-PCR反应从100个RAPD随机引物中筛选出10个能获得清晰多态性条带,反应稳定引物.RAPD-PCR扩增总体积为25 μL.包括10 ×PCR缓冲液2.5 μL,25 mmol $MgCl_2$ 2.5 μL,1 mmol/L dNTPs 2.5 μL,5 μmol/L引物1.5 μL,40 ng的

* 教育部科技重点项目(104105);福建省青年科技项目(2001J033)资助

原载于厦门大学学报(自然科学版),2004,431(Sup.):106-112

表1 4个纬度采样点桐花树种群的土壤条件①

Tab.1 The soil characteristics the populations of *Aegiceras corniculatum* in 4 sampling sites

样地	土壤含盐量	pH值	有机质/ %	全氮/ %	全磷/ %	纬度
海南东寨港	13.4	5.14	10.44	0.032	0.014	19°30′
广西山口	20.66	5.32	3.48	0.079	0.033	21°28′
深圳福田	19.08	7.4	4.69	0.247	0.122	22°32′
福建龙海	13.40	7.3	4.89	0.229	0.076	24°24′

表2 RAPD和ISSR引物号及序列

Tab.2 List of RAPD and ISSR primers and sequences

引物号	引物序列	引物号	引物序列
S50	GGTCTACACC	S148	TCACCACGGT
S53	GGGGTGACGA	S156	GGTGACTGTG
S55	CATCCGTGCT	ISSR4	$CCA(GTG)_4$
S124	GGTGATCAGG	ISSR6	$(AG)_8T$
S125	CCGAATTCCC	ISSR7	$(GA)_8T$
S129	CCAAGCTTCC	ISSR8	$(GA)_8GC$
S130	GGAAGCTTGG	ISSR9	$(AC)_8CA$
S141	CCCAAGGTCC		

DNA模板4 μL,5U Taq聚合酶0.2 μL,无菌水11.8 μL. PCR循环设置为:94 ℃变性1 min,36 ℃退火1 min,72 ℃延伸2 min,共40个循环,然后72 ℃延伸7 min,最后将结果置于4 ℃冰箱中.反应产物在含有EB的1.4%琼脂糖凝胶中电泳检测,电压为3 V/cm,2 h,电泳结束后,在紫外检测仪上观察,并在凝胶成像系统保存图像.

ISSR-PCR反应 按照与RAPD相同的筛选策略从30个ISSR引物中选出5个用于ISSR的PCR反应. ISSR-PCR扩增总体积为20 μL.包括10×PCR缓冲液2 μL,25 mmol $MgCl_2$ 1.6 μL,1 mmol/L dNTPs 4 μL,5 μmol/L引物1.6 μL,40 ng的DNA模板4 μL,5U Taq聚合酶0.2 μL,无菌水6.6 μL. PCR循环设置为:94 ℃预变性5 min后,94 ℃变性30 s,52 ℃退火45 s,72 ℃延伸2 min,共45个循环,然后72 ℃延伸7 min,最后将结果置于4 ℃冰箱中.反应产物在含有EB的2%琼脂糖凝胶中电泳检测,电压为3 V/cm,电泳结束后,在紫外检测仪上观察,并在凝胶成像系统保存图像.

① 黎中宝,红树植物桐花树和白骨壤的遗传变异与分化的生态遗传学研究,厦门大学博士论文,2000.

1.3 数据统计分析

在RAPD和ISSR扩增结果电泳图谱中,有带计为"1",无带计为"0",强带和弱带均计为"1".

遗传一致度、遗传距离及聚类分析:根据RAPD和ISSR扩增结果所统计的数据,遗传距离(D)和遗传一致度(F)的计算运用Nei指数法[10], $F = 2N_{xy}/(N_x + N_y)$,其中N_{xy}为两个个体共同享有的RAPD标记数,N_x,N_y分别为X和Y个体分别拥有的RAPD标记数,再经$D = 1 - F$计算相应的遗传距离.聚类分析采用UPGMA(unweighted pair group mean average)进行[11].

Shannon's信息多样性指数:利用公式$H = -\sum P_i \ln P_i$计算[12](P_i是种群内一条扩增产物存在的频率,或称表型频率).

H为表型多样性指数. H可以计算两种水平的多样性:H_{pop}和H_{sp}. H_{pop}是种群内平均遗传多样性的测度:$H_{pop} = 1/n \sum H$,n为亚种群数. H_{sp}是种群间总的遗传多样性的测度:$H_{sp} = -\sum P \ln P$,P为种群间1条扩增产物存在的频率. H_{pop}/H_{sp}是种群内多样性所占的比例;$(H_{sp} - H_{pop})/H_{sp}$为种群间多样性所占比例.

2 实验结果与分析

2.1 不同纬度桐花树种群的遗传分化

从100个RAPD引物中筛选出10个能获得清晰条带,反应稳定的引物进行研究. RAPD反应扩增情况见表3,片断大部分集中在0.35~3.5 kb范围内(如图1).10个引物在4个纬度桐花树种群间共扩增出217条带,多态性条带为47条,多态性条带百分率为21.66%.表明种群间的遗传变异小.

根据RAPD扩增结果,运用Nei指数计算4种不同纬度桐花树种群间的遗传距离,结果见表4.最大遗传距离是海南-福建为0.238,最小的是海南-广西为0.127,种群间平均遗传距离为0.188.说明种

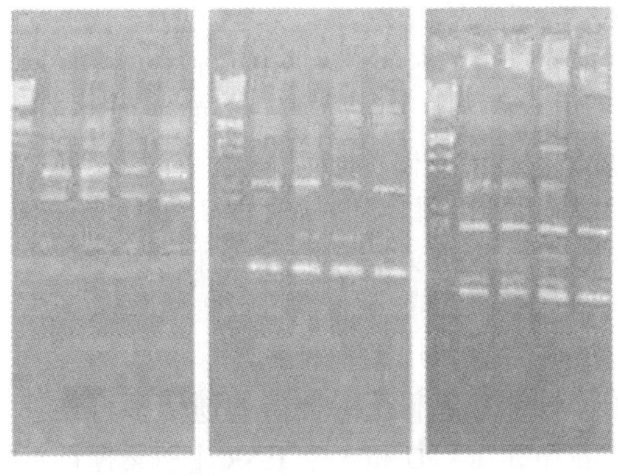

图 1 不同纬度桐花树种群间 RAPD 扩增图谱
1. 海南种群 2. 广西种群 3. 广东种群 4. 福建种群 M 为λDNA EcoR Ⅰ/ Hind Ⅲ分子量标记

Fig. 1 Genomic DNA fingerprints of populations in 4 latitudes by RAPD

表 3 RAPD 标记的遗传多态性

Tab. 3 Amplification of population by RAPD

	扩增条带	多态性条带	多态百分率/ %
种群间	217	47	21.66
种群内			
海南	680	200	33.3
广西	537	177	32.96
广东	646	146	22.6
福建	598	128	21.4

表 4 不同纬度桐花树种群间 RAPD 分析的遗传距离(下三角)及遗传一致度(上三角)

Tab. 4 Similarity matrix and genetic distance of populations in 4 various latitudes by RAPD

	海南	广西	广东	福建
海南	0	0.873	0.809	0.762
广西	0.127	0	0.809	0.794
广东	0.19	0.19	0	0.825
福建	0.238	0.206	0.175	0

群间的遗传分化小. 运用 UPGMA 统计对 4 个不同纬度桐花树种群进行聚类分析, 结果见图 2. 海南与广西聚为一个亚组, 广东与福建聚为另一个亚组. 根据遗传距离和聚类分析可看出, 空间距离与遗传距离有相关关系, 空间距离越大遗传距离也越大, 反之越小.

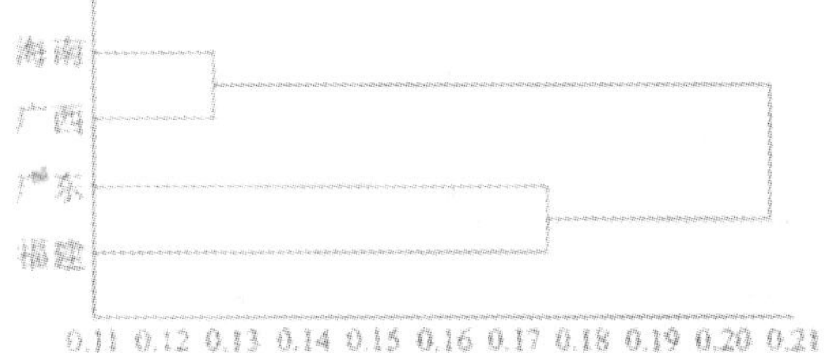

图 2 不同纬度桐花树种群间 RAPD 分析的遗传聚类图

Fig. 2 Molecular dendrogram of populations in 4 various latitudes by RAPD

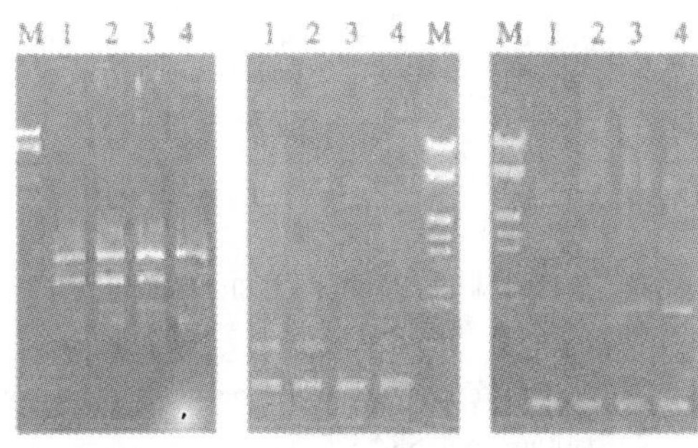

图 3 不同纬度桐花树种群间 ISSR 扩增图谱
1. 海南种群 2. 广西种群 3. 广东种群 4. 福建种群 M 为λDNA EcoR Ⅰ/ Hind Ⅲ 分子量标记

Fig. 3 Genomic DNA fingerprints of populations in 4 various latitudes by ISSR

ISSR 分析 本实验从 30 个 ISSR 引物中筛选出 5 个能获得清晰条带, 反应稳定的引物进行研究. ISSR 扩增情况见表 3, ISSR 反应扩增片断大部分集中在 0.2～2.0 kb 范围内, 如图 3, 5 个引物在 4 个纬度桐花树种群间共扩增出 71 条带, 多态性条带

表 5 ISSR 标记的遗传多态性

Tab. 5 Amplification of population by ISSR

	扩增条带	多态性条带	多态百分率/ %
种群间	71	19	26.77
种群内			
海南	315	65	20.63
广西	277	117	42.24
广东	295	55	18.64
福建	282	62	21.99

表 6 不同纬度桐花树种群间 ISSR 分析的遗传距离(下三角)及遗传一致度(上三角)

Tab. 6 Similarity matrix and genetic distance between populations of in 4 various latitude by ISSR

	海南	广西	广东	福建
海南	0	0.913	0.696	0.739
广西	0.087	0	0.696	0.739
广东	0.304	0.304	0	0.869
福建	0.261	0.261	0.13	0

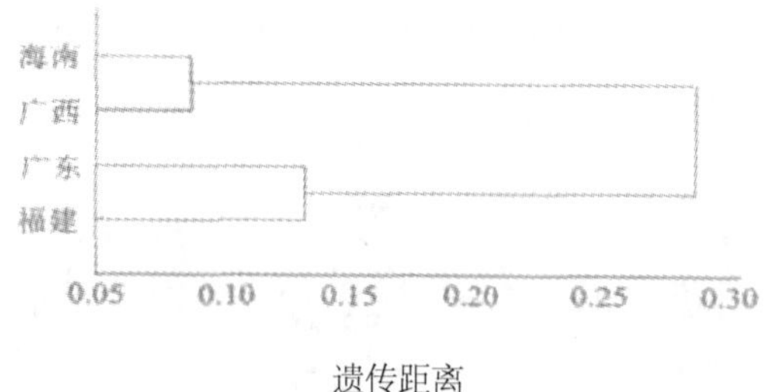

图 4 不同纬度桐花树种群间 ISSR 分析的遗传聚类图

Fig. 4 Molecular dendrogram of populations in 4 various latitude by ISSR

为 19,多态性条带百分率为 26.77 %. 也说明种群间遗传变异小.

根据 ISSR 扩增情况,运用 Nei 指数计算 4 种不同纬度桐花树种群间的遗传距离,结果见表 5. 最大遗传距离是海南-广东为 0.304,和广西-广东为 0.304. 最小的是海南-广西为 0.087,种群间平均遗传距离为 0.224. 说明种群间的遗传分化小. 运用 UPGMA 统计对 4 个不同纬度桐花树种群进行聚类分析,得到与 RAPD 分析相同的聚类图,结果见图 4,海南与广西聚为一个亚组,广东与福建聚为另一个亚组. 根据遗传距离和聚类分析可看出,空间距离小遗传距离也较小.

用 RAPD 和 ISSR 两种分子标记检测不同纬度桐花树种群遗传分化获得一致的遗传聚类图. RAPD 和 ISSR 技术对甜瓜[13]和黄麻[14]的种间、种内遗传关系的检测的相关系性很高. 说明经过优化的 RAPD 和 ISSR 方法两者对红树植物种群遗传多样性分析具有较高一致性和可行性. 但与应用等位酶方法研究结果则表现有较大的不同[15].

2.2 不同纬度桐花树种群的遗传多样性

用 10 个 RAPD 和 5 个 ISSR 个引物探讨不同纬度桐花树种群的遗传多样性分析(图 5,图 6). 各种群内的扩增情况见表 3、表 5. 从 RAPD 分析看,海南、广西种群的多态百分率较高且比较接近,分别为 33.3 %、32.96 %. 广东、福建种群的较低且也比较接近,分别为 22.6 %、21.4 %. 从 ISSR 分析看,广西种群的多态百分率最高为 42.24 %,广东种群的最低为 18.64 %. 从遗传多样性指数和多态百分率看,广西种群都较高. 可能与广西生境多样复杂、地理隔离特殊,等因素有关,有利于遗传变异的积累,适合桐花树生长.

Shannon 多样性指数估测各种群的遗传多样性见表 7,广西种群的多样性指数最高,为 0.331,福建种群多样性指数最低,为 0.199. 通过种群内与种群间的多样性分析比较见表 8,桐花树种群间的遗传分化程度低,4 个纬度桐花树种群总的遗传多样性指数 0.476,种群内的遗传多样性指数为 0.197,种群内的遗传多样性所占比率为 64 %,种群间的遗传多样性所占比率为 36 %. 说明有 64 %变异来自种群内,36 %变异来自种群间.

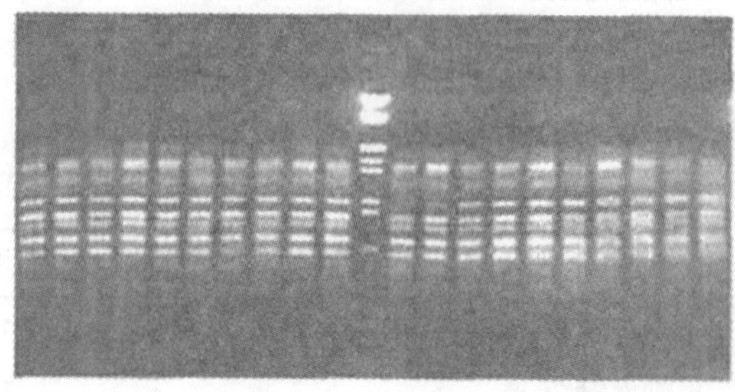

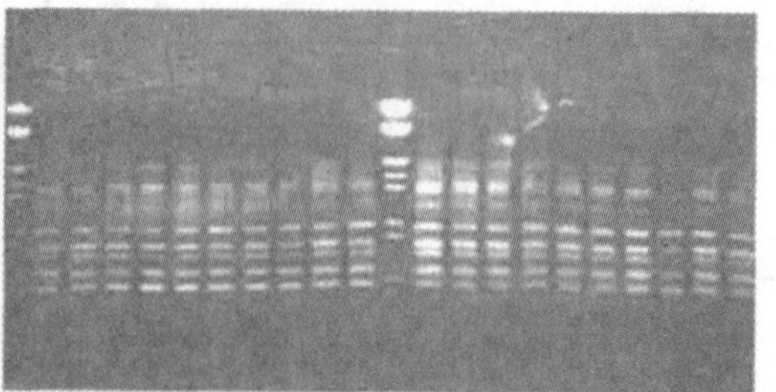

图 5 RAPD 引物 S50 对不同纬度桐花树种群的扩增

1~10 为海南种群, 11~20 为广西种群, 21~30 为广东种群, 31~40 为福建种群;M 为 λDNA EcoR Ⅰ/Hind Ⅲ 分子量标记

Fig. 5 Genomic DNA fingerprints of populations in 4 latitudes by RAPD S50 primer

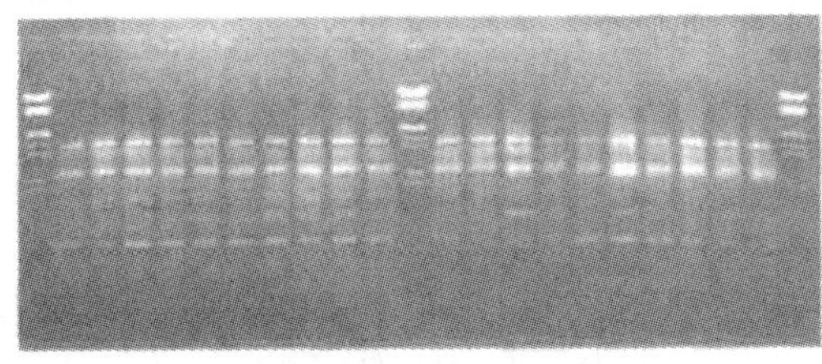

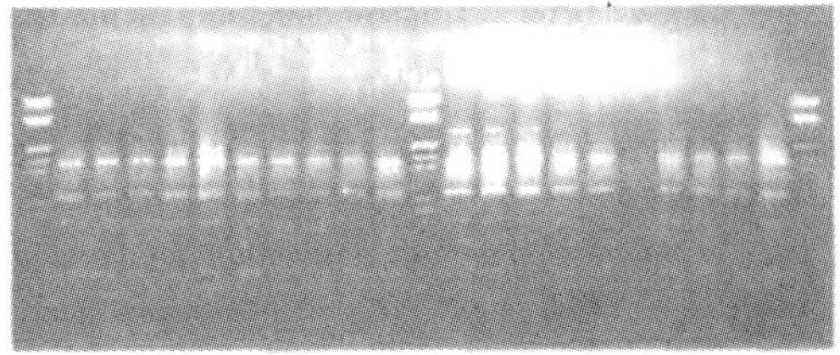

图 6 ISSR7 引物对不同纬度桐花树种群的扩增

1～10 为海南种群，11～20 为广西种群，21～30 为广东种群，31～40 为福建种群；M 为 λDNA EcoR Ⅰ/Hind Ⅲ 分子量标记

Fig.6 Genomic DNA fingerprints of populations in 4 latitudes by ISSR7 primer

表 7 4 个纬度桐花树种群内的遗传多样性(H)

Tab.7 Genetic diversity in of population in 4 various latitude

引物号	海南	广西	广东	福建
S50	0.253	0.29	0.302	0.23
S53	0.083	0.191	0.47	0.331
S55	0.365	0.61	0.534	0.29
S124	0.085	0.071	0.298	0.129
S125	0.347	0.199	0.222	0.191
S129	0.4	0.266	0.174	0.103
S130	0.588	0.833	0.219	0.23
S141	0.257	0.126	0.235	0.126
S148	0.043	0.043	0.012	0
S156	0	0.276	0	0.196
ISSR4	0.147	0.536	0.106	0.001
ISSR6	0.043	0.028	0.133	0.403
ISSR7	0.085	0.444	0	0.444
ISSR8	0.44	0.355	0.686	0.123
ISSR9	0.467	0.702	0.211	0.191
Average	0.24	0.331	0.24	0.199

3 讨 论

植物种群遗传变异的分布情形与该物种的地理分布情形、生态特征有关[16]. 有的学者认为遗传距离与空间距离的相关性很大，如 Kiang 和 Chiang[17] 的研究结果表明，遗传距离与纬度、物候农艺学性状有相关性. 顾少华[18]认为地理距离越远，基因频率的差异和遗传距离越大，反之，越小. 郎萍[19]对栗属的研究表明，地理距离和遗传距离有一定的相关性. 也有学者认为，遗传距离与空间距离没有相关性，胡志昂[20]等研究北京地区野生大豆天然群体遗传结构后认为，群体间的遗传距离与地理距离之间没有相关性.

许多研究表明，低温是限制红树植物向高纬度分布的重要因子[8]. 本研究 RAPD 分析，结果显示，不同纬度桐花树种群间的遗传距离与地理距离相关性明显，而 ISSR 分析的结果显示，遗传距离与空间距离相关性不明显. 黄生[21]等采用等位酶分析法对秋茄的遗传多样性进行分析也表明，遗传距离与纬度间也存在相关性. 赵萌莉[22]用 RAPD 分析法分析 5 个省 9 个不同纬度分布点秋茄种群的遗传分化，9 个种群间的遗传距离虽然较小，但种群间的遗传距离与纬度间也存在相关性. 黎中宝[15]用等位酶分析 4 个不同纬度桐花树种群遗传多样性的结果认为，4 个种群的平均遗传距离和空间距离无相关性.

多态位点百分率是衡量物种遗传变异水平高低的一个重要指标，是度量遗传多样性的重要参数. 本研究用 RAPD 和 ISSR 两种方法分析获得种群多态百分率分别为 21.66 %和 26.77 %. Shannon 多样性指数估测有 36 %变异来自种群间. 黎中宝[15]用等位酶分析 4 个不同纬度桐花树种群遗传多样性，种群的平均多态百分率为 44.28 %. 遗传分化度 $G_{ST}=0.215$，表明有 21.5 %的遗传变异来自种群间. Ge Xuejun 和 Sun Mei[23]用等位酶和 ISSR 两种方法研究桐花树的遗传多样性的结果显示，等位酶和 ISSR 分析的多态百分率分别为 4.76 %和 16.18 %. 遗传分化度分别为 0.106、0.178. 表明分别有10.6 %、17.8 %的遗传变异来自种群间. 许多研究表明，遗传变异与物种的地理分布范围有密切的关系[24]，广泛分布的种类比狭窄分布种群维持较多的遗传变异. 桐花树是一种广布红树植物，由于只分布与热带海岸潮间带，因此，与其它陆生植物相比，它的分布范围较小，加上海水传播，基因流较大，因而遗传变异也较小.

黎中宝[15]对4个纬度桐花树种群的聚类分析表明,海南东寨港、福建浮宫和广西英罗湾3个种群的遗传距离较近与深圳福田的遗传距离较远.与本研究的结果相差较大.本研究用RAPD和ISSR两种方法对4个纬度种群桐花树进行聚类分析,结果表明,4个种群明显分为南北两个类群,即海南与广西一个类群,广东与福建一个类群.并且,南部种群多态百分率和遗传多样性指数比北部种群高.桐花树可能起源于南部,也进一步说明红树植物是热带起源的,因此,桐花树的南部类群在研究其起源和演化中将显得更为重要.桐花树的遗传分化与其空间关系密切,因此在人工引种栽培时,可以就近采种或从其多样性分布中心地区采种.与其它植物相比,红树植物桐花树的遗传变异水平低,且大部分的遗传变异都存在于种群内,而种群间的遗传分化程度低.因此,局部的少数种群的消减不会导致其遗传多样性的迅速丧失.但长期的人为破坏及改变其生境,仍会导致其种群的衰退,必须注意保护.

表8 4个纬度桐花树种群内和种群间的遗传多样性

Tab.8 Genetic diversity within and among populations of in 4 various latitude

引物	种群内的遗传多样性(H_{pop})	总的遗传多样性(H_{sp})	种群内遗传多样性所占比率(H_{pop}/H_{sp})	种群间多样性所占比率($(H_{sp}-H_{pop})/H_{sp}$)
S50	0.269	0.414	0.65	0.35
S53	0.163	0.288	0.566	0.434
S55	0.442	0.79	0.56	0.441
S124	0.146	0.252	0.579	0.421
S125	0.24	0.241	0.996	0.004
S129	0.236	0.36	0.656	0.344
S130	0.467	0.863	0.541	0.459
S141	0.186	0.394	0.472	0.528
S148	0.024	0.031	0.774	0.226
S156	0.118	0.211	0.559	0.441
ISSR4	0.197	0.282	0.699	0.301
ISSR6	0.151	0.476	0.317	0.683
ISSR7	0.243	0.345	0.704	0.296
ISSR8	0.401	0.69	0.581	0.419
ISSR9	0.393	0.417	0.942	0.058
Average	0.245	0.404	0.64	0.36

参考文献:

[1] 阎文昭,蔡平钟,宣朴,等.RAPD技术在生物学研究中的应用[J].世界科技研究与发展,1999,2:83-84.

[2] Williams J G K, Kubelic A R, Livak KJ, et al. DNA polymorphisms amplified by arbitrary primers are useful as genetic markers[J]. Nuc. Aci. Res., 1990, 18(22):6 531-6 535.

[3] Ziekiewicz E, Rafalski A, Labuda D. Genome fingerprinting by simple sequence repeat (SSR)-anchored polymerase chain reaction amplification[J]. Genomica, 1994, 20:176-183.

[4] 王建波.ISSR分子标记及其在植物遗传学研究中的应用[J].遗传,2002,24(5):613-615.

[5] 杜金昆,姚颖垠,倪中福,等.普通小麦、斯卑尔脱小麦、密穗小麦和轮回小麦选择后代材料ISSR分子标记遗传差异研究[J].遗传学报,2002,29(5):44.

[6] 陈洪,朱立煌,徐吉臣,等.水稻RAPD分子连锁图谱的构建[J].植物学报,1995,37:677-684.

[7] Penner G A, Chong J, Levesque L M, et al. Identification of a RAPD marker linked to the oat stem rust gene Pg3[J]. Theor Appl Genet, 1993, 85:702-705.

[8] 林鹏.红树林[M].北京:海洋出版社,1984.

[9] 周涵韬,林鹏.中国红树科7种红树遗传多样性分析[J].水生生物学报,2001,25(4):362-369.

[10] Nei M, Li W. A mathematical model for studying genetic variation in the terms of restriction endonucleases[J].

Proc. Natl Acad Sci. USA ,1979 ,17 :5 269 - 5 273.

[11] Vierling R A ,Nguyen H T. Use of RAPD markers to determine the genetic diversity of diploid ,wheat genotyes[J]. Theor Appl Genet ,1992 ,84 :835 - 838.

[12] King L M ,Schall B A. Ribosomal DNA variation and distribution in Rudbeckia missouriensis [J]. Evolution , 1989 ,43 :1 117 - 1 119.

[13] 刘万勃 ,宋明等 ,刘富中 ,等. RAPD 和 ISSR 标记对甜瓜种质遗传多样性研究[J]. 农业生物技术学报 , 2002 ,10(3) :231 - 236.

[14] 周东新. 黄麻遗传多样性的 RAPD 和 ISSR 分析[D]. 福州:福建农林大学硕士学位论文. 2001.

[15] 黎中宝 ,林鹏. 不同纬度地区桐花树种群的遗传多样性研究[J]. 集美大学学报(自然科学板) ,2001 ,6(1) : 39 - 45.

[16] Loveless M D ,Hamrick J L. Ecological determinants of genetic structure in plant populations [J]. Annu. Rev. Ecol. Syst. ,1984 ,15 :65 - 95.

[17] King Y T ,Chiang Y C. Companing differentiation of wild soja bean (*Clycine soja Sieb and Zucc.*) population based on isozymes and quan titative traits[J]. Bot ball Acad Sin. ,1990 ,31 :129 - 142.

[18] 顾少华. 华东地区黑果蝇自然群体同工酶遗传多态的研究[J]. 遗传学报 ,1992 ,19 :228 - 235.

[19] 郎萍 ,黄宏文. 栗属中国特有种居群的遗传多样性及地域差异[J]. 植物学报 ,1999 ,41(6) :651 - 657.

[20] 胡志昂 ,王洪新. 北京地区野大豆天然群体遗传群体结构[J]. 植物学报 ,1985 ,27(6) :599 - 604.

[21] 黄生. 秋茄的区域性种群的遗传结构[J]. 生物多样性 ,1994 ,2(2) :68 - 75.

[22] 赵萌莉. 红树植物秋茄的遗传变异与分化的生态学研究[D]. 厦门:厦门大学博士论文 ,2000.

[23] Ge Xuejun ,Su Mei. Reproductive biology and genetic diversity of a cryptoviviparous mangrove *Aegiceras corniculatum* (Myrsinaceae) using allozyme and intersimple sequence repeat (ISSR) analysis [J]. Molecular Ecology , 1999 ,8 :2 061 - 2 069.

[24] Hamrick J K, Godt M J. Plant Population Genetics, Breeding and Genetic Resource [M]. Sundrlan Massachusetts:Sinauer Associates Inc ,1989. 43 - 63.

The Study on Genetic Diversity of *Aegiceras corniculatum* Populations in Different Areas

PAN Wen[1,2] ,ZHOU Han-tao[1,3] ,CHEN Pan[1] ,LIN Peng[1,2]

(1. School of Life Science ,Xiamen University ,2. Research Centre for wetlands and Ecological Engineering ,Xiamen University ,Xiamen 361005 ,China ;3. Biotechnology Centre , Fujian Academy of Agricultural Science ,Fuzhou 350003 ,China)

Abstract : RAPD and ISSR molecule markers were used to detect genetic diversity and genetic differentiation of *Aegiceras corniculatum* populations in different latitudes. The results showed that genetic variation of four populations was low. Percentage of polymorphic bands of RAPD and ISSR were 21. 66 % and 26. 77 %. Based on UPGMA cluster analysis ,it showed that four *Aegiceras corniculatum* populations were divided into two groups :the south and the north. There were some evidences showing noticeable correlation existing between the variations and the geographical latitudes. Shonnon diversity index indicated that 64 % variation existed within populations and 36 % variation existed among populations. This study discussed about genetic diversity protection of population and strategy of plant introduction.

Key words : RAPD ; ISSR genetic diversity ; *Aegiceras corniculatum*

厦门市红树植物白骨壤(*Avicennia marina*)两个种群的遗传变异及建立机制*

葛菁萍[1,2] 蔡柏岩[3] 林 鹏[1]

(1. 厦门大学生物学系,福建 厦门 361005;2. 黑龙江大学生命科学学院,黑龙江 哈尔滨 150080;
3. 黑龙江东方学院食品与环境工程学部,黑龙江 哈尔滨 150086)

摘 要:利用水平切片淀粉凝胶及聚丙烯酰胺凝胶分析了位于厦门市两个白骨壤(*Avicennia marina*)种群的遗传变异及遗传分化,进而分析了其中的游泳池种群的分布及建立机制。结果表明:游泳池白骨壤种群和东屿白骨壤种群相比较,种群内的遗传变异较大,两种群之间的遗传分化较大,游泳池种群建立时未受到建立者效应的影响。

关键词:白骨壤(*Avicennia marina*);种群建立;种群分布

中图分类号:Q94 **文献标识码**:A

白骨壤(*Avicennia marina*),马鞭草科(Verbenaceae),白骨壤属,是抗低温广布种,在国内福建、广东、广西、海南和台湾5省(区)均有自然分布[1]。它生长在泥沙滩上,常为群落先锋植物,但在不同滩位和潮带均可见,是一个多潮带广布种,能单独形成一个群落,也可与其它种类混生。白骨壤虽不是红树科植物,但它也具胎萌现象[2]:种子萌芽后,仍留在果皮内,把果皮填满。只有当果实掉入水中后,因果皮吸水胀破,幼苗才伸出果皮,插入泥滩,随即生根固定下来。

福建省厦门市白城有一处废弃多年不用的游泳池,池内荒草丛生,泥质松软,池水与外面海水相连通。在这个游泳池内,分布着一个小的白骨壤种群,数量不超过50株,同时混生有几株秋茄(*Kancelia candel*)和桐花树(*Aegiceras corniculatum* Blanco.)。该处白骨壤种群年龄较小,尚处于营养阶段,未见有繁殖体,但长势良好。很明显,这是一个刚建立的种群,尚处于种群增长阶段.那么游泳池这个白骨壤种群从它建立、发展到现在定居下来,是怎样一个过程呢?它会不会受到建立者效应(founder effect)的影响?基因多样性丰富吗?于是,我们选择了与游泳池白骨壤种群分布最近的厦门东屿白骨壤种群作为比较和参照对象,旨在通过等位酶分析技术,对这两个白骨壤种群遗传变异和遗传分化进行研究,了解游泳池白骨壤种群的分布及建立机制,为种群建立及分布机制研究提供参考。

1 材料和方法

1.1 取样地点及样地自然状况

白骨壤样地包括游泳池和东屿两个地点。游泳池样地位于厦门市白城(24°29′N, 118°01′E),泥质松软,无结构,该样地属南亚热带海洋性气候。据厦门市气象台资料,该地平均温20.8 ℃,极端最低温2 ℃,极端最高温39.4 ℃,年雨量1095mm,年相对温度77%。白骨壤种群生长在池内,散生有少量的秋茄桐花树,未见有繁殖体。东屿样地位于厦门市海域海沧东屿村附近海滩(24°28′N, 118°05′E),年均温21.1 ℃,年雨量1036mm,最低月均温12.3 ℃,年较差16.0 ℃,该群落土壤盐度为17.6‰(0-30cm)-19.91‰(30-60cm),pH呈微酸性

* 国家自然科学基金项目(39670135)资助
原载于黑龙江大学自然科学学报,2004,21(1):132-137

(0－30 cm 为 6.51)。该群落为白骨壤纯林，林冠整齐，外貌呈褐绿色，群落结构简单，高度 1.2 m，地表具指状呼吸根，高约 10 cm，密度 35 根/m^2。

1.2 样品采集方法及处理

在各样地内随机采集幼嫩的白骨壤叶片，株与株之间相隔 5 米以上．用棉花蘸取足够的水分包住茎枝条的基部，保持叶片新鲜不变质，迅速携至实验室内处理。也可以采集成熟的白骨壤果实，采集方法同叶片，砂培于盆内，待长出叶片后取样测定。由于红树植物叶片内富含单宁[2]，因此酶粗提液采用改进的 Tris－HCl 提取液[3]，冰浴中研成匀浆，备用。

1.3 电泳及酶谱分析

实验采用垂直板型不连续聚丙烯酰胺凝胶(PGE)和水平切片淀粉凝胶(SGE)两种凝胶类型(SUSAN, 1990)。共测定了 5 个酶系统 13 个位点。其中 PGE 浓缩胶和分离胶的浓度分别为 2.5% 和 7.0%，pH 分别为 6.7 和 8.9；SGE 采用美国 Sigma 公司的水解淀粉(S－5691)，淀粉胶浓度为 12%，相应凝胶缓冲液为 TBE(pH 8.6)(#10)。用于本次实验的酶系统、代码、位点数目及缓冲系统等详见表 1。酶组织化学染色方法见文献[3－4]。

1.4 计算方法

1.4.1 遗传变异的计算

多态位点百分率(P)：$P=(k/n)\times 100\%$，其中，k－ 多态酶位点的数目，n－ 所测定酶位点的总和。多态位点的标准按 Nei 氏(1975)的 0.99 划分，即最常见的等位基因出现的频率小于或等于 0.99。

杂合度(H)：即杂合位点的百分数。期望杂合度 $He=\sum_{i=1}^{n}(1-\sum_{j=1}^{mi}q_{ij}^2)/n$，观察杂合度 $Ho=\sum_{i=1}^{n}(1-\sum_{j=1}^{mi}p_{ij})/n$．其中 q_{ij} 为第 i 个位点上第 j 个等位基因的频率，p_{ij} 为第 i 个位点上第 j 个等位基因观察到的纯合基因型频率。

平均每位点有效等位基因数(Ae)：$Ae=\sum_{i=1}^{r}(1/\sum_{j=1}^{m}q_{ij}^2)/n$，$q_{ij}$ 为第 i 个位点上第 j 个等位基因的频率。

固定指数(F)：$F=1-Ho/He$

1.4.2 遗传分化的测定

基因分化系数(G_{ST})：$G_{ST}=D_{ST}/H_T$，D_{ST} 为种群间的基因多样度；H_T 为基因多样度总量。

遗传距离采用 Nei 氏(1987)[5] 的标准遗传距离。

$$I=\sum_k\sum_i XiYi/\sqrt{\sum\sum_{k\ i}X i^2\cdot\sum\sum_{k\ i}Yi^2}$$

$$D=-\ln I$$

表 1 白骨壤种群的酶系统、E.C. 代码、位点数目及缓冲系统

Table 1 The enzyme systems E.C.No numbers of loci and buffer systems of *Avicennia marina* populations

酶系统(括号内为缩写)	酶委 E.C. 代码	凝胶类型(括号内缓冲液)	位点数目(个)
乙醇脱氢酶(ADH)	E.C.1.1.1.1	SGE(#10)	1
酯酶(EST)	E.C.3.3.3－	PGE	3
过氧化物酶(POD)	E.C.1.11.1.7	PGE	3
天冬氨酸转氨酶(AAT)	E.C.2.6.1.1	SGE(#10)	2
超氧物歧化酶(SOD)	E.C.1.15.1.1	PGE	4

基因流采用 Wright(1965)的 F_{ST} 法统计出的 F_{ST} 值来计算：$Nm=(1-F_{ST})/4F_{ST}$

2 结 果

2.1 白骨壤种群的遗传变异

表 2 是两个白骨壤种群的等位基因频率及标准差。在检测到的 13 个位点中，有 5 个位点在游泳池和东屿种群中完全相同，其余 9 个位点差别较大。在 *Est*－1 中，*Est*－1A 只存在于东屿种群中，而 *Est*－1C 只存在于游泳池种群中，只有 *Est*－1B 为两个种群共有；*Est*－2 位点中，*Est*－2C 只存在于游泳池种群中，*Est*－2A,*Est*－2B 为两个种群共有，但 *Est*－2A 的频率在两个种群中相差很大，在其它几个位点中，*Adh*－1A、*Pod*－1B、*Pod*－2C、*Pod*－2D、*Sod*－2A、*Sod*－2B、*Sod*－2D、*Sod*－2E、*Sod*－4B、*Sod*－4C 这 10 个等位基因均只存在于游泳池种群中，*Pod*－1A 只存在于东屿种群中(表 3)，即使是两个种群共有的等位基因，其频率也相差很大。

由表 2 可以看出，两个白骨壤种群无论在等位基因频率上，还是在等位基因分布上，都有很大差别。共有等位基因个数只占全部等位基因个数的 54.5%，39.4% 的等位基因只存在于游泳池种群中，而仅有 6.1% 的等位基因只存在于东屿种群中。由此可见，虽然两地同属于九龙江口地区，直线距离不到 20 km，但作为一个刚建立的种群，游泳池白骨壤种群的等位基因并不是或并不全是来源于东屿种群，因为它们有 45.5% 的等位基因为各自独有,有可能共有的 54.5% 的等位基因也来源于其它种群,或者说,两个种群各自经过遗传漂变、突变、选择和迁移后才发展到目前这种状况。因此,游泳池白骨壤种群的建立可能来源于几个种群,是几个种群等位基因的重新组合。由于我们只把东屿种群作为一个参比对象,这就显得很不充分,但由此也可以对种群分布及建立机制窥见一斑。

表 2　两个白骨壤种群的等位基因频率(括号内为标准差)

Table 2　Allelic frequencies of *Avicennia marina* populations (standard errors in parentheses)

位点	等位基因	游泳池	东屿	位点	等位基因	游泳池	东屿
Aat－1	A	1.000(0.000)	1.000(0.000)	*Pod*－4	A	0.386(0.073)	0.400(0.077)
Aat－2	A	1.000(0.000)	1.000(0.000)		B	0.068(0.038)	0.000(0.000)
Est－1	A	0.000(0.000)	0.310(0.071)		C	0.545(0.075)	0.600(0.077)
	B	0.545(0.075)	0.690(0.071)	*Sod*－1	A	0.500(0.075)	0.500(0.075)
	C	0.455(0.075)	0.000(0.000)		B	0.500(0.075)	0.500(0.075)
Est－2	A	0.095(0.045)	0.500(0.075)	*Sod*－2	A	0.205(0.061)	0.000(0.000)
	B	0.595(0.076)	0.500(0.075)		B	0.114(0.027)	0.000(0.000)
	C	0.310(0.071)	0.000(0.000)		C	0.386(0.041)	1.000(0.000)
Est－3	A	0.500(0.112)	0.500(0.112)		D	0.114(0.027)	0.000(0.000)
	B	0.500(0.112)	0.500(0.112)		E	0.182(0.058)	0.000(0.000)
Adh－1	A	0.370(0.066)	0.000(0.000)	*Sod*－3	A	1.000(0.000)	1.000(0.000)
	B	0.630(0.066)	1.000(0.000)	*Sod*－4	A	0.455(0.075)	1.000(0.000)
Pod－1	A	0.000(0.000)	0.500(0.075)		B	0.500(0.075)	0.000(0.000)
	B	0.500(0.075)	0.000(0.000)		C	0.045(0.031)	0.000(0.000)
	C	0.500(0.075)	0.500(0.075)				
Pod－2	A	0.125(0.052)	0.500(0.075)				
	B	0.250(0.068)	0.500(0.075)				
	C	0.425(0.078)	0.000(0.000)				
	D	0.200(0.063)	0.000(0.000)				

受等位基因频率的影响,两个白骨壤种群在各项遗传变异参数上有很大区别(表 4)。

两个白骨壤种群的遗传变异水平都较高,但比较而言,无论是各位点平均值,还是多态位点平均值,游泳池种群都比东屿种群高出很多(表 4),表明游泳池种群的遗传变异性比东屿种群高。

表 3　游泳池和东屿白骨壤种群等位基因的分布

Table 3　The distribution of alleles in *Avicennia marina* populations

	只存在于游泳池	只存在于东屿	两个种群都存在
等位基因数(个)	13	2	18
百分比(%)	39.4	6.06	54.5

总的来看,游泳池种群的遗传多样性高于东屿种群,说明游泳池的等位基因并不全来自东屿,在种群建立初期,也并没有受到建立者效应的影响。游泳池种群的基因可能来自多个种群,是各地点种群综合作用的结果。

2.2　白骨壤种群的遗传分化

对白骨壤种群各项遗传分化指标的检测表明(表 5 和 6),两个白骨壤种群间的遗传分化偏高。

两个白骨壤种群的遗传一致度仅为 0.8409(表 5),小于一般种内种群间的平均值 0.9000[8],这可能是由于游泳池白骨壤种群刚刚建立,等位基因异常丰富,有些等位基因还没固定下来,使它与东屿种群的遗传距离加大(表 5)。各位点遗传一致度(表 6)除 *Aat*－1、*Aat*－3、*Est*－3、*Pod*－4、*Sod*－1、*Sod*－3 均为 1.000 外,其它各位点的均很小,其中以 *Pod*－2 为最小,仅为 0.506。

根据 Nei 氏的分化度 G_{ST} 和Wright 的 F－ 统计量方法计算得到的 2 个白骨壤种群的分化度很大(表 7),G_{ST} 高达 0.156,F_{ST} 也高达 0.136(表 8),说明二个白骨壤种群有 15.6% 的遗传变异保持在种群之间,有84.4%的遗传变异保持在种群内。根据 F_{ST} 计算出的基因流很小,仅为 1.588。但各位点之间的基因流却相差很大,其中以 *Pod*－4 最大,为 62.25,*Est*－3 和 *Sod*－1 最小,为 0.000。

表 4 两个白骨壤种群的遗传变异性

Table 4 Genetic diversity of *Avicennia marina* populations (standard errors in parentheses)

位点	游泳池				东屿			
	A	*Ae*	*He*	*Ho*	*A*	*Ae*	*He*	*Ho*
Aat－1	1.000	1.000	0.000(0.000)	0.000	1.000	1.000	0.000(0.000)	0.000
Aat－2	1.000	1.000	0.000(0.000)	0.000	1.000	1.000	0.000(0.000)	0.000
Est－1	2.000	1.984	0.496(0.250)	0.909	2.000	1.748	0.428(0.247)	0.619
Est－2	3.000	2.178	0.541(0.214)	0.810	2.000	2.000	0.500(0.250)	1.000
Est－3	2.000	2.000	0.500(0.250)	0.000	2.000	2.000	0.500(0.250)	0.000
Adh－1	2.000	1.873	0.466(0.249)	0.000	1.000	1.000	0.000(0.000)	0.000
Pod－1	2.000	2.000	0.500(0.250)	0.909	2.000	2.000	0.500(0.250)	1.000
Pod－2	4.000	3.347	0.701(0.157)	0.650	2.000	2.000	0.500(0.250)	1.000
Pod－4	3.000	2.219	0.549(0.204)	0.909	2.000	1.923	0.480(0.250)	0.800
Sod－1	2.000	2.000	0.500(0.250)	1.000	2.000	2.000	0.500(0.250)	1.000
Sod－2	5.000	3.998	0.750(0.130)	1.000	1.000	1.000	0.000(0.000)	0.000
Sod－3	1.000	1.000	0.000(0.000)	0.000	1.000	1.000	0.000(0.000)	0.000
Sod－4	3.000	2.178	0.514(0.198)	0.000	1.000	1.000	0.000(0.000)	0.000
位点平均	2.385	1.983	0.426(0.071)	0.476	1.538	1.426	0.262(0.071)	0.417
多态位点平均	2.800	2.378	0.554(0.077)	(0.129)	2.000	1.953	0.487(0.000)	(0.133)
P(%)			76.9				53.8	

二个白骨壤种群之间的 *Nm* 较小，说明它们之间的基因交换量很少，在很大程度上，会导致它们之间存在着一定的生殖隔离。

3 讨 论

游泳池和东屿这两个白骨壤种群的遗传变异水平较高。一般植物种群水平的期望杂

表 5 白骨壤种群的最大、最小和标准遗传距离（括号内为标准差）

Table 5 The maximum、minimum and standard genetic distance betwwen A. *marina* populations (Standard errors parentheses)

遗传一致度 *I*	最大遗传距离 *Dv*	最小遗传距离 *Dm*	标准遗传距离 *D*
0.8409	0.1892(0.1388)	0.1087(0.0310)	0.1733(0.0605)

表 6 白骨壤种群各位点的遗传一致度

Table 6 The genetic identity of each locus between A. *marina* populations

位点	*Aat*－1	*Aat*－2	*Est*－1	*Est*－2	*Est*－3	*Adh*－1	*Pod*－1	*Pod*－2
I	1.000	1.000	0.716	0.740	1.000	0.869	0.512	0.506
位点	*Pod*－4	*Sod*－1	*Sod*－2	*Sod*－3	*Sod*－4			
I	1.000	1.000	0.801	1.000	0.681			

表 7 白骨壤种群的分化度

Table 7 Coefficient of gene differentiation between A. *marina* populations

位点	*Hs*	*HT*	D_{ST}	G_{ST}
Est－1	0.462	0.543	0.081	0.149
Est－2	0.520	0.541	0.021	0.038
Adh－1	0.233	0.466	0.233	0.500
Pod－1	0.500	0.625	0.125	0.200
Pod－2	0.601	0.707	0.106	0.150
Pod－4	0.515	0.517	0.002	0.004
Sod－1	0.500	0.500	0.000	0.000
Sod－2	0.375	0.494	0.119	0.242
Sod－4	0.270	0.408	0.138	0.337
全部位点	0.344	0.408	0.064	0.156

表 8 二个白骨壤种群的 F－统计量

Table 8 F－statistics of A. *marina* populations

位点	F_{IS}	F_{IT}	*Nm*
Est－1	－0.655	－0.408	1.43
Est－2	－0.739	－0.540	1.94
Est－3	1.000	1.000	0.000
Adh－1	1.000	1.000	0.851
Pod－1	－0.909	－0.527	1.000
Pod－2	－0.374	－0.168	1.417
Pod－4	－0.662	－0.655	62.25
Sod－1	－1.000	－1.000	0.000
Sod－2	－0.333	－0.111	0.787
Sod－4	1.000	1.000	0.492
平均	－0.297	－0.120	1.588

合度(He)、多态位点百分比(P)和平均每位点等位基因数目(A)分别为0.113、35%和1.52[9]。在白骨壤种群中,期望杂合度平均为0.344,多态位点百分比平均为65.4%,平均每位点等位基因数为1.96,均高于一般植物种群水平的平均值。白骨壤种群具有较高的等位酶变异水平是地理范围、交配系统共同作用的结果。广泛分布的植物种类比狭窄分布种维持较多的遗传变异[9]。同时,以杂交为主的物种,其遗传变异性远高于自交为主的物种[9]。白骨壤是完全花,由于雄蕊先熟,加强了有性隔离,因此它是杂交种[10]。由风传粉的物种,其遗传变异性也高于动物传粉和自花传粉的物种。白骨壤为虫媒传粉,这在一定程度上会降低其遗传多样性。抛除刚建立的游泳池白骨壤种群来看,东屿白骨壤种群的各项变异水平不是很高。而游泳池白骨壤种群刚刚建立,种群还不是很稳定,其等位基因还处于选择和被选择阶段。由于具有丰富的等位基因,因此导致它的各项变异水平很高。从其等位基因数目和遗传多样性看出,该种群在建立初期尚未受到建立者效应的影响。

生物种群的分布决定于生态条件,类群的移动性以及历史上曾促进或限制过种群扩大的气候因素和地质因素等的联合作用。游泳池白骨壤种群的建立也受到生态条件、种的移动性以及人为条件的影响。红树林的土壤一般是较初生的土壤,在沉积下来之前已被河水分选过。它们多是精细的颗粒,通常是半流体,不坚固的,含有丰富的腐殖质,pH值3.5－7.5,大多pH在5以下。游泳池自荒废以来,池内淤泥逐渐增加,在白骨壤生长的池内,还可见到人为倾倒的沙堆。池内淤泥无结构,有机质含量较多,适合于红树植物的生长。同时,游泳池四周围以围墙,仅有墙底部与海水相通,大潮时,海水可涌到池内。这样形成了一个类似隐蔽海岸的生态环境,可抵御风浪的冲击,也适于红树植物的生长、定居。有了适合生长的环境后,关于游泳池白骨壤种群的建立作如下三种假设:一是受种的迁移性的控制;二是人为因素的控制;三是种的迁移性和人为因素共同作用的结果。白骨壤为隐胎生[2]植物,当果实掉落海水中后,可能一部分受海水流动的影响,被带到游泳池中,生根,发芽,从而导致该地种群的建立。另外,也不排除是人为引种或者是人为引种和自然散布共同作用的结果。但有一点可以确信,仅将东屿白骨壤种群作为假想的种群起源中心是远远不够的。从东屿和游泳池白骨壤种群等位基因的丰富程度来看,至少有39.4%的等位基因在东屿种群中不存在,那么这39.4%的等位基因也就不是来源于东屿种群。具体游泳池白骨壤种群和其它地点的白骨壤种群具有怎样的亲缘关系,以后还需作进一步的探讨。

游泳池和东屿白骨壤种群之间的基因流为1.588,这与同是广布种的木榄的基因流(3.85)相比,有些偏小。这就会在一定程度上,产生生殖隔离。这种较小的基因流与两个白骨壤种群所处的地理位置有关:海沧镇东屿村位于厦门西海域河口的内侧,果实成熟后向外界的散布受到一定程度的阻碍;而厦门白城游泳池位于厦门港河口的外侧,可直接与外海相连通。这样,外来种苗到达游泳池的机会会更多一些,同时受东屿海岸隐蔽性的影响,两地的基因交换也减少,这也说明了为什么游泳池白骨壤种群的遗传变异性会更高一些。

参考文献:

[1] 林 鹏.中国红树林生态系[M].北京:科学出版社,1997.18－20.

[2] 林 鹏.红树林[M].北京:海洋出版社,1984.

[3] 王中仁.植物等位酶分析[M].北京:科学出版社,1996.

[4] VALLEJOS C E. Enzyme activity staining[A]. TANKSLEY S D, ORTON T J. Isozyme in plant gentics and breeding [C]. Amsterdam: Elsevier Press, 1983.

[5] NEI M. Molecular evolutionary genetics [M]. New York: Columbia Univ Press, 1987.

[6] WRIGHT S. The interpretation of population structure by F－statistics with special regard to systems of mating [J]. Evolution, 1965, 19: 395－420.

[7] NEI M. Molecular population genetic and evolution [M]. Amsterdam and New York: North Holland Publ Co, 1975.

[8] SUSAN R K. Starch gel electrophoresis of plant isozymes: A comparative analysis of techniques [J]. American Journal of Botany, 1990, 77(5): 693－712.

[9] HAMRICK J L, GODT M J. Plant population genetics, breeding and genetic resource [M]. Sunderland Massachusetts: Sinauer Associates Inc, 1989. 43－63.

[10] TOMLINSON P B. The botany of mangroves [M]. London: Cambridge University Press, 1986.

Study on distribution and foundation mechanism of two *Avicennia marina* populations in Xiamen

GE Jing-ping[1,2], CAI Bai-yan[3], LIN Peng[1]

(1. Department of Biology, Xiamen University, Xiamen 361005, China; 2. College of Life Science, Heilongjiang University, Harbin 150080, China; 3. Department of Food and Science, Dongfang College, Harbin 150080, China)

Abstract: Using horizontal sliceable gel electrophoresis and polyacrylamide gel electrophoresis to examine the genetic diversity of two *Avicennia marina* populations in Xiamen, and to study the distribution and foundation mechanism of *Avicennia marina* population of Youyongchi. The results show that the genetic diversity of *Avicennia marina* between Youyongchi and Dongyu is quite large, and the genetic differentiation of these two populations is also large. The foundation of *Avicennia marina* population in Youyongchi is not affected by founder effect.

Key words: *Avicennia marina*; population foundation; population distribution

Mating system and population genetic structure of *Bruguiera gymnorrhiza* (Rhizophoraceae), a viviparous mangrove species in China*

Jing Ping Ge[a] Baiyan Cai[a,b] Wenxiang Ping[a] Gang Song[a] Hongzhi Ling[a] Peng Lin[c]

(*a. Key Laboratory of Microbiology, College of Life Science, Heilongjiang University, No. 74 Xuefu Road, Harbin* 150080, *PR China*; *b. Department of Food Science and Technology, Heilongjiang Dongfang College, No.* 334 *Xuefu Road, Harbin* 150080, *PR China*; *c. College of Life Science, Xiamen University, Xiamen* 361005, *PR China*)

Abstract

In order to explain the diversity patterns and develop the conservation strategies, the population genetic structures and the mating systems of *Bruguiera gymnorrhiza* from the coastlines of south China were investigated in this study. The mating system parameters were analyzed using progeny arrays for allozyme markers. The multilocus outcrossing rates (*tm*) ranged from 0.845 (Fugong) to 0.267 (Dongzhai harbor). High allozyme variations within the five collected populations were determined and compared with the published data of other plant species with the mixed mating systems. At species level, the percentage of polymorphic loci (P) was 80%, the average number of alleles per locus (A) was 2.440, and the heterozygosity (He) was 0.293. The total gene diversity within each population (H_S=0.2782) and the coefficient of genetic differentiation (G_{ST}=0.0579) among the populations were estimated. On the basis of this population genetic structure, it is suggested that the gene flow (Nm=3.85) is quite high, which is possibly related to its water-dispersed hypocotyls. It is also suggested that the mating system of this species is of mixed mating.

Keywords: Bruguiera gymnorrhiza; Mangrove; Mating system; Genetic structure

1. Introduction

Mangrove species are a characteristic of tropical and subtropical coastlines. Approximately 83 species from 24 plant families have been recognized, consisting primarily of trees and shrubs that normally grow above mean sea level in the intertidal zone of marine coastal environments or estuarine margins (Duke, 1992). Overexploitation in many parts of the tropics is depleting mangrove resources, resulting in the loss in many tropical nations of >1% of mangrove forest area per year (Umali et al., 1987; Hatcher et al., 1989). Sound scientific knowledge of the basic biology of

* From Journal of Experimental Marine Biology and Ecology, 2005, 326: 48—55

these species is essential to provide guidance for their conservation planning and management.

The mating system is believed to be a critical factor in determining patterns of genetic variation both within and among plant populations (Brown, 1989; Brown et al., 1989; Hamrick and Godt, 1989). Polymorphic allozyme loci have proven to be extremely useful in developing quantitative estimates of plant mating system parameters (Brown, 1989; Hamrick, 1989), and have also provided a rapid and effective means for studying patterns of genetic variation in plant populations (Soltis and Soltis, 1989). A method was used to investigate the mating system in *Kandelia candel*, another mangrove species (Sun et al., 1998) and outcrossing as the mating system was suggested.

The mixed mating model (Fyfe and Bailey, 1951) gives a robust and frequently used procedure for estimating mating system parameters. However, several new approaches have been developed (Brown, 1989), which enable further investigations of demographic and genetic causes of mating patterns. For example, the effective selfing model (Ritland, 1984, 1986) calculates mating system parameters in spatially structured populations. However, the mixed mating model is limited by certain assumptions, in which case the maternal plants are random samples from the population producing pollen.

This study was carried out to investigate the mating system of a typical species of mangrove, *Bruguiera gymnorrhiza* (Rhizophoraceae) in the coastlines of south China. The relationships between the mating system and population structure or patterns of genetic variation among the populations were estimated. *B.gymnorrhiza* is one of the dominant species of mangroves along the Chinese coast (Li and Lee, 1997). The information on the mating system parameters is helpful to explain the patterns of genetic diversity and enable to develop appropriate conservation strategies for this species.

2. Materials and methods

2.1. Population sampling and site descriptions

B. gymnorrhiza (Rhizophoraceae) is a kind of shrub or arbor and widely distributed in the south of China except Zhejiang province. Five populations were selected as sampling sites from the coastlines of south China (Fig. 1). Each population included over a hundred of individuals. The heights of the plants ranged from ~1 m to ~8 m. The collection areas were from 24°24′N, 117°55′E (Fugong site) to 19°51′N, 110°24′E (Dongzhai harbor site). Plant samples of the five populations were collected in 2001, which were used for population genetic structure. The young leaves were from about 50 individuals per population (see Fig. 1 for population names and locations).

For estimating mating system, the samples were collected from four populations located in Fugong, Shenzhen, Shankou and Dongzhai harbor, in which the abundant propagules were largely collectable at the time. About 20 trees per population and 10 propagules per tree in average were selected to provide the materials for progeny arrays for mating system analyses. The 10 propagules collected from the same tree were kept separately as a family basis and planted in sands in the condition same as the one at the collection sites until the completion of allozyme electrophoresis.

2.2. Allozyme electrophoresis

The young leaves were collected and kept in moisture until electrophoresis. The fresh materials were ground with slightly modified extraction buffer (Wang, 1996; Chen, 1997). Two electrophoresis buffers (Tris–Boric Acid–EDTANa_4, pH 8.6 and Tris–Boric Acid–EDTANa_2, pH 8.0) were used in the eight enzyme systems with 12.5% starch gels (Wang, 1996). The eight enzyme systems were: Malate dehydrogenase (MDH, E.C.1.1.37), Malic enzyme (ME, E.C.1.1.40), Esterase (EST, E.C.3.3.3), Peroxidase (POD, E.C.1.11.1.7), Aspartate aminotransferase (AAT, E.C.2.6.1.1), Alkaline phosphatase (ALP, E.C.3.1.3.1), alcohol dehydrogenase (ADH, E.C.1.1.1.1) and superoxide dismutase (SOD, E.C.1.15.1.1), respectively. The enzyme activity staining was performed according to the protocols of Wang (1996), Vallejos (1983) or Zhu et al. (1990). The genetic interpretation of the band patterns was estimated based on Wang (1996).

2.3. Data analysis

The genetic variation values such as polymorphism, allelic diversity and heterozygosity were mea-

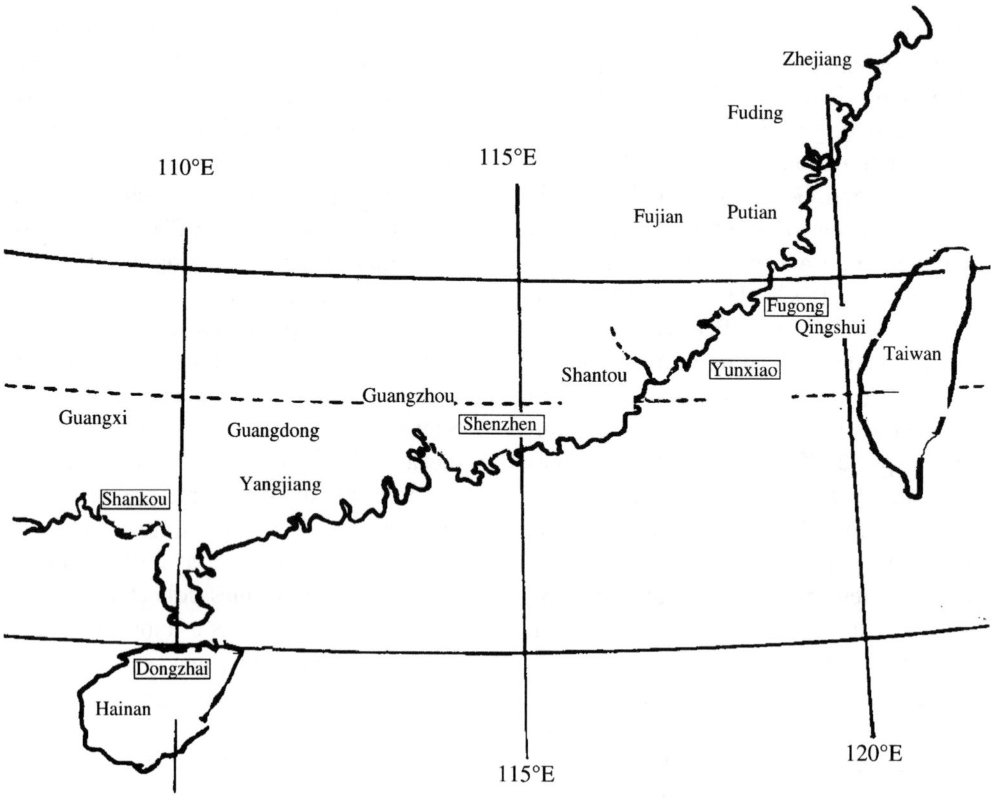

Fig. 1. Map showing location of the populations of *B. gymnorrhiza* sampled.

sured within each population. The percentage of polymorphic loci (P) was calculated based on the 99% criterion (Nei, 1972). The allelic diversity was measured based on the average number of alleles per locus (A) and the average effective number of alleles per locus (Ae). The gene diversity (He) was estimated based on $Hei = 1 - \sum_{j=1}^{m} P_j^2$, where P_j was the frequency of the jth allele at the ith locus, m was the allele numbers of ith locus, and $He = \sum_{i=1}^{n} Hei/n$, where n was the total number of allozyme loci surveyed. The mean of observed heterozygosity per locus (Ho) was calculated as $Ho = \sum_{i=1}^{n} Hoi/n = \sum_{i=1}^{n}\left(1 - \sum_{j=1}^{mi} P_{ij}\right)/n$, where P_{ij} was the frequency of pure genotype of jth allele at the ith locus, and mi was the numbers of pure genotype at ith locus.

The genetic variations within (H_S) and between populations (D_{ST}) were measured using Nei's (1973) gene diversity statistics. The coefficient of genetic differentiation among the populations, G_{ST}, was used to estimate the level of gene flow, Nm (the number of migrants exchanged between local populations per generation), based on the relationship $G_{ST}=1/(4Nm+1)$ where G_{ST} was Nei's (1972) estimator of F_{ST} (Wright, 1951). The UPGMA (unweighted pair-group arithmetic averaging) dendrogram of Nei's (1972) genetic distance was constructed.

For the estimation of outcrossing rates, three polymorphic allozyme loci, *Mdh*-1, *Mdh*-2 and *Me*-1, were used as gene markers in Fugong, Shankou and Dongzhai harbor, and four polymorphic allozyme loci, *Mdh*-1, *Mdh*-2, *Aat*-1 and *Aat*-2, were used in Shenzhen. Single locus Mendelian inheritance with these loci was verified by the conformity of progeny array patterns to Mendelian prediction. Mating system parameters were estimated using Ritland's MLT-1 program (1990). The estimations of mating system parameters included: (a) the multilocus outcrossing rate tm; (b) the average of single-locus outcrossing

Table 1
Allelic frequencies of the hypocotyls, maternal parents and pollen pools of four *B. gymnorrhiza* populations (standard errors in parentheses)

Locus	Allele	Fugong			Shenzhen			Shankou			Dongzhai harbor		
		Hypocotyls	Maternal parents	Pollen pool	Hypocotyls	Maternal parents	Pollen pool	Hypocotyls	Maternal parents	Pollen pool	Hypocotyls	Maternal parents	Pollen pool
Mdh-1	A	0.971	0.917	0.984	0.934	0.750	0.930	0.974	0.900	0.989	0.489	0.583	0.351
		(0.020)	(0.079)	(0.013)	(0.028)	(0.125)	(0.025)	(0.018)	(0.095)	(0.009)	(0.052)	(0.142)	(0.135)
	B	0.029	0.083	0.016	0.066	0.250	0.070	0.026	0.100	0.011	0.511	0.417	0.649
		(0.020)	(0.079)	(0.013)	(0.028)	(0.125)	(0.025)	(0.018)	(0.095)	(0.009)	(0.052)	(0.142)	(0.135)
Mdh-2	A	0.914	0.833	0.903	0.934	0.750	0.955	0.950	0.800	0.971	0.957	0.833	0.979
		(0.034)	(0.108)	(0.041)	(0.028)	(0.354)	(0.021)	(0.024)	(0.126)	(0.009)	(0.021)	(0.108)	(0.007)
	B	0.086	0.167	0.097	0.066	0.250	0.045	0.050	0.200	0.029	0.043	0.167	0.021
		(0.034)	(0.108)	(0.041)	(0.028)	(0.354)	(0.021)	(0.024)	(0.126)	(0.009)	(0.021)	(0.108)	(0.007)
Me-1	A	0.529	0.583	0.487				0.064	0.200	0.064	0.404	0.333	0.586
		(0.060)	(0.142)	(0.085)				(0.028)	(0.126)	(0.030)	(0.051)	(0.136)	(0.138)
	B	0.471	0.417	0.513				0.462	0.300	0.581	0.574	0.417	0.404
		(0.060)	(0.142)	(0.085)				(0.056)	(0.145)	(0.107)	(0.051)	(0.142)	(0.138)
	C	0.000	0.000	0.000				0.474	0.500	0.355	0.021	0.083	0.010
		(0.000)	(0.000)	(0.000)				(0.057)	(0.158)	(0.104)	(0.015)	(0.080)	(0.004)
Aat-1	A				0.263	0.333	0.233						
					(0.051)	(0.136)	(0.102)						
	B				0.579	0.500	0.639						
					(0.057)	(0.144)	(0.113)						
	C				0.158	0.167	0.128						
					(0.042)	(0.108)	(0.049)						
Aat-2	A				0.382	0.417	0.380						
					(0.056)	(0.142)	(0.136)						
	B				0.618	0.583	0.620						
					(0.056)	(0.142)	(0.136)						

Table 2
Mating system estimates in four populations of *B. gymnorrhiza* (standard errors in parentheses)

Statistic	Fugong	Shenzhen	Shankou	Dongzhai harbor
tm	0.845	0.561	0.820	0.267
	(0.276)	(0.154)	(0.100)	(0.010)
ts	0.740	0.530	0.835	0.242
	(0.162)	(0.131)	(0.109)	(0.100)
tsa	0.812	0.816	0.817	0.502
tm–ts	0.104	0.032	−0.015	0.025
	(0.138)	(0.063)	(0.025)	(0.025)

rate *ts*; (c) the outcrossing rate of each single-locus using MLT-1 *tsa*; (d) the value of *tm–ts* (if the value was over zero, it showed the parent inbreeding).

3. Results

3.1. The hypocotyl, parent and pollen allele frequencies

In order to estimate the mating systems of *B. gymnorrhiza*, four populations from Fugong, Shenzhen, Shankou and Dongzhai harbor, and five loci were analyzed. The data of allele frequencies for the hypocotyls, maternal parents and pollen pools at the loci in the mating system were presented in Table 1. Allele frequencies differences among the populations were significant. And all five loci appeared to be differentiated. In each four population, there were no significant differences between pollen and hypocotyl frequencies, but significant differences between maternal parent and pollen and between maternal parent and hypocotyl were detected. However, there were no significant differences between maternal parent and pollen at *Aat*-1C in Shenzhen, *Mdh*-1A in Shankou, *Me*-1B in Dongzhai harbor and *Mdh*-1A in Fugong, indicating that the maternal parents contributed equally to the pollen pool at these loci.

Outcrossing rates estimated with single locus and multilocus were given in Table 2. In these four populations, the differences of the multilocus outcrossing rates (*tm*) ranged from 0.267 (Dongzhai harbor) to 0.845 (Fugong). The values of (*tm–ts*) of the populations from Fugong, Shenzhen and Dongzhai harbor were positive, except the one from Shankou, suggesting that the mating systems in the three populations were slightly parental inbreeding but the system in Shankou was of random mating. The mean of single locus outcrossing rates (*overlinetsa*) in Fugong, Shenzhen and Shankou were 0.812, 0.816 and 0.817, respectively. However, the mean was 0.502 in Dongzhai harbor. The differences of the outcrossing rates were significant (Table 2).

3.2. Population genetic structure

High genetic diversity was found at both species and population levels in all five populations. Of 10 allozyme loci analyzed, the percentage of polymorphic loci (*P*) at species level was 80%. The average number of alleles per locus (*A*) was 2.440, and the heterozygosity (He) was 0.293. The estimations varied with averages of $P=70\%$, $A=1.960$ and He=0.277 at population level (Table 3). However, the estimations of Ap (*A* of polymorphic loci), Aep (Ae of polymorphic loci) and Hep (He of polymorphic loci) at population level were higher than at species level. The fixation indexes were positive in

Table 3
Allozyme variation within populations of *B. gymnorrhiza*

Index	Fugong	Yunxiao	Shenzhen	Shankou	Dongzhai harbor	Population level	Species level
P (%)	90	70	70	60	60	70	80
A	2.100	1.800	2.000	2.000	1.900	1.960	2.440
Ap	2.222	2.143	2.429	2.667	2.500	2.392	2.294
Ae	1.591	1.587	1.586	1.589	1.539	1.579	1.630
Aep	1.657	1.839	1.837	1.982	1.899	1.843	1.708
He	0.278 (0.079)	0.292 (0.076)	0.268 (0.087)	0.284 (0.083)	0.264 (0.084)	0.277	0.293
Hep	0.309 (0.081)	0.416 (0.062)	0.382 (0.094)	0.473 (0.054)	0.440 (0.075)	0.404	0.365
Ho	0.375 (0.130)	0.305 (0.138)	0.386 (0.139)	0.379 (0.145)	0.268 (0.139)	0.343	0.344
F	0.029	0.104	−0.076	−0.248	0.132		

Table 4
Genetic identity (below diagonal) and standard genetic distance (above diagonal) of *B. gymnorrhiza* populations (standard errors in parentheses)

	Fugong	Yunxiao	Shenzhen	Shankou	Dongzhai harbor
Fugong	–	0.0198 (0.0106)	0.0140 (0.0101)	0.0336 (0.0247)	0.0474 (0.0227)
Yunxiao	0.9804	–	0.0175 (0.1901)	0.0149 (0.0073)	0.0269 (0.0297)
Shenzhen	0.9861	0.9827	–	0.0410 (0.0280)	0.0444 (0.0253)
Shankou	0.9670	0.9852	0.9599	–	0.0534 (0.0361)
Dongzhai	0.9537	0.9734	0.9566	0.9479	–

Fugong, Yunxiao and Dongzhai harbor, indicating the reduced heterozygosity.

The total gene diversity (H_T=0.2953) was primarily distributed within populations (H_S=0.2782, and D_{ST}=0.0171). The coefficient of gene differentiation G_{ST} (=D_{ST}/H_T=0.0579) was relatively low. Based on the G_{ST} value, the number of migrants Nm exchanged between local populations was estimated to be 3.85 per generation. Nei's standard genetic distance (*D*) and Nei's genetic identity (*I*) were calculated as $D=-\ln(I)$ (Table 4). An average of *D*'s was 0.0313 and *I*'s was 0.9693. The UPGMA was showed in Fig. 2 according to the *I*'s between populations. It was apparent that the population of Dongzhai harbor was significantly differentiated compared with the other four populations.

4. Discussion

4.1. Mating system

The highest outcrossing rates (0.854 in Fugong) and the lowest (0.267 in Dongzhai harbor) were obtained, indicating the outcrossing rates of different populations were significantly different. The mating systems in Dongzhai harbor, Shenzhen and Fugong were of parental inbreeding while the system in Shankou was of random mating. It was suggested that *B. gymnorrhiza* as entomophilou plant was of mixed mating and mainly of outcrossing. Kondo et al. (1987) also found both of outcrossing and inbreeding patterns appeared in *B. gymnorrhiza* and *Rhizophora mucronata*, which was found by Tomlison (1986) as well. The observation by Tomlison (1986) on the floral biology of species of the Rhizophoraceae revealed a wide variety of pollination mechanism, such as bird pollination in large-flowered and butterfly pollination in small-flowered *Bruguiera* species. The outcrossing was also found in entomophilou plants (Barrett and Eckert, 1990) and in *K. candel* (Sun et al., 1998; Chen, 1997).

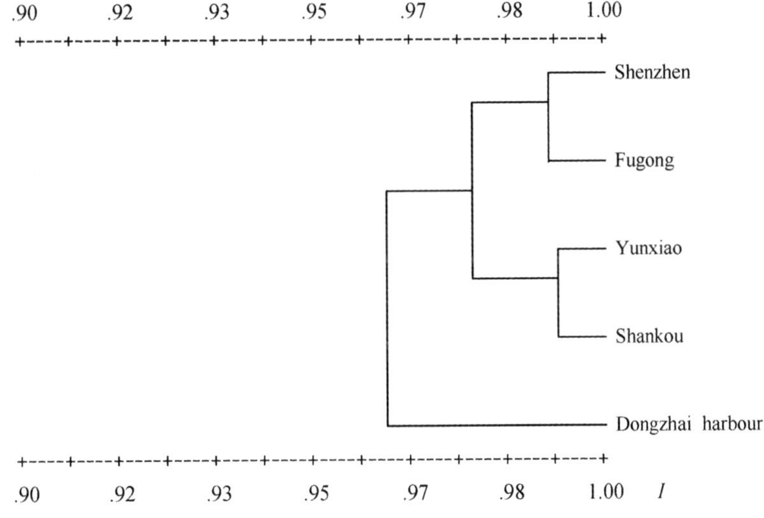

Fig. 2. The UPGMA of five *B. gymnorrhiza* populations.

The multilocus outcrossing rates among the populations of the *B. gymnorrhiza* varied on the basis of our studies. Schoen (1982) also found that this was a common phenomenon. Lande and Schemiske (1985) demonstrated that significant outcrossing or inbreeding pattern existed in balanced populations. The environmental factors and statistic errors could either break such balance or produce different outcrossing rates. Generally, the genetic factors such as pollination types, pollinator adaptabilities, corolla structures, population sizes/densities and genetic controls could have certain effects.

The outcrossing rates with big population density would increase because of the increase of relationship complexity among plants (Watkins and Levin, 1990). *B. gymnorrhiza* was of selfing compatibility species (Tomlison, 1986). When low population density was in such species, the opportunities of inbreeding and selfing would increase and the outcrossing rates would decrease. The low outcrossing rates of *B. gymnorrhiza* in Shenzhen possibly resulted from its individuals scatteredly distributed with low density.

The pollinator adaptabilities and activities would have effect on the mating systems, especially on the systems of entomophilou plants. The low outcrossing rates in Dongzhai harbor might be caused by bird activities, such as the food searching, the alteration in the population structure, etc.

4.2. Population genetic structure

The high genetic diversity at both species and population levels was observed in *B. gyumnorrhiza* of Fugong, Yunxiao, Shenzhen, Shankou and Dongzhai harbor. Of the 10 allozyme loci recorded, 8 (80%) were polymorphic at species level. The average number of alleles per locus (*A*) was 2.44, the average heterozygosity (He) was 0.293, and the average effective number of alleles per locus (Ae) was 1.63. These estimates were even lower at population level, with an average of P=70%, A=1.96, He=0.277, and Ae=1.58, compared with the common plants values P=50.5%, A=1.96, He=0.149, and Ae=1.21 at species levels and P=34.0%, A=1.53, He=0.113, and Ae=1.15 at population levels (Hamrick and Godt, 1989). Such high genetic diversity could be caused by species' biological characteristics and the life history. *B. gymnorrhiza* was widely spread species with dicotyledon, entomophilous pollination and outcrossing (Kondo et al., 1987; Tomlison, 1986). Its vivipary was capable of long-distance dispersal by ocean currents and is in the anaphase of succession. Such characteristics sustained the high diversity at both species and population levels. The high gene flow also could decrease the differentiation among the five populations and sustained the diversity within each population.

Population differentiation in *B. gymnorrhiza* (G_{ST}=0.0579) was somewhat lower than that in other plants observed (G_{ST}=0.224) by Hamrick (1989). But the differentiation was similar with that of mangrove species, such as *K. candel* (F_{ST}=0.043) (Huang, 1994) or (G_{ST}=0.064) (Sun et al., 1998) and *Rhizophora stylosa* (F_{ST}=0.023) (Tomlison, 1986). Compared with the species that have the similar biological characteristics and life history, the G_{ST} of *B. gymnorrhiza* was relatively low. The low G_{ST} of *B. gymnorrhiza* probably resulted from the dispersal pattern of vivipary, and could result in high gene flow estimated (Nm=3.85).

The genetic distance between each two populations in *B. gymnorrhiza* was quite small, which ranged from 0.0534 in Shankou and Dongzhai harbor to 0.0140 in Fugong and Shenzhen, at an average of 0.0313, which indicated that the genetic identity of five populations was quite high.

The distance isolation prominently effected the genetic differentiation of largely separated populations. The farther the distance isolation was, the larger the genetic differentiation. However the correlation between genetic distance and spatial distance was not quite clear. Alpert et al. (1993) pointed that the genetic distance was highly correlated with spatial distance (P<0.01) in *Frogaria chiloensis*, and Kiang and Chiang (1990) found that genetic distance was correlated with latitude, biotemperature and agriculture traits. However, we did not find obvious correlation between genetic distance and spatial distance in *B. gymnorrhiza*.

In conclusion, the genetic diversity within each population is significant and the differentiation among the populations is low in *B. gymnorrhiza*. High gene flow occurred in *B. gymnorrhiza*, possibly resulting from the particular vivipary dispersal pattern and the bird pollination. The mating system of this species is of mix-mating and mainly of outcrossing.

Acknowledgements

The article was financially supported by Outstanding Teacher Funds of Heilongjiang Province (1055G034); Outstanding Teacher Funds of Heilongjiang University (JC200310). Thanks to Professor Zhuangwei Lou who helped revise the English problems. [SS]

References

Alpert, P., Lumaret, R., Giusto, F.D., 1993. Population structure inferred from allozyme analysis in the clonal herb *Fragaria obiloensis* (Rosaceae). Am. J. Bot. 80 (9), 1002–1006.

Barrett, S.C.H., Eckert, C.G., 1990. Variation and evolution of mating systems in seed plants. In: Kawano, S. (Ed.), Biological Approaches and Evolutionary Trends in Plants. Academic Press, London, pp. 229–254.

Brown, A.H.D., 1989. Genetic characterization of plant mating systems. In: Brown, A.H.D., Clegg, M.T., Kahler, A.L., Weir, B.S. (Eds.), Plant Population Genetics, Breeding and Genetic Resources. Sinauer Associates, Sunderland, pp. 145–162.

Brown, A.H.D., Burdon, J.J., Jarosz, A.M., 1989. Isozyme analysis of plant mating system. In: Soltis, D.E., Soltis, P.S. (Eds.), Isozymes in Plant Biology. Dioscorides Press, Portland, pp. 73–86.

Chen, X.Y., 1997. Mating system and inferred inbreeding depression of a Cyclobal anopsis glauca population in Diaoqiao, Huangshan. Acta Ecol. Sin. 17 (5), 463–468 (in Chinese).

Duke, N.C., 1992. Mangrove floristics and biogengraphy. In: Robertson, A. (Ed.), Tropical Mangrove Ecosystem. American Geographical Union, Washington, DC, pp. 63–100.

Fyfe, J.L., Bailey, N.T.J., 1951. Plant breeding studies in leguminous forge crops: 1. Natural cross breeding in winter beans. J. Agric. Sci. 41, 371–378.

Hamrick, J.L., 1989. Isozymes and the analysis of genetic structure in plant populations. In: Soltis, D.E., Soltis, P.S. (Eds.), Isozymes in Plant Biology. Dioscorides Press, Portland, pp. 86–92.

Hamrick, J.L., Godt, M.J., 1989. Allozyme diversity in plant species. In: Brown, A.H.D., Clegg, M.T., Kahler, A.L., Weir, B.S. (Eds.), Plant Population Genetics, Breeding and Genetic Resources. Sinauer Associations, Sunderland, pp. 43–63.

Hatcher, B.G., Johannes, R.E., Robertson, A.I., 1989. Review of research relevance to conservation of shallow tropical marine ecosystems. Oceanogr. Mar. Biol. An Annu. Rev. 27, 337–414.

Huang, S., 1994. Genetic structure of *Kandelia candel* (L.) Druce in Taiwan. Chin. Biodivers. 2 (2), 68–75.

Kiang, Y.T., Chiang, Y.C., 1990. Comparing differentiation of wild soybean (*Glycine soja* Sieb. and Zucc.) populations based on isozymes and quantitative traits. Bot. Bull. Acad. Sin. 31, 129–142.

Kondo, K., Nakamura, T., Tsuruda, K., 1987. Pollination in *Bruguiera gymnorrhiza* and *Rhizophora mucronata* in Ishigaki Island, the Ryukyu Islands, Japan. Brotropica 19, 377–380.

Lande, R., Schemiske, D.W., 1985. The evolution of self-fertilization and inbreeding: 1. Genetic Models. Evolution 39, 24–40.

Li, M.S., Lee, S.Y., 1997. Mangroves of China: a brief review. For. Ecol. Manag. 96, 241–259.

Nei, M., 1972. Genetic distance between populations. Am. Nat. 106, 283–292.

Nei, M., 1973. Analysis of gene diversity in subdivided population. Proc. Natl. Acad. Sci. U. S. A. 12, 3321–3323.

Ritland, K., 1984. The effective proportion of self fertilization with consanguineous matings in inbred populations. Genetics 106, 139–152.

Ritland, K., 1986. Joint maximum likelihood estimation of genetic and mating structure using open pollinated progenies. Biometrics 42, 25–43.

Schoen, D.J., 1982. The breeding system of *Gilia achilleifolia*: variation in floral characteristics and outcrossing rate. Evolution 36, 596–613.

Soltis, D.E., Soltis, P.S. (Eds.), 1989. Isozyme in Plant Biology, Advances in Plant Science, vol. 4. Dioscorides Press, Portland, OR.

Sun, M., Wong, K.C., Lee, J.S.Y., 1998. Reproductive biology and population genetic structure of *Kandelia candel* (Rhizophoraceae), a viviparous mangrove species. Am. J. Bot. 85 (1), 1631–1637.

Tomlison, P.B., 1986. The Botany of Mangroves. Cambridge University press, Cambridge.

Umali, R.M., Zamora, P.M., Gaiera, R.R., Jara, R.S. et al., 1987. Mangrove of Asia and the pacific status and management, Technical report of the UNDP/UNESCO Research and Training Pilot programme on mangrove ecosystems in Asia and the Pacific (RAS/79/002). Quezon City, The Philippines.

Vallejos, C.E., 1983. Enzyme activity staining. In: Tankley, S.D., Orton, T.J. (Eds.), Isozyme in Plant Genetics and Breeding. Elsevier Science Publishers B.V., Amsterdam.

Wang, Z.R., 1996. Plant Allozyme Analysis. Science Press, Beijing. (in Chinese).

Watkins, L., Levin, D.A., 1990. Outcrossing rates as related to plant density in *Phlox drummordii*. Heredity 65, 81–90.

Wright, S., 1951. The genetical structure of population. Ann. Eugen. 15, 323–354.

Zhu, G.L., Zhong, H.W., Zhang, A.Q., 1990. Instruction to Plant Physiology. Beijing University Press, Beijing. (in Chinese).

六 红树林保护与生态恢复

PART VI MANGROVE CONSERVATION AND ECOLOGICAL REHABILITATION

自然保护区群网生态建设的几点思考*

林 鹏[1] 张宜辉[1] 杨志伟[1] 李振基[1] 郑达贤[2]

（1. 厦门大学生命科学学院，福建 厦门 361005；2. 福建师范大学地理科学学院，福州 350007）

摘要：论述了自然保护区在生态建设中的作用、我国国家级自然保护区的分布、福建国家级自然保护区的类型及其分布，提出了自然保护区群网建设的必要性．针对保护区建设的现状，提出了现阶段的工作重点：(1)建立自然保护区之间的群网关系；(2)开展自然保护区群网建设与研究工作；(3)保护区群网建设与森林涵养水源的重大意义和保护区域水的生态安全和生态效益；(4)保护区群网建设与发展循环经济；(5)主导资源保护与合理利用；(6)统一落实经费筹措；(7)加强部门、人员协调；(8)配合生态省建设工作．从而促进自然保护区的生态保护、生态修复和生态工程的协调发展．

关键词：自然保护区；群网；生态建设；发展

中图分类号：S759.9 **文献标识码**：A **文章编号**：1673-7105（2006）01-0067-07

Considerations about the Ecological Construction of Nature Reserve Network

LIN Peng[1], ZHANG Yi-hui[1], YANG Zhi-wei[1], LI Zhen-ji[1], ZHENG Da-xian[2]

(1. School of Life Sciences, Xiamen University, Xiamen, Fujian 361005, China;

2. School of Geographical Sciences, Fujian Normal University, Fuzhou 350007, China)

Abstract: Using national, especially Fujian provincial data, this paper deals with ecological conservation and restoration regarding nature reserves. It is pointed out that the nature reserves in China are highly scattered and managed by different administrative units, causing various investment and management handicaps. To effectively manage them, it is key to establish nature reserve network within a region, such as a province or an autonomous region. The authors firstly discusses the basic functions of nature reserves, main types and distribution of national level nature reserves in Fujian as well as in China, and the necessity of building up nature reserve network. Then major countermeasures and recommendations are presented: (1) to combine nature reserves in a certain region like Fujian province into a closely related network; (2) to carry out further scientific research on the network; (3) to achieve reasonable relations between water conservation and use within nature reserves; (4) to develop a circling economy related to nature reserve developments; (5) to advance the conservation and suitable utilization of dominant resources depending on nature reserves; (6) to implement an unified financial policy and plan; (7) to harmonize relations among different managerial institutions, and (8) to embed nature reserve related work into the strategy of constructing Fujian into an ecological province (the Eco-province Project).

Key word: nature reserve; network; ecological construction; development

截止2003年底，全国（不含香港、澳门特别行政区和台湾省）共建立各种类型、不同级别的自然保护区1 999个．至2006年，已批准建立的国家级自然保护区有265个，福建也有11个．但随着保护区在数量和规模上快速发展，经费投入不足，管护能力有限，技术支撑滞后，在建设与发展中还存在不少不容忽视的问题，特别是保护区分散建设，布局不够合理，保护区网络尚未完善．因此，如何利用景

* 国家自然科学基金项目(30370275)资助

原载于亚热带资源与环境学报，2006，1(1)：67－73.

观生态学的理论，在保护区与保护区之间建立廊道，通过保护区群网把区域的保护区联系起来，更好地对区域的生物多样性起着保护作用，已成为当务之急．福建省已在武夷山脉和戴云山脉建立了多个保护区．为此，试图以福建为例来探讨保护区群网建设的意义．

1 概况

1.1 自然保护区在生态建设中的作用

自然资源、生态与环境是人类赖以生存和发展的基本条件．人类在长期的社会实践中，认识到保护好自然资源、生态与环境，保护好生物多样性，对人类的生存和发展具有极为重要的意义．保护自然资源、生态与环境和生物多样性的最重要和最有效措施是建立自然保护区，自然保护区建设已成为衡量一个国家进步和文明的标准之一，是维护生态安全，促进生态文明，实现社会经济可持续发展和人与自然和谐共存的重要保障．

中国是世界上自然资源和生物多样性最丰富的国家之一，中国生物多样性保护对世界生物多样性保护具有十分重要的意义．根据1994年中华人民共和国国务院发布的《中华人民共和国自然保护区条例》，自然保护区是指对有代表性的自然生态系统，珍稀濒危野生动植物物种的天然集中分布区，有特殊意义的自然遗迹等保护对象所在的陆地、陆地水体或者海域，依法划出一定面积，给以特殊保护和管理的区域[1]．

我国的自然保护区事业，作为保护自然资源与生物多样性的重要手段，从无到有，蓬勃发展．1956年我国建立了第一个自然保护区——广东肇庆鼎湖山自然保护区．经过50多年的努力，中国的自然保护区建设取得了显著的成绩．特别是20世纪80年代以来，自然保护区事业发展很快，在全国初步建成一个类型较齐全、分布较合理的自然保护区网络．中国的自然保护区分森林、湿地、荒漠、野生动物、野生植物、草原、海洋、自然遗迹、古生物遗迹等9大类型，前5种属林业部门主管，数量最多，后4种类型中，草原归属农业部，海洋属于国家海洋局，而自然遗迹、古生物遗迹的行政管理权则属于地矿部门．四大行政管理部门之外，又由国家环保总局进行综合协调[2~4]．

根据《全国自然保护区名录(2003)》统计，截止2003年底，全国(不含香港、澳门特别行政区和台湾省)共建立各种类型、不同级别的自然保护区1 999个，保护区总面积143.98万 km^2 (陆地面积137.95万 km^2，海域面积6.03万 km^2)，陆地自然保护区面积约占陆地国土面积的14.37 %[5]．目前，中国70 %的陆地生态系统种类、80 %的野生动物种类和60 %的高等植物种类，特别是国家重点保护的珍稀濒危动植物绝大多数种类都在自然保护区里得到较好的保护．同时，这些自然保护区还起到了涵养水源、保持水土、防风固沙、稳定地区小气候等重要作用．中国自然保护区作为宣传教育的基地，通过对国家有关自然保护的法律法规和方针政策及自然保护科普知识的宣传，极大地提高了我国公民的自然保护意识．

1.2 中国国家级自然保护区的分布

国家环保总局于2006年2月23日宣布，经国务院批准，我国新建了22处国家级自然保护区[6]．至此，我国(不含香港、澳门特别行政区和台湾省)的国家级自然保护区已达265个，总面积91.85万 km^2．31个省、自治区、直辖市(不含香港、澳门特别行政区和台湾省)都拥有各自的国家级自然保护区(表1)，其中，数量达到10个以上的有内蒙古、四川、云南、黑龙江、甘肃、广西、福建、湖南、辽宁等9省(自治区)．

表1 中国国家级自然保护区数量分布(不含香港、澳门特别行政区和台湾省)

Table 1 Quantitative distribution of national level nature reserves in China (Hong Kong, Macao and Taiwan not included)

地区	数量/个	地区	数量/个	地区	数量/个	地区	数量/个
内蒙古	21	辽宁	11	新疆	8	山东	5
四川	20	河南	10	贵州	7	山西	5
云南	16	广东	9	湖北	7	江苏	3
黑龙江	15	吉林	9	陕西	7	天津	3
甘肃	13	西藏	9	安徽	6	重庆	3
广西	12	浙江	9	宁夏	6	上海	2
福建	11	海南	8	江西	5	北京	1
湖南	11	河北	8	青海	5		

说明：表中数据为2006年本文收集整理．

从总体上看，我国现有自然保护区的空间布局不尽合理，大部分保护区集中在中西部，特别是西部少人或无人地区．据统计，西部12个省、自治区、直辖市自然保护区面积占到全国自然保护区总面积的 84.5 %，而其中西藏羌塘、青海可可西里、三江源和新疆阿尔金山 4 个国家级自然保护区总面积达 54 万 km^2，占到国家级自然保护区总面积的 60 %以上．东部地区自然保护区的数量和面积远低于全国平均水平．在当前经济快速发展的情况下，东中部地区不少重要物种和生态系统的保护已迫在眉睫．同时，当前我国自然保护区以森林生态类型、珍稀濒危动物类型和荒漠化类型偏多，海洋类、湿地类、地质遗迹类等偏少．有必要加强部门合作，在科学考察、充分论证的基础上，合理布局不同区域、地区、类型自然保护区的建设．

2 福建国家级自然保护区的分布和分散建设的困难

福建省有良好的自然环境，是比较完整且相对独立的地理单元，山多海阔、水系纵横交错，森林覆盖率达 60.5 %，居全国首位．从 1979 年福建省建立第一个国家级自然保护区——武夷山自然保护区开始，到 2000 年陆续建成 5 个国家级自然保护区．2000 年之后，福建省自然保护区建设进入了一个新的重要发展时期，至 2006 年，又新增国家级自然保护区 6 个．这 11 个国家级自然保护区总面积为 18.40 万 hm^2，涵盖了森林（7 个）、湿地（1 个）、野生动物（1 个）、野生植物（1 个）、古生物遗迹（1 个）等 5 个类型，分属林业、海洋和环保部门管理（图 1，表 2）．

总体看，随着生态省建设的开展，福建省自然保护区建设力度逐步加大，发展势头良好．但是，在数量和规模快速发展的新形势下，由于经费投入不足，管护能力不强，技术支撑滞后，福建省国家级自然保护区的建设与发展还存在不少不容忽视的问题，特别是保护区分散建设，布局不够合理，全省自然保护区网络尚待建设．

图 1 福建省国家级自然保护区分布示意图

Figure 1 A sketch map of spatial distribution of national level nature reserves in Fujian Province

3 自然保护区群网建设的几点思考

针对目前福建省国家级保护区建设与发展中存在的问题，就自然保护区群网的建设提出以下几点建议，以实现保护区的和谐发展．

3.1 建立自然保护区之间的群网关系

随着保护区数量和规模的快速增长，应该把自然保护的重点转移到景观尺度，把对孤立的自然保护区的传统管理，转变为对景观动态斑块的管理[7~11]．福建省内有我国所特有的中亚热带生态系统自然保护区群，这一区群的各个保护区之间，既有共有的保护对象，也有各具特色的重点对象和这些保护对象所依托的典型生境，特别是有许多高价值种质资源及其特殊的生境，这是该区域自然保护区的最大特色[12]．

自然保护区的规划建设仅考虑单个保护区是不可取的，应该在更大的范围内采用节点—网络—斑块—廊道模式来考虑与设计自然保护区群网．节点相当于自然保护区，但是节点不可能维持和保护所有的生物多样性．所以，必须建设保护区网来连接各种节点，允许物种、基因、能量、物质在走廊中流动．生境走廊作为适应于生物移动的通道，把不同地方的保护区构成了大范围的保护区群网．

通过建立自然保护区群网，首先可以促进生物多样性的迁徙与传布；其次可以解决目前许多“岛屿”式的孤立保护区所不能解决的问题，将有效地提高自然保护区内的顶级森林生态系统的稳定性和抗干扰能力，扩大保护对象的活动空间和生境范围，大大提高生物多样性和种质资源的保护效果；再者可以把一些孤立的天然林景观斑块通过廊道连接起来，进而为恢复大片的稳定亚热带森林生态系统创造条件，从而很好地保护省内重要的常绿阔叶林生态系统，保护闽江流域的水系和水资源，极大地改善全省生态与环境．在全省开展自然保护区群网的研究与建立在保护区研究、规划、建设和可持续适应性管理上，将是一个新的突破，具有重大理论研究价值及长远经济社会环境可持续发展实践意义，它的建成必将对绿色海峡西岸的建设起到重要的促进作用．

表 2 福建省国家级自然保护区一览表 (2006)

Table 2 List of national level nature reserves in Fujian province

保护区名称	类型	面积/hm^2	主要保护对象	所属行政区	建立时间/年	主管部门
武夷山	森林生态	56 528	中亚热带森林生态系统及珍稀动植物	武夷山市、建阳市、光泽县、邵武市	1979	林业
梅花山	森林生态	22 169	华南虎、中亚热带森林生态系统	上杭县、新罗区、连城县	1985	林业
深沪湾	古生物遗迹	2 700	海底古森林遗迹	晋江市	1992	海洋
龙栖山	森林生态	12 600	中亚热带常绿阔叶林	将乐县	1998	林业
厦门	野生动物	33 088	中华白海豚、白鹭、文昌鱼	厦门市	2000	环保
虎伯寮	森林生态	2 650	南亚热带雨林	南靖县	2001	林业
天宝岩	森林生态	11 015	中亚热带森林生态系统	永安市	2003	林业
漳江口	湿地生态	2 360	红树林、湿地鸟类	云霄县	2003	林业
梁野山	野生植物	14 365	南方红豆杉及常绿阔叶林	武平县	2003	林业
戴云山	森林生态	13 472	森林及珍稀动植物	德化县	2005	林业
闽江源	森林生态	13 022	森林生态系统	建宁县	2006	林业

3.2 开展自然保护区群网建设与研究工作

随着自然保护区群网建设理念的提出，具体的建设与研究工作也提到了议事日程上．对自然保护区群网的建设工作可以从以下方面开展：

(1) 已有的国家、省级和市县级自然保护区的基础设施的建设；

(2) 在自然保护区群网规划的基础上，开展保护区之间生态廊道的建设，并在管理制度上加以创新．主要包括以下方面：①廊道的设置与建设及相关经济、社会、技术问题的解决；②廊道涉及到的生态公益林的保护和人工促进自然生态系统的修复；③划入廊道区域范围内的土地使用权的改变和森林生态系统的建设；

(3) 根据保护生物学研究在重要节点增设新的自然保护区；

(4) 保护区群网生态功能动态监测站网的建设．

此外，在自然保护区群网规划、建设和管理过程急需开展以下重点研究工作：

(1) 构建保护区群网监测、监护体系，对具有典型性、原生性、稀有性的森林生态系统，珍稀、濒危、特有、重要经济价值的生物资源进行生物学、生态学研究；

(2) 景观破碎化对生物多样性的影响分析与对策研究；

(3) 中亚热带常绿阔叶林生态系统的保育和重建研究；

(4) 维持生态系统水平的稳定性 (特别是正常的能量流动、物质循环和信息传递) 对于重要物种保护的意义和作用的研究；

(5) 珍稀、濒危和特有植物特定生境制图与适宜性分析；

(6) GIS 支持下的栖息地斑块连接性评估和栖息地选址；

(7) 保护区群网的构建在水源涵养、水系生态安全中的作用；

(8) 保护区群网的构建对区域经济可持续发展的作用的研究．

3.3 保护区群网建设与区域水的生态效益

水是生命的源泉、农业的命脉、生态与环境的支柱．森林生态系统健康，才会水源充足，才会保证水文生态健康，单纯强调节水而忽视养水，就会损害水文生态健康．

3.3.1 通过保护区蓄水达到“三水”的矛盾统一

“三水”即天上水 (大气降水)、地面水 (江、河、湖、海、冰川、雪水) 和地下水 (包括地下泉和土壤水）始终在运动转化之中[13]．应该自觉地运用对立统一规律，遵循自然生态系统的物质流循环原则来开展经济活动，而不能随意提口号、定政策．例如：忽视地面水与地下水的转化，一味地引用地面水灌溉，就会引起地下水上升，在滨海与干旱地区就会招致土壤盐渍化；在一些河段过量堵截引水，造成其下游干枯少水，破坏了区域水循环过程，引起不良后果，甚至生态灾害．

在自然保护区生态系统中，要以自然辩证法观点为指导，通过保护森林，利用森林蓄水，永续不断地保障整个流域的工农业用水，确保水源的生态安全．

3.3.2 保护水源，永续利用，达到养水与用水的辩证统一

水是生物圈的血液，是环境可持续发展的决定因素．2000 年以色列的 Uri Shamir 等提出了“蓝色水”（地表和地下水资源）、“绿色水”（一般也称“虚拟水”，围绕农产品运动的水）、“金色水”（货币和资金意义的水）和“灰色水”（管理水）的新概念[14]．需要保护更多的蓝色水（天上水、地表水和地下水资源），保护、涵养更多的绿色水（森林、农作物等生物水），建立绿色水库（特别是自然保护区水源涵养水），改善生态与环境，同时产生更多的金色水（生产更多货币与资金意义的淡水资源）．重视保护水源和生态效益，以及森林水源的生态补偿问题．

3.4 保护区群网建设与发展循环经济

所谓循环经济，本质上是一种生态保护型经济，它要求运用生态学规律而不是机械论规律来指导人类社会的经济活动．循环经济倡导的是一种与环境和谐的经济发展模式．它要求把经济活动组织成一个“资源 –产品 –再生资源”的反馈式流程，其特征是低开采、高利用、低排放．所有的物质和能量要能在这个不断进行的循环中得到合理和持久的利用，以把经济活动对自然环境的影响降低到尽可能小的程度[15]．简单来说，循环经济是一种“促进人与自然的协调与和谐”，即自然 –社会 –经济复合生态系统和谐发展的经济发展模式．

从大的区域来看，自然保护区如一颗颗绿色明珠，镶嵌在大地上，在区域气候调节、水源涵养等方面起着重要作用，保护区的存在确保了人与自然的协调与和谐．因此自然保护区群网建设是区域循环经济的基础．

3.5 主导资源保护与合理利用

自然保护区是“以保护为根本、以科技为支撑、以改革为动力、以可持续发展为最终目标”．如何设计、建设和管理好自然保护区，促进环境保护与生态建设和社会经济发展的双赢，是当前自然保护区管理研究与实践的一个焦点，特别是对于肩负着生态系统保护与区域社会经济可持续发展双重任务，以自然资源管理为主的综合功能自然保护区，这一问题显得更加突出．目前，国际上对于综合功能自然保护区如何开展保护区管理研究与实践已经基本达成了共识：即必须走环境保护与社会经济协调发展的双赢之路．自然保护区必须走可持续发展之路，在保护的前提下，进行合理的开发与利用，这样才能符合保护的要求和经济发展的需要，才能体现自然保护区的职能．

福建省各自然保护区蕴藏着极丰富的生物种质资源，是中国乃至世界少见的生物多样性富集区域，可以利用这一优势，找出几个保护区共有的功能性资源如药用植物资源，在全省规划设计与建立药用植物种质资源为主的种质资源种子库，分析药用植物种质资源遗传多样性，开展药用植物的开发利用，高质量药用植物的保护、选育、适宜地规模化种植及高附加值产业链的形成和延伸，都将会创造巨大的社会、经济和环境效益，服务于福建省社会、经济和环境和谐发展．

3.6 统一落实经费筹措

经费投入严重不足，是制约自然保护区事业发展的重要因素．自然保护区建设是经济社会持续发展的基础性工作．但长期以来自然保护区建设和管理经费主要依靠各级行政主管部门从部门经费中解决，大多自然保护区缺乏应有的投入．

自然保护区群网的建设，需要各级政府和部门加大财力支持，尽快把相关资金列入财政专项预算予以保障，特别是省科技厅、发改委应加大投入．

同时，充分调动全社会的力量参与建设，努力形成多元化的投融资机制．

3.7 加强部门、人员协调

《自然保护区条例》第二十一条规定“国家级自然保护区，由其所在地的省、自治区、直辖市人民政府有关自然保护区行政主管部门或者国务院有关自然保护区行政主管部门管理”. 现实的情况并非都按照《条例》来管理自然保护区，较为普遍的管理体制是业务由上级主管部门管理，行政由县级以上地方政府管理，实行业务与行政分离的管理体制. 这种管理体制存在着职责不清、权利不明的弊病. 此外，保护区各自为政，相互之间互不了解. 这种“各自为政”的管理不利于自然保护区的质量升级. 这些都表明现行的行政管理体制不适应自然保护区的发展.

针对上述实际情况，基于“自然保护区群网”的建设平台，有必要通过协调，理顺多头管理的现况，建立健全协调统一的管理体系. 如以环保部门对自然保护区实行统一综合监管，林业、国土资源、文化等相关行业部门协同管理，充分发挥各部门的主观能动作用，强化协作意识. 同时，依托大专院校、科研单位和有关部门，如厦门大学、福建师范大学、福建农林大学、林业厅自然保护中心等相关单位的技术力量，将自然保护区群网建设与优良药用植物生态学、生物学研究及种质资源库开发的规划设计和建立多学科相结合，为实现福建省森林生态保护与建设、生物多样性保护、研究保护区森林涵养水的规律、以及退化生态系统的修复和生物资源的高附加值开发加工提供科技支撑，实现产学研一体化的目标.

3.8 配合生态省建设工作

2002 年 10 月，《福建生态省建设总体规划纲要》在北京通过了国家环保总局和省政府组织的专家论证，福建成为继海南、吉林、黑龙江之后的第四个全国生态省建设试点省份. 在中共福建省委，福建省人民政府颁发的闽委发 [2004] 15 号文件《福建生态省建设总体规划纲要》中，提出了：“加强动植物物种、湿地保护和自然保护区建设. 重点保护亚热带天然常绿阔叶林、滨海天然湿地等典型生态系统，建立类型齐全、布局合理、等级完善、功能齐全的自然保护区 (小区、点) 和物种拯救基地网络，有效拯救、保护珍稀濒危野生动植物和优良地方畜禽品种”. 总体纲要计划至 2010 年，划定和建设国家级和省级自然保护区 60 个，全省自然保护区、小区面积达 122 万 hm^2，对 34 种珍贵濒危物种实施拯救.

此外，2004 年 11 月中共福建省委颁发的闽委发 [2004] 12 号文件《海峡西岸经济区建设纲要 (试行)》提出了未来福建省社会经济与环境可持续发展的若干重要战略目标，其中包括：加强生态与环境的保护和建设，构建可持续的生态支撑体系，建设绿色海峡西岸，促进人与自然的和谐. 主要的实施措施就是继续推进自然保护区和生态示范区建设，加强动植物资源保护，充分发挥福建省动植物物种资源十分丰富的特点，开发利用其中的药用物种资源，服务于福建省经济社会发展和生态与环境的保护.

可以看出，研究、规划、建立和管理可持续的自然保护区群网与研究、开发和利用其中某个功能性资源，如珍贵的药用植物种质资源两者相辅相成、互为依托. 建立和管理可持续的自然保护区群网可以更科学合理地保护其中的生物资源，而研究开发利用其中具有重大经济价值的药用植物资源既可以促进当地乃至整个福建省社会经济的健康快速发展，反过来又可以为保护区的建设和管理提供资金保证. 从而实现保护区的和谐发展，为建立人与自然和谐相处的社会作出更大贡献.

4 结语

本文根据福建省自然保护区建设与发展的现状，提出自然保护区群网研究、规划和建设的几点思路. 建立和管理自然保护区群网可以更科学合理地保护其中的生物资源，其管理模式与对策能够通过可持续的景观管理达到整合保护与发展的目标，达到人与自然的和谐发展；而研究开发利用具有重大经济价值的药用植物资源，既可以促进当地乃至整个福建省社会经济的健康快速发展，又可以对保护区的建设和管理提供资金保证，实现自然保护区的资源、环境和管理的生态整合. 保护区群网建设是以福建生态省建设总体规划纲要为指导，涉及经济、社会、技术各个层面，让更多的人来参与、关心、支持自然保护区事业，促进自然保护区的生态修复、生态工程协调发展. 从而达到天蓝、地绿、水碧、气新、人富的和谐社会.

参考文献 (References):

[1] Decree No. 167 of the State Council of the People's Republic of China. *Regulations on Nature Reserve Protection in the People's Republic of China*, 1994. [中华人民共和国国务院令第 167 号. 中华人民共和国自然保护区条例. 1994.]

[2] Wang X P, Cui G F. *The constructionand management of nature reserve* [M]. Beijing: Chemical Industry Press, 2003. [王献溥, 崔国发. 自然保护区建设与管理 [M]. 北京: 化学工业出版社, 2003.]

[3] Wang K eds. *The national nature reserves in China* [M]. Hefei: Anhui Science & Technology Press, 2003. [王恺 (主编). 中国国家级自然保护区 [M]. 合肥: 安徽科学技术出版社, 2003.]

[4] Department of Wildlife Conservation, State Forestry Administration ed. *Management handbook for nature reserves in China* [M]. Beijing: China Forestry Publishing House, 2004. [国家林业局野生动物保护司 (编). 中国自然保护区管理手册 [M]. 北京: 中国林业出版社, 2004.]

[5] Department of Nature and Ecology Conservation, State Environmental Protection Administration ed. *Introduction to the nature reserves in China* (2003) [M]. Beijing: China Environmental Sciences Press, 2004. [国家环境保护总局自然生态保护司. 全国自然保护区名录 (2003) [M]. 北京: 中国环境科学出版社, 2004.]

[6] State Environmental Protection Administration. Notice on promulgation of area, range and functional division of 22 new national nature reserves [EB/OL]. http://www.zhb.gov.cn. [国家环境保护总局. 关于发布山西五鹿山等 22 处新建国家级自然保护区面积、范围及功能分区等有关事项的通知 [EB/OL]. http://www.zhb.gov.cn/. 2006-04-05/2006-05-20.]

[7] Harris L D. *The fragmented forest: island biogeography theory and preservation of biotic diversity* [M]. Chicago IL: University of Chicago Press, 1984.

[8] Kingsland S. Designing nature reserves: adapting ecology to real-world problems [J]. *Endeavour*, 2002, 26 (1): 9-14.

[9] Margules C and Pressey R L. Systematic conservation planning [J]. *Nature*, 2000, 405: 243-253.

[10] Margules C, Higgs A J and Rafe R W. Modern biogeographic theory: are there any lessons for nature reserve design? [J]. *Biological Conservation*, 1982, 24: 115-128.

[11] Soule M E and Simberloff D. What do genetics and ecology tell us about the design of nature reserves? [J]. *Biological Conservation*, 1986, 35: 19-40.

[12] Special Group on the STDREEA. No.9 of Special Reports on Medium-Long Term Development Plan of Science and Technology of Fujian: Special Reports on STDREEA [R]. 2005. [资源、生态与环境科技问题研究专题组. 福建省中长期科技发展规划战略研究专题报告之九 ——资源、生态与环境科技问题研究专题报告 [R]. 2005.]

[13] Li P C. On the new thought concerning water-related human activities [J]. *Engineering Science*, 2000, 2 (2): 5-9. [李佩成. 试论人类水事活动的新思维 [J]. 中国工程科学, 2000, 2 (2): 5-9.]

[14] Zhang Z B. The great potential of biological water saving [N]. *Scientific Times*, 2006-04-10 (B3). [张正斌. 生物节水潜力巨大 [N]. 科学时报, 2006-04-10 (B3).]

[15] Qu G P. To develop circular economy is the general trend in the 21st century [J]. *Contemporary Eco-Agriculture*, 2002, (21) (Supplement): 18. [曲格平. 发展循环经济是 21 世纪的大趋势 [J]. 当代生态农业, 2002, (21) (增刊): 18.]

人工红树林幼林藤壶危害及防治研究进展*

向 平[1] 杨志伟[1,2] 林 鹏[1,2]

(1. 厦门大学生命科学学院，厦门 361005；2. 厦门大学湿地与生态工程研究中心，厦门 361005)

【摘要】 随着近年来红树林恢复性造林面积的扩大,海洋污损生物藤壶对红树林幼林的危害问题日益突出.文中综述了藤壶附着的生物化学、藤壶在红树林附着的生态学、藤壶对人工红树林幼林的危害和国内所采用的化学药物防治措施等方面的研究进展,以及今后的研究方向.藤壶在红树林的附着和分布模式受海水盐度、浸淹深度、林分郁闭度、水文条件等环境因素和生物因素的影响.而藤壶胶粘蛋白的氨基酸组成、一维结构,胶粘蛋白在水下的交联、组装和胶粘的过程与机制,以及藤壶危害红树幼苗的机制和危害权重尚需要深入探讨.研究红树植物对藤壶附着的响应和长期适应机制将为藤壶的防治提供更多的启示.

关键词 红树林 藤壶 红树林恢复

文章编号 1001 - 9332(2006) 08 - 1526 - 04 **中图分类号** Q948.1 **文献标识码** A

Barnacle damage and its control in young mangrove plantations: A research review. XIANG Ping[1], YANG Zhiwei[1,2], LIN Peng[1,2] ([1]*School of Life Science, Xiamen University, Xiamen 361005, China;* [2] *Research Center for Wetlands and Ecological Engineering, Xiamen University, Xiamen 361005, China*). -*Chin. J. Appl. Ecol.*, 2006, 17(8): 1526～1529.

With the increasing area of restored mangrove vegetation, marine-fouling organisms, barnacle in particular, are suggested to be an important factor affecting the survival and growth of mangrove seedlings. This paper reviewed the biochemical and ecological studies on the settlement of barnacle, its damage on mangrove seedlings, and its chemical control. The settlement and distribution model of barnacle on mangroves is significantly affected by the environmental factors such as seawater salinity, tide inundating depth, canopy density, hydrographical regime, and some biotic factors, but few are known about the amino acid composition and one-dimension structure of barnacle's adhesive proteins, especially their processes and mechanisms of cross-link, aggregation, and adhesion. More attention should be paid on understanding the damage mechanisms and its weight of barnacle on mangrove seedlings, and the study on the response and adaptation models of individual plant in nature mangrove ecosystem to barnacle disturbance should be strengthened, which are potentially valuable for the research of barnacle control.

Key words Mangrove, Barnacle, Restoration of mangrove vegetation.

1 引 言

红树林 (mangroves)是自然分布于热带、亚热带海岸和河口潮间带的木本植物群落[23,25],通常生长在港湾与河口地区的淤泥质滩涂上,是海滩上特有的森林类型.我国红树林自然分布在海南、广东、广西、台湾和福建等省,以及香港和澳门地区,浙江省有人工引种的红树林.红树林除了具有促淤保滩、巩固堤岸、抵抗风浪袭击、过滤有机物和污染物等功能外,还是鸟类、贝类、鱼、虾和蟹等海洋生物栖息繁衍的良好场所,在保护海滨湿地生态系统和生物多样性以及维护海湾、河口地区生态平衡等方面均有着不可替代的重要作用[24].同时,红树林也是进行旅游和社会教育的自然和人文景观.2004年 12月 26日印度洋发生海啸后,当印度洋各国正讨论建立海啸预警机制时,各国环保科学家们及其环保组织提出了一个业已存在并可以最大程度地降低自然灾害破坏力的天然武器 ——红树林.

多年来,由于围海造田、围塘养殖、填海造陆和修建公路码头等人为的因素,红树林的面积大幅度减少[21],甚至处于濒危状态.因此,建立红树林自然保护区和对红树林进行恢复性造林势在必行[24].在红树林恢复性造林实践、残次林改造和优良树种驯化研究中,红树林幼苗受到多种海洋底栖生物的影响,藤壶是人工红树林幼林最为常见的危害生物,甚至在某些地方成为红树幼苗正常生长发育的关键胁迫因子,成为危害红树林面积最大、程度最高的污损生物[22,37].本文主要介绍了藤壶附着的生物化学、藤壶在红树林附着的生态学、藤壶对人工红树林幼林的危害和国内所采用的化学药物防治措施等方面的研究进展,以及今后的研究方向.

2 藤壶附着的生物化学

藤壶属于节肢动物门甲壳纲,蔓足亚纲,完胸目,藤壶亚目,共记录有 8科约 559种[17].藤壶的幼虫时期需要经历无节幼体和金星幼体 (腺介幼体)的变态发育过程.金星幼体是一种特殊的幼体形态,无需摄食,仅是为了选择附着、变态的适宜基底[16].金星幼虫被流动的海水牵引到基底上,进行

* 国家自然科学基金项目(30270272)、厦门市海洋与渔业局、翔安农林水利局资助项目(2005－2008)

原载于应用生态学报,2006,17(8):1526－1529

暂时性附着并对所附着的基底进行探查.当幼虫找到合适的基底后,便从其第1触角第3节的附着吸盘开口处分泌胶体开始营固着生活.金星幼虫常选择粗糙不平或有凹陷及小沟的基底,特别是在有同种成体密集的基底表面附着[46].

藤壶通过分泌蛋白质类的藤壶胶在水下基底上附着营固着生活.根据收集方法和藤壶胶聚集的先后,藤壶胶分为初生胶和次生胶[36,43],这两种胶在氨基酸组成上是相似的[28].藤壶胶在水下的交联、组装和有效粘附,其过程必须完成胶凝、转移基底表面的水分子、建立界面接触和与各种基底间的分子吸附等功能.对这一系列功能和胶粘机制的研究均有赖于对胶粘蛋白的充分研究.由于藤壶胶不溶于盐、稀酸、稀碱等溶液中,故对藤壶胶的研究是随着胶粘蛋白溶解方法的改进而逐渐深入的.近年来,Kamino等[18,19]通过一种非水解的方法溶解出了90%以上的藤壶胶粘蛋白,并在此基础上,从 *Megabalanus rosa* 的胶粘蛋白中分离鉴别出3种主要的蛋白质 Mrcp-100K、Mrcp-52K和 Mrcp-68K. Mrcp-100K和 Mrcp-52K具有高疏水性和分泌到附着位置后的难溶特性.而 Mrcp-68K则含有独特的氨基酸组成,其中57%的组分是由丝氨酸(Ser, 16%)、苏氨酸(Thr, 14%)、甘氨酸(Gly, 15%)和丙氨酸(Ala, 12%)构成.同时,藤壶胶中另一种含量较少的蛋白 Mrcp-20K则含有丰富的 Cys等带电荷的氨基酸,它包括半胱氨酸(Cys 17.5%)、天门冬氨酸(Asp 11.5%)、谷氨酸(Glu 10.4%)和组氨酸(His 10.4%)等最主要的氨基酸[17].关于藤壶胶粘蛋白在水下的交联、组装和胶粘过程与机制的研究有赖于对胶粘蛋白的氨基酸组成和胶粘蛋白质一维结构的清楚认识.

藤壶附着在水下各种基底的表面,藤壶胶底板的形态、组分和胶粘蛋白的构成等均随着基质的结构和表面物化性质的不同而各异. Berglin等[1]对附着在不同基底上的藤壶(*Balanus improvisus*)底板研究发现,附着在低表面能和低模数基底(PDMS)上的藤壶合成厚的、有弹性的粒状附着底板,而在较高表面能和较高模数的基底(PMMA)上的藤壶则合成硬的、连续的附着底板.在较高表面能基底上附着的底板中含有方解石形态的钙质,而在低表面能的基底上附着的底板则没有检测到钙质.而在低表面能基底上附着的底板中,用0.5M的DTT和7M的盐酸胍可以溶解出在较高表面能基底上附着的底板中没有的3种蛋白质和一些小肽.

3 藤壶在红树林附着的生态学

藤壶在红树林的附着及其分布模式受多种因素的影响[14,20,22,38,39,42,47].主要包括以下几个方面:1)盐度因素.盐度对藤壶的分布影响较大,从数量上基本成正相关,同时,盐度的改变还引起优势种的变化;2)浸淹深度.随着浸淹深度的增加,在红树植物茎干和枝叶上附着的藤壶数量增加,对红树的危害程度增大,即便是同一样地中微地形的差异都会呈现这种相关性,同时,浸淹深度也引起优势种的变化;3)郁闭度(林分密度).在九龙江口红树林区,当林分郁闭度达到0.5时就基本没有藤壶附着,但李云等[22]报道,在广东红树林幼林的密度与藤壶附着没有正相关性,这可能与红树幼苗还没形成有效的郁闭度有关;4)水文条件.水流畅通程度是影响藤壶纵深分布的主要因子,开阔的海域藤壶对红树植物的危害程度较封闭的港湾严重,向海林缘较林内和向陆林缘附着严重;5)生物因素.影响藤壶附着和分布的生物因素包括同种其他个体的影响、特殊的食物源、细菌粘膜、其他生物如藻类和红树植物通过活性物质或通过对微生境的改变而对藤壶附着产生的抑制或诱导的影响.

在开放生态系统中(如水生生境),繁殖体或幼体的丰度和分布模式对成体生物的丰度和分布模式有决定性的影响[6,9,44],幼体在水层的运动和不同基底上的附着行为也产生较大影响[32].因此,研究引起幼体丰度和分布模式差异的原因是认识海洋动物种群动态的关键[38].特定区域幼体数量和分布模式受多种物理和生物因子制约,其中包括幼体的繁殖[11]、幼体食物丰度[29]、生理应力[3]和捕食[7]等都会影响浮游阶段的幼体的数量.与其他海洋幼体一样,藤壶的幼体在大尺度上的运动都是靠海水被动地牵引而完成的.因此,水动力因子如海流、潮汐和海浪都会影响幼体在特定位置的输送和丰度[2,31].这种被动的迁移有助于把幼体牵引到特定的附着位置,同时也可能通过水平的迁移使幼体远离合适的附着基质而导致死亡.在微观空间尺度上,幼体的行为对其附着模式起重要作用[33].如在水体平静的红树林区或者河口区域,幼体可能不再是作为被动的"颗粒"而受到上述水文动力因子的影响,其通过在水层的游动和垂直运动来改变在水体中的位置,从而影响幼体的附着和成体的分布[12,39,41].在微观尺度上,幼体的行为同样受到多种因子的影响,其中的物理因子包括光照、海水盐度、基底的物理特征(微地形)和微尺度的水流影响[8,10,40,42],生物因子则主要有基底上的细菌粘膜和同种个体的诱导作用[20,42].

在澳大利亚东南部的红树林白骨壤(*Avicennia marina*)林区,藤壶 *Elminius covertus* 金星幼体的丰度和金星幼体的行为都是影响红树林中藤壶附着模式的重要因子[34,37,39].在包括红树林在内的潮间带,在时间上,藤壶金星幼体密度的高峰出现在夜间的高潮水层中;在空间分布上,水层中的藤壶金星幼体密度随向海林缘到向陆林缘的方向逐渐减少,在向海林缘密度最大、林内较少,而在向陆林缘则几乎没有或很少有金星幼体分布.同时,向海林缘较林内和向陆林缘有更长的淹水时间.这种淹水时间的差异扩大了幼体密度差异对红树林内不同潮带幼体丰度的影响,使幼体的丰度在向海林缘、林内和向陆林缘产生3倍的差异.而在紧邻红树林向海林缘没有红树植物的光滩上,幼体密度和淹水时间没有差异,导致金星幼体丰度也没有显著差异.金星幼体的行为对藤壶在向海林缘、林内和向陆林缘的附着及其在垂直方向上的分布模式产生较大影响.同时,生物因子如细菌粘膜和附着藻类对金星幼体在红树林区基底上的附着也产生了显著影响.在水平方向和垂直方向上,金星幼体的丰度、附着行为和附着后的死亡过程对红树林区藤壶分布模式的影响不同.在水平方向上,幼体丰度和包括细菌粘膜及附着藻类等产生的诱导作用对藤壶的附着模式起主要作用;而在垂直方向上,藤壶附着后 1个月内的附着后死亡过程决定了藤壶附着的模式.红树林区的藤壶金星幼体对藤壶附着模式的影响随着种类的不同而不同.藤壶 *Elminius covertus*幼体在水层中的密度高峰在季节上出现在冬季,而藤壶 *Hexaminius popeiana* 的幼体密度高峰则出现在春夏季节.同时,附着密度所依赖的因素也存在差异,*E. covertus*金星幼体的附着密度与其丰度成正相关,而 *H. popeiana*幼体则没有这种相关性[34].

4 藤壶对红树林和人工红树林幼林的影响

在 20世纪 80年代初,林鹏等[23]研究了藤壶对天然红树林的影响.生长于潮间带的红树植物茎干、小枝、叶片、支柱根和气生根都能成为在潮间带生长的藤壶附着的基底[35,39,47].因此,只要影响藤壶生长的条件合适,藤壶便大量附着于红树植株的各个部位上.藤壶的大量附着除了增加植株地上部分的重量和潮水对植株冲击的受力面积,增大了潮水对红树植物正常生长的干扰外,还通过藤壶在叶片上的附着堵塞叶片上的气孔和减少叶片的光合面积,进而影响植株的正常生长.在藤壶附着特别严重的地方,重重叠叠的藤壶附着使得整个群落中约 1/3的植株死亡,并有约 1/3的植株处于濒死状态[23].

近年来,红树林恢复性造林引起了各国、各种组织的广泛关注和积极投入.在红树林人工营造过程中,藤壶附着对红树林幼林植株的危害也日益得到重视[5,15,22,30].我国在红树林造林实践中,研究者先后在海南东寨港红树林自然保护区、广东湛江红树林自然保护区和深圳福田红树林自然保护区就藤壶对人工红树林的危害进行了较为系统的研究,发现在幼苗茎干和枝叶上附着的藤壶主要有网纹藤壶 (*Balanus reticulatus*)和中华小藤壶 (*Chthamalus sinensis*)两种,并进一步证实藤壶是严重地影响人工红树林幼林正常生长发育的关键胁迫因子之一,是危害红树林面积最大、程度最高的污损生物.幼林单一植株茎叶上的附着藤壶一般为数个至数百个不等,有的多达四层[22].何斌源等[15,27]也发现,在较低潮位生长的幼苗上附着的潮间藤壶 (*Balanus littoralis*)和白条小藤壶 (*Chthamalus withersi*)是严重影响红海榄、桐花树等幼苗正常生长发育的关键胁迫因子.大量藤壶固着于幼苗的茎叶上,造成幼苗呼吸作用和光合作用受阻,生长缓慢,过厚、过重的藤壶负载甚至造成幼苗折断而死亡[15,27].笔者在红树林生态恢复实践过程中观察到秋茄幼苗的生长也受到大量藤壶附着的影响,在藤壶危害严重的地方,植株倒伏率达到 30%以上.通常认为藤壶在红树幼苗植株茎干和叶片上的附着会危害红树植物的正常生长,但 Satumanatpan等[39]对去除附着在白骨壤幼苗茎干和叶片上的藤壶的对比实验发现,藤壶在植株上的附着对幼苗的存活和生长没有显著影响,而是其他胁迫因子如藻类和海草、层积物的堵塞和哺乳动物的破坏、气候条件等严重地影响了幼苗的生长和存活.

红树幼苗的正常生长和存活率受到多种物理因子和生物因子的胁迫和制约.物理因子包括海水盐度、滩面潮位、温度、沉积物土质、潮流速度、淤泥在植株上的沉积和光照强度等,均能导致幼苗生长不良甚至死亡[4,24,26,45];生物因子如大型藻类和海草、哺乳动物和螃蟹等的干扰和啃食也会导致幼苗存活率降低[30].在上述因子中,海水盐度、滩面潮位和潮流速度等对红树植物生长有负影响的物理因子,同时对藤壶在红树林植株上的附着和藤壶的危害程度有正相关性.这种多因子的交互作用使得有关藤壶危害红树幼苗生长机制和危害权重的分析和研究较为困难和混乱.由此,清楚认识和精确评估藤壶危害方式和危害权重的实验设计和分析方法很有意义.目前,在天然的红树林中,红树植物对藤壶附着的响应模式和适应策略的研究尚未见报道.对此的深入研究,将为开展藤壶防治的研究提供更多的有用信息.

5 人工红树林幼林藤壶的防除措施

在红树林恢复性造林实践中,对幼林中过量附着的藤壶采取防治措施十分必要.韩维栋等[13]研究发现,采用多种农药混合的防治效果较佳,藤壶的死亡率略大于 50%.李云等[22]采用掺有农药的油漆对人工红树林幼苗茎干进行涂层,对藤壶的清除达 100%.其防治机理可能是:油漆紧紧地粘在藤壶上,使油漆中的农药不易流失,当潮水上涨,藤壶张开盖板摄取食物时易受药害;或者由于油漆的附着,藤壶难以打开盖板而饥饿致死.

采用化学药物对红树林幼林附着藤壶的防除措施,主要是由红树林幼林所处的特殊的潮间带环境和藤壶的生物学特性决定的.在红树林幼林上喷施的药物容易在每天周期性的海水浸淹过程中流失,降低了药物的局部浓度,从而减少了药物对藤壶的杀伤效果.同时,药物随着海水在海区的扩散,会造成整个海区的一定污染,特别是在城市边缘造林,较大剂量的药物可能有更大的潜在风险.藤壶的生物学特性也决定了对藤壶防除用药的困难,由于藤壶在退潮后盖板总是处于关闭状态,特别是在外界刺激的情况下,因此,化学药剂往往很难进入藤壶体内,使得对一般农业害虫有效的农药难以发挥作用,并会污染环境.同时,藤壶死亡后,其底板和壳板往往不会在短时间内脱落,并伴随有大量淤泥沉积和其他生物如黑荞麦蛤 (*Vignadula atrata*)的侵填,减小药物防治的效果.因此,在恢复性造林实践中,建立一套经济、安全高效、系统的藤壶防治措施十分必要.

6 展　望

近年来,随着红树林恢复性造林面积的扩大,藤壶对红树林幼林的影响问题日益突出,并引起广泛的关注.为提高红树存活率和确保幼苗的正常生长,建立一套经济、安全高效的藤壶防治措施是当前造林实践中所急需的.对藤壶胶粘蛋白的氨基酸组成、一维结构,胶粘蛋白在水下的交联、组装和胶粘的过程与机制,多种环境因素和生物因素对藤壶在红树林中的附着和分布模式的影响等进行深入研究,必将为开发一套经济、安全高效的藤壶防治技术提供重要的理论依据.针对多种环境因子对红树植物幼苗生长和藤壶危害的影响，开展多因子的交互作用实验设计和数据分析，以清楚地认识和精确评估藤壶危害方式和危害权重，是开发藤壶防治措施的重要前提. 同时，积极开展天然红树林中红树植物对藤壶附着的响应和长期适应机制的研究，不仅具有重要的理论意义，还将为藤壶的防治提供更多的启示.

参考文献

1 Berglin M, Gatenholm P. 2003. The barnacle adhesive plaque: Morphological and chemical differences as a response to substrate properties. *Colloid Surf B-Biointerfaces*, **28**: 107～117

2 Bertness MD, Gaines SD, Wahle RA. 1996. Wind-driven settlement patterns in the acorn barnacle *Semibalanus balanodies*. *Mar Ecol Prog Ser*, **137**: 103～110

3 Bhatnagar KM, Crisp DJ. 1965. The salinity tolerance of nauplius larvae of cirripedes. *J Anim Ecol*, **34**: 412～428

4 Chen LZ, Wang WQ, Lin P. 2004. Influence of water logging time on the growth of *Kandelia candel* seedlings. *Acta Oceanol Sin*, **23**: 149～158

5 Clarke PJ, Myerscough PJ. 1993. The intertidal distribution of the grey mangrove *Avicennia marina* in southeastern Australia: The effects of physical conditions, interspecific competition, and predation on propagule establishment and survival. *Aust J Ecol*, **18**: 307～315

6 Connell JH. 1985. The consequences of variation in initial settlement vs post-settlement mortality in rocky intertidal communities. *J Exp Mar Biol Ecol*, **93**: 11～46

7 Cowden C, Young CM, Chia FS. 1984. Differential predation on marine invertebrate larvae by two benthic predators. *Mar Ecol Prog Ser*, **14**: 145～149

8 Crisp DJ, Ritz DA. 1973. Responses of cirripede larvae to light I. Experiments with white light. *Mar Biol*, **23**: 327～335

9 Denley EJ, Underwood AJ. 1979. Experiments on factors influencing settlement, survival and growth in two species of barnacles in New South Wales. *J Exp Mar Biol Ecol*, **36**: 269～294

10 Dineen JF, Hines AH. 1994. Larval settlement of the polyhaline barnacle *Balanus eburneus* (Gould): Cue interactions and comparisons with two estuarine congeners. *J Exp Mar Biol Ecol*, **179**: 223～234

11 Geraci S, Romairone V. 1982. Barnacle larvae and their settlement in Genoa harbour (North Tyrrhenian Sea). *Mar Ecol*, **3**: 225～232

12 Grosberg PK. 1982. Intertidal zonation of barnacles: The influence of planktonic zonation of larvae on vertical distribution of adults. *Ecology*, **63**: 894～899

13 Hang W-D (韩维栋), Chen L (陈 亮), Yuan M-J (袁梦婕). 2004. The barnacle control on the planted young mangle trees. *J Fujian For Sci Technol* (福建林业科技), **31** (1): 57～62 (in Chinese)

14 He B-Y (何滨源), Cai Y-H (赖廷和). 2001. Study on the distribution characteristic of *Euraphia withersi* attached to the stems of different-aged *Aigiceras corniculatum*. *Mar Sci Bull* (海洋通报), **20** (1): 40～45 (in Chinese)

15 He B-Y (何滨源), Mo Z-C (莫竹承). 1995. Study on the growth and damage factors during the afforestation with atificial seedlings of *Rhizophora stylosa* in a bare tidal flat in Guangxi. *J Guangxi Acad Sci* (广西科学院学报), **11** (3): 37～42 (in Chinese)

16 Huang Y (黄 英), Ke C-H (柯才焕), Zhou S-Q (周时强). 2001. Advancements in research on settlement of barnacles larvae. *Mar Sci* (海洋科学), **25** (3): 30～32 (in Chinese)

17 Huang Z-G (黄宗国), Cai R-X (蔡如星). 1984. Marine Biofouling and Its Prevention. Beijing: Ocean Press (in Chinese)

18 Kamino K. 2001. Novel barnacle underwater adhesive protein is a charged amino acid-rich protein constituted by a Cys-rich repetitive sequence. *Biochem J*, **356**: 503～507

19 Kamino K, Inoue K, Maruyama T, *et al*. 2000. Barnacle cement proteins: Importance of disulfide bounds in their insolubility. *J Biol Chem*, **275**: 27360～27365

20 Keough MJ, Raimondi PT. 1995. Responses of settling invertebrate larvae to bioorganic films: Effects of different types of films. *J Exp Mar Biol Ecol*, **185**: 235～253

21 Li MS, Lee SY. 1997. Mangroves of China: A brief review. *For Ecol Manage*, **96**: 241～259

22 Li Y (李 云), Zheng D-Z (郑德璋), Zheng S-F (郑松发), *et al*. 1998. Barnacles harm to artificial mangroves and their chemical control. *For Res* (林业科学研究), **11** (4): 370～376 (in Chinese)

23 Lin P (林 鹏). 1997. Mangrove Ecosystem in China. Beijing: Science Press. 1～10 (in Chinese)

24 Lin P (林 鹏). 2001. A review on the mangrove research in China. *J Xiamen Univ* (Nat Sci) (厦门大学学报 · 自然科学版), **40** (2): 592～603 (in Chinese)

25 Lin P (林 鹏), Wei X-M (韦信敏). 1981. The ecological studies of the subtropical mangroves in Fujian, China. *Acta Phytoecol Sin* (植物生态学与地植物学丛刊), **5** (3): 177～186 (in Chinese)

26 Macnae W. 1966. Mangrove in eastern and southern Australia. *Aust J Bot*, **14**: 67～104

27 Mo Z-C (莫竹承), Huang H-Q (范航清), He B-Y (何滨源). 2003. Distributional characters of barnacles on artificial *Rhizophora stylosa* seedlings. *J Trop Oceanol* (热带海洋学报), **22** (1): 50～54 (in Chinese)

28 Naldrett MJ. 1993. The importance of sulphur cross-links and hydrophobic interactions in the polymerization of barnacle cement. *J Mar Biol Ass UK*, **73**: 689～702

29 Olson RR, Olson MH. 1989. Food limitation of planktotrophic marine larvae: Does it control recruitment success? *Annu Rev Ecol Syst*, **20**: 225～247

30 Perry MD. 1988. Effects of associated fauna on growth and productivity in the red mangrove. *Ecology*, **69**: 1064～1075

31 Pinda J. 1991. Predictable upwelling and the shoreward transport of planktonic larvae by internal bores. *Science*, **253**: 548～551

32 Raimondi PT. 1988. Settlement cues and determination of the vertical limit of an intertidal barnacle. *Ecology*, **69**: 400～407

33 Raimondi PT. 1991. Settlement behaviour of *Chthamalus anisopoma* larvae largely determines the adult distribution. *Oecologia*, **85**: 349～360

34 Ross PM. 2001. Larval supply, settlement and survival of barnacles in a temperate mangrove forest. *Mar Ecol Prog Ser*, **215**: 237～249

35 Ross PM, Underwood AJ. 1997. The distribution and abundance of barnacles in a mangrove forest. *Aust J Ecol*, **22**: 37～47

36 Saroyan JR, Linder E, Dooley CA. 1970. Repair and reattachment in the balanidae as related to their cementing mechanism. *Biol Bull*, **139**: 333～350

37 Satumanatpan S, Keough MJ. 1999. Effect of barnacles on the survival and growth of temperate mangrove seedlings. *Mar Ecol Prog Ser*, **181**: 189～199

38 Satumanatpan S, Keough MJ. 2001. Roles of larval supply and behavior in determining settlement of barnacles in temperate mangrove forest. *J Exp Mar Biol Ecol*, **260**: 133～153

39 Satumanatpan S, Keough MJ, Watson GF. 1999. Role of settlement in determining the distribution and abundance of barnacles in a temperate mangrove forest. *J Exp Mar Biol Ecol*, **241**: 45～66

40 Wethey DS. 1986. Ranking of settlement cues by barnacle larvae: influence of surface contour. *Bull Mar Sci*, **39**: 393～400

41 Woodin SA. 1986. Settlement of infauna: Larval choice? *Bull Mar Sci*, **39**: 401～407

42 Wright J, Boxshall AJ. 1999. The influence of small-scale flow and chemical cues on the settlement of two congeneric barnacle species. *Mar Ecol Prog Ser*, **183**: 179～187

43 Yan W-X (严文侠), Dong Y (董 钰), Yi F (尹 芬). 1983. Comparison between the primary and secondary cements of *Balanus reticulatus* utnomi. *Trop Oceanol* (热带海洋学报), **2** (3): 231～239 (in Chinese)

44 Yoshioka PM. 1982. Role of planktonic and benthic factors in the population dynamics of the bryozoan *Membranipora membranacea*. *Ecology*, **63**: 457～468

45 Zhang Q-M (张乔民), Sui S-Z (隋淑珍), Zhang Y-C (张叶春), *et al*. 2001. Marine environment indexes related to mangrove growth. *Acta Ecol Sin* (生态学报), **21** (9): 1427～1437 (in Chinese)

46 Zheng Z (郑 重). 1993. Ecological study on attachment and metamorphosis of marine planktonic larvae. *Chin J Ecol* (生态学杂志), **12** (3): 36～38 (in Chinese)

47 Zhou S-Q (周时强), Li F-X (李复雪), Hong R-F (洪荣发). 1993. Ecological studies on mangrove fouling animals in Jiulong River Estuary, Fujian. *J Oceanol Taiwan Strait* (台湾海峡), **12** (4): 335～341 (in Chinese)

厦门地区秋茄幼苗生长的宜林临界线探讨*

陈鹭真[1,2] 杨志伟[1,2] 王文卿[1,2] 林 鹏[1,2]

(1. 厦门大学生命科学学院,厦门 361005;2. 厦门大学湿地与生态工程研究中心,厦门 361005)

【摘要】 2003年5月在厦门大屿岛白鹭自然保护区西面滩涂上试种秋茄幼苗,研究秋茄的宜林临界线.结果表明,滩涂高程为黄零0.99 m处,每个潮水周期的平均淹水时间高达8 h,幼苗成活率低于50%,生长缓慢,不适合用秋茄造林;在滩涂高程为黄零1.62 m处,秋茄幼苗成活率达90%,生物量积累最大,光合同化作用较高,生长良好,为厦门沿海秋茄的最适生长区;而在高程为黄零1.31 m处,秋茄幼苗仍能正常生长.故厦门地区秋茄造林的宜林临界线应不低于黄零1.31 m(即厦零4.55 m),平均每个潮水周期淹水不高于5.6 h.

关键词 红树林 秋茄 造林 滩面高程 宜林临界线

文章编号 1001-9332(2006)02-0177-05 **中图分类号** S728.6 **文献标识码** A

Critical tidal level for planting Kandelia candel **seedlings in Xiamen.** CHEN Luzhen[1,2], YANG Zhiwei[1,2], WANG Wenqin[1,2], LIN Peng[1,2] ([1] *School of Life Sciences, Xiamen University, Xiamen 361005, China;* [2] *Research Centre for Wetlands and Ecological Engineering, Xiamen University, Xiamen 361005, China*). -*Chin. J. Appl. Ecol.*, 2006, **17**(2):177~181.

Plantable tidal flat is one of the most important factors affecting the survival rate of mangroves seedlings in forestation. In this paper, an experiment was conducted in the tidal zones of Umbrette Natural Reserve in the Dayu Island of Xiamen in May 2003 to investigate the critical tidal level for *Kandelia candel* forestation. The results showed that the tidal level of 0.99 m above the zero tidal level of the Huang Ocean was not suitable for planting *K. candel* seedlings, because the waterlogging time at this tidal level was longer than 8 h per-tide-cycle, and the survival rate was lower than 50%. At 1.62 m above the zero tidal level of Huang Ocean, *K. candel* seedlings had the best growth and the highest photosynthetic assimilation, with a survival rate of 90%. At 1.31 m above the zero tidal level of Huang Ocean, *K. candel* seedlings could still grow well. It could be concluded that the tidal level of 1.62 m was optimal for planting *K. candel* seedlings, and the critical tidal level of *K. candel* seedlings in the coastal areas of Xiamen was not lower than 1.31 m above the zero tidal level of Huang Ocean, where the waterlogging time was not longer than 5.6 h per-tide-cycle.

Key words Mangroves, *Kandelia candel*, Forestation, Tidal flat, Critical tidal level for planting.

1 引 言

秋茄(*Kandelia candel*)是我国境内天然分布最广且分布纬度最高的红树植物,也是我国东南沿海的主要红树林造林树种[7].以往对秋茄造林做了较多的研究探讨[6,8,11,12,18].由于红树林的生境受到破坏[5],以及全球海平面上升[4,15,17],导致了红树林宜林滩涂的高程下降.滩面高程和宜林水位线的确定在红树林造林工作中倍受关注[3,10~11,21].广西属于全日潮区,莫竹承等[12]根据广西沿海群众对潮水的计算方法给出了每流水中的小半眼水、半眼水和一眼水的低潮期最高水位线作为红树林的宜林临界水位线[12];深圳赤湾的不规则半日潮区,秋茄造林的潮滩基面高程应该大于130 cm,即不低于当地平均海面以下22 cm[6];海南东寨港也属于不正规半日潮区,当地秋茄宜林滩涂的潮汐基面高程应高于105 cm,不低于当地海平面25 cm[6].然而福建沿海特别是厦门地区属于正规半日潮区,也是强潮差海区,适应于福建红树林造林的宜林水位线还未见报道.

Chen等[1,2]用模拟潮汐系统对秋茄幼苗在没顶海水浸淹情况下的生长和生理反应进行研究,把潮间带按淹水时间不同平均划分为7个淹水梯度,认为适合秋茄幼苗生长的临界淹水时间为每个潮水周期淹水8 h.国内外学者在秋茄等红树植物淹水胁迫反应方面也做了不少工作,其中包括生长、生理指标、氧气运送和根系通气组织等方面[1,4,13,14,16,19].本研究在

* 国家自然科学基金项目(30200031)、教育部博士点基金项目(20030384007)和厦门市科技基金资助项目(3502Z20021046)

原载于应用生态学报,2006,17(2):177-181

潮汐淹水对秋茄幼苗生长和生理特性[1,2]的影响基础上，通过野外试种实验，给出适合厦门地区潮汐特点的秋茄造林临界淹水时间和确切的宜林临界线，为秋茄造林宜林临界线的确定提供科学依据。

2 研究地区与研究方法

2.1 自然概况

厦门大屿岛白鹭自然保护区位于厦门岛西部、九龙江出海口，地理坐标 24°27′30″～24°27′57″N，118°02′32″～118°02′51″E，总面积 1.8 km²。大屿岛距厦门岛 1.9 km、鼓浪屿 1.1 km、距嵩屿仅 300 m。全岛海岸线长 2.3 km。岛屿东南面陡峭，西面有山坳，退潮时此处露出一片滩涂，是本实验营造秋茄林的滩涂。厦门港属正规半日潮港，海水盐度高、潮差大。按厦门理论基面计算，多年平均高潮位 5.49 m、平均低潮位 1.50 m，平均潮差 3.98 m，最大潮差 6.92 m，最小潮差 0.99 m。厦门年均气温 20.7 ℃。黄海零点海面（黄零）相当于厦门潮位 3.24 m，即厦零 3.24 m。

2.2 研究方法

2.2.1 样地设置 2003 年 5 月开始，在大屿岛西面滩涂上，由高潮线向低潮线延伸，选择 Ⅰ～Ⅴ 5 个样带种植秋茄胚轴。每个样带设有 0.8 m ×1.4 m 3 个重复的平行样地，每个样地种 8 ×5 个秋茄胚轴，株行距 20 cm ×20 cm。2004 年 5 月补种样带 Ⅵ，种植胚轴数与前 5 个样带相同。每个样带均平行于高潮水位线，6 个样带在滩涂上的分布位置如图 1 所示。样地面积、种植情况、盐度与基质情况如表 1 所示。

同时，每次采样过程中，采集各样方内的土壤以及间隙水带回实验室用于测定。间隙水测定其盐度。土壤样品测定容重（环刀法）、pH 值（酸度计法）、盐度（电导法）、有机质含量（含盐量较高，用 375 ℃干烧法）以及土壤质地（我国土壤质地分类法）等指标[9]。土壤背景值如表 1 所示。

2.2.2 样品采集和测定 种植方法均采用胚轴直接插植，为预防潮水涨落过程中，潮汐力的冲击，胚轴的 1/2 插入土中。幼苗生长过程中，前 3 个月每月定期测定每个样方幼苗的成活数量，每株株高，并在每个样方随机取 2 株带回实验室测量生长指标。以后每季度测定一次，为期 1 年。样带 Ⅵ目前已经测定 3 个月。测定的生长指标包括生物量、基茎直径、全株叶片面积以及叶片的肉质化程度(多汁度)[1]。

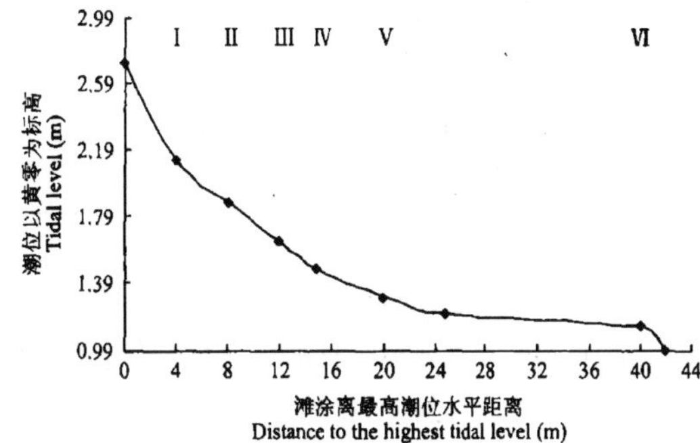

图 1 大屿岛滩涂秋茄各高程样带分布图

Fig. 1 Distribution of sample plots in the intertidal zone of Dayu Island.

2.2.3 秋茄幼苗光合作用测定 在植株生长 5 个月后，用便携式光合作用仪（Model CIRAS-1，UK）测定成熟叶片的光合速率和蒸腾速率，应用自然光源和内置 CO_2 提供 CO_2。水分利用效率（WUE）＝光合速率（*Pn*）/蒸腾速率（*Tr*）。

3 结果与分析

3.1 秋茄幼苗成活率

由表 2 可见，3 个月后，幼苗已经完全萌发（萌芽、萌根），统计其成活率。Ⅲ、Ⅳ、Ⅴ 3 个样地的幼苗成活率较为一致，而且成活率均高达 90 %。Ⅰ与Ⅱ两个样地属于砂壤土，胚轴不易固定，因此存活率低于 Ⅲ、Ⅳ、Ⅴ 3 个样带；Ⅰ与 Ⅱ两个样地原处于高潮带，而砂壤土的土壤质地严重影响了秋茄胚轴在滩涂上的固定和萌发，影响了其成活率。而样地 Ⅵ由于滩面高程最低，潮水浸淹时间最长，土壤受海水浸淹时间过长，含水量过高，胚轴不仅不易固定，而且

表 1 大屿岛滩涂各秋茄样方的土壤背景值

Table 1 Soil characters in different sample plots in the intertidal zone of Dayu Island (2003.5)

土壤背景值 Soil characters	样带 Sample plots					
	Ⅰ	Ⅱ	Ⅲ	Ⅳ	Ⅴ	Ⅵ
幼苗数量(苗) Seedling number	120	120	120	120	120	120
样地离岸距离 Distance to the highest tidal level (m)	4	8	12	15	20	42
滩面高程(黄零) Tidal level, upper the zero tidal level of Huang Ocean (m)	2.11	1.86	1.62	1.45	1.31	0.99
每个潮水周期平均浸淹时间 Waterlogging time per-tide-cycle (h)	2.7	3.9	5.0	5.6	6.5	8
水深 Depth of waterlogging (cm)	57	82	106	123	137	169
土壤间隙水盐度 Salinity of soil interstitial water (‰)	31	31	31	31	31	31
土壤 pH Soil pH	7.30 ±0.04	7.06 ±0.08	7.01 ±0.09	7.06 ±0.03	7.16 ±0.05	7.17 ±0.06
土壤盐度 Soil salinity (‰)	10.8 ±1.6	19.8 ±3.2	30.2 ±2.3	32.6 ±1.8	35.1 ±1.4	29.5 ±0.4
土壤容重 Soil bulk density($g\ cm^{-3}$)	1.66 ±0.05	1.52 ±0.07	1.21 ±0.05	1.09 ±0.05	1.13 ±0.06	1.11 ±0.05
土壤有机质 Soil organic matter (%)	1.51 ±0.37	2.57 ±0.43	3.81 ±0.29	3.79 ±0.30	3.61 ±0.15	3.24 ±0.08
土壤质地 Soil texture	砂壤1)	砂壤	粉粘土2)	粉粘土	粉粘土	粉粘土

土壤层次采样于 0～20 cm Soil sampled on the depth between 0 and 20 cm. 1) Sandy loam; 2) Silt clay.

表 2　大屿岛滩涂各样方秋茄幼苗的成活率和树高

Table 2 Survival rate and plant height of Kandelia candel seedlings in the tidal zone of Dayu Island

样带 Sample plot	成活率 Survival rate (%)			平均株高(包括原胚轴部分) Height of seedlings including the hypocotyls (cm)					
	3 个月 3 months	6 个月 6 months	12 个月 12 months	1 个月 1 month	2 个月 2 months	3 个月 3 months	6 个月 6 months	9 个月 9 months	12 个月 12 months
Ⅰ	70.0 ±4.3	72.5 ±8.7*	70.8 ±5.8	11.8 ±0.4	17.6 ±0.7	20.2 ±0.1	27.5 ±0.2	29.5 ±1.0	36.1 ±1.7
Ⅱ	67.5 ±4.3	67.5 ±4.3	65.0 ±2.5	11.9 ±0.1	19.2 ±1.1	23.2 ±1.5	31.6 ±1.1	33.7 ±2.6	40.4 ±4.2
Ⅲ	90.0 ±2.5	88.3 ±5.2	86.7 ±6.3	12.4 ±0.1	20.0 ±0.4	25.2 ±0.4	37.4 ±1.3	40.6 ±1.2	50.5 ±2.7
Ⅳ	90.0 ±5.0	90.0 ±5.0	85.8 ±5.2	12.2 ±0.4	20.8 ±0.7	25.9 ±1.2	38.4 ±1.9	42.8 ±1.2	53.6 ±2.8
Ⅴ	90.0 ±9.0	88.3 ±6.3	86.7 ±7.2	12.3 ±0.3	20.4 ±0.2	24.8 ±1.1	35.6 ±1.9	39.3 ±3.1	50.3 ±3.1
Ⅵ	43.3 ±3.8	-	-	0	13.1 ±0.3	20.2 ±0.1	-	-	-

*3 个月时统计为枯死的幼苗,但胚轴还活着,在 6 个月后长出小叶 Seedlings is regarded dead in 3 months, but leaves grow after 6 months.

受海水浸淹时间长,胚轴萌发缓慢,甚至不萌发,在胚轴种植 1 个月后,基本没有萌芽,少数萌根;3 个月后统计其存活率仅 43.3 % ±3.8 %,低于 50 %. 样地 Ⅵ的高程比样地 Ⅴ仅低了 0.32 m,而秋茄幼苗的成活率却显著下降了 51.9 %($P<0.001$). 而滩面高程相差 0.31 cm 的 Ⅲ和 Ⅴ两个样带的秋茄幼苗成活率却无显著的变化($P>0.05$). 可见,样地 Ⅵ的滩面高程已经低于宜林临界线,不适合在此高程造林. 可以将宜林临界线定义为红树植物幼苗生长 6~12 月内,成活率达 65 %以上,株高达 35 cm 以上的最低滩涂高程.

3.2　秋茄幼苗平均株高

如表 2 所示,秋茄幼苗种植后 1 个月、2 个月、3 个月、6 个月、9 个月和 1 年时间内,对样地上所有秋茄幼苗地面株高(包括原胚轴的地上部分)的测定和统计结果表明,Ⅲ、Ⅳ、Ⅴ 3 个样带的幼苗树高增长趋势基本一致. 一年后,样带 Ⅳ的幼苗树高最高,分别比样地 Ⅲ和 Ⅴ增高了 106.4 %和 106.2 % (* $P=0.023$, * * $P=0.001$) ,而样地 Ⅰ和 Ⅵ的幼苗平均树高较低.

3.3　秋茄幼苗生物量

生物量反映了各样地上幼苗生长的物质积累情况. 如图 2 和表 3 所示,总生物量随时间和滩面高程的变化而发生极显著地变化($P<0.001$),除了第一个月各高程样地上的总生物量没有显著的差异外,其它各月均有显著变化,即 Ⅲ、Ⅳ、Ⅴ 3 个样地的总生物量较高,样地 Ⅵ的总生物量最低,随后是样地 Ⅰ 和样地 Ⅱ. 茎生物量的变化趋势与总生物量和树高基本一致 (图 3,表 3). 秋茄胚轴种植一年后,样地 Ⅲ的总生物量、茎生物量和叶片生物量均显著高于样地 Ⅳ、Ⅴ($P<0.05$). 根系生物量(图 4,表 3) 的变化不显著($P>0.05$).

3.4　秋茄幼苗的茎叶特征

秋茄幼苗的基茎直径变化趋势与茎生物量和树高的变化较为一致(图 3,表 3),前 6 个月,各样带植株的基茎直径变化不大,但种植胚轴 9 个月后,Ⅲ、Ⅳ、Ⅴ 3 个样带的基茎直径比其它样带的基茎直径有显著提高($P<0.05$). Ⅲ、Ⅳ、Ⅴ3 个样带的植株不仅茎生物量和树高增加,同时基茎加粗,以此来提高植株抵抗潮汐力冲击的能力,从而更有效地提高幼苗的成活率. 在胚轴种植 2 个月后,各样带幼苗全株叶片面积无显著变化($P>0.05$);胚轴种植 3 个月后,其变化趋势与叶片生物量的变化较为一致,均表现为 Ⅲ、Ⅳ、Ⅴ 3 个样带的叶片面积较大. 叶片多汁度随样带的变化趋势不显著.

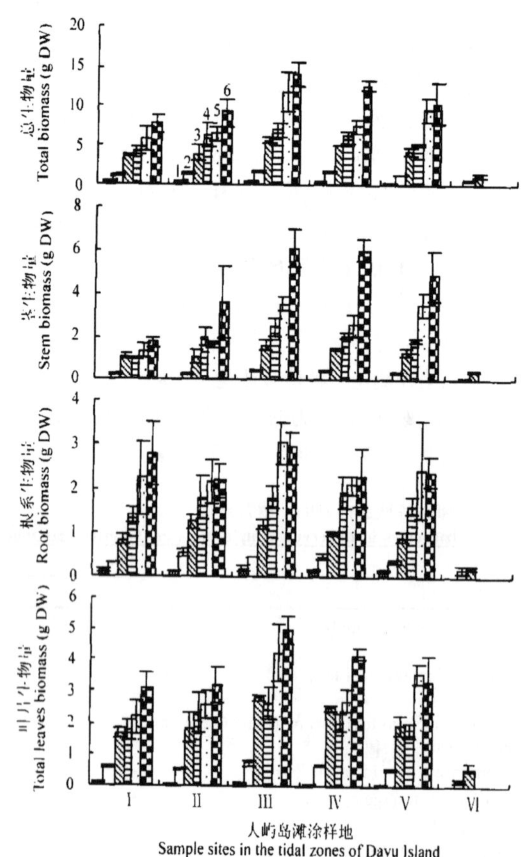

图 2　大屿岛滩涂秋茄幼苗单株生物量变化

Fig. 2 Individual plant biomass of *Kandelia candel* seedlings in the tidal zone of Dayu Island.

1) 1 个月 1 month; 2) 2 个月 2 months; 3) 3 个月 3 months; 4) 6 个月 6 months; 5) 9 个月 9 months; 6) 12 个月 12 months. 下同 The same below.

表3 采样时间和滩面高程对大屿岛滩涂秋茄幼苗影响的显著程度(F值)

Table 3 Results of ANOVA (F-values) for the effects of different cultivation time and tidal flat on growth of Kandelia candel **seedlings in Dayu Island**

参数 Parameter	II-way ANOVA			幼苗在不同滩面高程(各样带)的变化的显著性(I-way ANOVA) I-way ANOVA for the growth of *Kandelia candel* seedlings in different tidal flats					
	采样时间 Month	滩面高程 Elevation	M × E	1个月 1 month	2个月 2 months	3个月 3 months	6个月 6 months	9个月 9 months	12个月 12 months
苗高 Seedling height (cm)	870.213***	71.249***	8.605***	1.932	10.247**	15.454***	29.769***	21.827***	18.642***
总生物量 Total biomass (g)	232.434***	21.154***	4.332***	1.240	29.640***	7.565**	3.973*	7.165**	6.839**
茎生物量 Stem biomass (g)	201.786***	26.349***	8.002***	1.315	11.643**	4.326*	13.043**	22.219***	9.926**
根系生物量 Root biomass (g)	102.670***	3.625**	0.989	0.833	6.619**	7.045**	1.244	0.624	1.089
叶片生物量 Leaf biomass (g)	206.556***	22.385***	3.938***	1.952	6.223*	8.708**	2.455	6.254*	5.390*
茎基径 Diameter of root base (cm)	137.011***	8.622***	2.220**	0.152	0.782	0.899	1.122	5.328*	5.381*
全株叶面积 Leaf area of a seedling (cm)	178.109***	31.029***	4.673***	0.759	2.218	10.301**	2.965	10.062**	9.697**
叶片多汁度 Leave succulence (g·dm⁻²)	78.683***	4.672**	3.903***	0.387	9.241**	3.222	1.454	4.882*	4.092*

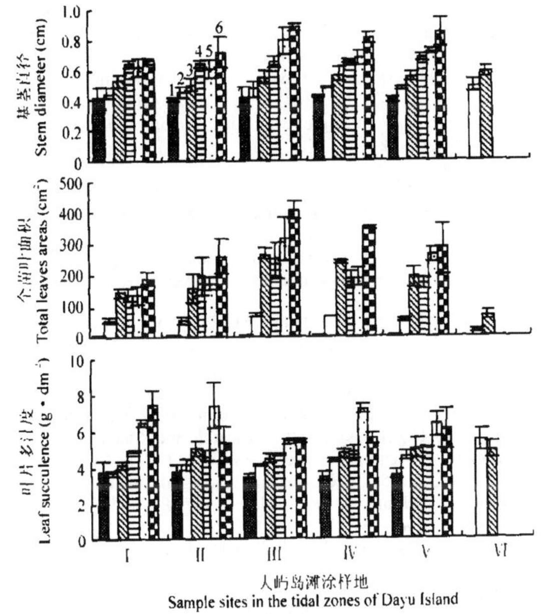

图3 大屿岛滩涂秋茄幼苗茎叶特征

Fig. 3 Stem and leaf characters of *Kandelia candel* seedlings in the tidal zone of Dayu Island.

表4 大屿岛滩涂秋茄幼苗的光合、蒸腾速率和水分利用率

Table 4 Pn, Tr and WUE of Kandelia candel **seedlings in the tidal zone of Dayu Island**

样带 Sample plot	光合速率 Photosynthesis rate (μmol·m^{-2}·s^{-1})	蒸腾速率 Transpiration rate (μmol·m^{-2}·s^{-1})	水分利用效率 Water use efficiency
I	9.72 ±1.01	5.39 ±0.44	1.97 ±0.15
II	10.94 ±1.09	5.93 ±0.10	1.77 ±0.11
III	11.30 ±0.84	5.78 ±0.11	2.08 ±0.11
IV	12.72 ±0.99	5.98 ±0.31	2.17 ±0.08
V	14.11 ±1.36	4.96 ±0.55	3.32 ±0.44

3.5 叶片光合和蒸腾速率及水分利用率

叶片的光合速率和蒸腾速率与叶片面积和叶片生物量有较大关系,特别是叶片光合速率,可以反映植株的光合同化能力和有机物的积累速率.由表4可知,Ⅲ、Ⅳ、Ⅴ3个样带的光合速率随滩面高程下降而有所升高($P<0.001$),特别是Ⅴ样带光合速率达到最大值14.11 ±1.36 μmol·m^{-2}·s^{-1}.各样地之间,蒸腾速率变化不显著($P>0.05$).由于水分利用效率(WUE)是光合速率和蒸腾速率的比值,因此,在样地Ⅰ~Ⅳ之间由于蒸腾速率变化不大,植株叶片的水分利用效率变化不显著($P>0.05$),但样带Ⅴ的水分利用效率比其它4个样带有极显著提高($P<0.001$, $df=29$).

4 讨 论

从造林成活率的结果来看,样地Ⅵ在潮间带上处于中低潮带,土壤条件与中高潮带Ⅲ~Ⅴ3个样地没有较大差异,但仅3个月成活率就低于50%,明显不适合秋茄造林.秋茄幼苗的生长受高程影响,而样地Ⅵ在6个样地中,相对滩面高程最低,淹水时间最长,秋茄幼苗生长不良.由此可以推知,样地Ⅵ的高程(黄零0.99 m)低于厦门地区宜林临界线的高程.

在属于高潮位的样地Ⅰ和Ⅱ(高程分别是黄零2.11 m和1.86 m)上,秋茄幼苗虽然可以生长,但成活率、树高和生物量均较低,生长不如样地Ⅲ~Ⅴ好.这可能与当地的土壤条件有关,从土壤质地、有机质含量和盐度等条件看(表1),贫瘠和坚硬的砂壤土不适合秋茄胚轴的固定、伸根和营养需求.因此,在厦门沿海的红树林造林过程中,高程高于黄零1.86 m的滩涂是有机物含量低的砂壤土,虽然其淹水时间较短,但也不适于秋茄幼苗的造林,应该避开.而Ⅲ~Ⅴ3个样地均处于中高潮带,不管从秋茄成活率、树高或生物量积累,还是从光合速率上来衡量,秋茄幼苗都生长良好,叶面积也较大,能进行正常的光合同化和营养积累,特别是样地Ⅲ(黄零1.62 m)上的秋茄幼苗,在种植1年后,其总生物量、茎生物量和叶片生物量均高于其它样地上的幼苗.

样地 Ⅲ的平均淹水时间为每个潮汐循环 5 h,这是最适合秋茄幼苗生长的淹水时间.这一结果比相关模拟实验[12]得到的秋茄幼苗适应于 2～4 h 的淹水结果高,但仍与秋茄在潮间带的分布较为一致[7].

受各地不同的潮汐特点制约,国内红树林造林宜林临界线的规定不尽相同.总体上认为,红树林主要分布于中潮滩面以上,个别延伸至低潮带,如白骨壤等[7].张乔民等[20]认为,红树林主要分布在平均海平面(或稍上)与回归潮平均高高潮位(或大潮高潮位,或最高天文潮位)之间的潮滩上.在广西的造林过程中,通常认为低潮期的海水最高水位线为低潮线,这条低潮线是潮滩的宜林临界线[12].根据各地秋茄造林的结果,深圳赤湾的秋茄林潮滩基面高程应高于 130 cm,即不低于当地平均海面以下 22 cm[6];而海南东寨港的秋茄宜林滩涂的潮汐基面高程应大于 105 cm,不低于当地平均海面以上 25 cm[6].而本研究的结果表明在厦门大屿岛试种秋茄,当高程达到黄零 0.99 m (样地 Ⅵ) 时,即每个潮水周期秋茄幼苗被海水浸淹长达 8 h 时,幼苗的成活率降低、生物量下降、生长不良,特别是胚轴仅种植 3 个月,其成活率就低于 50 %,说明这个高程不适合秋茄的造林,宜林临界线应高于这个高程,即临界淹水时间应低于 8 h.这一结果比 Chen 等[1]的淹水时间临界值为每个潮水周期浸淹 8 h 的时间短.说明人工模拟潮汐的淹水条件比自然条件下的宜林地淹水时间有所延长.当滩面高程不低于黄零 1.31 m 时 (即样地 Ⅲ～ Ⅴ),秋茄幼苗生长最好,此处每个潮水周期浸淹 5.0～6.5 h,因此认为,在厦门地区秋茄造林过程中,滩面高程应不低于黄零 1.31 m,淹水时间不高于 6.5 h.

结合潮汐模拟实验[1,2]的结果,认为在厦门地区进行秋茄造林最适的生长潮滩是高程为黄零 1.62 m,平均淹水时间约在每个潮水周期浸淹 5 h 的淤泥质潮滩;而宜林临界线高于黄零 1.31 m 时,平均淹水时间约在每个潮水周期浸淹 6.5 h 以内.这条宜林临界线的划定不仅为厦门地区秋茄造林提供科学依据,同时也是福建省红树林造林中第一条宜林临界线,将为福建省红树林造林中宜林地的选择提供借鉴,也有助于提高福建省红树林造林的成活率.

致谢 感谢陈长平博士、王龙和揭祖亮同学在野外调查和样品处理过程中给予帮助,感谢大屿岛白鹭自然保护区的王博和朱开建同志在调查过程中给予帮助.

参考文献

1 Chen L Z, Wang WQ, Lin P. 2005. Photosynthetic and physiological responses of *Kandelia candel* (L.) Druce seedlings to duration of tidal immersion in artificial seawater. *Environ Exp Bot*, **54**:256～266

2 Chen L Z, Wang WQ, Lin P. 2004. Influence of waterlogging time on the growth of *Kandelia candel* seedlings. *Acta Oceanol Sin*, **23** (1):149～158

3 Chen Y-J (陈玉军), Chen W-P (陈文沛), Zheng S-F (郑松发), *et al*. 2001. Researches on the mangrove plantation in Panyu, Guangdong. *Ecol Sci* (生态科学), **20**:25～31 (in Chinese)

4 Ellison AM, Farnsworth EJ. 1997. Simulated sea level change alters anatomy, physiology, growth, and reproduction of red mangrove (*Rhizophora mangle* L.). *Oecologia*, **112**:435～446

5 Fan H-Q (范航清), Li G-Z (黎广钊). 1997. Effect of sea dike on the quantity, community characteristic and restoration of mangroves forest along Guangxi coast. *Chin J Appl Ecol* (应用生态学报), **8**: 240～244 (in Chinese)

6 Liao B-W (廖宝文), Zheng D-Z (郑德璋), Zheng S-F (郑松发), *et al*. 1997. Studies on the cultivation techniques of mangrove *Kandelia candel*. In: Wong YS, Tam NFY, eds. Mangrove Research of Guangdong, China. Guangzhou: South China University of Technology Press. 479～486 (in Chinese)

7 Lin P. 1999. Mangrove Ecosystem in China. Beijing: Science Press.

8 Lu C-Y (卢昌义), Lin P (林 鹏). 1990. Afforesting techniques of *Kandelia candel* mangrove and their ecological principle. *J Xiamen Univ* (Nat Sci) (厦门大学学报·自然科学版), **29** (6):694～698 (in Chinese)

9 Lu R-K (鲁如坤). 2000. Methods of Agriculture and Chemistry Analysis of Soil. Beijing: China Agriculture Science and Technique Press. (in Chinese)

10 McKee, KL. 1996. Growth and physiological respones of neotropical mangroves seedlings to root zone hypoxia. *Tree Physiol*, **16**:883～889

11 Mo Z-C (莫竹承), Fan H-Q (范航清). 2001. Comparison of mangroves forestation methods. *Guangxi For Sci* (广西林业科学), **30** (2):73～75 (in Chinese)

12 Mo Z-C (莫竹承), Liang S-C (梁士楚), Fan H-Q (范航清). 1995. A preliminary study on planting techniques of Guangxi Mangroves. In: Fang HQ, Liang C, eds. Research and Management on China Mangroves. Beijing: Science Press. 164～172 (in Chinese)

13 Naidoo G. 1985. Effects of waterlogging and salinity on plant-water relations and on the accumulation of solutes in three mangrove species. *Aquat Bot*, **22**:133～143

14 Naidoo G, Rogalla H, Von-Willert DJ. 1997. Gas exchange responses of a mangrove species, *Avicennia marina*, to waterlogged and drained conditions. *Hydrobiologia*, **352**:39～47

15 Semeniuk V. 1994. Predicting the effect of sea-level rise on mangroves in northwestern Australia. *J Coast Res*, **10**:1050～1076

16 Skelton NJ, Allaways WG. 1996. Oxygen and pressure changes measured in situ during flooding in roots of the grey mangrove *Avicennia marina* (Forssk.) Vierh. *Aquat Bot*, **54**:165～175

17 Snedaker SC, Meeder JF, Ross MS, *et al*. 1994. Mangrove ecosystem collapse during predicted sea-level rise-Holocene analogues and implications-discussion. *J Coast Res*, **10**:497～498

18 Wang B-S (王伯荪), Liao B-W (廖宝文), Wang Y-J (王勇军), *et al*. 2002. Mangrove Forest Ecosystem and its Sustainable Development in Shenzhen Bay. Beijing: Science Press. 194～211 (in Chinese)

19 Ye Y, Tam NFY, Wong YS, *et al*. 2003. Growth and physiological responses of two mangrove species (*Bruguiera gymnorrhiza* and *Kandelia candel*) to waterlogging. *Environ Exp Bot*, **49**:209～221

20 Zhang Q-M (张乔民), Yu H-B (于红兵), Chen X-S (陈欣树), *et al*. 1997. The relationship between mangrove zone on tidal flats and tidal levels. *Acta Ecol Sin* (生态学报), **17** (3):258～265 (in Chinese)

21 Zheng D-Z (郑德璋), Li M (李 玫), Zheng S-F (郑松发), *et al*. 2003. Headway of study on mangroves recovery and development in China. *Guangdong For Sci Technol* (广东林业科技), **19** (1):10～14 (in Chinese)

厦门海岸红树林的保护与生态恢复*

林 鹏 张宜辉 杨志伟

(厦门大学生命科学学院,湿地与生态工程研究中心,福建 厦门 361005)

摘要: 红树林生态系统在维持海湾河口生态系统的稳定和平衡中起着特殊的作用.由于许多港湾围海造田、围滩(塘)养殖、填滩造陆和码头与道路的建设,厦门海岸红树林面积从60年代初的320 hm^2 降至现有的21 hm^2.红树林的消失严重影响了厦门海湾的生态系统,使得生物多样性和滨海环境质量下降,同时导致外来物种的入侵,红树林的保护和恢复种植已是当前的重要任务.本文在红树林引种及栽培技术研究工作以及厦门东西海域红树林宜林地现状调查的基础上,总结出一套红树林宜林地选择标准,提出厦门红树林造林主要影响因素.并论述厦门海岸红树林的环境保护与生态恢复存在的问题及解决办法.

关键词: 厦门;红树林;保护;生态恢复

中图分类号: S 728.6 **文献标识码**:A **文章编号**:0438-0479(2005)Sup-0001-06

1 红树林的特点及其生态效益

1.1 红树林种类概况

红树林是指热带海岸潮间带的木本植物群落.但是,由于温暖洋流的影响,有的可以分布到亚热带,有的因潮汐影响,在最高潮边缘而具有水陆两栖现象.红树林中生长的木本植物叫做红树植物,一般都没有包括群落内外的草本植物或藤本植物.由于各地学者对红树植物定义的理解不同或地区本身的认识不统一,因而有很多差异.我国现有红树植物12科15属27种(变种).关于红树植物界定标准.可参阅前文[1].

1.2 红树林特点

红树林生态系统处于海洋、陆地和大气的动态交界面,周期性遭受海水浸淹的潮间带环境,使其在结构和功能上既不同于海洋生态系统,也不同于陆地生态系统,作为独特的海陆边缘生态系统在维持海湾河口生态系统的稳定和平衡中起着特殊的作用.由于潮间带生境的高度盐渍化、土壤(沉积物)的缺氧、高光辐射及周期性的海水浸淹,经长期的自然选择和进化适应,红树植物逐渐形成了一套独特的形态及生理生化适应特征[1~6].

1.2.1 根系多样性

1) 支柱根和板状根:适应泥泞的红树植物,从茎基伸出拱形下弯的支柱根或宽厚的板状根,以抗御风浪.如秋茄在外海呈支柱根、在内河口海岸呈板状根;红海榄具有巨大的支柱根;角果木具小支柱根;在陆岸过渡带的海漆则只有板状根.

2) 呼吸根:白骨壤具有匍匐的缆状根,向上伸出地面的指状呼吸根;海桑有粗大的笋状呼吸根;木榄具有膝状呼吸根;木果楝的气根比较特别,呈蛇状呼吸根.

1.2.2 胎生现象

胎生现象是红树植物重要特点之一,它可分显胎生和隐胎生两种[5].红树植物中,红树科的果实成熟时仍留在树上,种子在母树的果实内发芽后,从果中伸出,形成一个下垂的胚轴,为显胎生,如木榄、角果木、秋茄等具有纺锤形或棍棒形的胚轴,长可达15~30 cm.成熟掉落后,胚根插入泥中,即可成苗;若掉入潮水中,由于胚轴有气道,可远漂传播.而白骨壤、桐花树等非红树科红树植物,种子萌芽后,仍留在果皮内,把果皮填满.当果实掉入水中,果皮吸水胀破后,幼苗才伸出果皮,插入泥中,即开始生根固着下来,为隐胎生.

还有一类,如银叶树、水椰等,并非胎生,而是果皮具木栓纤维层,也可浮于水面,远漂传播.

1.2.3 富含单宁

富含单宁是红树植物在化学成分上的显著特征,对红树植物具有重要的生态意义.红树植物树皮和果皮的单宁含量高,形成一个有效的保护层.由于单宁有涩味,避免或减轻了动物对植物活体的直接啃食.同时单宁有抑制微生物活动、杀灭病原菌的效能,增强了红树植物抗病能力和抗海水腐蚀的能力.

1.2.4 特殊的次生木质部结构

* 高等学校博士点基金(20030384002)和厦门市重点基金(3502Z20021046)资助

原载于厦门大学学报(自然科学版),2005,44(Sup.):1-6.

次生木质部结构中，导管直径小，导管分子长，导管分布频率多[6]：红树植物生长在海滨滩涂含盐量高，渗透压大，植物吸水需要更多的负压，导管直径小能提高植物的负压，增加植物的吸水能力。窄导管的输导效率虽低，但抗负压强，不易倒塌，且窄导管单位面积上的数量多，即使有部分导管被气泡堵塞，也不会导致整个输导系统丧失功能，这样可保证水分运输的安全性.

1.3 红树林的效益

大多数生态学家认为红树林有6个主要作用[2]：(1)通过网罗碎屑的方式促进土壤在林，内沉积、防止水土流失，并可抵抗海啸、风暴潮和洪水的冲击，保护堤岸、滨海良田和村庄；(2)过滤陆地迳流和内陆带出的有机物质和净化污染物；(3)为许多海洋动物(包括渔业、水产生物和鸟类)提供栖息、育苗滋生地和觅食的理想生境；(4)大量凋落物是为近海生产力提供有机碎屑的主要生产者；(5)植物本身的生产物，包括木材、薪炭、食物、药材和其它化工原料等；(6)红树林可以做为社会环境教育和旅游的自然和人文景观.

2005年1月在印度新德里举行的海啸专家会议上，科学家认为，在海岸地区种植红树林可以有效减轻海啸的灾难程度.与会科学家认为，红树林可以起到生物"盾牌"的作用，减缓海浪的速度，同时还可以减轻海岸地区遭受飓风、海岸暴风雨袭击的程度[7].这场灾难启示我们，要加强对红树林海岸生态系统的保护力度以及红树林沿海防护林体系的建设，以预防台风、风暴潮和海啸的侵袭，减轻自然灾害的危害程度.

2 厦门红树林的历史变迁

2.1 红树林种类及引种驯化

厦门大学红树林科研组历来就十分重视红树林引种及栽培技术的研究，早在80年代中期就开展了红树林的引种工作，并成功地从海南引种了若干红树林优良品种.红树植物种类从6种(秋茄、木榄、桐花树、白骨壤、老鼠簕、海漆)增加到现在10种左右(增加了海莲、尖瓣海莲、红海榄、无瓣海桑).后又发现厦门老鼠簕和卤蕨，共达12种.

从1997年开始，在海沧东屿村石塘码头东侧滩涂通过人工填土开辟了红树林试验园，进行了较大规模的红树林引种及育苗工作，并在青礁、东屿凤林美、山亭海滩和大屿、小兔屿等岛屿上进行了红树植物的试种工作.这些工作为厦门的红树林恢复种植提供了科学依据和实践经验.

2.2 红树林面积

厦门海岸线长达254 km，滩涂面积广阔，加上又位于九龙江出海口，有淡水补充，适宜于红树林的生长，曾经分布有大面积的红树林.在1960年前后，厦门约有320 hm^2的天然红树林.由于许多港湾围海造田、围滩(塘)养殖、填滩造陆和码头与道路的建设，使得厦门的红树林面积迅速下降.到1979年，厦门天然红树林面积下降为106.7 hm^2；到2000年，厦门红树林面积仅有32.6 hm^2[8]，和1960年相比，90%以上的天然红树林已经消失；近期调查结果表明，到2005年4月，厦门市现有天然红树林面积仅有21 hm^2，加上人工造林达43.4 hm^2(表1).红树林的消失严重影响了厦门海湾的生态系统，使得生物多样性和滨海环境质量下降，据中国红树林生态系统生态价值评估[9]，以2000年市场评估，其生态价值达17.3万元/hm^2 ·a，厦门海岸红树林面积从60年代初320 hm^2降至现有21 hm^2，其年生态价值从5 536万元/a降至363.3万元/a，年损失5 173万元/a，因而就此项而言滨海人民的生活环境质量下降，与此同时导致外来物种(如：互花

表1 厦门市红树林变迁及分布情况

Tab.1 Mangroves variance and their distributions in Xiamen

时间	红树林面积/hm^2	分布情况
1960年前后	320	主要分布于海沧青礁86.7 hm^2，海沧东屿湾80 hm^2，同安湾80 hm^2，马銮湾20 hm^2，杏林湾22.7 hm^2，筼筜港13.3 hm^2，高崎13.3 hm^2，其他小岛合计4 hm^2.
1979年前后	106.7	主要分布于海沧青礁6.7 hm^2，海沧码头3.3 hm^2，海沧东屿湾73.3 hm^2，高崎2 hm^2，杏林湾2 hm^2，同安湾16.7 hm^2，其他小岛合计2.7 hm^2亩.
2000年	32.6	主要分布于海沧青礁0.3 hm^2，海沧码头0.3 hm^2，海沧镇后井村0.1 hm^2，海沧东屿23.3 hm^2，同安鳄鱼岛0.1 hm^2，杏林湾0.4 hm^2，集美凤林7.3 hm^2，高崎0.1 hm^2.此外还有其他小岛0.7 hm^2.
2004年	21 (43.4)*	主要分布于海沧青礁1.3 hm^2，海沧东屿13.3 hm^2，同安鳄鱼岛0.1 hm^2，集美凤林5.4 hm^2，高崎0.2 hm^2，其他小岛0.7 hm^2.此外，近年来厦门海岸带人工种植22.4 hm^2，包括海沧青礁6 hm^2，海沧东屿1.3 hm^2，集美凤林8.4 hm^2，翔安山亭6.7 hm^2.

*天然林与人工林之和.

米草)的入侵,红树林的保护和恢复种植已迫在眉睫.

3 厦门东西海域红树林宜林地现状调查

在2002~2004年期间,厦门大学湿地与生态工程研究中心组织人员,沿厦门市的海岸线调查了海沧区、集美区、同安区和翔安区沿海滩涂.调查过程中,采用GPS定位样点,同时测定样点附近海水的盐度.各样点具体的方位和现况见表2.根据各样点在厦门海域分布的具体位置,我们可以将他们划分到下述3块海域:

(1) 厦门西海域:包括海沧嵩屿码头至海沧大桥、吴冠、马銮湾沿岸、杏林高浦等样点;

(2) 同安湾:包括集美区龙舟池、凤林美;同安区后田、丙洲、浦头;翔安区琼头、鳄鱼屿、刘五店等村镇的样点;

(3) 厦门东部海域和大嶝海域:包括翔安区澳头、蔡厝、莲河、大嶝等村镇的样点.

同一海域内各样点滩涂的状况较为相近,而不同海域之间差别较大,我们从滩涂高程、海水盐度和周边环境等方面详述如下.

3.1 厦门西海域

(1) 海沧区嵩屿码头至海沧大桥

海沧滨海大道位于厦门西海域西部的广阔浅滩上.滩涂外侧,从南到东北依次有大屿岛、白兔屿、小兔屿、大兔屿和火烧屿等.滩地的西北侧有京口围堤、东屿围堤及水头围堤,堤内有大面积的鱼塘和虾池.土壤属粉壤土、粘壤土类型,均属壤土.高潮时海水盐度处于27.9‰~29.7‰之间,平均为28.8‰,土壤盐度12‰~23‰.滨海大道外侧滩涂适当填土整地后可以恢复营造红树林.

(2) 海沧区吴冠

位于马銮湾口南侧,其土壤、水文情况和邻近的马銮湾以及高浦相近.从东方高尔夫球场外侧到吴冠,沿岸大部分滩涂中,养殖池潮位较低,经整治和填土工程后,大多数地段适宜恢复种植红树林.

(3) 马銮湾生态重构示范区

马銮湾位于厦门岛西海域西岸中部,原为天然海湾,由于湾口兴建海堤,隔断了马銮湾水域和厦门西海域水体的自然连通.周边汇入溪流有9条,现存6条.湖水变淡,但污染较严重.土壤类型基本为粘土或砂壤.马銮湾综合整治工程为西海域综合整治工程之一,根据"两湾三区"规划以及"厦门市马銮湾湿地生态重构示范区"规划,在示范区东北面计划建设红树林湿地公园区.

(4) 集美区高浦

附近海域有来自西面的马銮湾和北面的杏林湾水库的淡水补充,海水盐度为20‰~22‰.距该样点数公里处有一片白骨壤林分布,但已经被围到海堤公路内侧.由此可以看出该处适合红树植物先锋树种的生长,但该适合红树林生长的中高潮带滩涂已围海造公路,现存的海岸前沿滩涂绝大部分处于中低潮带,高程相对较低,需采用回填淤泥抬高滩涂高程后方能顺利种植红树植物,马銮湾开口后,潮汐水流急,宜注意防冲刷措施等.此外,该处有两个沙场,对滩涂土质有一定的影响.

3.2 同安湾

该海域的淡水补充主要来源于同安西溪,河口附近的海水盐度低(12‰~16‰),较适合红树植物的生长,现场勘察中有看到残留的秋茄.从港湾内侧到港湾出口,各样点的海水盐度逐步上升,最高为刘五店的28.7‰.此外,该海域处于内湾,风浪较小.

由于围海筑堤造田、围滩养殖以及人工挖沙,大片适合红树林生长的中高潮带滩涂被改变功能和用途.仅在同安湾西侧的集美凤林美,集美和同安交界处的后安;同安湾东侧的鳄鱼屿、琼头村白兔山到下李之间等地还有部分宜林滩涂.

(1) 同安湾西侧

集美区龙舟池外侧滩涂位于厦门大桥边,集美龙舟池外侧.顺岸方向长585 m,离岸方向宽100 m.现有滩涂较平整,除用于隔离的埂外,没有明显的突出物.目前主要利用方式是贝类养殖,滩涂高程为3.1 m(厦门理论最低潮面),属于低潮带滩涂.访问集美当地的老人得知,现在的龙舟池位置附近曾经有红树林生长,具体种类已经无从查考.根据目前测得的水体盐度及底质情况分析,白骨壤的可能性较大,但不排除也有秋茄.

根据对集美龙舟池外侧滩涂水文、海水盐度、土壤理化性质、潮位、有害生物等的测定,结合厦门的气象条件和临近地区红树林的调查,在该地块进行红树林的恢复种植的不利因素有:滩涂潮位低、生境含盐量高、土壤以泥沙质土壤为主,含沙量偏高、部分区段潮水流速快和有害生物(藤壶、藻类、鼠类及蟹类)危害、人为活动频繁、地块功能定位不明朗等.

因此,集美龙舟池滩涂土壤基本上是适合红树植物的生长的.但是,生境盐度过高尤其是海水盐度过高和土壤含沙量过高,潮位偏低,对移植后红树林苗木的成活及生长较为不利.

表 2 厦门沿海红树林恢复种植宜林滩涂现状

Tab. 2 Status of intertidal zones suitable for planting mangroves along coast of Xiamen

所在区	所在镇村	具体地点	现况	海水盐度/ ‰	宜林滩涂面积 (hm^2,估测)	备注
海沧区	东屿村	嵩屿码头至海沧大桥	滩涂和养殖	22～24(低潮时) 27～29(高潮时)	10～20 (滨海大道沿岸); 2～3(小岛)	需提高滩涂高程(填土或整治)
	吴冠村	吴冠	滩涂和养殖	20～25	13.3	需填土
集美和海沧区	西滨、陈井	马銮湾生态重构示范区	养殖	现在未开口,水较淡	33～40	海堤开口方式未确定,红树林重建计划尚无法确定
集美区	杏林镇	高浦	滩涂	20～22	13.3	流速大,需填土
	集美街道	龙舟池外侧	养殖	20～26	5～7	需填土
	凤林美村	污水厂西南侧	已进行人工种植	21～26	8.4	
		污水厂东侧	养殖	21～26	1～2	
同安区	西柯镇后田村	和集美交界处	养殖	20～25	4～5.3	
	西柯镇丙洲村	海堤纪念碑外侧	养殖	18	2～3	边上修船、有沙场,需改造
	西柯镇浦头村	策槽海堤外侧	养殖	12～16	4～5.3	需适当填土
		同安西溪西岸	滩涂	12～16	1～1.7	附近残留几株秋茄大树,适宜种植
翔安区	马巷镇琼头村	上礁码头东北侧	养殖	24	1～2	少量白骨壤生长,适宜种植
		白兔山和下李之间	养殖	23.5	6～66.6	需适当填土和整地
	新店镇鳄鱼屿		围网	26.0	1.3～2	已有白骨壤林 0.3 hm^2
	新店镇刘五店	刘五店和内安之间	港口,养殖	28.7	3～5.5	需适当填土和整地
	新店镇澳头		滩涂	30.8	无宜林地	风速大,盐度高
	新店镇蔡厝	大嶝海堤西南侧	盐田,养殖	31.0	难造林	风速大,盐度高
	新店镇莲河	码头	盐田,养殖	30.0	难造林	风速大,盐度高
	大嶝镇		港口,养殖	31.0	难造林	风速大,盐度高
总计	宜林滩涂面积 108.3～198.4 hm^2,其中 30～35.3 hm^2 仅需整地即可种植,另外 78.3～163.1 hm^2 需填土后才可种.					

从 2004 年 4 月上旬开始,集美凤林美污水厂西南侧约 126 亩的滩涂已经种上秋茄、无瓣海桑、海桑、老鼠簕和红海榄等红树植物. 其中,秋茄为当年成熟的胚轴,现长势良好. 其他种类的红树植物均为移栽一年生的小苗,经过近一年的生长适应,除少量个体因受机械伤害和高盐度胁迫而死亡外,大部分植株已经逐步适应该滩涂生境,萌发新的枝叶,高度分别达 50～100 cm.

集美凤林美污水厂外侧、同安后田等处的滩涂生境和集美凤林美污水厂西南侧滩涂基本相似,只要稍加整理就可种植红树林.

(2) 同安湾东侧鳄鱼屿上分布有 0.1 hm^2 左右的天然白骨壤林,琼头村上礁码头东北侧的滩涂上也有少量白骨壤生长,由此可以看出该处适合先锋红树植物的生长. 但适合红树林生长的中高潮带滩涂大部分已被用于养殖,现存的海岸前沿滩涂绝大部分处于中低潮带,高程相对较低,需采用回填淤泥抬高滩涂高程后方能种植红树植物. 其中,琼头-下李之间的滩涂位于厦门东部翔安区海岸,与厦门岛东面隔海相望,位于同安东西溪出海口. 滩涂面积约有 200 hm^2,地势宽阔而平坦,坡度一般 1 %～3 %. 其土壤质地为壤土或砂壤土,是典型的河口淤积地. 高潮区基本被围垦,中低潮区以淤泥为主,淤泥深厚达 80 cm 以上,人行下陷约 50 cm 以上. 该地段位于翔安规划中的滨海旅游区圈内,滩涂属水产限养区,早期是红树林生长带,具备红树林的生长条件,因此,在该地段先对较低的滩涂进行适当的海泥回填,然后种植红树林,在技术上是可行的. 工程方案可充分利用现有的养殖围埂,同时根据滩涂地形特征,顺应海岸线自然变化布置红树林的宽度,既达到了护岸的目的,又形成了比较自然的绿色红树林海岸景观.

3.3 厦门东部海域和大嶝海

该海域的淡水补充较少,海水盐度较高(30 ‰左右),并且该海域的风浪较大. 由于围海筑堤造田、围滩养殖,目前该海域的大片中高潮带滩涂被用来养殖和晒盐. 在大嶝海堤两头的滩涂上,厦门同安有关部门于 1997~1998 年人工种植了秋茄和白骨壤. 2004 年 4 月在实地勘察过程中发现秋茄的长势较差,植株矮小呈灌丛状,叶片肉质化程度大(叶片小而厚),并且死亡个体较多,部分植株受风浪影响而倾斜生长. 此外该处的白骨壤也较为矮小. 可以看出,在高盐度、风浪大等逆境胁迫下,红树植物生长受抑制且死亡率高.

4 生态恢复存在问题及解决办法的建议

随着公众生态环保意识的提高,目前国内红树林受到空前重视. 国家林业局决定 2001 年启动红树林保护工程,这是继湿地生态系统保护工程以后,我国生态建设事业的又一重大举措. 从 20 世纪 90 年代中期以来,在努力保护好原有的红树林的同时,全国华南沿海各地(包括福建)掀起了一股红树林造林热潮. 2002 年初,国家林业总局在深圳专门召开了一次会议,计划在未来 10 年内在全国营造 60 000 hm^2 红树林. 厦门市政府对恢复红树林也相当重视,将恢复红树林工作写入 2004 年政府工作报告. 红树林造林对于厦门的城市生态建设有特殊的意义,现在厦门市陆地的宜林地基本上都已绿化,进一步提高城市绿化率和人均绿地面积潜力有限,如果能将广阔的滩涂造林绿化,不仅可以提高绿地面积,还将大幅度提高城市的品位,这对厦门建设海湾型城市具有重要意义. 从红树林资源日趋衰退和大量适宜种植红树林滩涂急待绿化的现状出发,以及沿海防护林体系建设工程的需要,开展大规模的红树林造林已是当前的重要任务.

但是,红树林对生长环境有特殊要求,只能生长在平均海平面(或稍上)与回归潮平均高高潮位(或大潮高潮位)之间的潮滩面(图 1),潮水浸淹频率过高或过低均会导致红树林退化、死亡或难自然更新[10]. 由于修筑海堤与人工围垦时,为了多占地造陆,多围到中低潮带(图 1,A),使得原先适合红树林生长的中高潮带滩涂损失殆尽,目前的红树林造林多为中低潮带滩涂造林,人工造林恢复难度很大[8]. 如果仅围到中潮中带(图 1,B),就有较大滩涂作生态恢复使用,让自然拦淤红树林可以恢复. 据我们对福建沿海各县市的调查统计,红树林造林成功率不超过 50 %,许多地方甚至不到 20 %,全军覆灭的情况也时有发生. 造成这种情况的原因是多方面的,除了缺乏规划,未经宜林地可行性研究就盲目造林和管理方面的问题,以及滩涂高程过低外,还存在红树林生境破坏严重,基础研究滞后,尤其是缺乏一套红树林宜林地选择标准.

大量的造林实践表明,水淹时间成了影响红树林造林的重要的限制因子,红树林宜林临界线的确定,是红树林造林成败的关键. 张乔民等[11]对目前国内外有关红树林生长带的内外边界与潮汐水位的关系作了很好的总结,明确指出红树林生长带位于平均海平面以上的潮滩,这一结论对研究红树林宜林地的选择具有重要的指导意义. 但是,该结论只能告诉我们红树林在滩涂上的大概分布情况,事实上由于不同种红树植物耐水淹能力的不同,不同物种的宜林临界线是不同的. 因此,必须经过室内人工控制淹水时间的实验并结合野外实地观测,才有可能确定各红树林种类的宜林临界线.

针对上述问题,厦门大学湿地与生态工程研究中心研究人员设计了独特的淹水实验装置,研究了厦门主要红树林造林乡土树种:秋茄、桐花树、白骨壤的耐淹水生理生态学特性,并结合野外试种及跟踪测定,明确了各树种的临界淹水时间. 结合滩涂的水文、波浪、潮汐、土壤等的实地观测和红树植物种苗试种实验,分析了影响厦门红树林造林成活率的各种环境因素,明确提出红树林造林主要影响因素有 6 个方面:(1) 潮位要中高潮带以上;(2) 盐度在 10 ‰~20 ‰左右;(3) 潮

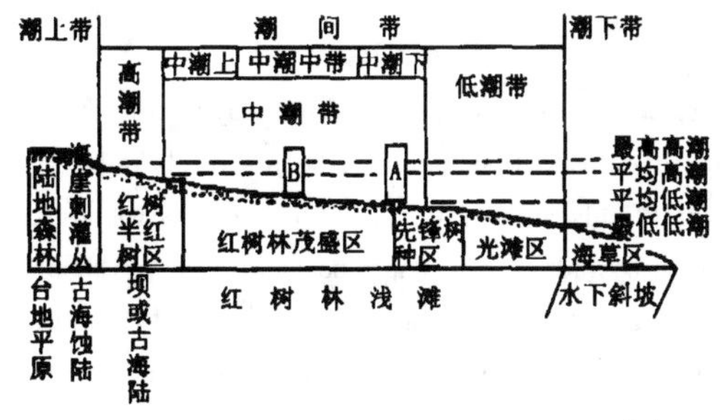

图 1 红树林生长潮位图(A、B 指围海造地位置)[11]

Fig. 1 The tidal level for mangroves growth (A, B indicate the places of reclamation of mudflat)

流速度不宜太大;(4)滩涂以泥质为主;(5)防治动物危害;(6)加强管理,防止人为干扰.在上述研究基础上,总结出一套红树林宜林地选择标准,为厦门恢复红树林提供必须的技术准备.对改变厦门红树林造林粗放经营的现状,提高科技含量、降低造林成本、提高造林存活率有很大的指导意义,其研究成果还可以辐射到福建省的其它地区.

此外,在恢复造林过程中,主要存在以下几个主要问题:

(1) 经过多年的围海造地(机场、港口、码头、道路和围湾造地等建设)及围塘养殖,目前厦门适合红树林生长的中高潮带滩涂多数已被改变功能和用途,现存的滩涂大部分处于中低潮带,因海水浸淹时间过长是限制厦门红树林恢复造林的主要因素,恢复红树林有相当的难度.必须结合海域整治,提高宜林地的潮滩高程.

(2) 目前,厦门的滩涂大部分已被开发,蛏田及泥蚶田密布.滩涂权限尚无法确定.从而影响红树林恢复种植工程的实施.考虑到当地群众的实际情况及红树林恢复种植的正常进行,必须取得当地政府及相关部门的支持才可进行.

(3) 围垦养殖和农民对红树林生态效益认识不足而造成的破坏,或讨小海时挖土造成幼苗死亡.必须取得当地政府和相关部门以及当地居民的支持,加强管理,责任到位.此外,开展对群众特别是海岸沿线村民群众、养殖业主的宣传教育,使社会形成自觉保护红树林的良好氛围.

(4) 由于滩涂环境因子对大范围推广红树林种植的资金筹措将有一定的难度,红树林的恢复工作要由政府有计划地投入一定资金,保障经费到位,让专门机构负责,积极开展优选良种,有计划地扩大滩涂红树林生态恢复工作.

参考文献:

[1] 林鹏,傅勤.中国红树林的环境生态及经济利用[M].北京:高等教育出版社,1997(中文版),2000(英文版).

[2] 林鹏.中国红树林生态系[M].北京:科学出版社,1997(中文版),1999(英文版).

[3] Field C D. Journey Amongst Mangroves[M]. Okinawa: ITTO and ISME Publication,1995. 16 - 63.

[4] Robertson A I,Alongi D M (eds.). Tropical Mangrove Ecosystems[M]. Washington DC: American Geophysical Union,1992. 63 - 100.

[5] Tomlinson P B. The Botany of Mangroves[M]. Cambridge: Cambridge University Press,1986.

[6] 林鹏,林益明,林建辉.红树植物次生木质部的结构与进化[J].海洋学报,1998,20(4):97 - 102.

[7] 张保平.海岸植树可减轻海啸危害[N].北京:科技日报,2005-01-25.

[8] 王文卿,赵萌莉,邓传远,等.福建沿岸地区红树林的种类与分布[J].台湾海峡,2000,19(4):534 - 540.

[9] 韩维栋,高秀梅,卢昌义,等.中国红树林生态系统价值评估[J].生态科学,2000,19(1):40 - 46.

[10] 范航清,黎广钊.海堤对广西沿海红树林的数量、群落特征和恢复的影响[J].应用生态学报,1997,8(3):240 - 244.

[11] 张乔民,于红兵,陈欣树,等.红树林生长带与潮汐水位关系的研究[J].生态学报,1997,17(3):258 - 265.

Protection and Restoration of Mangroves Along the Coast of Xiamen

LIN Peng, ZHANG Yi-hui, YANG Zhi-wei

(School of Life Sciences, Research Center for Wetlands and Ecological Engineering, Xiamen University, Xiamen 361005, China)

Abstract: Mangrove ecosystems play irreplaceable and important roles for the stabilization and equilibrium of coastal estuary. The mangrove areas of Xiamen have been severely reduced from 320 hm^2 to 21 hm^2 in the past tens years for destroying mangroves as agricultural fields, aquaculture ponds, urban construction and so on. It's an important task to protect and restore mangroves along coast of Xiamen nowadays. This paper deals with the problems of protection and restoration of mangroves in Xiamen. Base on the studies of introducing and planting mangrove species, and the status of intertidal zones along coast of Xiamen, we summed up a series of standards for selecting intertidal zones suitable for planting mangroves, and we also put forward the main factors that influenced the afforestation of mangroves in Xiamen.

Key words: Xiamen; mangroves; protection; restoration

Recent progresses in mangrove conservation, restoration and research in China*

Luzhen Chen[1,2,3] Wenqing Wang[1,2,3] Yihui Zhang[1,2] Guanghui Lin[1,2,3]

(1. *Key Laboratory of Ministry of Education for Coastal and Wetland Ecosystems, School of Life Sciences, Xiamen University, Xiamen* 361005, *China*; 2. *Department of Biology, School of Life Sciences, Xiamen University, Xiamen* 361005, *China*; 3. *State Key Laboratory of Marine Environmental Science, Xiamen University, Xiamen* 361005, *China*)

Abstract

Aims

In this paper, we highlighted some key progresses in mangrove conservation, restoration and research in China during last two decades.

Methods

Based on intensive literature review, we compared the distribution and areas of existing mangroves among selected provinces of China, discussed the issues associated with mangrove conservation and restoration and highlighted major progresses on mangrove research conducted by key institutions or universities in mainland China, Hong Kong, Taiwan and Macao.

Important findings

The population boom and rapid economic developments have greatly reduced mangrove areas in China since 1980s, leaving only 22 700 ha mangroves in mainland China in 2001. Chinese government has launched a series of programs to protect mangroves since 1980s and has established mangrove ecosystems as high-priority areas for improving environmental and living resource management. During last three decades, a total of 34 natural mangrove conservation areas have been established, which accounts for 80% of the total existing mangroves areas in China. Mangrove restoration areas in Mainland China accounted for <7% of the total mangroves areas in 2002. A great deal of research papers on Chinese mangroves has been published in international journals. However, more systematic protection strategies and active restoration measurements are still urgently needed in order to preserve these valuable resources in China.

Keywords:

China • conservation • mangrove • research • restoration

Received: 14 January 2009 Revised: 7 May 2009 Accepted: 7 May 2009

INTRODUCTION

Mangroves are the characteristic intertidal plants distributed in tropical and subtropical coastlines (Chapman 1976; Lin 1984; Tomlinson 1986). On the global scale, mangrove areas are becoming smaller or fragmented and their long-term survival is at great risk (Duke *et al.* 2007). Mangrove species in China belong to the Indo-Malaysia Northeast subgroup of East group and covered >50 000 ha in 1950s (Lin 1997; Wang and Wang 2007). Before their important ecological and economic values were recognized by Chinese publics in early 1990s, mangroves had been degraded seriously and the areas greatly reduced, with only 22 752 ha remained (Wang and Wang 2007). Issued in 1995, China's Biodiversity Conservation Action Plan included the action plans for marine biodiversity protection in China, which called for increasing mangrove conservation areas. As a result, the majority of natural mangroves have been protected as part of the national wide mangrove nature reserves.

Chinese scientists have conducted a great deal of research on mangroves since 1950s (Lin 1997a). An in-depth review on Chinese mangrove research was conducted by Li and Lee (1997). Since then, increased government investments have greatly improved the research on mangroves in China.

* From Journal of Plant and Ecology, 2009, 2(2): 45—54

Nearly 1 500 papers have been published in Chinese and international journals since 1990, which provided useful information for the conservation and utilization of mangroves in China. In addition, a series of books about Chinese mangroves were published during last decade (e.g. Chen and Miao 2000; Fan 2000; Liao 2009; Lin 1997; Lin and Fu 1995; Lu and Ye 2006; Tam and Wong 2000; Wang and Wang 2007; Wang *et al.* 2002; Zheng 1999). However, >75% of research papers and most books were written in Chinese, which are not accessible by the international research community. In this paper, we reviewed the rapid developments in the mangrove conservation, restoration and researches in China after 1990s and provide some perspectives and suggestions on future research areas.

MANGROVE CONSERVATION, AFFORESTRATION AND RESTORATION

In China, mangroves distributed naturally in Hainan, Guangdong, Guangxi, Fujian, Hong Kong, Macao and Taiwan (Fig. 1). In 1950s, one mangrove species was successfully transplanted to Yueqing County in Zhejiang Province, where the latitude was ~28°25′N (Zhejiang Forestry Bureau 1961). During the last few decades, the population boom and rapid economic development in agriculture, aquaculture, industry and urban construction have reduced mangrove areas greatly in China. According to the latest survey by State Forestry Administration in 2001 (Table 1), the total mangrove area in mainland China was ~22 000 ha, which was only 44% of that in 1950s (State Forestry Administration 2002). The existing mangrove area in Hong Kong, Macao and Taiwan was ~727 ha in total according to Chen (1997), Leung (1998) and Hsueh and Lee (2000) (Table 1). Thus, the total area of existing mangroves in China was ~22 700 ha in 2001.

Mangrove forests occur mainly in three southern provinces of China (Hainan, Guangdong and Guangxi), which account for 94% of the total mangrove area in China. Among them, Guangdong Province has the largest existing mangroves, followed by Guangxi and Hainan (Table 1). However, mangroves also naturally occupy the intertidal areas of the higher latitudes, such as Fujian Province and Taiwan. In addition, there are some small areas of natural mangroves in Hong Kong and Macao. Although not naturally distributed, there are some mangrove stands in Zhejiang Province, which was transplanted from Fujian during 1950s.

Although accounting only ~0.1% of world mangroves, the mangroves in China have some unique features and important values. Firstly, there are 24 true mangrove species in China, accounting for about one-third of the total true mangrove species worldwide. *Kandelia candel* (L.) Druce, one species in the family of Rhizophoraceae, has long been recognized as a monotypic mangrove genus and occurs in all mangrove distribution zones in China. Recent studies in chromosome number, molecular phylogeography, physiological adaptation and leaf anatomy realized that it should be recognized as *Kandelia obovata* Sheue, Liu and Yong (Sheue *et al.* 2003). Secondly, *K. obovata*, the northernmost distribution species in China, is a good model species for studying mangrove cold resistance and

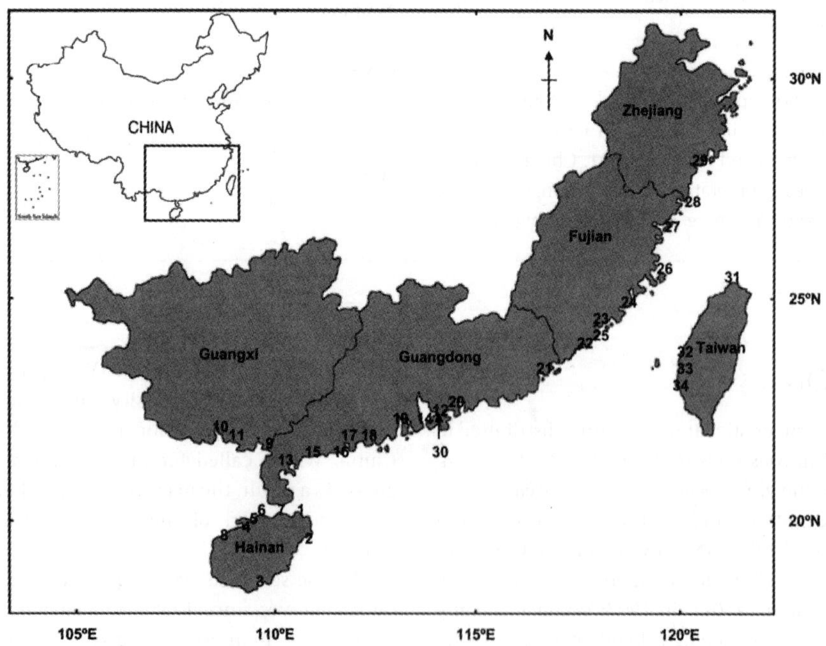

Figure 1: mangrove distribution in the coast along Hainan, Guangxi, Guangdong, Fujian, Zhejiang Provinces, Hong Kong and Taiwan. See Table 1 for the detail names of locations 1–34.

Table 1: surface areas of existing mangroves in China by province

Location	Total existing mangrove area (ha) 2002	Reforestation area[a] (ha) Up to 2002	Restoration area[a] (ha) Up to 2002
Hainan Province	3 930[a]	60	60
Guangdong Province	9 084[a]	672	298
Guangxi Province	8 375[a]	1 093	783
Fujian Province	615[a]	596	369
Zhejiang Province	21[a]	257	21
Hong Kong	380[b]	ND	ND
Macau	60[c]	ND	ND
Taiwan	287[d]	ND	ND
Total	22 752	2 678	1 531

[a] Reports of mangrove resources survey in China, State Forestry Administration (2002)
[b] Chen (1997)
[c] Leung (1998)
[d] Hsueh and Lee (2000); ND = no data.

possible responses to global warming. Thirdly, all true mangrove species in China except *Lumnitzera littorea* can be found in the Qinglan Reserve of Hainan, with a mangrove area of only 1 233 ha (Wang and Wang 2007). Such high species diversity of mangroves is rare in other mangrove regions, which makes the Qinglan Reserve an ideal location for studying the relationship between biodiversity and ecosystem functions. Furthermore, most mangroves in China are distributed in the areas where the population density is among the highest in the world and the local economic growth has increased dramatically during last three decades. Thus, studies and practices in mangrove conservation and restoration in China will provide a good model for other developing countries with similar high intensity of human disturbances to coastal wetlands. Finally, mangroves in China can serve as core components of windshield forests along the coastlines for reducing damage from high frequency of typhoons (State Forestry Administration and State Oceanic Administration 2006).

Mai Po Wetland in Hong Kong is the first mangrove reserve in China, which was established in 1976 and listed as Ramsar Site in 1995 (Li and Lee 1997). Since then, China has made rapid progress in mangrove conservation. Up to date, a total of 34 mangrove nature reserves have been established in different locations of China, and the total protected area was >18 000 ha, accounting for >80% of the mangrove area in China (Table 2). According to the Engineering Programs of Mangroves Conservation and Development in China (2006–16) (State Forestry Administration and State Oceanic Administration 2006), more mangrove natural reserves will be established in the coming decades.

In addition to the establishment of mangrove natural reserves, Chinese government has made great efforts in mangrove reforestation since the early 1990s (Zheng *et al.* 2003). Up to 2002, ~2 678 ha mangroves have been replanted, while only 57% of them were successful restored (Table 1). According to the statistics conducted by State Forestry Administration and State Oceanic Administration (2006), there are ~65 600 ha of intertidal zones suitable for mangrove afforestration in China, indicating great potential for the expansion of mangrove areas by silviculture.

ISSUES ASSOCIATED WITH MANGROVE CONSERVATION AND RESTORATION

Despite of the apparent success in mangrove conservation and reforestation during last two decades, there are still many threats to Chinese mangroves. Urban and aquaculture wastewater discharge (Wang *et al.* 2002), oil pollution (Liu *et al.* 2006; Wang *et al.* 2002), biological invasion (Lin 2003; Wang *et al.* 2002), insect outbreak (Jia *et al.* 2001a, 2001b; Qiang and Lin 2004) and the influence of water transportation (Wang and Wang 2007; Wang *et al.* 2002) remain serious threats to mangroves in China. For a long period of time, wastewater from the upstream and landfill pollution discharged directly into the mangrove wetland without proper treatments, which were popular in the coastlines of Southern China (Lin 1997). Although the self-purification functions of mangrove wetlands were reported (Dwivedi and Padmakumar 1983; Huang *et al.* 2000; Tam and Wong 1995; Wong *et al.* 1997), pollution still adversely changed the ecosystem functions and the biodiversity of the mangrove ecosystem (Wang *et al.* 2002). For example, the quantities and densities of benthic animals, birds or fishes declined in several polluted mangrove forests (Cai *et al.* 2000; Lin *et al.* 2007; Ma *et al.* 2003; Wang *et al.* 2002).

Biological invasion is a global problem for its great threats to native species and local ecosystems (Drake *et al.* 1989; Higgins and Richardson 1996), which is also common to the mangroves in China. For example, *Spartina alterniflora*, a C_4 grass native to the east coast of USA and was first introduced to Fujian Province in China as a windproof and beach-protecting plant. Because of its significant roles for these purposes, this species was later planted in large areas in other provinces, such as Jiangsu, Zhejiang and Guangdong. By 2000, the coverage of *S. alterniflora* was >112 000 ha in China (Zhu and Qin 2003). The strong dispersal and reproductive capacities of the seeds or new ramets from rhizome segments of *S. alterniflora* made it a very invasive species, which has brought serious threats to the native mangroves (Qian and Ma 1995). For example, >167 ha of *S. alterniflora* were found in Shankou mangrove nature reserve in 2005 and most of the mangrove tidal flats in Qi'ao mangrove nature reserve were covered entirely by *S. alterniflora* now. The *S. alterniflora* invasion caused hundred millions of dollar loss per year in China (Chen 1998; Chen *et al.* 2004b). For example, >100 million ¥ lost in the aquaculture was reported each year along in the six counties in Fujian Province (Li and Xie 2002). Several control methods have been developed to prevent and mitigate the spread of *S. alterniflora*, including burning, harvesting, herbicide application, freshwater flooding and replacement with an exotic

Table 2: detail information for current mangrove natural reserves in China (Up to 2007)

No.[a]	Name of reserve	Location	Conservation area (ha)	Mangrove area (ha)	Time of established	Class	No. of mangrove species[b]
Hainan Province							
1	Dongzhai Harbor, Haikou	19°54′N, 110°20′E	3 337	1 733	1980	National[c]	14 + 9
2	Qinglan Harbor, Wenchang	19°34′N, 110°45′E	2 948	1 233.3	1981	Provincial	23 + 12
3	Sanya	18°11′N, 109°33′E	923.5	59.7	1989	Local	18 + 9
4	Xinying, Danzhou	19°44′N, 109°16′N	115	79.1	1992	Local	10 + 6
5	Dongchang, Danzhou	19°51′N, 109°33′E	696	478.4	1986	Local	ND
6	Xinying, Lingao	19°51′N, 109°33′E	67	ND	1983	Local	7 + 5
7	Huachang Bay, Chengmai	19°54′N, 109°57′E	150	150	1995	Local	7 + 5
8	Dongfang	19°14′N, 108°39′E	1 429	123.6	2006	Local	ND
Guangxi Province							
9	Shankou, Hepu	21°28′N, 109°43′E	8 000	806.2	1990	National[c]	9 + 6
10	Beilunhe, Fangchenggang	21°30′N, 108°09′E	2 680	1 131.3	2000	National[c]	10 + 8
11	Maowei Bay, Qinzhou	21°43′N, 118°38′E	2 784	1 892.7	2005	Provincial	8 + 7
Guangdong Province							
12	Futian-Neilingding, Shenzhen	22°32′N, 114°05′E	921.6	70.0	1984	National	5 + 7
13	Zhanjiang	20°15′-21°55′N, 109°40′-110 °55′E	20 278.8	7 256.5	1997	National[c]	10 + 9
14	Qi'ao-Dangan Island, Zhuhai	22°26′N, 113°38′E	7 363	193.3	2000	Provincial	8 + 7
15	Shuidong Bay, Maoming	21°29′N, 111°03′E	1 950.0	150.9	1999	Local	ND
16	Chengcun, Yangxi	21°45′N, 111°44′E	1 320.0	ND	ND	Local	ND
17	Yangjiang, Yangxi	21°42′N, 111°55′E	1 060.0	ND	2000	Local	ND
18	Zhenhai Bay, Enping	22°00′N, 112°22′E	666.7	134.3	2000	Local	ND
19	Jiangmen	22°13′N, 113°05′E	520	133	2004	Local	10 + 7
20	Huidong, Huizhou	22°48′N, 114°48′E	533.3	136	2000	Local	8 + 3
21	Shantou	23°18′N, 116°45′E	10 333.3	1 644.6	2001	Local	2 + 1
Fujian							
22	Zhangjiang Estuary, Yunxiao	23°54′N, 117°27′E	2 360	83.3	1997	National[c]	7 + 3
23	Jiulongjiang Estuary, Longhai	24°26′N, 117°54′E	600	197.3	1988	Provincial	4 + 2
24	Quanzhou Bay, Quanzhou	24°47′N, 118°38′E	7 039	17	2003	Provincial	3 + 2
25	Xiamei, Zhangpu	23°58′N, 117°42′E	400	ND	1997	Local	ND
26	Changle, Fuzhou	26°01′N, 119°38′E	2 921	34.3	2003	Local	1 + 1
27	Sandu Bay, Ningde	26°41′N, 119°46′E	39 981	ND	1997	Local	1 + 1
28	Shacheng Harbor, Fuding	27°16′N, 120°18′E	2 174	7	1997	Local	1 + 0
Zhejiang							
29	Yueqing	28°20′N, 121°10′E	2 000	21.5	2005	Local	1 + 0
Hong Kong							
30	Mai Po	22°30′N, 114°02′E	380	120	1976	National[c]	8 + 7
Taiwan							
31	Tanshui, Taibei	25°09′N, 121°26′E	190	77.8	1986	Local	1 + 1
32	Haomeiliao, Chiayi	23°22′N, 121°07′E	1 171	31.7	1987	Local	2 + 1
33	Peimeng, Tainan	23° 16′N, 120°06′E	2 447	28.5	1987	Local	2 + 0
34	Sutsao, Tainan	23°02′N, 120°07′E	547	8.7	1994	Local	4 + 1
	Total			>18 033			

[a] The numbers are denoted to the location number in Figure 1,
[b] mangrove species is shown as true mangrove species + semi-mangrove species; the classification of true mangrove species and semi-mangrove species follows Wang and Wang (2007). *Excoecaria agallocha* belongs to true mangrove species (Wang and Wang 2007) and
[c] listed as Ramsar Wetland; ND = no data and local = county level or city level (Hsueh 1995; State Oceanic Administration 1996; Lin 1997a; Lin and Fu 2005; Zhang and Sui 2001; Chen *et al.* 2001; State Forestry Administration and State Oceanic Administration 2006).

fast-growing mangrove species, *Sonneratia apetala* (An *et al.* 2007; Lin 1997; Liu and Huang 2000; Tang *et al.* 2007). However, the effectiveness of all these methods appears to be very limited since the invasion of this exotic grass occurred in more areas during last few years.

On the other hand, there are still some great challenges in replanting mangroves on locations where mangroves have been destroyed. First, the survival rates in mangrove afforestation are quite low. As reported by State Forestry Administration (2002), the survival rate of replanting mangroves in Guangdong Province was <44% in 2001. Environmental factors, such as tidal inundation periods, seawater salinity and air temperature can affect the survival rate of mangrove reforestation. Selecting suitable tidal zones for mangrove replantion (i.e. the plantable tidal flats, which refer to the tidal flats where natural mangroves distributed and the planted mangrove seedlings can survive (Zhang *et al.* 1997)) is essential in any mangrove restoration project (Chen *et al.* 2004a). In Panyu County, Guangdong Province, for example, ~90% of the replanting mangrove seedlings survived in the plantable tidal flats, but >80% of those planted in the tidal flats of 0.8 m's lower than the plantable tidal zone died after 12–15 months (Chen *et al.* 2001). In the areas with higher latitude, such as Fuding in Fujian (27°20′N) and Yueqing in Zhejiang (28°15′N), low air temperature (−2.2 to −4.2°C) in winter is a major threat to the survival of mangrove seedlings. Only 20 ha of the total 256-ha planted mangroves survived in Zhejiang during the period from 1980 to 2001 (Table 1), which was mainly due to the relatively low air temperature in the winters.

Secondly, mono-species or exotic species are often used in the mangrove reforestation in China, which reduces the biodiversity of replanted forests. Although it has long been known that reduced biodiversity is sensitive to be easily subjected to insect outbreak and has low ecological values, a few species of native mangroves (*K. obovata*, *Sonneratia caseolaris*, *Rhizophora stylosa*) were frequently planted in monoculture for most of the reforestation projects. This is because most of these reforestation projects are aimed mainly for the appearance of the planted trees and for the high survival rates. On the other hand, some fast-growing exotic mangrove species were introduced and intensively used in many mangrove afforestation projects in China during last two decades (Liao *et al.* 2003; Wang *et al.* 2002). For example, *S. apetala* from Bangladesh were planted in many locations along the coastline of Southern China, such as the Dongzhaigang Mangrove Nature Reserve of Hainan, the Zhanjiang, Qi'ao and Futian Mangrove Nature Reserves of Guangdong, Beilunhe Mangrove Nature Reserve of Guangxi and several locations in Fujian. The total area of *S. apetala* forests was estimated as ~3 800 ha (Li and Liao 2008), about half of the total replanted mangrove areas so far. Ironically, *S. apetala* was also used to control the invasive *S. alterniflora* (Tang *et al.* 2007), even though the invasiveness of this exotic mangrove species was not yet fully studied (Chen *et al.* 2008b). It may be a good strategy to control invasive grasses using fast-growing pioneer mangroves such as *S. apetala* and then the exotic mangroves are replaced with native species. However, more studies are needed to test this approach before it is implemented at large scale.

PROGRESSES ON MANGROVE RESEARCH

Mangrove research in China began in early 1950s and has been well developed during last five decades (Lin 1997). Early work focused mainly on mangrove floristics (He 1951), taxonomy (Hou and He 1953), population ecology (He 1957), community and vegetation distributions (He 1957; Zhang *et al.* 1957). The research progresses on Chinese mangroves before 1990 were reviewed by Li and Lee (1997). Since then, significant amount of books and scientific papers have been published, indicating the rapid development in this research field. In total, there were 1 473 papers published in domestic and international journals from 1990 to 2007, according to the records in the Weipu Database of China and the Web of Science Database (Table 3). The number of officially published papers on Chinese mangrove (included in mainland China, Hong Kong, Macao and Taiwan) since 2000 was >2-fold of that during 1990–99 (Table 3). In fact, the number of annual published papers on mangroves by Chinese scientists increased exponentially from 1990 to 2007 (Fig. 2). The proportion of the peer review journal papers written by Chinese scientists to the world mangrove publications after 2000 was 10.2%, which was lower than that in 1990s (Table 3). Although research papers published by Chinese scientists were double after 2000, the world mangrove researches developed more rapidly based on the total number of published Science Citation Index (SCI) papers.

Between 1990 and 2007, the mangrove research in China focused on a dozen areas (Fig. 3), which included remote sense and modeling, aquaculture, global ecology, geography and hydrography, energy flow, morphology and anatomy, molecular ecology, pharmaceutics and active material exploitation, silviculture, community and population ecology, biodiversity, pollution ecology, ecophysiology, conservation and management. Among them, five research areas increased most rapidly, including molecular ecology, pollution ecology, biodiversity, conservation and management, silviculture and pharmaceutics and active material exploitation.

Extensive researches on mangrove ecosystem structure and function revealed extremely high biomass and primary production for the mangrove forests in China. The highest biomass among Chinese mangrove forests was found in the *Bruguiera sexangula* forest in Dongzhai Nature Reserve of Hainan, with the biomass of 248.5 t/ha, followed by the *R. stylosa* forest in Shankou nature reserve in Guangxi (196.2 t/ha) and the *K. obovata* forest in Hong Kong (129.6 t/ha) (Lin *et al.* 1990, 1992; Lee 1990). High litter production and litter decomposition rates were also found in the Chinese mangrove communities (Fan and Lin 1995; Lin and Fan 1992; Yin and Lin 1992). Based on these results, a "Three-High" or "3-H" theory on mangrove

Table 3: number of papers on China's mangroves published from 1990 to 2007. Officially published papers of Mainland China, Hong Kong, Taiwan and Macao mangroves were included. Abstracts submitted to conferences and book chapters were not included

Year	No. of publications in Weipu database of domestic journals in Mainland China	No. of publications in SCI journals of Web of Science database	Total No. of world mangrove publications in SCI journals of Web of Science database	Proportion of China's SCI publications to world mangrove publications (%)
1990–99	307	104	679	15.3
2000–07	780	282	2 752	10.2
Total (1990–2007)	1 087	386	3 431	11.3

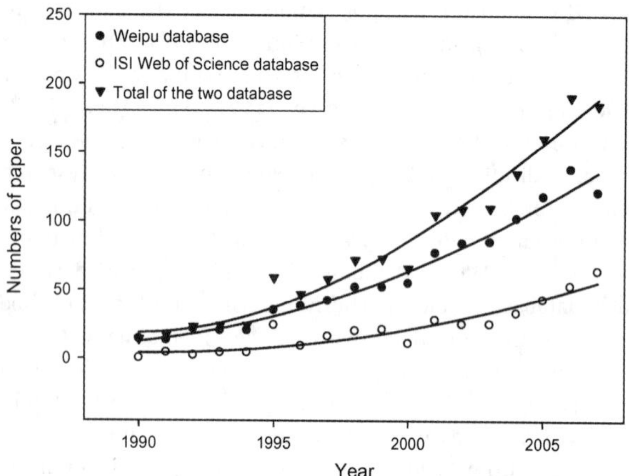

Figure 2: number of papers on China's mangroves from Weipu database and ISI Web of Science database from 1990 to 2007, including Mainland China, Hong Kong, Macau and Taiwan. These papers do not include abstracts submitted to conferences.

communities, i.e. high productivity, high return ratio and high decomposition ratio, was later proposed (Lin 1997).

There were increasing interests in studies on the interactions between Chinese mangrove ecosystems and global change. However, so far the work focused mainly on methane dynamics in mangrove wetlands (Lu *et al.* 1999; Ye *et al.* 1997) and the responses of mangroves to tidal flooding associated with sea level rise (e.g. Ye *et al.* 2003, 2004). Little is known about how mangroves and their ecosystems in China respond to elevated CO_2, global warming or nitrogen deposition. In China, sea walls were constructed in many locations along the coastline to prevent damages from frequent typhoons and hide tides, which not only increased the difficulty in mangrove reforestation (Fan 1995) but also limited mangrove landward migration in response to sea level rise. Thus, more studies in this field are urgently needed to assess potential impact of global change on the mangroves in China. Long-term ecological research stations are now established in Hainan's Donzhaigang Mangrove Nature Reserve (Liao 2009), Guanngdong's Zhanjiang Mangrove Reserve and Fujian Zhangjiangkou Mangrove Reserve (by our group), which will greatly enhance this field research in the coming decades.

A great deal of field and greenhouse studies pointed to great challenges in selecting plantable tidal flats for the mangrove afforestation efforts in China (Fan and Li 1997; Mo and Fan 2001). According to Zhang *et al.* (1997), mangrove can only occupy the tidal flats between the mean sea level (of slightly above) and the highest tidal level in the tropical region. Plantable tidal flats for *K. obovata* have been established for Guangxi (Mo *et al.* 1995), Shenzhen (Liao *et al.* 1997a) and Xiamen (Chen *et al.* 2006). Physiological studies showed that *B. gymnorrhiza* had lower tolerance to soil flooding than *K. obovata* (Ye *et al.* 2003), while the optimal tidal inundation period for *K. obovata* growth and photosynthesis was 2–4 hr per tidal cycle (Chen *et al.* 2004a, 2005). Studies on mangrove management and new techniques in silviculture developed rapidly after 2000 (Fig. 3). The exotic *S. apetala* was shown to have good tolerance of high tide and chilling conditions (Li *et al.* 1997; Liao *et al.* 2004), which may explain why it was considered as one of the best mangrove afforestation species and used almost in all mangrove afforestation projects in China.

Researches on the potentials of mangrove wetlands for wastewater treatments and pollutant degradation have been also greatly promoted in China since 1990s. Mangrove wetland was regarded as an effective ecological system for the removal of nutrients and other anthropogenic pollutants (Tam and Wong 1997; Wong *et al.* 1997). More than 70% of dissolved organic carbon, ammonia and total Kjeldahl nitrogen and ~50% of inorganic nitrogen could be removed by a constructed wetland of *K. ovobata* in a greenhouse study in Hong Kong (Wu *et al.* 2008a). Using a constructed wetland in Futian Mangrove Nature Reserve in Guangdong, Yang *et al.* (2008) found higher treatment efficiencies of *Aegiceras corniculatum* and *S. caseolaris* communities for sewage wastewater than the *K. obovata* communities. Furthermore, the bacterial consortium enriched in mangrove sediments was also shown to be very effective in facilitating the degradation of many polycyclic aromatic hydrocarbons (Luan *et al.* 2006; Zhong *et al.* 2007).

Medicinal applications of Chinese mangrove plants were known for a long period of time (Lin and Fu 1995), which stimulated great interests in the studies on the sources, compound structures and bioactivities of natural products from mangrove materials after 2000 (Wu *et al.* 2008b). Several natural products, such as xylocarpins A and B and scyphiphorins A and B, were isolated from the fruits and seeds of *Xylocarpus granatum* and *Scyphiphora hydrophyllacea*, and their configurations and bioactivities were fully evaluated (Li *et al.* 2007; Tao *et al.* 2007; Wu *et al.* 2008c). *Ceriops tagal* was another mangrove

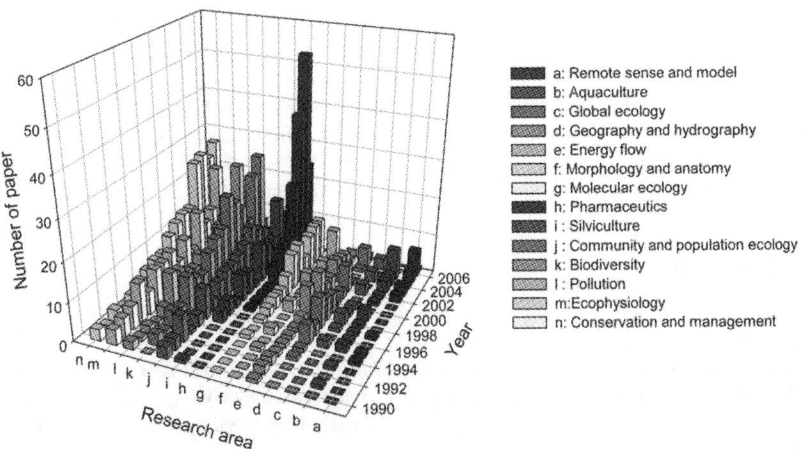

Figure 3: distribution of published papers in 14 major research areas from 1990 to 2007.

species with high medicinal values and three new dimeric diterpenes have been isolated from its roots recently (Chen *et al.* 2008a). However, the direct utilization of mangrove materials for medicine production will likely reduce mangrove resources and should be avoided. A better way for this application is to formulate new medicines through chemical synthesis base on the compound configurations of related compounds found in certain mangrove materials.

In the field of molecular ecology, great progresses have been made since 2000, especially in the areas on the geographical distances and species relationships of Chinese mangroves. The genomic basis of the adaptive evolution and speciation in mangrove was established (Zhong *et al.* 2002; Shi *et al.* 2005; Zhou *et al.* 2007), and several molecular biomarker, including chloroplast DNA, mitochondrial DNA and inter-simple sequence repeat, were used to identify the gene flows between South China Sea and nearby regions (Ge and Sun 2001; Su *et al.* 2006, 2007; Tan *et al.* 2005). Molecular markers were also used as taxonomic evidences for several mangrove species classifications (Qiu *et al.* 2008; Zhou *et al.* 2005). For example, two *Sonneratia* species in Hainan Province, *Sonneratia* × *hainanensis* and *Sonneratia* × *gulngai*, have been speculated as natural hybrids derived from hybridization between *Sonneratia alba* and *Sonneratia ovata* and *Sonneratia alba* and *Sonneratia caseolaris*, respectively (Duke 1984; Wang *et al.* 1999), but a recent molecular study suggested that they were not true hybrid species (Zhou *et al.* 2005). These studies illustrated the values of using modern technologies in resolving long-standing ecological or evolutionary issues in mangroves.

FUTURE PERSPECTIVES

Over the past two decades, a large number of case studies have significantly increased our understanding of the structure and function of the mangrove ecosystems as well as the values of mangroves. However, there are still many areas needed to be strengthened in the future.

Firstly, although mangrove ecosystem functions in China have been studied intensively, which mangrove species is the key stone species of Chinese mangrove ecosystems is still not resolved. More controlled experiments on the relationships between species diversity and ecosystem functions of mangroves should be conducted to resolve this issue.

Secondly, many studies showed that mangroves would migrate landward and expand laterally into areas of higher elevations in response to sea level rise (see review by Gilman *et al.* 2007). As pointed out earlier, the construction of sea walls, plus many skyscrapers behind natural mangrove wetlands, may prevent such migration from occurring, so we need to evaluate the fate of mangroves in China under rising sea levels in coming decades or century.

Thirdly, biological invasions such as those of *S. alterniflora* may jeopardize mangrove habitats. We still lack good understandings of their invasive mechanisms and the efficient measures for controlling such invasions. More field and greenhouse studies are needed in this field.

Fourthly, great efforts and achievements have been made in mangrove afforestation restoration in China, but there is still a lack of a universal standard system for evaluating such efforts and achievements. Collaborations among governmental agencies (such as State Administration of Forestry), research institutions and local communities are strongly encouraged in establishing such evaluation standard system for mangrove afforestation and restoration.

Finally, cooperation among related mangrove research institutions in mainland China, Hong Kong and Taiwan is essential to ensure more successful conservation, restoration and research of mangroves in China. There has been continuous cooperation on mangrove researches between mainland China and Hong Kong since 1990s, but collaboration between mainland China and Taiwan just began recently. We recognize that restoring and protecting mangrove wetlands in all three areas of China require collaborative efforts from all parties.

FUNDING

National Natural Science Foundation of China (30700092 to L. Chen and G. Lin and 30671646 to P. Lin); the "Minjiang Scholar" program of Fujian (to G. Lin); a China Postdoctoral Science Foundation award (20060400529 to L. Chen).

ACKNOWLEDGEMENTS

The authors are grateful to ML Hsueh in Taiwan Endemic Species Research Institute for providing the information on Taiwan mangroves and Mr WB Su in Hainan Forest Bureau and Mr L Wei in Guangdong Forest Research Institute for the information on Hainan and Guangdong mangroves.

Conflict of interest statement. None declared.

REFERENCES

An SQ, Gu BH, Zhou CF, *et al.* (2007) *Spartina* invasion in China: implications for invasive species management and future research. *Weed Res* **47**:183–91.

Cai LZ, Lin P, Liu JJ (2000) Quantitative dynamics of three species of large individual polychaete and environmental analysis on mudflat in Shenzhen Estuary. *Acta Oceanol Sin* **22**:110–6.

Chapman VJ (1976) *Mangrove Vegetation*. Vaduz: Cramer.

Chen CD (1998) The effect of exotic species on Chinese biodiversity. In: Chen CD (ed). *China's Biodiversity: A Country Studies*. Beijing, China: Environmental Science Press, 58–62.

Chen GZ, Miao SY (2000) *Mangrove Species Kandelia candel and Wetland Ecosystem Researches*. Guangzhou: Sun Yat-Sen University Press (in Chinese).

Chen JD, Qiu Y, Yang ZW, *et al.* (2008*a*) Dimeric diterpenes from the roots of the mangrove plant *Ceriops tagal*. *Helv Chim Acta* **91**: 2292–8.

Chen LZ, Tam NFY, Huang JH, *et al.* (2008*b*) Comparison of ecophysiological characteristic between introduced and indigenous mangrove species in China. *Estuar Coast Shelf Sci* **79**:644–52.

Chen LZ, Wang WQ, Lin P (2004*a*) Influence of waterlogging time on the growth of *Kandelia candel* seedlings. *Acta Oceanol Sin* **23**:149–58.

Chen LZ, Wang WQ, Lin P (2005) Photosynthetic and physiological responses of *Kandelia candel* L. Durce seedlings to duration of tidal immersion in artificial seawater. *Environ Exp Bot* **54**:256–66.

Chen LZ, Yang ZW, Wang WQ, *et al.* (2006) Critical tidal level for planting *Kandelia candel* seedlings in Xiamen. *Chin J Appl Ecol* **17**:177–81(in Chinese with English abstract).

Chen SP (1997) The mangrove of Hong Kong. *Trop Geogr* **17**:184–90(in Chinese with English abstract).

Chen YJ, Chen WP, Zheng SF, *et al.* (2001) Researches on the mangrove plantation in Panyu, Guangdong. *Ecol Sci* **20**:25–31.

Chen ZY, Li B, Zhong Y, Chen JK (2004*b*) Local competitive effects of introduced *Spartina alterniflora* on *Scirpus mariqueter* at Dongtan of Chongming Island, the Yangtze River estuary and their potential ecological consequences. *Hydrobiologia* **528**:99–106.

Drake JA, Mooney HA, di Castri F, *et al.* (1989) *Biological Invasions. A Global Perspective*. New York: Wiley.

Duke NC (1984) A mangrove hybrid, *Sonneratia* × *gulngai* (Sonneratiaceae) from north-eastern Australia. *Austrobaileya* **2**:103–5.

Duke NC, Meynecke JO, Dittmann S, *et al.* (2007) A world without mangroves? *Science* **317**:41–2.

Dwivedi SN, Padmakumar KG (1983) Ecology of a mangrove swamp near Juhu Beach, Bombay with reference to sewage pollution. In: Teas HJ (ed). *Biology and Ecology of Mangrove. TV: S*, Vol. 8. Lancaster: Dr W. Junk Publishers, 163–79.

Fan HQ (1995) An ecological pattern of sea dyke maintenance by mangroves and assessment of its benefits along Guangxi coast. *Guangxi Sci* **2**:48–52.

Fan HQ (2000) *Mangroves: The Environmental Guard of Coastline*. Nanning: Guangxi Science and Technology Press (in Chinese).

Fan HQ, Li GZ (1997) Effect of sea dike on the quantity, community characteristic and restoration of mangroves forest along Guangxi coast. *Chin J Appl Ecol* **8**:240–244(in Chinese with English abstract).

Fan HQ, Lin P (1995) Potential role of leaching in weight loss during the decomposition of mangrove *Kandelia candel* leaf litter. *Oceanol Limnol Sin* **26**:28–33(in Chinese with English abstract).

Ge XJ, Sun M (2001) Population genetic structure of *Ceriops tagal* (Rhizophoraceae) in Thailand and China. *Wetl Ecol Manag* **9**:203–9.

Gilman EH, Ellison J, Coleman R (2007) Assessment of mangrove response to projected relative sea-level rise and recent historical reconstruction of shoreline position. *Environ Monit Assess* **124**: 105–30.

He J (1951) Plant region and vegetation in Fujian Province. *Sci China* **2**:198–213(in Chinese).

He J (1957) Mangrove ecology. *Bull Biol* **8**:1–5(in Chinese).

Higgins SI, Richardson DM (1996) A review of models of alien plant spread. *Ecol Model* **87**:249–65.

Hou KZ, He CN (1953) Families of China's mangrove plants. *Acta Phytotaxon Sin* **2**:133–57(in Chinese).

Hsueh ML (1995) *The disappearing wetland forest: mangroves in Taiwan. Taiwan:* Taiwan Endemic Species Research Institution. (in Chinese).

Hsueh ML, Lee HH (2000) Diversity and distribution of the mangrove forests in Taiwan. *Wetl Ecol Manag* **8**:233–42.

Huang LN, Lan CY, Shu WS (2000) Sewage discharge on soil and plants of the mangrove wetland ecosystem. *Chin J Ecol* **19**:13–19(in Chinese with English abstract).

Jia FL, Chen HD, Wang YJ, *et al.* (2001*a*) The pest insects and analysis of its outbreaks cause in Futian Mangrove, Shenzhen. *Acta Sci Nat Univ Sunyatsen* **40**:88–91(in Chinese with English abstract).

Jia FL, Wang YJ, Zan QJ (2001*b*) The efficacy test of "Dimilin-III" and B.T. to *Oligochroa cantonelia* (Lepidoptera: Pyralidae). *Nat Enem Insect* **23**:86–89(in Chinese with English abstract).

Lee SY (1990) Primary productivity and particulate organic matter flow in a estuarine mangrove wetland in Hong Kong. *Mar Biol* **106**:453–63.

Leung V (1998) The distribution pattern of mangrove plant populations and its species composition in Macau. *Ecol Sci* **17**:25–31(in Chinese with English abstract).

Li M, Liao BW (2008) Introduction and ecological impact of *Sonneratia apetala*. *Prot For Sci Technol* **3**:100–102(in Chinese with English abstract).

Li MS, Lee SY (1997) Mangroves of China: a brief review. *For Ecol Manag* **96**:241–59.

Li MY, Wu J, Zhang S, *et al.* (2007) Xylocarpins A and B, two new mexicanolides from the seeds of a Chinese mangrove *Xylocarpus granatum*: NMR investigation in mixture. *Magn Res Chem* **45**:705–9.

Li Y, Zheng DZ, Chen HX, *et al.* (1997) Preliminary study on introduction of Mangrove *Sonneratia apetala* Buch-Ham. *For Res* **11**:39–44(in Chinese with English abstract).

Li ZY, Xie Y (2002) *Invasive Alien Species in China*. Beijing, China: China Forestry Publishing House (in Chinese), p. 211.

Liao BW (2009) *Studies on Dongzhai Harbor Mangrove Wetland Ecosystem on Hainan Island in China*. Qingdao:: China Ocean University Press (in Chinese).

Liao BW, Li M, Zheng SF, *et al.* (2003) Study on intraspecific and interspecific competition in exotic species *Sonneratia apetala*. *For Res* **16**:418–22 (in Chinese with English abstract).

Liao BW, Zheng DZ, Zheng SF, *et al.* (1997*a*) Studies on the cultivation techniques of mangrove *Kandelia candel*. In: Wong YS, Tam NFY (eds). *Mangrove Research of Guangdong, China*. Guangzhou: South China University of Technology Press, 479–86 (in Chinese).

Liao BW, Zheng DZ, Zheng SF, *et al.* (1997*b*) Studies on seedling nursing techniques of *Sonneratia caseolaris* and its seedling growth Rhythm. *For Res* **10**:295–302 (in Chinese with English abstract).

Liao BW, Zheng SF, Chen YJ, *et al.* (2004) Biological characteristics of ecological adaptability for nonindigenous mangroves species *Sonneratia apetala*. *Chin J Ecol* **23**:10–15 (in Chinese with English abstract).

Lin P (1984) *Mangrove Vegetation*. Beijing, China: Ocean Press (in Chinese).

Lin P (1997*a*) *Mangrove Ecosystem in China*. Beijing: Science Press (in Chinese).

Lin P (2003) The characteristics of mangrove wetlands and some ecological engineering questions in China. *Eng Sci* **5**:33–8 (in Chinese with English abstract).

Lin P, Fan HQ (1992) Seasonal model of the decomposition rates of *Kandelia candel* fallen leaves in Jiulongjiang River estuary. *J Xiamen Univ (Nat Sci)* **31**:430–34 (in Chinese with English abstract).

Lin P, Fu Q (1995) *Environmental Ecology and Economic Utilization of Mangroves in China*. Beijing: Higher education Press (in Chinese).

Lin P, Lu CY, Wang GL, *et al.* (1990) Biomass and productivity of *Bruguiera sexangula* mangrove forests in Hainan Island, China. *J Xiamen Univ (Nat Sci)* **29**:209–13 (in Chinese with English abstract).

Lin P, Yin Y, Lu CY (1992) Biomass and productivity of *Rhizophora stylosa* community in Yingluo Bay of Guangxi, China. *J Xiamen Univ (Nat Sci)* **31**:199–202 (in Chinese with English abstract).

Lin QR (1997*b*) Damage and management of *Spartina anglica* and *Spartina alterniflora*. *Fujian Geogr* **12**:16–20.

Lin QX, Chen XL, Lin PJ (2007) The effect of environment changing on birds at Dongyu mangrove area, Xiamen. *J Xiamen Univ (Nat Sci)* **46**:104–8 (in Chinese with English abstract).

Liu J, Huang JH (2000) Management of *Spartina alterniflora*. *Bull Oceanogr* **10**:68–72 (in Chinese with English abstract).

Liu JC, Yan CL, Macnair MR (2006) Distribution and speciation of some metals in mangrove sediments from Jiulong River estuary, People's Republic of China. *Bull Environ Contam Toxicol* **76**:815–22.

Lu CY, Wong YS, Tam NFY, *et al.* (1999) Methane flux and production from sediments of a mangrove wetland on Hainan Island, China. *Mangroves Salt Marshes* **3**:41–9.

Lu CY, Ye Y (2006) *Wetland Ecology and Engineering. Mangroves as Example*. Xiamen: Xiamen University Press (in Chinese).

Luan TG, Yu KSH, Zhong Y, *et al.* (2006) Study of metabolites from the degradation of polycyclic aromatic hydrocarbons (PAHs) by bacterial consortium enriched from mangrove sediments. *Chemosphere* **65**:2289–96.

Ma L, Cai LZ, Yuan DX (2003) Advances of studies on mangrove benthic fauna pollution ecology. *J Oceanogr Taiwan Strait* **22**:113–9 (in Chinese with English abstract).

Mo ZC, Fan HQ (2001) Comparison of mangrove forestation methods. *Guangxi For Sci* **30**:73–581 (in Chinese with English abstract).

Mo ZC, Liang SC, Fan HQ (1995) A preliminary study on planting techniques of Guangxi mangroves. In: Fan HQ, Liang SC (eds). *Research and Management on China Mangroves*. Beijing, China: Science Press, 164–72.

Qian YQ, Ma KP (1995) Bio-techniques and bio-safety. *J Nat Res* **10**:322–34.

Qiang N, Lin YG (2004) Characteristics of quantitative distribution and species composition of fouling fauna in Dongwan mangrove stands of Fangchenggang, Guangxi. *J Trop Oceanogr* **23**:64–8 (in Chinese, with English abstract).

Qiu S, Zhou R, Li Y, *et al.* (2008) Molecular evidence for natural hybridization between *Sonneratia alba* and *S. griffithii*. *J Syst Evol* **46**:391–5.

Sheue CR, Liu HY, Yong JWH (2003) *Kandelia obovata* (Rhizophoraceae), a new mangrove species from Eastern Asia. *Taxon* **52**: 287–94.

Shi S, Huang Y, Zeng K, *et al.* (2005) Molecular phylogenetic analysis of mangroves: independent evolutionary origins of vivipary and salt secretion. *Mol Phylogenet Evol* **34**:159–66.

State Forestry Administration (2002) *Report of Mangroves Survey in China*.

State Forestry Administration and State Oceanic Administration (2006) *Engineering Programs of Mangroves Conservation and Development in China (2006–2015)*.

State Oceanic Administration (1996) *The Plan of Action on the Marine Agenda of China in the 21st Century*. Beijing, China: Ocean Press.

Su G, Huang Y, Tan F, *et al.* (2006) Genetic variation in *Lumnitzera racemosa*, a mangrove species from the Indo-West Pacific. *Aquat Bot* **84**:341–6.

Su G, Huang Y, Tan F, *et al.* (2007) Conservation genetics of *Lumnitzera littorea* (Combretaceae), an endangered mangrove from the Indo-West Pacific. *Mar Biol* **150**:321–8.

Tam NFY, Wong YS (1995) Mangroves sediments as sinks for wastewater-borne pollutants. *Hydrobiologia* **295**:231–41.

Tam NFY, Wong YS (2000) *Hong Kong Mangroves*. Hong Kong: City University of Hong Kong Press.

Tam NFY, Wong YS (1997) Accumulation and distribution of heavy metals in a simulated mangrove system treated with sewage. *Hydrobiologia* **352**:67–75.

Tan F, Huang Y, Ge X, *et al.* (2005) Population genetic structure and conservation implications of *Ceriops decandra* in Malay Peninsula and North Australia. *Aquat Bot* **81**:175–88.

Tang GL, Shen LH, Weng WH, *et al.* (2007) Effects of using *Sonneratia apetala* to control the growth of *Spartina alterniflora* Loisel. *J South China Agri* **28**:10–3 (in Chinese with English abstract).

Tao SH, Wu J, Qi SH, *et al.* (2007) Scyphiphorins A and B, two new iridoid glycosides from the stem bark of a Chinese mangrove *Scyphiphora hydrophyllacea*. *Helv Chim Acta* **90**:1718–22.

Tomlinson PB (1986) *The Botany of Mangroves*. London: Cambridge University Press.

Wang BS, Liao BW, Wang YJ, *et al.* (2002) *Mangrove Forest Ecosystem and Its Sustaninable Development in Shenzhen Bay*. Beijing, China: Science Press (in Chinese).

Wang RJ, Chen ZY, Chen EY, *et al.* (1999) Two hybrids of the genus *Sonneratia* (Sonneratiaceae) from China. *Guihaia* **19**:199–204 (in Chinese with English abstract).

Wang WQ, Wang M (2007) *The Mangroves of China*. Beijing: Science Press (in Chinese).

Wong YS, Tam NFY, Lan CY (1997) Mangrove wetlands as wastewater treatment facility: a field trial. *Hydrobiologia* **352**:49–59.

Wu Y, Chung A, Tam NFY, *et al.* (2008*a*) Constructed mangrove wetland as secondary treatment system for municipal wastewater. *Ecol Eng* **34**:137–46.

Wu J, Xiao Q, Xu J, *et al.* (2008*b*) Natural products from true mangrove flora: source, chemistry and bioactivities. *Nat Prod Rep* **25**:955–81.

Wu J, Zhang S, Bruhn T, *et al.* (2008*c*) Bringmann G. Xylogranatins F-R: antifeedants from the Chinese mangrove, *Xylocarpus granatum*, a new biogenetic pathway to tetranortriterpenoids. *Chem Eur J* **14**:1129–44.

Yang Q, Tam NFY, Wong YS, *et al.* (2008) Potential use of mangroves as constructed wetland for municipal sewage treatment in Futian, Shenzhen, China. *Mar Pollut Bull* **57**:735–43.

Ye Y, Lu CY, Wong YS, *et al.* (1997) Methane fluxes from sediments of Bruguiera sexangula mangroves during different diuranal periods and in different flat zones. *J Xiamen Univ (Nat Sci)* **36**:925–30 (in Chinese with English abstract).

Ye Y, Tam NFY, Wong YS, *et al.* (2003) Growth and physiological responses of two mangrove species (*Bruguiera gymnorrhiza* and *Kandelia candel*) to waterlogging. *Environ Exp Bot* **49**:209–21.

Ye Y, Tam NFY, Wong YS, *et al.* (2004) Does sea level rise influence propagule establishment, early growth and physiology of *Kandelia candel* and *Bruguiera gymnorrhiza*? *J Exp Mar Biol Ecol* **306**:197–215.

Yin Y, Lin P (1992) Study on the litter fall of *Rhizophora stylosa* community in Yinluo Bay, Guangxi. *Guihaia* **12**:359–63 (in Chinese with English abstract).

Zhang HT, Zhang CS, Wang BS (1957) The mangrove community in Leizhou peninsular, Guangdong Province. *J Sun Yatsen Univ (Nat Sci)* **1**:122–45 (in Chinese with English abstract).

Zhang QM, Sui SZ (2001) The mangrove wetland resources and their conservation in China. *J Nat Resour* **16**:28–36. (in Chinese, with English abstract).

Zhang QM, Yu HB, Chen XS, *et al.* (1997) The relationship between mangrove zone on tidal flats and tidal levels. *Acta Ecol Sin* **17**:258–65 (in Chinese with English abstract).

Zhejiang Forestry Bureau (1961) *Report of Windshield Forest Investigation* (in Chinese).

Zheng DZ (1999) *Major Mangrove Species for Afforestation and the Research of Planting Technology*. Beijing: Science Press (in Chinese).

Zheng DZ, Li M, Zheng SF, *et al.* (2003) Headway of study on mangrove recovery and development in China. *For Sci Technol Guangdong Prov* **19**:10–4 (in Chinese with English abstract).

Zhong Y, Luan T, Wang X, *et al.* (2007) Influence of growth medium on cometabolic degradation of polycyclic aromatic hydrocarbons by *Sphingomonas* sp. strain PheB4. *Appl Microbiol Biotechnol* **75**:175–86.

Zhong Y, Zhao Q, Shi SH, *et al.* (2002) Detecting evolutionary rate heterogeneity among mangroves and their close terrestrial relatives. *Ecol Lett* **5**:427–32.

Zhou R, Shi S, Wu CI (2005) Molecular criteria for determining new hybrid species—an application to the *Sonneratia* hybrids. *Mol Phylogenet Evol* **35**:595–601.

Zhou R, Zeng K, Wu W, *et al.* (2007) Contrasting ancestral and extant polymorphisms in non-model organisms: I. Mangroves. *Mol Biol Evol* **24**:2746–54.

Zhu XJ, Qin P (2003) The alien species *Spartina alterniflora* and *Spartina* eco-engineering. *Mar Sci* **27**:14–9. (in Chinese with English abstract).

海处有片红树林*

□撰文/供图 王文卿

“红树林，它是红色的吗？”

也许在你的心底也曾出现过类似的疑问，然而事实上，大部分红树植物并没有红色的树皮，在它们扎根的地方，更是一片郁郁葱葱的绿。

红树林的名称来源于一种红树科植物——木榄，这种树的木材、树干、枝条和花朵都是红色的。马来西亚人在砍伐这种植物的时候，发现不仅裸露的木材显红色，砍刀的刀口也变成红色，他们利用这种植物的树皮提取物制作红色染料，而红树植物，也开始了这个富于传奇色彩的名字。

涨潮时仅露出树冠的红树植物

木榄红色的花及红褐色幼枝

那么，在植物学家的眼中，红树林又是指什么呢？那是热带与亚热带地区，海岸潮间带滩涂上生长的木本植物的总称，白骨壤、海莲、老鼠簕、角果木……这些名字奇特的植物构成了红树林复杂的家族。涨潮时它们被海水淹没或仅露出树冠，退潮时又安然无恙地显露出来，因此也被称为“海上森林”或“海底森林”。

全世界的红树林大致分布于南、北回归线之间，尤其集中在印度洋和西太平洋沿岸，113个国家和地区的海岸都有分布。如果以子午线为界，则恰好可以将世界红树林分为东方和西方两大中心：位于亚洲、大洋洲和非洲东海岸的东方群系，以苏门答腊和马来半岛的西海岸为中心，拥有丰富的植物种类；位于北美洲、西印度群岛和非洲西海岸的西方群系，植物种类贫乏。印度－马来半岛被认为是世界红树植物生物多样性最丰富的地区，澳大利亚则是第二大多样性中心。

中国的海岸线虽然辽阔，但是红树林面积仅占世界红树林面积的1.3‰，集中在海南、广东、广西、福建、香港、台湾等省（区），尤其是北部湾海岸和海南东海岸，占全国红树林面积的82%以上。由于中国处于热带北缘，无论红树植物的种类和高度都不如赤道附近，且受到严重的人为干扰，但它们对研究世界红树林的起源、分布和演化都具有特殊价值。

* 原载于生命世界，2008，8：10－17

生长于最外缘的桐花树

退潮时的红树林（福建福鼎）

生长于最外缘的杯萼海桑

割开后的木榄树皮

真红树和半红树

虽然红树林的名称来源于红树科植物，但它的家族其实涉及21科26属。我国的红树林专家林鹏院士在1995年提出了红树植物类型与鉴别标准，把专一性生长在潮间带的木本植物称为真红树植物，真红树只能在潮间带环境生长繁殖，在陆地环境不能繁殖。而既能生长于潮间带，有时成为优势种，但也能在陆地非盐渍土生长的两栖木本植物则被称为半红树植物。

我国有真红树植物24种，半红树植物12种。

木果楝表面根
木果楝果实

木果楝：国内唯一具有复叶的真红树植物，直径超过10厘米球形蒴果，有不甚发达的板根或蛇形表面根。

杯萼海桑果实

杯萼海桑花

杯萼海桑：广泛分布的杯萼海桑是海桑科植物，红树植物中耐盐能力最强的物种之一，常见于海滩外缘。

红海榄：红树科的红海榄具有非常发达的支柱根，生长于红树林的中内缘，属演替中后期树种。

秋茄：秋茄是红树科的常绿灌木或小乔木，板根或支柱根并不怎么发达。秋茄多分布于群落外缘，是最耐寒的红树植物。

红海榄胚轴

秋茄胚轴

角果木：角果木是红树科角果木属的常绿灌木或小乔木，有不甚发达的膝状根，多见于潮间带中上部，有时可分布到只有特大潮才淹及的高潮带上缘，常成纯林。

角果木纯林

红榄李花
红榄李:开着红色艳丽花朵的使君子科植物红榄李,有着细长的膝状呼吸根,纤维质果皮使得果实能够随水流传播。
白骨壤指状呼吸根
白骨壤:马鞭草科白骨壤的耐盐和耐淹水能力最强,也是我国分布面积最大的红树植物,它拥有发达的指状呼吸根和隐胎生硕果。
白骨壤隐胎生果实
桐花树花蕾(叶片上白色斑点为盐腺分泌的盐的结晶)
老鼠簕:爵床科老鼠簕多生长于红树林内缘、潮沟两侧,它的叶片形状受到叶龄、含盐量、光照的影响,变化极大。
桐花树花
老鼠簕花(注意其全缘的叶片)
老鼠簕果实
桐花树:紫金牛科桐花树属常绿灌木或小乔木,叶片表面有盐腺,能够泌盐,有牛角形的隐胎生果实,耐寒能力仅次于秋茄,常出现于红树林外缘,属演替的中前期树种。
水椰:棕榈科植物水椰生长在有淡水注入的隐蔽海湾河口,常在红树林中后缘出现。
水椰果序
水椰地下茎

黄槿花
玉蕊：玉蕊科常绿小乔木，花序顶生，下垂，长达60厘米，花芳香，晚间开放，天亮即谢。中果皮纤维质，质轻，果实借此随水流传播，生长于受潮汐影响的河流两岸或有淡水输入的红树林内缘。
黄槿果实
黄槿：锦葵科木槿属常绿灌木或乔木，海岸沙地、泥沙地及淤泥质滩涂均能够生长，常在红树林内、林缘、堤岸及不受潮汐影响的高地出现，广泛应用于城市园林绿化。
莲叶桐果实
莲叶桐：莲叶桐科植物莲叶桐生长在滨海高潮带以上疏林中，黑色核果包在黄色的腊质总苞内，悬挂在树上好像一串串铃铛，十分优美。
玉蕊果实
玉蕊花晚上开放
银叶树果实
银叶树：梧桐科的银叶树有非常发达的板根，叶背密被银白色鳞秕，银叶树由此得名。坚果木质，背部有龙骨状突起；中果皮有厚的木栓状纤维层，外种皮与内果皮间有孔隙，果实能借此随水流传播。
海檬果果实
海檬果：夹竹桃科海檬果全株具丰富乳汁，核果阔卵形或球形，未成熟时绿色，成熟时橙红色（剧毒）。具疏松而质轻的纤维质中果皮，果实借此随水流传播。
海檬果果实（示纤维质的果皮）

板根
支柱根
生存法宝之多样根系
呼吸根
膝状呼吸根
根系表面的皮孔
红树植物生长在海岸潮间带，经常遭受海水的浸渍和风浪的冲击，海水盐度较高，淤泥质滩涂通气不良，并且时有毒气(硫化氢)产生，在如此恶劣的环境中扎根生存，红树植物也练就了一身生存法宝，形态多样的根系就是其中之一。
多而发达的根系帮助红树植物支撑躯干、抵抗风浪，它们衍生出支柱根、膝状根、表面根、板状根、笋状根等类型。根和树干的表面还密布了许多类似于人类鼻孔一样的皮孔，帮助这些植物进行呼吸。

封 面 故 事

生存法宝之抗旱代谢

虽然生活在潮湿的海岸，但高盐度的海水和土壤都限制了红树植物对水分的吸收。于是，几乎所有的红树植物都采取了开源节流的策略：高渗透势的叶片强化了它们吸收水分的能力；有的红树叶片肉质化，可以保存更多的水分；有的叶片具有复表皮，有的叶背密生绒毛，减少水分蒸发；还有的植物能够调节叶片生长角度，使它们与光线平行，避免强光直射。

红树植物的叶片表面有厚的角质层

生存法宝之泌盐特性

高盐环境中生长的红树植物，在吸收水分的同时也要及时排出多余的盐分。红树植物具备了超强的过滤系统，在根系吸收水分的时候将水中大部分盐分过滤掉，对于不得不进入体内的多余盐分，则集中起来通过叶片表面的盐腺排出去。还有的红树林能把多余的盐分集中在老叶，落叶时一并排除。

白骨壤叶片盐腺泌盐

老鼠簕叶片盐腺泌盐

桐花树叶片盐腺泌盐

生存法宝之胎生繁殖

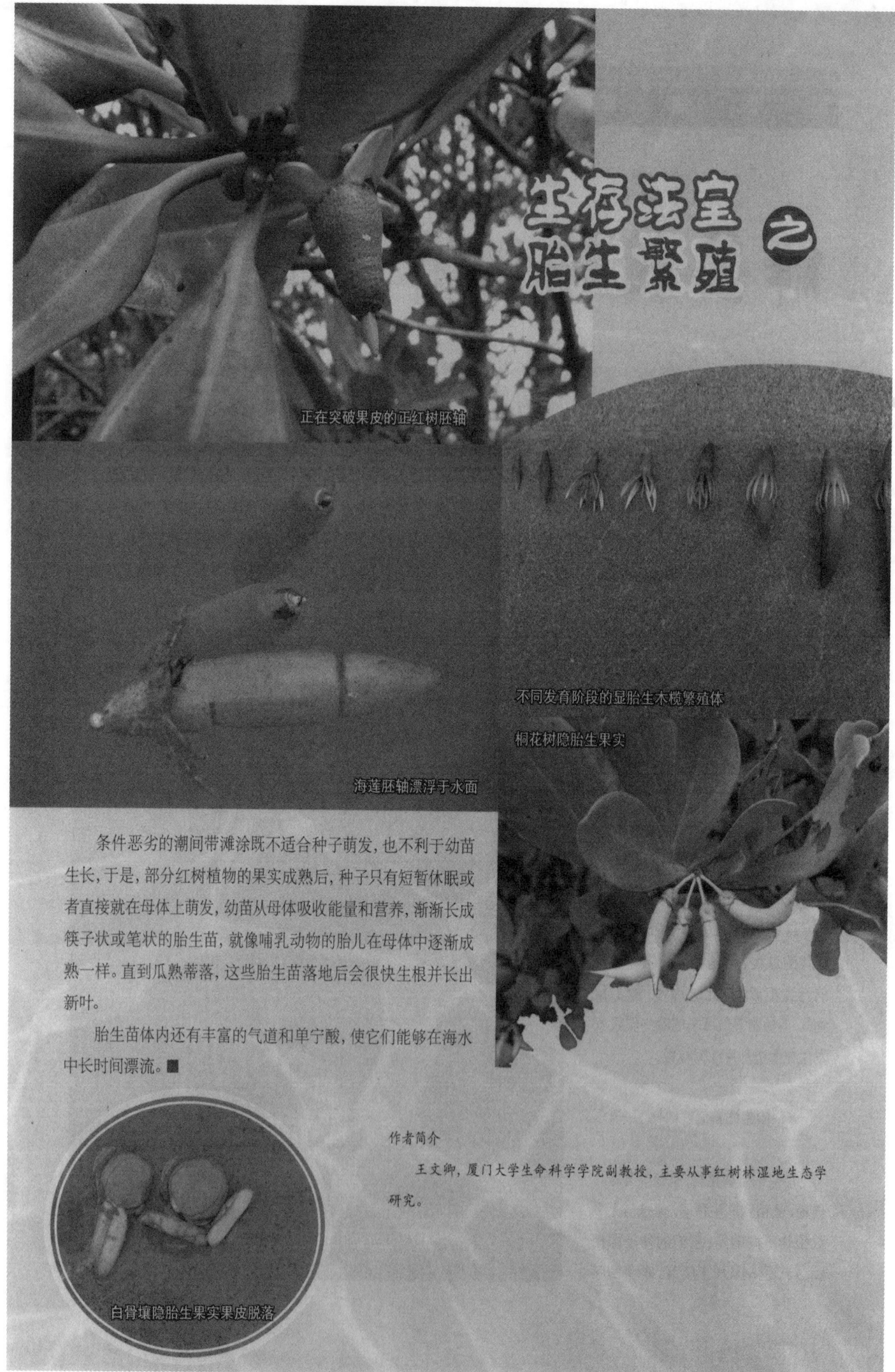

正在突破果皮的正红树胚轴

不同发育阶段的显胎生木榄繁殖体

桐花树隐胎生果实

海莲胚轴漂浮于水面

条件恶劣的潮间带滩涂既不适合种子萌发，也不利于幼苗生长，于是，部分红树植物的果实成熟后，种子只有短暂休眠或者直接就在母体上萌发，幼苗从母体吸收能量和营养，渐渐长成筷子状或笔状的胎生苗，就像哺乳动物的胎儿在母体中逐渐成熟一样。直到瓜熟蒂落，这些胎生苗落地后会很快生根并长出新叶。

胎生苗体内还有丰富的气道和单宁酸，使它们能够在海水中长时间漂流。■

白骨壤隐胎生果实果皮脱落

作者简介

王文卿，厦门大学生命科学学院副教授，主要从事红树林湿地生态学研究。

图书在版编目(CIP)数据

红树林研究论文集.第7集(2005—2009)/林光辉,林鹏主编.—厦门:厦门大学出版社,2010.10
ISBN 978-7-5615-3098-6
Ⅰ.①红… Ⅱ.①林…②林… Ⅲ.①红树林-文集 Ⅳ.①S796-53
中国版本图书馆CIP数据核字(2010)第206753号

厦门大学出版社出版发行
(地址:厦门市软件园二期望海路39号 邮编:361008)
http://www.xmupress.com
xmup@public.xm.fj.cn
厦门集大印刷厂印刷
2010年10月第1版 2010年10月第1次印刷
开本:889×1194 1/16 印张:27 插页:4
字数:750千字 印数:1～1 000册
定价:58.00元